S0-APC-896

Methods in Enzymology

Volume 153

RECOMBINANT DNA

Part D

METHODS IN ENZYMOLOGY

EDITORS-IN-CHIEF

John N. Abelson   Melvin I. Simon

Methods in Enzymology

Volume 153

Recombinant DNA

Part D

EDITED BY

Ray Wu

SECTION OF BIOCHEMISTRY
MOLECULAR AND CELL BIOLOGY
CORNELL UNIVERSITY
ITHACA, NEW YORK

Lawrence Grossman

DEPARTMENT OF BIOCHEMISTRY
THE JOHNS HOPKINS UNIVERSITY
SCHOOL OF HYGIENE AND PUBLIC HEALTH
BALTIMORE, MARYLAND

ACADEMIC PRESS, INC.
Harcourt Brace Jovanovich, Publishers
San Diego New York Berkeley Boston
London Sydney Tokyo Toronto

ACADEMIC PRESS, INC.
1250 Sixth Avenue
San Diego, California 92101

United Kingdom Edition published by
ACADEMIC PRESS INC. (LONDON) LTD.
24-28 Oval Road, London NW1 7DX

LIBRARY OF CONGRESS CATALOG CARD NUMBER: 54-9110

ISBN 0-12-182054-8 (alk. paper)

PRINTED IN THE UNITED STATES OF AMERICA
87 88 89 90   9 8 7 6 5 4 3 2 1

Table of Contents

CONTRIBUTORS TO VOLUME 153 ix

PREFACE xiii

NATHAN O. KAPLAN xv

VOLUMES IN SERIES xvii

Section I. Vectors for Cloning DNA

1. Production of Single-Stranded Plasmid DNA — JEFFREY VIEIRA AND JOACHIM MESSING — 3

2. pKUN, Vectors for the Separate Production of Both DNA Strands of Recombinant Plasmids — RUUD N. H. KONINGS, ELS J. M. VERHOEVEN, AND BEN P. H. PEETERS — 12

3. Restriction Site Bank Vectors for Cloning in Gram-Negative Bacteria and Yeast — JOHN DAVISON, MICHEL HEUSTERSPREUTE, AND FRANÇOISE BRUNEL — 34

4. Plasmids for the Selection and Analysis of Prokaryotic Promoters — JÜRGEN BROSIUS AND JAMES R. LUPSKI — 54

5. A λ DNA Protocol Based on Purification of Phage on DEAE–Cellulose — CYNTHIA HELMS, JAMES E. DUTCHIK, AND MAYNARD V. OLSON — 69

6. Double *cos* Site Vectors: Simplified Cosmid Cloning — PAUL BATES — 82

7. Improved *in Vitro* Packaging of λ DNA — SUSAN M. ROSENBERG — 95

8. λ Phage Vectors—EMBL Series — A. M. FRISCHAUF, N. MURRAY, AND H. LEHRACH — 103

9. Plasmid and Phage Vectors for Gene Cloning and Analysis in *Streptomyces* — DAVID A. HOPWOOD, MERVYN J. BIBB, KEITH F. CHATER, AND TOBIAS KIESER — 116

10. Cosmid Shuttle Vectors for Cloning and Analysis of *Streptomyces* DNA — R. NAGARAJA RAO, M. A. RICHARDSON, AND S. KUHSTOSS — 166

11. Host–Vector Systems for Gene Cloning in Cyanobacteria — C. J. KUHLEMEIER AND G. A. VAN ARKEL — 199

12. Genetic Engineering of the Cyanobacterial Chromosome — SUSAN S. GOLDEN, JUDY BRUSSLAN, AND ROBERT HASELKORN — 215

13. Conjugal Transfer of Plasmids to Cyanobacteria — TERESA THIEL AND C. PETER WOLK — 232

14. Site-Directed Chromosomal Rearrangements in Yeast — RICHARD T. SUROSKY AND BIK-KWOON TYE — 243

15. Improved Vectors for Plant Transformation: Expression Cassette Vectors and New Selectable Markers — S. G. ROGERS, H. J. KLEE, R. B. HORSCH, AND R. T. FRALEY — 253

16. Vectors for Cloning in Plant Cells — R. DEBLAERE, A. REYNAERTS, H. HÖFTE, J.-P. HERNALSTEENS, J. LEEMANS, AND M. VAN MONTAGU — 277

17. Binary Ti Vectors for Plant Transformation and Promoter Analysis — GYNHEUNG AN — 292

18. Detection of Monocot Transformation via *Agrobacterium tumefaciens* — PAUL J. J. HOOYKAAS AND ROB A. SCHILPEROORT — 305

19. Direct Gene Transfer to Protoplasts of Dicotyledonous and Monocotyledonous Plants by a Number of Methods, Including Electroporation — RAYMOND D. SHILLITO AND INGO POTRYKUS — 313

20. Uptake of DNA and RNA into Cells Mediated by Electroporation — W. H. R. LANGRIDGE, B. J. LI, AND A. A. SZALAY — 336

21. Electroporation of DNA and RNA into Plant Protoplasts — MICHAEL FROMM, JUDY CALLIS, LOVERINE P. TAYLOR, AND VIRGINIA WALBOT — 351

22. Cloning Vectors of Mitochondrial Origin for Eukaryotes — CHRISTINE LANG-HINRICHS AND ULF STAHL — 366

Section II. Vectors for Expression of Cloned Genes

23. Short Homopeptide Leader Sequences Enhanced Production of Human Proinsulin in *Escherichia coli* — WING L. SUNG, FEI-L. YAO, AND SARAN A. NARANG — 385

24. Expression of Bovine Growth Hormone Derivatives in *Escherichia coli* and the Use of the Derivatives to Produce Natural Sequence Growth Hormone by Cathepsin C Cleavage — HANSEN M. HSIUNG AND WARREN C. MACKELLAR — 390

25. Expression of Eukaryotic Genes in *Escherichia coli* with a Synthetic Two-Cistron System — BRIGITTE E. SCHONER, RAMA M. BELAGAJE, AND RONALD G. SCHONER — 401

26. Expression of Heterologous Unfused Protein in *Escherichia coli* — ERIK REMAUT, ANNE MARMENOUT, GUUS SIMONS, AND WALTER FIERS — 416

27. Directing Ribosomes to a Single mRNA Species: A Method to Study Ribosomal RNA Mutations and Their Effects on Translation of a Single Messenger in *Escherichia coli* — ANNA HUI, PARKASH JHURANI, AND HERMAN A. DE BOER — 432

28. New Expression Vectors for Identifying and Testing Signal Structures for Initiation and Termination of Transcription — KLAUS SCHNEIDER AND CHRISTOPH F. BECK — 452

29. Synthesis and Sequence-Specific Proteolysis of Hybrid Proteins Produced in *Escherichia coli* — KIYOSHI NAGAI AND HANS CHRISTIAN THØGERSEN — 461

30. Expression Plasmid Containing the λ P_L Promoter and cI857 Repressor — GREGORY MILMAN — 482

31. Expression and Secretion of Foreign Proteins in *Escherichia coli* — GUY D. DUFFAUD, PAUL E. MARCH, AND MASAYORI INOUYE — 492

32. Engineering for Protein Secretion in Gram-Positive Bacteria — SHING CHANG — 507

33. Expression and Secretion Vectors for Yeast — GRANT A. BITTER, KEVIN M. EGAN, RAYMOND A. KOSKI, MATTHEW O. JONES, STEVEN G. ELLIOTT, AND JAMES C. GIFFIN — 516

34. Vaccinia Virus as an Expression Vector — ANTONIA PICCINI, MARION E. PERKUS, AND ENZO PAOLETTI — 545

AUTHOR INDEX 565

SUBJECT INDEX 593

Contributors to Volume 153

Article numbers are in parentheses following the names of contributors.
Affiliations listed are current.

GYNHEUNG AN (17), *Institute of Biological Chemistry, Washington State University, Pullman, Washington 99164*

PAUL BATES (6), *Department of Microbiology, University of California, San Francisco, San Francisco, California 94143*

CHRISTOPH F. BECK (28), *Institut für Biologie III, Albert-Ludwigs-Universität, D-7800 Freiburg i. Br., Federal Republic of Germany*

RAMA M. BELAGAJE (25), *Department of Molecular Biology, Lilly Research Laboratories, A Division of Eli Lilly and Company, Lilly Corporate Center, Indianapolis, Indiana 46285*

MERVYN J. BIBB (9), *Department of Genetics, John Innes Institute, Norwich NR4 7UH, England*

GRANT A. BITTER (33), *AMGen, Thousand Oaks, California 91320*

JÜRGEN BROSIUS (4), *Department of Genetics and Development and Center for Neurobiology and Behavior, Columbia University, New York, New York 10032*

FRANÇOISE BRUNEL (3), *Unit of Molecular Biology, International Institute of Cellular and Molecular Pathology, B-1200 Brussels, Belgium*

JUDY BRUSSLAN (12), *Department of Molecular Genetics and Cell Biology, The University of Chicago, Chicago, Illinois 60637*

JUDY CALLIS (21), *Horticulture Department, University of Wisconsin, Madison, Wisconsin 53706*

SHING CHANG (32), *Microbial Genetics, Cetus Corporation, Emeryville, California 94608*

KEITH F. CHATER (9), *Department of Genetics, John Innes Institute, Norwich NR4 7UH, England*

JOHN DAVISON (3), *Unit of Molecular Biology, International Institute of Cellular and Molecular Pathology, B-1200 Brussels, Belgium*

R. DEBLAERE (16), *Laboratorium voor Genetica, Rijksuniversiteit Gent, B-9000 Gent, Belgium*

HERMAN A. DE BOER (27), *Department of Biochemistry of the Gorlaeus Laboratory, University of Leiden, 2300 RA Leiden, The Netherlands*

GUY D. DUFFAUD (31), *Department of Biochemistry, State University of New York at Stony Brook, Stony Brook, New York 11794*

JAMES E. DUTCHIK (5), *Department of Genetics, Washington University School of Medicine, St. Louis, Missouri 63110*

KEVIN M. EGAN (33), *AMGen, Thousand Oaks, California 91320*

STEVEN G. ELLIOTT (33), *AMGen, Thousand Oaks, California 91320*

WALTER FIERS (26), *Laboratory of Molecular Biology, State University of Ghent, B-9000 Ghent, Belgium*

R. T. FRALEY (15), *Plant Molecular Biology Group, Biological Sciences Department, Corporate Research and Development Staff, Monsanto Company, Chesterfield, Missouri 63198*

A. M. FRISCHAUF (8), *European Molecular Biology Laboratory, D-6900 Heidelberg, Federal Republic of Germany*

MICHAEL FROMM (21), *United States Department of Agriculture, Agricultural Research Service, Pacific Basin Area Plant Gene Expression Center, Albany, California 94710*

JAMES C. GIFFIN (33), *AMGen, Thousand Oaks, California 91320*

SUSAN S. GOLDEN (12), *Department of Biology, Texas A&M University, College Station, Texas 77843*

ROBERT HASELKORN (12), *Department of Molecular Genetics and Cell Biology, The University of Chicago, Chicago, Illinois 60637*

CYNTHIA HELMS (5), *Collaborative Research, Inc., Lexington, Massachusetts 02173*

J.-P. HERNALSTEENS (16), *Laboratorium Genetische Virologie, Vrije Universiteit Brussel, B-1640 Sint-Genesius-Rode, Belgium*

MICHEL HEUSTERSPREUTE (3), *Unit of Molecular Biology, International Institute of Cellular and Molecular Pathology, B-1200 Brussels, Belgium*

H. HÖFTE (16), *Plant Genetic Systems, Inc., B-9000 Ghent, Belgium*

PAUL J. J. HOOYKAAS (18), *Department of Plant Molecular Biology, Biochemistry Laboratory, University of Leiden, 2333 AL Leiden, The Netherlands*

DAVID A. HOPWOOD (9), *Department of Genetics, John Innes Institute, Norwich NR4 7UH, England*

R. B. HORSCH (15), *Plant Molecular Biology Group, Biological Sciences Department, Corporate Research and Development Staff, Monsanto Company, Chesterfield, Missouri 63198*

HANSEN M. HSIUNG (24), *Lilly Research Laboratories, A Division of Eli Lilly and Company, Lilly Corporate Center, Indianapolis, Indiana 46285*

ANNA HUI (27), *Department of Cell Genetics, Genentech, Inc., South San Francisco, California 94080*

MASAYORI INOUYE (31), *Department of Biochemistry, University of Medicine and Dentistry of New Jersey at Rutgers, Robert Wood Johnson Medical School, Piscataway, New Jersey 08854*

PARKASH JHURANI (27), *Department of Organic Chemistry, Genentech, Inc., South San Francisco, California 94080*

MATTHEW O. JONES (33), *AMGen, Thousand Oaks, California 91320*

TOBIAS KIESER (9), *Department of Genetics, John Innes Institute, Norwich NR4 7UH, England*

H. J. KLEE (15), *Plant Molecular Biology Group, Biological Sciences Department, Corporate Research and Development Staff, Monsanto Company, Chesterfield, Missouri 63198*

RUUD N. H. KONINGS (2), *Laboratory of Molecular Biology, Faculty of Science, University of Nijmegen, Toernooiveld, 6525 ED Nijmegen, The Netherlands*

RAYMOND A. KOSKI (33), *AMGen, Thousand Oaks, California 91320*

C. J. KUHLEMEIER (11), *Laboratory of Plant Molecular Biology, The Rockefeller University, New York, New York 10021*

S. KUHSTOSS (10), *Molecular Genetics Research, Lilly Research Laboratories, A Division of Eli Lilly and Company, Lilly Corporate Center, Indianapolis, Indiana 46285*

CHRISTINE LANG-HINRICHS (22), *Institut für Mikrobiologie, Institut für Gärungsgewerbe und Biotechnologie, D-1000 Berlin 65, Federal Republic of Germany*

W. H. R. LANGRIDGE (20), *Boyce Thompson Institute for Plant Research, Cornell University, Ithaca, New York 14853*

J. LEEMANS (16), *Plant Genetic Systems, Inc., B-9000 Ghent, Belgium*

H. LEHRACH (8), *The Imperial Cancer Research Fund, London WC2A 3PX, England*

B. J. LI (20), *Department of Biology, Chungshan University, Kwangchou, Kwangdong, People's Republic of China*

JAMES R. LUPSKI (4), *Department of Pediatrics and Institute for Molecular Genetics, Baylor College of Medicine, Texas Medical Center, Houston, Texas 77030*

WARREN C. MACKELLAR (24), *Lilly Research Laboratories, A Division of Eli Lilly and Company, Lilly Corporate Center, Indianapolis, Indiana 46285*

PAUL E. MARCH (31), *Department of Biochemistry, University of Medicine and Dentistry of New Jersey at Rutgers, Robert Wood Johnson Medical School, Piscataway, New Jersey 08854*

ANNE MARMENOUT (26), *Innogenetics, Zwijnaarde, Belgium*

JOACHIM MESSING (1), *Waksman Institute of Microbiology, Rutgers, The State University of New Jersey, Piscataway, New Jersey 08855*

GREGORY MILMAN (30), *Department of Biochemistry, The Johns Hopkins University, School of Hygiene and Public Health, Baltimore, Maryland 21205*

N. MURRAY (8), *Department of Molecular Biology, University of Edinburgh, Edinburgh EH9 3JR, Scotland*

KIYOSHI NAGAI (29), *Medical Research Council Laboratory of Molecular Biology, Cambridge CB2 2QH, England*

SARAN A. NARANG (23), *Division of Biological Sciences, National Research Council of Canada, Ottawa, Ontario, Canada K1A 0R6*

MAYNARD V. OLSON (5), *Department of Genetics, Washington University School of Medicine, St. Louis, Missouri 63110*

ENZO PAOLETTI (34), *Laboratory of Immunology, Wadsworth Center for Laboratories and Research, New York State Department of Health, Albany, New York 12201*

BEN P. H. PEETERS (2), *Department of Genetics, University of Groningen, 9751 NN Haren (GR), The Netherlands*

MARION E. PERKUS (34), *Laboratory of Immunology, Wadsworth Center for Laboratories and Research, New York State Department of Health, Albany, New York 12201*

ANTONIA PICCINI (34), *Laboratory of Immunology, Wadsworth Center for Laboratories and Research, New York State Department of Health, Albany, New York 12201*

INGO POTRYKUS (19), *Institute for Plant Sciences, CH-1892 Zurich, Switzerland*

R. NAGARAJA RAO (10), *Molecular Genetics Research, Lilly Research Laboratories, A Division of Eli Lilly and Company, Lilly Corporate Center, Indianapolis, Indiana 46285*

ERIK REMAUT (26), *Laboratory of Molecular Biology, State University of Ghent, B-9000 Ghent, Belgium*

A. REYNAERTS (16), *Plant Genetic Systems, Inc., B-9000 Ghent, Belgium*

M. A. RICHARDSON (10), *Molecular Genetics Research, Lilly Research Laboratories, A Division of Eli Lilly and Company, Lilly Corporate Center, Indianapolis, Indiana 46285*

S. G. ROGERS (15), *Plant Molecular Biology Group, Biological Sciences Department, Corporate Research and Development Staff, Monsanto Company, Chesterfield, Missouri 63198*

SUSAN M. ROSENBERG (7), *Institute of Molecular Biology, University of Oregon, Eugene, Oregon 97403*

ROB A. SCHILPEROORT (18), *Department of Plant Molecular Biology, Biochemistry Laboratory, University of Leiden, 2333 AL Leiden, The Netherlands*

KLAUS SCHNEIDER (28), *Institut für Biologie III, Albert-Ludwigs-Universität, D-7800 Freiburg i. Br., Federal Republic of Germany*

BRIGITTE E. SCHONER (25), *Department of Molecular Genetics, Lilly Research Laboratories, A Division of Eli Lilly and Company, Lilly Corporate Center, Indianapolis, Indiana 46285*

RONALD G. SCHONER (25), *Department of Molecular Genetics, Lilly Research Laboratories, A Division of Eli Lilly and Company, Lilly Corporate Center, Indianapolis, Indiana 46285*

RAYMOND D. SHILLITO (19), *Biotechnology Research, CIBA-GEIGY Corporation, Research Triangle Park, North Carolina 27709*

GUUS SIMONS (26), *N.I.Z.O., 6710 Ede, The Netherlands*

ULF STAHL (22), *Fachgebiet Mikrobiologie,*

Technische Universität Berlin, D-1000 Berlin 65, Federal Republic of Germany

WING L. SUNG (23), *Division of Biological Sciences, National Research Council of Canada, Ottawa, Ontario, Canada K1A 0R6*

RICHARD T. SUROSKY (14), *Department of Molecular Genetics and Cell Biology, The University of Chicago, Chicago, Illinois 60637*

A. A. SZALAY (20), *Boyce Thompson Institute for Plant Research, Cornell University, Ithaca, New York 14853*

LOVERINE P. TAYLOR (21), *Carnagie Institution of Washington, Stanford, California 94305*

TERESA THIEL (13), *Department of Biology, University of Missouri-St. Louis, St. Louis, Missouri 63121*

HANS CHRISTIAN THØGERSEN (29), *Biostruktur Afdeling, Kemisk Institut, Århus Universitet, 8200 Århus N, Denmark*

BIK-KWOON TYE (14), *Section of Biochemistry, Molecular and Cell Biology, Division of Biological Sciences, Cornell University, Ithaca, New York 14853*

G. A. VAN ARKEL (11), *Department of Molecular Cell Biology, University of Utrecht, 3584 CH Utrecht, The Netherlands*

M. VAN MONTAGU (16), *Laboratorium Genetische Virologie, Vrije Universiteit Brussel, B-1640 Sint-Genesius-Rode, Belgium, and Laboratorium voor Genetica, Rijksuniversiteit Gent, B-9000 Gent, Belgium*

ELS J. M. VERHOEVEN (2), *Department of Biology, Antoni van Leeuwenhoekhuis, 1066 CX Amsterdam, The Netherlands*

JEFFREY VIEIRA (1), *Waksman Institute of Microbiology, Rutgers, The State University of New Jersey, Piscataway, New Jersey 08855*

VIRGINIA WALBOT (21), *Department of Biological Sciences, Stanford University, Stanford, California 94305*

C. PETER WOLK (13), *MSU-DOE Plant Research Laboratory, Michigan State University, East Lansing, Michigan 48824*

FEI-L. YAO (23), *Division of Biological Sciences, National Research Council of Canada, Ottawa, Ontario, Canada K1A 0R6*

NATHAN O. KAPLAN
June 25, 1917–April 15, 1986

Preface

Recombinant DNA methods are powerful, revolutionary techniques for at least two reasons. First, they allow the isolation of single genes in large amounts from a pool of thousands or millions of genes. Second, the isolated genes or their regulatory regions can be modified at will and reintroduced into cells for expression at the RNA or protein levels. These attributes allow us to solve complex biological problems and to produce new and better products in the areas of health, agriculture, and industry.

Volumes 153, 154, and 155 supplement Volumes 68, 100, and 101 of *Methods in Enzymology*. During the past few years, many new or improved recombinant DNA methods have appeared, and a number of them are included in these three new volumes. Volume 153 covers methods related to new vectors for cloning DNA and for expression of cloned genes. Volume 154 includes methods for cloning cDNA, identification of cloned genes and mapping of genes, chemical synthesis and analysis of oligodeoxynucleotides, site-specific mutagenesis, and protein engineering. Volume 155 includes the description of several useful new restriction enzymes, detail of rapid methods for DNA sequence analysis, and a number of other useful methods.

RAY WU
LAWRENCE GROSSMAN

Nathan O. Kaplan

In the past half century, knowledge in the natural sciences has progressed at a rate unmatched in previous history. Biochemistry appears closer than ever to the attainment of its ultimate objective: creation of a body of knowledge rationalized in a conceptual structure which provides a solid basis for understanding life processes. In these fabulous times, there have been fabulous people among whom may be included Nathan ("Nate") Kaplan. His many, varied and massive contributions to crucially important areas of biochemical research added to his creative activities as an editor, scholar, and academic statesman have left a lasting impression on the history of these exciting times. We are fortunate in having an account of his life philosophy and experiences which he himself provided in "Selected Topics in the History of Biochemistry" (edited by G. Semenga; Vol. 30, p. 255 *et seq.*; Elsevier Science Publishers).

His potential was manifest early in his career at Berkeley where he collaborated with Barker, Hassid, and Doudoroff in the late 1930s, providing biochemical expertise crucial for the demonstration that in the phosphorolysis of sucrose the phosphate ester formed was glucose 1-phosphate. His first scientific publication on sucrose phosphorylase included an account of these seminal researches. His full potential was realized when, under the watchful eye of Fritz Lipmann, his great mentor and life-long admirer and friend, he made essential contributions in collaboration with Lipmann and Dave Novelli to the isolation and characterization of coenzyme A, work which later formed part of the basis for the Nobel Prize to Lipmann.

Nate followed his unerring intuition in continuing his career at the McCollum-Pratt Institute under the aegis of W. D. McElroy. He built a body of research on NAD, NAD analogs, and associated dehydrogenases to earn a leading position as an international authority on the pyridine nucleotide coenzymes. In the course of these investigations he began a life-long collaboration with another "biochemist's biochemist"—Sidney Colowick—which resulted in the creation of the monumental series *Methods in Enzymology,* which was to become the definitive source of methodology in the biochemical sciences.

Nate, as he so vividly detailed in the account I have referred to above, stressed the importance of following research wherever it led, even if assured results might not be immediately evident. As an example, one notes that his investigations of the pyridine nucleotide cofactors ignited an interest in comparative biochemistry, elaborated in many researches of major significance for biochemical evolution.

Nate's intuitive insights into things biochemical also extended to an uncanny ability to assess potential in budding biochemists. His success in finding and recruiting talent was never better shown than in the creation of the Graduate Department of Biochemistry at Brandeis in the late 1950s. Those in the remarkable group he assembled—which included W. Jencks, L. Grossman, G. Sato, M. E. Jones, L. Levine, H. Van Vunakis, and J. Lowenstein—owed their start in large part to his unstinting guidance and encouragement.

He found time to serve on a multitude of policy-making committees and was always available, however hard pressed, to take over editorial chores, however onerous. I recall the many hours he spent helping to organize and edit a Festschrift and symposium celebrating the fact I had survived to age 65. And then there was the salvage and rebuilding operation he so unselfishly initiated to revive the ailing *Analytical Biochemistry* journal when his old friend, Al Nason, its Editor-in-Chief, fell seriously ill.

No project engaged Nate's attention and devotion more than his labors with Colowick to oversee and assure the publication and excellence of the many volumes which make up the *Methods in Enzymology* series, now numbering more than a hundred, which will stand as a lasting monument to his memory. Certainly nothing could be more appropriate than the present dedication.

MARTIN D. KAMEN

METHODS IN ENZYMOLOGY

EDITED BY

Sidney P. Colowick and Nathan O. Kaplan

VANDERBILT UNIVERSITY
SCHOOL OF MEDICINE
NASHVILLE, TENNESSEE

DEPARTMENT OF CHEMISTRY
UNIVERSITY OF CALIFORNIA
AT SAN DIEGO
LA JOLLA, CALIFORNIA

I. Preparation and Assay of Enzymes
II. Preparation and Assay of Enzymes
III. Preparation and Assay of Substrates
IV. Special Techniques for the Enzymologist
V. Preparation and Assay of Enzymes
VI. Preparation and Assay of Enzymes (*Continued*)
Preparation and Assay of Substrates
Special Techniques
VII. Cumulative Subject Index

METHODS IN ENZYMOLOGY

EDITORS-IN-CHIEF

Sidney P. Colowick and Nathan O. Kaplan

VOLUME VIII. Complex Carbohydrates
Edited by ELIZABETH F. NEUFELD AND VICTOR GINSBURG

VOLUME IX. Carbohydrate Metabolism
Edited by WILLIS A. WOOD

VOLUME X. Oxidation and Phosphorylation
Edited by RONALD W. ESTABROOK AND MAYNARD E. PULLMAN

VOLUME XI. Enzyme Structure
Edited by C. H. W. HIRS

VOLUME XII. Nucleic Acids (Parts A and B)
Edited by LAWRENCE GROSSMAN AND KIVIE MOLDAVE

VOLUME XIII. Citric Acid Cycle
Edited by J. M. LOWENSTEIN

VOLUME XIV. Lipids
Edited by J. M. LOWENSTEIN

VOLUME XV. Steroids and Terpenoids
Edited by RAYMOND B. CLAYTON

VOLUME XVI. Fast Reactions
Edited by KENNETH KUSTIN

VOLUME XVII. Metabolism of Amino Acids and Amines (Parts A and B)
Edited by HERBERT TABOR AND CELIA WHITE TABOR

VOLUME XXXII. Biomembranes (Part B)
Edited by SIDNEY FLEISCHER AND LESTER PACKER

VOLUME XXXIII. Cumulative Subject Index Volumes I–XXX
Edited by MARTHA G. DENNIS AND EDWARD A. DENNIS

VOLUME XXXIV. Affinity Techniques (Enzyme Purification: Part B)
Edited by WILLIAM B. JAKOBY AND MEIR WILCHEK

VOLUME XXXV. Lipids (Part B)
Edited by JOHN M. LOWENSTEIN

VOLUME XXXVI. Hormone Action (Part A: Steroid Hormones)
Edited by BERT W. O'MALLEY AND JOEL G. HARDMAN

VOLUME XXXVII. Hormone Action (Part B: Peptide Hormones)
Edited by BERT W. O'MALLEY AND JOEL G. HARDMAN

VOLUME XXXVIII. Hormone Action (Part C: Cyclic Nucleotides)
Edited by JOEL G. HARDMAN AND BERT W. O'MALLEY

VOLUME XXXIX. Hormone Action (Part D: Isolated Cells, Tissues, and Organ Systems)
Edited by JOEL G. HARDMAN AND BERT W. O'MALLEY

VOLUME XL. Hormone Action (Part E: Nuclear Structure and Function)
Edited by BERT W. O'MALLEY AND JOEL G. HARDMAN

VOLUME XLI. Carbohydrate Metabolism (Part B)
Edited by W. A. WOOD

VOLUME XLII. Carbohydrate Metabolism (Part C)
Edited by W. A. WOOD

VOLUME XLIII. Antibiotics
Edited by JOHN H. HASH

VOLUME XLIV. Immobilized Enzymes
Edited by KLAUS MOSBACH

VOLUME XLV. Proteolytic Enzymes (Part B)
Edited by LASZLO LORAND

VOLUME LX. Nucleic Acids and Protein Synthesis (Part H)
Edited by KIVIE MOLDAVE AND LAWRENCE GROSSMAN

VOLUME 61. Enzyme Structure (Part H)
Edited by C. H. W. HIRS AND SERGE N. TIMASHEFF

VOLUME 62. Vitamins and Coenzymes (Part D)
Edited by DONALD B. MCCORMICK AND LEMUEL D. WRIGHT

VOLUME 63. Enzyme Kinetics and Mechanism (Part A: Initial Rate and Inhibitor Methods)
Edited by DANIEL L. PURICH

VOLUME 64. Enzyme Kinetics and Mechanism (Part B: Isotopic Probes and Complex Enzyme Systems)
Edited by DANIEL L. PURICH

VOLUME 65. Nucleic Acids (Part I)
Edited by LAWRENCE GROSSMAN AND KIVIE MOLDAVE

VOLUME 66. Vitamins and Coenzymes (Part E)
Edited by DONALD B. MCCORMICK AND LEMUEL D. WRIGHT

VOLUME 67. Vitamins and Coenzymes (Part F)
Edited by DONALD B. MCCORMICK AND LEMUEL D. WRIGHT

VOLUME 68. Recombinant DNA
Edited by RAY WU

VOLUME 69. Photosynthesis and Nitrogen Fixation (Part C)
Edited by ANTHONY SAN PIETRO

VOLUME 70. Immunochemical Techniques (Part A)
Edited by HELEN VAN VUNAKIS AND JOHN J. LANGONE

VOLUME 71. Lipids (Part C)
Edited by JOHN M. LOWENSTEIN

VOLUME 72. Lipids (Part D)
Edited by JOHN M. LOWENSTEIN

VOLUME 85. Structural and Contractile Proteins (Part B: The Contractile Apparatus and the Cytoskeleton)
Edited by DIXIE W. FREDERIKSEN AND LEON W. CUNNINGHAM

VOLUME 86. Prostaglandins and Arachidonate Metabolites
Edited by WILLIAM E. M. LANDS AND WILLIAM L. SMITH

VOLUME 87. Enzyme Kinetics and Mechanism (Part C: Intermediates, Stereochemistry, and Rate Studies)
Edited by DANIEL L. PURICH

VOLUME 88. Biomembranes (Part I: Visual Pigments and Purple Membranes, II)
Edited by LESTER PACKER

VOLUME 89. Carbohydrate Metabolism (Part D)
Edited by WILLIS A. WOOD

VOLUME 90. Carbohydrate Metabolism (Part E)
Edited by WILLIS A. WOOD

VOLUME 91. Enzyme Structure (Part I)
Edited by C. H. W. HIRS AND SERGE N. TIMASHEFF

VOLUME 92. Immunochemical Techniques (Part E: Monoclonal Antibodies and General Immunoassay Methods)
Edited by JOHN J. LANGONE AND HELEN VAN VUNAKIS

VOLUME 93. Immunochemical Techniques (Part F: Conventional Antibodies, Fc Receptors, and Cytotoxicity)
Edited by JOHN J. LANGONE AND HELEN VAN VUNAKIS

VOLUME 94. Polyamines
Edited by HERBERT TABOR AND CELIA WHITE TABOR

VOLUME 95. Cumulative Subject Index Volumes 61–74, 76–80
Edited by EDWARD A. DENNIS AND MARTHA G. DENNIS

VOLUME 96. Biomembranes [Part J: Membrane Biogenesis: Assembly and Targeting (General Methods; Eukaryotes)]
Edited by SIDNEY FLEISCHER AND BECCA FLEISCHER

VOLUME 97. Biomembranes [Part K: Membrane Biogenesis: Assembly and Targeting (Prokaryotes, Mitochondria, and Chloroplasts)]
Edited by SIDNEY FLEISCHER AND BECCA FLEISCHER

VOLUME 98. Biomembranes (Part L: Membrane Biogenesis: Processing and Recycling)
Edited by SIDNEY FLEISCHER AND BECCA FLEISCHER

VOLUME 99. Hormone Action (Part F: Protein Kinases)
Edited by JACKIE D. CORBIN AND JOEL G. HARDMAN

VOLUME 100. Recombinant DNA (Part B)
Edited by RAY WU, LAWRENCE GROSSMAN, AND KIVIE MOLDAVE

VOLUME 101. Recombinant DNA (Part C)
Edited by RAY WU, LAWRENCE GROSSMAN, AND KIVIE MOLDAVE

VOLUME 102. Hormone Action (Part G: Calmodulin and Calcium-Binding Proteins)
Edited by ANTHONY R. MEANS AND BERT W. O'MALLEY

VOLUME 103. Hormone Action (Part H: Neuroendocrine Peptides)
Edited by P. MICHAEL CONN

VOLUME 104. Enzyme Purification and Related Techniques (Part C)
Edited by WILLIAM B. JAKOBY

VOLUME 105. Oxygen Radicals in Biological Systems
Edited by LESTER PACKER

VOLUME 106. Posttranslational Modifications (Part A)
Edited by FINN WOLD AND KIVIE MOLDAVE

VOLUME 107. Posttranslational Modifications (Part B)
Edited by FINN WOLD AND KIVIE MOLDAVE

VOLUME 108. Immunochemical Techniques (Part G: Separation and Characterization of Lymphoid Cells)
Edited by GIOVANNI DI SABATO, JOHN J. LANGONE, AND HELEN VAN VUNAKIS

VOLUME 109. Hormone Action (Part I: Peptide Hormones)
Edited by LUTZ BIRNBAUMER AND BERT W. O'MALLEY

VOLUME 110. Steroids and Isoprenoids (Part A)
Edited by JOHN H. LAW AND HANS C. RILLING

VOLUME 111. Steroids and Isoprenoids (Part B)
Edited by JOHN H. LAW AND HANS C. RILLING

VOLUME 112. Drug and Enzyme Targeting (Part A)
Edited by KENNETH J. WIDDER AND RALPH GREEN

VOLUME 113. Glutamate, Glutamine, Glutathione, and Related Compounds
Edited by ALTON MEISTER

VOLUME 114. Diffraction Methods for Biological Macromolecules (Part A)
Edited by HAROLD W. WYCKOFF, C. H. W. HIRS, AND SERGE N. TIMASHEFF

VOLUME 115. Diffraction Methods for Biological Macromolecules (Part B)
Edited by HAROLD W. WYCKOFF, C. H. W. HIRS, AND SERGE N. TIMASHEFF

VOLUME 116. Immunochemical Techniques (Part H: Effectors and Mediators of Lymphoid Cell Functions)
Edited by GIOVANNI DI SABATO, JOHN J. LANGONE, AND HELEN VAN VUNAKIS

VOLUME 117. Enzyme Structure (Part J)
Edited by C. H. W. HIRS AND SERGE N. TIMASHEFF

VOLUME 118. Plant Molecular Biology
Edited by ARTHUR WEISSBACH AND HERBERT WEISSBACH

VOLUME 119. Interferons (Part C)
Edited by SIDNEY PESTKA

VOLUME 120. Cumulative Subject Index Volumes 81–94, 96–101

VOLUME 121. Immunochemical Techniques (Part I: Hybridoma Technology and Monoclonal Antibodies)
Edited by JOHN J. LANGONE AND HELEN VAN VUNAKIS

VOLUME 122. Vitamins and Coenzymes (Part G)
Edited by FRANK CHYTIL AND DONALD B. MCCORMICK

VOLUME 123. Vitamins and Coenzymes (Part H)
Edited by FRANK CHYTIL AND DONALD B. MCCORMICK

VOLUME 124. Hormone Action (Part J: Neuroendocrine Peptides)
Edited by P. MICHAEL CONN

VOLUME 125. Biomembranes (Part M: Transport in Bacteria, Mitochondria, and Chloroplasts: General Approaches and Transport Systems)
Edited by SIDNEY FLEISCHER AND BECCA FLEISCHER

VOLUME 126. Biomembranes (Part N: Transport in Bacteria, Mitochondria, and Chloroplasts: Protonmotive Force)
Edited by SIDNEY FLEISCHER AND BECCA FLEISCHER

VOLUME 127. Biomembranes (Part O: Protons and Water: Structure and Translocation)
Edited by LESTER PACKER

VOLUME 128. Plasma Lipoproteins (Part A: Preparation, Structure, and Molecular Biology)
Edited by JERE P. SEGREST AND JOHN J. ALBERS

VOLUME 129. Plasma Lipoproteins (Part B: Characterization, Cell Biology, and Metabolism)
Edited by JOHN J. ALBERS AND JERE P. SEGREST

VOLUME 130. Enzyme Structure (Part K)
Edited by C. H. W. HIRS AND SERGE N. TIMASHEFF

VOLUME 131. Enzyme Structure (Part L)
Edited by C. H. W. HIRS AND SERGE N. TIMASHEFF

VOLUME 132. Immunochemical Techniques (Part J: Phagocytosis and Cell-Mediated Cytotoxicity)
Edited by GIOVANNI DI SABATO AND JOHANNES EVERSE

VOLUME 133. Bioluminescence and Chemiluminescence (Part B)
Edited by MARLENE DELUCA AND WILLIAM D. MCELROY

VOLUME 134. Structural and Contractile Proteins (Part C: The Contractile Apparatus and the Cytoskeleton)
Edited by RICHARD B. VALLEE

VOLUME 135. Immobilized Enzymes and Cells (Part B)
Edited by KLAUS MOSBACH

VOLUME 136. Immobilized Enzymes and Cells (Part C)
Edited by KLAUS MOSBACH

VOLUME 137. Immobilized Enzymes and Cells (Part D) (in preparation)
Edited by KLAUS MOSBACH

VOLUME 138. Complex Carbohydrates (Part E)
Edited by VICTOR GINSBURG

VOLUME 139. Cellular Regulators (Part A: Calcium- and Calmodulin-Binding Proteins)
Edited by ANTHONY R. MEANS AND P. MICHAEL CONN

VOLUME 140. Cumulative Subject Index Volumes 102–119, 121–134

VOLUME 141. Cellular Regulators (Part B: Calcium and Lipids)
Edited by P. MICHAEL CONN AND ANTHONY R. MEANS

VOLUME 142. Metabolism of Aromatic Amino Acids and Amines
Edited by SEYMOUR KAUFMAN

VOLUME 143. Sulfur and Sulfur Amino Acids
Edited by WILLIAM B. JAKOBY AND OWEN GRIFFITH

VOLUME 144. Structural and Contractile Proteins (Part D: Extracellular Matrix)
Edited by LEON W. CUNNINGHAM

VOLUME 145. Structural and Contractile Proteins (Part E: Extracellular Matrix)
Edited by LEON W. CUNNINGHAM

VOLUME 146. Peptide Growth Factors (Part A)
Edited by DAVID BARNES AND DAVID A. SIRBASKU

VOLUME 147. Peptide Growth Factors (Part B)
Edited by DAVID BARNES AND DAVID A. SIRBASKU

VOLUME 148. Plant Cell Membranes
Edited by LESTER PACKER AND ROLAND DOUCE

VOLUME 149. Drug and Enzyme Targeting (Part B)
Edited by RALPH GREEN AND KENNETH J. WIDDER

VOLUME 150. Immunochemical Techniques (Part K: *In Vitro* Models of B and T Cell Functions and Lymphoid Cell Receptors)
Edited by GIOVANNI DI SABATO

VOLUME 151. Molecular Genetics of Mammalian Cells
Edited by MICHAEL M. GOTTESMAN

VOLUME 152. Guide to Molecular Cloning Techniques
Edited by SHELBY L. BERGER AND ALAN R. KIMMEL

VOLUME 153. Recombinant DNA (Part D)
Edited by RAY WU AND LAWRENCE GROSSMAN

VOLUME 154. Recombinant DNA (Part E)
Edited by RAY WU AND LAWRENCE GROSSMAN

VOLUME 155. Recombinant DNA (Part F)
Edited by RAY WU

VOLUME 156. Biomembranes (Part P: ATP-Driven Pumps and Related Transport: The Na,K-Pump) (in preparation)
Edited by SIDNEY FLEISCHER AND BECCA FLEISCHER

VOLUME 157. Biomembranes (Part Q: ATP-Driven Pumps and Related Transport: Calcium, Proton, and Potassium Pumps) (in preparation)
Edited by SIDNEY FLEISCHER AND BECCA FLEISCHER

VOLUME 158. Metalloproteins (Part A) (in preparation)
Edited by JAMES F. RIORDAN AND BERT L. VALLEE

Section I

Vectors for Cloning DNA

[1] Production of Single-Stranded Plasmid DNA

By JEFFREY VIEIRA and JOACHIM MESSING

Introduction

In the study of gene structure and function, the techniques of DNA analysis that are efficiently carried out on single-strand (ss) DNA templates, such as DNA sequencing and site-specific *in vitro* mutagenesis, have been of great importance. Because of this, the vectors developed from the ssDNA bacteriophages M13, fd, or f1, which allow the easy isolation of strand-specific templates, have been widely used. While these vectors are very valuable for the production of ssDNA, they have certain negative aspects in comparison to plasmid vectors (e.g., increased instability of some inserts, the minimum size of phage vectors). Work from the laboratory of N. Zinder showed that a plasmid carrying the intergenic region (IG) of f1 could be packaged as ssDNA into a viral particle by a helper phage.[1] This led to the construction of vectors that could combine the advantages of both plasmid and phage vectors.[2] Since that time a number of plasmids carrying the intergenic region of M13 or f1 have been constructed with a variety of features.[3]

A problem that has been encountered in the use of these plasmid/phage chimeric vectors (plage) is the significant reduction in the amount of ssDNA that is produced as compared to phage vectors. Phage vectors can have titers of plaque-forming units (pfu) of 10^{12}/ml and give yields of a few micrograms per milliliter of ssDNA. It might then be expected that cells carrying both a plage and helper phage would give titers of 5×10^{11}/ml for each of the two. However, this is not the case due to interference by the plage with the replication of the phage.[4] This results in a reduction in the phage copy number and, therefore, reduces the phage gene products necessary for production of ssDNA. This interference results in a 10- to 100-fold reduction in the phage titer and a level of ss plasmid DNA particles of about 10^{10} colony forming units (cfu) per milliliter.[1] Phage mutants that show interference resistance have been isolated.[4,5] These mutants can increase the yield of ss plasmid by 10-fold and concurrently

[1] G. P. Dotto, V. Enea, and N. D. Zinder, *Virology* **114,** 463 (1981).
[2] N. D. Zinder and J. D. Boeke, *Gene* **19,** 1 (1982).
[3] D. Mead and B. Kemper, *in* "Vectors: A Survey of Molecular Cloning Vectors and Their Uses." Butterworth, Massachusetts, 1986.
[4] V. Enea and N. D. Zinder, *Virology* **122,** 222 (1982).
[5] A. Levinson, D. Silver, and B. Seed, *J. Mol. Appl. Genet.* **2,** 507 (1984).

increase the level of phage by a similar amount. Whether wild-type (wt) phage or an interference-resistant mutant is used as helper the yield of plasmid ssDNA is usually about equal to that of the phage,[3] and as the plasmid size increases the ratio shifts to favor the phage.[5] In order to increase both the quantitative and qualitative yield of the plasmid ssDNA, a helper phage, M13KO7, has been constructed that preferentially packages plasmid DNA over phage DNA. In this chapter, M13KO7 will be described and its uses discussed.

M13 Biology

Certain aspects of M13 biology and M13 mutants play an important role in the functioning of M13KO7, so a short review of its biology is appropriate.[6,7] M13 is a phage that contains a circular ssDNA molecule of 6407 bases packaged in a filamentous virion which is extruded from the cell without lysis. It can infect only cells having an F pili, to which it binds for entering the cell. The phage genome consists of 9 genes encoding 10 proteins and contains an intergenic region of 508 bases. The proteins expressed by the phage are involved in the following processes: I and IV are involved in phage morphogenesis, III, VI, VII, VIII, and IX are virion proteins, V is an ssDNA binding protein, X is probably involved in replication, and II creates a site-specific (+) strand nick within the IG region of the double-stranded replicative form (RF) of the phage DNA molecule at which DNA synthesis is initiated.

Phage replication consists of three phases: (1) ss–ds, (2) ds–ds, and (3) ds–ss. The ss–ds phase is carried out entirely by host enzymes. For phases 2 and 3, gene II, which encodes both proteins II and X, is required for initiating DNA synthesis; all other functions necessary for synthesis are supplied by the host. The DNA synthesis initiated by the action of the gene II protein (gIIp) leads to both the replication of the ds molecule and the production of the ssDNA that is to be packaged in the mature virion. The phage is replicated by a rolling circle mechanism that is terminated by gIIp cleaving the displaced (+) strand at the same site and resealing it to create a circular ssDNA molecule. Early in the phage life cycle this ssDNA molecule is converted to the ds RF but later in the phage life cycle gVp binds to the (+) strand, preventing it from being converted to dsDNA and resulting in it being packaged into viral particles. The assembly of the virion occurs in the cell membrane where the gVp is replaced by the

[6] D. T. Denhardt, D. Dressler, and D. S. Ray (eds.), "The Single-Stranded DNA Phages." Cold Spring Harbor Lab., Cold Spring Harbor, New York, 1978.

[7] N. D. Zinder and K. Horiuchi, *Microbiol. Rev.* **49,** 101 (1985).

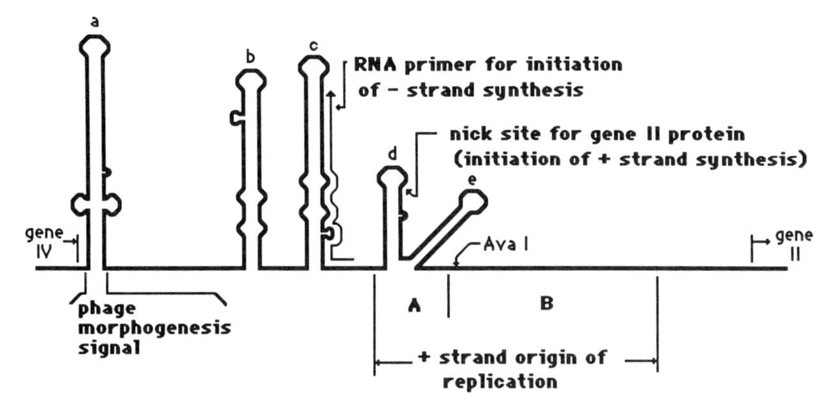

FIG. 1. The M13 intergenic region is schematically presented. It is 508 nucleotides long and is situated between genes II and IV. Potential secondary structure is represented by hairpin structures a–e.[8] Important functional regions are also shown.

gVIIIp and the other virion proteins as the phage particle is extruded from the cell.

The IG structure contains regions important for four phage processes[8–10]: (1) The sequences necessary for the recognition of an ssDNA by phage proteins for its efficient packaging into viral particles; (2) the site of synthesis of an RNA primer that is used to initiate (−) strand synthesis; (3) the initiation; and (4) the termination of (+) strand synthesis. In Fig. 1 the IG, which has the potential to form five hairpin structures, is represented schematically and important regions designated. Most important to the functioning of M13KO7 is the origin of replication of the (+) strand. The origin consists of 140 bp and can be divided into two domains. Domain A, about 40 bp, is essential for replication and contains the recognition sequence for gIIp to create the nick that initiates and terminates replication of the RF. Domain B is about 100 bp long and acts as an enhancer for gIIp to function at domain A. The effect of domain B can be demonstrated by the fact that a disruption or deletion of it will decrease phage yield by 100-fold.[9] Two types of mutants, a qualitative mutation from M13mp1[11] and two quantitative ones from R218 and R325,[12] that compensate for the loss of a functional domain B have been analyzed. The qualitative mutant from mp1, which has an 800-bp insertion within B,

[8] H. Schaller, *Cold Spring Harbor Symp. Quant. Biol.* **45,** 177 (1978).

[9] G. P. Dotto, K. Horiuchi, and N. D. Zinder, *J. Mol. Biol.* **172,** 507 (1984).

[10] G. P. Dotto and N. D. Zinder, *Virology* **130,** 252 (1983).

[11] J. Messing, B. Gronenborn, B. Muller-Hill, and P. H. Hofschneider, *Proc. Natl. Acad. Sci. U.S.A.* **74,** 3642 (1977).

[12] G. P. Dotto and N. D. Zinder, *Proc. Natl. Acad. Sci. U.S.A.* **81,** 1336 (1984).

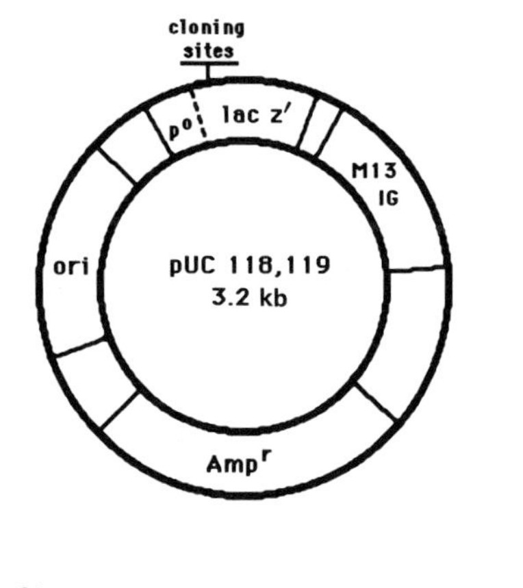

Cloning Sites

pUC 118

Xma I
met-->lac z′ Sst I Sma I Xba I Pst I HindIII
ATGACCATGATTACGAATTCGAGCTCGGTACCCGGGGATCCTCTAGAGTCGACCTGCAGGCATGCAAGCTTGCA
EcoRI Kpn I BamHI Sal I Sph I
Asp718 Acc I
HincII

pUC 119

HincII
Acc I Asp718
met-->lac z′ Sph I Sal I BamHI Kpn I EcoRI
ATGACCATGATTACGCCAAGCTTGCATGCCTGCAGGTCGACTCTAGAGGATCCCCGGGTACCGAGCTCGAATTC
HindIII Pst I Xba I Sma I Sst I
Xma I

FIG. 2. Structure of pUC 118 and 119 and the DNA sequence of the unique restriction enzyme sites within the sequence encoding the *lacZ* peptide.

consists of a single G-to-T substitution that changes a methionine (codon 40) to an isoleucine within the gIIp.[13] This change allows the mp1gIIp to function efficiently enough on an origin consisting of only domain A to give wild-type levels of phage. In R218 and R325 the loss of a functional domain B is compensated for by mutations that cause the overproduction of a normal gIIp at 10-fold normal levels.[12,13] Even though a wild-type gIIp works very poorly on a domain B-deficient origin, the excess level of gIIp achieves enough initiation of replication to give normal levels of phage.

pUC 118 and 119

All ss plasmid DNA vectors carry a phage intergenic region. The entire complement of functions necessary for the packaging of ssDNA

[13] G. P. Dotto, K. Horiuchi, and N. D. Zinder, *Nature* (*London*) **311,** 279 (1984).

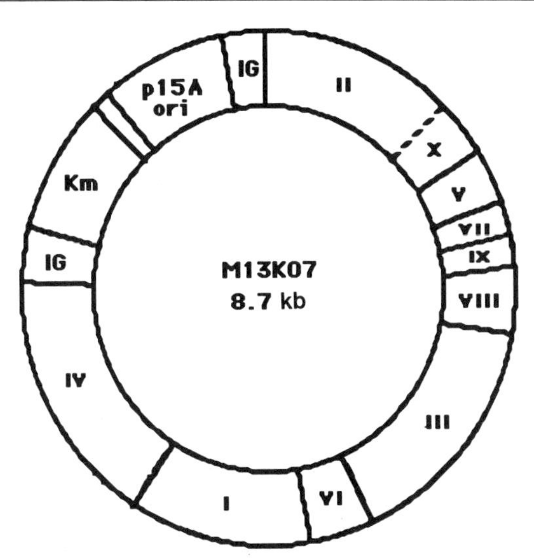

FIG. 3. Structure of M13KO7.

into viral particles will work *in trans* on an IG region. The vectors used in the experiments described here are pUC 118 and 119 (Fig. 2). They are pUC 18 and 19,[14] respectively, with the IG region of M13 from the *Hgi*AI site (5465) to the *Dra*I site (5941) inserted at the unique *Nde*I site (2499) of pUC. The orientation of the M13 IG region is such that the strand of the *lac* region that is packaged as ssDNA is the same as in the M13mp vectors.

M13KO7

M13KO7 (Fig. 3) is an M13 phage that has the gene II of M13mp1 and the insertion of the origin of replication from p15A[15] and the kanamycin-resistance gene from Tn 903[16] at the *Ava*I site (5825) of M13. With the p15A origin, the phage is able to replicate independent of gIIp. This allows the phage to overcome the effects of interference and maintain adequate genome levels for the expression of proteins needed for ssDNA production when it is growing in the presence of a plage. The effect of the addition of the plasmid origin is shown in Fig. 4B. The insertion of the p15A origin and the kanamycin-resistance gene separates the A and B

[14] J. Norrander, T. Kempe, and J. Messing, *Gene* **26,** 101 (1983).

[15] G. Selzer, T. Som, T. Itoh, and J. Tomizawa, *Cell* **32,** 119 (1983).

[16] N. D. F. Grindley and C. M. Joyce, *Proc. Natl. Acad. Sci. U.S.A.* **77,** 7176 (1980).

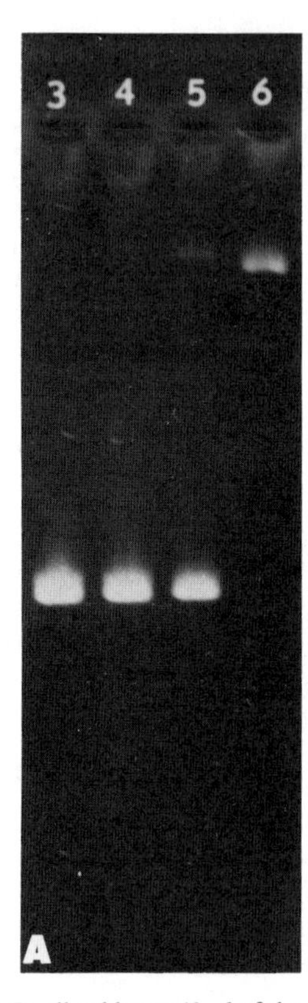

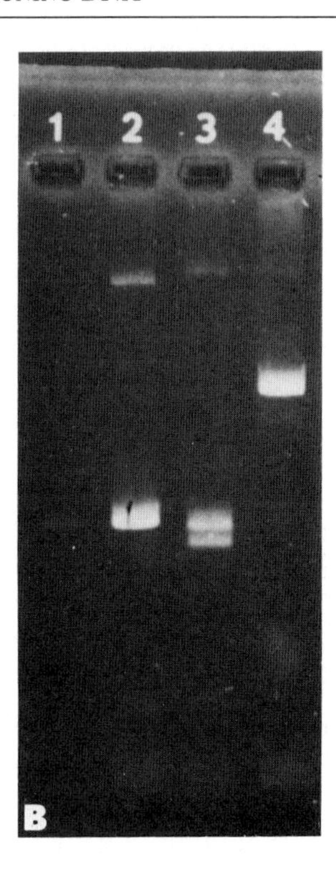

FIG. 4. In all gel lanes 40 μl of the supernatant fraction after centrifugation of the culture was mixed with 6 μl of SDS gel-loading buffer and loaded on the gel. (A) Lane 3: pUC 118 with M13KO7 as helper phage. Plasmid titer is 5×10^{11} cfu/ml, phage titer is 8×10^9 pfu/ml. Lane 4: pUC 119 with M13KO7 as helper phage. Plasmid titer is 6×10^{11} cfu/ml, phage titer is 8×10^9 pfu/ml. Lane 5: pUC 119 with M13KO19 (similar to KO7, but with a deletion of domain B of the phage origin of replication) as helper phage. Lane 6: M13KO7. (B) Lane 1: pUC 119 with an M13mp8 phage carrying the kanamycin gene, but no plasmid origin of replication, as helper phage. Lane 2: pUC 119 with M13KO19 as helper phage. Lane 3: pUC 19 with the M13 IG region in the same location as 119, but in the opposite orientation. Lane 4: pUC 118 with 2.5-kb insert.

domains of the phage origin of replication, creating an origin that is less efficient for the functioning of the mp1 gIIp than the wild-type origin carried by the plage. This, plus the high copy number of pUC, leads to the preferential packaging of plasmid DNA into viral particles. The mp1 gIIp functions well enough on the altered origin when M13KO7 is grown by itself to produce a high titer of phage for use as inoculum for the production of ss plasmid.

Materials and Reagents

Strains

MV1184: *ara*,Δ(*lac–pro*), *strA*, thi, (ϕ80Δ*lacIZ*ΔM15),Δ(*srl–recA*) 306::Tn*10*(tet^r); F′: *traD*36, *proAB*, *lacI*^{q}ZΔm15)

Media

2× YT (per liter): 16 g Difco Bacto tryptone, 10 g Difco Bacto yeast extract, 5 g NaCl, 10 m*M* KPO_4, pH 7.5

2× YT plates: 15 g Difco Bacto agar added to 1 liter of 2× YT

YT soft agar (per liter): 8 g Difco Bacto tryptone, 5 g yeast extract, 5 g NaCl, 7 g agar

M9 plates: For 1 liter of 10× M9 salts: combine 60 g Na_2HPO_4, 30 g KH_2PO_4, 0.5 g NaCl, 10 g NH_4Cl dissolved in H_2O to a final volume of 970 ml and autoclave. After autoclaving add 10 ml of a sterile 1 *M* $MgSO_4$ solution and 20 ml of a sterile 0.05 *M* $CaCl_2$ solution. For 1 liter of plates autoclave 15 g of agar in 890 ml. After autoclaving add 100 ml 10× M9 salts, 10 ml of a 20% glucose solution, and 1 ml of a 1% thiamin solution

Solutions

SDS gel loading buffer: 0.05% bromphenol blue, 0.2 *M* EDTA, pH 8.0, 50% glycerol, 1% SDS

TE buffer: 10 m*M* Tris–HCl, pH 8.0, 1 m*M* EDTA, pH 8.0

Growth of M13KO7

M13KO7 exhibits some instability of the insert during growth, but this does not create a problem if it is propagated correctly. The procedure for the production of M13KO7 is the following. M13KO7 supernatant is streaked on a YT agar plate and then 4 ml of soft agar, to which 0.5 ml of a culture of MV 1184 ($OD_{600} > 0.8$) has been added, is poured across the plate from the dilute side of the streak toward the more concentrated side. After 6–12 hr of incubation at 37° single plaques are picked and grown individually in 2–3 ml of YT containing kanamycin (70 μg/ml) overnight. The cells are then pelleted by centrifugation, and the supernatant is used as inoculum of M13KO7. The phage in the supernatant will remain viable for months when stored at 4°.

Production of ss Plasmid DNA

For the production of ss plasmid DNA it is important that a low-density culture of plage-containing cells, infected with M13KO7, be grown for 14–18 hr with very good aeration. The medium that is used is

2× YT supplemented with 0.001% thiamin, 150 μg/ml ampicillin, and, when appropriate, 70 μg/ml kanamycin. Commonly used methods are the following:

1. A culture of MV1184 (pUC 118/119) in early log phase is infected with M13KO7 at a multiplicity of infection (moi) of 2–10 and incubated at 37° for 1 hr and 15 min. The infection should be carried out on a roller or a shaker at low rpm. After this time the cells are diluted, if necessary, to an $OD_{600} < 0.2$ and kanamycin is added to a final concentration of 70 μg/ml. The culture is then grown for 14–18 hr at 37°. Culture conditions are usually 2–3 ml in an 18-mm culture tube on a roller or 5–10 ml in a 125-ml culture flask on a shaker at 300 rpm. Pellet the cells by centrifugation (8000 *g*, 10 min) and remove the supernatant to a fresh tube. Add one-ninth of the supernatant volume of 40% PEG and of 5 *M* sodium acetate and mix well. Place on ice 30 min and pellet the viral particles by centrifugation (8000 *g*, 10 min) and pour off the supernatant. Remove the remaining supernatant with a sterile cotton swab. Resuspend the pellet in 200 μl TE buffer by vortexing. Add 150 μl of TE-saturated phenol (pH 7) and vortex for 30 sec. Add 50 μl of $CHCl_3$, vortex, and centrifuge for 5 min (Brinkman Eppendorf centrifuge). Remove the aqueous layer to a fresh tube and repeat phenol/$CHCl_3$ extraction. Remove the aqueous layer to a fresh tube and add an equal volume of $CHCl_3$, vortex, and centrifuge for 5 min. Remove the aqueous layer to another tube and add 3 vol of ether. Vortex well and centrifuge briefly. Remove the ether, add one-twentieth the volume of 3 *M* sodium acetate (pH 7), and precipitate the DNA with 2.5 vol of ethanol at −70° for 30 min and then pellet by centrifugation. Once the pellet is dry it can be resuspended in TE and used in the same manner as has been previously described for the use of M13 ssDNA templates.[17]

2. For the screening of plasmid for inserts a colony selected from a plate is added to 2–3 ml of medium containing M13KO7 ($\sim 10^7$/ml) and grown at 37° for a few hours. Kanamycin is then added and the cultures are incubated for 14–18 hr at 37°. The cells are then pelleted and 40 μl of supernatant is mixed with 6 μl of loading buffer and electrophoresed on a 1% agarose gel, stained with ethidium bromide, and viewed with UV illumination.

Discussion

The use of M13KO7 for the production of ss plasmid DNA normally gives titers of cfu of 10^{11}–5×10^{11}/ml and phage titers 10- to 100-fold lower

[17] J. Messing, this series, Vol. 101, p. 20.

(Fig. 4A). Plasmids containing inserts as large as 9 kb have been packaged as ssDNA without a significant loss in yield (M. McMullen and P. Das, personal communication) and instability has not been a problem. It has been observed that some clones, irregardless of size, give reduced levels of ssDNA. This reduction in yield has been both dependent (M. McMullen, personal communication) and independent (J. Braam, personal communication) of the orientation of the insert. M13KO7 has given high yields of ssDNA from pUC-derived vectors, but when it was used as a helper phage with pZ150,[19] a vector constructed from pBR 322, the yield of ssDNA was not significantly different from the yield given by other helper phages. Whether this is due to the lower copy number of pBR as compared to pUC or to some effect of the vector structure is not known. It has been noted that the position and orientation of the IG region within the plasmid can affect its packaging as ssDNA. An example is shown in Fig. 4B (lane 3). This plasmid has the IG region inserted in the same position but the opposite orientation as compared to pUC 119/118, and always gives two bands. However, if the IG region, in the opposite orientation of 118/119, is inserted within the polycloning sites of a pUC vector, the resulting plasmid yields a single band after gel electrophoresis (data not shown). A large variation in the yield of ss plasmid DNA has been seen between different bacterial strains. MV 1184 (derived from JM 83) and MV 1190 (derived from JM 101) have given satisfactory yields. MV 1304 (derived from JM 105) gives much reduced yields and JM 109 undergoes significant lysis when it contains both plasmid and phage.

Acknowledgments

We would like to thank B. McClure, R. Zagursky, M. Berman, and D. Mead for valuable discussions. We also thank M. Volkert for the MV bacterial strains and Claudia Dembinski for help in preparing this manuscript. This work was supported by the Department of Energy, Grant #DE-FG05-85ER13367.

[18] M. Zoller and M. Smith, this series, Vol. 100, p. 468.

[19] R. J. Zagursky and M. L. Berman, *Gene* **27,** 183 (1984).

[2] pKUN, Vectors for the Separate Production of Both DNA Strands of Recombinant Plasmids

By RUUD N. H. KONINGS, ELS J. M. VERHOEVEN, and BEN P. H. PEETERS

Introduction

In the past few years the advent of rapid DNA sequencing,[1] *in vitro* mutagenesis,[2-4] hybridization,[5,6] DNA shuttling,[7] and S1 nuclease mapping[8,9] techniques has been paralleled by the development of cloning vehicles which make it possible to obtain one of the strands of a recombinant DNA molecule in a single-stranded form.[10-20a]

Until recently the only vectors available for this purpose were the genomes of the F-specific filamentous single-stranded (ss) DNA phages M13, f1, or fd.[10-12] The use of these genomes as cloning vectors is due to

[1] F. Sanger, S. Niclen, and A. R. Coulson, *Proc. Natl. Acad. Sci. U.S.A.* **74,** 5463 (1977).

[2] M. J. Zoller and M. Smith, this series, Vol. 100, p. 468.

[3] J. Norrander, T. Kempe, and J. Messing, *Gene* **26,** 101 (1983).

[4] R. M. Myers, L. S. Lerman, and T. Maniatis, *Science* **229,** 242 (1985).

[5] N.-T. Hu and J. Messing, *Gene* **17,** 271 (1982).

[6] F. Thierry and O. Danos, *Nucleic Acids Res.* **10,** 2925 (1982).

[7] S. Artz, D. Holzschu, P. Blum, and R. Shand, *Gene* **26,** 147 (1983).

[8] A. J. Berk and P. A. Sharp, *Cell* **12,** 721 (1977).

[9] J. F. Burke, *Gene* **30,** 63 (1984).

[10] J. Messing, this series, Vol. 101, p. 20.

[11] N. D. Zinder and J. D. Boeke, *Gene* **19,** 1 (1982).

[12] C. Yanisch-Perron, J. Vieira, and J. Messing, *Gene* **33,** 103 (1985).

[13] J. Vieira and J. Messing, this volume [1].

[14] G. P. Dotto, V. Enea, and N. D. Zinder, *Virology* **114,** 463 (1981).

[15] L. Dente, G. Cesareni, and R. Cortese, *Nucleic Acids Res.* **11,** 1645 (1983).

[16] R. J. Zagursky and M. L. Berman, *Gene* **27,** 183 (1984).

[17] D. A. Mead, E. Szczesna-Skapura, and B. Kemper, *Nucleic Acids Res.* **14,** 1103 (1985).

[18] C. Baldari and G. Cesareni, *Gene* **35,** 27 (1985).

[19] K. Geider, C. Hohmeyer, R. Haas, and T. Meyer, *Gene* **33,** 341 (1985).

[20] B. P. H. Peeters, J. G. G. Schoenmakers, and R. N. H. Konings, *Gene* **41,** 39 (1986).

[20a] R. N. H. Konings, B. P. H. Peeters, and R. G. M. Luiten, *Gene* **46,** 269 (1986).

the unique biological properties of these viruses.[21,22] A few of these properties are the following:

1. Infection of *Escherichia coli* by filamentous phages does not result in cell lysis or cell killing; instead the infected cells continue to grow and divide although at a slower rate than uninfected cells.
2. After infection the ss phage genome is replicated via a double-stranded (ds) intermediate (replicative form or RF DNA). This RF DNA, which can be manipulated as if it were a plasmid,[10–12] is eventually replicated asymmetrically, resulting in the biosynthesis of large amounts of progeny ssDNA. Packaging and extrusion of this DNA results in the production of 10^{11} to 10^{12} phage particles/ml of culture medium, thus allowing the easy isolation of copious amounts of (recombinant) ssDNA.
3. Almost certainly because of their unique filamentous morphology there is little constraint on the size of DNA that can be packaged into filamentous particles.[10–12]

Although these properties make filamentous phages very attractive tools for cloning, a number of disadvantages have also been encountered: (1) large inserts cloned in filamentous phage vectors are often unstable[11,12,23]; (2) only one of the (recombinant) DNA strands is synthesized in an ss form and subsequently packaged into phage particles[10–12]; and (3) because of the alteration of the physiology of the host cell after phage infection, a plasmid rather than a phage vector is preferred for functional studies of cloned fragments.

Our studies[24–26a] on the similarities and differences between the replication mechanisms of the filamentous *E. coli* phages M13 and IKe have given some clues as to how their replication properties can be used to advantage in the construction of new cloning vectors, i.e., the pKUN plasmids. These plasmids allow the separate biosynthesis of both DNA strands of a recombinant plasmid in an ss form and thus overcome the drawbacks of the filamentous phage vectors described above.[20,20a]

[21] D. Denhardt, D. Dressler, and D. S. Ray (eds.), "The Single-Stranded DNA Phages." Cold Spring Harbor Lab., Cold Spring Harbor, New York, 1978.

[22] N. D. Zinder and K. Horiuchi, *Microbiol. Rev.* **49,** 101 (1985).

[23] R. Hermann, K. Neugebauer, E. Pirkl, H. Zentgraf, and H. Schaller, *Mol. Gen. Genet.* **177,** 231 (1980).

[24] B. P. H. Peeters, R. Peters, J. G. G. Schoenmakers, and R. N. H. Konings, *J. Mol. Biol.* **181,** 27 (1985).

[25] B. P. H. Peeters, Ph.D. thesis. Univ. of Nijmegen, Nijmegen, The Netherlands, 1985.

[26] B. P. H. Peeters, J. G. G. Schoenmakers, and R. N. H. Konings, *Nucleic Acids Res.* **14,** 5067 (1986).

[26a] B. P. H. Peeters, J. G. G. Schoenmaker, and R. N. H. Konings, *DNA* **6,** 139 (1987).

Before presentation of the properties of these cloning vectors, a short survey of the filamentous phages will be given, because some basic knowledge of their biology, and particularly of their DNA replication mechanism, is a prerequisite for a proper understanding of the versatile characteristics of the pKUN plasmids.

Biology and Replication of Filamentous Phages

Filamentous Phages

Filamentous phages consist of a circular, covalently closed ssDNA genome encapsulated in a long slender protein coat which consists of at least two but at most of five different subunits.[21,24,27] One of the smallest subunits (major coat protein, M_r ~5000) is present in the virion in about 3000 copies whereas of the largest subunit (M_r ~45,000) about 5 copies are present. For adsorption and penetration, the filamentous phages are dependent on the presence of specific pili at the surface of the host cell.[21,28–30] These pili are generally encoded by conjugative plasmids. Following attachment to the tip of the pilus the phage genome is brought into the host cell by a mechanism that is not understood. After replication of the phage genome, the progeny virions are assembled concomitantly with extrusion of the virion through the inner and outer cell membrane. Because infected cells continue to grow and divide at a reduced rate, cells can be infected or transformed to yield either turbid plaques or recombinant phage-producing colonies.

On the basis of their host and/or pilus specificity filamentous phages can be divided into different classes.[21,28–30] The best studied filamentous phages are those which have *E. coli* as host.[21,22,24,25] Genetic studies as well as nucleotide sequence analyses have demonstrated that the F plasmid-specific phages, i.e., M13, f1, and fd, are almost identical and thus can be considered as natural variants of the same phage,[31–33] in this chapter further called Ff. The *E. coli* phages with different plasmid specificity

[27] R. G. M. Luiten, J. G. G. Schoenmakers, and R. N. H. Konings, *Nucleic Acids Res.* **11,** 8073 (1983).

[28] V. A. Stanisich, *J. Gen. Microbiol.* **84,** 332 (1974).

[29] D. E. Bradley, *Plasmid* **2,** 632 (1979).

[30] D. E. Bradley, J. N. Coetzee, and R. W. Hedges, *J. Bacteriol.* **154,** 505 (1983).

[31] E. Beck, R. Sommer, E. A. Auerswald, C. Kurz, B. Zink, G. Osterburg, H. Schaller, K. Sugimoto, H. Sugisaki, T. Okamoto, and M. Takanami, *Nucleic Acids Res.* **5,** 4495 (1978).

[32] P. M. G. F. van Wezenbeek, T. J. M. Hulsebos, and J. G. G. Schoenmakers, *Gene* **11,** 229 (1980).

[33] D. F. Hill and G. P. Petersen, *J. Virol.* **44,** 32 (1982).

are, however, less homologous.[24,34,35] For example the genome of bacteriophage IKe, a phage specific for the broad-host-range plasmids of the N-incompatibility group (IncN),[36] is only 55% homologous to that of Ff; both genomes have, however, an identical gene order (Fig. 1A).[24]

The genomes of Ff and IKe contain 10 genes which are functionally clustered (Fig. 1A). One cluster consists of genes VII, IX, VIII, III, and VI, which code for structural phage proteins.[21,22,24,37–40] Another cluster (genes I and IV) encodes proteins involved in phage morphogenesis,[21,41] whereas a third cluster (genes II, X, and V) specifies proteins important for DNA replication.[21,22,24–26a,42–44] Besides these gene clusters, the filamentous genome contains a relatively large intergenic region (IR) in which cis-acting DNA elements, involved in DNA replication and phage morphogenesis, are located (Fig. 1B).[21,22,24–26a,45–47] As one moves from gene IV to gene II, one first meets a sequence required for phage morphogenesis, which overlaps a *rho*-dependent transcription termination signal. Then follows a sequence [complementary strand or (−) origin] required for the conversion of the viral strands into dsDNA, which in turn is followed by a sequence [viral strand or (+) origin] required for the asymmetric synthesis of the viral strands.

Filamentous Phage DNA Replication

After penetration of the host the dismantled viral strand is replicated in three stages (Fig. 1C):

1. First the parental DNA strand is converted into a ds replicative form (ss to RF IV). This complementary (minus) strand synthesis is

[34] R. G. M. Luiten and R. N. H. Konings, unpublished results.

[35] D. F. Hill, personal communication.

[36] H. Kathoon, R. V. Iyer, and V. Iyer, *Virology* **48,** 145 (1972).

[37] C. A. van den Hondel, A. Weyers, R. N. H. Konings, and J. G. G. Schoenmakers, *Eur. J. Biochem.* **53,** 559 (1975).

[38] G. F. M. Simons, G. H. Veeneman, R. N. H. Konings, J. H. van Boom, and J. G. G. Schoenmakers, *Nucleic Acids Res.* **10,** 821 (1982).

[39] T. C. Lin, R. E. Webster, and W. Konigsberg, *J. Biol. Chem.* **255,** 10331 (1980).

[40] G. F. M. Simons, R. N. H. Konings, and J. G. G. Schoenmakers, *Proc. Natl. Acad. Sci. U.S.A.* **78,** 4194 (1981).

[41] R. E. Webster and J. Lopez, *in* "Virus Structure and Assembly" (S. Casjens, ed.). Jones and Bartlett, Boston, 1985.

[42] D. Pratt, H. Tzagoloff, and W. S. Erdahl, *Virology* **30,** 397 (1966).

[43] D. Pratt and W. S. Erdahl, *J. Mol. Biol.* **37,** 181 (1968).

[44] W. Fulford and P. Model, *J. Mol. Biol.* **178,** 137 (1984).

[45] W. Wickner, D. Brutlag, R. Schekman, and A. Kornberg, *Proc. Natl. Acad. Sci. U.S.A.* **69,** 965 (1972).

[46] G. P. Dotto and N. D. Zinder, *Virology* **130,** 252 (1983).

[47] R. A. Grant and R. E. Webster, *Virology* **133,** 329 (1984).

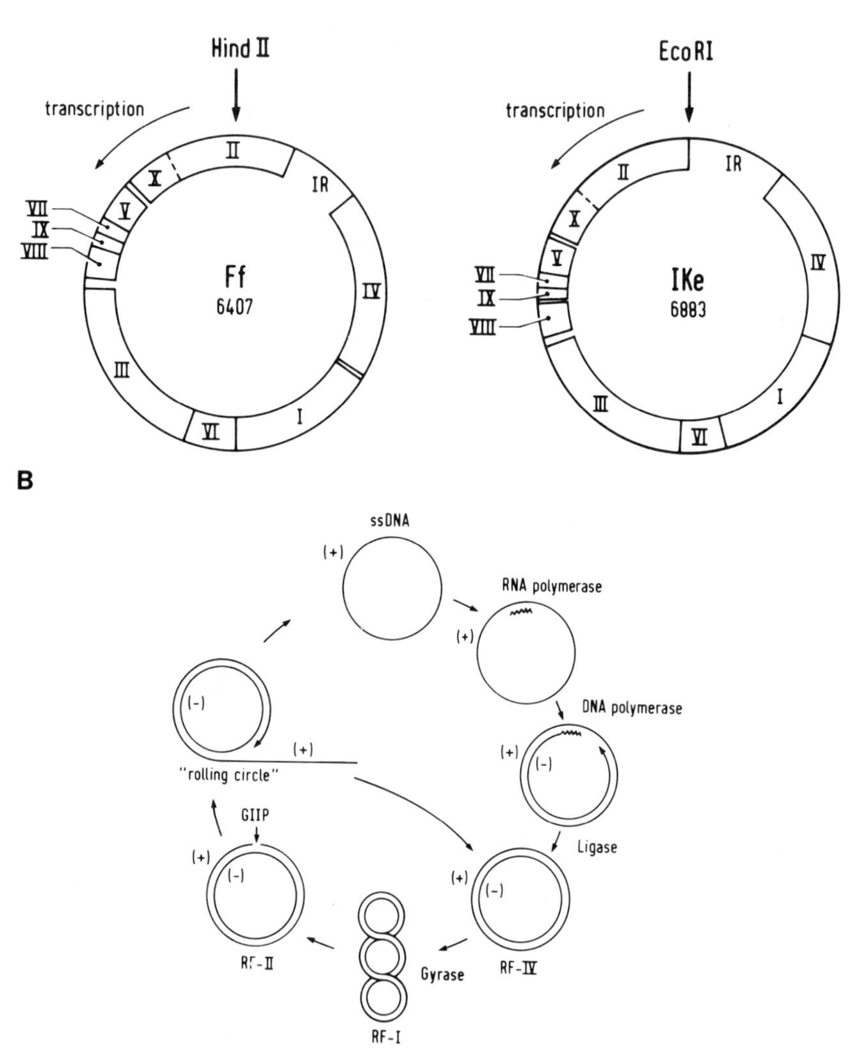

FIG. 1. (A) Circular genetic maps of the genomes of the bacteriophages Ff and IKe. Genes are indicated by Roman numerals and the direction of transcription (i.e., 5′–3′ polarity of the viral strand) is indicated. Note that gene X is located within the 3′-terminal region of gene II and that the 3′-terminal end of gene I overlaps in Ff the 5′-terminal end of gene IV. (B) Mechanism of replication of the single-stranded DNA genome of the filamentous bacteriophages Ff and IKe. For explanation see text. (C) Schematic representation of the location of the morphogenetic signal (M) and the complementary (−) and viral strand (+) replication origins in the intergenic region (IR) of the genomes of the filamentous phages Ff and IKe. The two IR's are drawn to scale.

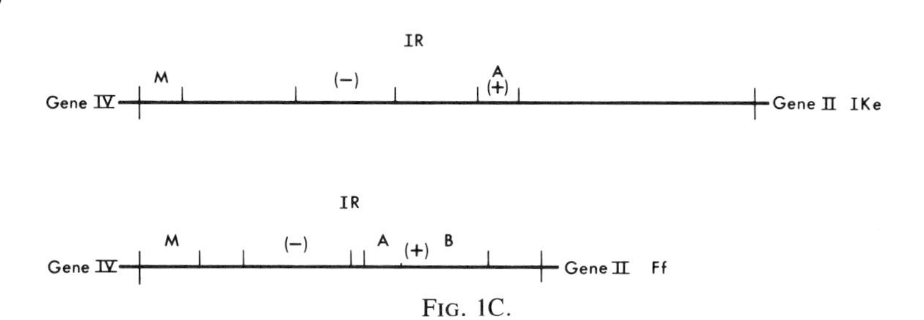

FIG. 1C.

entirely dependent on host-encoded functions and is initiated at the (−) origin present in the IR (Fig. 1B).[45,48,49]

2. After conversion of the parental RF into a supercoil by DNA gyrase (RF IV to RF I),[50] it is replicated according to a rolling circle mechanism,[51] thereby yielding a pool of about 100 progeny RF molecules (RF to RF). Besides host-encoded functions the phage-encoded gene II protein is absolutely required for this process.[21,22,25–26a,52–56] This protein is a site-specific topoisomerase which introduces a nick in the replication origin of the viral strand [(+) origin; Fig. 1B] of RF I, thereby creating a free 3′-OH end which serves as a primer for the further DNA replication. Gene II protein is also involved in the termination of viral strand replication.[22,26a,54–57] After one round of replication gene II protein again cleaves the displaced viral strand at exactly the same position and seals the resulting molecules, yielding a covalently closed ss viral DNA and a ds RF IV molecule, both of which can undergo the replication processes described above.

3. Late in infection when sufficient gene V protein molecules have accumulated, phage DNA synthesis becomes highly asymmetric, producing almost exclusively viral strands which eventually are incorporated

[48] K. Geider, E. Beck, and H. Schaller, *Proc. Natl. Acad. Sci. U.S.A.* **75,** 645 (1978).

[49] C. P. Gray, R. Sommer, C. Polke, E. Beck, and H. Schaller, *Proc. Natl. Acad. Sci. U.S.A.* **75,** 50 (1978).

[50] K. Horiuchi, J. V. Ravetch, and N. D. Zinder, *Cold Spring Harbor Symp. Quant. Biol.* **43,** 389 (1979).

[51] W. Gilbert and D. Dressler, *Cold Spring Harbor Symp. Quant. Biol.* **33,** 437 (1968).

[52] T. F. Meyer and K. Geider, *J. Biol. Chem.* **254,** 12642 (1979).

[53] T. F. Meyer, K. Geider, C. Kurz, and H. Schaller, *Nature (London)* **278,** 365 (1979).

[54] K. Horiuchi, *Proc. Natl. Acad. Sci. U.S.A.* **77,** 5226 (1980).

[55] G. P. Dotto, V. Enea, and N. D. Zinder, *Proc. Natl. Acad. Sci. U.S.A.* **78,** 5421 (1981).

[56] G. P. Dotto, K. Horiuchi, and N. D. Zinder, *Proc. Natl. Acad. Sci. U.S.A.* **79,** 7122 (1982).

[57] G. P. Dotto, K. Horiuchi, K. S. Jakes, and N. D. Zinder, *J. Mol. Biol.* **162,** 335 (1982).

into mature filamentous particles (RF to SS).[21,22,58,59] Gene V protein is a phage-encoded ssDNA binding protein that, by binding to the viral strands, prevents the synthesis of complementary strands.

After formation, the rod-shaped nucleoprotein complex of gene V protein and viral DNA moves to the host cell membrane where, concomitant with the substitution of gene V protein by the coat proteins, extrusion of the virus particle takes place. For efficient packaging of the ssDNA molecules a specific nucleotide sequence (morphogenetic signal) located in the IR immediately distal to gene IV (Fig. 1B) is required.[14,22,25–26a,46,47] To act this morphogenetic signal must have the same orientation as the viral strand (+) origin but need not contiguous with it.

The Viral Strand Origins of Ff and IKe

The viral strand or (+) origin of Ff consists of two domains (A and B; Fig. 1C),[14,22,58,60–64] whereas that of IKe consists of only one domain (A),[24–26a] whose nucleotide sequence strongly resembles that of domain A of Ff.[23] Domain A of Ff and IKe is about 45 nucleotides long. It can be subdivided into three distinct but partially overlapping sequences: a sequence required for nicking of the viral strand by gene II protein, a sequence required for initiation, and a sequence required for termination of viral strand synthesis.

Domain B, which is located in Ff immediately distal to domain A, is about 100 nucleotides long. Its function is to increase, according to a mechanism still unknown, the efficiency of viral strand replication. Domain B thus is not absolutely required but rather facilitates the initiation of viral strand replication.

Additional functional differences between the (+) origins of IKe and Ff are located in domain A.[24–26a] In particular we have observed that the nucleotide sequence which is responsible for initiation of viral strand replication, and which is located at the 3′-side of the gene II protein cleavage site, is highly phage specific. This means that this sequence is only recognized by its cognate gene II protein and, consequently, that the domains A of IKe and Ff, and *mutatis mutandi* their gene II proteins, are not interchangeable.

[58] B. J. Mazur and P. Model, *J. Mol. Biol.* **78,** 285 (1973).
[59] N. J. Mazur and N. D. Zinder, *Virology* **68,** 490 (1975).
[60] S. Johnston and D. S. Ray, *J. Mol. Biol.* **177,** 685 (1984).
[61] M. H. Kim, J. C. Hines, and D. S. Ray, *Proc. Natl. Acad. Sci. U.S.A.* **78,** 6784 (1981).
[62] G. P. Dotto and N. D. Zinder, *Nature (London)* **311,** 279 (1984).
[63] G. P. Dotto and N. D. Zinder, *J. Mol. Biol.* **172,** 507 (1984).
[64] J. M. Cleary and D. S. Ray, *Proc. Natl. Acad. Sci. U.S.A.* **77,** 4638 (1980).

Principle of the Method

Plasmids for the Production of ssDNA

The unique replication process of the filamentous phages M13 and IKe can also be exploited for the production of ssDNA of (recombinant) plasmids. Cloning of the viral strand (+) origin plus morphogenetic signal of either Ff or IKc into a plasmid diverts the plasmids upon superinfection to the Ff or IKe mode of replication.[13–20a,25–26a] Because the superinfecting phage supplies *in trans* the gene products for asymmetric DNA synthesis, as well as the proteins required for phage assembly and extrusion, both filamentous phages and filamentous particles containing ss plasmid DNA will bud from the cell. These particles can easily be concentrated and purified (see Materials, Reagents, and Procedures) and used, for example, for sequence analysis,[1,10,13] mutagenesis,[2,4,10] DNA recombination studies,[65] DNA shuttling experiments,[7,18] and S1 nuclease mapping.[8–10] For most experiments the genome of the helper phage does not have to be purified away. However, when necessary, preparative agarose gel electrophoresis can be used to separate the two classes of DNA.

It is of paramount importance to realize that as a result of asymmetric DNA replication, only the DNA strand of the plasmid on which the cleavage site for the cognate gene II protein is located will eventually be incorporated into phagelike particles. Separate packaging of both plasmid strands is possible, however, if the same plasmid carries, in opposite orientation, the viral strand replication origin plus morphogenetic signal of both IKe and Ff.[20,25–26a] A vector in which these properties are incorporated is plasmid pKUN.[20,20a,26a]

Construction of pKUN9 and pKUN19

In the pKUN vectors, which are derivatives of the pUC plasmids,[10,13,66] the following five properties are combined (Fig. 2A):

1. The ColE1 replication origin which, in the absence of helper phage, enables the vector to replicate as a high copy number plasmid
2. The *bla* gene encoding β-lactamase, which confers ampicillin resistance to cells harboring these plasmids and which thus can be used as a selectable marker for transformation
3. A fragment of the *E. coli lac* operon containing the regulatory region and a short fragment encoding the first 77 amino acids (α-peptide)

[65] R. H. Hoes and K. Abrenski, *J. Mol. Biol.* **181,** 351 (1985).
[66] J. Vieira and J. Messing, *Gene* **19,** 259 (1982).

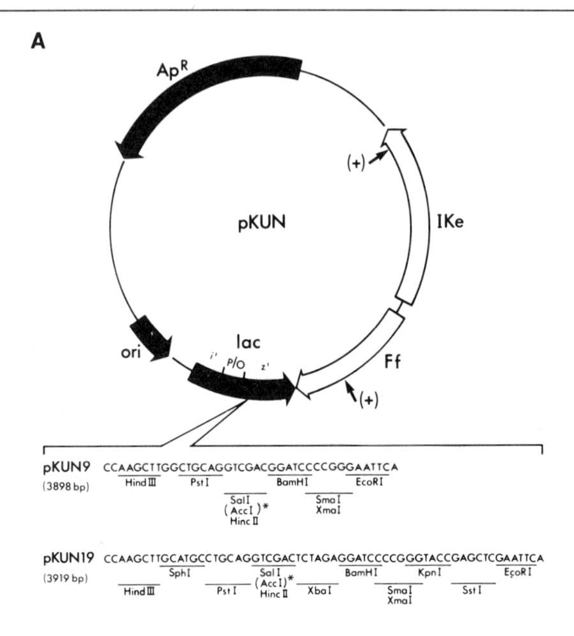

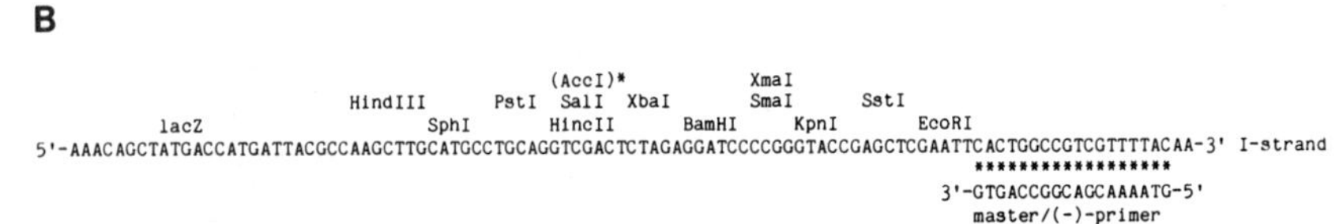

FIG. 2. (A) Genetic organization of plasmid pKUN. pKUN contains the morphogenetic signals plus the viral and complementary strand origins of IKe and Ff in opposite orientations. The nucleotide sequence of the multiple cloning sites linker present in the *lacZ*-gene fragment is given. The size (in base pairs) of pKUN9 and pKUN19 is indicated. *, *Acc*I cleavage site not unique. (B) Nucleotide sequence of the 5′-terminal end of the *lacZ′* gene present in plasmid pKUN19 showing the multiple cloning sites and the position of the master and reverse primer. The I and F strands are defined as the DNA strands which are packaged under instruction of the helper bacteriophages IKe (IKe-9, *Mike,* and *Mike*Δ) and Ff (IR1, R4O8, and M13KO7), respectively.

of β-galactosidase and containing a multiple-cloning sites linker.[10,13,66,67] The *lacZ* fragment confers to the plasmid a selectable marker for cloning. The α-peptide complements a defective β-galactosidase gene present on the F′-plasmid of the host cell (*E. coli* JM101 or JM101[pCU53]; see below).[10,12,13] This complementation gives rise to blue colonies when the

[67] J. Messing, R. Crea, and P. Seeburg, *Nucleic Acids Res.* **9,** 309 (1981).

cells are plated in the presence of the inducer IPTG and the chromogenic substrate X gal (see Materials, Reagents, and Procedures). Insertion of DNA into the polylinker region of the *lacZ* gene destroys the complementation ability and thus plasmids that contain inserts give rise to white (colorless) colonies when grown in the presence of IPTG and Xgal[10,12,13]

4. A fragment containing the morphogenetic signal and the viral (+) and complementary (−) strand replication origins of Ff (nucleotides 5488–6001 of f1)[33] cloned in the unique *Nar*I site of pUC9 or pUC19

5. A fragment containing the morphogenetic signal and the viral (+) and complementary (−) strand replication origins of IKe (nucleotides 5921–6621)[24] inserted in the unique *Nde*I site of pUC9, or pUC19, in an orientation opposite to that of Ff

Thus upon superinfection of cells, harboring (recombinant) pKUN plasmids, with IKe, the strand containing the recognition sequence for IKe gene II protein (the I strand in Fig. 2B) will be packaged, while the complementary strand (F strand and containing the recognition sequence for Ff gene II protein; Fig. 2B) will be packaged into phagelike particles upon superinfection with Ff.

One should realize that, because of the complementarity of the strands packaged by IKe and Ff, different primers must be used for sequence analysis, hybridization studies, or site-directed mutagenesis of the DNA inserts. The ssDNA packaged under the direction of IKe should be sequenced with the aid of the master primer; on the other hand, for sequence analysis of the ssDNA packaged under the direction of Ff, the reverse primer should be used (Fig. 2B) (see Materials, Reagents, and Procedures). Plasmids pKUN9 and pKUN19 contain the same multiple cloning sites as are present in the *lacZ* gene of plasmid pUC9 and pUC19, respectively.[10,12,13] The unique restriction enzyme cleavage sites present in pKUN9 and pKUN19 are shown in Fig. 2. The complete nucleotide sequence of pKUN has been compiled from the known sequences of pUC19,[12] Ff,[32,33] and IKe.[24] Sequence fusions generated during the construction of pKUN have been verified by nucleotide sequence analysis. The nucleotide sequences of pKUN9 and pKUN19 are available upon request.

Helper Phages for the Production of ss Plasmid DNA

Plating of wild-type (wt) Ff or IKe on *E. coli* strains harboring plasmids on which their cognate viral strand replication origin is located (such as pKUN) leads to superinfection interference.[14,22,25–26a,60,68] This means that the plating efficiency and phage yield is lower than when the plasmid

[68] V. Enea and N. D. Zinder, *Virology* **122,** 222 (1982).

does not carry a viral strand origin. This interference is most likely the result of competition for the phage-specific replication proteins between the viral strand origins on the multicopy plasmid and on the superinfecting helper phage genome. Higher plating efficiencies and phage yields are obtained when interference-resistant mutants of Ff and IKe are used.[14,20,20a,22,25,68] These mutants can be selected for by serial passage of the wt phage through cells that carry recombinant plasmids with a wt viral strand origin. They grow as well as wt phages in a strain not harboring a plasmid, but differ both in the level and the ratio of production of ss plasmid DNA to ss phage DNA.

For the selective packaging of the I strand of pKUN (Fig. 2b), either one of the phages IKe-9,[20] *Mike*,[20a] or *Mike*Δ[69] (see below) can be used. On the contrary, for the synthesis and packaging of the F strand, either one of the Ff phages IR1,[68] M13KO7,[13,70] or R4O8[70a] should be used. The efficiency of packaging of the plasmid DNA strands by either IKe-9, *Mike,* or IR1 equals that of the phage genome.[20,20a] On the other hand, using the helper phages, *Mike*Δ, M13KO7, or R4O8, the efficiency of packaging of the (recombinant) plasmid strand is generally significantly higher than that of the phage genome[13,69–70a] (see Examples of Packaging and Rapid Dideoxy Sequencing).

The bacteriophages *Mike* and *Mike*Δ are interference-resistant chimeric phages that have been constructed in our laboratory from the genomes of Ff and IKe.[20a,69] Their genomes consist of the replication functions of IKe (i.e., genes II, X, and the viral strand origin) and of the genes V, VII, IX, VIII, III, VI, I, and IV and of the complementary strand origin and morphogenetic signal of Ff. As a result of this construction, these phages can only infect cells which contain a F or a F′-plasmid. They replicate, however, according to the replication mode of IKe. *Mike*Δ is a derivative of *Mike* that contains a deletion in the nucleotide sequence required for phage morphogenesis. Its chimeric genome has been constructed via *in vitro* recombination of the largest *Ban*I fragment of RF of the Ff phage R4O8[70a] and the smallest *Ban*I fragment of *Mike* RF.[69] The major advantage of the introduction of this deletion in the morphogenetic signal of the filamentous genome is that these phages generally package and secrete the plasmid strand at a higher efficiency than their own ss genome.[70a]

[69] R. N. H. Konings, unpublished results.

[70] M13KO7 is an M13 phage with a missense mutation (*met* → *ile*) at position 40 in the gene II protein. It also contains the origin of replication of p15A and the gene for kanamycin resistance from Tn*903*.[13]

[70a] M. Russel, S. Kidd, and M. R. Kelley, *Gene* **45,** 333 (1986).

Host Strains and Biological Containment

The NIH guidelines currently place filamentous phage vectors and plasmids, which are specifically designed for the production of ss plasmid DNA via helper phage infection, in the exempt category (EKO). When performing other than self-cloning experiments (i.e., cloning of *E. coli* DNA) certain *E. coli* strains deficient in conjugation must be used as host. For Ff cloning vectors these requirements are met by using *E. coli* strains that contain F-plasmids defective in transfer (*traD* and/or *traI*).[10-12] Thus, although these plasmids make the host conjugation deficient they still confer to the host phage sensitivity. For IKe this requirement now is met either by the use of *E. coli* strains which contain conjugation-deficient plasmids of the N-incompatibility group, e.g., pCU53,[71] or by using the chimeric phages *Mike* or *Mike*Δ as a helper.

Plasmid pCU53 (ca. 20 kb) has been constructed by cloning of the largest *Bgl*II fragment of the wt IncN plasmid pCU1 in the unique *Bam*HI site of plasmid pACYC184.[72] Except for being *tra*$^-$, pCU53 has the additional advantage over the wt IncN plasmids that it gives rise to a higher number of N-pili at the surface of the host cell.[73] Because the conjugative plasmids F and pCU53 (or plasmid N3, which we previously used[20]) are members of different incompatibility groups,[74] by definition these plasmids can coexist in the same host, thus rendering the cell sensitive to both Ff and IKe.

Materials, Reagents, and Procedures

Preliminary Remarks

All reagents are made from sterilized stock solutions. All glassware and related equipment are either heated at 200° for 2 hr, or autoclaved. All manipulations are carried out at low temperatures (0–4°) unless otherwise specified. All *in vitro* reactions are carried out in sterile plastic tubes (Eppendorf), and all solutions are handled with sterile pipets and/or sterile pipet tips (Oxford Laboratories). Dialysis tubing is boiled in 5% sodium bicarbonate containing 0.1 *M* EDTA and washed in distilled water. The pH of all buffers is measured at 20°.

[71] R. N. H. Konings and E. Verhoeven, unpublished results.

[72] V. Thatte and V. N. Iyer, *Gene* **21,** 227 (1983).

[73] V. Thatte, D. E. Bradley, and V. N. Iyer, *J. Bacteriol.* **163,** 1229 (1985).

[74] A. I. Bukhari, J. A. Shapiro, and S. L. Adhya, *in* "DNA," p. 601. Cold Spring Harbor Lab., Cold Spring Harbor, New York, 1977.

Bacteria and Bacteriophages

Each of the bacteriophages IR1, M13KO7, R4O8, IKe-9, *Mike*, or *Mike*Δ can be used as a helper for the selective synthesis and secretion of the (recombinant) DNA strands of plasmid pKUN9 or pKUN19 (Fig. 2A; see Principle of the Method). The use of M13KO7 as a helper has the advantage that infected cells can be selected with the aid of kanamycin.[13] For the preparation of high-titer phage stocks, *Mike* and *Mike*Δ and the Ff bacteriophages IR1, R4O8, or M13KO7 are propagated on *E. coli* JM101 (*supE,thi,*Δ(*lacproAB*)[F′,*traD*36,*proAB,lacI*^{q}*Z*ΔM15]), whereas IKe-9 is propagated on *E. coli* JE2571 (*fla,gal,lac,leu,mal,str,thr,xyl*)[pCU53, *tra, cam*]. For the separate packaging of the strands of plasmid pKUN9 or pKUN19, we have constructed an *E. coli* strain which is permissive for all helper phages i.e., JM101 [pCU53,*tra,cam*][71] (see Principle of the Method). Because the phages *Mike*, *Mike*Δ, IR1, R4O8, and M13KO7 are F-plasmid-specific phages, *E. coli* JM101 can be used as host for these helper phages as well. To assure retention of plasmid pCU53, cells should be maintained on plates containing chloramphenicol. Maintenance of the F′ plasmid in JM101 is based on its ability to confer proline prototrophy. Therefore JM101 and JM101[pCU53] should be streaked out on minimal glucose plates (see below). Always use colonies from these plates to grow overnight cultures for transformation. All strains can be kept frozen at −70° after the addition of glycerol to 15% (v/v).

Enzymes

Enzymes can be purchased from Boehringer, Pharmacia, New England BioLabs, or Bethesda Research Laboratories, and should be used as recommended by the supplier.

Media, Nutritional Supplements, and Buffers

Use distilled water for media and buffers (dH_2O) and double-distilled water (ddH_2O) for enzyme reactions.

All biochemicals can be purchased from Merck (Darmstadt, FRG) unless stated otherwise.

Minimal glucose plates: Mix after autoclaving (20 min at 15 lb): 100 ml of 10× M9 salts (Na_2HPO_4, 60 g; KH_2PO_4, 30 g; NaCl, 5 g; NH_4Cl, 10 g; per liter of dH_2O), 1 ml of 1 *M* $MgSO_4 \cdot 7H_2O$, 10 ml of 20% (w/v) glucose, 1 ml of 1% (w/v) thiamin (BDH; sterilized by filtration), 10 ml of 0.01 *M* $CaCl_2$, and 900 ml of 2% (w/v) agar (Difco), cooled to 60°. When appropriate add antibiotics (see below) before pouring the plates

2× YT medium: Bacto tryptone (Difco), 16 g; yeast extract (Difco), 10 g; NaCl, 5 g; per liter of dH_2O. Autoclave for 20 min at 15 lb

2× YT agar: Same as 2× YT medium but including 15 g agar/liter. After autoclaving cool to 55° in a water bath and add, when appropriate, 1 ml of Xgal, 1 ml of IPTG, and antibiotics (see below) before pouring the plates

CW agar: Casamino acids (Difco), 30 g; glycerol, 20 g; yeast extract (Difco), 1 g; $MgSO_4 \cdot 7H_2O$, 1 g; agar, 10 g (Difco); per liter of dH_2O. Autoclave for 20 min at 15 lb

TE: 10 m*M* Tris–HCl (pH 8.0), 1 m*M* EDTA

NaOAc: 3 *M* sodium acetate adjusted to pH 6.0 with glacial acetic acid

IPTG: 100 m*M* solution of isopropylthiogalactoside (Sigma) in dH_2O. Sterilize by filtration through a 0.22-μm Millipore filter. Store at −20°

Xgal: 4% solution of 5-dibromo-4-chloro-3-indolylgalactoside (Sigma) in dimethylformamide. Store in the dark at −20°

Antibiotics

- Ampicillin (Ap, Serva), 100 mg/ml stock solution in dH_2O; sterilize by filtration through 0.22-μm Millipore filter
- Chloramphenicol (Cm, Sigma), 30 mg/ml stock solution in 70% ethanol
- Kanamycin (Km, Boehringer), 10 mg/ml stock solution in dH_2O; sterilize by filtration

Working concentrations of the respective antibiotics are 100, 30, and 70 μg/ml in rich media, and 50, 15, and 35 μg/ml in minimal media and media for transformation experiments

Phenol: To 1 kg of freshly distilled phenol add 8-hydroxyquinoline (Sigma) to a final concentration of 0.1% (w/v), extract once with an equal volume of unbuffered 1 *M* Tris, followed by two extractions with an equal volume of TE, containing 0.2% (v/v) 2-mercaptoethanol. Store working solution under equilibration buffer at 4° in the dark, and the rest in the dark at −20°

Phenol/chloroform: Mix equal volumes of TE-saturated phenol and a 24 : 1 (v/v) mixture of chloroform and isoamyl alcohol. Mix well and store at 4° in the dark

PEG/NH_4OAc: 3.5 *M* NH_4OAc and 20% (w/v) PEG 6000 (BDH) in double-distilled H_2O

$CaCl_2$: For the preparation of competent *E. coli* cells use a 50 m*M* solution of $CaCl_2$, prepared freshly before use and sterilized by filtration through a 0.22-μm Millipore filter

Ethidium bromide: 5 mg/ml stock solution in dH_2O. Keep in the dark at 4°. Avoid skin contact or inhalation of powder; treat as a mutagen. Caution: in the presence of light ethidium bromide introduces nicks into DNA

Agarose (BRL): 0.8–1% (w/v) solution in TBE (see below). Dissolve by autoclaving at 15 lb for 20 min

10× TBE: Tris, 121 g; boric acid, 61.8 g; EDTA, 9.3 g; per liter of dH_2O

2-Propanol: Add 2-propanol to a saturated solution of CsCl in TE. Mix well and use the upper 2-propanol phase for removal of ethidium bromide from DNA

Phage dilution buffer: KH_2PO_4, 3 g; Na_2HPO_4, 7 g; NaCl, 5 g; per liter of dH_2O. Sterilize by autoclaving for 20 min at 15 lb

RNase A: 10 mg/ml of pancreatic ribonuclease A (BDH) in TE. Heat at 90° for 10 min, to inactivate residual DNase activity

Loading buffer: 0.5% (w/v) bromphenol blue (BDH), 0.2 *M* EDTA (pH 8.3) and 50% (v/v) glycerol

Sequencing and hybridization primers: For sequence analysis and for the preparation of hybridization probes two deoxynucleotide primers are used.

5′-dG-T-A-A-A-A-C-G-A-C-G-G-C-C-A-G-T-G-3′ (master primer)
5′-dA-A-C-A-G-C-T-A-T-G-A-C-C-A-T-3′ (reverse primer)

They are complementary to the I and F strands of pKUN, respectively (see Fig. 2B). Primers may be obtained from one of the following companies: Pharmacia, P-L Biochemicals, Bethesda Research Laboratories, Amersham, or Promega Biotec.

Preparation of High-Titer Phage Stocks

To obtain good yields of ss (recombinant) plasmid DNA it is important to infect at a multiplicity of infection (moi) of 10 to 20. This ensures that all cells are infected. It is therefore necessary to have a high-titer phage stock at the time of infection. The following procedure is used for the preparation of high-titer phage stocks required for the separate packaging of the (recombinant) DNA strands of plasmid pKUN harbored in *E. coli* JM101[pCU53] (or JM101, if for packaging either one of the phages *Mike*, *Mike*Δ, IR1, R4O8, or M13KO7 is used). The procedure is given for a 1-liter culture.

Grow *E. coli* JM101 (for *Mike*, *Mike*Δ, IR1, R4O8, or M13KO7) or *E. coli* JE2571[pCU53] (for IKe-9) at 37° in 50 ml of well-aerated 2× YT medium until a density of approximately 10^7 cells/ml is reached. Add one single plaque (see below) of either one of the bacteriophages to the appropriate culture and continue incubation for another 6 hr. Especially in case of IKe-9 and M13KO7 it is advisable to infect the culture with a fresh plaque and not with phages from a previously prepared phage stock.

The rationale behind this procedure is that, for reasons still unknown, during serial propagation of IKe-9 and M13KO7 deletion mutants are generated much faster than during serial propagation of the other phages.

Centrifuge the culture for 10 min at 10,000 rpm at 4° in the Beckmann JA-14 rotor (or its equivalent). Pour the supernatant in a sterile flask and store at 4°. The phage titer of this culture supernatant (see below) should be around 10^{11} pfu/ml. The next day inoculate 1 liter of 2× YT medium with 10 ml of a fresh overnight culture of JM101 or JE2571/[pCU53] and grow at 37° until a density of 10^8 cells/ml is reached. Infect the culture by adding the supernatant of the first culture and leave for 10 min without shaking. In the case of M13KO7 infection, infected cells are selected by the addition of kanamycin (70 μg/ml) to the culture medium 30 min after phage infection.

Subsequently, shaking is continued under good aeration for another 6 hr (or overnight) at 37°. After that time centrifuge the culture for 10 min at 8000 rpm in the Beckman JA-10 rotor (or its equivalent) and pour the supernatant into a sterile flask. Precipitate the phage by the addition of solid NaCl and PEG 6000 (Serva) to a final concentration of 0.5 *M* and 4% (w/v), respectively. Mix well until all PEG 6000 is dissolved and leave at 4° overnight. Recover the precipitate by centrifugation for 20 min at 8000 rpm in the Beckman JA-10 rotor (or its equivalent) and pour off the supernatant. Resuspend the pellet thoroughly in 20 ml of TE containing 0.1% (w/v) Sarcosyl (Ciba-Geigy). Again precipitate the phage by the addition of 5 ml of PEG/NH_4OAc solution. Mix well and leave at room temperature for 30 min. Recover the precipitate by centrifugation at 20,000 rpm for 15 min in a Beckman JA-20 rotor (or its equivalent) and carefully remove the supernatant.

Dissolve the pellet in 1.5 ml of TE, transfer the suspension to an Eppendorf tube, heat for 10 min at 60° to kill the remaining bacteria, and store at 4°. At this stage the isolated phages can already be used for packaging. If desired, a further purification of the phages can be achieved by cesium chloride density gradient centrifugation.[75] The phage titer (plaque forming units, pfu) is established by pipetting 100 μl of serial dilutions (up to 10^{-10}, in phage dilution buffer, or YT medium) and 0.3 ml of an overnight culture of JM101 (for *Mike, Mike*Δ, IR1, R4O8, or M13KO7) or JE2571/[pCU53] (for IKe-9) in an empty Petri dish (no bottom agar required). Then add 5 ml of CW agar (cooled to 50°), mix well by

[75] K. R. Yamamoto, B. M. Alberts, R. Benzinger, L. Lawhorne, and G. Treiber, *Virology* **40,** 734 (1970).

rotating the plate, and incubate overnight at 37°. The titer of the phage suspension should be 10^{13} to 10^{14} pfu/ml.

Isolation of Plasmids pKUN9 and pKUN19

Plasmid pKUN, or its recombinants, can be isolated from *E. coli* JM101 or *E. coli* JM101[pCU53] by a number of methods designed for the isolation of plasmid DNA. The method routinely used in your laboratory should work well. We usually use the alkaline lysis method of Birnboim and Doly[76] as described by Maniatis *et al.*,[77] with the exception that lysozyme is omitted from solution I. After isolation, the plasmid DNA is further purified by cesium chloride centrifugation. To 8 ml of plasmid DNA in TE buffer add 8.2 g of CsCl in a Ti50 centrifuge tube, mix gently to dissolve the CsCl, and subsequently add 0.5 ml of ethidium bromide. Overlay the solution with light mineral oil and centrifuge (Beckman rotor Ti50) at 15° for 60 hr at 34,000 rpm. Following centrifugation the lower band, containing exclusively covalently closed circular DNA, is collected as described by Messing.[10] Remove the ethidium bromide by four to five extractions with an equal volume of 2-propanol. Dialyze the aqueous phase against several changes of 300 vol of TE buffer. Precipitate the DNA by the addition of 0.1 vol of 3 *M* NaOAc and 3 vol of ethanol. After standing for 1 hr at −20°, the DNA is recovered by centrifugation, washed once with 70% ethanol, and dried under reduced pressure. For a 500-ml culture the DNA is dissolved in 0.5 ml of TE buffer. Inspect the quality of the DNA preparation by horizontal agarose gel electrophoresis using a 0.8–1.0% submarine minigel (apparatus GNA-100; Pharmacia) in TBE (include 0.1 μg/ml of ethidium bromide in the gel and the running buffer). Mix 1 μl of the DNA sample with 8 μl of TE buffer and 1 μl of loading buffer and run the gel at 30 mA for 2 hr. Visualize the DNA bands by illumination with a long-wave UV lamp. The yield should be 1–2 μg of plasmid/ml of culture medium. In case there is still some RNA present in the preparation (a smear of fluorescent material at the bottom of the gel), add 10 μl of RNase A, and incubate at 37° for 30 min, extract twice with an equal volume of phenol/chloroform, and reprecipitate the DNA as described above.

Cloning in pKUN

Standard techniques can be used for cloning in plasmid pKUN9 or pKUN19. A comprehensive description of these techniques is given by

[76] H. C. Birnboim and J. Doly, *Nucleic Acids Res.* **7,** 1513 (1979).

[77] T. Maniatis, E. F. Fritsch, and J. Sambrook, "Molecular Cloning: A Laboratory Manual," p. 90. Cold Spring Harbor Lab., Cold Spring Harbor, New York, 1982.

Messing in Vol. 101 of this series.[10] The unique restriction enzyme cleavage sites, present in the multiple cloning sites linker of pKUN9 and pKUN19, are shown in Fig. 2.

Competent cells are prepared by inoculation of the appropriate volume of 2× YT medium with a colony of JM101 or JM101[pCU53]. Grow under aeration at 37° to a density of 5×10^7 cells/ml.

Collect cells by centrifugation for 5 min at 5000 rpm and 4° in a Beckman JA-14 rotor (or its equivalent). Resuspend the pellet gently into one-half of its growth volume of ice-cold 50 m*M* $CaCl_2$. Keep the cells on ice for 40 min and centrifuge again. Gently resuspend the cells in one-tenth of the original culture volume of ice-cold 50 m*M* $CaCl_2$ and keep on ice for an additional 1–2 hr. Before use gently swirl the tubes to suspend the cells evenly.

Transformation is carried out by the addition of appropriate amounts of recombinant DNA to 0.2 ml of competent cells in a sterile Eppendorf tube and incubation on ice for 40 min. Heat the mixture at 42° for 2 min to induce uptake of DNA and subsequently put it back on ice for 1 min. Add 0.8 ml of 2× YT medium and incubate at 37° for 60–90 min. Spread 0.2-ml portions on freshly prepared and dried 2× YT agar plates containing Xgal, IPTG, ampicillin, and, in case of JM101[pCU53], chloramphenicol. Incubate the plates overnight at 37°. Recombinants give rise to white (colorless) or light blue colonies, whereas self-ligated vector molecules give rise to deep blue-colored colonies. To enhance the blue color, place the plates at 4° for 2 hr. If you have any doubt about whether your white colony represents a recombinant or an F′-missing clone, streak it onto a minimal medium plate. An F'^--cell will not grow, an F'^+ will.

Colonies for DNA hybridizations should be on plates lacking Xgal and IPTG and also if there is some suspicion about whether the product of the insert might be toxic. Store plates of transformed cells at 4°. Colonies may also be screened with antibodies if open reading frames have been inserted. This may be especially useful for screening cDNA libraries. Alternatively, colonies may be analyzed by restriction mapping of DNA isolated from minipreparations.[10,12] Using freshly plated cells and storing cells in the cold is important for obtaining good yields of ssDNA.

Selective Packaging of the (Recombinant) DNA Strands of pKUN

The production of ss recombinant DNA with this system in many respects resembles that for obtaining recombinant M13 DNA.[10,12] Therefore experience with M13 should help to identify potential problems. To obtain the maximum yield of ssDNA it is important to infect the cells with a high multiplicity of infection. This ensures that all cells will be infected.

Prepare a small overnight culture (in 2× YT medium containing ampicillin and, when appropriate, chloramphenicol) of the colonies from which you want to isolate ss plasmid DNA. Inoculate 15 ml of 2× YT medium with 0.02 vol of the overnight culture and grow the cells for 2–3 hr at 37° (cell density should be around 2×10^8 cells/ml). Divide the culture into four 3-ml portions and infect each portion with 1.2×10^{10} pfu of the helper phage (IR1, R4O8, or M13KO7 for packaging of the F strand; *Mike, Mike*Δ, or IKe-g for packaging of the I strand). To allow phage adsorption, leave the cultures for 5 min without shaking, then continue shaking under very good aeration for 8 hr or overnight at 37°. After that time centrifuge 1.5 ml of culture for 5 min at maximum speed in an Eppendorf centrifuge and transfer 1.0 ml of supernatant to another Eppendorf tube containing 250 μl of PEG/NH_4OAc solution. Mix well and leave at room temperature for 15 min. Centrifuge at maximum speed for 10 min and carefully remove the supernatant with a Pasteur pipet. Recentrifuge for a few seconds and remove the last traces of fluid with a glass capillary drawn from a Pasteur pipet. Resuspend the pellet in 100 μl of TE buffer and add 50 μl of buffer-saturated phenol. Vortex for 1 min, leave at room temperature or alternatively at 65° for 5 min, vortex again for 30 sec, and centrifuge for 2 min. Remove the top aqueous phase and repeat the extraction twice with phenol/chloroform; any interphase should be left behind. Transfer the aqueous phase to a new tube and precipitate the ssDNA by adding 0.5 vol of 7.5 *M* NH_4OAc, pH 7.0, and 3 vol of ethanol. Leave at −80° for 30 min. Recover the precipitate by centrifugation, wash once with 70% ethanol, dry briefly in a vacuum desiccator, and dissolve the pellet in 25 μl of TE buffer. The efficiency of packaging of plasmid DNA can be analyzed by electrophoresis of 5 μl of the DNA solution on a 0.8–1.0% agarose gel in TBE buffer as described above. Two DNA bands should be visible on the gel, one representing ss phage DNA and the other ss plasmid DNA (see Fig. 3A). Sometimes you might also see traces of high-molecular-weight DNA. In case of packaging by IKe-9 another, fast migrating, (miniphage DNA) band might be visible. These DNA's do not interfere with most techniques (sequencing, S1 nuclease mapping, hybridization, etc.) for which pKUN has been developed. When required the vector DNA strand can be purified away from the other bands by preparative electrophoresis on agarose gels, followed by electroelution. There have been a few cases where no ss plasmid DNA was observed although phage DNA was produced. To ensure optimal production of ssDNA we recommend initially trying infections with several helper phages. Even if the amount of plasmid DNA is as little as 5% of the total ssDNA, it can still be sequenced.

Another method by which the packaging efficiency can be established is by determining the number of phage particles (pfu) and transducing particles (cfu) present in the culture supernatant. The phage titer is determined as described above, whereas the number of transducing particles is determined by heating the culture supernatant for 3 min at 60°, followed by adding 10 μl of serial dilutions in YT medium to 100 μl of exponentially growing cells (JM101 for *Mike, Mike*Δ, IR1, R4O8, and M13KO7 packaged particles and JE2571[pCU53] for IKe-9 packaged particles), followed by incubation at 37° for 20 min to allow infection. Subsequently the mixtures are spread on dry 2× YT plates containing 100 μg/ml Ap. The number of colonies can then be used to calculate the number of transducing particles (cfu) present in the supernatant. The best yield of phage and transducing particles one can expect is about 10^{11} pfu or cfu/ml, respectively. We have noticed that yields of ssDNA are not significantly affected by the presence of ampicillin and/or kanamycin in the culture medium, but it ensures that all cells harbor a plasmid and, in case of M13KO7, are infected.

Preparation of Hybridization Probes, Mutagenesis, and Nucleotide Sequence Analysis

The ss plasmid DNA prepared as described above is usually sufficiently pure for the preparation of hybridization probes, mutagenesis, and sequencing.[10,12] Due to the presence of ss helper phage DNA, it is advisable to perform the primer hybridization step more stringently than is needed using filamentous phage vectors.

For example, the annealing of the master primer (18-mer) is carried out for 1–2 hr in a waterbath of 54°, whereas the hybridization of the reverse primer (15-mer) is carried out at 40°. When using other primers, hybridization should be carried out at a temperature which lies 2° below the T_m of the oligonucleotide. The T_m can be calculated with the following equation: $T_m = 4\times$ (number of GC base pairs) + $2\times$ (number of AT base pairs).[78] Five to 10 μl of the ssDNA, prepared as described above, is sufficient for sequencing. This corresponds to an amount of ssDNA obtained from approximately 0.20–0.40 ml of culture supernatant. Standard techniques can be used for sequencing (or for the preparation of hybridization probes, mutagenesis, etc.) of the differentially packaged DNA strands of pKUN.[10,12] Be sure, however, to use the right primer with the right template: IR1-, R4O8-, or M13KO7-packaged DNA should be sequenced with the reverse primer, whereas for sequence analysis, or hybridization probe labeling, of the DNA strands packaged by *Mike, Mike*Δ, or IKe-9, the master primer should be used (Fig. 2B). The choice of other

[78] S. V. Suggs, T. Hirose, T. Miyake, E. H. Kawashima, M. J. Johnson, K. Itakura, and R. B. Wallace, *in* "Developmental Biology Using Purified Genes" (D. D. Brown, ed.), p. 683. Academic Press, New York, 1981.

primers depends on the desired starting point in the DNA strand which is packaged.

Examples of Packaging and Rapid Dideoxy Sequencing

The efficiency of cloning in pKUN9 and pKUN19 is comparable to that of cloning in the pUC vectors, and cells harboring a recombinant

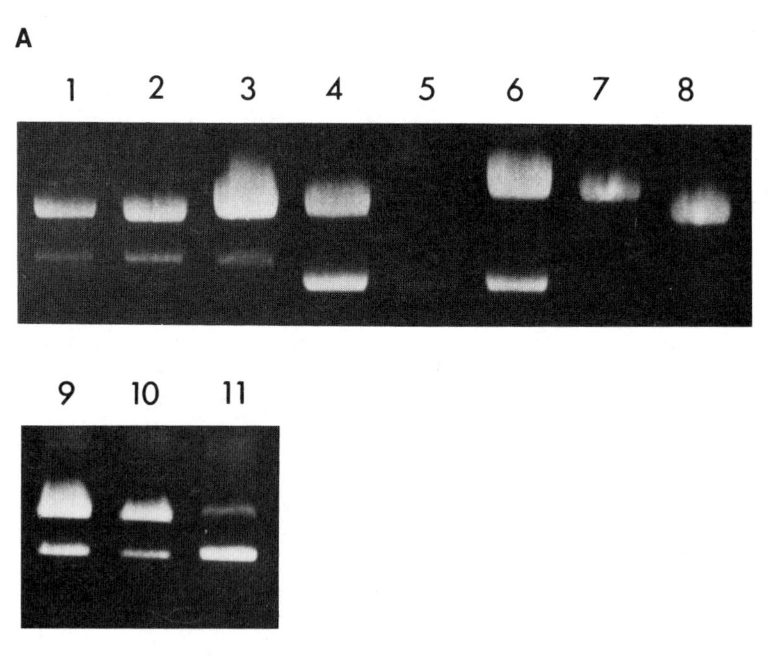

FIG. 3. Agarose gel analyses of the single-stranded (ss) DNA isolated from filamentous particles produced upon superinfection of cells harboring recombinant pKUN plasmids. Shown in each lane is the ethidium bromide-stained DNA, equivalent to 0.2 ml of culture supernatant (see the Materials section). The top bands are the ss helper phage DNAs and the bottom bands are the ss plasmid DNAs. For packaging three different recombinant pKUN plasmids were used, containing inserts of 1000 (lanes 1–3), 400 (lanes 4–6), and 600 (lanes 9–11) bp, respectively. Lanes 1, 3, and 9: ssDNA packaged with IKe-9 (host: JE2571[pCU53], lane 1; JM101[pCU53], lanes 3 and 9). Lanes 2, 4, and 10: ssDNA packaged with IR1 (host: JM101, lanes 2 and 10; JM101[pCU53], lane 4). Lanes 5 and 11: ssDNA packaged with M13KO7 (host: JM101, lane 5; JM101[pCU53], lane 11). Lane 6: ssDNA packaged with *Mike* (host: JM101). Lanes 7 and 8: ssDNA of the helper phages *Mike* and IKe-9 produced by cells lacking pKUN plasmids. (B) Autoradiogram of dideoxy sequence analysis from the 5′- and 3′-terminal end of the *Hae*III C fragment (531 bp) of the filamentous bacteriophage Pf3 inserted in the *Sma*I site of pKUN9. The DNA strand packaged with the aid of IKe-9 was sequenced with the master primer (left) while its complementary strand, packaged with IR1, was sequenced with the reverse primer. The arabic numerals indicate the positions of the respective nucleotides on the genetic map [R. G. M. Luiten, D. G. Putterman, J. G. G. Schoenmakers, R. N. H. Konings, and L. A. Day, *J. Virol.* **56,** 268 (1985)].

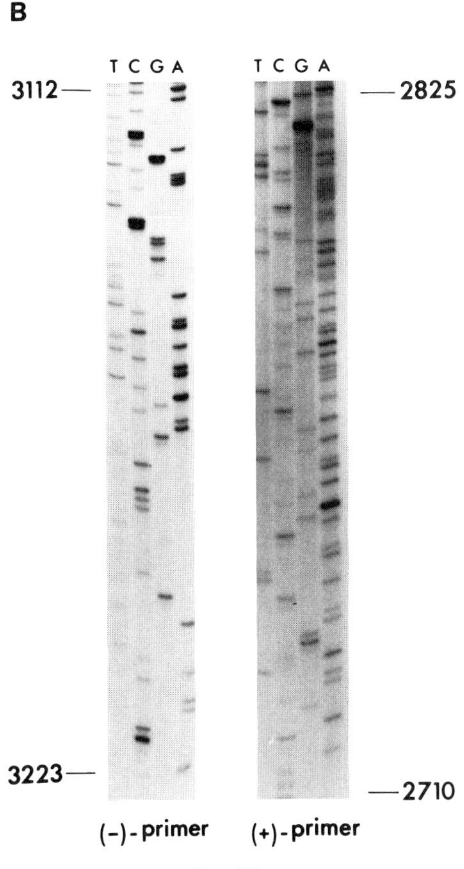

FIG. 3B.

pKUN plasmid grow as well as nontransformed cells. The largest fragment we have cloned thus far is 6.3 kb long. Up to now we have not encountered severe problems in cloning fragments from either eukaryotic or prokaryotic origin.

Some examples of packaging of ss recombinant plasmid DNA are shown in Fig. 3A. The level and ratio of ss phage DNA (upper band in Fig. 3A) to ss plasmid DNA (lower band) vary for unknown reasons with the helper phage used and/or with the recombinant plasmid to be packaged. Therefore it is advisable, but not absolutely necessary, to test all helper phages for production of ssDNA and then choose the helper phage that produces upon superinfection the largest amount of ss plasmid DNA. An example of rapid sequence analysis is presented in Fig. 3B. The complementary plasmid strands were recovered via superinfection of JM101[pCU53] with the helper phages IKe-9 and IR1 (see Materials, Reagents, and Procedures). After isolation of the ss plasmid DNA's the nucleotide sequence of the 5'- and 3'-terminal regions of the cloned DNA fragment was established with the aid of the master and reverse sequencing primers, respectively (see Materials, Reagents, and Procedures). Thus although both ss phage and plasmid DNA are present in the reaction mixture, there is no interference by the phage DNA when copying the plasmid DNA using a plasmid-specific primer.

Acknowledgments

The authors would like to thank Dr. N. Lubsen for critical reading and Dr. R. Luiten for both critical reading and great help in the preparation of the manuscript. Dr. Marjorie Russel, Dr. Jeff Vieira, and Dr. Joachim Messing are gratefully acknowledged for communication of results prior to publication.

[3] Restriction Site Bank Vectors for Cloning in Gram-Negative Bacteria and Yeast

By JOHN DAVISON, MICHEL HEUSTERSPREUTE, and FRANÇOISE BRUNEL

The dramatic advances of genetic engineering are in large part due to our ability to cleave DNA molecules at specific locations with the use of restriction enzymes. Over the last few years the number of identified restriction enzymes has increased enormously so that a recent review[1] recorded several hundred from many different species of bacteria and algae, while new restriction endonucleases are still being found. For the efficient exploitation of this impressive collection of restriction endonucleases, new cloning vectors having the appropriate restriction recognition sites are needed.

Historically, vectors have been constructed by attaching two or three antibiotic resistance genes to a suitable replicon so that cloning into a unique site located in one of the antibiotic genes eliminates resistance to that antibiotic (insertional inactivation) and facilitates recombinant detec-

[1] C. Kessler, *Gene* **47,** 1 (1987).

tion.[2] While extremely useful, this concept has serious limitations. The number and nature of the restriction sites in the antibiotic resistance gene are determined by the coding requirements of the corresponding protein and can be changed only with difficulty. It is possible to alter the sequence without losing the resistance marker, though it has only rarely been done.[3,4] The increased use of synthetic DNA and site-directed mutagenesis may make it simpler in the future. The alternative is to add more antibiotic resistance genes,[5] but this leads to an undesirable size increase and often to duplication of existing sites. A different approach to the insertional inactivation philosophy of vector construction has been the development of the synthetic polylinker system as found in the M13 vectors and pUC and pIC series.[6-8] In this case, cloning into the polylinker destroys the reading frame of the small α peptide of β-galactosidase so that the plaque or colony is white, instead of blue.

However, there are many situations where an insertion inactivation approach is of no advantage. For example, when the ideal restriction site is absent from the indicator gene, it may be necessary to use a different site and to modify the ends of the DNA with DNA polymerase or an exonuclease prior to blunt end cloning. Thus, both recircularized and recombinant molecules would carry an inactive indicator gene. Similarly, insertional inactivation is of little use when screening for rare specific recombinants in a "shotgun" cloning experiment. In this case, different techniques are needed to screen for hybrid colonies, e.g., using DNA or RNA hybridization,[9-11] or antibody probes,[12-14] by complementation with genetically marked host strains,[15] or by simply screening large numbers of potential recombinants in small rapid DNA preparations.[16] The propor-

[2] F. Bolivar, R. L. Rodriguez, M. C. Betlach, and H. W. Boyer, *Gene* **2,** 75 (1977); F. Bolivar, R. L. Rodriguez, P. J. Greene, M. C. Betlach, H. L. Heyneker, H. W. Boyer, J. H. Crosa, and S. Falkow, *Gene* **2,** 95 (1977).

[3] K. Talmage and W. Gilbert, *Gene* **12,** 235 (1980).

[4] J. Davison, M. Heusterspreute, M. Merchez, and F. Brunel, *Gene* **28,** 311 (1984).

[5] X. Soberon, L. Covarrubias, and F. Bolivar, *Gene* **9,** 287 (1980).

[6] J. Vieira and J. Messing, this volume [1].

[7] J. Messing, this series, Vol. 101, p. 20.

[8] J. L. Marsh, M. Erfle, and E. J. Wykes, *Gene* **32,** 481 (1984).

[9] M. Grunstein and J. Wallis, this series, Vol. 68, p. 379.

[10] J. W. Szostak, J. I. Stiles, B.-K. Tye, P. Chiu, F. Sherman, and R. Wu, this series, Vol. 68, p. 419.

[11] D. Hanahan and M. Meselson, this series, Vol. 100, p. 333.

[12] L. Clarke, R. Hitzeman, and J. Carbon, this series, Vol. 68, p. 436.

[13] D. Anderson, L. Shapiro, and A. M. Skalka, this series, Vol. 68, p. 428.

[14] D. A. Kaplan, L. Greenfield, and R. J. Collier, this series, Vol. 100, p. 342.

[15] L. Clarke and J. Carbon, this series, Vol. 68, p. 396.

[16] H. C. Birnboim, this series, Vol. 100, p. 243.

tion of recombinant clones can be greatly increased by treatment of the linearized vector DNA with alkaline phosphatase prior to ligation[17] or by cleaving both the donor and vector DNA with restriction enzymes, generating different cohesive ends and separating the respective fragments by electrophoresis prior to ligation.

For the above reasons, it is difficult, and often unnecessary, to design a versatile cloning vector based on insertional inactivation and we have evolved a different concept of vector construction, called the "restriction site bank," in which a great many unique cloning sites are assembled together on the smallest possible DNA fragment.[4,18–21a] This chapter describes the application of these restriction site banks to construction of *Escherichia coli* vectors, wide-host-range cosmid vectors (able to grow in most gram-negative bacteria), and to *E. coli*–yeast shuttle vectors.

Rationale for Construction of Restriction Site Bank Vectors

Minimum requirements for a cloning vector are an origin of replication, a selective marker (often antibiotic resistance), and at least one restriction site at which insertion of foreign DNA does not prevent either replication or selection. If a plasmid has only one resistance gene then none of the restriction sites located within it can normally be used for cloning since the resulting plasmid can no longer be selected following transformation of the host. Thus, most vectors described below contain two antibiotic resistance markers, thereby enabling a choice of restriction sites located in either one or the other. Certain combinations of antibiotic resistance genes do not lead to duplication of restriction sites (e.g., Tc^R and Ap^R, Km^R and Sm^R) and are therefore preferred. In addition to the above requirements, an ideal general purpose cloning vector would combine two often contradictory features: a minimum size and a maximum number of unique restriction sites. The small size facilitates all subsequent analyses of recombinants and allows the complete DNA sequence to be determined. The large number of unique restriction sites allows use

[17] A. Ullrich, J. Shine, J. Chirgwin, R. Pictet, E. Tischer, W. J. Rutter, and H. M. Goodman, *Science* **196,** 1313 (1977).

[18] M. Heusterspreute and J. Davison, *DNA* **3,** 259 (1984).

[19] M. Heusterspreute, J. Oberto, V. Ha-Thi, and J. Davison, *Gene* **34,** 363 (1985).

[20] M. Heusterspreute, V. Ha-Thi, S. Emery, S. Tournis-Gamble, N. Kennedy, and J. Davison, *Gene* **39,** 299 (1985).

[20a] M. Heusterspreute, V. Ha-Thi, S. Emery, S. Tournis-Gamble, N. Kennedy, and J. Davison, *Gene* **53,** 299 (1987).

[21] J. Davison, M. Heusterspreute, N. Chevalier, V. Ha-Thi, and F. Brunel, *Gene* **51,** 273 (1987).

[21a] F. Brunel, N. Chevalier, and J. Davison, manuscript in preparation (1987).

to be made of the vast choice of restriction enzymes now commercially available. For example, of the 64 possible 6-bp palindromic restriction sites (probably the most useful for everyday cloning), 48 are cleaved by known restriction endonucleases.[1] However, pBR322,[2] the most popular vector at present, has unique sites for only 20 of these restriction enzymes (Table I), thereby limiting its usefulness. To make a more versatile vector we have constructed a series of "restriction site banks" (defined as a collection of unique restriction sites on a small DNA fragment) and have inserted these into a suitably modified pBR322 derivative. The construction of such a bank of sites does not have the limitations imposed by the need to maintain an open reading frame, as in the case of the M13mp and pUC series of cloning vectors,[6] so that it is easier to collect a very large number of restriction sites. The most recent of this series is pJRD184, a 3793-bp plasmid carrying 39 unique 6-bp palindromes recognized by known restriction enzymes.[20,20a]

Application of a similar rationale to other organisms has allowed the construction of *E. coli*–yeast shuttle vectors[19] and of wide-host-range gram-negative vectors equipped with restriction site banks.[21]

Materials and Techniques

Strains

Escherichia coli strain MM294 (*endo*I$^-$, r_k^-, m_k^-, *B1*$^-$, *pro*$^-$) was used as the host for all plasmids.[22] An exception to this was the use of strain GM48 for the preparation of plasmid DNA lacking the *dam*$^-$- and *dcm*$^-$-dependent methylation, which is necessary when it is desired to cleave the DNA with certain restriction enzymes (e.g., *Bcl*I, *Mbo*I, *Taq*I, etc.).[22] Conjugation was performed using *E. coli* S17-1,[23] in which the RP4 conjugative plasmid is integrated into the host chromosome so that, while it can mobilize RSF1010 derivatives (such as pJRD203 and pJRD215), it does not transfer itself during conjugation. Furthermore, in S17-1, the *Km*^{R} gene of the integrated RP4 plasmid is inactivated due to a Tn*7* insertion, thus facilitating the detection of the *Km*^{R} following transformation or cosmid infection. However, this strain is not absolutely essential and one of several different conjugative plasmids may be used in the donor strain.[24] Conjugation is particularly useful for the transfer of RSF1010-

[22] K. Backman, *Gene* **11,** 169 (1980).

[23] R. Simon, U. Priefer, and A. Pühler, *Biotechnology* **1,** 784 (1983).

[24] P. T. Barth, L. Tobin, and G. S. Sharpe, *in* "Molecular Biology, Pathogenicity and Ecology of Bacterial Plasmids" (S. B. Levy, R. C. Clowes, and E. L. Koenig, eds.), p. 439. Plenum, New York, 1981.

based recombinant DNA derivatives to bacterial species that cannot be transformed by naked DNA. Experiments with *Saccharomyces cerevisiae* used strain GRF18 (α, *leu*2–3, 2–212; *his*3–11, 3–15; *can*^{R}) or MC333 (*trp*1) according to whether the selective marker on the plasmid was *LEU*2 or *TRP*1.[19] *Pseudomonas* sp. ATCC19151 was isolated by its ability to use sodium dodecyl sulfate as sole carbon and sulfur source.[25] It is also able to utilize vanillate as carbon source and is used here for the cloning of these genes.

Plasmids

E. coli plasmids pJRD158, RSF1010, and pEMBL8 have been described previously.[4,26,27] Yeast vectors pJO158 and pMH158 are derived, respectively, from YEp7[28] and pJDB207[29] by the addition of the restriction site bank of pJRD158.[19] pJRD203 is a preliminary version[21] of pJRD215 described below. It is based on the RSF1010 replicon and contains *Km*^{R}, λ *cos*, and the pEMBL8 polylinker but lacks the restriction site bank of pJRD184.

DNA Techniques

Restriction endonucleases and DNA-modifying enzymes were obtained from New England BioLabs or Boehringer–Mannheim and used according to the manufacturer's instructions. Methods for the preparation of DNA, conversion of cohesive ends to blunt ends, removal of phosphate groups using bacterial alkaline phosphatase, addition of synthetic linkers, ligation of DNA, and transformation of *E. coli* and *S. cerevisiae* have been described previously.[30] DNA sequencing followed the method of Maxam and Gilbert,[31] and the sequences were analyzed on an Apple IIe microcomputer using the programs of Larson and Messing.[32]

DNA fragments were separated in 1% agarose for fragment sizes 1 to 20 kb or in 2% agarose for fragments of 0.1 to 1 kb. Specific DNA fragments were transferred by electrophoresis onto DEAE–cellulose paper placed just in front of the migrating band, followed by elution from the

[25] Y. C. Hsu, *Nature* (*London*) **200,** 1091 (1960).

[26] K. Nagahari and K. Sakaguchi, *J. Bacteriol.* **134,** 1527 (1978).

[27] L. Dente, G. Cesareni, and R. Cortese, *Nucleic Acids Res.* **11,** 1645 (1983).

[28] D. T. Stinchcomb, K. Struhl, and R. W. Davis, *Nature* (*London*) **282,** 39 (1979).

[29] J. R. Broach, this series, Vol. 101, p. 307.

[30] T. Maniatis, E. F. Fritsch, and J. Sambrook, "Molecular Cloning: A Laboratory Manual." Cold Spring Harbor Lab., Cold Spring Harbor, New York, 1982.

[31] A. M. Maxam and W. Gilbert, *Proc. Natl. Acad. Sci. U.S.A.* **74,** 560 (1977).

[32] R. Larson and J. Messing, *Nucleic Acids Res.* **10,** 39 (1982).

DEAE–cellulose paper in a high salt buffer, and ethanol precipitation according to Dretzen *et al.*[33] DNA isolated in this manner is a good substrate for restriction enzymes, DNA polymerase Klenow fragment, and T4 DNA ligase. It can also be used directly with alkaline phosphatase and T4 polynucleotide kinase for DNA sequencing.

Construction of a Pseudomonas Gene Bank

Large *Pseudomonas* DNA fragments (for cloning in wide-host-range cosmid vectors) were prepared by partial digestion with *Mbo*I. The conditions of reaction were determined in advance on small samples by monitoring the digestion products by electrophoresis. The large-scale digestion products (100–200 μg/500 μl) were subjected to sucrose gradient (10–30%) centrifugation in a Beckman SW27 rotor at 20,000 rpm for 18 hr. The gradient was collected as 0.8-ml fractions and the size distribution determined by electrophoresis. Fractions in the 30- to 35-kb range were pooled, dialyzed against TE buffer (10 m*M* Tris–HCl, 1 m*M* EDTA, pH 7.4), concentrated by extraction with unsaturated butanol, and precipitated with ethanol.

The cosmid vector pJRD203 was cleaved at its unique *Bam*HI site and dephosphorylated with bacterial alkaline phosphatase[17] to prevent formation of polycosmids. A 3-μg aliquot was then ligated to 3 μg of 30- to 35-kb *Pseudomonas* DNA fragments (at a concentration of 300 μg/ml), this being sufficient to make a gene bank representing the entire *Pseudomonas* genome. The ligation was then packaged into bacteriophage λ particles using a DNA packaging kit (Amersham, Ltd.). Although λ packaging extracts[34] were previously prepared in the laboratory,[35] we now consider it more efficient, more reproducible, and less labor intensive to use a commercial packaging mixture. The packaged cosmid recombinants were infected into *E. coli* S17-1 followed by selection for Km^R colonies.

Identification of Cosmid Recombinants by Complementation

The method of selection of recombinant clones by complementation of a mutant is of general application to any gram-negative bacterium able to support the growth of the pJRD203, pJRD205, and pJRD215 vectors. A second requirement is that it must be possible to mutate the strain to obtain the appropriate mutant phenotype. This problem is not trivial; for example, certain obligate methylotrophs have proved extremely difficult

[33] G. Dretzen, M. Bellard, P. Sassone-Corsi, and P. Chambon, *Anal. Biochem.* **12,** 295 (1981).

[34] L. Enquist and N. Sternberg, this series, Vol. 68, p. 281.

[35] J. Davison, F. Brunel, M. Merchez, and V. Ha-Thai, *Gene* **17,** 101 (1982).

to mutate. The example given below describes the cloning of the genes concerned with vanillate utilization in *Pseudomonas* sp. ATCC19151. In this case mutants were easily isolated following mutagenesis by *N*-methyl-*N'*-nitro-*N*-nitrosoguanidine[36] by replica plating on M9 minimal agar[36] plates containing either vanillate or protocatechuate as carbon source. This choice of substrate assumes that the first step of vanillate degradation in *Pseudomonas* is its demethylation by vanillate *O*-demethylase to protocatechuate as has been suggested by Cartwright and Smith.[37] However, caution should be exercised, since these authors did not demonstrate protocatechuate to be the reaction product, possibly because of high protocatechuate oxygenase activity in the cell-free extracts.

The cosmid recombinants in *E. coli* S17-1 were grown individually in Nunc microwell plates and then transferred using a multipronged replicating tool to an L broth mating plate spread with a vanillate-negative mutant (*van5*) of the parent strain ATCC19151. After overnight growth and mating at 37° on these plates, the mating areas were transferred to an M9 plate with vanillate as sole carbon source and kanamycin as a selective marker. On this plate, the *E. coli* S17-1 cannot grow due to the inability of *E. coli* to use vanillate as a carbon source. Similarly, the ATCC19151 *van*$^-$ mutant cannot grow unless it receives a recombinant plasmid containing the intact vanillate gene able to complement its mutational defect. Due to the large size of the inserts, positive clones are rather frequent, representing about 0.2–0.5% of the total gene bank. Large cosmid clones of this type vary in their stability, often giving rise to deletion derivatives. Thus, plasmid DNAs from different isolates of the same clone may have different but related restriction patterns. For this reason, it is advisable to maintain the recombinants as DNA preparations rather than as bacterial cultures. The stability usually improves when the gene of interest is subcloned on a smaller DNA fragment.

Procedures and Results

E. coli Restriction Site Bank Vector pJRD158

Plasmid pJRD158 is an early version of the pBR322-based restriction site bank vectors. It contains 35 unique restriction cleavage sites of which

[36] J. H. Miller, "Experiments in Molecular Genetics." Cold Spring Harbor Lab., Cold Spring Harbor, New York, 1972.

[37] N. J. Cartwright and A. R. W. Smith, *Biochem. J.* **102,** 826 (1967).

30 are of the 6-bp palindromic type[4,18] (the original reports described only 28 unique sites but new restriction enzymes have since been discovered). Figure 1 shows the positions of these sites on the plasmid. The regions concerned with antibiotic resistance and DNA replication are largely unchanged except for certain point mutations introduced at the time that the *Pst*I and *Hin*cII sites were removed from the Ap^R gene.[4,18] It will be noted that the restriction site bank of the pJRD158 plasmid, represented by the DNA between *Sac*I and *Xho*I, has been added between the end of the Tc^R gene and the origin of replication, replacing the corresponding region of pBR322, which is known to be nonessential for plasmid replication and maintainance. As a result of this exchange pJRD158 is considerably smaller than pBR322 (3903 bp compared to 4363) while retaining both resistance markers and having many more useful cloning sites.

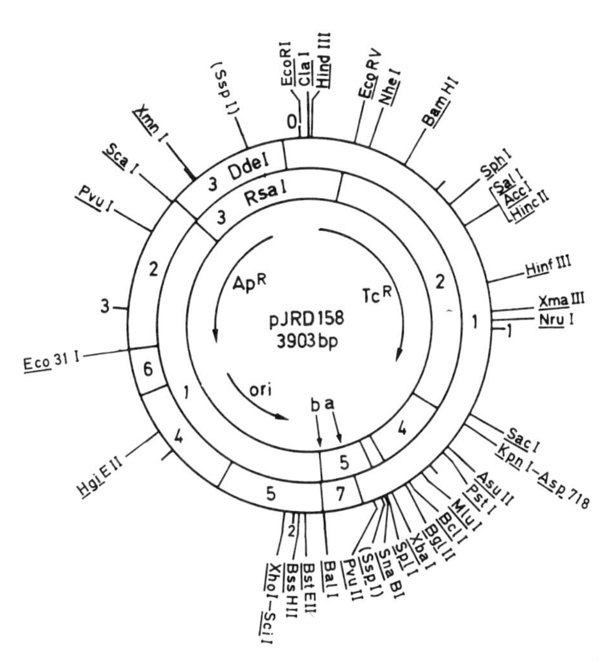

FIG. 1. Restriction map of *E. coli* vector pJRD158. The restriction sites on the exterior of the circle are all unique with the exception of *Ssp*I, which is present twice, but is shown because it was used as the site of insertion of an 83-bp fragment (see text). The inner circles contain the *Dde*I and *Rsa*I restriction sites. The region between the *Dde*I marked (a) and the *Rsa*I site marked (b) was removed by deletion (see text) to give pJRD182. Abbreviations: Ap^R, resistance to ampicillin; Tc^R, resistance to tetracycline; ori, the origin of replication of pMB1.

Addition and Deletion of Small DNA Segments

In the construction of the restriction site banks it was often necessary to add small DNA fragments containing new restriction sites. Such fragments were frequently identified by computer scanning of DNA sequences determined either in this laboratory or from the published data of others. Most DNA fragments were taken from phage T5 DNA, the Km^R region of the Tn*5* transposon, or the 2μ plasmid of yeast. It would often have been easier to use DNA from phage λ (which has been completely sequenced), but this has been avoided to prevent cross-homology between the pJRD plasmid series and existing λ vectors, since it is often the case that a cDNA clone isolated in a plasmid will be used directly as a hybridization probe to identify a genomic clone in a λ gene bank.

Addition of DNA segments implies an increase in size, which would be a conflict with our aim that the final vector should be small, preferably smaller than the parent plasmid pBR322. To accomplish this, it has been necessary to evolve methods for both the insertion and the deletion of small DNA fragments in such a way that preexisting unique sites are neither duplicated nor removed. It is not possible, in the space of this chapter, to describe every insertion and deletion event. Instead, one example of each used in the construction of pJRD184 from pJRD158 will be given.

Deletion of a 121-bp DdeI–RsaI Fragment of pJRD158

In order to add additional DNA fragments containing unique sites to pJRD158 without greatly increasing its size it was desirable to remove nonessential DNA from the plasmid. Inspection of the physical map (Fig. 1) shows a 170-bp region (from *Pvu*II to *Bal*I sites at coordinate 1740 and 1910, respectively) without any useful restriction sites. A fusion between the *Pvu*II and *Bal*I sites would have removed this entire region but was undesirable since the unique *Pvu*II and *Bal*I sites would both be destroyed in the process. Instead, an alternate procedure was used, based on the method of Heusterspreute and Davison,[38] for generating predetermined deletions. The pJRD158 DNA was cleaved with *Dde*I and the large fragment (1845 bp) isolated. The 5′ protruding ends were repaired with DNA polymerase Klenow fragment to give blunt ends and the molecule was then cleaved with *Bam*HI. In a second reaction, the pJRD158 molecule was cleaved with both *Rsa*I and *Pvu*I and the large 1370-bp fragment isolated. In a third reaction the pJRD158 vector was cleaved with *Pvu*I

[38] M. Heusterspreute and J. Davison, *Gene* **23,** 35 (1983).

and *Bam*HI and the small 1009-bp fragment containing the 5′ ends of the Ap^R and Tc^R genes isolated. Finally, the fragments from these different restriction reactions were ligated together and used to transform *E. coli* MM294, selecting for Ap^R and Tc^R. Only those molecules which have correctly reconstituted the *Bam*HI cohesive ends in Tc^R and the *Pvu*I cohesive ends in Ap^R are able to confer resistance to both antibiotics. The third ligation necessary for circle formation is the successful junction of the *Rsa*I site at coordinate 1897 to the repaired *Dde*I site at position 1776. This ligation has the effect of deleting the DNA between these positions resulting in the plasmid pJRD182. It is always advisable to verify the junction of blunt-end ligations by DNA sequencing, as was done in this case.

Addition of an 83-bp Segment to pJRD182

Plasmid pJRD158 (and its deletion derivative pJRD182) described above contain two sites for *Ssp*I, one of these sites being in the restriction site bank (pJRD158 coordinate 1698) and the other just clockwise from the Ap^R gene (coordinate 3710). This latter site is present in pBR322 and its doubling in the restriction site bank of pJRD158 is a historical accident, due to the fact that the restriction endonuclease *Ssp*I was still undiscovered at the time pJRD158 was constructed. It was decided to clone a blunt-ended DNA fragment into the *Ssp*I site at coordinate 1698, thus destroying this site (thereby rendering the other unique) while introducing additional sites carried by the new DNA fragment. The new DNA fragment was identified as being interesting by computer analysis,[32] which showed an 83-bp *Eco*RI–*Stu*I fragment of the yeast 2μ plasmid carrying unique *Sna*I, *Ppu*MI, and *Spe*I sites,[39] neither of these being present on pJRD182. The 83-bp *Eco*RI–*Stu*I fragment was repaired at the *Eco*RI end with DNA polymerase Klenow fragment (the *Stu*I end already being a blunt end) and the repaired product was inserted into a partial (linearized) *Ssp*I digest of pJRD182. No selection was available for this insertion but it was easily found by using a 4-fold molar excess of the fragment in the ligation and screening rapid DNA preparations[16] for their cleavage pattern with the enzyme *Taq*I, the pattern of the desired recombinant being predicted by computer analysis. Of the total of 24 samples screened, 4 of the rapid DNA preparations contained the correct insert. The new plasmid pJRD183 was checked by cleavage with *Ssp*I and the new junctions verified by DNA sequencing.

[39] J. L. Hartley and J. E. Donelson, *Nature (London)* **286,** 860 (1980).

TABLE I
RESTRICTION ENZYMES RECOGNIZING PALINDROMIC SEQUENCES IN pJRD184[a] AND pBR322

Restriction enzyme	Recognition site	Coordinates	
		pJRD184	pBR322
*Aat*II	GACGTC	3334, 3716	4286
*Afl*II	CTTAAG	1575	—
*Apa*I	GGGCCC	1549	—
*Apa*LI	GTGCAC	2219, 3465	2219, 2789, 4035
*Asu*II	TTCGAA	1413	—
*Avr*II	CCTAGG	1582	—
*Bal*I	TGGCCA	1800	1444
*Bam*HI	GGATCC	375	375
*Bbe*I	GGCGCC	413, 434, 548, 1205	413, 434, 548, 1205
*Bcl*I	TGATCA	1566	—
*Bgl*II	AGATCT	1571	—
*Bsp*MII	TCCGGA	—	1664
*Bss*HII	GCGCGC	1883	—
*Cla*I	ATCGAT	23	23
*Dra*I	TTTAAA	2662, 2681, 3373	3232, 3251, 3943
*Eco*47III	AGCGCT	232, 494, 775	232, 494, 775, 1727
*Eco*RI	GAATTC	3791	4361
*Eco*RV	GATATC	185	185
*Hin*dIII	AAGCTT	29	29
*Hpa*I	GTTAAC	1541	—
*Kpn*I	GGTACC	1323	—
*Mlu*I	ACGCGT	1537	—
*Mst*I	TGCGCA	260, 3018	260, 1356, 1454, 3588
*Nae*I	GCCGGC	401, 769, 929, 1902	401, 769, 929, 1283
*Nco*I	CCATGG	1528	—
*Nde*I	CATATG	1562	2297
*Nhe*I	GCTAGC	229	229
*Nru*I	TCGCGA	972	972
*Nsi*I	ATGCAT	1559	—
*Pst*I	CTGCAG	1443	3609
*Pvu*I	CGATCG	3165	3735
*Pvu*II	CAGCTG	1751	2066
*Sac*I	GAGCTC	1289	—
*Sac*II	CCGCGG	1553	—
*Sal*I	GTCGAC	651	651
*Sac*I	AGTACT	3276	3846
*Sma*I	CCCGGG	1546	—
*Sna*I	GTATAC	1680	2246

TABLE I (*continued*)

Restriction enzyme	Recognition site	Coordinates	
		pJRD184	pBR322
*Sna*BI	TACGTA	1618	—
*Spe*I	ACTAGT	1684	—
*Sph*I	GCATGC	562	562
*Spl*I	CGTACG	1616	—
*Ssp*I	AATATT	3600	4170
*Stu*I	AGGCCT	1579	—
*Xba*I	TCTAGA	1588	—
*Xho*I	CTCGAG	1938	—
*Xma*III	CGGCCG	939	939

[a] Restriction endonucleases recognizing unique nonpalindromic sites are not included in this list but are shown in Fig. 2 (*Bst*EII, 1858; *Bsm*I, 1879; *Eco*31I, 2865; *Hin*fIIIB, 847; *Hgi*EII, 2486; *Ppu*MI, 1644; *Xmn*I, 3393).

E. coli Restriction Site Bank Vector pJRD184

Plasmid pJRD183 was similarly improved by the addition of unique *Mlu*I, *Nsi*I, and *Nde*I sites and the final version obtained after addition of two synthetic polylinkers in collaboration with Dr. N. Kennedy and S. Emery of Biosyntech GmbH.[20,20a] In the final plasmid pJRD184, 55 of the 64 possible 6-bp palindromic recognition sites are present, 46 of these are unique, and 39 are recognized by restriction enzymes already identified to date. The positions of these sites (and of other unique cleavage sites on the molecule) relative to the origin of replication and the antibiotic resistance markers are shown in Fig. 2. In Table I, the number and exact coordinates of the sites are shown and compared to those on pBR322 to demonstrate the increased versatility of the vector. The *Aat*II site is interesting since it is unique in pBR322 but double in pJRD158 and pJRD184. This is a consequence of the fortuitous generation of a second *Aat*II site during the mutagenesis to remove the *Hin*cII and *Pst*I sites from the Ap^R gene and has been described previously.[4,18] No attempt has been made to make this site unique since the removal of the intervening region during cloning inactivates only the Ap^R gene. Some of the other nonunique recognition sites may also be used for cloning where loss of the intervening DNA does not prevent replication and selection (*Eco*47III, *Nae*I, *Bbe*I). Finally, there are several unique interrupted palindromic and nonpalindromic restriction sites (Fig. 2) that may be useful in special circumstances.[20,20a]

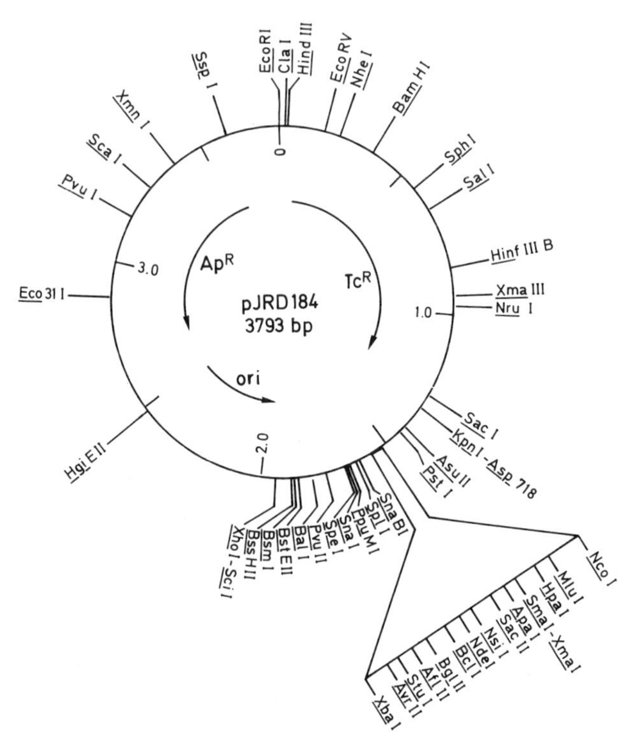

FIG. 2. Restriction map of *E. coli* vector pJRD184. Only unique restriction sites are indicated. Abbreviations as in Fig. 1.

The original restriction site bank vector pJRD158 has a very high copy number (120/cell) compared to pBR322 (30/cell). This is for two independent reasons: first, it lacks the *rop* gene concerned with replication control; and second, it is mutant in the RNA I regulator, which controls copy number and incompatibility.[4,18,40] This very high copy number is useful for DNA preparations and is suitable for most cloning experiments. However, we have noticed that it poses a burden on the host cells, causing them to grow more slowly and when coupled to excessive gene expression may be lethal. For this reason, we constructed an alternate variant pJRD158b which differs from pJRD158 by only a single base pair corresponding to the mutation in RNA I.[4,18] This plasmid has a copy number intermediate between that of pJRD158 and pBR322 (due to the effect of the *rop* deletion) and forms the basis of pJRD184.

[40] J. Davison, *Gene* **28,** 1 (1984).

E. coli–Yeast Shuttle Restriction Site Bank Vectors pMH158 and pJO184

E. coli–S. cerevisiae shuttle vectors are dual replicon plasmids which replicate in *E. coli* via a pBR322-based replicon and in yeast either via the origin of replication of the 2μ yeast plasmid or a chromosomal autonomous replication segment (*ARS*). They are usually selected in yeast by complementation of a nonreverting auxotrophic mutant due to a functional gene carried by the plasmid. For example, vector YRp7 replicates in yeast via the *ARS*1 replication origin which is located on the same 1453-bp *Eco*RI–*Eco*RI fragment as the selective marker *TRP*1.[28] In the case of pJDB207, the replication origin is that of the 2μ plasmid and the selection is via the *LEU*2 gene in *leu*2 auxotrophic strain GRF18.[29] Restriction site bank variants have been constructed for both of the plasmids by replacing the pBR322 replicon with the corresponding region from pJRD158. The number of unique restriction sites was thus raised from 21 in pJDB207 to 28 in the plasmid pMH158 (Fig. 3) and from 14 in YRp7 to 29 in pJO158,[19] thereby considerably enhancing the versatility of these vectors.

The availability of the new restriction site bank in pJRD184 gave the possibility of further increasing the usefulness of these vectors. In the case of the pJDB207 derivative pMH158, substitution of the restriction

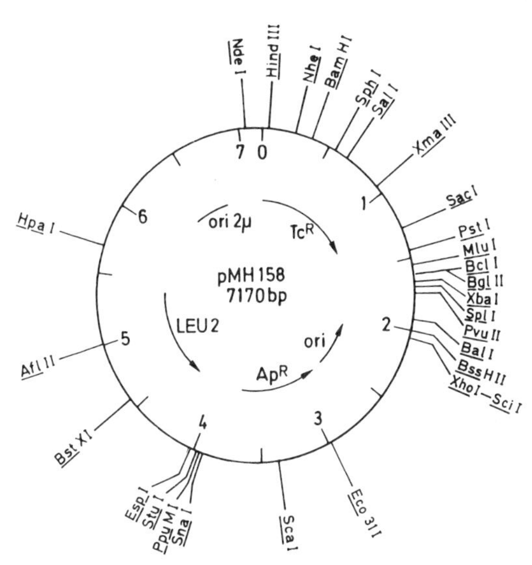

FIG. 3. Restriction map of *E. coli–S. cerevisiae* shuttle vector pMH158. Only unique restriction sites are indicated. Abbreviations: as in Fig. 1; ori 2μ, the origin of replication of the yeast 2μ plasmid: LEU2, the yeast *LEU*2 gene.

site bank of pJRD158 by that of pJRD184 would have made relatively little difference due to duplication of preexisting sites, and was therefore not attempted. In the case of YEp7 derivative pJO158, a considerable improvement could easily be achieved simply by replacing the *Bam*HI–*Xho*I region (containing the restriction site bank) by that of pJRD184. Figure 4 shows the result of this construction: pJO184, which has all of the properties of YEp7 but now has 42 unique restriction sites.

Wide-Host-Range Restriction Site Bank Cosmid Vector pJRD215

Plasmids based on the pMB1 replicon such as pBR322 and pJRD184 replicate only in *E. coli* and related enteric bacteria. It is often desirable to be able to transfer recombinant plasmids between species and one method of doing this is the double-replicon shuttle vector of the type described above for *S. cerevisiae*. However, some gram-negative plasmids are naturally able to replicate in many different species. Many of these, such as RK2, are large self-transferable conjugative plasmids which are not useful for cloning experiments. Some miniderivatives of these have been constructed, such as pRK290 and pLAFR1, but tend to be large and have few useful cloning sites.[41,42]

Another type of wide-host-range plasmid includes the related (possibly identical) plasmids RSF1010, R300B, and R1162 (*E. coli* incompatibility group Q), which have been reported to replicate in *Acetobacter xylinum, Acinetobacter calcoaceticus, Agrobacterium tumefaciens, Alcaligenes eutrophus, Azotobacter vinelandii, E. coli, Klebsiella aerogenes, Methylophilus methylotrophus, Providencia* sp., *Pseudomonas* sp., *Rhizobium melitoti, Rhodopseudomonas spheroides, Salmonella* sp., and *Serratia marcescens*.[24,26,43] These plasmids are small (8.7 kb) and non-self-transmissible but nonetheless can be transferred between species in the presence of several groups of conjugative plasmids such as RP4.[24] The physical and genetic map of RSF1010 is shown in Fig. 5. The plasmid carries genes coding for resistance to two antibiotics, sulfonamide and streptomycin, and three genes essential for DNA replication.[44,45] It is un-

[41] M. Ditta, S. Stanfield, D. Corbin, and D. R. Helsinki, *Proc. Natl. Acad. Sci. U.S.A.* **77,** 7347 (1980).

[42] A. M. Friedman, S. R. Long, S. E. Brown, W. J. Buikema, and F. M. Ausubel, *Gene* **18,** 289 (1982).

[43] M. Bagdasarian, R. Lurz, B. Rückert, F. C. H. Franklin, M. M. Bagdasarian, J. Frey, and K. N. Timmis, *Gene* **16,** 237 (1981).

[44] K. Kim and R. J. Meyer, *J. Mol. Biol.* **185,** 755 (1985).

[45] V. Haring, P. Scholz, E. Scherzinger, J. Frey, K. Derbyshire, G. Hatfull, N. S. Willett, and M. Bagdasarian, *Proc. Natl. Acad. Sci. U.S.A.* **82,** 6090 (1985).

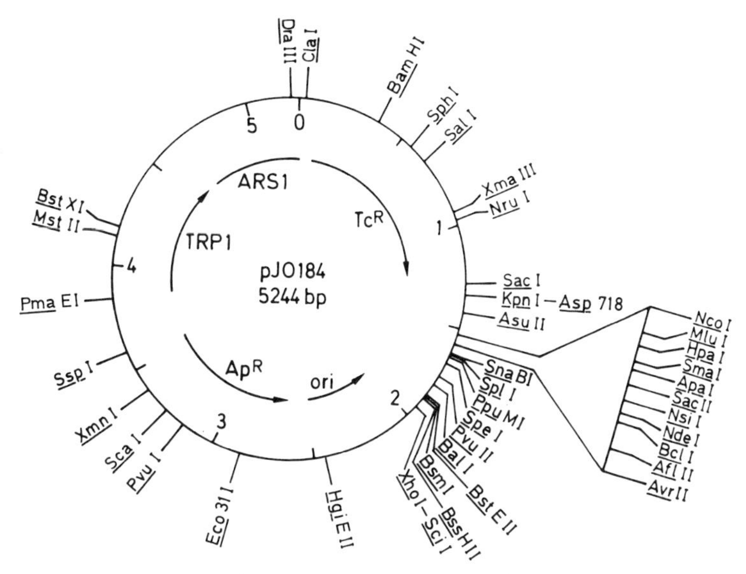

FIG. 4. Restriction map of *E. coli–S. cerevisiae* shuttle vector pJO184. Only unique restriction sites are indicated. Abbreviations: as in Fig. 1; ARS1, yeast autonomous replication sequence; TRP1, the yeast *trp*1 gene.

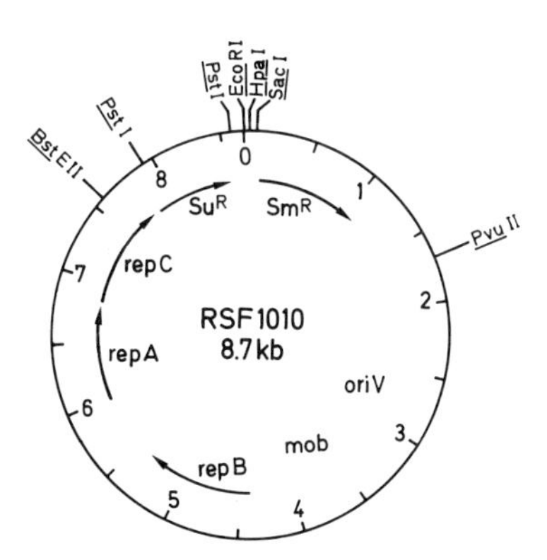

FIG. 5. Restriction map of wide-host-range plasmid RSF1010. Abbreviations: Su^R, resistance to sulfonamide: Sm^R, resistance to streptomycin; oriV, origin of replication; mob, mobilization functions; rep, genes concerned with replication.

satisfactory as a cloning vector, having very few unique restriction sites that can be used for cloning, and only *Eco*RI, *Hpa*I, *Sst*I, and *Pst*I (double) are in dispensible regions.[43,46] Despite this, the advantages of wide host range and mobilization make RSF1010 (R300B, R1162) potentially useful as a cloning vehicle and several improved vectors have been reported.[43,47,48]

The remarkable scarcity of unique restriction sites of RSF1010 can be turned to advantage for the construction of a wide-host-range restriction site bank vector since the addition of a restriction site bank duplicates very few preexisting sites. A useful restriction site bank, pJRD215 (derived from RSF1010), is shown in Fig. 6. The construction of this plasmid is very complicated and this will be described in detail elsewhere.[21] In clockwise orientation starting from the *Eco*RI site (defined as zero) it contains a 187-bp *Eco*RI–*Eco*RV region from pBR322, carrying the Tc^R promoter; the *Hpa*I (0.01) to *Pst*I (7.9) region of RSF1010, carrying the Sm^R gene, the origin of replication, the replication genes, and the mobilization region; the Km^R gene of Tn*5* derived from pKC7[49]; an *Eco*RI–*Hin*dIII polylinker from pEMBL8[27] carrying a 400-bp λ cos region from MUA10[50]; part of a polylinker from pIC20R,[8] and finally the *Kpn*I–*Pvu*II from a pJRD184 (Fig. 2) variant that had been modified to remove the *Bgl*II restriction site. In addition to the above an *Asu*II site in the region 3′ to the Km^R gene was replaced by an *Xho*I linker and a *Hin*dIII site upstream of the Km^R gene was converted to *Nhe*I by repair with DNA polymerase I Klenow fragment. As already described above for other restriction site bank vectors, these operations are designed to maximize the number of unique cloning sites without significant increase in size. In fact, pJRD215 is both smaller (10.2 kb) and contains many more cloning sites (23 unique restriction sites) than other wide-host-range vectors (e.g., pLAFR1: 21.6 kb, 1 unique site[42]; pKT230: 11.9 kb, 6 unique sites[43]; pGSS33: 13.4 kb, 9 unique sites[48]).

pJRD215 retains its wide-host-range properties and can be transferred to many different species of gram-negative bacteria in the presence of a self-transmissible plasmid in the donor strain. Conjugation is an efficient process and this may facilitate transfer to species which carry hostile restriction systems. In this context it is interesting to speculate that the statistically abnormal scarcity of 6-bp palindromic restriction sites in

[46] R. Meyer, M. Hinds, and M. Brasch, *J. Bacteriol.* **150,** 552 (1982).

[47] J. Frey, M. Bagdasarian, D. Feiss, F. C. H. Franklin, and J. Deshusses, *Gene* **24,** 299 (1983).

[48] G. S. Sharpe, *Gene* **29,** 93 (1984).

[49] R. N. Rao and S. G. Rogers, *Gene* **7,** 79 (1979).

[50] E. M. Meyerowitz, G. M. Guild, L. S. Prestidge, and D. S. Hogness, *Gene* **11,** 271 (1980).

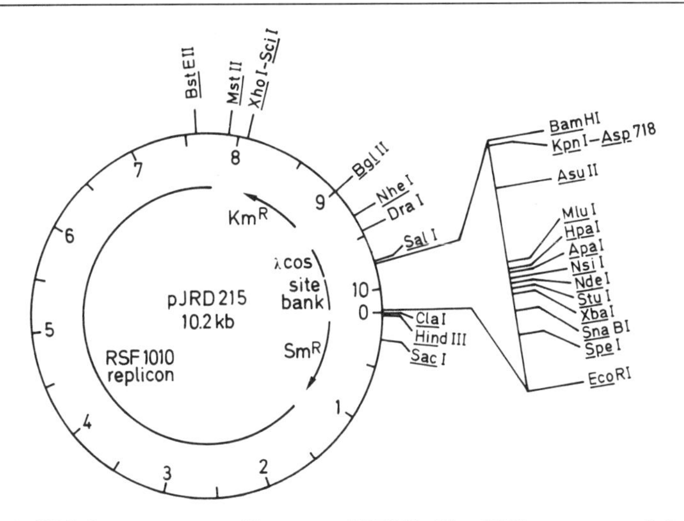

FIG. 6. Wide-host-range cosmid vector pJRD215. The DNA sequence of the region between coordinates 7.9 and 0.185 is known, and the sequencing of the RSF1010 replication region has nearly been completed (P. Scholz, personal communication). Thus only sites known to be unique have been listed, though other unique sites may be present. Sites for restriction enzymes: *Aat*II, *Bal*I, *Bbe*I, *Bcl*I, *Bss*HII, *Eco*RV, *Nae*I, *Nco*I, *Nru*I, *Pst*I, *Pvu*II, *Sma*I, *Sph*I, *Ssp*I, *Tth*111I, and *Xma*III are known to be multiple. Abbreviations: as in Fig. 5; Km^R, kanamycin resistance; λcos, cohesive ends DNA sequence of phage λ recognized by λ terminase.

RSF1010 may be an evolutionary response to avoid restriction systems present in its various hosts.

Construction and Screening of a Pseudomonas Cosmid Gene Bank

The inclusion in pJRD215 of the 400-bp λ *cos* region containing the sealed λ sticky ends enables the plasmid to be used as a cosmid when the total size of the inserted DNA together with that of the vector is in the range 35–50 kb. Packaging in phage λ is a highly effective process, being 100–1000 times more efficient than transformation, so that a complete gene bank can be made from only a few micrograms of DNA. Similarly, the large size of the DNA insert ensures that relatively few recombinant colonies have to be screened to isolate the desired gene.[51] In practice, cosmid clones able to complement the *Pseudomonas* vanillate utilization mutant *van*5 (Materials and Techniques) were obtained at a frequency of 0.2–0.5%, a number easily handled in microwell plates. Similar results have been obtained with other *Pseudomonas* genes. The use of very large

[51] J. Collins, this series, Vol. 68, p. 309.

35- to 40-kb partial *Mbo*I digests as the source of donor DNA avoids problems due to the nonstatistical distribution of restriction sites around the gene of interest. For example, enzymes such as *Bam*HI that cut the DNA on the average every 4 kb may cleave the region of interest many times so that the intact gene or operon may be rarely obtained. However, the large size of the partial *Mbo*I insert poses problems for the subsequent deletion analysis or subcloning of the gene necessary for further investigation. In practice, we have found it advisable to isolate several independent positive clones. All of these differ in the exact end point of the insert of the *Mbo*I partial digest but since all complement the same mutant they must contain a region in common. Comparison of the restriction patterns of independent clones, using restriction enzymes that cleave infrequently, usually shows a common fragment for at least one of the enzymes used. This fragment is then subcloned into pJRD215, the task being facilitated by the wide choice of unique restriction sites available in this vector. Thus, comparison of clones able to complement the *van*5 mutant of *Pseudomonas* showed a common 3.7-kb *Bam*HI fragment. Subcloning of this fragment led to the isolation of the complementing gene.[21a] Some caution should be exercised in this approach since the genes isolated may turn out to be part of an operon or cluster of related genes. To obtain the complete operon it is advisable to isolate several different mutants having the same phenotype. For example, in the case of vanillate, it was subsequently found that not all van^- mutants were complemented by the 3.7-kb *Bam*HI fragment, but all were complemented by a 4.5-kb *Asu*II fragment which included this *Bam*HI fragment.[21a]

This result demonstrates the necessity of having a good collection of phenotypically identical mutants for screening of the clones. If some of the mutants in the collection are not complemented by any of the clones able to complement the mutant initially used then it is possible that the respective genes are unlinked on the *Pseudomonas* chromosome and that a rescreening of the *Pseudomonas* cosmid bank against other mutants is necessary.

More recently, the same cloning procedure has been used to isolate the aspartase and asparaginase genes from an industrial strain of *Pseudomonas*.[52]

Discussion

This chapter summarizes the restriction site bank series of vectors that have been constructed as tools for the genetic manipulation of DNA. The

[52] D. Ursi, D. Prozzi, J. Davison, and F. Brunel, *J. Biotech.* **5**, 221 (1987).

advantage of the restriction site bank vectors is that they maximize the choice of restriction enzymes that can be used for cloning and facilitate subsequent subcloning and deletion analysis. Thus they permit, in a single plasmid, a variety of operations that could otherwise be done only with a combination of several different vector systems. Of the known restriction endonucleases that recognize 6-bp palindromes, 39 have unique recognition sites in *E. coli* vector pJRD184 and 4 others, while not unique, can also be used for cloning without disturbing essential vector functions. Only two known restriction enzymes of this class lack a recognition site on pJRD184 (*Bsp*MII and *Pma*EI, enzymes discovered since the completion of the vector). On the other hand, there are seven additional unique 6-bp palindromes in pJRD184 which may represent cleavage sites for restriction enzymes yet to be discovered. Finally, pJRD184 contains several cloning sites for the less useful restriction enzymes that recognize interrupted palindromes or nonpalindromic sequences. Modified restriction site banks have also been applied to the improvement of vectors for *Saccharomyces cerevisiae,* resulting in a large increase in the number of cloning sites.

The RSF1010-based pJRD215 vector is of particular interest, combining the advantages of wide host range among gram-negative bacteria, transferability by conjugation (thereby avoiding problems with nontransformable species and host restriction systems), cosmid packaging into phage λ particles, and cloning versatility due to the restriction site bank containing 23 unique sites. This vector should facilitate the cloning of genes from many interesting but genetically undefined organisms.

The example given involves the cloning of the genes for vanillate utilization from an uncharacterized pseudomonad ATCC19151. Vanillate is a key intermediate in the degradation of the lignin molecule, one of the world's most abundant carbon sources and a potential source of industrial raw material and energy. This system is also a useful choice to demonstrate the generality of the cosmid cloning, conjugal transfer, complementation method of gene isolation. Neither the bacterium used, nor the genetics and biochemistry of vanillate biodegradation, are well documented in the scientific literature and it is doubtful that these genes could have been identified by any other method. Efforts to purify the vanillate demethylase have not been successful[37] and even the reaction catalyzed (vanillate to protocatechuate) has not been rigorously demonstrated; possibly due to the presence, in *Pseudomonas* lysates, of protocatechuate oxygenase that degrades the reaction product.[37] We hope that the eventual expression of the *van* genes in *E. coli* (which does not degrade protocatechuate) will provide a cleaner biochemical background for the determination of the corresponding enzymatic activities. Such an approach, if successful, would represent a kind of "reverse biochemistry," where the enzymatic activity is determined only after the gene has been cloned and overexpressed in a new environment.

In the future, it should be possible to apply modified restriction banks to the improvement of vector systems in other organisms (*Bacillus, Streptomyces, Corynebacterium,* higher eukaryotes, etc.). Similarly, we are presently constructing expression vectors having restriction site banks downstream of the promoter to facilitate insertion and manipulation of the expressed gene.

Acknowledgments

We thank N. Chevalier, V. Ha Thi, N. Kennedy, M. Merchez, J. Oberto, M. F. Pilaete, D. Prozzi, S. Tournis-Gamble, D. Ursi, all of whom actively participated in various parts of this project. This research was partially financed by Research Contract GBI-3-016-B of the Biomolecular Engineering Programme and BAP-0048-B (GDF) of the Biotechnology Action Program of the Commission of the European Communities, and partially by the Labofina Co. (Belgium).

[4] Plasmids for the Selection and Analysis of Prokaryotic Promoters

By JÜRGEN BROSIUS and JAMES R. LUPSKI[1]

Present information about transcription initiation stems mainly from the comparative analysis of wild-type and mutant promoters.[1a–5] Plasmid vectors that can select or screen for transcription promoters are important tools for the isolation and characterization of DNA fragments containing such regulatory sequences. A number of these vectors are available.[6–9] In this chapter two vectors (pKK175-6 and pKK232-8) are described which

[1] Abbreviations: Ap^r, ampicillin resistance; bp, base pair; Cm^r, chloramphenicol resistant; Cm^s, chloramphenicol sensitive; CAT, chloramphenicol acetyltransferae; kb, kilobase; LB, Luria-Bertani; RBS, ribosome-binding site (Shine–Dalgarno sequence); Tc^r, tetracycline resistant; Tc^s, tetracycline sensitive.

[1a] D. Pribnow, *Biol. Regul. Dev.* **1,** 219 (1980).

[2] M. Rosenberg and D. Court, *Annu. Rev. Genet.* **13,** 319 (1979).

[3] U. Siebenlist, R. B. Simpson, and W. Gilbert, *Cell* **20,** 269 (1980).

[4] D. K. Hawley and W. R. McClure, *Nucleic Acids Res.* **11,** 2237 (1983).

[5] W. R. McClure, *Annu. Rev. Biochem.* **54,** 171 (1985).

[6] G. Widera, F. Gautier, W. Lindenmaier, and J. Collins, *Mol. Gen. Genet.* **163,** 301 (1978).

[7] R. Neve, R. W. West, and R. L. Rodriguez, *Nature (London)* **277,** 324 (1979).

[8] R. W. West and R. L. Rodriguez, *Gene* **20,** 289 (1982).

[9] M. Rosenberg, K. McKenney, and D. Schumperli, *in* "Promoters—Structure and Function" (R. L. Rodriguez and M. J. Chamberlin, eds.), p. 387. Praeger, New York, 1982.

utilize antibiotic resistance structural genes whose natural promoters have been deleted. These promoter probe vectors are particularly useful because they have very low background of antibiotic resistance when no promoter is present and they are constructed to protect against the creation of artifactual promoters.[10] Plasmids pKK175-6 (Ap^r, Tc^s) and pKK232-8 (Ap^r, Cm^s) enable the fusion of a transcription initiation signal to a structural gene whose product can be assayed readily. Procedures for assaying promoter activity by measuring the drug resistance gene product are described as well as applications of these vectors to the study of various gene systems.

Vectors for the Selection and Characterization of Promoters

The two featured plasmids possess promoter-deficient antibiotic resistance genes flanked by transcription terminators. The purpose of the terminators is (1) to block transcription initiated from other plasmid-borne promoters from continuing into the antibiotic resistance structural genes (this minimizes the background levels of antibiotic resistance gene expression) and (2) to prevent transcription initiating at cloned strong promoters from interfering with transcription of plasmid genes which in turn might result in a destabilization of the host–vector system. Thus the terminators isolate the transcription unit of the gene whose product is to be assayed. In addition, in one vector (pKK232-8) there are translational stop codons in all three reading frames upstream from the Cm gene, which prevents translational readthrough from any ATG start codons that might be introduced with the selected promoter fragment. This ensures that translation will proceed only from the Cm^r start codon through the entire structural gene.

Plasmid pKK175-6 (Fig. 1) is a derivative of pBR322[10a] but lacks the region between the Tc gene and the origin of replication between the *Ava*I site (which is filled in and inactivated) and the *Pvu*II site (that has been regenerated). This region is important in the copy number control of ColE1 plasmids.[11,11a] This 639-bp fragment has been replaced with the 498-bp *Alu*I segment that carries the 5 S rRNA gene and the two transcription terminators T_1, T_2 from the *rrn*B ribosomal RNA operon.[12] The sec-

[10] J. Brosius, *Gene* **27,** 151 (1984).

[10a] F. Bolivar, R. L. Rodriguez, P. J. Greene, M. C. Betlach, H. L. Heynecker, H. W. Boyer, J. H. Crosa, and S. Falcow, *Gene* **2,** 95 (1977).

[11] A. J. Twigg and D. Sherratt, *Nature (London)* **283,** 216 (1980).

[11a] J. R. Lupski, S. J. Projan, L. S. Ozaki, and G. N. Godson, *Proc. Natl. Acad. Sci. U.S.A.* **83,** 7381 (1986).

[12] J. Brosius, T. J. Dull, D. D. Sleeter, and H. F. Noller, *J. Mol. Biol.* **148,** 107 (1981).

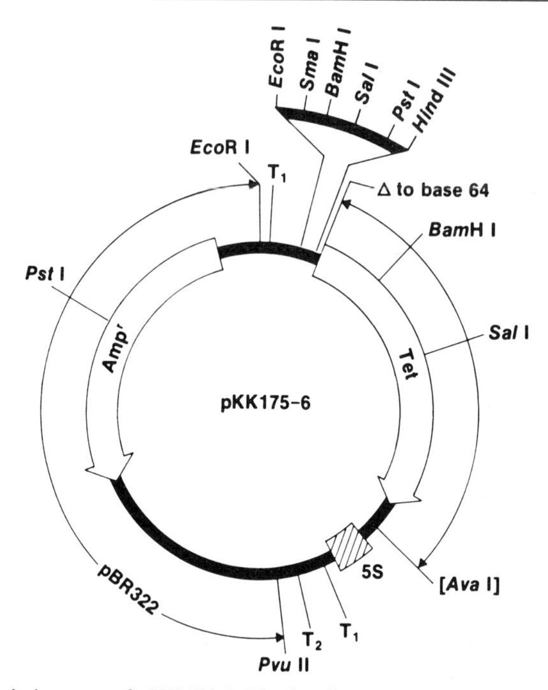

FIG. 1. Restriction map of pKK175-6. The fine lines on the outer circle (from *Eco*RI to *Pvu*II and from *Ava*I to base 64) depict pBR322 sequences. The sequence between the inactivated *Ava*I site and the *Pvu*II site is a 498-bp *Alu*I fragment derived from *E. coli rrn*B ribosomal RNA operon (position 6416–6913[12]) and contains the 5 S rRNA gene and the two transcription terminators T_1 and T_2. The terminators function in both orientations but most efficiently in their natural clockwise orientation. A shorter fragment containing *rrn*B T_1 only (position 6610–6778[12,15]) is located (terminating most efficiently in the clockwise orientation) upstream from the *Eco*RI–*Sma*I–*Bam*HI–*Sal*I–*Pst*I–*Hin*dIII polylinker (the region magnified outside the circle). The Tc^r promoter of pBR322 is deleted since this portion of the plasmid is deleted. The −35 region of the "antitet" promoter[20] pointing in the counterclockwise orientation is still present. The plasmid confers resistance to Ap but is (without promoter insert in the polylinker) sensitive to Tc.

ond portion of pBR322 that is deleted is from the *Eco*RI site at position 1 to base 64 of pBR322.[13,14] This eliminates the promoter for the Tc^r gene. The deletion of the entire Tc^r promoter reduces the chance of fortuitous promoter regeneration involving either the remaining −10 or −35 region of the *tet* promoter. In pKK175-6 this area is replaced by *rrn*B transcription terminator T_1 (position 6610–6778 of *rrn*B[12,15]) and the polylinker

[13] J. G. Sutcliffe, *Cold Spring Harbor Symp. Quant. Biol.* **43,** 77 (1979).

[14] K. Peden, *Gene* **22,** 277 (1983).

[15] J. Brosius, *Gene* **27,** 161 (1984).

```
ACCTGACGTCTAAGAAACCATTATTATCATGACATTAACCTATAAAAATAGGCGTATCAC

GAGGCCCTTTCGTCTTCAAGGAATTCCAGGCATCAAATAAAACGAAAGGCTCAGTCGAAA

GACTGGGCCTTTCGTTTTATCTGTTGTTTGTCGGTGAACGCTCTCCTGAGTAGGACAAAT

TCCGCCGGGAGCGGATTTGAACGTTGCGAAGCAACGGCCCGGAGGGTGGCGGGCAGGACG

CCCGCCATAAACTGCCAGGAATTCCCGGGGATCCGTCGACCTGCAGCCAAGCTTGGGCTA

AACGCAGTCAGGCACCGTGTATGAAATCTAACAATGCGCTCATCGTCATCCTCGGCACCG
```

FIG. 2. Nucleotide sequence (inferred) of the cloning region of pKK175-6. The *Aat*II and *Eco*RI restriction sites, as well as the polylinker (*Eco*RI, *Sma*I, *Bam*HI, *Sal*I, *Pst*I, and *Hin*dIII), are underlined. The region 5′ to the proximal *Eco*RI site corresponds to pBR322 sequence (position 4281–4361[13]). The region between the two *Eco*RI sites contains *rrn*B transcription terminator T_1 (dyad symmetry shown under- and overscored). The area distal to the polylinker contains pBR322 sequences from position 64–123,[13] including the putative ATG start codon (boxed) of the Tc structural gene.

from pUC8.[16] The *rrn*B T_1 terminator (on a 170-bp *Eco*RI fragment) prevents clockwise transcription from plasmid promoters into the Tc^r gene. The polylinker located upstream from the Tc^r gene allows for the insertion of fragments into unique *Sma*I and *Hin*dIII sites. A limited digest is necessary for insertion into the *Eco*RI, *Bam*HI, *Sal*I, and *Pst*I sites since they also occur elsewhere on the plasmid. The *Sma*I site should be most useful since it will accept any blunt-end fragment generated by other blunt-end cutters, fill-in reactions with DNA polymerase I large fragment, or exonuclease treatment. Functional promoters located on DNA fragments cloned into one of designated sites of pKK175-6 should confer Tc^r at various levels (depending, in part, on promoter strength) to the host strain. Inserted sequences can be cut out via the flanking restriction sites. A partial sequence of the relevant area around the cloning region of pKK175-6 is shown in Fig. 2.

The second promoter selection vector, pKK232-8 (Fig. 3) is similar in design to the previous vector. The main difference is the Tc^r structural gene has been replaced with the structural gene for Cm^r. Insertion of a promoter in one of the polylinker sites confers chloramphenicol resistance to the host. An additional feature in pKK232-8 is the insertion of a DNA fragment that places translational stop codons in all three reading frames between the polylinker and the ATG start codon of the CAT gene. The *rrn*B transcription terminator protects from clockwise transcription into the Cm^r structural gene and since the terminator functions to some extent in the opposite orientation it will also reduce counterclockwise transcription from an inserted divergent promoter into the Ap^r structural

[16] J. Vieira and J. Messing, *Gene* **19,** 259 (1982).

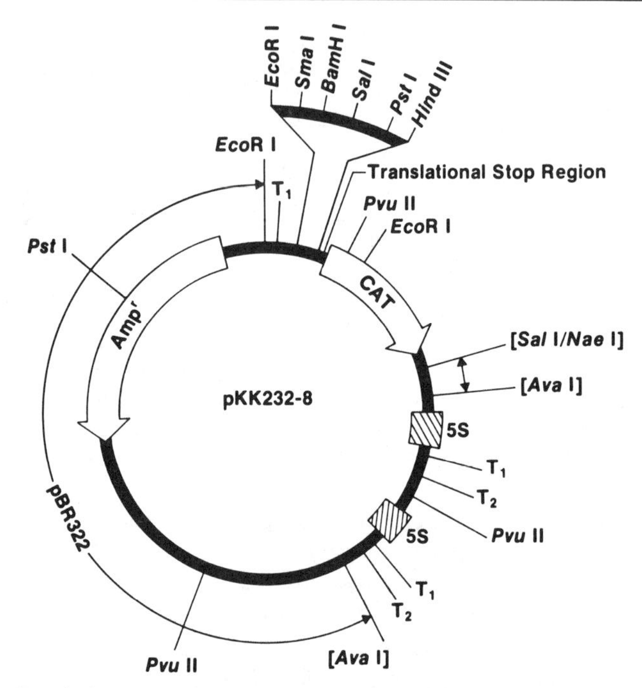

FIG. 3. Restriction map of pKK232-8. The fine lines on the outer circle flanked by arrowheads depict pBR322 sequences. Two 498-bp *Alu*I fragments from *rrn*B (position 6416–6913[12]) containing the 5 S rRNA gene and transcription terminators T_1 and T_2 are tandemly inserted into the filled-in and thus inactivated *Ava*I site. The terminators function in both orientations but most efficiently in their natural—in this plasmid clockwise—orientation. Another fragment containing *rrn*B T_1 only (position 6610–6778) is located upstream from the polylinker sequence (enhanced segment above the outer circle). This terminator is able to block plasmid-originated clockwise transcription from reaching the CAT marker gene and to a lesser extent eventual counterclockwise transcription from a cloned DNA fragment that contains a divergent double promoter. Between the pUC8 polylinker[16] and the CAT cartridge[17,36] a short DNA fragment providing translational stop codons in all three reading frames has been inserted. The CAT cartridge from pMC71[17] was fused with its proximal *Hin*dIII site to the *Hin*dIII site of the polylinker and with its distal filled-in *Sal*I site (leading to its inactivation) to one of the *Nae*I sites in the Tc^r gene of pBR322 (probably position 1285[13]), which is also inactivated and therefore shown in brackets.

gene whose product β-lactamase serves as an internal control in the assay described below. The T_1 fragment, polylinker, and CAT cartridge[17] replace pBR322 sequences from the *Eco*RI site up to one of the *Nae*I sites in the Tc^r gene. Two terminator fragments (T_1 and T_2 from *rrn*B) are inserted

[17] T. J. Close and R. L. Rodriguez, *Gene* **20,** 305 (1982).

```
ACCTGACGTCTAAGAAACCATTATTATCATGACATTAACCTATAAAAATAGGCGTATCAC

GAGGCCCTTTCGTCTTCAAGGAATTCCAGGCATCAAATAAAACGAAAGGCTCAGTCGAAA

GACTGGGCCTTTCGTTTTATCTGTTGTTTGTCGGTGAACGCTCTCCTGAGTAGGACAAAT

CCGCCGGGAGCGGATTTGAACGTTGCGAAGCAACGGCCCGGAGGGTGGCGGGCAGGACGC

CCGCCATAAACTGCCAGGAATTCCCGGGGATCCGTCGACCTGCAGCCAAGCTTGAGTAGG

ACAAATCCGCCGAGCTTCGACGAGATTTTCAGGAGCTAAGGAAGCTAAAATGGAGAAAAA

AATCACTGGATATACCACCGTTGATATATCCCAATGGCATCGTAAAGAACATTTTGAGGC

ATTTCAGTCAGTTGCTCAATGTACCTATAACCAGACCGTTCAGCTGGATATTACGGCCTT

TTTAAAGACCGTAAAGAAAAATAAGCACAAGTTTTATCCGGCCTTTATTCACATTCTTGC

CCGCCTGATGAATGCTCATCCGGAATTC
```

FIG. 4. Nucleotide sequence of the cloning region of pKK232-8. The *Aat*I, *Eco*RI, *Pvu*II, and polylinker sites are underlined. The region 5′ to the proximal *Eco*RI site corresponds to pBR322 sequence (position 4281–4361[13]). The segment between the two proximal *Eco*RI sites contains *rrn*B transcription terminator T_1 (region of dyad symmetry shown under- and overscored). The area 3′ to the *Hin*dIII site contains a filled-in 20-bp *Dde*I–*Hpa*II fragment from pKK3535 [J. Brosius, U. Ullrich, M. A. Raker, A. Gray, T. J. Dull, R. R. Gutell, and H. F. Noller, *Plasmid* **6,** 112 (1981)] containing nucleotides 6689–6708 from the *rrn*B operon,[12] providing translational stop codons (overscored) in two reading frames. This region is followed by the CAT cartridge,[17] providing two more stop codons in the remaining reading frame (overscored). The ATG start codon of the CAT gene is boxed. A portion containing the CAT coding sequence (up to the distal *Eco*RI site) is included.

into the filled in *Ava*I site. The pBR322 sequences between the destroyed *Ava*I site and the *Pvu*II site remain intact, resulting in a copy number similar to pBR322. The nucleotide sequence of the relevant portion of pKK232-8 is shown in Fig. 4. The unique *Sma*I, *Bam*HI, *Sal*I, and *Hin*dIII restriction sites of the polylinker in pKK232-8 can be used for insertion of fragments, while if insertion into the *Eco*RI or *Pst*I sites is required the plasmid has to be linearized by limited digestion since these sites also occur elsewhere in the vector.

Since it has been observed that the correlation between promoter strength and tetracycline resistance was not always linear in pKK175-6, its precursors, and related plasmids, the use of this plasmid is more limited except in procedures to select for promoter down mutations. Therefore all further discussions and procedures are directed primarily at pKK232-8, which shows, within the limitations discussed below, an excellent correlation of promoter activity and levels of chloramphenicol resistance or CAT activity.

Transcriptional Gene Fusions Using Promoter Probe Vectors

Plasmid pKK175-6 (Ap^r, Tc^s) and pKK232-8 (Ap^r, Cm^s) enable the fusion of a transcription initiation signal to a structural gene whose product can be assayed readily. If a DNA sequence which is cloned into the polylinker contains a promoter it will activate the drug resistance structural gene. Some caveats to this method of analyzing transcription initiation by assaying a protein product are as follows: (1) DNA sequences between the promoter and the end of the polylinker contained within the cloned insert may interfere with the assay results.[18] For example, if those sequences have homologies with sequences within the structural gene to be assayed, RNA secondary structures may alter the assay results. It is also possible that these sequences may contain a terminator or attenuator. (2) The assay results reflect the turnover of the protein being measured. Results may vary somewhat depending on strain background, growth conditions, and the time in the cell cycle of the host in which the assay is performed. (3) Plasmid copy numbers may vary as a function of inserted DNA. Thus in an optimized experiment to determine the relative strengths of different promoters using these gene fusion systems one should use isogenic host strains, and protein extracts should be prepared under the same conditions. In addition, the gene fusions should contain identical sequences between the promoter start site and the gene whose product is to be assayed. Finally, there should also be an internal control to normalize results.

Plasmid pKK232-8 can be used for a number of applications.

1. The cloning of composite promoters that are assembled from multiple DNA components (including synthetic DNA): Only the correct assembly of a functional promoter will result in the expression of chloramphenicol resistance in the host.

2. Subcloning of promoter fragments; for example, in a determination of minimal size of functional promoter sequences and cis-acting DNA regulatory sequences following digestion with exonuclease or for nuclease S1 mapping of the transcription start sites.[19,20]

3. Subcloning of strong promoters (e.g., ribosomal RNA promoters from *Escherichia coli* or other organisms): This would not be possible in many vectors because the altered level of transcription would interfere with other plasmid functions, i.e., plasmid replication functions or overexpression of a detrimental protein.

4. Selection of up and down mutations in cloned promoters: This

[18] J. Brosius and A. Holy, *Proc. Natl. Acad. Sci. U.S.A.* **81,** 6929 (1984).
[19] A. J. Berk and P. A. Sharp, *Cell* **12,** 721 (1977).
[20] J. Brosius, R. L. Cate, and A. P. Perlmutter, *J. Biol. Chem.* **257,** 9205 (1982).

should be possible by selecting for overexpression of the Cm^r structural gene using chloramphenicol gradient plates[21,22] to isolate cells harboring pKK232-8 with up mutations in a cloned promoter. For the selection of down mutations in promoters Tc^s colonies would be identified by a positive selection procedure,[23] with cells harboring pKK175-6 including a cloned promoter insert. In addition, cloned promoter segments can be subjected to site-specific or random mutagenesis and the effects of mutations screened and assayed by using the drug resistance marker. The nucleotide sequence of the mutations can be determined by the direct plasmid sequencing method.[24]

5. Shotgun cloning of promoters from plasmids, phages, or genomic DNA from *E. coli* and other organisms for the selection of promoters: By adjusting the concentration of chloramphenicol during plating it is possible to select for promoters of variable strength.

6. Quantifying promoter activity by (1) plating the bacterial strains harboring recombinant plasmid DNA on media containing different concentrations of antibiotic; (2) direct measurements of CAT protein levels[18,20] in sodium dodecyl sulfate polyacrylamide gels,[25] or (3) assaying the CAT activity enzymatically; β-lactamase activity may be used as an internal control.[26–27a]

The manipulations for applications (1)–(3) are standard and are described in other volumes of *Methods in Enzymology* and by Maniatis *et al.*[28] Application (4) has not been implemented to date while (5) and (6) will be discussed in detail.

Shotgun Cloning of Promoters

Preparation of pKK232-8 Vector. Plasmid DNA is prepared from a cesium chloride gradient[28] or can be purchased from Pharmacia (Piscataway, NJ). The amount of linear plasmid DNA to be prepared will depend on the number and type of experiments to be performed. Twenty micro-

[21] V. Bryson and W. Szybalski, *Science* **116,** 45 (1952).
[22] W. Szybalski and V. Bryson, *J. Bacteriol.* **64,** 489 (1952).
[23] B. R. Bochner, H. C. Huang, G. L. Schieven, and B. N. Ames, *J. Bacteriol.* **143,** 926 (1980).
[24] E. Y. Chen and P. Seeburg, *DNA* **4,** 165 (1985).
[25] U. K. Laemmli, *Nature (London)* **227,** 680 (1970).
[26] J. R. Lupski, A. A. Ruiz, and G. N. Godson, *Mol. Gen. Genet.* **195,** 391 (1984).
[27] S. C. Li, C. L. Squires, and C. Squires, *Cell* **38,** 851 (1984).
[27a] R.-A. Klotsky and I. Schwartz, *Gene,* in press (1987).
[28] T. Maniatis, E. F. Fritsch, and J. Sambrook, "Molecular Cloning." Cold Spring Harbor Lab., Cold Spring Harbor, New York, 1982.

grams of plasmid is digested with 40 U of restriction endonuclease *Sma*I (New England BioLabs, Beverly, MA) in a 100-μl reaction volume of 20 m*M* KCl, 6 m*M* Tris–HCl (pH 8.0), 6 m*M* $MgCl_2$, 6 m*M* 2-mercaptoethanol, for 1 hr at 37°. About 1 μl of the digest should be run on a 1% agarose gel to confirm complete digestion of plasmid DNA. To digested DNA 100 μl of distilled water and 4 μl of 0.25 *M* EDTA (pH 8.0) are added. The aqueous phase is extracted twice with 200 μl of a mixture of 25 parts liquified phenol (equilibrated to pH 7.5[28]), 24 parts chloroform, and 1 part isoamyl alcohol. After addition of 50 μl 10 *M* NH_4OAc (final aqueous concentration 2 *M*) and two parts (= 500 μl) of ethanol the mixture is chilled for 5 min at −70° and spun in a microcentrifuge at 15,000 *g* for 5 min in the cold room. The pellet is resuspended in 250 μl of 0.3 *M* NaOAc and reprecipitated with 700 μl of EtOH. After washing the pellet with 80% EtOH and drying, the DNA is dissolved in 100 μl of water at a final concentraton of 0.2 μg/μl. Treatment of the linearized vector with phosphatase[29] is optional but not essential since transformed cells containing recircularized plasmids which do not contain an insert will not grow on selection medium.

Preparation of Inserts to Be Shotgunned. The amounts of DNA which serve as source for the inserts are variable. For plasmid DNA, 0.1–0.3 μg should be sufficient. Bacteriophage λ sized phage DNA requires about 1.5 μg and genomic DNA requires at least 10 μg of DNA. The DNA is digested with restriction enzymes that cut frequently and leave blunt ends [e.g., *Alu*I, *Dpn*I (for *dam*-methylated DNA), *Fnu*DII, *Hae*III, *Mnl*I, *Nla*IV, or *Rsa*I], extracted with phenol/chloroform, and EtOH precipitated as above.

When enzymes, which have 5′ or 3′ protruding ends, are used in order to generate fragments for cloning the ends can be treated with DNA polymerase I, large fragment (Klenow fragment), in the following way: DNA fragments or fragment mixtures (0.2–5.0 μg) are incubated in a total volume of 50 μl in a reaction mixture containing 10 m*M* Tris–HCl (pH 7.5), 50 m*M* NaCl, 10 m*M* $MgCl_2$, 1 m*M* dithiotreitol, 50 μ*M* dATP, dCTP, dGTP, dTTP each (from a stock solution adjusted to pH 7.0 with NaOH[28]), and 5 U of *E. coli* DNA polymerase I, large fragment (Pharmacia), at room temperature for 1 hr. DNA fragments with 3′ overhangs are treated identically except that only one dNTP is added (the first base-paired 3′ nucleotide; for example in a *Pst*I fragment dCTP, or in a *Pvu*I fragment dGTP).

Alternatively, restriction fragments can be made blunt ended with

[29] A. Ullrich, J. Shine, J. Chirgwin, R. Pictet, E. Tisher, W. J. Rutter, and H. M. Goodman, *Science* **196,** 1313 (1977).

nuclease S1, mung bean nuclease, or especially when a shortening of the fragment is acceptable or intended with exonuclease BAL-31[30]: 0.2–5.0 μg of DNA fragments are treated in a mixture of 20 m*M* Tris–HCl (pH 8.0), 600 m*M* NaCl, 12 m*M* $CaCl_2$, 12 m*M* $MgCl_2$, 1 m*M* EDTA, and 0.1 U of BAL-31 at room temperature. The reaction is stopped after a given time by pipetting into phenol/chloroform and the DNA is extracted as described above. The option exists, of course, to stop the reaction, e.g., in samples (20 μl) at different times. For example, in one experiment by terminating the reaction after 0.5, 1, 2, and 4 min and using the conditions described above we deleted approximately 5–50 bp from each side of the fragment. Of course, the length of DNA removed varies with the amount and activity of the enzyme. For critical applications it is recommended to calibrate the enzyme by monitoring the decrease in size of DNA fragments (e.g., pBR322 *Hin*fI digest) of known chain lengths. Another important consideration is the purity of DNA to be treated with BAL-31 exonuclease: We have noted that large amounts of RNA (e.g., from alkali preparations without CsCl gradient purification) inhibit the enzyme and that RNase treatment does not eliminate this problem. As a further option linkers can be added to blunt-ended fragments and as a result they can be inserted into a particular restriction site.

Ligation. Typically when using bacteriophage λ transducing phage DNA 2 μg of *Sma*I linearized plasmid was ligated with 1.5 μl of digested λ DNA. If one is interested in selecting specific promoters it is recommended to ligate a mixture of two or three different separate digests, since cleavage of a specific restriction site may inactivate the desired promoter. Furthermore, the ratio of insert to vector should be kept low to avoid multiple inserts which might lead to confusion in the analysis of the promoter.

Transformation and Selection. Competent *E. coli* HB101[31] or DH1[32] cells are transformed with an aliquot of the ligation mix by standard procedures[28] or to increase efficiency by the method of Hanahan.[32] Cells are plated on LB plates containing from 5 to 500 μg chloramphenicol/ml.

Assays of Promoter Strength Using Transcriptional Gene Fusions

In general three methods are used to assay promoter activity utilizing the promoter probe vector pKK232-8. All rely on constructing a transcriptional gene fusion by cloning a DNA fragment containing a promoter

[30] H. B. Gray, D. A. Ostrander, J. L. Hodnett, R. J. Legerski, and D. L. Robberson, *Nucleic Acids Res.* **2,** 1459 (1975).

[31] H. W. Boyer and D. Roulland-Dussoix, *J. Mol. Biol.* **41,** 459 (1969).

[32] D. Hanahan, *J. Mol. Biol.* **166,** 557 (1983).

sequence into the polylinker of pKK232-8 and then assaying the product of the Cm^r structural gene—CAT. The first is a semiquantitative assay which utilizes LB plates (10 g Bacto tryptone, 5 g yeast extract, 5 g NaCl, 15 g Bacto agar per liter of water[28]) supplemented with different concentrations of antibiotics. The growth of bacterial cells harboring recombinant plasmids containing a promoter cloned into the polylinker of the promoter probe plasmid is then assessed. This method works for both pKK175-6 and pKK232-8. The other two methods are quantitative assays which rely on measuring the Cm^r gene product either directly on sodium dodecyl sulfate–polyacrylamide gels or indirectly measuring CAT activity by a colorimetric assay.

Semiquantitative Assay of Promoter Strength. Cells harboring recombinant plasmids containing a promoter fragment are plated on LB plates supplemented with ampicillin (50 μg/ml) to select for the recombinant plasmid and either tetracycline (5 μg/ml) or chloramphenicol (5 μg/ml). The 5 μg/ml concentration of antibiotic appears to be the minimum concentration which will destroy all cells harboring a recombinant plasmid which does not contain a cloned insert yet enable selection for weak promoters.

As a first approximation to determining the strength of a cloned promoter, *E. coli* cells harboring gene fusions in pKK175-6 or pKK232-8 are streaked on plates containing increasing concentrations of antibiotics[33,34] (see Fig. 5). Plates are prepared freshly by placing 1.5 g agar in 100 ml LB broth, autoclaving and cooling the agar to 50°, then adding the appropriate concentration of antibiotics. This allows one to make four plates (25 ml each) at each drug concentration. Ampicillin (50 μg/ml) is added to all plates to ensure presence of the recombinant plasmid. Figure 5 shows a typical experiment where the control, *E. coli* HB101 harboring pKK232-8 (A), and cells harboring eight different gene fusions from the macromolecular synthesis operon[26,35] (B through I), are streaked on plates containing 50 μg/ml ampicillin plus increasing concentrations of chloramphenicol (5–600 μg/ml). Note the control cells (A) grow on ampicillin but are unable to grow on 5 μg/ml chloramphenicol. Cells harboring the different fusions with promoters of varying strengths stop growing at different concentrations of chloramphenicol. This same methodology has been correlated with analysis of transcripts from the cloned promoter sequences.[34]

Quantitative Assays of Promoter Strength. If the promoter upstream from the CAT gene is of sufficient strength, the protein from crude cell

[33] J. R. Lupski, B. L. Smiley, and G. N. Godson, *Mol. Gen. Genet.* **189,** 48 (1983).
[34] A. A. Ruiz, L. S. Ozaki, F. Zavala, and G. N. Godson, *Gene* **41,** 135 (1986).
[35] J. R. Lupski and G. N. Godson, *Cell* **39,** 251 (1984).

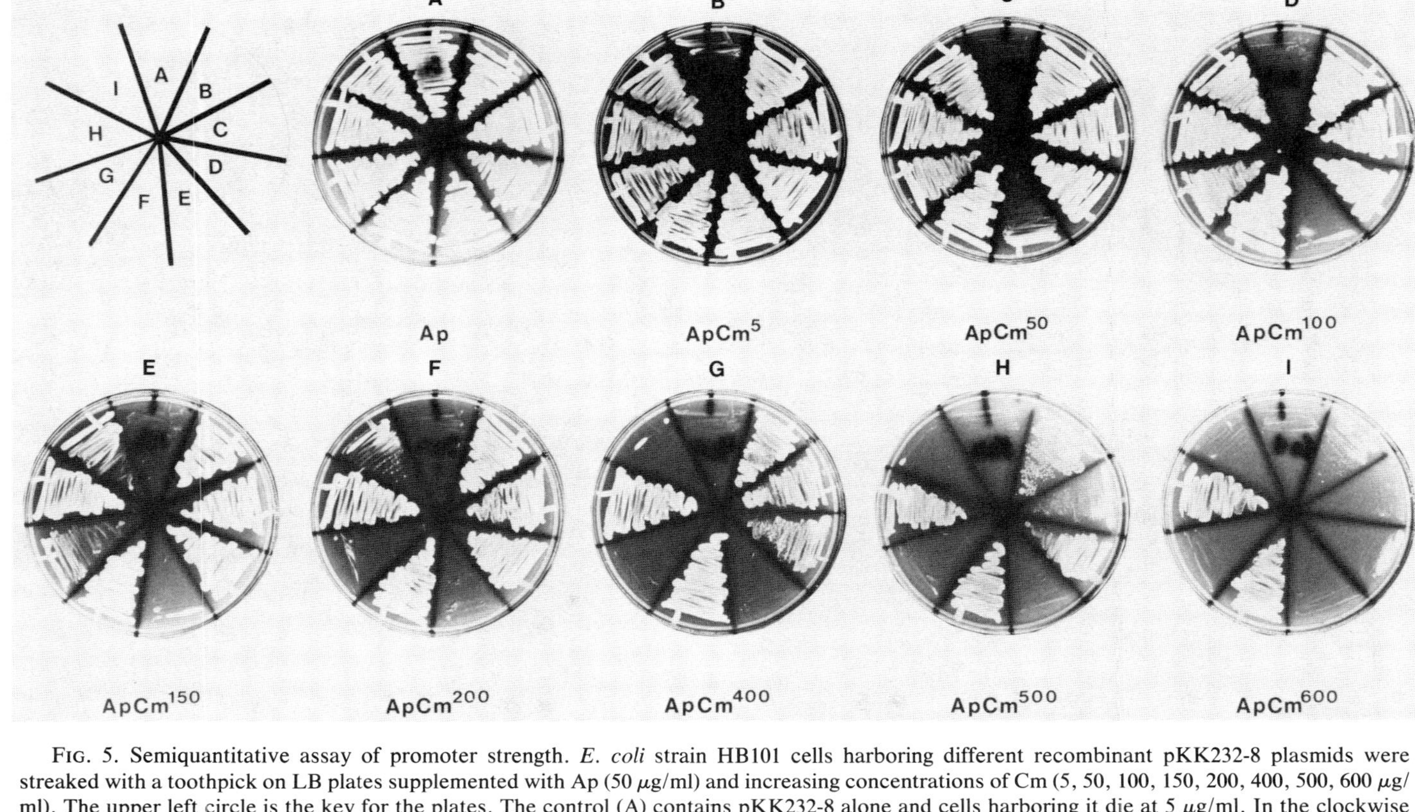

FIG. 5. Semiquantitative assay of promoter strength. *E. coli* strain HB101 cells harboring different recombinant pKK232-8 plasmids were streaked with a toothpick on LB plates supplemented with Ap (50 μg/ml) and increasing concentrations of Cm (5, 50, 100, 150, 200, 400, 500, 600 μg/ml). The upper left circle is the key for the plates. The control (A) contains pKK232-8 alone and cells harboring it die at 5 μg/ml. In the clockwise direction (B–I) are different promoter fusions to the Cm^r gene in the vector. Cells harboring these constructions are not viable at different concentrations of Cm.[26]

extracts can be directly visualized on SDS–polyacrylamide gels after staining with Coomassie blue. One-milliliter samples of bacterial cultures grown in LB medium are harvested at an optical density of $A_{550} = 0.8$ and collected by centrifugation in a micro-test tube at 15,000 *g* for 20 sec. The pellet is boiled for 5 min in loading buffer (10% glycerol, 5% 2-mercaptoethanol, 2.3% sodium dodecyl sulfate, 62.5 m*M* Tris–HCl, pH 6.8[25]) and 10 μl is loaded on a 12% acrylamide gel. The gel is stained with Coomassie Brilliant Blue, destained, and scanned. The CAT protein has a molecular weight of approximately 25,660.[36] Bands from other *E. coli* proteins can be used as internal standards. However, the copy number of the plasmids could vary upon insertion of different DNA fragments. Therefore it is advisable to estimate the plasmid copy numbers, e.g., by performing alkaline minipreparations and agarose gel electrophoresis[37] or using fluorescence densitometry to determine the plasmid copy number per chromosomal equivalent.[11a,38]

Chloramphenicol acetyltransferase can be indirectly assayed *in vitro*. The constitutively expressed β-lactamase activity from the Ap^r gene located elsewhere on the plasmid provides a selectable marker and an internal positive control for normalization. That is, the ratio of the CAT activity to the β-lactamase activity provides a measure of the strength of the promoter inserted into the pKK232-8 polylinker irrespective of plasmid copy number or overall metabolic activity of the cell.[26,27]

To prepare extracts, cells harboring recombinant pKK232-8 DNA are inoculated into 7 ml LB-broth supplemented with ampicillin (50 μg/ml) and grown overnight. The next day 7 ml LB Ap is inoculated with 70 μl from the overnight culture and grown to an $OD_{600} = 0.6$. The CAT specific activity appears to vary the least, between $OD_{600} = 0.5$ and 0.8. Cells (6 ml) are collected in a 15-ml Corex tube by centrifugation at 7000 rpm for 10 min in a Beckman JA17 rotor. The cells are washed in 2 ml of 50 m*M* Tris–HCl (pH 7.8), 30 m*M* dithiothreitol, collected, and resuspended in 600 μl of the same buffer. The cells are then transferred to Eppendorf tubes, placed at −70° for 1 hr, thawed, and sonicated with 10-sec pulses, with at least a 15-sec incubation in ice between pulses, two or three times or until clear, using a special stepped microtip and a Heat Systems Ultrasonics (Farmingdale, NY) sonicator. The cell extracts are centrifuged at 4° for 15 min to remove the cell debris. The supernatant is placed in another Eppendorf tube and used for subsequent assays: CAT, β-lactamase, and protein. The assays are performed at least in triplicate and the mean ± standard deviations is calculated.

[36] N. K. Alton and D. Vapnek, *Nature* (*London*) **282,** 864 (1979).

[37] H. C. Birnboim, this series, Vol. 100, p. 243.

[38] S. R. Projan, S. Carlton, and R. P. Novick, *Plasmid* **9,** 182 (1983).

The assay for CAT is a modification of the spectrophotometric method of Shaw,[39] which takes advantage of the generation of a free CoA sulfhydryl group coincident with the transfer of the acetyl group to chloramphenicol by CAT. The reduced CoA reacts with 5,5′-dithiobis-2-nitrobenzoic acid (DTNB) to yield the mixed disulfide of CoA and thionitrobenzoic acid plus a molar equivalent of free 5-thio-2-nitrobenzoate. The generation of the latter forms the basis of the assay since it has a molar extinction coefficient of 13,600 at 412 nm.

The reaction mixture is freshly prepared by dissolving 8 mg of DTNB in 2.0 ml 1.0 *M* Tris–HCl (pH 7.8), adding 0.4 ml acetyl-CoA from a 5 m*M* frozen stock solution, and adjusting the total volume to 20 ml with distilled water. Acetyl-CoA is very unstable and must be kept frozen. The reaction mixture (600 μl) is placed in a water-jacketed cuvette equilibrated at 37° placed in a Beckman double beam recording spectrophotometer (model 30) and 20 μl of the enzyme extract, prepared as described above from cells harboring the recombinant pKK232-8 plasmid DNA, is added to the reaction mixture. The A_{412} is adjusted to zero then measured for 30 sec using the spectrophotometer recorder to obtain the background CAT activity. To the enzyme containing reaction mixture 12 μl of 5 m*M* chloramphenicol (2-threo, dissolved in 70% EtOH, final concentration 0.1 m*M*) is added to start the reaction. The increase in absorbance (A_{412}) is recorded for about 5 min. The CAT enzyme specific activity (in nmol/min/mg) is calculated as A_{412}/0.0136/mg protein. An alternate assay to the spectrophotometric method uses ^{14}C-radiolabeled chloramphenicol and thin-layer chromatography to separate its acetylated forms, in order to measure CAT activity.[40,41] The CAT assay employing radioactivity is used mostly in eukaryotic systems where small amounts of CAT enzyme activity are present. When this method is used with *E. coli,* cell extracts have to be diluted several-fold to avoid saturation kinetics of the CAT enzyme.[41a]

The β-lactamase enzyme is assayed using the same extracts as above by a modification of Ross and O'Callahan.[42] The assay takes advantage of the fact that the β-lactam antibiotic cephaloridine has a maximum absorption associated with the β-lactam ring at 255 nm. An aliquot of a reaction mixture (600 μl) consisting of 41.5 μg/ml cephaloridine (Sigma) in 0.1 *M* potassium phosphate buffer, pH 7.0, is placed in a water-jacketed cuvette

[39] W. V. Shaw, this series, Vol. 43, p. 737.

[40] C. M. Gorman, L. F. Moffat, and B. H. Howard, *Mol. Cell. Biol.* **2,** 1044 (1982).

[41] C. M. Gorman, M. C. Willingham, I. Pastan, and B. H. Howard, *Proc. Natl. Acad. Sci. U.S.A.* **79,** 677 (1982).

[41a] M. Nesan, N. Almond, J. R. Lupski, and G. N. Godson, in preparation.

[42] G. W. Ross and C. H. O'Callahan, this series, Vol. 43, p. 69.

and equilibrated at 37°. This is blanked against 0.1 *M* potassium phosphate buffer, pH 7.0, and the A_{255} is recorded for 30 sec to get a baseline level. The absorbance of the cephaloridine mixture (A_{255}) is usually between 1.2–1.4. To start the reaction 20 μl of enzyme extract is added to the reaction mixture and the A_{255} recorded for about 5 min. The specific activity for β-lactamase (nmol/min/mg protein) is given by (A_{255}/min) (375/mg protein).

The amount of protein is assayed by the method of Bradford,[43] which is based on the fact that binding of Coomassie Brilliant Blue G-250 to protein causes a shift in the absorption maximum of the dye from 465 to 595 nm. The increased absorption at 595 nm is monitored. This is easily performed using the Bio-Rad (protein assay kit) dye reagent concentrate. For a calibration curve we have obtained the best results using bovine γ-globulin (Bio-Rad; 20–140 μg in 20-μg increments). From the extract as prepared above 100 μl usually yields approximately 20–40 μg protein. This amount must be divided by 5 in the final calculations as only 20 μl of extract is used in the assay.

Conclusion

Promoter–probe plasmid vectors allow one to construct transcriptional gene fusions between cloned promoters and drug resistance structural genes. These are, therefore, useful tools for analyzing transcription initiation signals because the protein products of these constructions are readily assayed. The vectors enable direct selection of promoter sequences and an easy method to evaluate the effects of mutations in cloned promoters. There are, however, limitations to the method of measuring a translational product to determine effects on a transcription initiation signal. Many of these limitations can be overcome by assaying a stable RNA product which is transcribed by a cloned promoter.[18] One must keep in mind that plasmid DNA copy number and DNA topology may introduce artifacts that do not truly represent events occurring *in vivo* on a single-copy chromosome. Nevertheless, this system allows comparative analysis of promoters, especially from genes which are vital to cell integrity and so not amenable to study by other means.

Acknowledgments

We would like to thank Neil Almond for critically reviewing the manuscript and Pharmacia, Molecular Biology Division, for permission to use their prints of Figs. 1 and 3.

[43] M. M. Bradford, *Anal. Biochem.* **72,** 248 (1976).

[5] A λ DNA Protocol Based on Purification of Phage on DEAE–Cellulose

By CYNTHIA HELMS, JAMES E. DUTCHIK, and MAYNARD V. OLSON

Introduction

Applications of λ cloning frequently require the purification of small quantities of DNA from large numbers of individual clones. We describe here a protocol for going from a phage plaque to a DNA sample that is well suited to such applications. In our laboratory, we have used this protocol to analyze restriction digests of over 5000 λ clones, which were processed in sets of 100 or more. The key step in the protocol is the purification of λ virions from either plate stocks or liquid lysates using small DEAE–cellulose columns.[1] Careful attention is also given to the growth of the phage, which is often the limiting step in preparing λ DNA by this or other protocols.

Protocol for DNA Preparation

Most applications of the DEAE–cellulose method are likely to involve isolating DNA from phage that are grown in a small liquid culture or on a single Petri plate. In describing the basic protocol, we will assume the availability of a 5-ml plate overlay or liquid lysate with a titer of at least 5×10^9 pfu/ml (plaque-forming units/milliliter). Detailed protocols for growing λ are presented in a later section, but the importance of obtaining good phage growth should be emphasized at the outset. Efforts to isolate DNA from fewer than 10^{10} pfu are likely to be futile (2×10^{10} pfu = 1 μg λ DNA).

The required materials are listed below in the approximate order of their first appearance in the protocol. Reagents that are normally autoclaved after preparation are marked with an asterisk:

*1. L broth, agar, and soft agar: 1% tryptone (Difco), 0.5% yeast extract (Difco), 0.5% NaCl, 1 m*M* $MgCl_2$ (final pH of broth adjusted to 7.5 with NaOH); add 1.5% Bacto agar (Difco) for bottom agar, 0.7% for top agar

*2. λ Diluent (required only for plate stocks): 10 m*M* Tris–HCl (pH 8), 2 m*M* $MgCl_2$

[1] C. Helms, M. Y. Graham, J. E. Dutchik, and M. V. Olson, *DNA* **4,** 39 (1985).

3. DNase I (required only for liquid lysates, see Notes): 1 mg/ml in 0.15 *M* NaCl, stored at −20° in small aliquots
4. DEAE–cellulose: preswollen DEAE–cellulose (Whatman, DE-52). The cellulose is washed in 0.1 *M* HCl until the pH of the decanted wash is <2, rinsed with 0.1 *M* Tris–HCl (pH 8) until the pH of the decanted wash is >7, and then rinsed twice with 10 m*M* Tris–HCl (pH 8); it can be stored indefinitely at 4° in a 1 : 1 slurry with the final equilibration buffer
5. Columns: 8-mm i.d., bed volumes 2 ml, spontaneous flow rates ~0.2 ml/min (e.g., Bio-Rad Econocolumns)

*6. Chase buffer: 10 m*M* Tris–HCl (pH 8), 10 m*M* $Mg(OAc)_2$, 60 m*M* NaOAc (final pH of mixed reagents not adjusted, but is expected to be 8–9; OAc = acetate)

*7. Elution buffer: 10 m*M* Tris–HCl (pH 8), 50 m*M* $Mg(OAc)_2$ (final pH of mixed reagents not adjusted, but is expected to be 8–9)

8. Proteinase K (Boehringer–Mannheim): 0.1 mg/ml in H_2O, stored at −20° in small aliquots
9. $NaDodSO_4$: 10% (w/v) in H_2O, clarified through a membrane filter

*10. 3 *M* KOAc: prepared in H_2O (pH is not adjusted but is expected to be 8–9)

11. Mussel glycogen (Sigma, type VII, required only when total yields of <2 μg are expected): 1 mg/ml in H_2O, clarified through a membrane filter. If necessary to obtain a clean pellet that redissolves readily, the glycogen can be extracted with phenol and chloroform before filtration
12. Alcohol: 100% 2-propanol and ethanol

*13. TE8: 10 m*M* Tris–HCl (pH 8), 1 m*M* EDTA (pH 8)

The basic protocol for processing 5 ml of liquid lysate or a similar volume λ–diluent overlay from a lysed plate is summarized below:

1. A. Plate lysates: Normally, no preprocessing of the sample is required (see, however, the discussion of lysate dilution in Trouble Shooting). Store 1 ml of the lysate at 4° over a few drops of chloroform in a tightly sealed vial for a permanent phage stock.

B. Liquid lysates: Add 50 μl of a fresh 1 : 1000 dilution in H_2O of the 1 mg/ml DNase I stock (i.e., 50 ng of DNase I) and incubate at 37° for 30 min. Store 1 ml of the treated lysate for a permanent phage stock as described for plate lysates in 1(A). Dilute the remaining 4 ml of lysate by adding 2 ml of H_2O (see discussion of lysate dilution in the Trouble Shooting section). Centrifuge to remove insoluble debris (40 min at 1300 *g*, as in a typical table-top centrifuge at ~2000 rpm, or 5 min at 3000 *g*, as in a typical high-speed centrifuge at ~6000 rpm).

2. Pour lysate over the column, discarding the flowthrough.

3. Pour 5 ml of chase solution over the column, taking some care not to disturb the top of the column; discard the flowthrough.

4. Pour 1 ml of elution buffer over the column, discarding the flowthrough.

5. Pour 0.6 ml of elution buffer over the column, collecting the eluted phage in a 1.5-ml microcentrifuge tube.

6. Add 10 μl of 0.1 mg/ml proteinase K and 25 μl of 10% $NaDodSO_4$.

7. Mix and let sit at room temperature for 5 min.

8. Add 100 μl of 3 *M* KOAc (precipitate will form).

9. Heat at 88° for 20 min (precipitate will dissolve).

10. Allow to cool to room temperature and then chill at 0° for 30 min (precipitate will reform).

11. Centrifuge for 15 min in a microcentrifuge (12,000 *g*), at or below room temperature.

12. Transfer the supernatant to a new microcentrifuge tube.

13. If yields of <2 μg are anticipated, add 20 μl of 1 mg/ml mussel glycogen as carrier.

14. Add 700 μl of 2-propanol (1 vol).

15. Cool to $-20°$ for 30 min or more.

16. Spin in a microcentrifuge for 5–30 min (see Notes: shorter time appropriate if a visible, fibrous precipitate formed immediately in step 14, longer time if not).

17. Pour off the supernatant and add 1.5 ml of ethanol to the tube.

18. Spin in a microcentrifuge for 3–15 min (see Notes: choose shorter or longer time, as in step 16).

19. Dry the pellet under vacuum for 5 min or until visible wetness disappears, whichever is longer.

20. Resuspend the pellet in TE8. Aim for a concentration no higher than 0.2 μg/μl and use at least 25 μl.

21. Store the DNA indefinitely at 4°.

Notes on DNA Preparation

Step 1. While we have found no dilution of plate stocks to be required and the recommended 50% dilution of liquid lysates to be adequate, some workers have found that higher dilutions are required before the phage will adsorb to DEAE–cellulose (see Trouble Shooting). In the liquid lysates, the only purpose of the light DNase treatment is to reduce the viscosity of the lysates so that they do not clog the columns. If column clogging is not a problem, the treatment may be safely omitted. Cellular DNA and RNA are efficiently removed by the column step. Vigorous DNase treatment (or any RNase treatment) of the lysates is not recommended since the cellular nucleic acids are more readily removed if they are left at high molecular weight instead of being degraded to oligonucleotides. Finally, the widespread practice of growing phage on agarose rather than agar plates is discouraged; in our hands, phage grow somewhat better on agar, and the column step efficiently removes the sulfated polysaccharides that are extracted from agar and inhibit many restriction enzymes.

Steps 2–5. The DEAE–cellulose columns do not develop air pockets when allowed to drain freely so they should simply be left until they stop dripping after each addition of a solution to the top of the column. The columns can actually be left for days at room temperature at any stage of the procedure without adverse consequences. The elution regime recommended in steps 4 and 5 is appropriate for columns with a 2-ml bed volume and no significant dead volume below the packing. If columns with a different geometry are employed, the elution profile of the λ DNA should be checked (see Trouble Shooting).

Step 5. After this step, residual phage on the columns are killed by passing 10 ml of 1 *M* NaOH through the columns, and the DEAE–cellulose is then discarded; the columns themselves are autoclaved and repacked with fresh DEAE–cellulose. We have not discovered a satisfactory regeneration protocol.

Step 6. The proteinase K is unnecessary for phage disruption or DNA recovery, but is critical for inactivating a nuclease activity from *Escherichia coli* that is resistant to heat and detergent treatments.

Step 7. When processing large numbers of samples, it is convenient to add the proteinase K (step 6), the $NaDodSO_4$ (step 6), and the KOAc (step 8) sequentially and then mix and heat (step 9). This variation has worked well in our hands, but if problems with DNA degradation are encountered, it may be prudent to include step 7 to provide time for the protease to act before carrying out the protein precipitation.

Steps 13–18. The use of 2-propanol for the initial precipitation allows the peak sample volume to be kept under 1.5 ml; DNA precipitation occurs at a lower alcohol concentration when 2-propanol is used instead of ethanol. The switch to ethanol for the final rinse facilitates drying of the samples.

In this and other small-scale procedures for DNA purification, DNA is often lost at the alcohol precipitation step. The most important consideration in avoiding losses at this stage is an appropriate choice of centrifugation conditions. The physical state of alcohol precipitates of DNA changes abruptly at a concentration of approximately 1–3 μg/ml (for molecules in the size range of lambda or typical plasmids and restriction fragments). Above this critical concentration, fibrous precipitates are obtained that have high sedimentation coefficients. These precipitates will actually settle out without centrifugation, and the only purpose of spinning them is to attach the precipitate to the wall of the tube. Excessively long or vigorous centrifugation of these precipitates is counterproductive since it makes them difficult to resuspend. Below the critical concentration, the precipitate is colloidal and has a low sedimentation coefficient. For such precipitates, long, high-speed spins are essential for good recovery. In contrast to the fibrous precipitates, the colloidal precipitates resuspend readily. Essentially the same considerations apply for the initial spin (step 16) and the spin that follows the alcohol rinse (step 18), although the former is the more critical of the two. Unfortunately, different samples may be on one side or the other of the colloidal/fibrous discontinuity. In processing large batches of clones with highly variable yields, it is best to err on the side of using conditions appropriate to low-yield clones. If yields are consistently above 5 μg, the protocol can be speeded up and recoveries improved by using the shorter centrifugation times.

As originally recommended by Tracy,[2] we find that glycogen is an excellent carrier for DNA precipitations when yields are low. We are unaware of any common use of DNA samples with which glycogen interferes. If yields are consistently above 5 μg, the glycogen can be safely omitted.

λ *Growth*

Although the problem of obtaining adequate phage growth is not specific to this protocol, it will be discussed in some detail since poor phage yields are the main cause of failure when λ DNA is prepared by any method. The only reliable way to monitor phage growth is to titer typical lysates. If the average titer for a series of clones is $<5 \times 10^9$ pfu/ml, initial efforts should go into improving the phage titers rather than optimizing the DNA extraction procedure.

Phage growth is inherently an erratic process. Only vigorously growing cells can support large phage bursts; optimum yields are obtained when the final rounds of phage amplification, which lead to lysis of the culture, occur at cell densities that are relatively high, but not so high that the cells are starting to enter stationary phase. If the final rounds of phage growth occur too early in the growth of the infected culture, as will be the case if the phage input is too high or the host input is too low, clearing of the culture will be rapid and complete, but phage yields will be small. On the other hand, if the phage input is too low or the host input is too high, most cells will enter stationary phase without ever becoming infected. In

[2] S. Tracy, *Prep. Biochem.* **11,** 251 (1981).

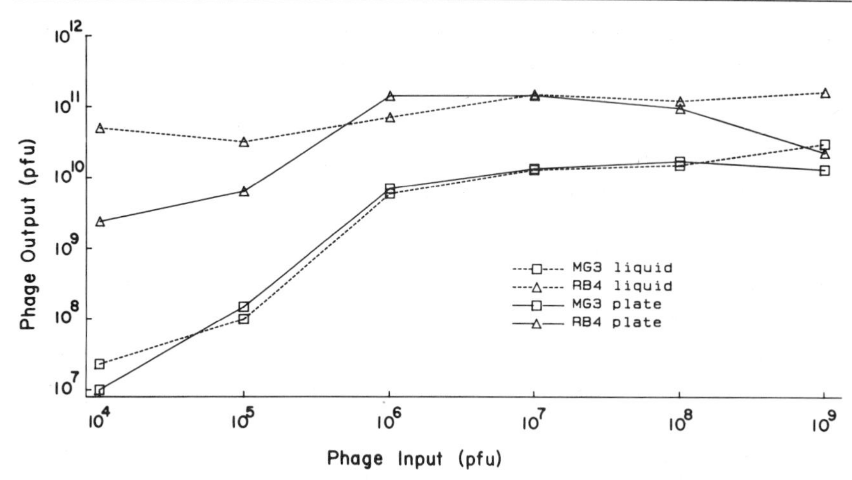

FIG. 1. Effects of varying the phage input on the titers of phage stocks. Phage were grown according to the protocols in this chapter for plate stocks and liquid lysates. The values on the ordinate are the total number of pfu's obtained from one plate or one 5-ml liquid lysate. For λMG3, the host was a C600 rk^-mk^+ strain, while for λRB4, the host was Y1088.[4]

growing any large series of λ clones, it is impractical to control the phage input precisely, and the burst sizes of the different clones will, in any event, vary considerably. Consequently, there is no hope of finding a general phage-growth protocol that will consistently produce excellent results.

We are convinced that the most common problem when growing λ is the use of inappropriate phage inputs. Until this cause of poor growth is excluded, little time should be wasted testing such variables as the growth medium or culture-to-culture variation of the common host strains (although one should check that one's host stocks are not overtly contaminated). Figure 1 shows the relationship between phage input and phage output for two λ clones, grown both on plates and in liquid cultures. The two clones illustrate the range of growth characteristics that is commonly encountered in cloning applications. λMG3 is a derivative of L47[1,3]; its growth properties are typical of clones in most high-capacity cloning vectors. λRB4, which is a λgt11 clone,[4] illustrates the much better growth that is often displayed by clones that have intact *red* and *gam* genes.[5]

[3] W. A. M. Loenen and W. J. Brammar, *Gene* **10,** 249 (1980).
[4] R. A. Young and R. W. Davis, *Science* **222,** 778 (1983).
[5] F. W. Stahl, *Annu. Rev. Genet.* **13,** 7 (1979).

The main lesson from the data in Fig. 1 is that it is important to use adequate phage inputs. Using the growth protocols summarized below, there is little risk of using too many phage, but experiments often fail because too few phage are used. Only the most vigorous clones show declines in phage output even at extreme phage inputs (e.g., λRB4 on plates). In the absence of detailed information about a particular clone, a standard phage input of 10^7 pfu is recommended, and even at this level it remains safer to err on the high side than the low.

Especially for the case of λ clones that grow poorly, the data in Fig. 1 pose an obvious problem. How does one get from a single plaque to a phage stock that contains the 10^7 phage that are required as the input for a "small-scale" growth protocol? We recommend one or two stages of amplification on grid plates, as described below. For λMG3 and λRB4, typical values for the number of phage that can be extracted from ordinary plaques and the macro-plaques of grid plates are as shown in the following tabulation:

	λMG3 (pfu)	λRB4 (pfu)
Primary plaque	3×10^4	5×10^5
First grid plate	4×10^7	6×10^8
Second grid plate	1×10^8	3×10^9

If the first grid plate gives a clear uniform zone of lysis, as was the case in the above experiment, there is generally only a moderate additional amplification (2–5×) on the second grid plate. However, if the first grid plate gives only spotty lysis, the second plate often has a major effect (10–50×). In routine practice, if the first amplification gives good lysis, we stop there and use a whole grid position as input for a liquid or plate lysate when growing a phage such as λMG3, while using 10% of that amount when growing a phage such as λRB4. If the first amplification gives mediocre lysis, we continue to the second grid plate and employ the same proportions of its phage yield.

A final point about the starting conditions for a phage lysate concerns the number of host cells used. The protocols summarized below employ relatively high inocula, and the growth phases extend for only 4–8 hr. Although the combination of lower inocula and longer growth phases allows lower phage inputs to be used and sometimes gives enormous one-step amplifications, we have found all such protocols to be too erratic to be suitable for growing large sets of clones. Higher host inocula (2–5×

those recommended) can be used safely only when growing vigorous phage; if employed, they should be combined with phage inputs at the high end of what we recommend.

Protocol for λ Growth

1. To obtain single plaques, mix 0.1 ml of λ–diluent containing 25–100 phage with 0.1 ml of a fresh L broth overnight of the host strain and incubate for 15–30 min at 37°.

2. Add 2.5 ml of melted top agar, maintained at 50°, mix briefly, and pour into a uniform layer on a standard 100 × 15 mm Petri plate. Incubate overnight at 37° with the agar surface down.

3. Prepare a grid plate by overlaying an L plate with 0.1 ml of a fresh L broth overnight of the host strain in 2.5 ml of melted top agar, as above. Allow the top agar to solidify for a few minutes, but do not delay excessively before using the plate.

4. Pick individual plaques from the plate in step 2 onto the plate prepared in step 3, employing an evenly spaced grid that can accommodate 30–40 plaques. To pick a plaque, touch the plaque once with the sharp end of a sterile toothpick, and then deposit the phage onto the surface of the grid plate by touching the soft agar gently several times over a zone that is a few millimeters in diameter. Incubate the grid plate for 6–8 hr at 37° with the agar surface down. After incubation, store the plate at 4°.

5. For phage that do not give a clear zone of lysis a few millimeters in diameter, repeat steps 3 and 4, transferring phage from the first grid plate to a second grid plate. When picking phage up from the first grid plate use the broad end of a sterile toothpick, touching the grid position repeatedly throughout the region of lysis while avoiding any gouging of the soft agar.

6. Excise each grid position by cutting out a plug that extends all the way through the bottom agar, using the large end of a standard Pasteur pipet (i.d. 5 mm). Transfer the plug to a tube containing 0.2 ml of λ–diluent and extract overnight at 4° with or without gentle agitation.

7. For a liquid lysate, add 0.15 ml of a fresh L broth overnight of the host strain, incubate for 15–30 min at 37°, add 5 ml of prewarmed L broth, and incubate on a culture tube rotor at 37° until lysed or for 6 hr, whichever comes first.

8. For a plate lysate, add 0.1 ml of a fresh L broth overnight of the host strain, incubate for 15–30 min at 37°, plate in 2.5 ml of top agar, and incubate for 6–7 hr at 37° in a humidified box with the agar surface up. At

the end of the incubation period, chill the plate and overlay with 5 ml of cold λ–diluent. Extract overnight at 4° with or without gentle agitation.

Notes on λ Growth

The protocol described here is a composite of numerous published techniques for growing λ.[1,6-9]

Step 1. Recipes for media and λ-diluent are given at the beginning of the DNA purification protocol. Slightly better phage growth is often observed with 2 m*M* $MgCl_2$ instead of the specified 1 m*M*, but the improvement is small and the higher $MgCl_2$ concentration interferes with phage adsorption to DEAE–cellulose.

Steps 3 and 8. For preamplification steps or final plate lysates, fresh L plates give the best results; the plates are normally poured the day before they are needed and allowed to set overnight at room temperature. A convenient method of preparing a few fresh plates at frequent intervals is to melt agar in L broth instead of starting from scratch. We have never experienced problems with contamination when using plates prepared in this way as long as they are used within 1 day and only incubated for the short periods required for phage growth.

Steps 4, 7, and 8. These cultures should not be incubated longer than the specified time or they will become overgrown with λ-resistant host cells, which can absorb substantial numbers of phage. Using the recommended host inocula and phage inputs, we have never seen phage yields improve when incubation was extended for longer than the specified times. For liquid lysates (step 7), there is little correlation between phage yields and the extent of clearing, while the correlation is better on plates (steps 4 and 8).

Step 6. Phage extraction from these plugs is quite rapid, and agitation has only a small effect. When using the plugs as a source of phage for liquid lysates, the step can be skipped altogether; in the case of plate

[6] R. W. Davis, D. Botstein, and J. R. Roth, "A Manual for Genetic Engineering: Advanced Bacterial Genetics." Cold Spring Harbor Lab., Cold Spring Harbor, New York, 1980.

[7] W. Arber, L. Enquist, B. Hohn, N. E. Murray, and K. Murray, *in* "Lambda II" (R. W. Hendrix, J. W. Roberts, F. W. Stahl, and R. A. Weisberg, eds.), p. 433. Cold Spring Harbor Lab., Cold Spring Harbor, New York, 1983.

[8] F. R. Blattner, B. G. Williams, A. E. Blechl, K. Denniston-Thompson, H. E. Faber, L.-A. Furlong, D. J. Grunwald, D. O. Kiefer, D. D. Moore, J. W. Schumm, E. L. Sheldon, and O. Smithies, *Science* **196,** 161 (1977).

[9] T. Maniatis, E. F. Fritsch, and J. Sambrook. "Molecular Cloning: A Laboratory Manual." Cold Spring Harbor Lab., Cold Spring Harbor, New York, 1982.

lysates, little useful extraction occurs during the growth phase so at least a 30-min preextraction is recommended.

Step 8. It is important for the plates to be thoroughly chilled before they are overlayed or excessive amounts of agar are extracted, thereby clogging the DEAE–cellulose columns. As is the case with plugs, agitation of the plates during the phage extraction has at most a small effect.

Troubleshooting

The only persistent failures of the DNA preparation that we are aware of have been due to poor adsorption of the phage to DEAE–cellulose. Additional dilution of the lysates with water invariably solves the problem. Even a small dilution can have dramatic effects. In one experiment, for example, we pooled several standard liquid lysates and then divided the pool into 4-ml aliquots that were processed after varying amounts of dilution. With no dilution, 97% of the recovered phage were in the run-through and the chase, while with a dilution of only 0.5 ml, 97% were in the first milliliter after phage collection normally begins. With our recommended dilution of 2 ml, <0.1% of the recovered phage eluted prematurely. We have seen no variation in the required dilution (2 ml for liquid lysates, none for plate lysates) in experiments spanning 3 years and employing many lots of DEAE–cellulose, the growth medium ingredients, and the other reagents. Nonetheless, the requirement for slightly higher dilutions has been observed independently in a number of laboratories, so if poor phage adsorption is observed, a dilution series should be analyzed. The profile of phage elution is readily monitored, either by titering the various fractions or by using a gel assay that has been described elsewhere[1]; regardless of whether viable phage or λ DNA are being monitored, recoveries from the column should be essentially 100%.

In routine practice, our overall success rate in going from a plaque to a usable lane of gel data is over 95%. Nearly all the failures are due to low yields, which usually arise because of poor phage growth. Figure 2 shows typical data, as well as some special cases, which were selected to illustrate the types of problems that arise when large sets of λ clones are analyzed. Most of these problems are general ones that will be encountered regardless of the DNA preparation protocol that is used. In most instances, the appropriate corrective action is obvious once the problem is diagnosed, but supplementary data will often be required before an accurate diagnosis can be made. Particularly when dealing with digests that contain nonstoichiometric bands, one should proceed with caution since there are several common causes for this phenomenon. Although under digestion is one obvious possibility, overdigestion is also common

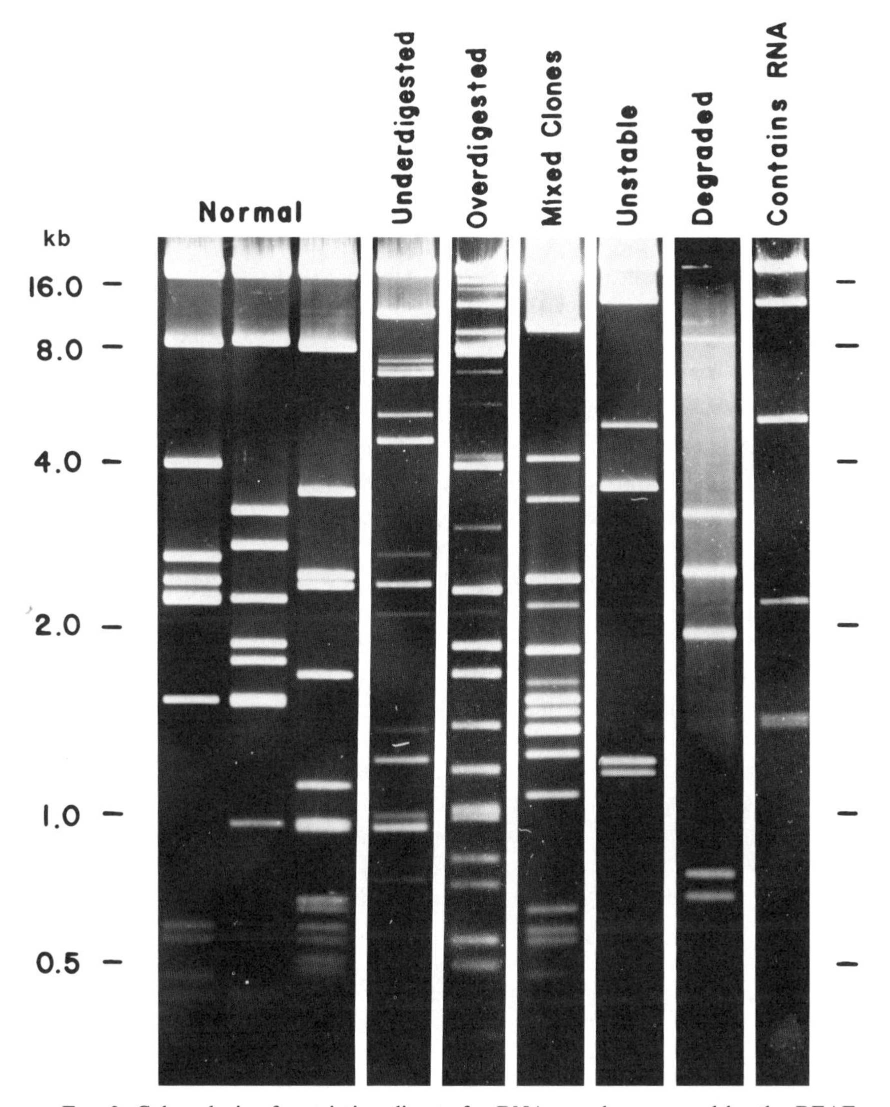

FIG. 2. Gel analysis of restriction digests for DNA samples prepared by the DEAE–cellulose method. All samples contained 0.5 μg of DNA, cleaved with a mixture of *Eco*RI and *Hin*dIII, and electrophoresed on 1.2% agarose gels, as described.[1] For a discussion of the atypical samples, see the text.

with some enzymes.[10,11] Other causes of nonstoichiometric bands include samples that contain mixtures of two unrelated clones, as well as clones that are unstable and generate deletions or duplications at a high frequency. The best method of distinguishing between the various causes of nonstoichiometric bands is to carry out a series of digestions at different enzyme concentrations. If the nonstoichiometric bands are unchanged over a wide range of enzyme concentrations, then the sample should be re-plaque-purified and several independent plaques analyzed. In the case of mixtures of unrelated clones, some plaques will give one pattern and some the other. Plaques derived from unstable clones may continue to generate the same nonstoichiometric bands during phage growth, or stable variants may become fixed.

More specific comments on the problems illustrated in Fig. 2 are presented below:

1. Underdigestion is usually characterized by multiple bands of erratic itensity, extending over most of the size range of the limit-digest fragments. Cases also occur, however, that are difficult to distinguish from limit digests because only one or two sites have been poorly cleaved while all the remaining sites have been cleaved quantitatively. We have consistently found that DNA prepared by the DEAE–cellulose protocol cleaves as readily as any highly purified source of λ DNA[1]; underdigestion is more likely to be symptomatic of problems with the restriction enzyme or the digestion conditions than the DNA preparation.

2. Overdigestion (cleavage at secondary sites) is primarily a problem with selected enzymes. It is usually characterized by a larger number of fainter bands than one observes in a typical underdigested sample. Although lowering the enzyme concentration is one obvious cure for overdigestion, a change in the digestion conditions is often even more effective.[10,11]

3. Imperfectly plaque-purified samples give two superimposed ladders of bands, each of which defines a stoichiometric series. Samples containing a mixture of two unrelated clones are encountered more often than might be expected even when the phage have been picked from sparsely populated plates. The phage should be taken through another round of plaque purification, and several isolates analyzed.

4. Unstable clones give rise to nonstoichiometric bands during phage growth. The DNA rearrangements usually all occur in one region of the clone so most of the bands are unaffected. Another round of plaque purification should be carried out, as above, to assess the severity and reproducibility of the phenomenon.

[10] E. Malyguine, P. Vannier, and P. Yot, *Gene* **8,** 163 (1980).

[11] J. George and J. G. Chirikjian, *Proc. Natl. Acad. Sci. U.S.A.* **79,** 2432 (1982).

5. Degraded samples show a diffuse background starting at the largest band, and the intensity of the bands diminishes in proportion to fragment size. In our experience, the source of the degradative activity is almost always the cells from which the DNA was extracted. A less common source, at least when commercial enzymes are being used, is the restriction enzyme. DNA-degrading activities are almost never introduced adventitiously during routine sample handling. Because the common degradative activities are Mg^{2+} dependent, the degradation often occurs during restriction digestion even though the source of the nuclease is the DNA sample. A good control is to carry out a mock digestion in restriction enzyme buffer without restriction enzyme. If the degradative activity is found to be in the DNA, the proteinase K step is suspect.

6. RNA-containing samples give rise to a diffuse band at the bottom of typical agarose gels (<500 bp in Fig. 2). RNA elutes from DEAE–cellulose columns well after the phage peak. If it is present in the DNA samples, the columns are being eluted too vigorously. The elution buffers and column geometry should be checked.

Discussion

The DEAE–cellulose method occupies an intermediate niche in the hierarchy of λ DNA preparations. Methods that bypass the purification of the phage can be carried out more rapidly and have higher yields, but they are also less reliable and produce DNA of lower quality (yields are higher because both packaged and unpackaged λ DNA is recovered from the lysates).[12] Classical techniques based on equilibrium sedimentation, like the DEAE–cellulose method, recover only packaged DNA; the DNA is of high purity, but the procedures are laborious and poorly suited to the processing of large numbers of samples.[6,9] In general, the DEAE–cellulose method is most likely to prove attractive in applications that place high demands on DNA quality and involve processing batches of at least 10–20 clones at regular intervals.

In applications of this type, it should be emphasized that the method of DNA preparation is only one aspect of the overall experimental design. DNA from lytic λ clones is relatively precious material, no matter how it is prepared. Although the DEAE–cellulose method can be scaled up,[1] large volumes of λ lysates are inherently more difficult to process than are comparable volumes of plasmid-containing cells; on a molar basis, they also contain much smaller quantities of DNA. The scale of the protocols that have been described here was designed to provide adequate DNA for

[12] R. W. Davis, M. Thomas, J. Cameron, T. P. St. John, S. Scherer, and R. A. Padgett, this series, Vol. 65, p. 404.

preliminary gel analysis and for subcloning into plasmid vectors; if larger amounts of DNA are required, subcloning should be considered as a serious alternative to scaling up the λ DNA preparations.

Another way to minimize the scale on which λ DNA must be prepared is to use sensitive analytical techniques. Given optimized protocols for running, staining, and photographing gels, a few tenths of a microgram of λ DNA is an adequate sample for a typical gel lane. Restriction fragments as small as 25 bp can be detected in such samples on polyacrylamide gels, and 200-bp fragments can be readily visualized on agarose gels. Small samples of λ DNA are also more effectively employed if their concentrations are known precisely. Particularly when working with large sets of clones, a reliable DNA assay is essential since the growth of the clones, and therefore the DNA yields, will often be highly variable. In this context, an advantage of the DEAE–cellulose method is that it produces DNA that is free of cellular nucleic acids, and consequently the concentration of λ DNA is readily determined fluorometrically. We have described elsewhere the particular solutions that our laboratory has adopted to the various analytical problems associated with characterizing large numbers of λ clones,[1] and excellent general treatments of the relevant methodology are also available.[6,9,13]

Acknowledgments

We would like to acknowledge M. Graham's contributions to the development of this protocol. We also thank P. Oeller for helpful comments about his experiences with the method and R. Barstead for supplying λRB4. The research was supported in part by a grant from the National Institutes of Health (GM28232).

[13] This series, Vols. 65 and 68.

[6] Double *cos* Site Vectors: Simplified Cosmid Cloning

By PAUL BATES

Recent studies of eukaryotic gene organization and expression demonstrate the need to be able to clone large segments of DNA. It is often desirable to isolate a gene or gene family within one recombinant molecule, thus increasing the likelihood of preserving any functional domains which may be important in the regulation of the genes. The presence of multiple introns in a single gene or of large intergenic regions in gene families often makes this impossible using the limited cloning capacity (approximately 23 kb) of λ vectors. Analysis of extensive regions of a eukaryotic genome by serial isolation of a number of overlapping clones (i.e., genome walking) is also facilitated by cloning fragments as large as possible. Again the limited capacity of λ vectors, thus the larger number of steps needed to span a specific region, can be a hindrance when performing these types of studies.

Cosmids are hybrid plasmid–bacteriophage vectors which have been designed for cloning large fragments of DNA.[1,2] These vectors are modified plasmids which contain a plasmid replicon, a selectable drug resistance marker, and the cohesive end or *cos* site necessary for packaging λ DNA. Cosmids accept DNA inserts of 30–45 kb and utilize the λ *in vitro* packaging system for efficient introduction of the DNA into bacterial cells. Because of the high efficiency of transduction into bacteria and the large inserts which can be accommodated, cosmids are ideal vectors for construction of eukaryotic genome libraries.

Despite these apparent advantages, various technical problems prevented cosmid cloning from being widely employed in the construction of genomic libraries. The three main problems encountered were vector concatamerization resulting in cosmids lacking inserted eukaryotic DNA, recombinational rearrangements of the eukaryotic DNA caused by ligation of multiple inserts into a single cosmid, and finally differential growth of cosmid clones causing misrepresentation of eukaryotic sequences in amplified cosmid libraries. Most of the presently available cosmid vectors contain a single *cos* site. Use of these single *cos* site vectors in a cloning scheme which avoids the problems outlined above, such as the procedure described by Ish-Horowicz and Burke, requires separate preparation of the two vector arms before insertion of the foreign DNA.[3] This step is not only laborious and time consuming, but requires that the two vector arm preparations be mixed in equal molar ratios for maximum cloning efficiency. Several years ago, we described a cosmid cloning system which overcame the problems described above, yet allowed rapid and efficient preparation of cosmid libraries.[4]

Principle of the Method

Our cloning system uses plasmids which contain two *cos* elements, thus avoiding the necessity of preparing the vector arms separately. A

[1] J. Collins, this series, Vol. 68, p. 309.
[2] J. Collins and B. Hohn, *Proc. Natl. Acad. Sci. U.S.A.* **75,** 4242 (1978).
[3] D. Ish-Horowicz and J. F. Burke, *Nucleic Acids Res.* **9,** 2989 (1981).
[4] P. F. Bates and R. A. Swift, *Gene* **26,** 137 (1983).

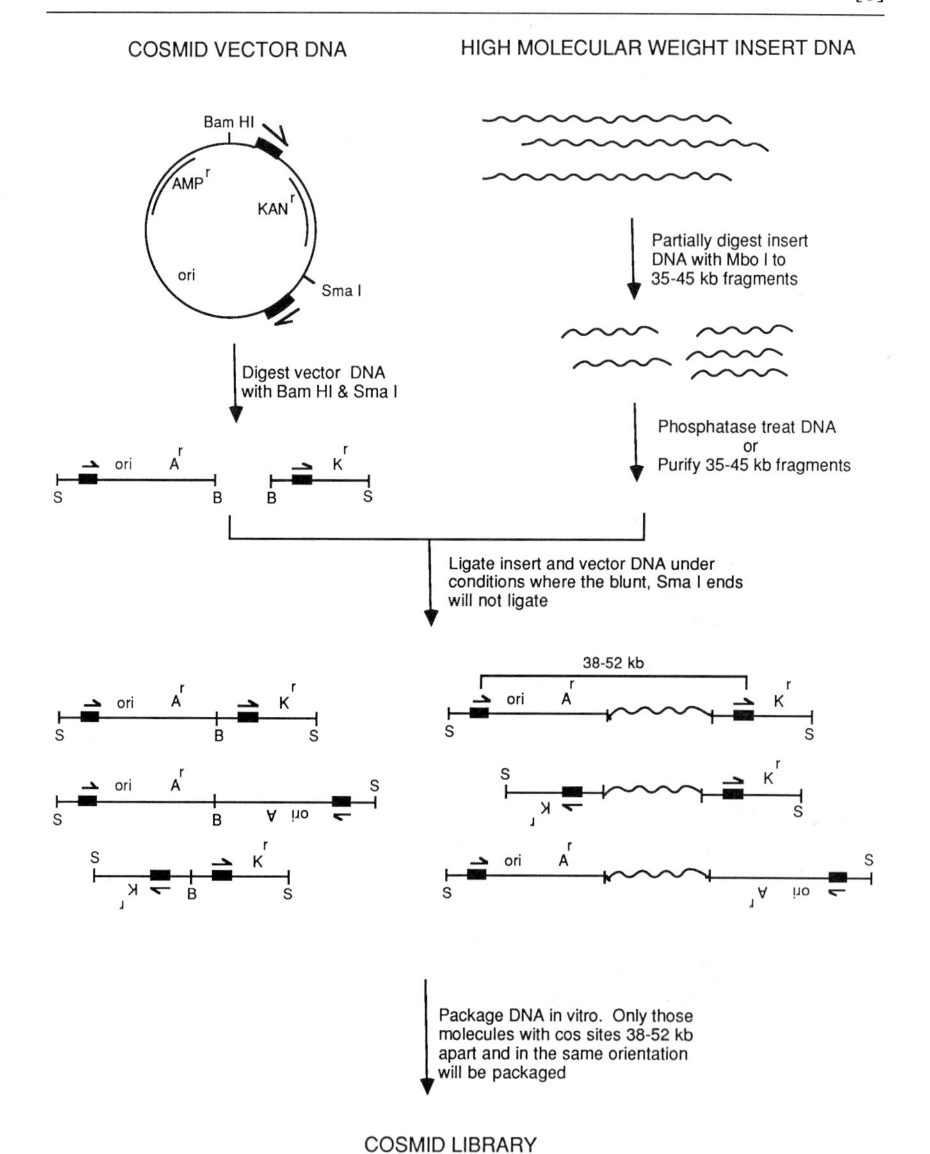

FIG. 1. Schematic diagram outlining the construction of cosmid libraries using the double *cos* site vectors. The closed boxes represent the *cos* sites with the arrows indicating the orientation of the *cos* sites. A^r and K^r denote the ampicillin- and kanamycin-resistance genes, respectively. The figure is not drawn to scale.

blunt-end restriction enzyme site (*Sma*I), placed between the two *cos* sites, prevents cosmid concatamerization during the ligation to the inserted DNA. A strategy for cloning in the double *cos* site vectors is shown in Fig. 1. The vector DNA is digested with *Bam*HI and *Sma*I to generate the two cosmid vector arms. Genomic DNA is partially digested with *Mbo*I to produce fragments which are primarily 35–45 kb in length. This insert DNA is treated with phosphatase to prevent self-ligation or alternatively the 35- to 45-kb DNA fragments are purified by size selection on NaCl gradients. The insert DNA fragments are then ligated to the vector arms using T4 DNA ligase under conditions which specifically inhibit blunt-end ligation. Only the complementary overlapping ends of the *Bam*HI and *Mbo*I will ligate. The ligated DNA is packaged *in vitro*. Only those molecules which contain the *cos* sites 38–52 kb apart and in which the *cos* sites are oriented in the same direction will be encapsidated into λ phage heads. Any dimers formed by ligation of two vector arms at the *Bam*HI site are excluded from packaging by the size specificity of the packaging system, thus eliminating any vector-alone background. The cloning efficiency of the system which is described here is very high, and generally ranges from 10^5 to 10^6 colonies/μg insert DNA. With this level of efficiency only 2–4 μg of insert DNA is needed to generate a complete cosmid library from most eukaryote genomes. This system avoids all the problems outlined above yet is simple to use, allowing rapid and efficient construction of cosmid libraries.

Structure of the Double *cos* Site Vectors

The family of double *cos* site vectors is shown in Fig. 2. All of these vectors are derived from the c2XB vector. The salient feature of this vector, and of all these vectors, is the arrangement of the two *cos* sites flanking the single *Sma*I site, with a single *Bam*HI site for cloning lying outside this region. It is this combination of sites that allows the rapid production of cosmid libraries with this family of vectors. The construction of the c2XB and c2RB vectors has been described.[4] All the other derivatives were constructed as described in the legend to Fig. 2. The c2RB derivative has been designed to facilitate genome walking experiments. In c2RB the *Bam*HI site is flanked by two *Eco*RI sites. The insert DNA can easily be separated from the vector DNA by digestion with *Eco*RI, thus allowing rapid isolation of the inserted DNA for rescreening the cosmid library to isolate overlapping clones. c2XBHC and c2RBHC are high-copy derivatives of c2XB and c2RB, respectively, and are generally used for constructing libraries which are to be screened with radiolabeled hybridization probes. Since they replicate at a higher copy number,

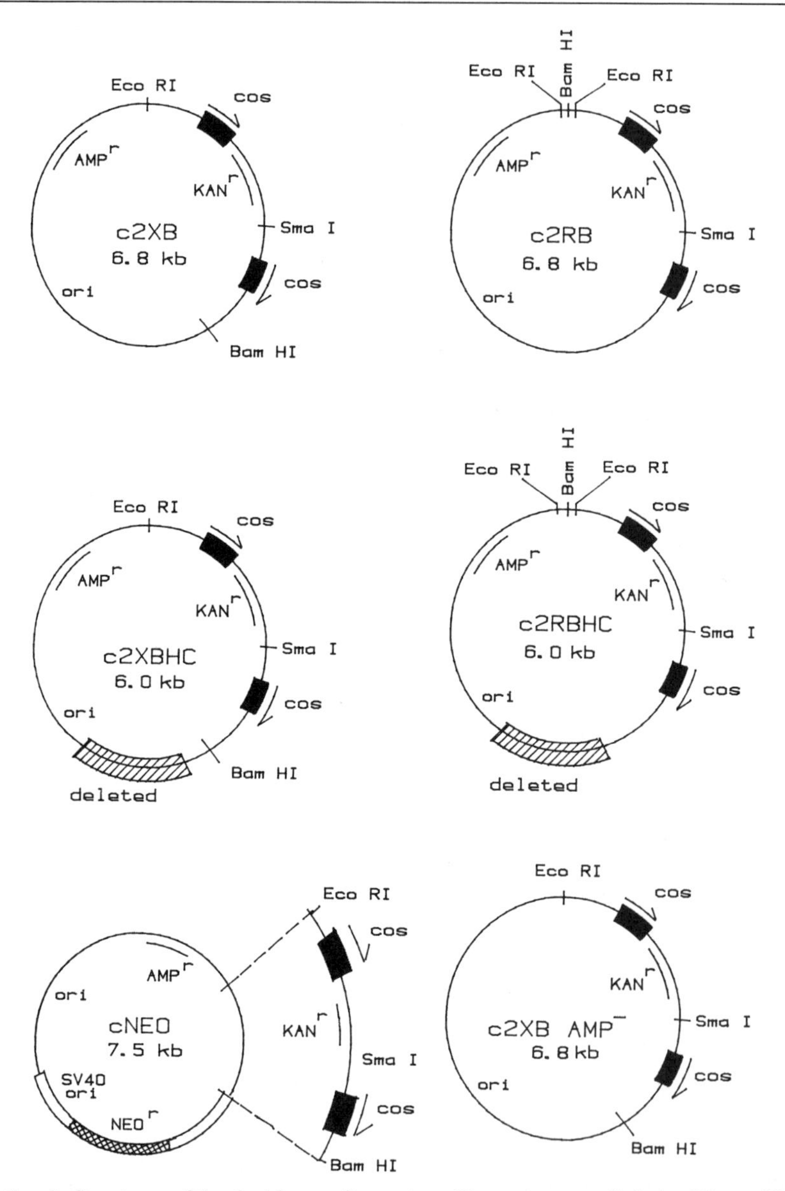

FIG. 2. Structures of the double *cos* site vectors. The vectors are all derived from c2XB. The high-copy vectors, c2XBHC and c2RBHC, were constructed by transferring the double *cos* region into pAT153. The hatched region denotes the sequences deleted in pAT153 relative to pBR322. The ampicillin-resistance gene of the c2XB AMP$^-$ vector was inactivated by cutting c2XB at the *Pvu*I site which lies in the ampicillin resistance gene, then filling in the overhanging ends with T4 DNA polymerase and recircularizing the plasmid. cNeo was

these vectors tend to give more intense hybridization signals than the c2XB and c2RB parents. The c2XB AMP^- vector has been designed for screening libraries by genetic selection, relying on expression of a drug-resistance marker acquired in the insert DNA for selection. When this vector is used the recombinant cosmid contains the plasmid replication origin but no drug resistance marker. The kanamycin-resistance gene used for the selective growth of the cosmid vector is lost during packaging into the λ headshell (see Fig. 1). The c2XB AMP^- vector has been used to rescue the sequences adjacent to the insertion site of a plasmid in the germline of transgenic mice by selection for the ampicillin-resistance marker of the plasmid.[5] Similarly, c2XB AMP^- has also been used to isolate sequences which have been tagged with the ampicillin-resistance gene prior to transfection into eukaryotic cells. Another derivative of c2XB has been constructed for use in transfection of cosmids into eukaryotic cells. This vector, cNEO, carries the region of the pSV2neo plasmid which confers G418 resistance to eukaryotic cells.[6] Cosmids constructed by using the cNEO vector can be used to shuttle genes from bacterial to eukaryotic cells using the ampicillin- and G418-resistance genes as selectable markers.

Materials and Reagents

Strains and Plasmids

Escherichia coli 1046 (*recA*$^-$, *supE*, *supF*, *hsdS*$^-$, *met*$^-$), the *in vitro* packaging extract strains, BHB2690 and BHB2688, were from Ed Fritsch. ED8767 (*supE*, *supF*, *hsdS*, *lacY*, *recA*56) was supplied by S. Guyourt. The plasmids c2XB and c2RB have been previously described. All derivatives of c2XB were constructed as described in the legend to Fig. 2. pSV2neo was from Dan Robinson. c2XB AMP^- was constructed by Rick Woychik. The packaging strain SMR10 was from Susan Rosenberg.

[5] R. P. Woychik, T. A. Stewart, L. G. Davis, P. D'Eustachio, and P. Leder, *Nature (London)* **318,** 36 (1985).

[6] P. J. Southern and P. Berg, *J. Mol. Appl. Genet.* **1,** 327 (1982).

constructed by replacing the *Bam*HI to *Eco*RI region of pSV2neo with the *Bam*HI to *Eco*RI sequences of c2XB. The cross-hatched region denotes the G418-resistance gene of pSV2neo. Only the relevant restriction enzyme sites are shown in all of the vectors. The sizes given are the sizes of the intact vector plasmid sizes. The amount of vector remaining in the recombinant cosmids, after loss of the sequences between the two *cos* sites, is 1.7 kb smaller than the intact vector size.

Enzymes

Restriction enzymes and T4 DNA ligase were purchased from New England BioLabs and were used according to the directions of the supplier. Calf intestinal alkaline phosphatase, special quality for molecular biology, was from Boehringer. Pancreatic RNase (Sigma) was dissolved at 10 mg/ml in 10 m*M* sodium acetate, pH 5.2, then boiled for 2 min to inactivate DNases. Aliquots were stored at $-20°$.

Solutions

TE: 10 m*M* Tris–HCl, 1 m*M* EDTA
Low TE: 10 m*M* Tris–HCl, 0.1 m*M* EDTA
TAE buffer for gel electrophoresis: 40 m*M* Tris–acetate, 2 m*M* EDTA, pH 8.2
10× ligase buffer: 660 m*M* Tris–HCl, pH 7.5, 50 m*M* $MgCl_2$, 50 m*M* DTT, 25 m*M* ATP
10× alkaline phosphatase buffer: 500 m*M* Tris–HCl pH 9.0, 10 m*M* $MgCl_2$
LB: 10 g Bacto tryptone, 5 g yeast extract, 10 g NaCl, pH adjusted to 7.5 with NaOH, autoclaved
Tris-saturated phenol: 500 g phenol dissolved at 60° and equilibrated with 300 ml of 1 *M* Tris–HCl, pH 7.5
1.25 *M* NaCl: 1.25 mol (72.5 g) NaCl dissolved in 1 liter of 10 m*M* Tris–HCl, pH 7.5, 5 m*M* EDTA, filter, then autoclave
5 *M* NaCl: 5 mol (290 g) NaCl dissolved in 1 liter 10 m*M* Tris–HCl, pH 7.5, 5 m*M* EDTA, filter, then autoclave
SM: 5.8 g NaCl, 2 g $MgSO_4 \cdot 7H_2O$, 50 ml 1 *M* Tris–HCl, pH 7.5, 5 ml 2% gelatin, dissolved to 1 liter, autoclaved

Methods

Preparation of Genomic DNA

Successful construction of cosmid libraries is dependent upon the starting size of the genomic DNA. The DNA should be ≥200 kb if cosmid cloning is to even be considered. I use a modification of the procedure described by Blin and Stafford to prepare very high-molecular-weight DNA.[7] This procedure avoids all pipetting and precipitation steps which may cause mechanical shearing of the DNA. Also, the phenol and chloroform extractions are performed with a large surface area between the

[7] N. Blin and D. W. Stafford, *Nucleic Acids Res.* **3,** 2903 (1976).

aqueous and organic phases so that vigorous mixing is not required to achieve efficient extraction. The procedure I use is as follows. A 1-cm^3 piece of tissue is homogenized in 10 ml of TE in a 50-ml screw-cap tube using three 1-min pulses of a polytron (Brinkman Instruments). Alternatively, 1×10^8 tissue culture cells are trypsanized and resuspended in 10 ml TE. The cells are lysed by the addition of 40 ml of a solution containing 0.5 *M* EDTA, 0.5% SDS, and 200 μg/ml proteinase K. Proteolytic digestion is carried out for 24–48 hr at 55°. The solution is gently mixed by inversion every few hours. After digestion, this viscous solution is poured into a 2-liter Erlenmeyer flask and extracted by gentle swirling with an equal volume of Tris-saturated phenol. This mixture is poured back into 50-ml tubes, and the phases are separated by centrifugation at 3000 rpm for 5 min. The upper, organic phase is removed (note that the phases are inverted due to the high EDTA concentration in the lysis buffer) and another phenol and two chloroform extractions are performed as described above. The aqueous solution is poured into dialysis tubing and dialyzed extensively against 50 m*M* Tris–HCl, pH 7.5, 50 m*M* EDTA, 200 m*M* NaCl. After dialysis, the solution is digested with 50 μg/ml RNase at 37° for 2 hr, then SDS (0.5%) and proteinase K (200 μg/ml) are added. Digestion is performed at 55° for 1 hr. The DNA solution is then phenol/chloroform extracted as described above and finally dialyzed against low TE. After dialysis, the DNA should be stored at 4°, not at −20°, since repeated freezing and thawing will shear the DNA.

Generally, the DNA obtained from this procedure is at least 200 kb in size and is usually at a concentration of 35 to 75 μg/ml. Even at this rather dilute concentration the solution is viscous due to the size of the DNA. The DNA size can be checked by analysis on a 0.2% agarose gel. These gels are very soft and are best run at 4° to help stabilize the gel. Also, the best resolution is obtained if the gel is run slowly at 1–2 V/cm for 18–20 hr. Multimers of λgt10 DNA, ligated at the *cos* site and then partially digested with *Eco*RI, make good size markers.

Cosmid Vector Preparation

One of the most attractive features of this cosmid cloning scheme is the ease of preparation of the vectors for cloning. The only important consideration to keep in mind is that the *Sma*I digest must be complete or there will be significant background colonies caused by vector concatamerization. To avoid this problem the vector DNA is digested with a 5-fold excess of *Sma*I. Vector DNA (10μg) is digested with 50 U *Sma*I, then phenol extracted and ethanol precipitated. The DNA is then digested with *Bam*HI. Many preparations of *Sma*I seem to have nuclease contamination, therefore the *Bam*HI and *Sma*I digests should be done sequentially. After digestion, the DNA should be phenol extracted, precipitated, and then resuspended in 20 μl low TE (500 μg/ml DNA concentration). Store the prepared vector DNA at 4°.

Insert DNA Preparation

There are two methods for preparing the insert DNA which are regularly employed. The choice of which method is used depends mainly on the amount of genomic DNA available and also on the efficiency of the packaging extracts available. When the amount of DNA available is limited or when the extracts used are highly efficient ($\geq 5 \times 10^8$ pfu/μg λ DNA), the insert DNA is used directly after partial digestion, without size selection. If the packaging extracts are significantly less efficient than 1×10^8 or if the amount DNA is not limiting, then the insert DNA is size selected on NaCl gradients. Size selection will increase the cloning efficiency (cfu/μg insert DNA) 10- to 20-fold; however, larger amounts of DNA and more preparation time are required. Generally, I use phosphatase-treated DNA to prepare most libraries, however, both methods will be presented. The details of the insert preparation are as follows.

Testing Partial Digest Conditions. Set up a digestion reaction with 10 μg of genomic DNA in 1× *Mbo*I buffer. Add one-tenth volume of 10× *Mbo*I buffer to the volume of DNA solution containing 10 μg. The final volume of the reaction is dependent upon the concentration of the genomic DNA to be digested. Since the DNA solution is very viscous it often takes some time to mix in the buffer solution. Be patient. Equilibrate the temperature of the solution to 37° for 5 min. Remove one-fifth of the reaction volume as uncut control. Add 0.8 U *Mbo*I (0.01 U/μg DNA). Mix gently by inverting the tube several times. Incubate at 37°. Remove $\frac{1}{5}$-vol aliquots at 5-, 10-, 20-, and 60-min time points. Stop the digest with EDTA at a final concentration of 50 m*M*. Hold these aliquots on ice till ready for gel analysis. Remove one-quarter of each aliquot (0.5 μg DNA) and load on a 0.3% agarose gel. Use various λ vector DNA's as size markers; e.g., Charon 28 (40 kb), λBF101 (46 kb), and λ cut with *Hin*dIII (28 and 23 kb). Bracket the region of interest. Run the gel at 2 V/cm for 16 hr. After ethidium bromide staining, evaluate the gel and choose the time which gives the most intense staining in the 35- to 50-kb range. Experience has shown that the best condition, based on the cloning efficiency, is usually a slightly shorter digestion time than the time which gives the most intense 35- to 50-kb staining. For example, if 10 min gives the most intense staining, then use 8 min for the large-scale digest.

Large-Scale Partial Digest. From the test digest conditions scale up exactly and digest 30 μg genomic DNA (for DNA which is to be size selected use 300 μg and scale up the rest of this protocol accordingly). Remove one-third of the sample at a time slightly shorter than the optimal conditions and allow one-third of the sample to digest slightly longer than the optimal time. For example, if the partial test conditions suggested an 8-min digest, then remove one-third aliquots at 6.5, 8, and 9.5 min. Stop the digests with EDTA, remove the enzyme by phenol extraction, and ethanol precipitate the DNA. At this point the DNA should only be of the order of 50 kb in length and shearing is not a major problem; however, some precautions to avoid shearing should still be used. After precipitation gently resuspend the DNA in 95 μl low TE. Allow the DNA to dissolve at 4° for 1 day, then proceed with dephosphorylation or size selection. Vortexing the DNA to get it back into solution usually decreases the cloning efficiency by about one-half.

Dephosphorylation of Insert DNA. DNA which is to be cloned directly without size selection must be dephosphorylated to avoid ligation of two or more separate insert fragments into a single recombinant, resulting in a rearrangement of the eukaryotic sequences. After the partially digested DNA samples have dissolved completely in low TE, add 10 μl 10× alkaline phosphatase buffer and 0.5 U calf intestinal alkaline phosphatase. Incubate at 37° for 30 min then add 0.5 U more phosphatase and incubate an additional 30 min. Heat to 68° for 45 min to kill the enzyme, then phenol extract and ethanol precipitate the DNA. Gently resuspend the DNA in 20 μl low TE (final DNA concentration 500 μg/ml). Store the DNA at 4°.

Size Selection of Insert DNA. Although it is not necessary to size select the insert DNA when using the vectors described here, size selection can be used with these vectors to increase the overall cloning efficiency (in colony forming units per microgram insert DNA) of this system. This increase in efficiency occurs for two reasons. First, more of the vector and insert fragments ligate to form packageable molecules with the *cos* sites positioned 38–51 kb apart when sized DNA is used. Second, the insert DNA is not phosphatase treated when size selection is employed, therefore this DNA ligates much more efficiently. This also increases the cloning efficiency. The increased cloning efficiency obtained with sized DNA can be used to offset the effects of an inefficient batch of packaging extracts, or alternatively the size-selected DNA may be used with highly efficient extracts to construct a very large cosmid library. Construction of such large libraries is often necessary when multiple screening steps are employed in genomic walking experiments. Since the insert DNA is stable for long periods of time at 4°, the sized DNA can be used to construct multiple libraries over a period of time.

DNA (300 μg) for size fractionation is partially digested as described above. The DNA from the three separate time points is pooled, then divided in half and layered onto two 11-ml 1.25–5 M NaCl gradients (150 μg DNA/gradient). The gradients should be poured immediately before use and kept at 4° till use. Heat the insert DNA at 65° for 10 min before layering the samples on the gradients. The partially digested DNA is sized by velocity sedimentation at 39K rpm in an SW41 rotor for 4 hr at 4°. After centrifugation, fractionate the gradient from the bottom, collecting 0.5-ml samples. Dilute each fraction 4-fold with TE and ethanol precipitate the DNA. Gently resuspend the fractions in 0.5 ml TE. Analyze 15 μl of each sample on a 0.3% agarose gel with λ DNA size markers. Pool the fractions containing the 35- to 50-kb fragments and ethanol precipitate the DNA. Gently resuspend the precipitated DNA in low TE to a final concentration of approximately 500 μg/ml. Store the samples at 4° till used in ligations.

Ligation of Vector and Insert DNA

The ligation conditions which are used in this protocol are designed to inhibit ligation of the blunt-ended *Sma*I sites while allowing the *Bam*HI and *Mbo*I ends to ligate. This is accomplished by raising the ATP concentration in the ligation reaction to 2.5 mM. At this concentration blunt-end ligation with T4 DNA ligase is effectively eliminated. If ATP concentrations below this are employed with these vectors, self-concatamerization will occur and significant vector-only background colonies will result. In these vectors the kanamycin-resistance gene lies outside the region normally packaged into the λ headshell in the recombinant cosmids (see Fig. 1), therefore the level of vector-only background can easily be checked by titering the cosmid library on kanamycin plates. Another condition of the vector/insert ligations which must be controlled is the ratio of insert to vector DNA fragments. To ensure that all insert fragments ligate to the vector DNA, the molar ratio of vector to insert is 10 : 1. This high vector-to-insert ratio is especially important when the insert DNA has not been phosphatase treated and insert fragments can ligate together to form recombinants which are not representative of the genome configuration.

The ligation reaction mixture should contain 4 μl *Sma*I- and *Bam*HI-cut vector DNA (500 μg/ml), 2 μl insert DNA (500 μg/ml), 1 μl 10× ligase salts, and water to bring the volume to 10 μl. Add 0.1 U of T4 DNA ligase and incubate the reaction for 12 hr at 16°. The ligation of vector and insert DNA is very difficult to assay by gel analysis since it is hard to detect a mobility shift of only 5 kb in the 35- to 50-kb insert fragments. A simple way to check the ligation is to remove 0.2-μl samples before and after the ligation reaction and to analyze these samples on a 0.7% agarose minigel (6.5 × 9 × 0.2 cm). Before ligation the DNA should be resolved into three bands; the two vector arms and a smear of high-molecular-weight insert

DNA. After successful ligation the two vector bands should disappear and be replaced by three bands corresponding to the three vector dimers which form (see Fig. 1). Since the insert DNA migrates in the nonlinear region of the 0.7% gel, it is not possible to detect any differences in the ligated and unligated insert DNA band. Generally in experiments where the vector DNA has ligated to form dimers, the vector to insert ligations are also successful. After ligation the DNA can be used directly for *in vitro* packaging.

In Vitro Packaging and Transduction of Cosmids into Bacteria

The packaging extracts were prepared by protocol II of Maniatis *et al.* using strains BHB2688 and BHB2690.[8,9] The single-strain packaging system using SMR10 described by Rosenberg has also been used for packaging cosmids.[10] The efficiency of either of these systems was $0.5-1.0 \times 10^8$ plaques/μg λ DNA. For packaging the cosmids, 5 μl of the ligation reaction (0.5 μg insert DNA) was packaged per reaction. After packaging for 1 hr, the packaging reaction was diluted with 500 μl of SM and 20 μl of chloroform was added. The packaged cosmids are transduced into *E. coli* ED8767 or 1046. A saturated culture is grown in LB + 0.2% maltose, then pelleted by centrifugation and resuspended in one-half volume 10 m*M* $MgSO_4$. For titering the packaged cosmids, 10 μl of a 10^{-1} dilution of the packaging reaction is adsorbed to 100 μl of bacteria. Adsorption of the packaged cosmids is carried out at 37° for 20 min, then 500 μl of LB is added and the incubation is continued for 45 min at 37°. The mixture is spread on an LB + AMP (50 μg/ml) plate and incubated at 37°. Colonies usually take 15–18 hr to grow, although very small colonies (0.1 mm) can be seen in 10–12 hr. For screening libraries, the transduction is scaled up as follows; for 9-cm plates use 100 μl bacteria and up to 100 μl packaged cosmids; for 15-cm plates use 300 μl bacteria and 300 μl cosmids; for a 23 × 30 cm Pyrex dish use 2 ml bacteria and up to 2 ml cosmids.

Screening of Cosmid Libraries

Hybridization Screening. Cosmids are screened using the same methods which have been designed for plasmids. The high-density screening method of Hanahan and Meselson works well with these vectors, especially the c2XBHC and c2RBHC derivatives.[11] Using this method a com-

[8] T. Maniatis, E. F. Fritsch, and J. Sambrook, "Molecular Cloning: A Laboratory Manual." Cold Spring Harbor Lab., Cold Spring Harbor, New York, 1982.

[9] B. Hohn, this series, Vol. 68, p. 299.

[10] S. Rosenberg, this volume [7].

[11] D. Hanahan and M. Meselson, this series, Vol. 100, p. 333.

plete human library of 2×10^5 cosmids can be screened on a single 23×30 cm Pyrex dish. Background hybridization is often a problem when screening cosmids. Addition of competitor nucleic acids to 50 μg/ml seems to help reduce this background. Generally the competitor is made from total nucleic acids extracted from *E. coli* containing the cosmid vector. Also, the hybridization probes for cosmid screening must be free of any plasmid sequences or screening will be impossible.

Genetic Selection. The c2XB AMP^- vector is designed for use in genetic selection systems. This vector has been used to rescue the ampicillin-resistance marker but should be applicable for any drug-resistance marker. For genetic selection, the cosmid library is transduced into bacteria, then grown in liquid culture without selection for 2–3 hr. The bacteria are collected by centrifugation, then resuspended in LB and plated on LB + AMP (50 μg/ml) plates. The 2- to 3-hr growth without selection usually results in isolation of some identical recombinants due to *E. coli* replication, but this growth period also seems to increase the probability that all the different types of recombinants are isolated. Generally, from a ligation and packaging reaction where 1 μg of size-selected insert DNA is used, 15–20 ampicillin-resistant colonies are recovered. Of these colonies, 2–3 will be different recombinants.

Comments

This cloning system provides a simple method for constructing cosmid libraries. With the procedures and vectors described here cosmid library construction is no more difficult than λ cloning, therefore cosmid cloning can now be considered as a reasonable alternative to λ cloning for construction of genomic libraries. The double *cos* vectors described here have been used to construct cosmid libraries from organisms ranging from plants to archaebacteria to humans. Unlike cosmid libraries made with some other vectors, these libraries are free of any vector-only background. In over 10^7 cosmids screened for kanamycin-resistant colonies, no vector background has been detected. Yet these double *cos* site vectors are extremely simple to use, requiring only a double restriction digest to prepare the vector for cloning.

Acknowledgments

I wish to thank Bob Swift for valuable help and discussion in the early phases of this work. Much of this work was done under the guidance of Jerry Dodgson, to whom I am very grateful. P.B. is supported by a Damon Runyon–Walter Winchell postdoctoral fellowship (DRG-865).

[7] Improved *in Vitro* Packaging of λ DNA

By SUSAN M. ROSENBERG

Apart from its usefulness for studying bacteriophage λ, *in vitro* packaging offers an efficient means of recovering DNA fragments cloned into either λ or cosmid vectors. For maximum usefulness, an *in vitro* packaging system should be simple to prepare, package added DNA efficiently, and be free from contaminating endogenous phage. Endogenous phage are plaque-forming units present in packaging extracts to which no exogenous λ DNA has been added. Standard *in vitro* packaging systems yield substantial numbers of endogenous phage. These particles can confound λ genetic experiments and limit the usefulness of some cloning strategies. My colleagues and I employed λ genetic trickery to produce an *in vitro* packaging system devoid of endogenous phage. Our system uses a sole bacterial strain which, when prepared by a simple procedure, yields high-efficiency packaging extract. I outline here our method for preparing the crude packaging mixture and review some of our previously published results.[1,2]

How *in Vitro* Packaging Works

The λ enzyme terminase performs packaging by first binding DNA at the packaging origin, *cos*, and also binding to an empty prohead (a structural protein assembly that is part of the phage coat). Terminase then cleaves DNA at *cos* and spools it into the prohead (see Feiss and Becker[3] for a review). Phage tails attach spontaneously to DNA-filled heads, yielding complete, infective particles.

For *in vitro* packaging, *Escherichia coli* cells (lysogens) carrying λ prophage are grown then induced for production of packaging proteins including terminase and proheads. Induced cells are broken. The resulting extract (used crude or after fractionation) plus additional ATP and polyamines (which facilitate DNA condensation) can package added λ DNA. The traditional genetic strategy[4–9] for packaging extract production (Fig. 1) uses two complementary λ lysogens, each mutant in a different

[1] S. M. Rosenberg, M. M. Stahl, I. Kobayashi, and F. W. Stahl, *Gene* **38,** 165 (1985).

[2] S. M. Rosenberg, *Gene* **39,** 313 (1985).

[3] M. Feiss and A. Becker, *in* "Lambda II" (R. Hendrix, J. Roberts, F. Stahl, and R. Weisberg, eds.), p. 305. Cold Spring Harbor Lab., Cold Spring Harbor, New York, 1983.

[4] D. Kaiser and T. Masuda, *Proc. Natl. Acad. Sci. U.S.A.* **70,** 260 (1973).

[5] B. Hohn and T. Hohn, *Proc. Natl. Acad. Sci. U.S.A.* **71,** 2372 (1974).

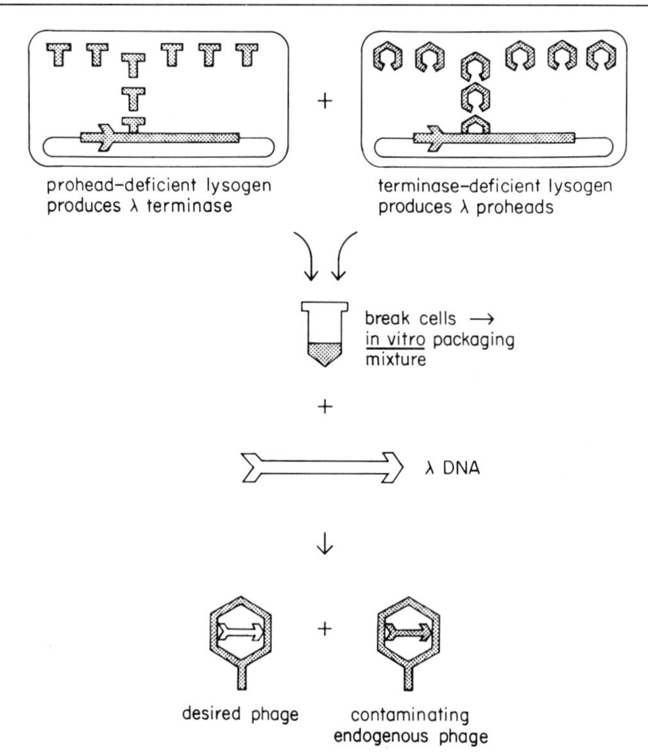

FIG. 1. Complementary two-strain *in vitro* packaging system. Lysogens are partnerships of λ DNA stably integrated into the *E. coli* chromosome. They can be induced to lytic growth, causing λ proteins to be made. The T shape represents λ terminase. Proheads are represented by open hexagons. The λ chromosome is represented by a thick line with an arrowhead (*cos,* the packaging origin). Reprinted from S. M. Rosenberg, M. M. Stahl, I. Kobayashi, and F. W. Stahl, *Am. Soc. Microbiol. News* **51,** 386 (1985).

packaging protein gene. The two strains are grown and induced separately. Neither endogenous λ can be packaged during growth for want of (for example, see Fig. 1) either terminase or proheads. Upon lysis and mixing of the extracts, added DNA is packaged, but so is the endogenous λ DNA. Endogenous phage result. Unsuccessful strategies for elimination of endogenous phage have been reviewed.[1]

---

[6] A. Becker and M. Gold, *Proc. Natl. Acad. Sci. U.S.A.* **72,** 581 (1975).

[7] F. R. Blattner, A. E. Blechl, K. Denniston-Thompson, H. E. Faber, J. E. Richards, J. L. Slightom, P. W. Tucker, and O. Smithies, *Science* **202,** 1279 (1978).

[8] B. Hohn, this series, Vol. 68, p. 299.

[9] T. Maniatis, E. F. Fritsch, and J. Sambrook, "Molecular Cloning: A Laboratory Manual." Cold Spring Habor Lab., Cold Spring Harbor, New York, 1982.

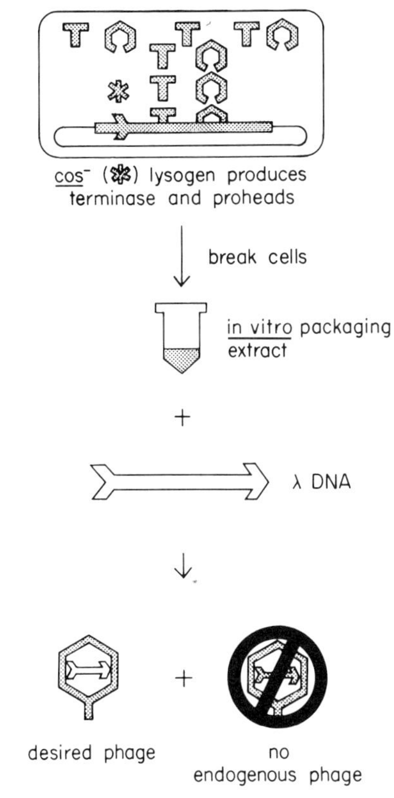

FIG. 2. One-strain *cos*-less *in vitro* packaging system. The symbols are the same as in Fig. 1. *, Mutant *cos*. Reprinted from S. M. Rosenberg, M. M. Stahl, I. Kobayashi, and F. W. Stahl, *Am. Soc. Microbiol. News* **51,** 386 (1985).

A Different Packaging Strategy

Our alternative design (Fig. 2) employs a single λ lysogen wild type for packaging proteins but unpackageable owing to a deletion mutation at *cos*. When induced, this strain produces all of the packaging proteins in one cell and cannot encapsidate its own DNA. Thus, *in vitro* packaging extracts of the *cos*-less lysogen produce virtually no endogenous phage.[10]

[10] This is true of cos^- lysogens of *E. coli* strain C but not strain K-12. Analysis of K-12-derived endogenous phage suggested that λ-homologous regions of the *E. coli* K-12 chromosome donated cos^+ to the λ prophage, restoring prophage packageability. *E. coli* C contains very little λ-homologous DNA and seems not to "convert" a cos^- prophage to cos^+.[1]

Preparation of *in Vitro* Packaging Extract

Bacterial Strain

SMR10 = *E. coli* C1a Su^- (λ *cos2* ΔB *xis1 red3 γam210 c*I*ts857 nin5 Sam7*)/λ. (See Ref. 1 for construction of cos^- lysogens and description of λ genetic elements.) Slants, streaks, overnights, and other cultures of this lysogen must be grown at or below 34°

Chemicals

1. ATP-2Na-3H_2O, M_r 605 (P-L Biochemicals)
2. Spermidine-3HCl, M_r 254.6 (Sigma)
3. Putrescine-2HCl, M_r 161 (Calbiochem)

Stock Solutions

1. ATP (0.1 *M*), 60.5 mg/ml, OD_{259} of 1/2000 should be 0.77. Adjust to pH 7 with 1 *N* NaOH, ~190 μl/ml
2. Spermidine (0.25 *M*), 64 mg/ml. Adjust to pH 7 with 1 *N* NaOH, ~15 μl/ml
3. Putrescine (0.25 *M*), 40 mg/ml. Adjust to pH 7 with 1 *N* NaOH, ~13 μl/ml

Buffers

1. SMC: Sterile double-distilled H_2O, 190 ml; SMC-A, 8 ml; SCM-B, 2 ml. Do not autoclave

   SMC-A: Anhydrous Na_2HPO_4, 52.5 g; KH_2PO_4, 22.5 g; double-distilled H_2O to adjust to 300 ml. Autoclave

   SMC-B: NaCl, 10 g; NH_4Cl, 20 g; 1 *M* $MgCl_2$, 20 ml; 1 *M* $CaCl_2$, 2 ml; double-distilled H_2O to adjust to 200 ml. Autoclave

2. TSP: Sterile double-distilled H_2O, 13.2 ml; 1 *M* Tris · HCl (pH 7.9), 0.6 ml; 0.25 *M* spermidine, 0.6 ml; 0.25 *M* putrescine, 0.6 ml

3. TEK (10× stock solution): Double-distilled H_2O, 850 ml; KCl, 37.3 g; 1 *M* Tris · HCl (pH 8.1 at room temperature), 100 ml; 0.2 *M* EDTA (pH 7.0), 50 ml. Autoclave

Medium

LBK: Double-distilled H_2O, 1 liter; bacto tryptone, 10 g; NaCl, 10 g; yeast extract, 5 g; 1 *N* NaOH, 4 ml

Preparation of Packaging Extract

Prepare in advance:

460 ml LBK divided among four 2-liter Fernbach flasks, prewarm to 34°
Bacterial overnight culture in LBK at 34°
Water bath shaker at 44°
Shakers at 34 and 37°
Liquid Nitrogen
Two 4-liter Erlenmeyer flasks

Chill:

Two 250-ml centrifuge bottles
One 12-ml centrifuge tube
Rotors for both

1. Inoculate 0.5 ml fresh overnight into each Fernbach flask at 34°. Shake at 34°.
2. At 0° make "cocktail." Allow DMSO and water mixture to cool on ice several minutes before adding ATP (DMSO + $H_2O \rightarrow$ heat).

Double-distilled H_2O, 255 μl
DMSO, 300 μl
ATP (0.1 *M*, pH 7.0); 45 μl

3. Aliquot 5 μl cocktail into each of 60 conical 1.5-ml microfuge tubes at 0°. Keep at 0–4°.
4. When the cells have grown to OD_{550} = 0.80 (2.5–3 hr later), pour the culture into two 4-liter Erlenmeyer flasks prewarmed in a 44° bath. Shake *vigorously* at 44° for 15 min. Excellent aeration is critical.
5. Shake *vigorously* at 37° for 90 min.
6. Chill in ice water; swirl for 5 min. From this stage onward, the cells must never warm up.
7. Centrifuge at 6000 rpm for 6 min at 4° in precooled 250-ml centrifuge bottles and a precooled rotor.
8. Carefully remove all supernatant and add 4.5 ml precooled TSP to each pellet. Suspend gently. Remove both suspensions into one precooled 12-ml centrifuge tube.
9. Centrifuge in a precooled rotor at 6000 rpm for 6 min at 4°.
10. Carefully remove all supernatant. To the pellet, add 0.35 ml of precooled TSP. Mix gently with a thin glass rod at 0°. The cell suspension is very thick and pasty, but not viscous. Do not thin the suspension.
11. Aliquot 20 μl of this suspension with a 50-μl-size glass pipet into

each microfuge tube with cocktail at 0°. Take care to dispense uniform 20-μl aliquots of cells in order to minimize tube-to-tube variation in packaging efficiency. Vortex and leave at 0° until step 12.

12. Freeze uncapped tubes in liquid nitrogen. Uncapped tubes are stored under liquid nitrogen. Under these conditions packaging efficiency of a given batch of extract remained constant for at least 4 months. When liquid nitrogen-frozen tubes were stored at −70°, packaging efficiency began to decrease after approximately 1 week.

Packaging

1. Dump liquid nitrogen from frozen tubes of packaging mixture and immediately add 10 μl DNA in TEK. Immediately, thaw tubes in 28° water for 30 sec. Mix carefully with thin glass rod. Avoid foaming.
2. Incubate at 28° for 90 min.
3. To each tube, add 0.5 ml SMC containing 50 μg/ml DNase I and also add 3 μl $CHCl_3$. Vortex until lysate in each tube is homogeneous with no solid material left visible.
4. Microfuge for 2 min. Collect supernatant.
5. Plate on appropriate bacterial indicator strains. Titers of *in vitro* packaged phage remained stable for at least 5 months when the lysates were stored capped at 4°.

Endogenous Phage

To assay endogenous phage from any *in vitro* packaging system, tubes of extract are "blank packaged" with buffer containing no DNA. The lysate is then plated on indicator bacteria appropriate for revealing particles with the genotype of the prophage(s) from which the extract was derived. Two types of endogenous phage may be observed. "Healthy" endogenous phage are capable of plaque formation on nonlysogenic bacteria and are therefore presumed to contain intact λ genomes. "Defective" endogenous phage plaque only on heteroimmune lysogens; therefore, one assumes that such particles are missing essential λ gene(s), which must be recovered by recombination with the heteroimmune prophage. Table I is a comparison of healthy and defective endogenous phage levels in two, two-strain complementary packaging systems and the SMR10 system. The *cos*-less packaging system produces no healthy and few defective endogenous phage,[10] in contrast with a crude or purified two-lysogen system.

TABLE I
ENDOGENOUS PHAGE LEVELS

In vitro packaging system	Lysogens	Preparation of extract	Number of endogenous phage per standard volume extract[a]	
			Healthy	Defective
1. Blattner *et al.* (1978),[7] Becker and Gold (1975)[6]	BHB2671 BHB2673	Sonic extract plus Freeze-thaw lysate	$(4.0 \pm 0.5) \times 10^3$	$(1.6 \pm 0.27) \times 10^4$
2. Kobayashi and Ikeda (1977),[13] modified from Hohn and Hohn (1974)[5]	BH2688 BHB2690	Crude, whole-cell extract	930 ± 140	$(5.5 \pm 1.9) \times 10^5$
3. Rosenberg *et al.* (1985)[1]	SMR10	Crude, whole-cell extract	$<1.6 \times 10^{-2}$[b]	850 ± 150

[a] To compare frequencies of endogenous phages in different *in vitro* packaging systems, titers are normalized to that found in whatever volume of packaging extract is used to package 1 μg exogenous λ DNA (standard volume extract). The bacteria and lysogens used to assay phage from each system are given in Ref. 1.

[b] No plaque formers were detected in 57 tubes of extract. Value is the upper limit of plaque formers expected within 95% confidence limits of the Poisson distribution.

Packaging Efficiency

Eleven different batches of *in vitro* packaging extract prepared from SMR10 yielded consistently high packaging efficiencies ranging from 1 to 5×10^{-3} plaque formers produced/λ chromosome added ($0.2–1.0 \times 10^{8}$ plaque formers/μg λ DNA). Packaging efficiency is independent of the amount of λ DNA added per tube, in the range of 1.0×10^{-5}–4.0 μg λ DNA added per tube.[1,11]

*Eco*K Restriction during *in Vitro* Packaging

The endogenous restriction system of *Escherichia coli* strain K works *in vitro*,[12] and only DNA grown in *E. coli* K is modified against *Eco*K restriction. Cloners of eukaryotic DNA and other non-*E. coli* K DNA should wonder whether insert DNA is restricted during *in vitro* packaging. Since *E. coli* C has no endogenous restriction system, *E. coli* C-derived packaging mixture can serve as a standard against which to measure the extent of restriction in K-12-derived packaging systems. A mixture of genetically marked modified and unmodified λ DNAs was packaged in the SMR10 (*E. coli* C) system and in various K-12-derived systems. The ratio of packaging efficiencies (unmodified λ DNA/modified λ DNA) of DNA packaged in the K-12-derived systems was compared with that of DNA packaged in SMR10 extract. Losses (2- to 7-fold) of the unmodified genotype were detected from packagings in the K-12 extracts.[2] No such losses occurred in control experiments in which DNAs of both genotypes were modified. Whereas the loss effect is ~100× smaller than the effect of *Eco*K restriction of λ *in vivo*, this restriction *in vitro* may specifically decrease packaging efficiency of hybrid molecules whose insert DNA contains the heptameric *Eco*K recognition sequence and thereby bias λ and cosmid genomic libraries. The loss effect may also vary widely between different K-12-derived packaging systems which have been optimized for packaging efficiency of K-modified λ DNA, not for lack of *Eco*K restriction. The SMR10 system should give a higher true packaging efficiency from ligations with unmodified insert DNA and relieve any such bias.

[11] In this range, plaque formers produced varied linearly with the number of λ chromosomes added when the λ DNA was annealed cohesive end to cohesive end or ligated at *cos*. To form linear monomers of λ, a DNA solution must be heated to 60°, then chilled on ice immediately before packaging. Under such conditions, numbers of plaques produced vary as the square of input λ molecules.[1] While the latter result has interesting implications for the mechanism of packaging,[1] it is probably not important for cloners.

[12] M. Meselson and R. Yuan, *Nature* (*London*) **217,** 1110 (1968).

Comments

SMR10 *in vitro* packaging mixture is free from endogenous phage and *Eco*K restriction. Additionally, SMR10 crude packaging extracts are as efficient as more purified (sonic extract plus freeze–thaw lysate) preparations of complementary, two-strain *in vitro* packaging mixture. (The sonic extract plus freeze–thaw lysate method of preparing two-strain packaging systems[6,9] yields higher efficiency extracts than does crude preparation of the same two strains.[5,13] Sonic extract of SMR10 added to freeze–thaw lysate of SMR10 may yield an even more efficient system than the crude extract outlined here, but I have not tried this.)

Acknowledgments

My friends Mary Stahl, Ichizo Kobayashi, Frank Stahl, and David Thaler are gratefully acknowledged for collaboration and interest. Oliver Smithies suggested an improvement to the procedure. Thanks to George Sprague for comments on the manuscript and to Janet Rosenberg for the illustrations. The author was supported by N.I.H. predoctoral fellowship No. 5-T32-GM07413. Work reviewed here was supported by N.I.H. Grant No. GM33677 and N.S.F. Grant No. PCM 8409843 to F. W. Stahl.

[13] I. Kobayashi and H. Ikeda, *Mol. Gen. Genet.* **153,** 237 (1977).

[8] λ Phage Vectors—EMBL Series

By A. M. FRISCHAUF, N. MURRAY, and H. LEHRACH

Introduction

To establish genomic libraries from large genomes and to identify specific genomic clones, either cosmids or λ replacement, vectors have been used. Both systems rely on the packaging of recombinant genomes into phage heads *in vitro* to give infective particles as a means of introducing DNA into the bacterial cell with high efficiency. Packaging requires the cutting of specific DNA sequences, termed *cos*, and infectious particles are formed efficiently only if the length of DNA between two *cos* sequences is between 78 and 105% of the length of wild-type λ (38–55 kb).[1] Hence, the size of the vector DNA determines the capacity remaining for the inserted DNA molecule. Approximately 30 kb of DNA is

[1] M. Feiss and A. Becker, *in* "Lambda II" (R. Hendrix, J. Roberts, F. Stahl, and R. Weisberg, eds.), p. 305. Cold Spring Harbor Lab., Cold Spring Harbor, New York, 1983.

necessary to encode the functions required for lytic growth, so the capacity of λ vectors has an inherent upper limit of approximately 22 to 24 kb. A vector comprising only the essential DNA regions (i.e., 30 kb in length) would be too small to be packaged; therefore, the theoretical cloning capacity can only be used in replacement vectors in which a nonessential region of the vector, flanked by the restriction sites used for cloning, is replaced by the inserted DNA.

Cosmids, in contrast, require only a few kilobases of DNA to encode the functions essential for their replication and selection, and the *cos* sequence, or sequences necessary for packaging, consequently they allow the cloning of DNA fragments of close to 50 kb in length. This larger capacity makes cosmids very attractive for those experiments in which the aim is either to analyze long regions by overlapping clones, or to clone genes too large to fit into λ vectors. λ replacement vectors, however, offer advantages in the efficiency of library construction and screening. Using the protocol described later, which allows cloning samples of analogous (or even identical) DNA preparations in either λ or cosmid vectors, we find cloning efficiencies of roughly 5×10^5 clones/μg of starting DNA for cloning in EMBL3[2] and approximately 5×10^4 clones/μg DNA for cloning in pcos2 EMBL.[3] Taking the different size of the insert into account, roughly five times more DNA is needed to construct cosmid libraries than λ libraries of equivalent coverage.

In addition, many laboratories find λ libraries somewhat easier to handle, to screen, and to amplify. Cloning in a λ replacement vector may be preferred if there are no special requirements for the larger capacity of cosmid vectors, if only small amounts of DNA are available, or if several libraries have to be constructed and screened.

EMBL3 and EMBL4

EMBL3 and EMBL4[2] are λ replacement vectors derived from λ 1059[4,5] which allow the cloning of DNA fragments with sizes between 8 and 23 kb. Polylinkers flanking the nonessential middle fragment contain symmetrically arranged sites for *Sal*I, *Bam*HI, and *Eco*RI (Fig. 1). Therefore fragments created by these enzymes, or a number of different restriction enzymes generating compatible ends, can be cloned [*Sal*I and *Xho*I in

[2] A.-M. Frischauf, H. Lehrach, A. Poustka, and N. Murray, *J. Mol. Biol.* **170,** 827 (1983).

[3] A. Poustka, H.-R. Rackwitz, A.-M. Frischauf, B. Hohn, and H. Lehrach, *Proc. Natl. Acad. Sci. U.S.A.* **81,** 4129 (1984).

[4] J. Karn, S. Brenner, L. Barnett, and G. Cesareni, *Proc. Natl. Acad. Sci. U.S.A.* **77,** 5172 (1980).

[5] J. Karn, S. Brenner, and L. Barnett, this series, Vol. 101, p. 3.

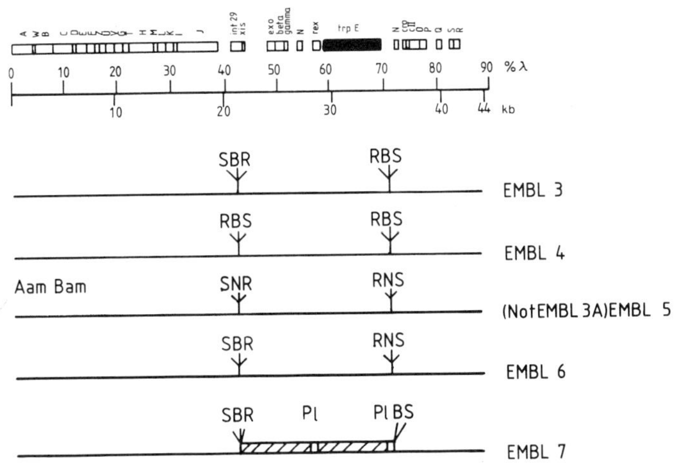

FIG. 1. Schematic structure of the λ vectors EMBL3, EMBL4, EMBL5 (NotEMBL3A), EMBL6, and EMBL7. EMBL7 contains as a middle fragment two head-to-tail copies of pEMBL18 [L. Dente, M. Sollazzo, C. Baldari, G. Cesareni, and R. Cortese, *in* "DNA Cloning" (D. M. Glover, ed.), Vol. 1. IRL, in press] with a modified polylinker sequence (M. Burmeister and F. Michiels, unpublished observations). The sequence of sites in the polylinker (P1) is *Hin*dIII, *Sph*I, *Pst*I, *Mlu*I, *Bam*HI, *Xma*I (*Sma*I, *Ava*I), *Kpn*I, *Sst*I, *Eco*RI. S, *Sal*I; B, *Bam*HI; R, *Eco*RI; N, *Not*I.

the *Sal*I site, *Bam*HI, *Bgl*II, *Bcl*I, *Xho*II, and *Sau*3A (or *Mbo*I) in the *Bam*HI site, and *Eco*RI or *Eco*RI* in the *Eco*RI site]. Of these enzymes, *Sau*3A and its isoschizomer, *Mbo*I, are most generally useful, since partial digestion with these enzymes, which recognize a 4-bp sequence, minimizes potential distortion of the sequence representation in the library due to either preferential cutting or irregular distribution of sites. When *Sau*3A fragments are cloned in a vector cut with *Bam*HI, three out of four ligation events will not recreate the *Bam*HI site. In most cases, therefore, the excision of the cloned fragment from the vector with this enzyme is not possible. However, the flanking sites of the polylinker (*Sal*I in EMBL3 or EMBL3A, *Eco*RI in EMBL4) can be used to excise the cloned fragment. *Sal*I is especially convenient in this regard since these sites are underrepresented in mammalian DNA so that in many cases, the insert can be recovered in a single *Sal*I fragment.

As in the vectors 1059[4,5] and 2001,[6] the replaceable middle fragment of the EMBL vector carries the *red* and *gam* genes of λ, and these can be transcribed, irrespective of their orientation, from a promoter included on this fragment. Expression of *red* and *gam*, particularly *gam*, prevents the

[6] J. Karn, H. W. D. Mathes, M. J. Gait, and S. Brenner, *Gene* **32,** 217 (1984).

growth of the vector phage on *Escherichia coli* lysogenic for phage P2.[7] This Spi$^+$ (sensitive to P2 interference) phenotype of the vector phage allows strong genetic selection against the vector phage, and therefore enrichment for ligation products in which the middle fragment has been replaced by inserted DNA.

There is, however, a disadvantage associated with this powerful selection system. In the absence of the phage *red* and *gam* genes, phage growth requires the host recombination function (for a more detailed discussion, see Refs. 8 and 9). This is a consequence of the fact that multimeric λ DNA, the substrate for packaging, can be produced in two ways, one of which is dependent on recombination, the other on the product of the *gam* gene. Following infection λ DNA is first replicated in the theta mode to give rise to monomeric circles, which can oligomerize via recombination. Later in the phage infection cycle replication proceeds by a rolling circle mode, giving long concatemeric molecules. This change of mode is dependent on inactivation of *Exo*V[10] by the phage *gam* gene product.

Growth of Spi$^-$ (*red*$^-$ *gam*$^-$) phages therefore depends on the host RecA function to generate oligomeric circles by recombination, a process that is only efficient if either the vector or the cloned insert contains a Chi sequence (see Ref. 8). Libraries constructed in EMBL phages therefore are generally propagated on a recA$^+$ strain which might enhance the potential instability of repetitive sequences and might make the occurrence of deletions and rearrangements more likely. Alternatives are the use of recBC$^-$ hosts, the provision of the *gam* gene transcribed from pR′ on a helper plasmid in a RecA$^-$ host,[11] or, at the cost of losing the genetic selection for chimeric molecules, the use of one of a number of *gam* replacement vectors of which Charon 34 and Charon 35 are the most versatile.[12]

As an alternative, or in addition to a genetic selection, religation of the original vector molecule can be avoided by cleaving the inner site in the polylinker molecule (*Eco*RI in EMBL3). The *Bam*HI end of the replaceable fragment will be released on a small connector fragment, which can be removed by the selective precipitation of the larger DNA fragments by 2-propanol.

[7] J. Zissler, E. Signer, and F. Schafer, *in* "The Bacteriophage Lambda" (A. D. Hershey, ed.), p. 469. Cold Spring Harbor Lab., Cold Spring Harbor, New York, 1971.

[8] G. Smith, *in* "Lambda II" (R. Hendrix, J. Roberts, F. Stahl, and R. Weisberg, eds.), p. 175. Cold Spring Harbor Lab., Cold Spring Harbor, New York, 1983.

[9] N. Murray, *in* "Lambda II" (R. Hendrix, J. Roberts, F. Stahl, and R. Weisberg, eds.), p. 395. Cold Spring Harbor Lab., Cold Spring Harbor, New York, 1983.

[10] R. C. Unger and A. J. Clark, *J. Mol. Biol.* **70,** 539 (1972).

[11] G. F. Crouse, *Gene* **40,** 151 (1985).

[12] W. A. M. Loenen and F. R. Blattner, *Gene* **26,** 171 (1983).

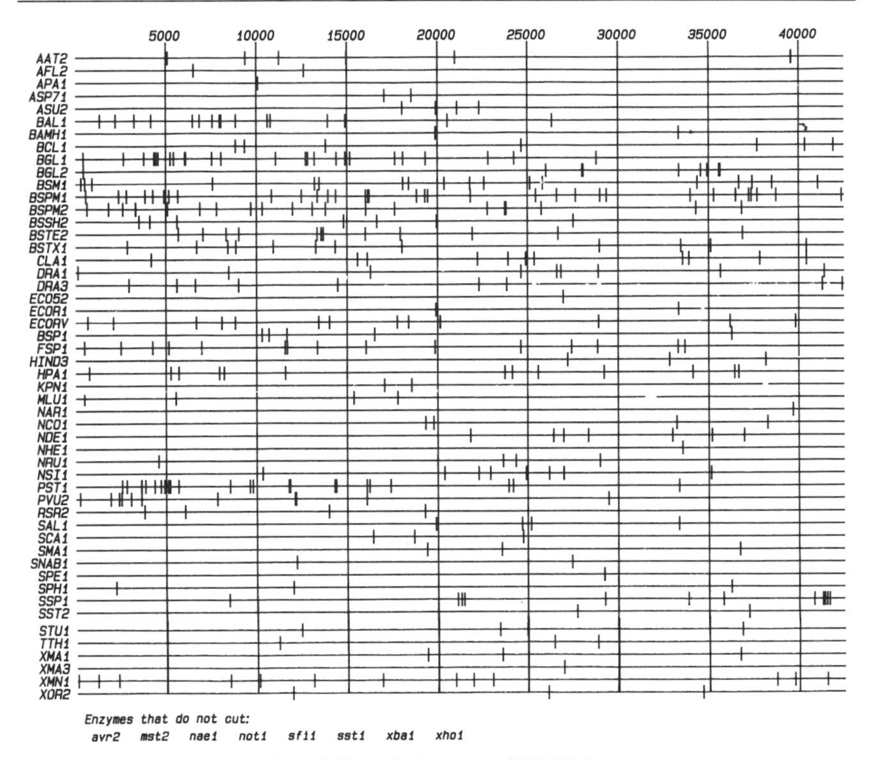

FIG. 2. Restriction map of EMBL3.

Though the EMBL3 vector itself has not been sequenced, the published sequence of λ and various mutants[13] can be used to predict high-resolution restriction maps. A map based on these data is shown in Fig. 2.

Amber Derivatives of EMBL Vectors

Amber derivatives of EMBL3 have been constructed. These include EMBL3A, carrying amber mutations in the *A* and *B* genes, EMBL3S, carrying a *Sam*7 mutation, and EMBL3AS with the *Aam Sam* mutations. Due to the proximity of the *Sam*7 mutation to the Chi site (ChiD) carried in the EMBL vectors, all *Sam*7 derivatives have lost the Chi mutation, and must therefore rely on Chi sites provided by the cloned sequence.

These vectors can be used to selectively clone sequences carrying

[13] D. L. Daniels, J. L. Schroeder, W. Szybalski, F. Sanger, A. R. Coulson, G. F. Hong, D. F. Hill, G. B. Peterson, and F. R. Blattner, *in* "Lambda II" (R. Hendrix, J. Roberts, F. Stahl, and R. Weisberg, eds.), p. 519. Cold Spring Harbor Lab., Cold Spring Harbor, New York, 1983.

suppressor genes, e.g., *supF*.[14] The use of two amber mutations eliminates the background due to reversion of the amber mutations. However, with the use of double amber mutations, the growth of suppressor-independent phages (probably formed by loss of the *Aam Bam* mutations by recombination with DNA from the *in vitro* packaging extract) has been observed. It is expected that the use of the *Sam* mutation, also present in the prophages used for preparation of packaging extracts, would reduce this background considerably.

An *orf* in the right arm of λ has been shown to be required for efficient phage plasmid recombination.[15] This *orf* is missing in EMBL3A and many other vectors, as the consequence of the *nin*5 deletions. EMBL3A, therefore, is not generally useful for the isolation of phage clones by homologous recombination with suppressor-carrying probe plasmids.[16] A derivative of EMBL3 containing the sequence responsible for a high rate of recombination has been described.

EMBL Vector Derivatives with Further Cloning Sites

Derivatives of the EMBL vectors have been constructed to provide additional cloning sites in the polylinker sequence. In particular, cloning sites for enzymes which cut a relatively small number of sites in the mammalian genome have been introduced. These enzymes either recognize an 8-bp sequence site (*Not*I, *Sfi*I) or recognition sites of 6-bp length containing the CG dinucleotide sequence underrepresented in mammalian DNA. Of these, *Not*I (GC′GGCCGC) is especially useful, since this enzyme creates fragments of millions of base pairs in length, has no recognition site in λ, and generates cohesive ends with 5′ extensions. Other enzymes that cut mammalian DNA infrequently and create ends which are easy to ligate are *Mlu*I (A′CGCGT), *Bss*HII (G′CGCGC), *Sac*II (CCGC′GG), and *Pvu*I (CGAT′CG).

Since the number of sites for these enzymes in the genome is quite small, fragments ending in or containing such rare restriction sites constitute a small, well-defined subset of the entire genome. A library of junction fragments therefore provides a group of easily identifiable reference points covering the whole genome. When such fragments are used as probes they can be mapped relative to each other by genomic restriction

[14] M. Goldfarb, K. Shimizu, M. Perucho, and M. Wigler, *Nature (London)* **296,** 404 (1982).

[15] H. V. Huang, P. F. R. Little, and B. Seed, *in* "Vectors: A Survey of Molecular Cloning Vectors and Their Uses" (R. Rodriguez, ed.). Butterworth, in press.

[16] B. Seed, *Nucleic Acids Res.* **11,** 2427 (1983).

analysis using pulsed field gradient gel electrophoresis techniques.[17] Cloning of purified fragments isolated from pulsed field gradient gels can be used to derive probes hundreds of kilobases from the start point. Various procedures making use of rare cutting enzymes should therefore be very helpful in coping with the large distances in mammalian genomes.

*Not*I-Linking Fragment Cloning

A derivative of EMBL3A in which the *Bam*HI sites of the polylinkers have been replaced by *Not*I sites (NotEMBL3A, Fig. 1) has been constructed. This is used for the selective cloning of fragments containing *Not*I sites (*Not*I-linking fragment clones).

In the cloning procedure *Mbo*I-digested, sized DNA (10–20 kb) is circularized at low concentration in the presence of a *Bam*HI-cleaved plasmid carrying a suppressor sequence.[18] Under appropriate conditions[19] a large fraction of the circles will contain a suppressor plasmid. The circles containing a *Not*I recognition site can be linearized by this enzyme and subsequently cloned in *Not*I-cut NotEMBL3A vector. By plating on a suppressor-free host phages containing *Not*I junction fragments can be selected. If DNA from interspecies hybrid cells has been used, clones from, e.g., the human chromosome(s) can be identified by hybridization with, in this case, human repetitive sequences.[20]

End Fragment Cloning

A derivative of EMBL3 has been constructed in which only one of the two *Bam*HI sites has been substituted by a *Not*I site (EMBL6, Fig. 1). This allows the cloning of the ends of large, gel-purified *Not*I fragments after partial digestion with *Mbo*I.

For a number of enzymes that cut rarely in mammalian DNA, the construction of specific cloning vectors is quite difficult, due to the large number of sites in λ DNA (e.g., *Mlu*I, *Bss*HII). Vectors allowing the cloning of the ends of fragments created by these enzymes have, however, been constructed (Michiels, unpublished) by cloning a *Mlu*I-containing polylinker at the position of the right *Eco*RI site in EMBL3 (EMBL7, Fig. 1). To clone either *Mlu*I or *Bss*HII fragment ends, the right

[17] D. C. Schwartz and C. R. Cantor, *Cell* **37,** 67 (1984).

[18] A. Levinson, D. Silver, and B. Seed, *J. Mol. Appl. Genet.* **2,** 507 (1984).

[19] F. S. Collins and S. M. Weissman, *Proc. Natl. Acad. Sci. U.S.A.* **81,** 6812 (1984).

[20] J. F. Gusella, C. Jones, F. T. Kao, D. Housman, and T. T. Puck, *Proc. Natl. Acad. Sci. U.S.A.* **79,** 7804 (1982).

arm from this vector has to be isolated and used in combination with a left vector arm created, e.g., by *Bam*HI digestion of EMBL3.

Additional Derivatives

To provide additional sites for *Xba*I and *Sac*I in the polylinker, Ernst Natt and Gerd Scherer (personal communication) have constructed EMBL3-12, in which the *Sal*I–*Eco*RI fragments of EMBL3 are replaced by the *Sal*I–*Eco*RI polylinker segment of the pUC12 plasmid.[21] The modified polylinker therefore contains sites for *Sal*I, *Xba*I, *Bam*HI, *Sam*I, *Sac*I, and *Eco*RI. This arrangement of sites in the polylinker allows the excision of inserts created by ligation of *Nhe*I and *Spe*I fragments to the ligation-compatible *Xba*I sites by the flanking *Sal*I sites.

DNA Analysis

Analysis of the partial digestion products of λ clones can be facilitated by tagging one *cos* sequence with a complementary radiolabeled oligonucleotide.[22] Since this analysis is complicated by the vector DNA between the labeled cohesive ends and the cloning site, an EMBL4 derivative in which *cos* is adjacent to one of the cloning sites has been constructed.

Construction of Libraries

Principle

The protocol used by us aims to minimize the manipulations involved in library construction, taking advantage of a combination of genetic and biochemical selection steps to rule out side reactions and to ensure the correct structure of the final clones. In our experience, this leads to high reproducibility, high yields, and the capability to handle very small amounts of material. The protocol described here for the construction of λ libraries is essentially identical to the protocol used by us in cosmid library construction. It is therefore possible, and in many cases advantageous, to construct both λ and cosmid libraries in parallel.

In constructing the library, the DNA to be cloned is first digested partially with *Sau*3A or *Mbo*I to a (number) average size of slightly above 20 kb. This DNA is then dephosphorylated using calf intestine alkaline phosphatase, and ligated to EMBL3 or EMBL4 cleaved with *Bam*HI.

[21] J. Vieira and J. Messing, *Gene* **19,** 259 (1982).

[22] H. R. Rackwitz, G. Zehetner, A. M. Frischauf, and H. Lehrach, *Gene* **30,** 195 (1984).

Due to the phosphatase treatment of the insert, ligation events between different originally unconnected insert fragments are ruled out, while ligation of inserts to vector can proceed over the 5′ phosphates carried by the vector arms.

As mentioned before, religation of the middle fragment to the vector arms can be prevented for EMBL3 by removal of the *Bam*HI-ligatable end as a small fragment by recleavage with *Eco*RI, followed by selective precipitation of the long DNA with 2-propanol. (Alternatively, or in addition, the Spi selection can be used to remove remaining uncut or religated vector from the library.)

After ligation, the DNA is packaged *in vitro*. The packaging product can be plated directly. Alternatively, if large amounts of packaged material have to be plated at high density, inhibitory material can be removed from the packaging product by a CsCl step gradient. In general we use either NM538 (*SupF,* rk^-, mk^+, the nonselective host) or NM539 [NM538 (P2cox3), the host used for Spi selection] for plating.

For EMBL3 and EMBL4, many other Rec^+, restriction strains can, however, be used. EMBL3A is best plated on *supF*-carrying strains, since it will form very small plaques on hosts providing only *supE*. To allow recovery of sequences carrying inverted repeats in the library, *recB recC sbcB* host strains can be used as alternatives to the strains mentioned above.[23,24]

Procedures

Vector Preparation

1. Plate Stock. To prepare a plate stock for vector preparation a well-separated single plaque is picked from a fresh BBL plate, and suspended in 300 μl λ diluent. After 1 hr or more for diffusion the phage is plated with DH1 or NM538 bacteria to give a slightly less than confluent plate. Five milliliters λ diluent is left on the plate for 3 hr at room temperature. The plate stock is harvested, treated with 0.2 ml $CHCl_3$, and titrated on NM538, NM539, and, for amber-containing phages, on a suppressor-free host (NM430, MC1061).

2. Preparation of Phage DNA. If the background of Spi^- (and suppressor independent) phages is sufficiently low, the plate stock is used to infect an exponentially growing culture of NM538 in L broth containing

[23] D. R. F. Leach and F. W. Stahl, *Nature (London)* **305,** 448 (1983).

[24] A. R. Wyman, L. B. Wolfe, and D. Botstein, *Proc. Natl. Acad. Sci. U.S.A.* **82,** 2880 (1985).

5 mM $MgSO_4$ (multiplicity of infection 1 : 10 for EMBL3 and EMBL4, 1 : 1 for *Aam Bam* derivatives). Phage is precipitated with polyethylene glycol (PEG 6000) and purified by two cycles of CsCl equilibrium density gradient centrifugation. DNA is extracted by standard procedures and stored in TE.

3. Testing of Vector DNA. To test the quality of the vector DNA preparations, samples are packaged *in vitro* directly, or cut with *Bam*HI, religated, and packaged. The material is plated on NM538 and NM539 to check for religatability and contamination with nonphage DNA. Religated material should give approximately 10% of the original number of plaques. The ratio of background as the result of the cloning of DNA on NM539/538 should not increase significantly.

4. Digestion. EMBL3 DNA (20 μg) is digested with 40 U *Bam*HI in 100 μl 100 mM NaCl, 10 mM $MgCl_2$, 10 mM Tris · HCl (pH 7.6) for 1 hr at 37°. *Eco*RI (100 U) is added and digestion continued for 30 min at 37°. EDTA is added to a final concentration of 15 mM and the mixture is heated 15 min at 68°. To 120 μl reaction mixture, 18 μl 3 M NaAc and 90 μl 2-propanol are added, the tube is left on ice for 5 min, then centrifuged for 5 min. The pellet is washed once with 100 μl 0.45 M NaAc plus 60 μl 2-propanol, air dried, and dissolved in TE buffer, to a final concentration of 0.25 μg/μl.

Comments

EMBL3 has the option of cutting with *Eco*RI after the *Bam*HI digestion and thus removing the *Bam*HI sticky ends from the middle fragment. (This is not convenient for EMBL4, where the *Sal*I site is very close to the *Eco*RI site and cutting is inefficient.) The small fragment can be removed by selective precipitation of the long DNA with 2-propanol. The efficiency of recutting and small fragment removal is checked by religating *in vitro,* packaging, and plating on NM538. The background should be no more than 1%.

The vector DNA may be phenol extracted and ethanol precipitated before use. Alternatively, heat inactivation of *Bam*HI and *Eco*RI is usually sufficient if the vector is used within a short time.

Preparation of Insert DNA

1. Partial Digestion. The appropriate enzyme concentration is established by partial digestion on an analytical scale. One microgram high-molecular-weight DNA (>100 kb) is digested with 0.1 U *Mbo*I in 10 mM Tris · HCl (pH 7.6), 10 mM $MgSO_4$, 1 mM DTT, 50 mM NaCl in a total volume of 30 μl; 5-μl aliquots are taken after 0, 5, 10, 20, 40, and 80 min. Four microliters of 0.1 M EDTA is added to each aliquot immediately, then the samples are incubated 10 min at 68° and separated overnight on a 0.3% agarose gel at 1 V/cm.

The appropriate enzyme concentration is chosen from the result of the analytical test. The reaction is scaled up 15-fold. Three time points are taken corresponding to 0.5, one, and two times the optimal enzyme concentration. EDTA is added to a final concentration of 15 m*M*. The samples are heated 15 min at 68°, and aliquots are analyzed on a 0.3% agarose gel. The remainder is ethanol precipitated, centrifuged 1 min in a table top centrifuge, and redissolved overnight in TE.

2. *Phosphatase Treatment*. To 20 μl *Mbo*I-digested DNA (0.25 μg/μl), 2.5 μl 500 m*M* Tris · HCl (pH 9.5), 10 m*M* spermidine, 1 m*M* EDTA, and 2 U alkaline phosphatase from calf intestine are added and the samples are incubated 30 min at 37°. After addition of 3 μl 100 m*M* trinitriloacetic acid (pH 8) the samples are heated 15 min at 68°. The DNA is precipitated with ethanol/NaAc, centrifuged, and dissolved overnight in TE to a concentration of 0.25 μg/μl.

Comments

For partial digest reactions it is important to have a completely dissolved homogeneous DNA solution. Otherwise part of the DNA is degraded rapidly while the rest is not accessible to the enzyme and its molecular weight remains high.

The chosen fractions should show the maximum ethidium bromide staining in the >20 kb range. Digesting too little rather than too much makes it more likely that larger inserts will be obtained. A partially overlapping, slightly larger size cut may be used for the parallel construction of cosmid libraries in a *Bam*HI vector (e.g., pcos2EMBL). If the starting DNA was somewhat degraded, better results are obtained with more digested samples, because the likelihood of a molecule having *Sau*3A ends on both sides is increased.

If only very small amounts of starting DNA are available, analytical digests can be carried out on 50 ng DNA. After electrophoresis, DNA is transferred onto nitrocellulose and visualized by hybridization to labeled repetitive DNA.

It is important that during construction of the library, fragments of DNA cannot be ligated to give packageable phage DNA containing inserts not contiguous in the genome. One way to assure this result is to size fractionate. An alternative is the use of phosphatase to prevent ligation of the DNA molecules to each other while leaving them able to be ligated to the unphosphatased vector DNA. To check the phosphatase reaction, aliquots of DNA treated with phosphatase or untreated are ligated to themselves and compared on a gel to unligated samples.

Ligation and Packaging

Eight micrograms of pooled, digested, phosphatase-treated DNA is ligated with 16 μg *Bam*HI- and *Eco*RI-cut EMBL3 in 80 μl 40 m*M* Tris · HCl (pH 7.6), 10 m*M* $MgCl_2$, 1 m*M* DTT, 0.2 m*M* ATP, and 400 U ligase for at least 16 hr at 15°.

To 80 μl ligation at room temperature first 80 μl of sonic extract and then 240 μl of freeze–thaw lysate[26] are added, mixed well, and left for 4 hr. λ diluent (800 μl) is added and the mixture is then kept at 4°. Alternatively, the packaged mixture can be applied to a CsCl step gradient [31, 42, 54% CsCl (w/w) in λ diluent].

The gradients are centrifuged in an SW60 rotor for 3 hr at 35,000 at 20°. Fractions are collected from the top with a Pasteur pipet, then tested by plating on NM538. The pooled fractions containing phage are dialyzed against λ diluent before plating.

Comments

A weight ratio of 1 : 2 or 1 : 3 of insert DNA to vector works well. If availability of insert DNA is limiting, a ratio of 1 : 4 or lower may be advisable. Even when vector cut with *Bam*HI only is used, excess of vector does not significantly increase the background.

Ligation is usually carried out at 10–100 μg vector DNA/ml overnight at 15°.

Commercial packaging extracts can be used. We prepare two-component packaging extract essentially as described by Scherer *et al.*[25]

A gradient concentration step is useful if the library construction is relatively inefficient. This can be the case because of unknown technical difficulties, or because there are few clones that should give rise to plaques in a very large background of other ligation products. More pfu's can be added to plating cells after this purification step, facilitating plating of the library at the desired density of plaques per plate.

Plating and Library Amplification

Before amplification the library is titrated on NM538 (both vector and recombinant grow), NM539 (recombinants only grow), and DH1 (vector only grows).

Plating cells are prepared from fresh, saturated overnight cultures by spinning down the cells and resuspending them in 0.5 vol 10 m*M* $MgSO_4$.

[25] G. Scherer, J. Telford, C. Baldari, and V. Pirotta, *Dev. Biol.* **86,** 438 (1981).

For a 22 × 22 cm plate, 2 ml plating cells is incubated with 1–2 × 10^5 pfu's for 15 min at 37°, 30 ml BBL top agar (or agarose for direct screening) plus 10 m*M* $MgSO_4$ are added, and the mixture is plated on BBL agar plates.

After incubation at 37° overnight the library is either screened directly or the top layer is scraped off, 30 ml λ diluent and 1 ml $CHCl_3$ are added, and the mixture is stirred for 20 min at room temperature. The agar is then removed by centrifugation and the library is stored over a drop of $CHCl_3$ at 4°.

Comments

In our experience, different preparations of plating cells give different plating efficiencies of *in vitro* packaged phage. It is therefore advisable to test different batches of cells before plating a library. Cells can be kept in 10 m*M* $MgSO_4$ at 4° for at least 2 weeks without loss of plating efficiency.

If it is not expected that the library will be rescreened a number of times over an extended period, a primary nonamplified library should be screened. Different phage clones grow at quite different rates and the complexity of the library decreases significantly on each amplification step. If the library is amplified on plates, it is advisable not to pool the plate stocks from different plates. This is especially important when screening with probes that occur a few times in the genome (e.g., a gene family with many pseudogenes).

Materials

L broth: 10 g Bacto tryptone (Difco, 0123-01), 5 g yeast extract (Difco, 0127-01), 5 g NaCl, water to 1 liter, adjust to pH 7.2

BBL agar: 10 g Baltimore Biological Laboratories trypticase, peptone (11921), 5 g NaCl, 10 g agar (Difco), add water to 1 liter, adjust to pH 7.2

BBL top layer agar: Like BBL agar, but 6.5 g agar/liter

BBL top layer agarose: Like BBL agar, but 5 g agarose/liter

λ diluent: 10 m*M* Tris/HCl (pH 7.6), 10 m*M* $MgSO_4$, 1 m*M* EDTA

TE: 10 m*M* Tris (pH 7.6), 1 m*M* EDTA

Enzymes: Restriction endonucleases were purchased from Bethesda Research Laboratories, New England BioLabs, or Boehringer–Mannheim; T4 DNA ligase was from New England BioLabs; alkaline phosphatase, intestinal, from Boehringer–Mannheim.

[9] Plasmid and Phage Vectors for Gene Cloning and Analysis in *Streptomyces*

By DAVID A. HOPWOOD, MERVYN J. BIBB, KEITH F. CHATER, and TOBIAS KIESER

The development of efficient gene cloning procedures for *Streptomyces* is leading to the detailed study of morphological and physiological differentiation in the colonies of these complex bacteria.[1] It also allows new approaches to the industrial exploitation of streptomycetes for increasing antibiotic yield,[2] generating new "hybrid" compounds,[3] and producing "foreign" gene products[4] from these nonpathogenic, industrially well-understood microorganisms.

Since the first full review of cloning in *Streptomyces*,[5] the subject has expanded considerably. Many genes and gene clusters have been cloned,[6] mostly using the plasmid and phage systems described in that review, and the further study of the cloned DNA has necessitated the development of vectors for specific analytical purposes. The range of vector families has also increased greatly. The field has been reviewed continually (in particular, a detailed laboratory manual has recently been published[7]). Against this background, the aims of this chapter are (1) to present a detailed bench-level account of typical shotgun cloning experiments using a plasmid and a phage vector, and (2) to catalog and discuss a representative selection of all the tried and tested vectors that we are aware of. We include some promoter–probe vectors as well as general cloning vectors,

[1] K. F. Chater, *in* "Microbial Development" (R. Losick and L. Shapiro, eds.), p. 89. Cold Spring Harbor, Lab., Cold Spring Harbor, New York, 1984.

[2] D. A. Hopwood, F. Malpartida, and K. F. Chater, *in* "Regulation of Secondary Metabolite Formation" (H. Kleinkauf, H. V. Döhren, H. Dornauer, and G. Neseman, eds.), p. 23. VCH, Weinheim, Federal Republic of Germany, 1986.

[3] D. A. Hopwood, F. Malpartida, H. M. Kieser, H. Ikeda, J. Duncan, I. Fujii, B. A. M. Rudd, H. G. Floss, and S. Ōmura, *Nature (London)* **314,** 642 (1985).

[4] G. S. Gray, G. Selzer, G. Buell, S. Escanez, P. Shaw, S. Hofer, P. Voegeli, and C. J. Thompson, *Gene* **32,** 21 (1984).

[5] K. F. Chater, D. A. Hopwood, T. Kieser, and C. J. Thompson, *Curr. Top. Microbiol. Immunol.* **96,** 69 (1982).

[6] D. A. Hopwood, T. Kieser, D. J. Lydiate, and M. J. Bibb, *in* "The Bacteria" (S. W. Queener and L. E. Day, eds.), Vol. 9, p. 159. Academic Press, New York, 1986.

[7] D. A. Hopwood, M. J. Bibb, K. F. Chater, T. Kieser, C. J. Bruton, H. M. Kieser, D. J. Lydiate, C. P. Smith, J. M. Ward, and H. Schrempf, "Genetic Manipulation of Streptomyces: A Laboratory Manual." John Innes Foundation, Norwich, England, 1985.

but space will not permit detailed coverage of vectors now being developed for more specific purposes, such as making translational fusions and analysing protein secretion.

The Host

Because commercially important *Streptomyces* antibiotics are produced by derivatives of different soil isolates, a wide range of hosts is used in genetic studies. However, one strain, *Streptomyces coelicolor* A3(2), which produces four antibiotics (none of them of commercial importance), is preeminent as a model system. A circular chromosomal linkage map with more than 150 nutritional, antibiotic resistance and biosynthesis and differentiation genes is available for this strain.[8,9] The genes for three chemically diverse antibiotics (actinorhodin, undecylprodigiosin, and CDA) reside in distinct clusters on the chromosome. The organism contains at least four native plasmids: the large, physically difficult to isolate SCP1, which carries genes for a fourth antibiotic (methylenomycin), and the 30-kb SCP2, both low-copy-number autonomous sex factors[10]; the 17-kb SLP1 sequence, which is integrated into the chromosome but capable of autonomous replication and sex factor activity when transferred into other species[11,12]; and the physically uncharacterized SLP4.[13] *Streptomyces coelicolor* A3(2) also contains at least two transposable elements, IS*110*[14] and a 2.6-kb minicircle,[15] and shows several genetically unstable (and sometimes reversible) phenotypes (e.g., resistance to chloramphenicol[16] or to phage ϕC31[17]). Unfortunately, DNA last propagated in *Escherichia coli* or in some other streptomycetes is subject to restriction by *S. coelicolor*. It is mainly for this reason that the closely related and apparently nonrestricting strain, *Streptomyces livi-*

[8] K. F. Chater and D. A. Hopwood, *in* "The Biology of the Actinomycetes" (M. Goodfellow, M. Mordarski, and S. T. Williams, eds.), p. 229. Academic Press, London, 1983.

[9] D. A. Hopwood and K. F. Chater, *in* "Genetics and Breeding of Industrial Microorganisms" (C. Ball, ed.), p. 7. CRC Press, Boca Raton, Florida, 1984

[10] M. J. Bibb and D. A. Hopwood, *J. Gen. Microbiol.* **126,** 427 (1981).

[11] M. J. Bibb, J. M. Ward, T. Kieser, S. N. Cohen, and D. A. Hopwood. *Mol. Gen. Genet.* **184,** 230 (1981).

[12] C. A. Omer and S. N. Cohen, *Mol. Gen. Genet.* **196,** 429 (1984).

[13] D. A. Hopwood, T. Kieser, H. M. Wright, and M. J. Bibb, *J. Gen. Microbiol.* **129,** 2257 (1983).

[14] K. F. Chater, C. J. Bruton, S. G. Foster, and I. Tobek, *Mol. Gen. Genet.* **200,** 235 (1985).

[15] D. J. Lydiate, H. Ikeda, and D. A. Hopwood, *Mol. Gen. Genet.* **203,** 79 (1986).

[16] R. J. Freeman, M. J. Bibb, and D. A. Hopwood, *J. Gen. Microbiol.* **98,** 453 (1977).

[17] T. A. Chinenova, N. M. Mkrtumian, and N. D. Lomovskaya, *Genetika* (*Moscow*) **18,** 1945 (1982).

dans 66,[18] has become the preferred host for initial cloning experiments and for the expression of mammalian genes in *Streptomyces*. This strain naturally contains two physically unisolable plasmids, SLP2 and SLP3, which have been eliminated from derivatives used in cloning and in the analysis of vector properties.[13] *Streptomyces lividans* 66 has the interesting characteristic of frequently giving rise to derivatives in which specific DNA segments are present in highly amplified tandem arrays.[19] DNA amplification has been observed in many other species,[20,21] including *S. coelicolor* A3(2), but it is unusually reproducible in *S. lividans,* and may be useful in recombinant DNA work.

Of course, some cloning strategies require *Streptomyces* hosts other than *S. lividans* (for example, for the complementation of particular mutants). However, irrespective of the choice of recipient strain, the basic procedure for introducing isolated DNA is the same; protoplasts, prepared by lysozyme treatment of mycelial cultures, are exposed to plasmid or phage DNA in the presence of polyethylene glycol, and then regenerated into mycelial colonies. With *S. lividans* or *S. coelicolor,* more than 10^7 transformation or transfection events can regularly be obtained with 1 μg of supercoiled plasmid or linear phage DNA (or 10^3–10^5 with 1 μg of a ligation mixture in a cloning experiment). For plasmid-mediated transformations, colonies are clearly visible after 2 days, though several more days may be necessary to allow spore formation to take place (usually a prerequisite for replica plating to diagnostic media and for further subculture to allow plasmid isolation). In transfection experiments with phage vectors, overnight incubation suffices for the appearance of plaques. [It is interesting to note that in liquid cultures at 30°, *S. coelicolor* A3(2) can attain mass doubling times only twice as long as those of *E. coli* grown at 37° in comparable media, despite its possession of a 1.5- to 2-fold more complex genome.]

Streptomyces Plasmids

A recent review[6] listed plasmids from *Streptomyces* strains and discussed their known properties in some detail. They range from very large (>100 kb) low-copy-number types such as SCP1 to small (<10 kb) very high-copy-number plasmids such as pIJ101. Most, if not all, are self-transmissible, with transfer determinants occupying short segments of

[18] N. D. Lomovskaya, N. M. Mkrtumian, N. L. Gostimskaya, and V. N. Danilenko, *J. Virol.* **9,** 258 (1972).

[19] J. Altenbuchner and J. Cullum, *Mol. Gen. Genet.* **195,** 134 (1984).

[20] S. E. Fishmann, P. R. Rosteck, and C. L. Hershberger, *J. Bacteriol.* **161,** 199 (1985).

[21] H. Schrempf, *J. Bacteriol.* **151,** 701 (1982).

DNA (e.g., not more than about 2 kb in pIJ101), in contrast to the much more complex transfer regions of plasmids of gram-negative bacteria. Self-transfer is usually associated with sex factor activity leading to chromosomal recombination, and also to "lethal zygosis" (the Ltz^+ phenotype), seen on plates as a zone of inhibition of growth and/or sporulation at the junction of $plasmid^+$ and $plasmid^-$ cultures. The Ltz^+ phenotype, best visualized as small circular zones of altered morphology (called "pocks") when plasmids spread through confluent lawns of plasmid-free bacteria, was invaluable as a marker in the development of transformation (it allows the detection of a single $plasmid^+$ transformant among 10^9 plasmid-free cells), and still can be exploited in some cloning experiments. However, many of the multicopy plasmid vectors are deleted for transfer functions and are therefore Ltz^-, to avoid the confusion that can arise when plasmids are transferred between adjacent clones on crowded transformation plates.

Pock formation as a detection system for transformants has usually been replaced by the inclusion in vectors of selectable genes for antibiotic resistance, mostly cloned from *Streptomyces* strains. In some vectors, insertional inactivation of marker genes can be used for clone recognition and for the rapid assessment of the efficiency of a cloning experiment. Recently, such markers have been augmented in plasmid vectors by systems for insertional *activation,* in which promoterless, and unexpressed, genes for antibiotic resistance acquire surrogate promoters from a significant fraction of inserted DNA fragments, giving rise to a readily scorable fraction of resistant colonies. The shotgun cloning experiment described below incorporates this approach.

Example of Shotgun Cloning Using a Plasmid Vector

The aim of this experiment is to isolate chromosomal genes of *S. coelicolor* involved in proline biosynthesis, but the same principles and methods apply to the isolation of other genes such as those for antibiotic biosynthesis. The recipient is a plasmid-free derivative of *S. lividans* 66 called 3104,[13] carrying the marker *pro-2* (for proline auxotrophy), with *spc-1* (for spectinomycin resistance) as an extra nonselected marker. The vector, pIJ486 (Fig. 1), is a broad-host-range, multicopy promoter–probe vector with a convenient polylinker region and carrying the thiostrepton-resistance gene (*tsr*) for selection of transformants.[22] The promoterless kanamycin resistance gene (*neo*) is expressed by inserting promoter-ac-

[22] J. M. Ward, G. R. Janssen, T. Kieser, M. J. Bibb, M. J. Buttner, and M. J. Bibb, *Mol. Gen. Genet.* **203,** 468 (1986).

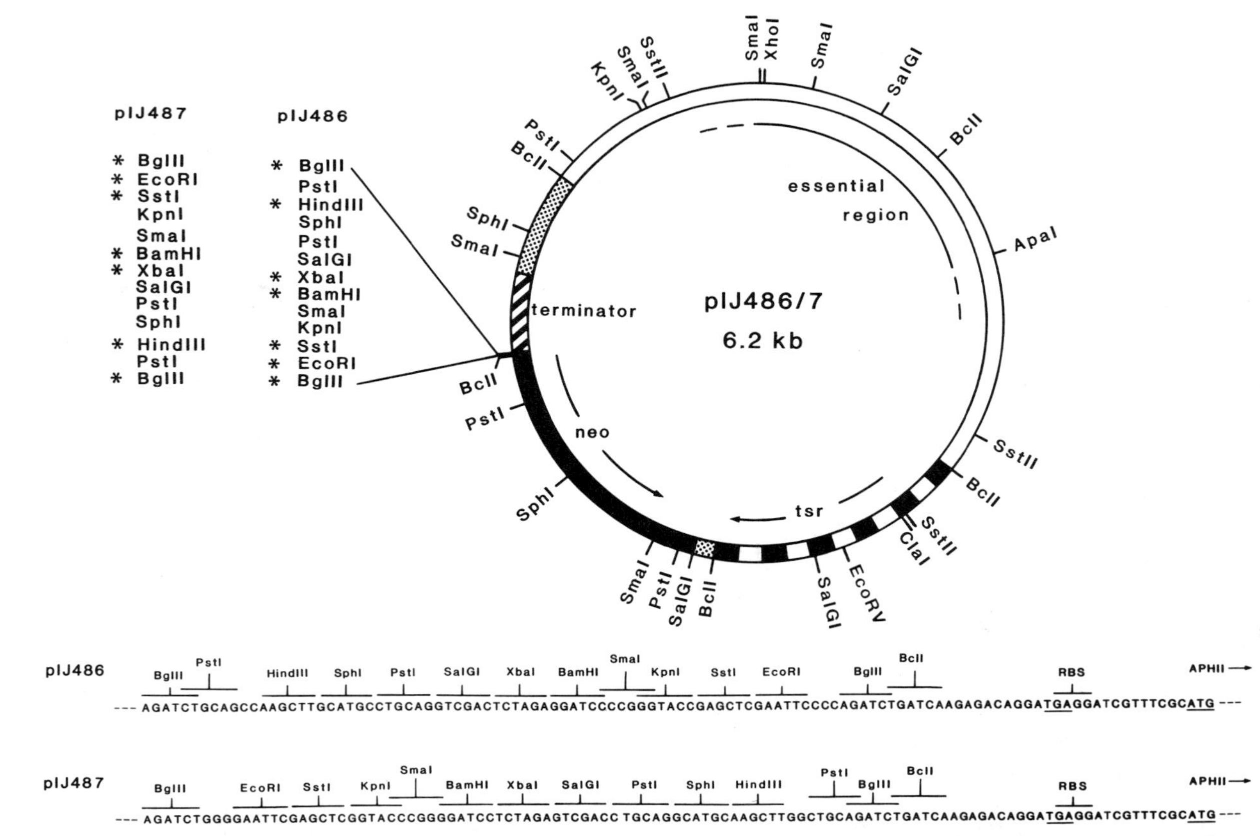

FIG. 1. Restriction map of pIJ486 (and pIJ487, which differs from it only in the orientation of the polylinker fragment, whose sequence is shown below). The unshaded segment represents pIJ101 sequences and the dotted regions are parts of the *mel* fragment of pIJ702.

tive fragments (in the correct orientation) into one of the cloning sites of the polylinker. Remarkably, more than 50% of all *Streptomyces* DNA inserts larger than 2 kb give rise to kanamycin resistance at 5 μg/ml and at least 15% of inserts give resistance at 15 μg/ml. This insertional activation can therefore be used to estimate the proportion of transformants containing hybrid plasmids. Ideally, the chromosomal DNA is cleaved partially with an enzyme (in this case *Mbo*I) that cuts *Streptomyces* DNA frequently (if using other restriction enzymes, remember that *Streptomyces* DNA has 73% G + C base pairs) and is then size fractionated (by sucrose or salt gradient centrifugation or gel electrophoresis) to yield fragments in the size range of 8–30 kb, thus ensuring that some contain the desired gene(s) in a genomic library of reasonable size. A DNA fragment containing the streptomycin phosphotransferase gene of *S. glaucescens*[23] is added to the cleaved sample of chromosomal DNA as an internal standard for the efficiency of cloning, a useful control when attempting to generate a genomic library for the first time.

Figure 2 summarizes the experiment. Each major step specific to *Streptomyces* is described in detail in a subsequent section. Other steps are explained by footnotes to Fig. 2. More complete details of some procedures and much background information about handling the organisms will be found in the manual of Hopwood *et al.*[7]

Media and Buffers

Yeast extract–malt extract medium (YEME) (per liter): 3 g of Difco yeast extract, 5 g of Difco Bacto peptone, 3 g of Oxoid malt extract, 10 g of glucose, and 340 g of sucrose. After autoclaving, add 2 ml of 2.5 *M* $MgCl_2$ and (for protoplast preparation) 25 ml of 20% glycine (some *Streptomyces* strains may need a different glycine concentration)

Minimal medium (MM) (per liter): 10 g of agar, 0.5 g of L-asparagine, 0.5 g of K_2HPO_4, 0.2 g of $MgSO_4 \cdot 7H_2O$, 0.01 g of $FeSO_4 \cdot 7H_2O$, and 10 g of glucose (added as a 50% solution after autoclaving separately)

R2YE agar: Make up the following solution in 800 ml water: 103 g of sucrose, 0.25 g of K_2SO_4, 10.12 g of $MgCl_2 \cdot 6H_2O$, 10 g of glucose, and 0.1 g of Difco casamino acids. In 250-ml Erlenmeyer flasks, place 2.2 g Difco Bacto agar and 80 ml of the solution. Autoclave. At time of use, remelt the agar and add to each flask: 1 ml of 0.5% KH_2PO_4, 8 ml of 3.68% $CaCl_2 \cdot 2H_2O$, 1.5 ml of 20% L-proline, 10 ml of 5.73% TES biological buffer adjusted to pH 7.2, 0.2 ml of trace element solution [per liter: $ZnCl_2$, 40 mg; $FeCl_3 \cdot 6H_2O$, 200 mg; $CuCl_2 \cdot 2H_2O$, 10 mg; $MnCl_2 \cdot 4H_2O$, 10 mg; $Na_2B_4O_7 \cdot 10\ H_2O$, 10 mg;

[23] G. Hintermann, R. Crameri, M. Vögtli, and R. Hütter, *Mol. Gen. Genet.* **196,** 513 (1984).

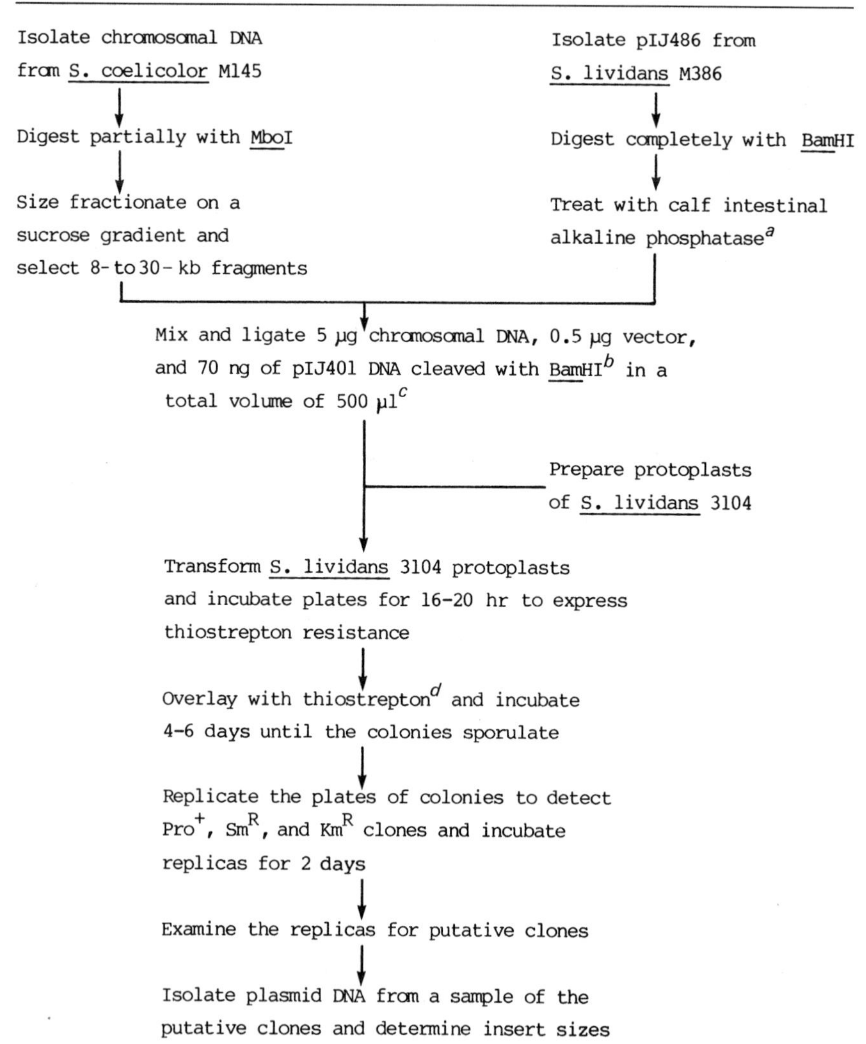

FIG. 2. Outline of the shotgun cloning experiment using *Streptomyces* plasmid vector pIJ486. [a]Alkaline phosphatase removes the 5′ terminal phosphates from the plasmid DNA and so prevents self-ligation. For large plasmid vectors (e.g., stable derivatives of SCP2* like pIJ941 and pIJ943) self-ligation is less than for small vectors (since the concentration of one end of the molecule in the proximity of the other is much lower) and phosphatase treatment is usually omitted since it seems to reduce recombinant yield. [b]Digestion of pIJ401[23] with *Bam*HI yields fragments of 5.4 kb (pBR325) and 10 kb (a segment of *S. glaucescens* DNA containing a streptomycin phosphotransferase gene). Thus, 70 ng of pIJ401 represents a copy number for the streptomycin resistance gene of ~10 per genome. [c]Ligation is carried out at a relatively low DNA concentration (11 μg/ml) to discourage the formation of concatamers. [d]Some growth of regenerated protoplasts is clearly visible at the time of overlaying.

$(NH_4)_6Mo_7O_{24} \cdot 4H_2O$, 10 mg], 0.5 ml of 1 *N* NaOH, and 5 ml of 10% Difco yeast extract

TE buffer: 10 m*M* Tris–HCl, 1 m*M* sodium EDTA, pH 8

P buffer: Make up the following basal solution in 800 ml water: 103 g of sucrose, 0.25 g of K_2SO_4, 2.02 g of $MgCl_2 \cdot 6H_2O$, 2 ml of trace element solution (see under R2YE agar). Autoclave in 80-ml aliquots. Before use, add to each aliquot: 1 ml of 0.5% KH_2PO_4, 10 ml of 3.68% $CaCl_2 \cdot 2H_2O$, and 10 ml of 5.73% TES biological buffer adjusted to pH 7.2

L (lysis) buffer: Mix the following sterile solutions: 100 ml of 10.3% sucrose, 10 ml of 5.73% TES biological buffer (pH 7.2), 1 ml of 2.5% K_2SO_4, 0.2 ml of trace element solution (see under R2YE agar), 1 ml of 0.5% KH_2PO_4, 0.1 ml of 2.5 *M* $MgCl_2$, and 1 ml of 0.25 *M* $CaCl_2$. Just before use, add 1 mg/ml lysozyme and filter sterilize

T (transformation) buffer: Mix the following sterile solutions: 25 ml of 10.3% sucrose, 75 ml of distilled water, 0.2 ml of trace element solution (see under R2YE agar), 1 ml of 2.5% K_2SO_4. To 9.3 ml of this solution, add 0.2 ml of 5 *M* $CaCl_2$, and 0.5 ml of Tris–maleic acid buffer (adjust 1 *M* Tris to pH 8.0 by adding maleic acid). For use, add 3 ml of the above solution to 1 g of molten PEG 1000 previously sterilized by autoclaving

Bacterial Strains

S. coelicolor M145: This is a genetically unmarked derivative of the A3(2) wild type, cured of the native plasmids SCP1 and SCP2[10]

S. lividans 3104: This is a derivative of strain 66, cured of SLP2 and SLP3 and carrying chromosomal markers *pro-2* and *spc-1*[13]

S. lividans M386: This is a derivative of strain 66 cured of SLP2 and SLP3 and carrying the chromosomal marker *str-6* and the vector pIJ486[22]

E. coli carrying pIJ401 (pBR325 with a 10-kb *Bam*HI insert of *S. glaucescens* DNA including the streptomycin phosphotransferase gene[23])

Isolation of Chromosomal DNA from S. coelicolor A3(2)

This procedure has been used without significant modification for many *Streptomyces* strains.

To 500 ml YEME in a 2-liter Erlenmeyer flask add 0.1–0.2 ml concentrated spore suspension of a derivative of *S. coelicolor* A3(2) such as M145. Incubate at 30° with vigorous shaking in an orbital incubator for ~48 hr. Harvest mycelium (usually 5–8 g wet weight) in a Büchner filter on two sheets of Whatman No. 1 filter paper[24] and wash the mycelium with 100 ml of 10% glycerol. The mycelium may now be indefinitely stored frozen as a paste at −20° or used immediately. Suspend 1 g mycelium in 5 ml TE buffer in a screw-cap bottle; add 10 mg lysozyme (i.e., to 2 mg/ml) and swirl to dissolve. Incubate at 30°, mixing by pipetting every 15 min, until a drop of suspension on a microscope slide is completely cleared by addition of a drop of 10% SDS.[25] Add 1.2 ml of 0.5 *M* EDTA (i.e., to 0.1 *M*) and mix gently. Add 0.13 ml of 10 mg/ml Pronase solution (i.e., to 0.2 mg/ml), mix gently, and incubate at 30° for 5 min. Add 0.7 ml of 10% SDS (i.e., to 1%); tilt immediately, then incubate at 37° for up to 2 hr.[26] *Protect your eyes and wear gloves.* Add 6 ml phenol (500 g melted phenol, 0.5 g 8-hydroxyquinoline, and 65 ml TE buffer containing 0.1 *M* NaCl) and shake thoroughly by hand for 10 min at room temperature. Add 6 ml chloroform and shake for 5 min at room temperature. Centrifuge on a bench centrifuge (~3000 rpm) for 10 min. Carefully transfer the upper (aqueous) phase with a shortened (wide bore) Pasteur pipet into a fresh screw-cap bottle.[27] To the phenol phase, add 5 ml TE buffer plus 100 m*M* NaCl; shake for 5 min and centrifuge for 10 min at 3000 rpm. Remove the upper phase, pool with the previous aqueous phase, and reextract with phenol and chloroform at least once (and if there is still a lot of interphase, twice). To remove phenol, add 6 ml chloroform and shake for 5 min at room temperature. Centrifuge at 3000 rpm for 5 min. Carefully remove the upper (aqueous) phase with a shortened Pasteur pipet into a fresh screw-cap bottle. Repeat the chloroform extraction (use a preweighed bottle for the last transfer). Weigh the bottle plus sample. Add RNase to 40 μg/g and incubate 1 hr at 37°. Add 0.25 vol of 5 *M* NaCl and mix. Add 30% PEG 6000 to 10% final concentration, and mix gently but thoroughly until DNA precipitates (small RNA remains in solution). Spool the DNA on a glass rod or sealed Pasteur pipet and transfer to a fresh screw-cap bottle, taking care to transfer as little PEG solution as possible. Dissolve the DNA in 5 ml TE buffer overnight at 4° (gentle swirling accelerates the process). If

[24] Check the culture for contamination before filtration, e.g., microscopically and by streaking the culture out for later examination.

[25] One hour is usually adequate; the solution becomes very viscous. Some streptomycetes lyse very quickly and their DNA is degraded (recognized by low viscosity of lysates). For these strains, it is best to add EDTA, Pronase, and SDS (premixed) after a very short lysozyme treatment.

[26] At this stage, the solution should clear completely and be very viscous; if not, the DNA has been degraded and it is best to start again with a fresh sample of mycelium.

[27] At each step, take care to leave behind the white interphase with the phenol and chloroform layer. Pipet gently to avoid shearing the DNA.

the DNA pellet is not completely dissolved transfer to a 30° orbital incubator. When the DNA is completely dissolved, add 0.6 ml of 3 *M* sodium acetate and 12 ml absolute ethanol and mix. The DNA will precipitate immediately. Spool the DNA on a sealed Pasteur pipet and transfer to a fresh screw-cap bottle. Wash the DNA with 2 ml of 70% ethanol. Remove the ethanol with a Pasteur pipet and finally dissolve the DNA in 1–5 ml sterile TE buffer depending on the amount of DNA (normally 0.5–2 mg). Store at 4°.

Size Fractionation of Partially Digested S. coelicolor A3(2) DNA

Digest 2 μg of *S. coelicolor* DNA with 1 μg of *Mbo*I at 37°. Take out 200 ng samples at 0, 5, 15, 30, 45, 60, 90, and 120 min and inactivate by heating at 65° for ~15 min. Assay the digestion on a 1% agarose gel (use, e.g., λ DNA digested with *Hin*dIII as size standards). Determine the appropriate degree of digestion (e.g., such that most of the fragments are 10–30 kb in length, i.e., such that most of the fluorescence is at 20–60 kb). Repeat digestion on 100 μg DNA (usually in a larger volume, but not necessarily 50 times larger!) for the predicted time interval and assay 200 ng as before. If digestion is appropriate, load the digested DNA in a volume of 100 μl onto a 12-ml 10–40% sucrose gradient in TE buffer containing 1 *M* NaCl.[28] Centrifuge at 35,000 rpm for 16 hr at 17° in a Beckman SW41 rotor or equivalent and collect 400-μl fractions. Assay 5-μl samples of each fraction on a 1% agarose minigel. Dilute the appropriate fractions by adding a half-volume (200 μl) of TE buffer and add carrier tRNA to a final concentration of 20 μg/ml. Add an equal volume of 2-propanol and stand at room temperature for 10 min. Pellet the DNA in a microcentrifuge for 5 min and discard the supernatant. Rinse the pellet in 75% and 100% ethanol, then dry and redissolve it according to needs.

Large-Scale Preparation of pIJ486 Plasmid DNA from Streptomyces lividans Strain M386

This procedure is discussed in detail by Kieser.[29] Prepare and wash mycelium of *S. lividans* M386 as for *S. coelicolor* A3(2) described earlier. Transfer the mycelium to a 250-ml centrifuge bottle and resuspend by

[28] Sucrose gradients can be made by several methods. A simple and reliable technique is carefully to layer 6 ml of 10% sucrose in TE buffer + 1 *M* NaCl over 6 ml of 40% sucrose in TE buffer + 1 *M* NaCl and seal the tube with a tightly fitting stopper. Slowly rotate the tube to the horizontal position and leave at room temperature for 3–4 hr. Slowly rotate back to the vertical position and remove the stopper.

[29] T. Kieser, *Plasmid* **12,** 19 (1984).

shaking in 45 ml of 25 m*M* Tris–HCl, 25 m*M* EDTA, 0.3 *M* sucrose (pH 8). Add 5 ml lysozyme solution (10 mg/ml in 25 m*M* Tris–HCl, 25 m*M* EDTA, 0.3 *M* sucrose). Add 250 μl RNase (10 mg/ml in 0.3 *M* sodium acetate, pH 4.8, preheated at 90° for 10 min to inactivate DNases). Incubate at 37° for 30–60 min, with occasional shaking. Add 30 ml of freshly made 0.3 *M* NaOH, 2% SDS. Mix thoroughly and incubate at 70° for 15 min.[30] Cool to room temperature. *Protect your eyes and wear gloves*. Add 20 ml phenol/chloroform (500 g phenol, 500 ml chloroform, 200 ml water, 0.5 g 8-hydroxyquinoline). Mix very thoroughly and centrifuge at 10,000 rpm for 15 min at room temperature.[31] Remove the upper (aqueous) layer to a new bottle. Add 7 ml unbuffered 3 *M* sodium acetate and 70 ml 2-propanol. Leave for 10 min at room temperature and centrifuge at 10,000 rpm for 10 min. Discard the supernatant and dissolve the precipitate in 10 ml TE buffer +50 m*M* NaCl. Transfer to a screw-cap bottle. Add 1 ml 3 *M* sodium acetate and 5 ml phenol/chloroform and shake vigorously. Centrifuge for 10 min and remove the upper (aqueous) phase. Add 5 ml chloroform; shake and centrifuge briefly. Remove the aqueous phase and repeat the chloroform extraction. Add 10 ml 2-propanol to the aqueous phase and invert the bottle to mix. Leave for 10 min at room temperature. Centrifuge for 10 min at room temperature and discard the supernatant. Rinse the DNA pellet with ~1 ml absolute ethanol (this helps to make the precipitate visible) and blow dry. Dissolve the pellet *completely* in 12 ml TE buffer and add 12.6 g CsCl. After dissolution add 0.6 ml ethidium bromide (10 mg/ml). Mix well. Distribute into two 12-ml ultracentrifuge tubes and spin at 55,000 rpm for 6–12 hr followed by 36,000 rpm for 6 hr at 20° (or 36,000 rpm for 48 hr). Remove the plasmid DNA with a hypodermic needle and syringe under long wave length (300–360 nm) UV illumination. Add an equal volume of 1-butanol or 2-propanol saturated with aqueous 5 *M* NaCl in TE buffer. Mix the phases by shaking vigorously. Allow the phases to separate (if necessary, centrifuge at 1500 *g* at room temperature for 3 min) and discard the upper layer. Repeat the extraction two to three times until the pink color disappears from the aqueous solution. Dialyze the aqueous phase against 1000 vol TE buffer at 4° at least three times each for 1 hr. (Alternatively, add 2 vol water to the aqueous phase, then 6 vol ethanol and keep at −20° for 1 hr; centrifuge and wash

[30] The NaOH/SDS mixture should be made with a freshly prepared solution of NaOH (not more than 1 week old). Incomplete mixing leads to incomplete removal of chromosomal DNA. The heat treatment at 70° may be longer (up to 30 min) to ensure complete lysis. At this stage, the linear DNA is fully denatured and most of the RNA is hydrolyzed.

[31] The acid phenol lowers the pH of the sample to around pH 10, allowing renaturation of the covalently closed circular DNA. Most single-stranded DNA and protein are precipitated. The supernatant after the spin should be clear.

the pellet with 70 and then 100% ethanol; dry and dissolve the pellet in TE buffer.) A 500-ml culture should yield 100–300 μg plasmid DNA.

Preparation of Protoplasts from Streptomyces lividans Strain 3104

This procedure is based on the work of Okanishi *et al.*[32] and was adapted and optimized for high transformation frequency by Bibb *et al.*[33] and Thompson *et al.*[34] Many different *Streptomyces* species can be protoplasted and regenerated using this procedure. Alternative procedures, additional information and references can be found in other reports.[35–38]

To a baffled flask add 25 ml YEME medium. Add ~0.1 ml of a dense spore suspension (~10^8 spores) of *S. lividans* 3104 and incubate 36–40 hr at 30° on an orbital shaker. Subsequent manipulations are done at room temperature unless stated otherwise. Pour the culture into a 20-ml screw-cap bottle and spin in the bench centrifuge (3000 rpm for 10 min).[24,39] Discard the supernatant and suspend the pellet in 15 ml 10.3% sucrose; spin as before and repeat this washing step. Discard the supernatant and resuspend the mycelium in 4 ml lysozyme solution (1 mg/ml in P or L buffer, filter sterilized); incubate at 30° for 15–60 min.[40] Mix by pipetting three times (this helps to free the protoplasts from the mycelium) and incubate for a further 15 min. Add 5 ml P buffer. Mix as before. Filter through cotton wool and transfer to a plastic tube.[41] Sediment the proto-

[32] M. Okanishi, K. Suzuki, and H. Umezawa, *J. Gen. Microbiol.* **80,** 389 (1974).

[33] M. J. Bibb, J. M. Ward, and D. A. Hopwood, *Nature (London)* **274,** 398 (1978).

[34] C. J. Thompson, J. M. Ward, and D. A. Hopwood, *J. Bacteriol.* **151,** 668 (1982).

[35] R. H. Baltz and P. Matsushima, *J. Gen. Microbiol.* **127,** 137 (1981).

[36] R. H. Baltz and P. Matsushima, *in* "Protoplasts 1983" (I. Potrykus, C. T. Harms, A. Hinnen, R. Hütter, P. J. King, and R. D. Shillito, eds.), p. 143. Birkhäuser Verlag, Basel, Switzerland, 1983.

[37] T. Furumai, T. Takeda, and M. Okanishi, *J. Antibiot.* **35,** 1367 (1982).

[38] T. Shirahama, T. Furumai, and M. Okanishi, *Agric. Biol. Chem.* **45,** 1271 (1981).

[39] Occasionally, mycelium will not pellet from the growth medium containing 34% sucrose on centrifugation; in this case, add water to reduce the density of the medium. Take care in pouring off the supernatant from the loose pellet.

[40] The mycelial pellet, without added liquid, can be frozen at −20° at this stage (before lysozyme treatment). For most purposes, lysozyme is dissolved in P buffer; use of L buffer gives more efficient transformation of *S. lividans* by plasmid DNA. Incubation times vary from strain to strain; e.g., 15 min is usually long enough for *S. lividans*, whereas *S. coelicolor* requires 60 min and other strains may take longer. Protoplast formation can be monitored by phase-contrast microscopy. For *S. lividans*, the time (15–60 min) is not critical, but other strains may lyse (especially in L buffer) if left too long. If this happens, P buffer plus lysozyme can be substituted for L buffer and protoplasts may be made at a lower temperature (e.g., room temperature) or for a shorter time.

[41] We use disposable sterile tubes (e.g., 16 × 100 mm, round-bottom, polystyrene test tubes, Cat. No. 142AS, made by Sterilin, Teddington, Middlesex, England). These are preferable to tubes which are washed and reused because traces of detergent are thereby avoided.

plasts by spinning in a bench centrifuge (3000 rpm for 7 min). Discard the supernatant and suspend the protoplasts in 10 ml P buffer.[42]

Count the protoplasts (OD measurements can be very unreliable if the protoplasts are pigmented) in a small sample of the suspension in a hemocytometer. (Use a 10-fold diluted sample: e.g., 0.01 ml added to 0.09 ml of P buffer.) Divide the suspension into aliquots each containing $\sim 4 \times 10^9$ protoplasts for immediate use, or for freezing and storage.[43]

Transformation of Streptomyces lividans Protoplasts

Spin down, separately, three aliquots of protoplasts in a benchtop centrifuge (3000 rpm for 7 min). (If the volume of protoplast suspension is very small, add a few milliliters of P buffer before spinning because this spin also serves as a wash.) Resuspend the protoplasts in the drop of buffer left after pouring off the supernatant. Add 20 μl of ligation mix (larger volumes may lyse the protoplasts) to one aliquot of protoplasts, 20 μl of uncut vector DNA (diluted to contain 10 ng) to a second, and TE buffer to a third aliquot as controls. Immediately (to avoid degradation of the DNA) add 0.5 ml T buffer (which contains PEG) or 25% PEG made in P buffer and mix by pipetting up and down once, using the same pipet. As soon as possible (within 3 min after adding the PEG) add 5 ml P buffer and spin in the benchtop centrifuge. Pour off the supernatant and resuspend in 1 ml P buffer. Plate out 0.1 ml on each of 10 R2YE plates for the cloning experiment (one each for the controls) and incubate at 30°.

Selection of Clones

The first step is to select transformants by overlaying the plates, after overnight incubation (16–20 hr), to allow the expression of thiostrepton resistance. Add thiostrepton, dissolved in DMSO at 50 mg/ml but not sterilized, to molten (45–50°) soft nutrient agar (Difco nutrient broth con-

[42] At this and any other steps, when pelleted protoplasts are to be resuspended, resuspend in the remaining drop of liquid by tapping the side of the tube with a finger until the protoplasts are dispersed to form a creamy suspension, then add the suspending P buffer (otherwise the protoplast pellet is difficult to disperse). To assess the proportion of nonprotoplasted units in the suspension, samples can be diluted in parallel in P buffer and in dilute detergent (e.g., 0.01% SDS) and plated on regeneration plates. Any colonies arising after dilution in detergent are likely to have arisen from nonprotoplasted units. For *S. lividans*, the proportion of nonprotoplasted units should be less than 0.1%.

[43] To freeze the protoplasts for storage, place aliquots of the protoplast suspension in small plastic tubes, close them, and place them in ice in a plastic beaker. Place the beaker at −70° overnight. Free the frozen protoplasts in their tubes from the ice and store at −70°. To thaw, shake the frozen tube under running warm water (i.e., freeze slowly, thaw quickly).

taining 0.3% agar) at a concentration of 500 μg/ml (the antibiotic forms a precipitate which should be dispersed by swirling) and pipet 2.5 ml of the agar over the surface of each transformation plate. [Alternatively, 1 ml of a suspension of thiostrepton (made by adding 0.1 ml of the stock solution in DMSO to 25 ml water) may be poured over the plate surface; with this procedure, the colonies will sporulate sooner.] After incubation of the plates to allow sporulation (usually for a further 4–8 days), many colonies should appear on both sets of transformation plates (several hundred for the ligation mixture and about 1000 for the uncut vector) and none (or very few) on the control plates from the protoplasts treated only with buffer. Now replica plate the colonies with velvet to plates containing kanamycin (say 15 μg/ml in MM) to assess the frequency of clones in general and to MM lacking proline to select Pro^+ clones. Also replicate to streptomycin (10 μg/ml in MM) to detect clones carrying the streptomycin resistance gene of *S. glaucescens* which was seeded into the ligation mixture as an internal cloning standard.

Minilysate Procedure for the Analysis of Streptomyces Plasmids

This method was devised for *Escherichia coli* and is a modification[44] of that of Ish-Horowicz and Burke.[45] Adjustment of incubation conditions[46] by H. A. Baylis and R. B. Sinclair has produced a rapid method for *Streptomyces* with a high yield of plasmid DNA that is easily digested by most restriction endonucleases. It is used here to confirm the presence of hybrid plasmids in transformants and to estimate the sizes of the inserts.

Pick a selection, say 24, of transformants (some kanamycin resistant) to plates of R2YE + thiostrepton (50 μg/ml) to give 1-cm^2 patches of growth. When well grown (3–4 days), inoculate 5-ml samples of Oxoid tryptone soya broth (CM129) in 20-ml screw-top bottles with mycelium and spores obtained from these patches (or use macerated colonies from the original transformation plates or their replicas, but they tend to give less consistent results) and incubate with shaking at 30° for 20–48 hr (5 μg/ml thiostrepton in liquid medium is suitable to maintain selection of potentially unstable clones). Either transfer mycelial clumps to a large Eppendorf tube using a cut-off 1-ml disposable tip or 5-ml pipet or, in the case of profuse and dispersed growth, transfer 1.5 ml directly to a 1.5-ml Eppendorf tube. Spin the Eppendorf tubes for 20 sec and remove all supernatant. Add 100 μl of 2 mg/ml lysozyme in 50 m*M* glucose, 25 m*M* Tris, pH 8, 10 m*M* EDTA and disperse the cells by vortex mixing. Incubate at 37°

[44] S. Y. Chang, personal communication.

[45] D. Ish-Horowicz and J. F. Burke, *Nucleic Acids Res.* **9**, 289 (1981).

[46] J. L. Schottel, M. J. Bibb, and S. N. Cohen, *J. Bacteriol.* **146**, 360 (1981).

for 30 min. Add 200 μl of freshly made 0.2 *N* NaOH, 1% SDS, and mix gently by inverting three or four times. Incubate on ice for at least 5 min and add 150 μl of ice-cold 3 *M* potassium acetate, pH 4.8. Mix gently by inverting three or four times and incubate on ice for 5 min. Centrifuge for at least 2 min and remove 400 μl to a clean 1.5-ml Eppendorf tube (try not to transfer any cell debris). Add 100 μl phenol : chloroform : isoamyl alcohol (25 : 24 : 1) and mix thoroughly by vortexing. Centrifuge for 30 sec and remove the aqueous phase to a clean 1.5-ml Eppendorf tube (do not carry over any interphase material). Add 800 μl cold (−20°) absolute ethanol; mix thoroughly by inverting three or four times and stand at room temperature for 2 min. Centrifuge at least 2 min; discard the supernatant and resuspend the pellet in 50 μl of 100 m*M* ammonium acetate. Add 150 μl cold (−20°) absolute ethanol; mix thoroughly by inverting three or four times and stand at room temperature for 2 min. Centrifuge for 2 min; remove the ethanol completely and dry the pellets. Resuspend in 50 μl TE buffer. Run 1 μl on a 1% agarose gel. Use this gel to determine the volumes of minilysate material required for digestion with *Bgl*II[47] (generally 0.5–2 μl). Confirm the presence and estimate the size of inserts by coelectrophoresis with *Bgl*II-cleaved pIJ486 and suitable size markers. Alternatively, the size of undigested plasmids can be estimated by agarose gel electrophoresis; for this purpose the method of Kieser[29] is especially efficient.

A Selection of Plasmid Cloning Vectors

Table I and Fig. 3 describe a selection of *Streptomyces* plasmid cloning vectors. The vectors differ in their origins of replication, copy number, host range, marker genes (Table II, p. 161), and cloning sites. Some vectors, although constructed to serve as promoter probes, are included here because they can also be useful as general cloning vectors, as is demonstrated in the model shotgun cloning experiment with pIJ486 (Fig. 1). Some of the plasmids have been made bifunctional by the incorporation of an *E. coli* replicon and selective marker. Of the constructs examined so far, bifunctional plasmids containing pACYC184 are usually more stable in *S. lividans* than similar plasmids containing a pBR322-derived origin of replication. The usefulness of such bifunctional plasmids may be limited because streptomycetes frequently have potent restriction systems which destroy DNA introduced from *E. coli* or other streptomycetes, but *S. lividans* 66 is, at least largely, nonrestricting. Our main

[47] The final preparation contains a high concentration of RNA, needed for the rapid ethanol precipitations; it is therefore usually necessary to add RNase (final concentration of 20–50 μg/ml) to the sample(s) either before or during restriction endonuclease digestion.

TABLE I
CLONING VECTORS

Vector and reference[a]	Copy number[b] (size in kb)	Markers[c]	Cloning sites[d] (capacity of phage vectors)	Parent replicon[e] (strain of origin)	Remarks[f]
Plasmid vectors					
pARCl[g]	5 (18.5)	Ltz *tsr* Pigment production	— — *Bam*HI	SLP1.2 (*S. lividans* 66[pp])	Promoter–probe vector: only strong promoters will induce pigment production
pBC6[h]	5 (22.1)	Ltz *tsr* *β-gal* *amp*	— — *Bam*HI —	SLP1.2 (*S. lividans* 66[pp]) pBR322	Bifunctional promoter–probe vector. β-Gal$^-$ *S. lividans* or *S. albus* G. are recommended hosts. Seven sites for *Stu*I. No sites for *Hap*I, *Xba*I
pBT37[i]	40–100 (8.2)	Sf^R Ak^R	*Cla*I, *Pvu*II *Bgl*II, *Sac*I, *Xho*I	pBT1 (*S. griseobrunneus* ISP5066)	No sites for *Bam*HI, *Eco*RI, *Eco*RV, *Hin*dIII, *Hin*cII, *Hpa*I, *Stu*I
pCAO170[j]	1 (20.6)	Ltz *tsr* *km* *amp* —	— — — — *Bam*HI, *Sac*I	SLP1 integrated sequence from *S. coelicolor* A3(2)	Bifunctional integration vector; autonomous plasmid only in *E. coli*. No sites for *Hpa*I, *Nde*I, *Xba*I
pEB11[k]	100 (9.2)	Ltz *tsr* *aphI* —	— *Cla*I, *Eco*RV — *Xba*I, *Xho*I	pSVH1 (*S. venezuelae* DSM40755)	Compatible with pIJ101, pSG2, pSG5 and SLP1.2. No sites for *Bcl*I, *Bgl*II, EcoRI, *Hin*dIII, *Sph*I

(*continued*)

TABLE I (*continued*)

Vector and reference[a]	Copy number[b] (size in kb)	Markers[c]	Cloning sites[d] (capacity of phage vectors)	Parent replicon[e] (strain of origin)	Remarks[f]
pEB102[k]	100 (11.2)	Ltz	—	pSVH1 (*S. venezuelae* DSM40755) pACYC184	Bifunctional vector. Same compatibility as pEB11. No sites for *Bam*HI
		tsr	*Eco*RV		
		neo	*Bgl*II		
		cat	*Eco*RI		
		—	*Hin*dIII, *Xho*I		
pFJ342[l]	500 (9.2)	*tsr*	—	pNM100 (*S. virginiae* A41030)	Copy number determined in *S. ambofaciens* ATCC 15154
		aphI	*Bam*HI		
		—	*Bcl*I, *Eco*RI, *Hin*dIII		
pGM4[k]	20 (4.8)	*tsr*	*Cla*I, *Eco*RV, *Pvu*II	pSG5 (*S. ghanaensis* DSM2932)	Compatible with pIJ101, pSG2, pSVH1, and SLP1.2. No sites for *Bgl*II, *Eco*RI, *Hin*dIII, *Xho*I
		aphI	*Bam*HI		
		—	*Kpn*I, *Xba*I		
pGM102[k]	20 (10.5)	*tsr*	*Eco*RV	pSG5 (*S. ghanaensis* DSM2932) pACYC184	Bifunctional vector. Same compatibility as pGM4. No sites for *Kpn*I
		neo	*Bgl*II		
		cat	*Eco*RI		
		—	*Bam*HI, *Hin*dIII, *Xho*I		
pHJL197[m]	1 (13.2)	*tsr*	*Eco*RV	SCP2* [*S. coelicolor* A3(2)] pBR322	Bifunctional vector without the *par* function of SCP2*. No sites for *Bgl*II
		*aphI**	*Bam*HI		
		amp	—		
		—	*Eco*RI, *Xba*I, *Xho*I		

pHJL210[n]	10 (11.2)	*tsr* *aphI** *amp* —	*Eco*RV *Bam*HI — *Eco*RI, *Xba*I, *Xho*I	SCP2* [*S. coelicolor* A3(2)] pBR322	Bifunctional vector without the *par* function of SCP2*. 2 *Nco*I sites. No sites for *Bgl*II, *Hpa*I
pHJL302[n]	1000 (5.1)	*tsr* *amp* *lac*α	— — *Bam*HI, *Eco*RI, *Hin*dIII, *Pst*I, *Sac*I, *Sma*I, *Xba*I	SCP2* [*S. coelicolor* A3(2)] pUC19	Bifunctional vector with very high copy number in *S. lividans* 66 (copy number lower in *S. ambofaciens* ATCC15154). No sites for *Bgl*II, *Hpa*I, *Xho*I
pHJL401[n]	10 (5.8)	As pHJL302	As pHJL302	As pHJL302	Bifunctional vector without the *par* function of SCP2*. Three *Acc*I, 3 *Aha*III, 5 *Ava*I, 2 *Bgl*I, 3 *Bst*EII, and 5 *Hinc*II sites. No sites for *Bgl*II, *Xho*I
pIJ61[o]	5 (14.8)	Ltz *tsr* *aphI* —	— *Eco*RV *Bam*HI, *Pst*I *Bcl*I, *Sph*I, *Xba*I, *Xho*I	SLP1.2 (*S. lividans* 66[pp])	Three *Pvu*II and 8 *Bst*EII sites
pIJ486/ pIJ487[p]	100 (6.2)	*tsr* *neo*	— *Bam*HI, *Eco*RI, *Hin*dIII, *Sst*I, *Xba*I	pIJ101 (*S. lividans* ISP 5434)	Promoter–probe vectors, differing only in orientation of the polylinker (Fig. 1)
pIJ680[q]	100 (5.3)	*tsr* *aphI* —	*Cla*I, *Eco*RV, *Pvu*II *Bam*HI *Xba*I	pIJ101 (*S. lividans* ISP5434)	Four *Bst*EII and 4 *Sma*I sites. No sites for *Bgl*II, *Eco*RI, *Hin*dIII, *Sph*I

(*continued*)

TABLE I (*continued*)

Vector and reference[a]	Copy number[b] (size in kb)	Markers[c]	Cloning sites[d] (capacity of phage vectors)	Parent replicon[e] (strain of origin)	Remarks[f]
pIJ702[r]	100 (5.8)	*tsr* *mel* —	— *Bgl*II, *Sph*I, *Sst*I *Pst*I, *Kpn*I	pIJ101 (*S. lividans* ISP5434)	See also pMT660. pMT676 is a derivative of pIJ702 lacking *Pst*I site[(1)] in Fig. 3 and instead having a unique *Pst*I site in the *mel* gene.[bb] Four *Bst*EII and 4 *Sma*I sites. No sites for *Eco*RI, *Hin*dIII, *Xba*I
pIJ860[s]	100 (10.3)	*tsr* *amp* *neo* —	*Eco*RV — *Bgl*II *Eco*RI, *Hin*dIII	pIJ101 (*S. lividans* ISP 5434) pBR322	Bifunctional vector. No sites for *Sst*I, *Xba*I
pIJ941[t]	1 (25.0)	Ltz *tsr* *hyg* —	— *Cla*I, *Eco*RV *Eco*RI *Bam*HI, *Bgl*II, *Pst*I, *Xho*I	SCP2* [*S. coelicolor* A3(2)]	Stable in *S. lividans* (contains *par* function). No sites for *Hin*dIII
pIJ943[q]	1 (20.6)	*tsr* *mel* —	— *Bgl*II *Eco*RI, *Xba*I, *Xho*I	SCP2* [*S. coelicolor* A3(2)]	Stable in *S. lividans* (contains *par* function). pIJ943 is Ltz$^-$ but still mobilizes the chromosome at a low frequency. No sites for *Hin*dIII

pJAS14[u]	5 (17.1)	*tsr*	—	SLP1.2 (*S. lividans* 66[pp])	Promoter–probe vector. No sites for *Sst*I, *Xba*I
		ampC	*Bam*HI		
pKC293[v]	300 (11.3)	*hph*	*Bam*HI	pFJ103 (*S. granuloruber* A39912.13)	Bifunctional vector. Copy number determined in *S. ambofaciens*
		vph	—		
		amp	—		
		—	*Cla*I, *Hin*dIII, *Kpn*I	pBR322	
pKST2[w]	100 (4.3)	*tsr*	*Cla*I, *Sal*GI	pTA4001 (*S. lavendulae* 1080)	Compatible with pIJ101, SCP2*, SLP1.2. No sites for *Hin*dIII, *Sac*I, *Pst*I, *Xho*I
		Sth[R]	*Eco*RI, *Sma*I		
pMCP10[x]	100 (8.0)	*tsr*	*Cla*I, *Pvu*II, *Sst*II	pSRCl-b (*S. roseochromogenes* S264)	Compatible with pIJ101 and pSL1. No sites for *Bam*HI, *Eco*RI, *Hin*dIII
		Ltz	*Mlu*I (replacement)		
pMCP28[y]	50 (6.9)	Km[R]	—	pSL1 (*S. lavendulae* KCC SO985)	Compatible with pIJ101. Contains part of the *mel* gene (inactive). No sites for *Bcl*I, *Hin*dIII, *Xba*I, *Xho*I
		—	*Mlu*I		
pMG312[z]	1–3 (10.9)	*tsr*	*Cla*I, *Eco*RV	pMG200 (*S. chrysomallus* ZIMET 43686)	Replicates in *S. noursei* G2 ZIMET 43760 and in *S. griseus* IMET JA 3933. Could not be introduced into *S. hygroscopicus* IMET JA 6599 by transformation. No sites for *Eco*RI, *Hin*dIII, *Sma*I
		vph	—		

(*continued*)

TABLE I (*continued*)

Vector and reference[a]	Copy number[b] (size in kb)	Markers[c]	Cloning sites[d] (capacity of phage vectors)	Parent replicon[e] (strain of origin)	Remarks[f]
pMS63[aa]	100 (5.0)	*tsr* *aph*	*Cla*I, *Eco*RV *Bgl*II, *Sst*I (replacement)	pIJ101 (*S. lividans* ISP5434)	The *Sst*I–*Bgl*II–*Sst*I linker is inserted in phase into the *aph* (ribostamycin resistance) gene which remains active. Further insertions into the linker will normally inactivate the gene
pMS75[aa]	10 (12.7)	*tsr* Km^{R}II	*Eco*RV *Bam*HI, *Eco*RI, *Hpa*I, *Sst*I	pSF765 (*S. fradiae* SF765)	Compatible with pIJ101. No sites for *Bgl*II, *Hin*dIII, *Xba*I
pMT660[bb]	100 (5.8)	*tsr* *mel*	— *Bgl*II, *Sph*I, *Sst*I	pIJ101 (*S. lividans* ISP5434)	Derivative of pIJ702 with temperature-sensitive replication
pOA154[cc]	20 (13.7)	Ltz *tetB* —	— — *Bam*HI, *Sma*I	pOA15 (unknown)	Replicates in *S. griseus* ATCC10137. Not tested for replication in *S. lividans*. Seven *Bst*EII sites. No sites for *Aat*I, *Bgl*II, *Cla*I, *Hin*dIII, *Hpa*I, *Kpn*I, *Sac*I, *Sca*I, *Xba*I, *Xho*I
pSK21-K3[dd]	>20 (8.0)	*tsr* —	— *Bam*HI, *Bcl*I, *Bgl*II	pSK2 (*S. kasugaensis* MB273)	
pSKO2[ee]	100 (8.5)	*tsr* *amp* *galK*	— — *Bgl*II, *Sac*I, *Xba*I	pIJ101 (*S. lividans* ISP 5434) ColE1	Bifunctional promoter–probe vector. GalK$^-$ strains needed as hosts. No sites for *Sph*I

pSKO3[ee]	1 (16.0)	*tsr*	—	SCP2* (*S. coelicolor* A3(2)]	Bifunctional low-copy-number promoter–probe vector. GalK⁻ strains needed as hosts
		amp	—		
		galK	*Bam*HI, *Hpa*I, *Xho*I		
pSW1[ff]	5 (16.6)	*tsr*	—	pSG2 (*S. ghanaensis* DSM2932) pACYC184	Bifunctional vector. Compatible with pIJ101, pSG5, pSVHl, and SLP1.2, No sites for *Bam*HI, *Bgl*II. (*Cla*I sites marked * are protected in DNA from dam⁺ *E. coli*)
		cat	*Eco*RI		
		—	*Sph*I		
pUC1112[gg]	100–200 (7.6)	*tsr*	*Cla*I	pIJ101 (*S. lividans* ISP 5434)	The three marker genes can be excised on a single *Bcl*I fragment. No sites for *Eco*RI, *Hin*dIII, *Xba*I
		mel	*Bgl*II		
		vph	—		
pUC1120[hh]	30–50 (14.8)	Ltz	—	pUC17 (*S. violaceus-ruber*)	Host range described as "moderately wide." No sites for *Eco*RI, *Hin*dIII
		tsr	*Cla*I		
		mel	—		
pVE30[ii]	High (11.2)	Ltz	*Bam*HI, *Bgl*II	pVE1 (*S. venezuelae* ATCC14585)	Four *Not*I, 5 *Mlu*I, 7 *Sst*II, and 7 *Nru*I sites. No sites for *Dra*I, *Hpa*I, *Kpn*I, *Nco*I, *Xba*I, *Xho*I, *Xmn*I
		tsr	—		
		—	*Eco*RI, *Hin*dIII, *Nde*I		
pVE138[ii]	High (7.7)	*tsr*	—	pVE1 (*S. venezuelae* ATCC14585) pBR322	Bifunctional vector; unstable in *S. lividans* (35% loss). Two *Mlu*I, 2 *Not*I, 2 *Pvu*II, 3 *Aha*III, 3 *Apa*I, 3 *Bst*EII, 3 *Nru*I, 3 *Pvu*I, 4 *Sst*II, and 5 *Sma*I sites. No sites for *Bcl*I, *Bst*XI, *Hpa*I, *Kpn*I, *Mst*II, *Nco*I, *Sfi*I, *Xho*I
		vph	—		
		amp	—		
		—	*Bal*I, *Bam*HI, *Eco*RI, *Hin*dIII, *Mst*I, *Nde*I, *Sca*I, *Xba*I, *Xmn*I		

(continued)

TABLE I (*continued*)

Vector and reference[a]	Copy number[b] (size in kb)	Markers[c]	Cloning sites[d] (capacity of phage vectors)	Parent replicon[e] (strain of origin)	Remarks[f]
pVE223[ii]	High (6.33)	*tsr* *mel* —	— *Bgl*II *Bal*I, *Nde*I	pVE1 (*S. venezuelae* ATCC14585)	Two *Bst*EII, 3 *Apa*I, 3 *Sma*I, 4 *Nru*I, and 5 *Sst*II sites. No sites for *Aha*III, *Bam*HI, *Bcl*I, *Bst*XI, *Eco*RI, *Hin*dIII, *Hpa*I, *Kpn*I, *Mst*II, *Nco*I, *Sca*I, *Sfi*I, *Xba*I, *Xho*I, *Xmn*I
pWOR126[jj]	150 (7.2)	*tsr* *mel*	— *Bgl*II, *Sph*I	pJV1 (*S. phaechromogenes* NRRL-3559)	Compatible with pIJ101 and SLP1.2, No sites for *Bcl*I, *Hin*dIII, *Hpa*I, *Pst*I, *Xho*I
Phage vectors					
KC304[kk]	(39.6)	*vph*, *tsr*	*Bam*HI replacement of *vph* (up to 5 kb) *Sst*I replacement in *vph* (up to 3.5 kb) *Pst*I insertion (up to 3 kb) *Pvu*II insertion in *tsr* (up to 3 kb)	ϕC31	The vector is c$^+$, attP$^+$ and can therefore efficiently lysogenize many strains, introducing cloned DNA stably at single copy

KC505[ll]	(40.7)	*vph* (*tsr*)	*Pst*I replacement of *tsr* (2 to 8 kb)	ϕC31	Lacks both *att*P and the *c* gene (the ΔC3, 1.8-kb deletion removes the *c* gene: Fig. 5). Use for complementation of ϕC31-lysogenic mutants or for plaque hybridization
KC515[ll] KC516	(38.6)	*vph*, *tsr*	*Bam*HI–*Bgl*II replacement of *tsr* (up to 6 kb) Insertion (up to 4 kb) into *Bam*HI, *Bgl*II, *Pst*I, or *Xho*I sites retaining *tsr* and *vph*. Insertion (up to 4 kb) into *Pvu*II site retaining *vph* *Sst*I replacement (up to 4.5 kb) in *vph*	ϕC31	Can be used for mutant complementation, using either insert-directed recombination into strains related to the DNA donor or recombination into ϕC31 lysogens. Effective in mutational cloning (but vector promoters may cause some clones to be nonmutagenic). Cloning in the *Bam*HI site of KC515 or the *Bgl*III or *Xho*I sites of KC516 may influence *vph* expression
KC518[ll]	(36.8)	*vph* (*tsr*)	*Bam*HI replacement of *tsr* (2 to 8 kb)	ϕC31	As KC505

(*continued*)

TABLE I (*continued*)

Vector and reference[a]	Copy number[b] (size in kb)	Markers[c]	Cloning sites[d] (capacity of phage vectors)	Parent replicon[e] (strain of origin)	Remarks[f]
KC684[mm]	(40.5)	*tsr, lacZ*	*Bgl*II, *Xho*I insertions (up to 2.5 kb)	ϕC31	c^+, *attP* deleted. Main use is mutational cloning, potentially leading to *in situ lacZ* transcriptional fusions
PM8[nn]	(39.3)	*tsr, hyg,* fd ter	*Bam*HI, *Bgl*II, *Pst*I, *Pvu*II *Xho*I or *Sst*I insertions (up to 3.5 kb)	ϕC31	c^+, *attP* deleted. Main use is mutational cloning. The *Bgl*II, *Bam*HI, and *Pst*I sites in the polylinker, next to the fd terminator, help mutation cloning
TG78[oo]	(38.8)	*tsr*	*Bam*HI, *Pst*I insertion (up to 4.5 kb)	TG1	c^+, attP$^+$. The host range of TG1, like that of ϕC31, is wide: the two host ranges are partially complementary. Eight *Bcl*I and 11 *Kpn*I sites

[a] See Figs. 1, 3, and 5 for restriction maps of cloning vectors. If there is no published reference to the vector plasmid, the most recent relevant reference (e.g., to the parent plasmid) is given. Addresses of authors who appear only as personal communications are also given; [b]Copy number determined in *S. lividans* unless otherwise specified; [c]See Table II for details; [d]Cloning sites shown on the same line as a marker either inactivate or (in promoter–probe vectors) activate the marker. Only unique cloning sites are listed for plasmid vectors. Unique sites and convenient replacement sites are shown for phage vectors; [e]SLP1.2-derived vectors have a narrow host range at least partially limited by the presence of integrated, incompatible elements. Most other vectors have wide host ranges, perhaps largely limited by restriction; [f]pIJ101-derived vectors have not been found to be suitable in chloramphenicol-producing *S. venezuelae* or erythomycin-

producing *S. erythreus*. Vectors derived from pIJ101, SLP1.2, and SCP2* are always compatible with each other. Extra restriction sites listed here either are unmapped or have been omitted to simplify presentation of the maps; [g]S. Horinouchi and T. Beppu, *J. Bacteriol.* **162,** 406 (1985); [h]T. Eckhardt and D. Stein, personal communication (Smith Kline & French Research and Development, Philadelphia, PA 19101); [i]N. Nakanishi, T. Oshida, S. Yano, K. Takeda, T. Yamaguchi, and Y. Ito, *Plasmid* **15,** 217 (1986); [j]C. A. Omer and S. N. Cohen, *Mol. Gen. Genet.* **196,** 429 (1984); C. A. Omer, personal communication; [k]W. Wohlleben, W. Muth, E. Birr, and A. Pühler, *Proc. Int. Symp. Biol. Actinomycetes, 6th,* p. 99 (1986); [l]J. T. Fayerman, M. D. Jones, and M. A. Richardson, *in* "Microbiology—1985" (L. Leive, ed.), p. 414. Am. Soc. Microbiol., Washington, D.C., 1985; [m]J. L. Larson and C. L. Hershberger, *J. Bacteriol.* **157,** 314 (1984); [n]J. L. Larson and C. L. Hershberger, *Plasmid* **15,** 199 (1986); C. L. Hershberger, personal communication; [o]C. J. Thompson, T. Kieser, J. M. Ward, and D. A. Hopwood, *Gene* **20,** 51 (1982); [p]J. M. Ward, G. R. Janssen, T. Kieser, M. J. Bibb, M. J. Buttner, and M. J. Bibb, *Mol. Gen. Genet.* **203,** 468 (1986); [q]D. A. Hopwood, M. J. Bibb, K. F. Chater, T. Kieser, C. J. Bruton, H. M. Kieser, D. J. Lydiate, C. P. Smith, J. M. Ward, and H. Schrempf, "Genetic Manipulation of Streptomyces: A Laboratory Manual," p. 314. John Innes Foundation, Norwich, England, 1985; [r]E. Katz, C. J. Thompson, and D. A. Hopwood, *J. Gen. Microbiol.* **129,** 2703 (1983); [s]J. F. Martin and J. A. Gil, *Biotechnology* **1,** 63 (1984); [t]D. J. Lydiate, F. Malpartida, and D. A. Hopwood, *Gene* **35,** 223 (1985); [u]B. Jaurin and S. N. Cohen, personal communication [National Defence Research Institute, ABC Research Department (FOA 4), S-901 82 Umeå, Sweden]; [v]S. Kuhstoss and N. R. Rao, *Gene* **26,** 295 (1984); [w]T. Kobayashi, H. Shimotsu, S. Horinouchi, T. Uozumi, and T. Beppu, *J. Antibiot.* **37,** 368 (1984); S. Horinouchi, personal communication; [x]Y. Shindoh, M. M. Nakano, and H. Ogawara, *J. Antibiot.* **37,** 512 (1984); H. Ogawara, personal communication; [y]M. M. Nakano, Y. Shindoh, and H. Ogawara, *FEMS Microbiol. Lett.* **13,** 279 (1982); M. M. Nakano, H. Mashiko, and H. Ogawara, *J. Bacteriol.* **157,** 79 (1984); H. Ogawara, personal communication; [z]H. Krügel and G. Fiedler, *Plasmid* **15,** 1 (1986); [aa]T. Murakami, C. Nojiri, H. Toyama, E. Hayashi, Y. Yamada, and K. Nagaoka, *J. Antibiot.* **36,** 429 (1983); T. Murakami, C. Nojiri, H. Toyama, E. Hayashi, K. Katumata, H. Anzai, Y. Matsuhashi, Y. Yamada, and K. Nagaoka, *J. Antibiot.* **36,** 1305 (1983); K. Nagaoka, personal communication; [bb]A. W. Birch and J. Cullum, *J. Gen. Microbiol.* **131,** 1299 (1985); [cc]T. Oknuki, T. Katoh, I. Tadayuki, and S. Aiba, *J. Bacteriol.* **161,** 1010 (1985); T. Ohnuki, personal communication; [dd]S. Nabeshima, Y. Hotta, and M. Okanishi, *J. Antibiot.* **37,** 1026 (1984); [ee]M. E. Brawner, J. I. Auerbach, J. A. Fornwald, M. Rosenberg, and D. P. Taylor, *Gene* **40,** 191 (1985); [ff]W. Wohlleben and A. Pühler, *Prepr.—Eur. Congr. Biotechnol., 3rd, 1984,* p. 219; [gg]D. Clemans and J. J. Manis, personal communication (The Upjohn Company, Kalamazoo, MI 49001); [hh]Y. Yagi and J. J. Manis, personal communication (The Upjohn Company, Kalamazoo, MI 49001); [ii]I. MacNeil, T. MacNeil, and P. Gibbons, personal communication (Merck Sharp & Dohme Research Laboratories, Rahway, NJ 07065); [jj]C. R. Bailey, C. J. Bruton, M. J. Butler, K. F. Chater, J. E. Harris, and D. A. Hopwood, *J. Gen. Microbiol.* **132,** 2071 (1986); [kk]Maureen J. Bibb, personal communication (John Innes Institute, Norwich NR4 7UH, England); [ll]M. R. Rodicio, C. J. Bruton, and K. F. Chater, *Gene* **34,** 283 (1985); [mm]A. A. King and K. F. Chater, *J. Gen. Microbiol.* **132,** 1739 (1986); [nn]F. Malpartida, personal communication (John Innes Institute, Norwich NR4 7UH, England); [oo]F. Foor, G. P. Roberts, N. Morin, L. Snyder, M. Hwang, P. H. Gibbons, M. J. Paradiso, R. L. Stotish, C. L. Ruby, B. Wolanski, and S. L. Streicher, *Gene* **39,** 11 (1985); F. Foor, personal communication; [pp]SLP1.2 was found in *S. lividans* 66 after conjugation with *S. coelicolor* A3(2), where the SLP1 sequence resides in the chromosome.[j]

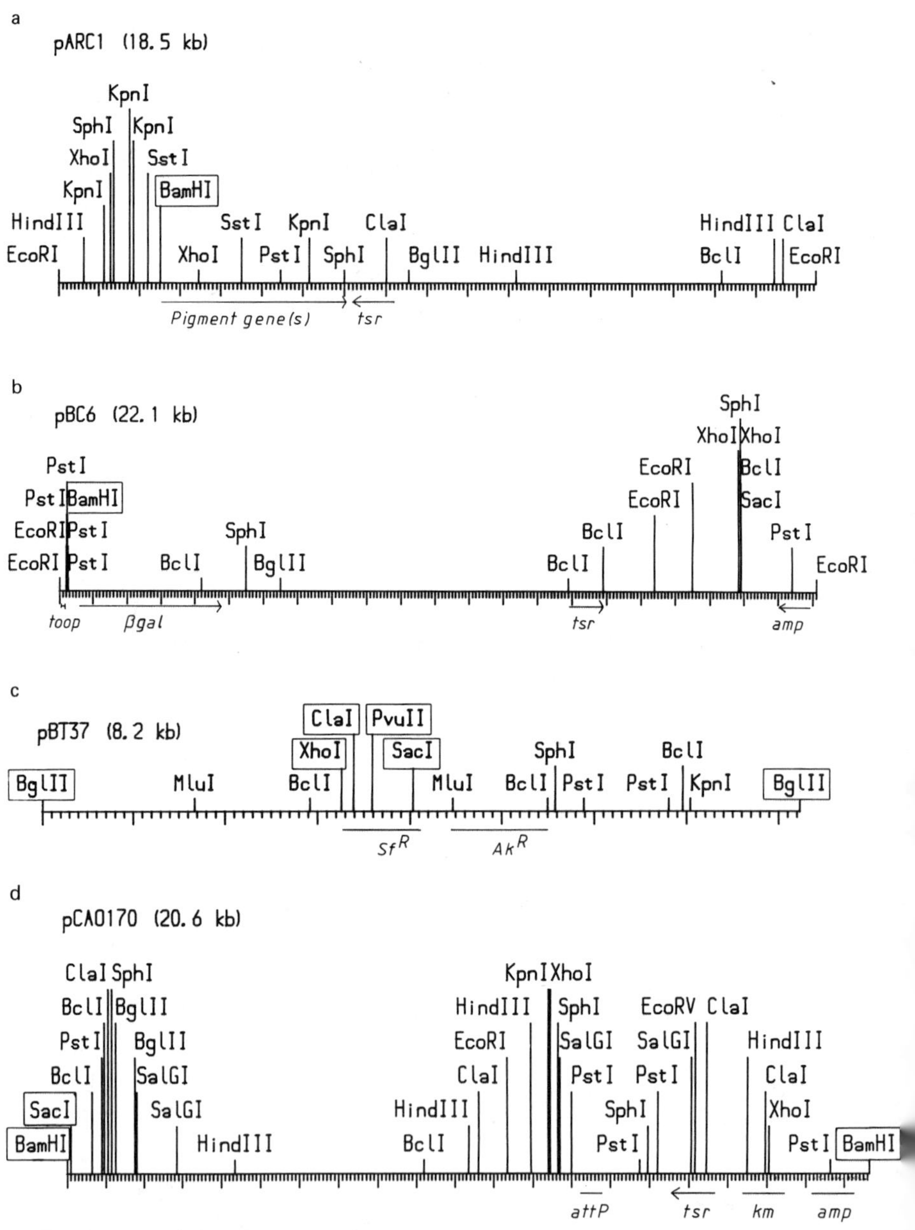

FIG. 3. Maps of plasmid vectors. See Table I for further information on each vector and Table II for details of the markers. Boxed restriction sites are unique cloning sites. All maps are drawn the same length, and therefore to different scales; each major scale division represents 1 kb.

e

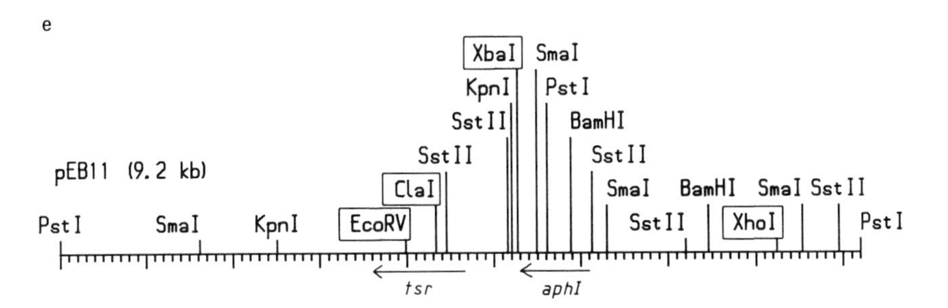

f

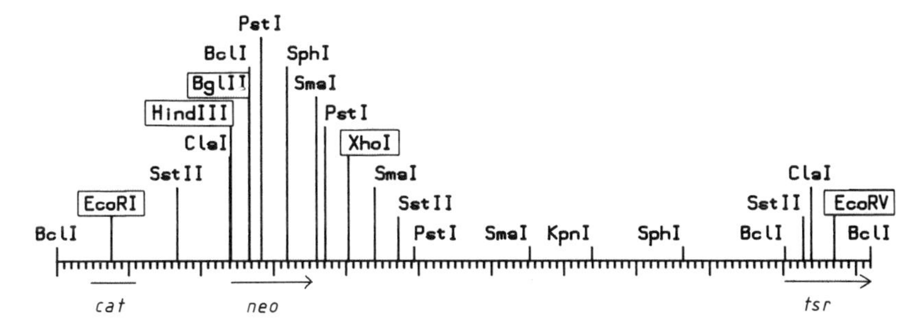

g

pFJ342 (9.2 kb)

ClaI PvuI PstI SalGI KpnI HindIII SalGI BamHI PvuII BclI ClaI ClaI KpnI PvuI PvuII KpnI PstI EcoRI SalGI PstI

tsr aphI

h

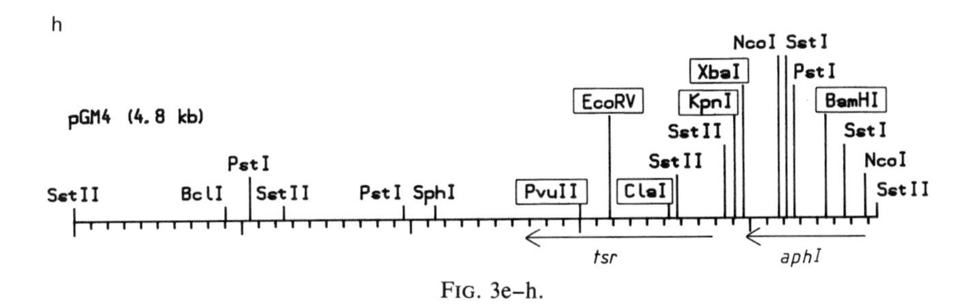

FIG. 3e–h.

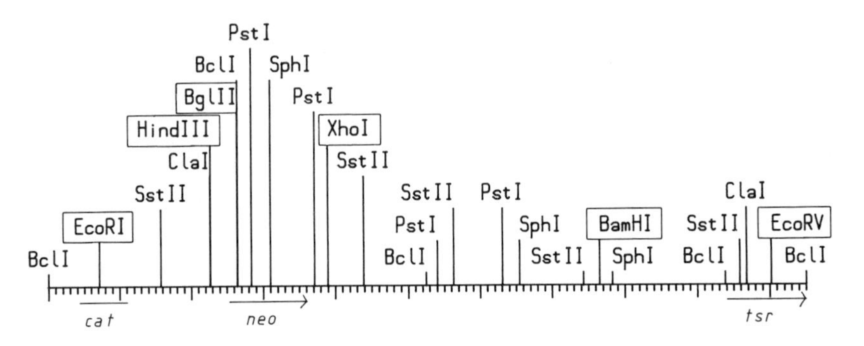

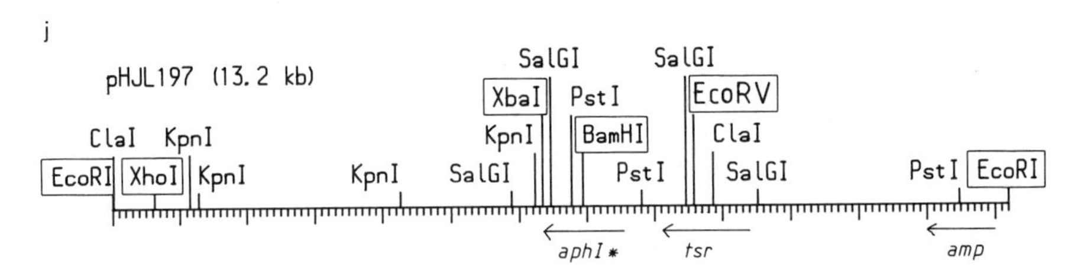

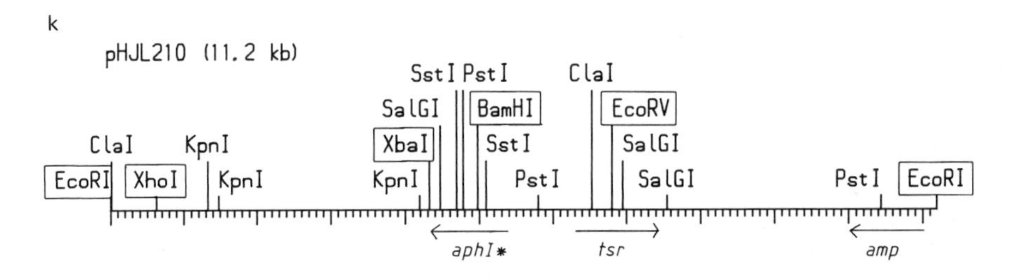

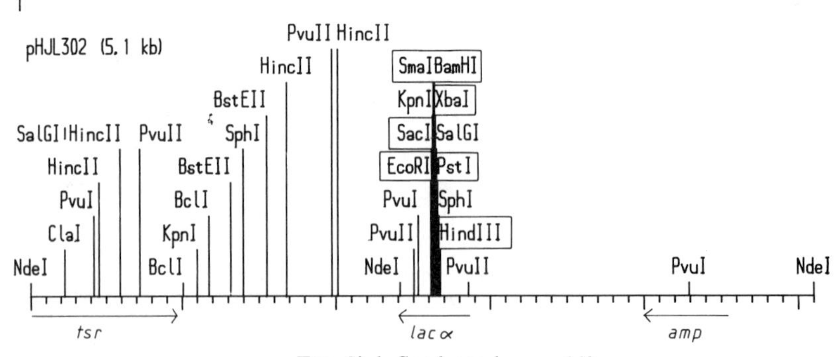

FIG. 3i–l. See legend on p. 142.

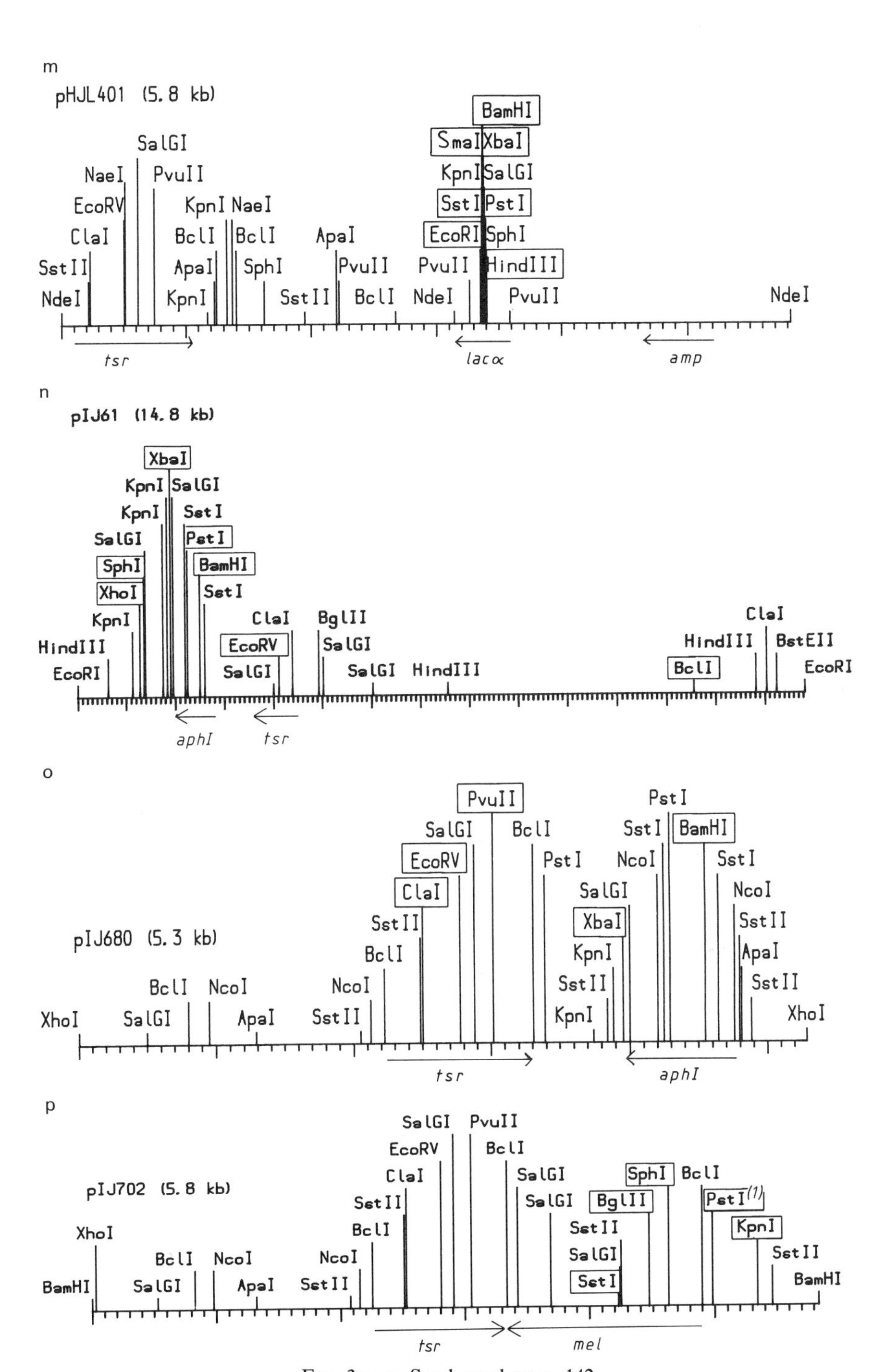

FIG. 3m–p. See legend on p. 142.

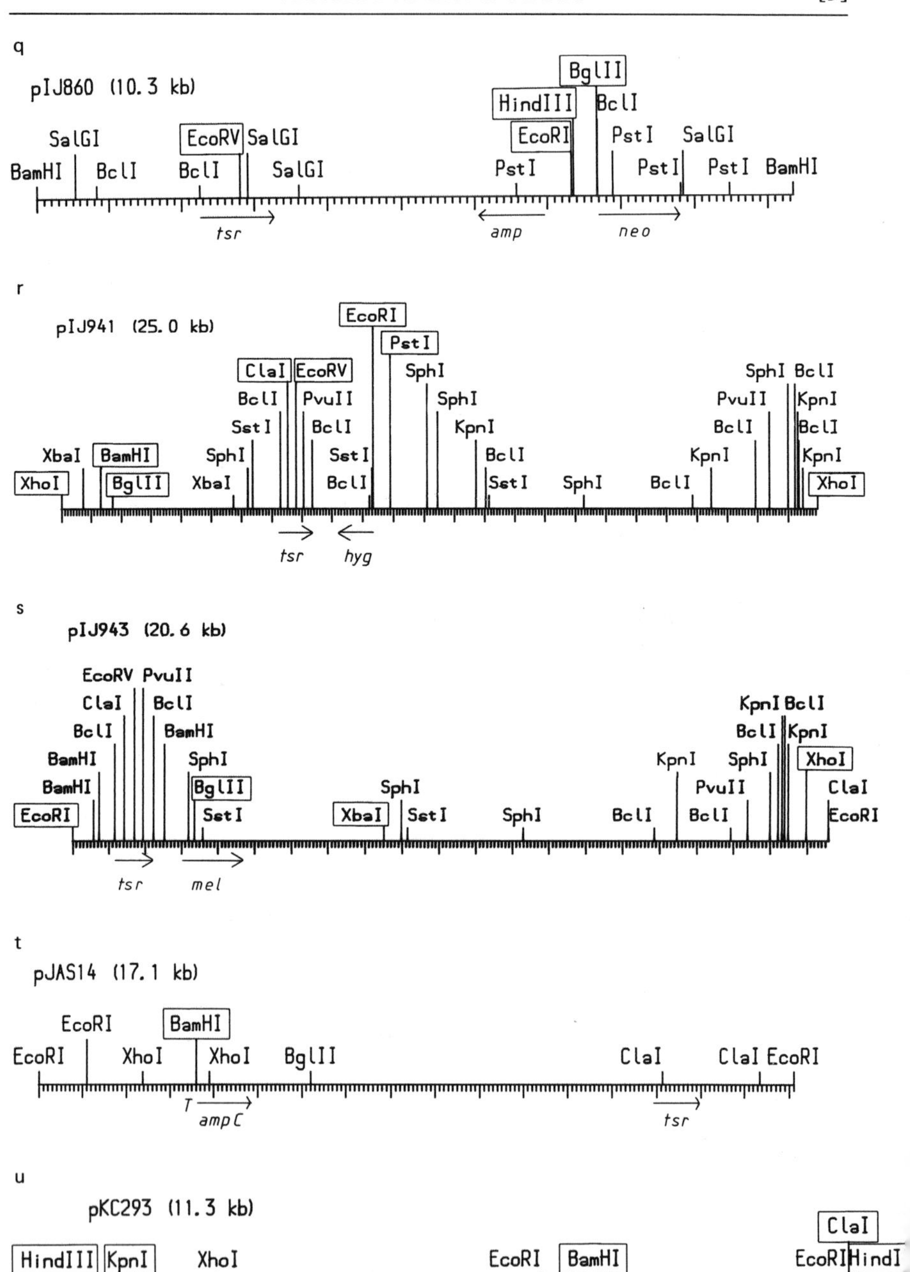

FIG. 3q–u. See legend on p. 142.

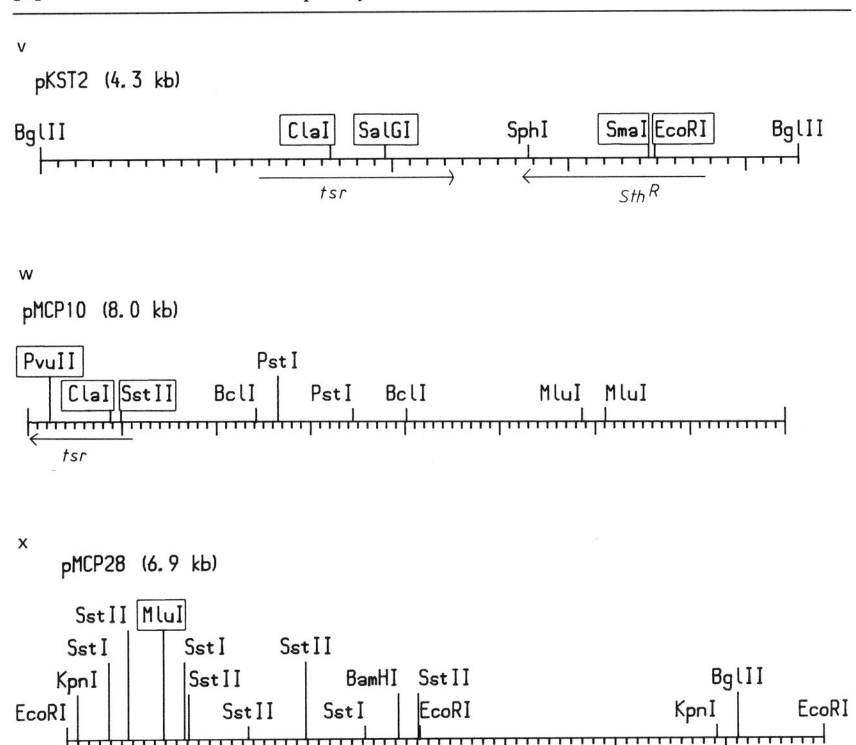

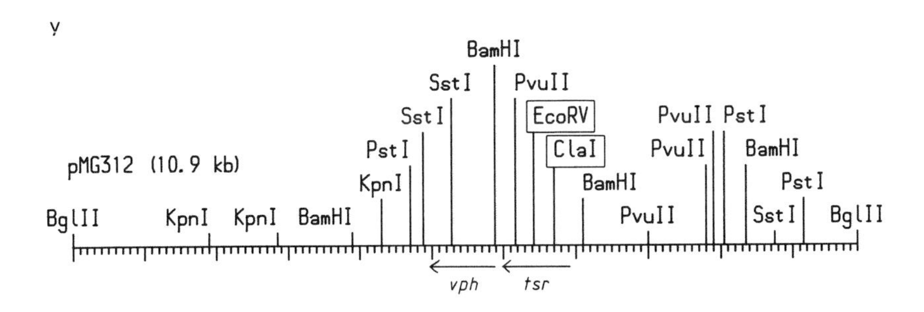

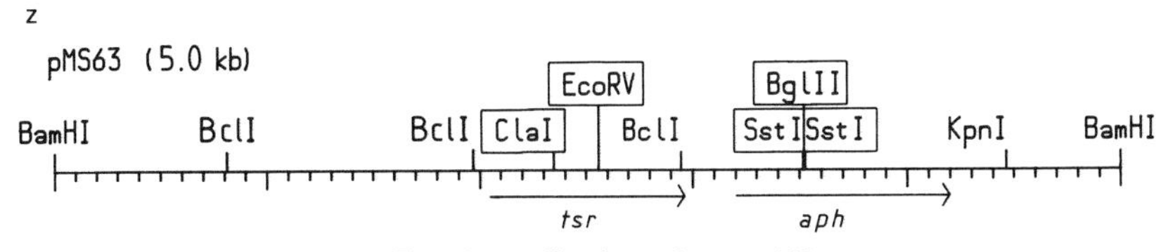

FIG. 3v–z. See legend on p. 142.

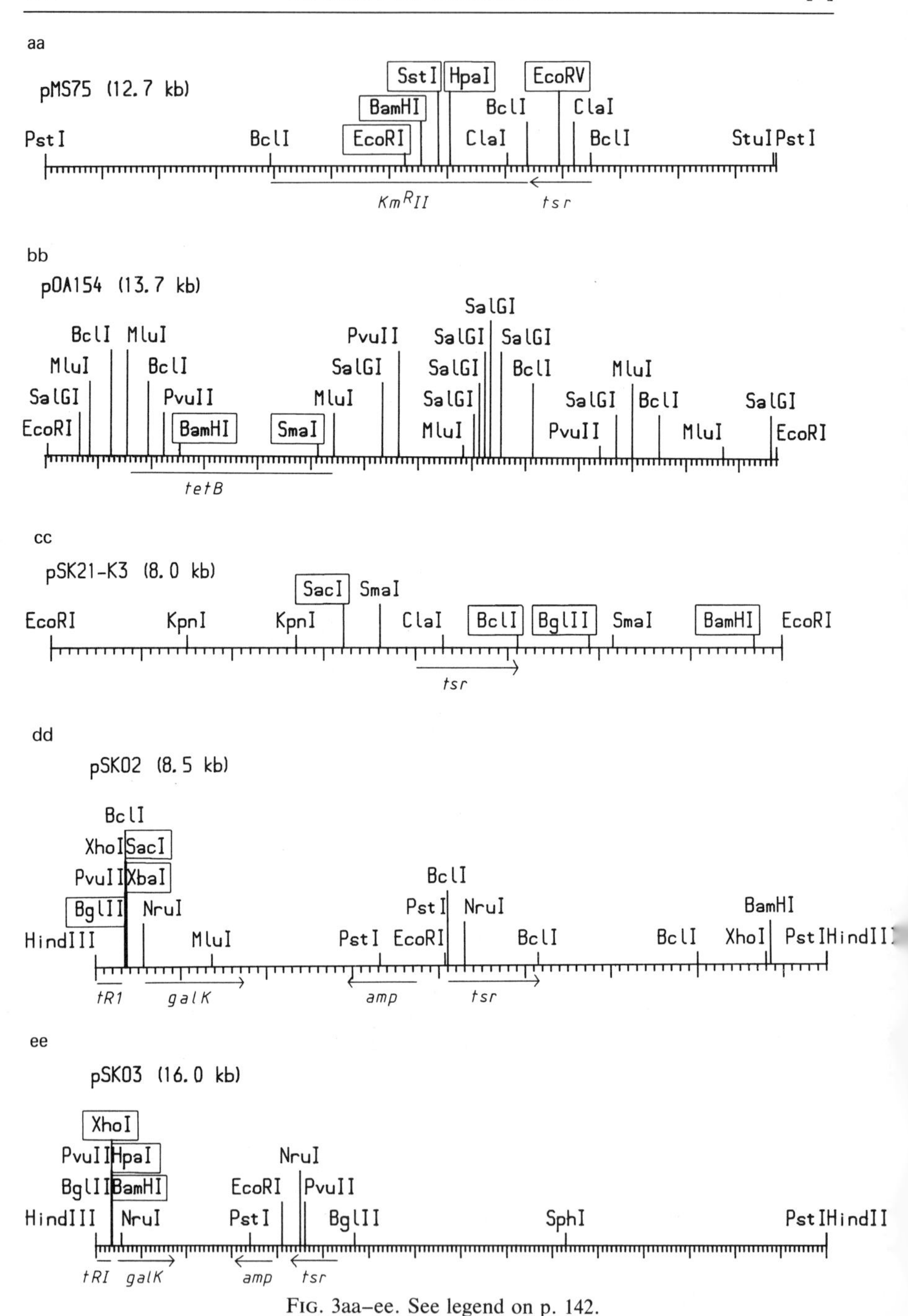

FIG. 3aa–ee. See legend on p. 142.

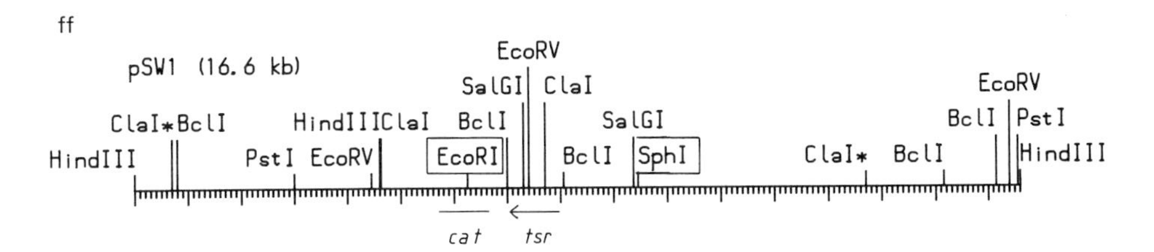

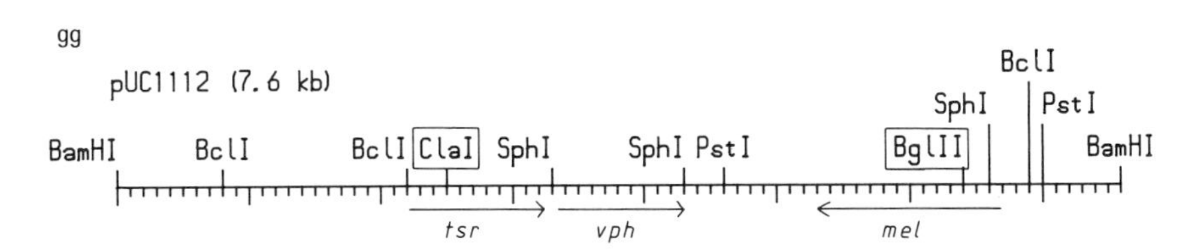

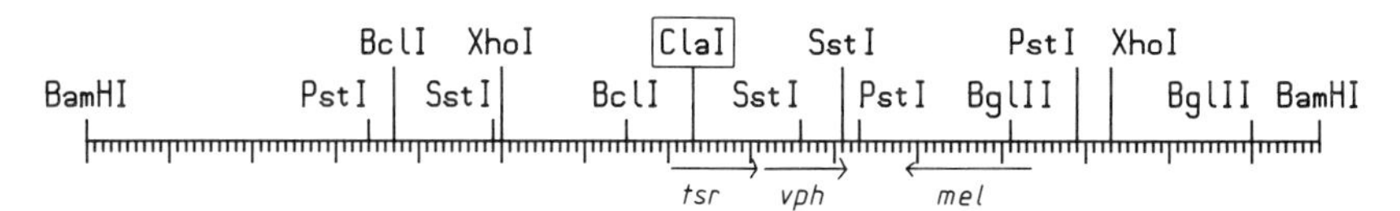

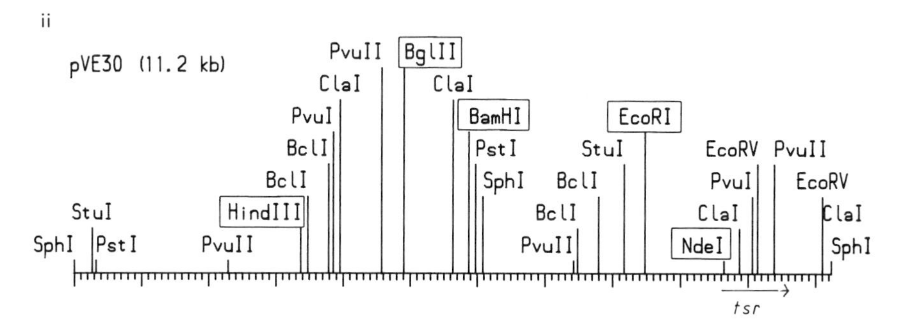

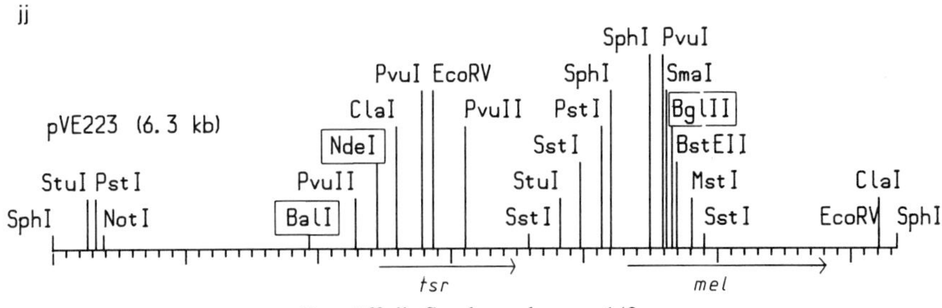

FIG. 3ff–jj. See legend on p. 142.

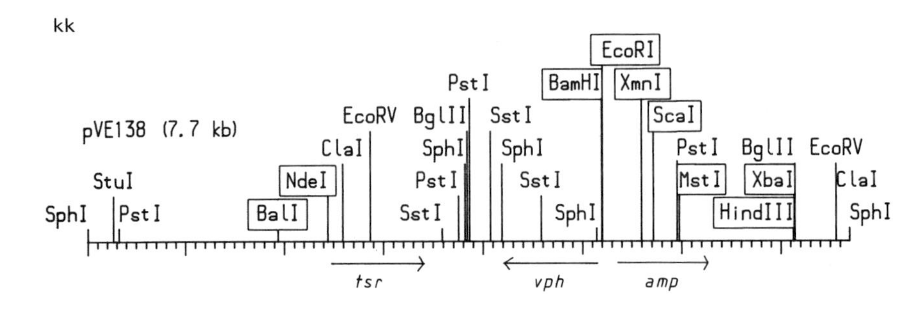

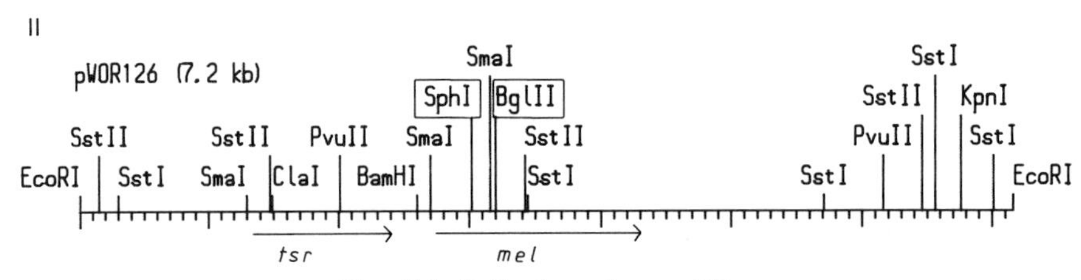

FIG. 3kk–ll. See legend on p. 142.

criterion for including vectors here is that they should have been used successfully for cloning experiments. Our selection is subjective and probably does not include many existing useful vectors. The fact that most of the vectors in Table I have been constructed very recently means that many more better developed and tested vectors will become available rapidly. We understand that all the vectors will be made available to academic researchers.

Streptomyces lividans 66 and its derivatives can probably support replication of all the listed plasmids. Other strains may be more selective (e.g., *S. coelicolor* does not support autonomous replication of SLP1.2-based plasmids). For this reason we have presented a rather wide range of vectors based on different replicons. Unless a selectable gene, e.g., one for antibiotic resistance, is to be cloned the vector needs at least one selectable marker. Almost all the listed vectors contain the *tsr* gene for thiostrepton resistance, useful in almost all streptomycetes. Most of the other selectable markers confer resistance to aminoglycoside antibiotics where natural resistance is widespread in streptomycetes. Additional features like insertional inactivation or activation of marker genes are useful for clone recognition. Probably more important are the copy number of the vector and the presence or absence of promoters near the cloning sites. Absence of promoters is usually desirable when the regulation of cloned genes is to be studied and for the same purpose it may be important to maintain a cloned gene at approximately its natural copy number.

Most of the listed vectors are of high copy number which facilitates DNA, and often gene product, isolation, and a great many genes have been cloned with such vectors, notably with pIJ702. In contrast, SCP2*-derived unit copy plasmids have been used successfully in situations where a multicopy vector failed, as in the stable cloning of large fragments of DNA carrying an entire set of antibiotic biosynthetic genes.[48] Some vectors have recently been developed from other replicons with low or intermediate copy number (Table I).

For studying the interaction of cloned genes with each other, compatible vectors are needed. The limited information suggests that most of the *Streptomyces* replicons isolated from nature are compatible with each other, but since most existing vectors rely on *tsr* as their selective marker, further derivatives of some of these replicons will be useful.

Streptomyces Phages as Cloning Vectors

A recent review[49] describes several families of *in vitro*-manipulated *Streptomyces* phages which have potential as cloning vectors. Of these, only ϕC31 derivatives have yet been put to extensive use. ϕC31 is genetically the most studied *Streptomyces* phage.[50] Like coliphage λ, it has a repressor gene (*c*) to establish and maintain lysogeny, a specific site (*attP*) in its DNA for integration into the host chromosome, and cohesive ends to its DNA. As a foundation for the use of ϕC31 as a vector, a 7.5-kb inessential region of its DNA was first replaced with pBR322,[51] which has since been the target for various internal substitutions with marker resistance genes in different vector derivatives. Vectors are available with unique sites (or replaceable fragments) within this region, for the restriction enzymes *Bam*HI, *Bgl*II, *Pst*I, *Pvu*II, *Sst*I, and *Xho*I. Most other potentially useful enzymes have several sites in essential regions of ϕC31 DNA.[52] An important consideration in using ϕC31 vectors is a limitation on the size of inserts, enforced by the need to package the hybrid DNA into phage particles. In practical terms, genomes of between 36 and 43 kb are packaged (the wild-type genome consists of 41.4 kb). The various vector derivatives now in use contain from 31 to 34 kb of ϕC31 DNA together with various DNA segments of diverse origin, so that in principle it should be possible to insert up to 12 kb of foreign DNA. In practice, for

[48] F. Malpartida and D. A. Hopwood, *Nature* (*London*) **309,** 462 (1984).

[49] K. F. Chater, *in* "The Bacteria" (S. W. Queener and L. E. Day, eds.), Vol. 9, p. 119. Academic Press, New York, 1986.

[50] N. D. Lomovskaya, K. F. Chater, and N. M. Mkrtumian, *Microbiol. Rev.* **44,** 206 (1980).

[51] J. E. Suarez and K. F. Chater, *Nature* (*London*) **286,** 527 (1980).

[52] M. R. Rodicio, C. J. Bruton, and K. F. Chater, *Gene* **34,** 283 (1985).

a variety of reasons the current upper limit is about 9.5 kb (e.g., replacing the *Bam*HI–*Pst*I segment of KC401).[53]

The general scheme for use of these vectors in cloning is to transfect *S. lividans* protoplasts with a "ligation mixture," and then, after ~16 hr, to replica plate from the plaques to plates spread with spores of a suitable indicator strain for infection of the indicator to occur. Lysogens of the indicator may be detected by subsequent replica plating to medium containing an antibiotic appropriate to the selective marker of the vector, and the phenotype of the lysogens for the desired character may then be determined. The specialized transducing phages obtained by this route can easily be used in conventional genetic tests. Some phage vectors lack the *attP* site, but they can lysogenize their host if they contain a cloned fragment of its genome, by using homologous recombination. Such insert-directed integration of the prophage into any desired region of the host genome gives rise to several interesting genetic possibilities, including the generation of mutants (see the section Mutational Cloning).

Example of Shotgun Cloning Using a Phage Vector

The experiment in Fig. 4 illustrates how a library of *Streptomyces* DNA can be constructed using a ϕC31 vector. Many of the media and methods already described for cloning with plasmids are also used in this experiment. Those which have not been described earlier are given in the next few sections. The vector chosen for this example is KC304 (Fig. 5; Table I) an att^+ c^+ phage with the *tsr* gene for selection and the *vph* fragment, flanked by *Bam*HI sites, for replacement by *Streptomyces* DNA (hence the use of size-fractionated DNA in the range 3–5 kb). Recently, some KC304 DNA preparations have given unexpectedly low frequencies of inserts.

However, when phages from single plaques were used to prepare high titer lysates for DNA extraction, no difficulties were encountered. Alternative vectors and strategies are given in a later section (ϕC31 Vectors Lacking the *attP* Site, and Mutational Cloning).

Additional Media, Buffers, and Reagents for Phage Cloning Experiments

Nutrient agar (per 200 ml): Place 4.6 g Difco nutrient agar in a 250-ml Erlenmeyer flask and add 200 ml distilled water. Close the flasks and autoclave. For use in ϕC31 experiments, melt, then add separately autoclaved solutions of glucose (2 ml of a 50% solution), $Ca(NO_3)_2$ (2 ml of a 0.8 *M* solution), and $MgSO_4$ (2 ml of a 1 *M* solution)

[53] J. E. Harris, K. F. Chater, C. J. Bruton, and J. M. Piret, *Gene* **22,** 167 (1983).

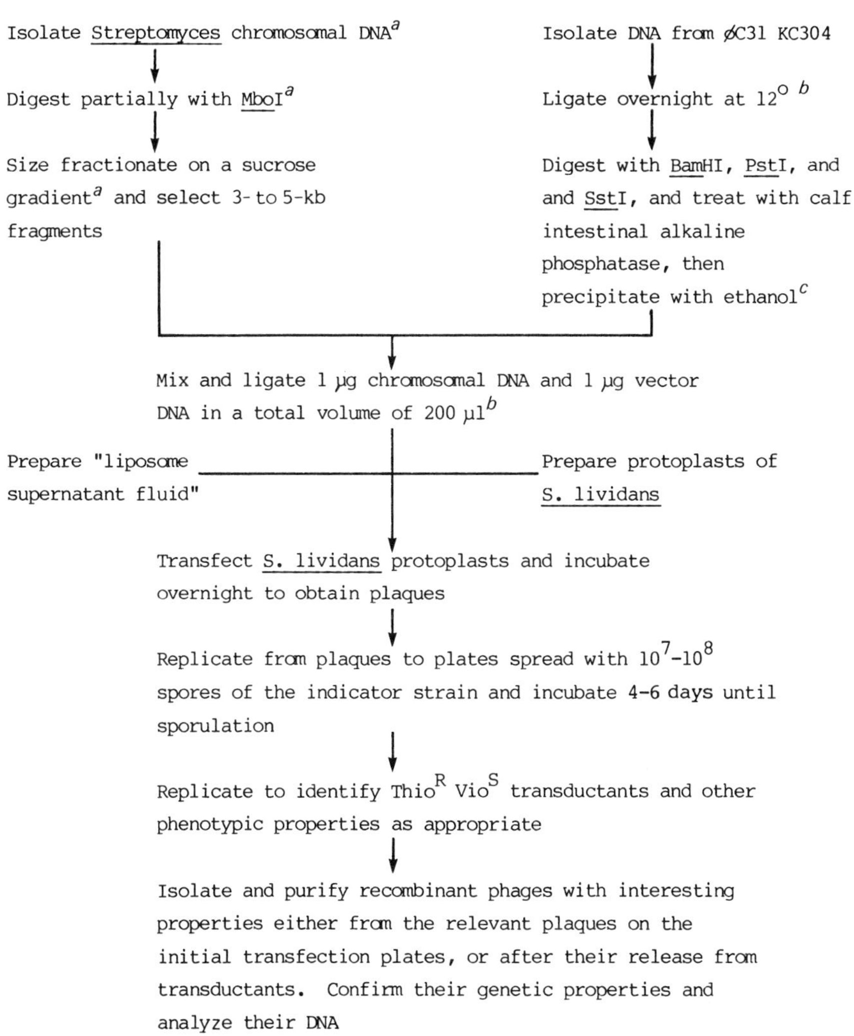

FIG. 4. Outline of the shotgun cloning experiment using *Streptomyces* phage vector ϕC31KC304. [a] See earlier sections for details. [b] Ligation of the vector DNA, usually at <100 µg/ml, joins the *cos* ends covalently. The cloning ligation then involves only two DNA molecules, as in cloning with plasmids (Fig. 2), and can be done at a low DNA concentration. In many other experiments, with a variety of ϕC31 vectors, we have not preligated the *cos* ends, nor used alkaline phosphatase. The cloning ligation is then most efficient at high (100–200 µg/ml) DNA concentration.[59] [c] *Bam*HI digestion cuts out the *vph*-containing fragment, which is to be replaced by up to 5 kb of *Streptomyces* chromosomal DNA. The further digestions with *Pst*I and *Sst*I cleave the *vph* fragment, effectively preventing it from being ligated back into the vector. Treatment with alkaline phosphatase, minimizing recircularization of the vector without inserts, results in at least 50% of transfectant phages having inserts (Maureen J. Bibb and R. E. Melton, personal communication).

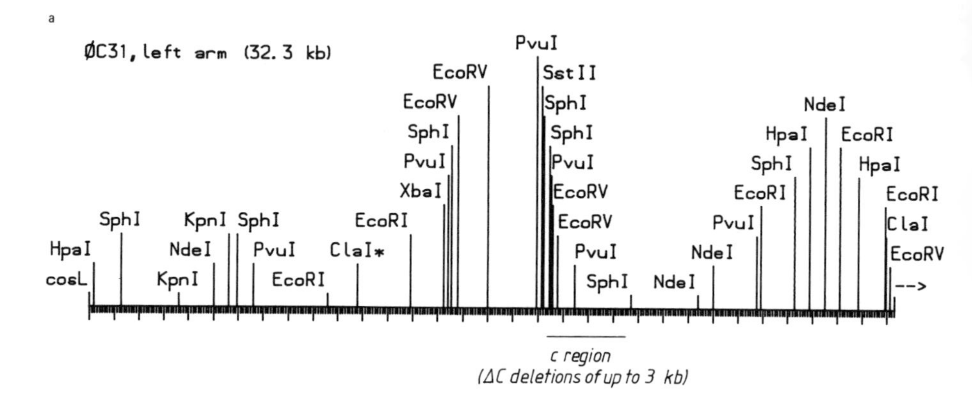

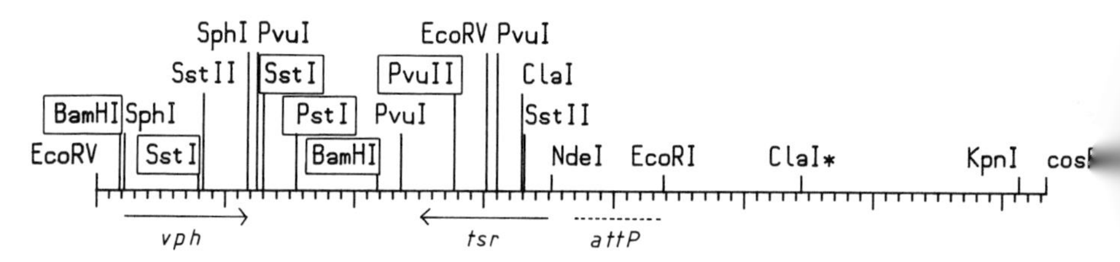

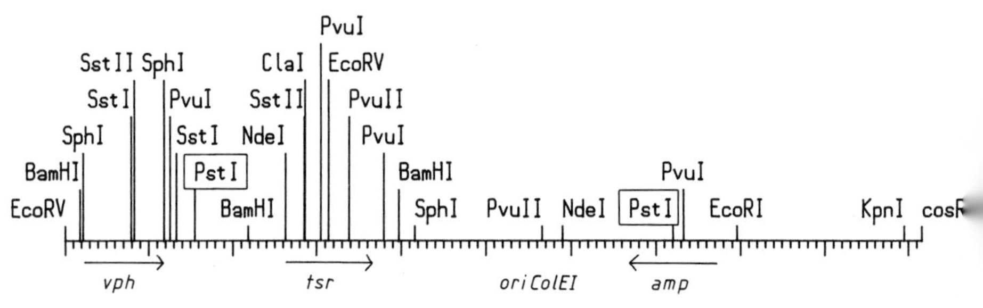

FIG. 5. Maps of phage vectors. Throughout, *cosL* and *cosR* represent the left and right cohesive ends. Vectors with KC and PM numbers are based on ϕC31. They all share a common left arm of 32.4 kb, except that 1.9 kb is deleted from the *c* region of ΔC3-containing derivatives such as KC518. The common left arm and the right arms of the vectors are therefore shown separately. See Tables I and II for additional information on the vectors and markers, and the caption to Fig. 3 for other conventions on the maps.

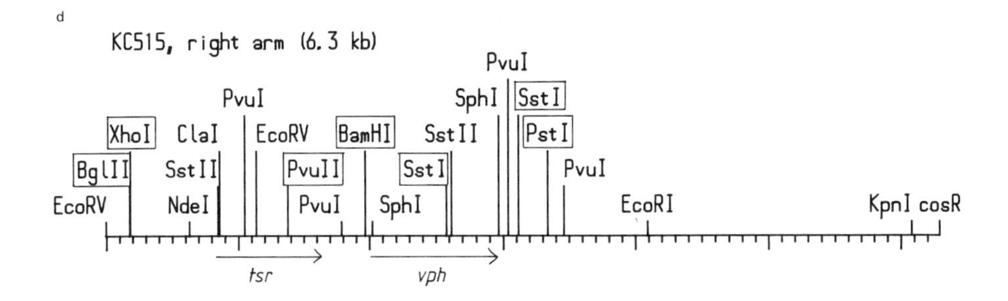
d
KC515, right arm (6.3 kb)
PvuI
PvuI
SphI
SstI
XhoI
ClaI
EcoRV
BamHI
SstII
PstI
BglII
SstII
PvuII
SstI
PvuI
EcoRV
NdeI
PvuI
SphI
EcoRI
KpnI
cosR
tsr
vph

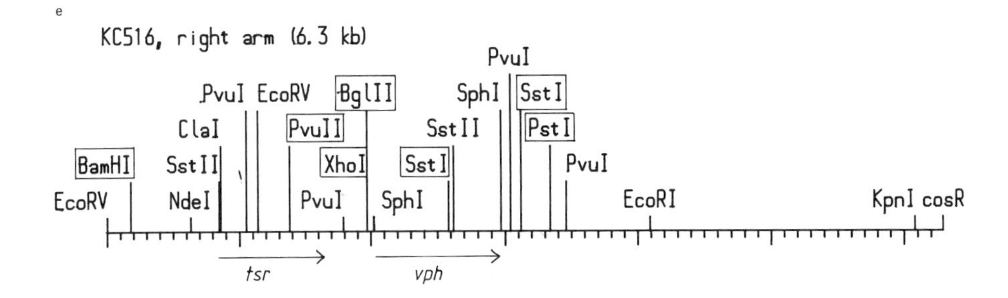
e
KC516, right arm (6.3 kb)
PvuI
PvuI
EcoRV
BglII
SphI
SstI
ClaI
PvuII
SstII
PstI
BamHI
SstII
XhoI
SstI
PvuI
EcoRV
NdeI
PvuI
SphI
EcoRI
KpnI
cosR
tsr
vph

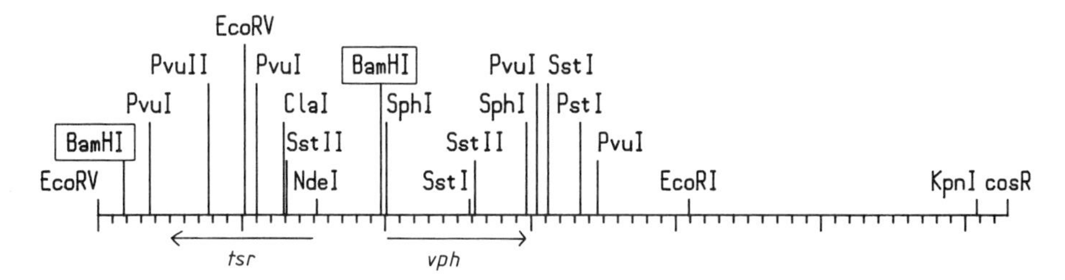
f
KC518, right arm (6.3 kb)
EcoRV
PvuII
PvuI
BamHI
PvuI
SstI
PvuI
ClaI
SphI
SphI
PstI
BamHI
SstII
SstII
PvuI
EcoRV
NdeI
SstI
EcoRI
KpnI
cosR
tsr
vph

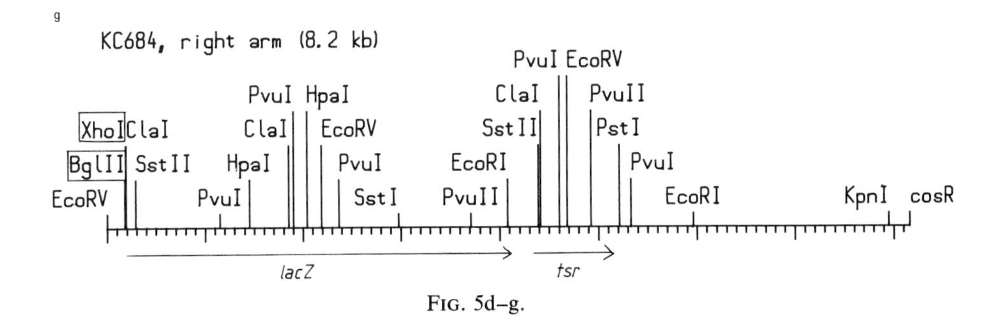
g
KC684, right arm (8.2 kb)
PvuI
EcoRV
PvuI
HpaI
ClaI
PvuII
XhoI
ClaI
ClaI
EcoRV
SstII
PstI
BglII
SstII
HpaI
PvuI
EcoRI
PvuI
EcoRV
PvuI
SstI
PvuII
EcoRI
KpnI
cosR
lacZ
tsr

FIG. 5d–g.

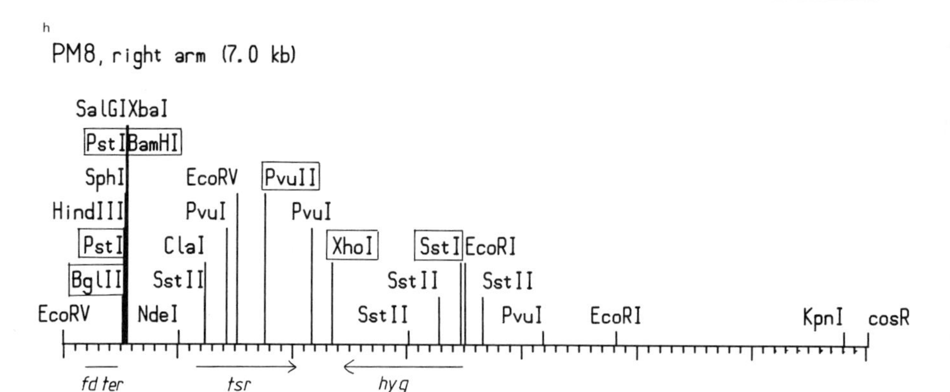

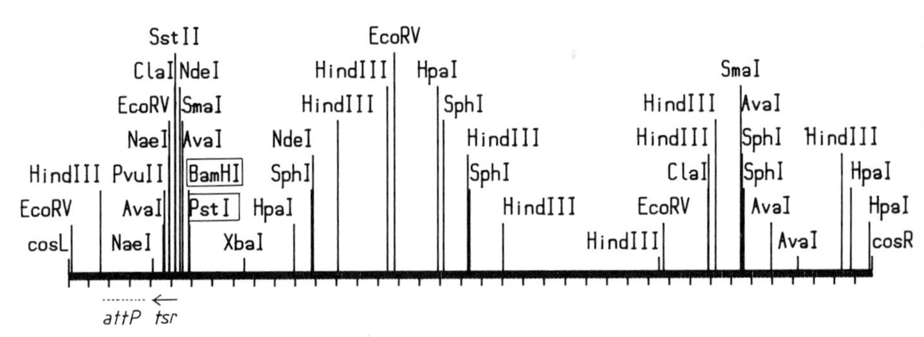

FIG. 5h–i. See legend on p. 154.

Soft nutrient agar (per liter): 8 g of Difco nutrient broth powder, and 3 g of agar

SM buffer (per liter): 20 ml of 1 *M* Tris–HCl (pH 7.5), 1 ml of 1 *M* $MgSO_4$, 20 ml of 5 *M* NaCl, and 1 g of gelatin. Dissolve the gelatin in 10 ml of water and add to the other components, making the volume up to 1000 ml with distilled water. Sterilize by autoclaving

Liposome supernatant fluid: Dissolve 2.5 mg of L-phosphatidylcholine (Sigma, type IX-E, No. P-8640) and 0.12 mg of stearylamine (PL Biochemicals, No. 6516) in 10 ml chloroform. (*Note:* At present PL Biochemicals has ceased production of stearylamine. We have not yet established an entirely satisfactory alternative source.) Deposit the lipids as a thin film in a scrupulously clean 100-ml round-bottom flask by rotary evaporation under vacuum at 55°. Resuspend the film in 0.5 ml G buffer (15 m*M* NaCl, 1.5 m*M* sodium citrate, 1 m*M* $CaCl_2$,

280 m*M* sucrose, 100 m*M* L-threonine, 100 m*M* L-histidine) by very vigorous manual shaking. Mix with 2 ml 5.2% KCl in 10% ethanol. Centrifuge (~6000 rpm, bench centrifuge, room temperature) and retain the supernatant. This "liposome supernatant fluid" significantly stimulates PEG-mediated transfection of protoplasts[54]

Large-Scale Isolation of DNA from ϕ*C31 Derivatives*

When a phage vector is to be used in cloning experiments, it is wise to have available a large amount of DNA from highly purified phage particles. The following procedure has proved reproducible in our hands. It involves the CsCl-gradient purification of phages soaked out from large areas of confluently-lysed agar overlays.

Mix 3×10^5 plaque-forming units of vector with about 2×10^9 spores of *S. lividans* in 150 ml molten soft nutrient agar at 45° and pour over a total of 0.3 m^2 (surface area) of freshly poured nutrient agar (e.g., in large autoclavable bowls). Incubate overnight at 30° and scrape the soft agar into Difco nutrient broth (200 ml). Shake and then stand at room temperature for 2 hr. Centrifuge (16,000 *g*, 10 min, 4°) to remove agar.

Transfer the supernatant(s) to a 250-ml Erlenmeyer flask, and keep them at 4° while reextracting each agar pellet with 50 ml Difco nutrient broth. Dispense the pooled supernatants into eight 50-ml polycarbonate centrifuge tubes. Centrifuge (34,000 *g*, overnight, 4° or 54,500 *g*, 75 min, 4°). Discard the supernatants and gently suspend each pellet in 0.5 ml SM buffer (either stand at 4° for 3 hr or agitate on a rotary shaker at 37° for 15–30 min). Pool the suspensions and adjust the volume to 7 ml with SM buffer. Add 5.95 g CsCl and dissolve; transfer the solution to a 12-ml centrifuge tube, overlay with liquid paraffin, and centrifuge (85,000 *g*, 18 hr, 20°). Open the tube and remove the blueish phage band through the side using a 2-ml syringe (this preparation can be stored at 4° if necessary). Dialyze (2 × 2 hrs against 100 ml SM buffer, 4°). Transfer to an 11-ml conical centrifuge tube. Add one-twentieth volume 20× SSC (3 *M* NaCl, 0.3 *M* sodium citrate), cover (e.g., with Parafilm), heat (60°, 10 min), and cool to room temperature. Add 5 *M* NaCl to 0.25 *M* final concentration. *Protecting your eyes and wearing gloves,* add an equal volume of phenol solution (as in "Isolation of Chromosomal DNA"), cover with Parafilm, and invert 50 times. Centrifuge (~3000 rpm, 10 min, room temperature). Using a shortened Pasteur pipet, transfer the upper (aqueous) phase to a fresh conical centrifuge tube. Reextract the aqueous phase twice with phenol. Also, reextract the phenol layers with an equal volume of TE buffer containing 0.25 *M* NaCl. Pool the aqueous layers and

[54] M. R. Rodicio and K. F. Chater, *J. Bacteriol.* **151,** 1078 (1982).

transfer to a 20-ml screw-cap bottle. Extract the phenol by shaking gently with an equal volume of ether (use a fume cupboard!). Discard the upper (ether) phase. Repeat the ether extraction twice. Precipitate the DNA by adding 2.5 vol of absolute ethanol, swirling the tube to allow formation of a fibrous clot which can be spooled on a sealed Pasteur pipet and transferred to a 5-ml screw-capped bottle. Wash the DNA with 1 ml 70% ethanol and centrifuge briefly. Remove all visible ethanol with a Pasteur pipet, dissolve the DNA (typical yield ~200 μg) in 0.5–1 ml sterile TE buffer, and store at 4°.

Transfection of Streptomyces lividans Protoplasts

Transfection may be done either directly with the DNA still in ligation buffer, or with ethanol-precipitated DNA redissolved in TE buffer. Protoplasts of *S. lividans* 66 (or its derivatives, such as 3104) are freshly prepared as for transformation from fresh or stored (frozen at −20°) mycelium, with the final wash immediately preceding transfection. (Protoplasts which have been prepared and then frozen give very low transfection frequencies.) For transfection, the protoplast pellet is resuspended in the drop of buffer left after centrifugation, and then a mixture of 100 μl of liposome supernatant fluid and half of the DNA (up to 20 μl) is added, followed immediately by 0.5 ml of 60% PEG (mix 3 g PEG 1000 with 2 ml P buffer). The remainder of the DNA is usually stored as a "back-up." The whole suspension is mixed by pipetting up and down a Pasteur pipet three times. After 1 min, samples are ready for plating. Typically, after removal of 50 μl for a dilution series, the undiluted mixture is mixed with 50 ml molten soft R2YE (molten R2YE mixed with an equal volume of P medium and held at 45°) to which about 5×10^8 spores of *S. lividans* have been freshly added, and distributed across the surface of 10 surface-dried R2YE plates. Samples of 0.1 ml of 10- and 100-fold dilutions are similarly plated, but using only two plates per dilution. The quality of the transfection conditions is usually checked by transfecting replicate samples of protoplasts with a standard preparation of ϕC31 DNA (50 ng/transfection). Usually, 0.1 ml of a 10^{-3} dilution of this control transfection will give about 100 plaques. The cloning transfection should give 200–1000 plaques on each 10^0 plate (2000–10,000 altogether).

Screening a Library of Streptomyces DNA Cloned in KC304

In the diagram (Fig. 4) the steps following transfection are those used when it is necessary for many clearly independent clones to be analyzed (i.e., where screening is limiting). If screening is easy, or if a permanently available library is sought, then it may be convenient to harvest the primary transfection plates directly in Difco nutrient broth, to be filter sterilized and stored at 4° for future use (e.g., as a source of phages that transduce particular markers). Chloroform is not usually used to preserve ϕC31 suspensions (though it can be used at low concentrations such as 0.1%).

If plaques from *S. lividans* are to be replicated, it is often helpful to kill the *S. lividans* host first by exposing plates to chloroform vapor for 20 min. This does not appear to reduce the viability of phages in the plaques. Lysogenization of indicator strains on an agar surface, to screen for desired phenotypic changes in lysogens, is done either by replicating plaques, as in Fig. 4, or by adding phage suspensions to plates spread with spores of the indicator strain ($\sim 10^7$–10^8 spores per plate). The medium is one which allows both phage infection and host sporulation, such as R2YE for most strains. After sporulation, the transduction plates are replicated: to plates (R2YE or suitable minimal medium) containing 50 μg/ml thiostrepton (to detect transductants), to plates containing 30 μg/ml viomycin (to identify Vio^s colonies), and, if appropriate, to plates selecting clones of the desired phenotype. Vectors such as KC304, which retain the *attP* site for chromosome integration, transduce many streptomycetes at high frequency, and usually give stable lysogens. It is therefore easy to test for "complementation" of several mutants. In libraries of 5000 clones with 4-kb inserts each gene is represented with about 90% probability.

Small-Scale Isolation and the Analysis of DNA Cloned in ϕC31-Based Vectors

When a transducing phage has been isolated, its DNA is usually analyzed. This first involves a small-scale DNA preparation. This procedure is adapted by C. J. Bruton from procedures used by Suarez and Chater[55]; an alternative version, using microcentrifuge tubes, is given by Hopwood *et al.*[7] From three or six 9-cm fresh nutrient agar plates, each containing 2×10^4 fresh plaques in a 2.5-ml *S. lividans* top layer, transfer the soft agar to Difco nutrient broth (12.5 ml in a 50-ml centrifuge tube per three plates), and mix the slurry by pipetting twice in a 10-ml pipet. After 2 hr at room temperature, centrifuge (16,000 *g*, 10 min, 4°), transfer the supernatant to a 30-ml centrifuge tube, and sediment the phage particles (55,000 *g*, 25 min, 4°). Use an orbital incubator (37°, 20 min) to resuspend the phages in 2 ml RNase solution (50 μg RNase/ml in 10 m*M* Tris–HCl, pH 7.5, 1 m*M* $MgCl_2$, solution preincubated at 95° for 5 min). Transfer the suspension to a 15-ml centrifuge tube, mix with 0.4 ml fresh SDS mix (1 vol 10%

[55] J. E. Suarez and K. F. Chater, *J. Bacteriol.* **142,** 8 (1980).

SDS, 1 vol 2 *M* Tris–HCl, pH 9.6, 2 vol 0.5 *M* NaEDTA, pH 7.4), and incubate at 70° for 30 min. Immediately add 0.5 ml 8 *M* potassium acetate and leave on ice for 15 min. Centrifuge (28,000 *g*, 30 min, 2°). Carefully transfer the supernatant to a 50-ml polycarbonate centrifuge tube, add 7 ml TE buffer, and mix. Add 25 ml absolute ethanol, and leave at −20° overnight. Centrifuge (7800 *g*, 10 min, room temperature), and pour off the supernatant. Invert the tube over a tissue for about 10 min to drain. Resuspend the pellet in 1 ml TE buffer at 37° in the orbital shaker for about 30 min. Add 0.1 ml of 3 *M* sodium acetate and mix, then add 2.5 ml absolute ethanol. Swirl the tube until the DNA precipitates as a fibrous clot. Spool the DNA on a sealed Pasteur pipet and transfer it to an Eppendorf tube. Dissolve it in 0.3 ml TE buffer (37°, 30 min). Add 30 μl 3 *M* sodium acetate (with mixing) and 1 ml absolute ethanol. The DNA should precipitate rapidly. Centrifuge briefly and remove all traces of liquid with a Pasteur pipet. Wash the precipitate with 1 ml 70% ethanol, centrifuge briefly, and remove all traces of liquid before finally dissolving the DNA in 50–100 μl sterile TE buffer. Store at 4°. About 10 μg of DNA is usually obtained. When used in restriction digests, it is often cleaved poorly by *Eco*RI and *Sph*I, but otherwise gives few problems. The region of the vector used for cloning is close to one end (*cosR*) of the DNA so that fragments extending from within the cloned fragment to *cosR* are in a measurable size range and are easily identified by their reduced intensity on gels (because part of the population of *cosR*-containing fragments anneals to *cosL*-containing fragments and migrates more slowly on the gel).

φC31 Vectors Lacking the attP Site, and Mutational Cloning

Considerable use has been made of φC31 vectors lacking *attP*. Useful examples of such vectors are KC515 and KC518 (Fig. 5; Table I). KC518 can accommodate 2- to 8-kb inserts, with loss of the *tsr* gene showing the presence of an insert. However, lysogenization of an indicator strain can proceed only by recombination of the incoming phage DNA with homologous sequences present in the host. The homology may be provided by a φC31 prophage previously introduced into the indicator strain, in which case either c^+ (e.g., KC515) or *c*-deleted (e.g., KC518) vectors can be used.[56] Alternatively, for c^+ vectors, a nonlysogenic derivative of the strain from which the cloned DNA originated can be used as the indicator, in which case the homology is provided by the cloned fragment.[57] Such "insert-directed integration" positively identifies plaques containing

[56] J. M. Piret and K. F. Chater, *J. Bacteriol.* **163,** 965 (1985).

[57] H. Ikeda, E. T. Seno, C. J. Bruton, and K. F. Chater, *Mol. Gen. Genet.* **196,** 501 (1984).

TABLE II
DESCRIPTION OF MARKERS ON MAPS IN FIGS. 1, 3, AND 5

Marker	Remarks
Ak^R	Amikacin resistance gene from *S. litmocidini* ISP5164. Confers resistance also to dibekacin, kanamycin A and B, ribostamycin, and tobramycin[a]
amp	Ampicillin resistance (β-lactamase) gene from pBR322.[b] Confers resistance in *E. coli* only, but detectable in *S. lividans* by use of nitrocefin
ampC	Promoterless β-lactamase gene from *E. coli*.[c] Detected in *S. lividans* by use of nitrocefin
aph	Ribostamycin resistance gene from *S. ribosidificus* SF733[d]
aphI	Aminoglycoside phosphotransferase gene from *S. fradiae* ATCC 10745.[e,f] Confers resistance to neomycin but not kanamycin. Not expressed in *E. coli* in vectors listed
*aphI**	High-resistance mutation of *aphI*[g]
attP	The attachment (integration) site of pCAO170, φC31, or TG1
βgal	Promoterless β-galactosidase gene from *S. lividans* 66[h]
c region	Region containing the φC31 repressor gene
cat	Chloramphenicol acetyltransferase gene from pACYC184.[i] Confers resistance in *E. coli* and in sensitive *Streptomyces* strains [*S. coelicolor* A3(2) and *S. lividans* 66 are naturally resistant but sensitive variants are available[j]]
galK	Promoterless galactokinase gene from *E. coli*[k]
hph	Hygromycin B phosphotransferase gene from *E. coli* plasmid. Confers resistance to *E. coli* and *Streptomyces*.[l,m] The *Bam*HI site is in the regulatory region of the gene
hyg	Hygromycin phosphotransferase gene from *S. hygroscopicus* NRRL 2387.[n] The *Eco*RI site is probably in the regulatory region
km	Kanamycin phosphotransferase gene from pACYC177 (derived from Tn903[i,o]). Confers resistance in *E. coli* and only low resistance in *Streptomyces*
Km^R	Kanamycin resistance (ribosomal) gene from *S. kanamyceticus* ISP5500.[p] Confers resistance to amikacin gentamicin, kanamycin A, sisomicin, and tobramycin
Km^{R}II	Kanamycin resistance gene from *S. kanamyceticus* M1164. Confers resistance to kanamycin A and B, dibekacin, gentamicin, but not neomycin and ribostamycin[d]
lacα	Modified *lacZ* gene, producing α-complementing fragment of β-galactosidase; part of pUC19.[q] Chromogenic only in special *E. coli* strains
lacZ	The *lacZ* gene of *E. coli*, encoding β-galactosidase. In KC604 (Table I, Fig. 5) the 5′ end of the sequence is actually the promoterless *Bam*HI–*Eco*RI fragment from pMC903[r] containing a *trp–lacZ* fusion. (The 3′ end of the sequence also contains part of the *lacY* gene)
Ltz	Lethal zygosis phenotype: ability of a *Streptomyces* plasmid to inhibit the development of neighboring plasmid-free mycelium.[s] "Pock" formation is an expression of the Ltz$^+$ function. Ltz$^+$ plasmids are self-transmissible and usually also mobilize the chromosome

(continued)

TABLE II (*continued*)

Marker	Remarks
mel	Melanin (tyrosinase) gene from *S. antibioticus* IMRU3720[t,u]
neo	Aminoglycoside phosphotransferase gene from Tn*5*. Confers resistance to neomycin and kanamycin in *E. coli* and in *Streptomyces*. The *Bgl*II site separates the structural gene from its promoter.[v,w]
Sf^{R}	Sulfomycin resistance gene from *S. viridochromogenes* ssp. *sulfomycini* ATCC29776.[a] Both Sf^{R} and *tsr* confer resistance to sulfomycin and thiostrepton. Most streptomycetes are sensitive to these antibiotics but gram-negative organisms are not affected. The *tsr* and Sf^{R} fragments cross-hybridize[a]
Sth^{R}	Streptothricin acetyltransferase gene from *S. lavendulae* 1080[x]
T, fd ter, terminator	Transcriptional terminator from phage fd[y]
tetB	Tetracycline resistance gene from *S. rimosus* ATCC10970. Reduces the uptake of tetracycline in *S. griseus*[z]
toop, tR1	Transcriptional terminators from phage λ[aa]
tsr	Thiostrepton resistance gene from *S. azureus* ATCC 14921.[d,bb] The *Nde*I site is not in the gene. (See also Sf^{R})
vph	Viomycin phosphotransferase gene from *S. vinaceus* NCIB8852.[d,bb] The gene lacks its own promoter in all the constructions listed; viomycin resistance is expressed by readthrough both in streptomycetes and in *E. coli* in all *vph*-containing bifunctional plasmids in Table I and Fig. 3, as well as in the ϕC31 derivatives in Table I and Fig. 5

[a] N. Nakanishi, T. Oshida, S. Yano, K. Takeda, T. Yamaguchi, and Y. Ito, *Plasmid* **15,** 217 (1986); [b]F. Bolivar, R. L. Rodriguez, P. J. Greene, M. C. Betlach, H. L. Heynecker, H. W. Boyer, J. H. Crosa, and S. Falcow, *Gene* **2,** 95 (1977); [c]B. Jaurin, T. Grundström, and S. Normark, *EMBO J.* **1,** 875 (1982); [d]T. Murakami, C. Nojiri, H. Toyama, E. Hayashi, K. Katumata, H. Anzai, Y. Matsuhashi, Y. Yamada, and K. Nagaoka, *J. Antibiot.* **36,** 1305 (1983); [e]C. J. Thompson, T. Kieser, J. M. Ward, and D. A. Hopwood, *Gene* **20,** 51 (1982); [f]C. J. Thompson and G. S. Gray, *Proc. Natl. Acad. Sci. U.S.A.* **80,** 5190 (1983); [g]J. L. Larson and C. L. Hershberger, *J. Bacteriol.* **157,** 314 (1984); [h]D. P. Taylor, T. Eckhardt, and L. R. Fare, *Ann. N.Y. Acad. Sci.* **413,** 47 (1983); [i]A. C. Y. Chang and S. N. Cohen, *J. Bacteriol.* **134,** 1141 (1978); N. K. Alton and D. Vapnek, *Nature (London)* **282,** 864 (1979); [j]J. L. Schottel, M. J. Bibb, and S. N. Cohen, *J. Bacteriol.* **146,** 360 (1981); [k]M. E. Brawner, J. I. Auerbach, J. A. Fornwald, M. Rosenberg, and D. P. Taylor, *Gene* **40,** 191 (1985); [l]S. Kuhstoss and N. R. Rao, *Gene* **26,** 295 (1984); [m]K. R. Kaster, S. G. Burgett, R. N. Rao, and T. D. Ingolia, *Nucleic Acids Res.* **11,** 6895 (1983); [n]F. Malpartida, M. Zalacain, A. Jimenez, and J. Davies, *Biochem. Biophys. Res. Commun.* **117,** 6 (1983); [o]N. D. F. Grindley and C. M. Joyce, *Proc. Natl. Acad. Sci. U.S.A.* **77,** 7176 (1980); [p]M. M. Nakano, H. Mashiko, and H. Ogawara, *J. Bacteriol.* **157,** 79 (1984); [q]C. Yanisch-Perron, J. Vieira, and J. Messing, *Gene* **33,** 103 (1985); [r]M. J. Casadaban, J. Chou, and S. N. Cohen, *J. Bacteriol.* **143,** 971 (1980); [s]M. J. Bibb, R. F. Freeman, and D. A. Hopwood, *Mol. Gen. Genet.* **154,** 155 (1977); [t]E. Katz, C. J. Thompson, and D. A. Hopwood, *J. Gen. Microbiol.* **129,** 2703 (1983); [u]V. Bernan, D. Filpula, W. Herber, M. Bibb, and E. Katz, *Gene* **37,** 101 (1985); [v]E. A. Auerswald, G. Ludwig, and H. Schaller, *Cold Spring*

References to TABLE II (*continued*)

Harbor Symp. Quant. Biol. **45,** 107 (1981); [w]E. Beck, G. Ludwig, E. A. Auerswald, B. Reiss, and H. Schaller, *Gene* **19,** 327 (1982); [x]T. Kobayashi, H. Shimotsu, S. Horinouchi, T. Uozumi, and T. Beppu, *J. Antibiot.* **37,** 368 (1984); [y]R. Gentz, A. Langner, A. C. Y. Chang, S. N. Cohen, and H. Bujard, *Proc. Natl. Acad. Sci. U.S.A.* **78,** 4936 (1981); [z]T. Ohnuki, T. Katoh, I. Tadayuki, and S. Aiba, *J. Bacteriol.* **161,** 1010 (1985); [aa]M. Rosenberg, B. de Crombrugghe, and R. Musso, *Proc. Natl. Acad. Sci. U.S.A.* **73,** 717 (1978); M. Rosenberg, D. Court, H. Shimatake, C. Brady, and D. L. Wulff, *Nature (London)* **272,** 414 (1978); [bb]M. J. Bibb, M. J. Bibb, J. M. Ward, and S. N. Cohen, *Mol. Gen. Genet.* **199,** 26 (1985).

DNA inserts, even if they are greatly outnumbered by plaques without inserts, because only they can generate lysogens.

Figure 5 and Table I also describe the attP$^-$ c$^+$ vector KC684.[58] Insert-directed integration of KC684 is expected to give rise to *in situ* fusions of *lacZ* to chromosomal transcription units in many cases. This should provide rapid access to studies on transcriptional regulation of the cloned DNA at single copy number.

When a cloned fragment is internal to a transcription unit, insert-directed integration of an *attP*-deleted vector leads to the disruption of the transcription unit and therefore often to a detectable mutant phenotype. The detection of mutants among the insert-directed transductants of a wild-type indicator strain, termed mutational cloning, has been used to clone genes involved in antibiotic biosynthesis,[59] and to analyze subclones of DNA cloned by other routes.[60,61] It has the potential benefit of bypassing the need for preexisting mutants. Shotgun mutational cloning is mainly applicable to situations where relatively large transcription units (of at least 2 kb) are expected, because insert-directed integration is very inefficient with inserts of less than 1 kb. The cloning of fragments of about 1–2.5 kb maximizes the probability that mutant transductants will be obtained. It has been calculated[62] that such libraries should usually generate nonproducers of an antibiotic at a frequency of about 1 per 1000 clones screened (i.e., comparable with the frequency at which mutants might be generated by efficient conventional mutagenesis). Mutants resulting from this procedure are not only potentially valuable in their own right for biochemical genetics; in addition, they release phages carrying the rele-

[58] A. A. King and K. F. Chater, *J. Gen. Microbiol.* **132,** 1739 (1986).

[59] K. F. Chater and C. J. Bruton, *Gene* **26,** 67 (1983).

[60] E. T. Seno, C. J. Bruton, and K. F. Chater, *Mol. Gen. Genet.* **193,** 119 (1984).

[61] K. F. Chater and C. J. Bruton, *EMBO J.* **4,** 1893 (1985).

[62] K. F. Chater, A. A. King, M. R. Rodicio, C. J. Bruton, S. H. Fisher, J. M. Piret, C. P. Smith, and S. G. Foster, *in* "Microbiology—1985" (L. Leive, ed.), p. 421. Am. Soc. Microbiol., Washington, D.C., 1985.

vant cloned DNA sequence which, with no further analysis, should directly provide a specific probe for the mRNA of interest.

The original description of mutational cloning with ϕC31[59] assumed that there was negligible readthrough transcription from vector promoters into the cloned DNA. In fact, there is some readthrough in several cloning sites in vectors such as KC515. In some cases, this results in cloned fragments being mutagenic in one orientation only.[63] To obviate this problem, some vectors have been constructed in which the fd terminator is located next to the restriction site used for cloning. In one on these, PM8, the cloning site is a polylinker and the phage carries the *hyg* gene for vector selection (Fig. 5.)[63]

Another Phage-Based Cloning System

Phage TGl[64] resembles ϕC31, and has been developed along very similar lines (but with pACYC177 being used in place of pBR322 to provide the main region for cloning). The host range of TG1 includes at least one species, *S. cattleya,* which is not easily infected by ϕC31. Both $attP^+$ and presumptive *attP*-deleted derivatives are available, some of which can also replicate in *E. coli.*[64] A particularly well-developed derivative, TG78, is given in Fig. 5 and Table I.

Beyond Shotgun Cloning

We have tried in this chapter to provide a detailed account of shotgun cloning in *Streptomyces*. Doubtless, vectors for this purpose will continue to be developed [for example, high- and low-copy-number plasmids with efficient selection for inserted DNA (positive selection vectors) are currently being tested[65]] even though primary shotgun cloning of *Streptomyces* genes no longer presents major problems.

The initial cloning of a gene is usually only the first stage in manipulations that will eventually lead to an understanding at the molecular level of its organization and functioning, and possibly to its modification (e.g., for overexpression). The identification of transcription initiation signals by the use of vectors with promoterless indicator genes (''promoter–probes'') has already been very successful. There are high-copy-number plasmids with the indicator genes *neo* from Tn*5* or *galK* from *E. coli* (e.g., pIJ486 and pSK02; Figs. 2 and 3, Table I) and low-copy-number vectors

[63] F. Malpartida, personal communication.

[64] F. Foor and N. Morin, U.S. Patent 4,460,689 (1984); F. Foor, personal communication.

[65] T. Kieser, unpublished results.

based on SCP2* or SLP1.2 with the indicator genes *galK* (pSKO3), *cat* from Tn*9* (pSLP114, pSLP124[66]), β-galactosidase (pBC6), and a pigment gene (pARC1). In addition, *attP*$^+$ ϕC31 promoter–probe derivatives containing *lacZ* (KC680[58]) and *neo* (KC310[62]) have been described. Insert-directed integration leading to *in situ* fusion (as mentioned earlier for ϕC31 KC684) is also a promising approach to analyzing transcriptional regulation. Translational fusions with a truncated *neo* gene (from Tn*5*) can be made efficiently using vectors now being developed; pIJ688 is an example.[67] The functional analysis of cloned DNA by mutagenic procedures such as mutational cloning[59] or deletion selection with ϕC31 vectors[68] is proving helpful in dissecting complex gene clusters. These procedures have so far used *attP*-deleted ϕC31 vectors. A recently described temperature-sensitive mutant of pIJ702 (pMT660; Table I) should provide further opportunities for investigating the applications of insert-directed vector integration.

The available data suggest that streptomycetes can readily transcribe and translate from heterologous expresion signals and may be more versatile in this respect than *E. coli*.[66,67] Special expression systems have not yet received extensive attention, but systems based on highly expressed, regulatable promoters such as the *gyl* operon promoter[69] have some potential: the *S. coelicolor* agarase gene has both attractive regulatory properties and the feature that its product is a secreted enzyme[70]; vectors are being constructed to exploit these aspects.[71] A quite different route to increased gene expression has been suggested by work on amplifiable chromosomal DNA sequences, which are of widespread occurrence in *Streptomyces*.[19–21] In a few cases, it is possible to select for amplification of particular DNA segments and genes present in (or introduced into) these segments may then be highly expressed.[72–74]

Recently, two systems other than temperate phages with the natural capability of stable integration into recipient chromosomes have been described. One of these, pCA0170 (Fig. 3 and Table I), is based on SLP1

[66] M. J. Bibb and S. N. Cohen, *Mol. Gen. Genet.* **187,** 265 (1982).

[67] T. Kieser, M. T. Moss, J. W. Dale, and D. A. Hopwood, *J. Bacteriol.* **168,** 72 (1986).

[68] S. H. Fisher, C. J. Bruton, and K. F. Chater, *Mol. Gen. Genet.* **206,** 35 (1987).

[69] C. P. Smith, Ph.D. Thesis, Univ. of East Anglia, Norwich, England, 1986.

[70] K. Kendall and J. Cullum, *Gene* **29,** 315 (1984).

[71] M. J. Bibb, G. H. Jones, R. Joseph, M. J. Buttner, and J. M. Ward, *J. Gen. Microbiol.* **133,** in press (1987).

[72] V. N. Danilenko, L. I. Stavodubstev, and S. M. Navashin, *Biotechnology* (*Russ.*) **2,** 62 (1985).

[73] K.-P. Koller, *in* "Biological, Biochemical and Biomedical Aspects of Actinomycetes" (G. Szabó, S. Biró, and M. Goodfellow, eds.), p. 177. Akadémiai Kiadó, Budapest, 1986.

[74] J. Altenbuchner and J. Cullum, *Bio/Technology,* in press (1987).

which, in its complete form, carries an *att* site which integrates into a preferred chromosomal site in an $SLP1^-$ *S. lividans* recipient.[12] The other system is based on a DNA species of *S. coelicolor* A3(2) (the 2.6-kb minicircle) with some transposon-like features, which also displays some preference for a particular integration site in *S. lividans*.[15]

With the range of vectors and approaches now becoming available we can confidently anticipate a burgeoning of knowledge of *Streptomyces* biology at the molecular level in the next few years. It will be surprising if this does not lead also to practical applications.

Acknowledgments

We are grateful to all those colleagues in other laboratories (and our own) who provided unpublished information on their vectors and tirelessly cooperated on checking the accuracy of our use of it. We thank Bob Findlay of the John Innes Institute for the development of computer software, without which the restriction maps could not have been produced, and Anne Williams for her skillful production of the typescript.

[10] Cosmid Shuttle Vectors for Cloning and Analysis of *Streptomyces* DNA

By R. NAGARAJA RAO, M. A. RICHARDSON, and S. KUHSTOSS

Introduction

The genus *Streptomyces* is well known for its ability to produce a variety of commercially useful therapeutic compounds.[1] (These compounds may be antimicrobial or nonantimicrobial in nature.) In addition, the genus produces a variety of extracellular degradative enzymes (proteases, nucleases, and enzymes degrading lignin[2,3]). The genus is capable of cellular differentiation.[4] Streptomycetes are gram-positive microorganisms whose DNA content is about three times that of *Escherichia coli*[5] (approximately 10^7 bp), and whose DNA is high in mole GC base compo-

[1] J. Berdy. *Adv. Appl. Microbiol.* **18,** 309 (1974).

[2] M. Goodfellow and S. T. Williams, *Annu. Rev. Microbiol.* **37,** 189 (1983).

[3] D. L. Crawford, A. L. Pometto, and R. L. Crawford, *Appl. Environ. Microbiol.* **45,** 898 (1983).

[4] K. F. Chater, "Microbial Development," p. 89. Cold Spring Harbor Lab., Cold Spring Harbor, New York, 1984.

[5] R. Benigni, P. A. Petrov, and A. Carere, *Appl. Microbiol.* **30,** 324 (1975).

sition[6,7] (~70%). Streptomycetes exhibit an unusual extent of genome plasticity,[8-14] whose relevance to the biology and evolution of *Streptomyces* is unclear at the moment. To study these phenomena a variety of plasmid[15-20] and phage[18,20,21] vectors and transformation systems[22-25] have been developed. Some of these vectors can shuttle between *Streptomyces* and *E. coli*.[18-20] These have the advantages of ease of structural analysis and manipulation[18-20] coupled with functional analysis in streptomycetes.

In this chapter, we discuss the use of one class of shuttle vectors that can be packaged in *E. coli* bacteriophage λ particles *in vitro*—shuttle cosmid vectors. These vectors make it possible to clone large inserts (up to a maximum of 40 kb). This allows the investigator to isolate genes for complex biosynthetic pathways in a single DNA clone.[26] Assuming random DNA cleavage and ideal representation of the chromosome in the clone library, a collection of 1200 clones, each with an average insert size of 40 kb, will include any genomic sequence from a 10^7-bp genome with greater than 99% probability.[27] When the insert size is reduced to 20 kb, 2400 clones[27] are needed.

[6] C. Frontali, L. R. Hill, and L. G. Silvestri, *J. Gen. Microbiol.* **38,** 243 (1965).
[7] E. M. Tewfik and S. G. Bradley, *J. Bacteriol.* **94,** 1994 (1967).
[8] M. Hasegawa, G. Hintermann, J.-M. Simonet, R. Crameri, and R. Hutter, *Mol. Gen. Genet.* **200,** 375 (1985).
[9] H. Schrempf, *Mol. Gen. Genet.* **189,** 501 (1983).
[10] S. E. Fishman and C. L. Hershberger, *J. Bacteriol.* **155,** 459 (1983).
[11] S. E. Fishman, P. R. Rosteck, and C. L. Hershberger, *J. Bacteriol.* **161,** 199 (1985).
[12] J. Altenbuchner and J. Cullum, *Mol. Gen. Genet.* **195,** 134 (1984).
[13] R. H. Baltz and J. Stonesifer, *J. Antibiot.* **38,** 1226 (1985).
[14] R. H. Baltz, *in* "The Bacteria" (L. Day and S. Queener, eds.), Vol. 9, pp. 61–94. Academic Press, New York, 1986.
[15] D. A. Hopwood, M. J. Bibb, K. F. Chater, and T. Kieser, this volume [9].
[16] I. S. Hunter, *DNA Cloning* **2,** 19 (1985).
[17] J. T. Fayerman, M. D. Jones, and M. A. Richardson, *Microbiology* **1985,** 414 (1985).
[18] K. F. Chater, D. A. Hopwood, T. Kieser, and C. J. Bruton, *Curr. Top. Microbiol. Immunol.* **96,** 96 (1981).
[19] J. L. Larson and C. L. Hershberger, *J. Bacteriol.* **157,** 314 (1984).
[20] D. A. Hopwood, M. J. Bibb, K. F. Chater, T. Kieser, C. J. Bruton, H. M. Kieser, D. J. Lydiate, C. P. Smith, and J. M. Ward, "Genetic Manipulation of *Streptomyces*: A Laboratory Manual." John Innes Foundation, Norwich, England, 1985.
[21] K. F. Chater, A. A. King, M. R. Rodicio, C. J. Bruton, S. H. Fischer, J. M. Piret, C. P. Smith, and S. G. Foster, *Microbiology* **1985,** 421 (1985).
[22] M. J. Bibb, J. M. Ward, and D. A. Hopwood, *Nature (London)* **274,** 398 (1978).
[23] C. J. Thompson, J. M. Ward, and D. A. Hopwood, *J. Bacteriol.* **151,** 668 (1982).
[24] P. Matsushima and R. Baltz, *J. Bacteriol.* **163,** 180 (1985).
[25] H. Yamamoto, K. H. Maurer, and C. R. Hutchinson, *J. Antibiot.* **39,** 1304 (1986).
[26] R. Stanzak, P. Matsushima, R. H. Baltz, and R. N. Rao, *Biotechnology* **4,** 229 (1986).
[27] L. Clarke and J. Carbon, *Cell* **9,** 91 (1976).

Early work on *in vitro* packaging of recombinant λ DNA,[28] and packaging of cosmids[29,30] has been discussed in this series. In these cosmid systems the vectors had only one *cos* site (cohesive sequence), the DNA molecules being ligated were not dephosphorylated, and there was no size fractionation of the insert DNA. This resulted in the packaging of vector molecules alone, or vector molecules with small inserts or vector molecules with rearranged inserts.[31–33] Subsequently, the size of cosmid vector was reduced,[32,34] and the cloning efficiency was increased by size fractionation of insert DNA and by the dephosphorylation of vector DNA.[34] The need for size fractionation of the insert was eliminated by an elegant scheme of generating two vector arms from two molecules of one *cos*-containing vector and by using dephosphorylated insert DNA.[35] Further refinements were made by introducing two *cos* sites into one vector,[36] thus allowing the generation of both the arms from one molecule of cosmid vector.

All the above-mentioned systems allow DNA to be cloned in *E. coli* but do not allow it to be transferred to *Streptomyces*. For functional analysis to be physiologically meaningful, it must be carried out in a homologous context. This requires the addition of a minimum of two modules to the cosmid vectors: (1) a replication origin or a gene transplacement system, and (2) a selectable marker, both of which are functional in the homologous context. The behavior of the system can be influenced by the specific modules added and their locations relative to the other components of the vector. The following description is mainly restricted to one shuttle cosmid vector, pKC505 (Fig. 1).[37–40] Structures of some other cosmid shuttle vectors, their construction, and characteristics are described in Figs. 2,[41–43] 3, 4,[44,45] and 5 and Table I.[46,47] The principles

[28] L. Enquist and N. Sternberg, this series, Vol. 68, p. 281.

[29] B. Hohn, this series, Vol. 68, p. 299.

[30] J. Collins, this series, Vol. 68, p. 309.

[31] J. Collins and H. J. Bruning, *Gene* **4,** 85 (1978).

[32] B. Hohn and J. Collins, *Gene* **11,** 291 (1980).

[33] F. G. Grosveld, H.-H. M. Dahl, E. deBoehr, and R. A. Flavell, *Gene* **13,** 227 (1981).

[34] E. M. Meyerowitz, G. M. Guild, L. S. Prestidge, and D. S. Hogness, *Gene* **11,** 271 (1980).

[35] D. Ish-Horowicz and J. F. Burke, *Nucleic Acids Res.* **9,** 2989 (1981).

[36] A. Poustka, H.-R. Rackwitz, A.-M. Frischauf, B. Hohn, and H. Lehrach, *Proc. Natl. Acad. Sci. U.S.A.* **81,** 4129 (1984).

[37] J. L. Larson and C. L. Hershberger, *Plasmid* **15,** 199 (1986).

[38] H. Steller and V. Pirotta, *EMBO J.* **3,** 165 (1984).

[39] M. Haenlin, H. Steller, V. Pirotta, and E. Mohier, *Cell* **40,** 827 (1985).

[40] R. N. Rao, N. E. Allen, J. N. Hobbs, Jr., W. E. Alborn, Jr., H. A. Kirst, and J. W. Paschal, *Antimicrob. Agents Chemother.* **24,** 689 (1983).

[41] D. J. Lydiate, F. Malpartida, and D. A. Hopwood, *Gene* **35,** 223 (1985).

[42] R. N. Rao, *Gene* **31,** 247 (1984).

[43] R. N. Rao and S. G. Rogers, *Gene* **7,** 79 (1979).

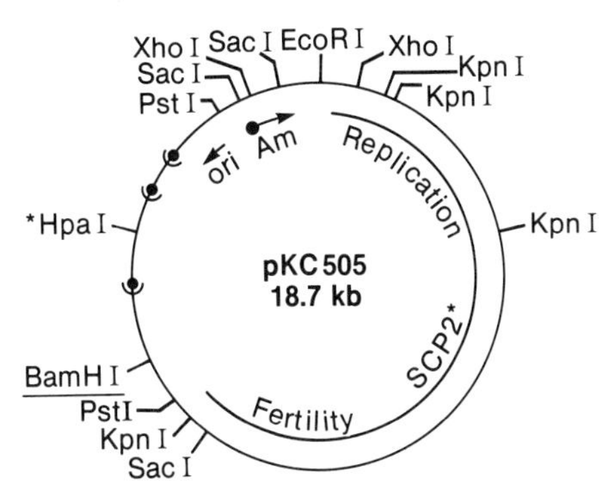

FIG. 1. This plasmid was constructed from SCP2* containing plasmid pHJL202,[19,37] *cos*4,[38,39] and apramycin resistance gene containing pKC222.[40] Clockwise from the unique *Eco*RI site, *Eco*RI–*Bam*HI fragment (12.9 kb) comes from plasmid pHJL202 and includes SCP2* replication and fertility functions. Plasmids that carry smaller SCP2* segments than that present in pKC505 are not stable in *S. lividans* without selective pressure. There is about 800 bp of *S. fradiae aph* gene sequence preceding the unique *Bam*HI (cloning) site. *Bam*HI–*Pst*I fragment (4.7 kb) comes from plasmid *cos*4 and includes three direct repeats of λ *cos* sites. A unique *Hpa*I site separates one *cos* site from the other two *cos* sites: (denotes left *cos* end and • denotes right *cos* end. *Pst*I–*Eco*RI fragment (1.1 kb) comes from the plasmid pKC222 and includes *E. coli* apramycin resistance gene that can be selected both in *E. coli* and in *Streptomyces* transformations. Digestion of the plasmid with *Hpa*I and *Bam*HI generates the two *cos* arms necessary for *in vitro* packaging reaction. Plasmid pKC505 has fertility functions and it was transferred at a frequency of 5×10^{-5} in matings between *S. lividans* strains. This vector and the recombinants derived from the vector can be easily and faithfully shuttled between *S. griseofuscus* and *E. coli*.

described in this chapter can be applied to a variety of other *E. coli* cosmid vectors in converting them to shuttle cosmid vectors for use in *Streptomyces*.

Principle

The substrate for normal *in vivo* packaging of λ is concatemeric DNA, while monomeric circles are not packaged.[48] Packaging requires *Nu*1[49]

[43a] S. Kuhstoss and R. N. Rao, *Gene* **26,** 295 (1983).

[44] M. A. Richardson, J. A. Mabe, N. E. Beerman, W. M. Nakatsukasa, and J. T. Fayerman, *Gene* **20,** 451 (1982).

[45] T. M. Roberts, S. L. Swanberg, A. Poteete, G. Riedel, and K. Backman, *Gene* **12,** 123 (1980).

[46] P. F. R. Little and S. H. Cross, *Proc. Natl. Acad. Sci. U.S.A.* **82,** 3159 (1985).

[47] A. E. Chambers and I. S. Hunter, *Biochem. Soc. Trans.* **12,** 644 (1984).

[48] M. Feiss and A. Becker, *in* "Lambda II" (R. W. Hendrix, J. W. Roberts, F. W. Stahl, and R. A. Weisberg, eds.), p. 305. Cold Spring Harbor Lab., Cold Spring Harbor, New York, 1983.

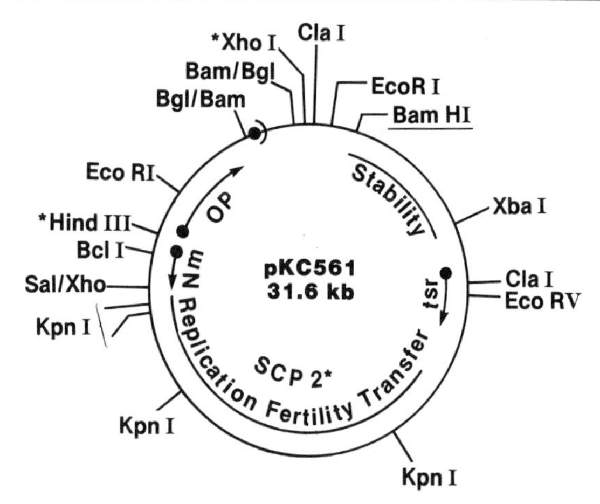

FIG. 2. Plasmid pKC561 was constructed from pIJ922,[41] pHC79,[32] and pKC31 [pKC31 is a λdv *Hin*dIII–*Bam*HI fragment[42] (4.1 kb) joined to a *Hin*dIII–*Bam*HI fragment[43] (1.9 kb) from Tn*5*]; 1.6-kb *Bgl*II fragment from plasmid pHC79 that includes *cos* was cloned into the unique *Bam*HI site present in plasmid pKC31 generating plasmid pKC521. This was digested with *Sal*I and *Xho*I and ligated with *Xho*I-digested pIJ922 plasmid (24.0 kb). pIJ922 includes stability, transfer, fertility, and replication functions of SCP2* and carries tsr^R marker that can be selected in *Streptomyces*. Tn*5* Nm^R gene can be selected both in *E. coli* and in *Streptomyces*.[43a] Small fragment from *Bam*HI and *Hin*dIII digestion and large fragment from *Bam*HI and *Xho*I digestion constitute the two arms necessary for *in vitro* packaging. Plasmid pKC561 has both fertility and transfer (pock-forming ability) functions and it was transferred at a frequency of 1 in matings between *S. lividans* strains. This vector can be easily shuttled between *S. lividans* and *E. coli*.

and *A*[50] proteins and a specific DNA sequence[51] (cohesive end sequences, *cos*). The *cos* sequence can be as short as 223 bp.[52] Packaging proteins bind to one *cos* sequence and packaging proceeds in a linear fashion until another *cos* sequence is encountered.[53] Terminase (protein coded by the *A* gene) introduces a 12-bp staggered break at the *cos* site.[49] For packaging to occur the DNA molecule must have two *cos* sites in the same orientation, and they must be separated by at least 37.8 kb and not more than 50.9 kb.[54,55] The development of *in vitro* λ packaging systems[56,57] played a

[49] J. C. Wang and A. D. Kaiser, *Nature* (*London*), *New Biol.* **241,** 16 (1973).
[50] A. D. Kaiser, M. Syvanen, and T. Masuda, *J. Mol. Biol.* **91,** 175 (1975).
[51] R. Wu and E. Taylor, *J. Mol. Biol.* **57,** 491 (1971).
[52] T. Miwa and K. Matsubara, *Gene* **20,** 267 (1982).
[53] B. Hohn, *J. Mol. Biol.* **98,** 93 (1979).
[54] M. Feiss and D. A. Siegele, *Virology* **92,** 190 (1979).
[55] M. Feiss, R. A. Fischer, M. A. Crayton, and C. Egner, *Virology* **77,** 281 (1977).
[56] B. Hohn and K. Murray, *Proc. Natl. Acad. Sci. U.S.A.* **74,** 3259 (1977).
[57] N. Sternberg, D. Tiemeir, and L. Enquist, *Gene* **1,** 255 (1977).

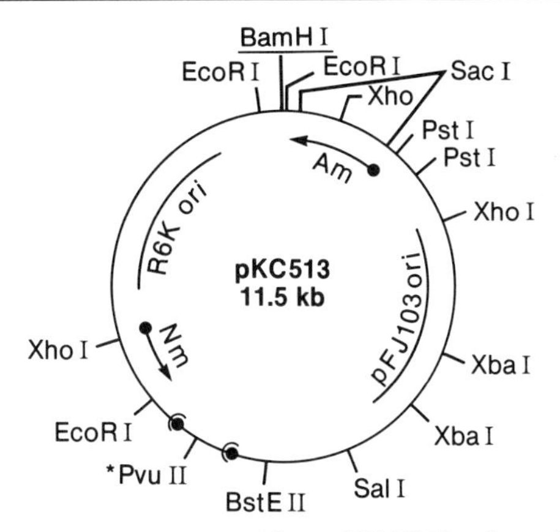

FIG. 3. Plasmid pKC513 was constructed from pKC462a[26] and p*cos*2 EMBL.[36] Clockwise from the unique *Bam*HI (cloning) site, *Bam*HI–*Sal*I fragment (5.4 kb) comes from plasmid pKC462a and includes Am^R gene and *Streptomyces* plasmid replication origin, pFJ103. *Sal*I–*Bam*HI fragment (6.1 kb) comes from plasmid p*cos*2EMBL and includes the two λ *cos* sites separated by a unique *Pvu*II site, Tn*903* Nm^R gene and plasmid R6K origin. Digestion of the plasmid with *Bam*HI and *Pvu*II generates the two *cos* arms necessary for *in vitro* packaging reaction. Plasmids isolated from *S. ambofaciens* transformed with this vector had altered structure.

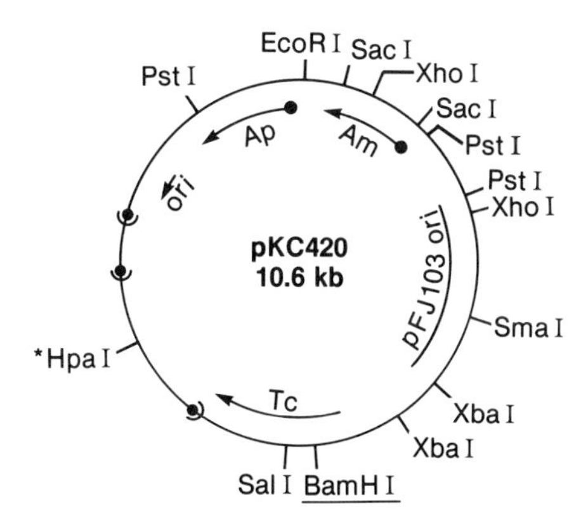

FIG. 4. Plasmid pKC420 was constructed from pKC222,[40] pFJ103,[44] pTR262,[45] and *cos*4.[38,39] Clockwise from the unique *Eco*RI site, *Eco*RI–*Pst*I fragment (1.1 kb) includes the apramycin resistance gene from plasmid pKC222. *Pst*I–*Bam*HI fragment includes 3′ sequence (346 bp) of λ c^+ gene, *Streptomyces* plasmid replication origin (2.9 kb) from pFJ103, 106 bp of internal sequence from λ c^+ gene, and *Hin*dIII–*Bam*HI segment from pBR322. *Bam*HI–*Eco*RI fragment (5.5 kb) comes from plasmid *cos*4 and includes three direct repeats of λ *cos* sites. A unique *Hpa*I site separates one *cos* site from the other two *cos* sites. Digestion of the plasmid with *Hpa*I and *Bam*HI generates the two *cos* arms necessary for *in vitro* packaging reaction. Plasmids isolated from *S. ambofaciens* transformed with this vector can be shuttled into *E. coli*. However, recombinants derived from this and derivative vectors, when transformed into *S. lividans* or *S. fradiae*, were difficult to recover.

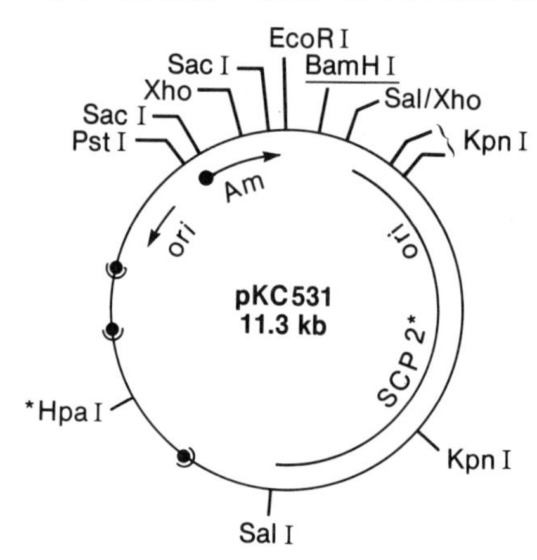

FIG. 5. Plasmid pKC531 was constructed from pKC222,[40] pHJL125,[19] and *cos*4.[38,39] Clockwise from the unique *Eco*RI site, *Eco*RI–*Sal*I fragment (0.7 kb) comes from the plasmid *cos*4 and includes the cloning site, a unique *Bam*HI site. *Xho*I–*Sal*I fragment (5.1 kb) comes from the plasmid pHJL125 and includes SCP2* replication functions. *Sal*I–*Pst*I fragment (4.4 kb) comes from plasmid *cos*4 and includes three direct repeats of λ *cos* sites. A unique *Hpa*I site separates one *cos* site from the other two *cos* sites. *Pst*I–*Eco*RI fragment (1.1 kb) comes from the plasmid pKC222 and includes *E. coli* apramycin resistance gene. Digestion of the plasmid with *Hpa*I and *Bam*HI generates the two *cos* arms necessary for *in vitro* packaging reaction. This vector could be easily shuttled between *S. griseofuscus* and *E. coli*.

very important role in the development of cosmid vectors. It was soon recognized that for packaging, the only λ sequences required are the *cos* sites and all the rest of λ is dispensable and can be replaced by donor sequences.[30,58]

The principles involved in cosmid cloning are illustrated in Fig. 6. The basic aim is to generate ligated recombinant DNA molecules that are suitable for *in vitro* packaging by the λ system. Such molecules (Fig. 6) involve insert DNA (as it was in the donor chromosome) flanked by two vector DNA fragments, each with a *cos* site in the same orientation and the two fragments together being able to generate a complete vector molecule. It is important to restrict all packageable molecules to such a configuration. This ensures that what is packaged contains inserts of donor DNA which have not been rearranged.

There are two ways to generate the flanking vector DNA segments with *cos* sites:

1. *One-cos system*. Both flanking segments are generated from the same cosmid vector but by two separate restriction enzyme cleavages.

[58] J. Collins and B. Hohn, *Proc. Natl. Acad. Sci. U.S.A.* **75,** 4242 (1978).

TABLE I
CHARACTERISTICS OF COSMID VECTORS[a]

Cosmid	Number of *cos* sites	Vector size (kb)	Allowed insert sizes (kb)	*E. coli*		*S. griseofuscus*		Reference
				Origin	Selection	Origin	Selection	
Loric	1	6.3	31.5–44.3	λdv	Nm	None	Nm	46
p*cos*2EMBL	2	6.1	31.7–44.8	R6K	Nm	None	None	36
*cos*4	3	6.0	31.8–44.9	pBR322	Ap, Tc	None	None	38, 39
pPZ74	1	12.5	25.3–38.4	pBR325	Tc, Cm	pIJ303	tsr	47
pIJ610	1	12.0	25.8–38.9	pBR322	Ap, vph	pIJ101	tsr, vph	T. Kieser (personal communication)
pKC575	1	13.3	24.5–37.6	λdv	Nm	SCP2*	Nm	This paper
pKC561	1	31.6	6.2–19.3	λdv	Nm	SCP2*	Nm, tsr	This paper
pKC513	2	11.5	26.3–39.4	R6K	Nm, Am	pFJ103	Am	This paper
pKC420	3	10.6	27.2–40.3	pBR322	Am, Ap	pFJ103	Am	This paper
pKC531	3	11.3	26.5–39.6	pBR322	Am	SCP2*	Am	This paper
pKC462a	3	11.5	26.3–39.4	pBR322	Ap, Am, Nm	pFJ103	Am, Nm	26
pKC505	3	18.7	19.1–32.2	pBR322	Am	SCP2*	Am	This paper

[a] Loric, p*cos*2EMBL, and *cos*4 are not shuttle vectors. They have been included in this table because either they or their design principles were used in the construction of shuttle vectors described in this chapter. We have no experience with the shuttle vectors pPZ74 or pIJ610. All the vectors that included pFJ103 *Streptomyces* replication functions (pKC513, pKC420, and pKC462a) behaved normally in *E. coli* but were difficult to work with in some strains of *Streptomyces*. In a number of instances, there were difficulties in shuttling the cosmid DNAs from *Streptomyces* into *E. coli*. We do not understand the reasons for this behavior. We do not think that this reflects an intrinsic property of the pFJ103 replicon fragment used in the construction of shuttle vector, as we did not experience difficulties with some other shuttle vectors that included the pFJ103 replicon. All the SCP2* replicon-based vectors (pKC575, pKC561, pKC531, and pKC505) could be shuttled easily from *Streptomyces* into *E. coli* and the shuttled plasmids retained their original structure. So far, we have used only pKC505 extensively in generating chromosomal libraries. We have not yet encountered any difficulties in shuttling clones made in the vector pKC505 from *Streptomyces* into *E. coli*. At the present time, this is the vector of choice. Note that the small insert capacity (19–32 kb) of this vector is compensated by its stability and by the presence of fertility functions.

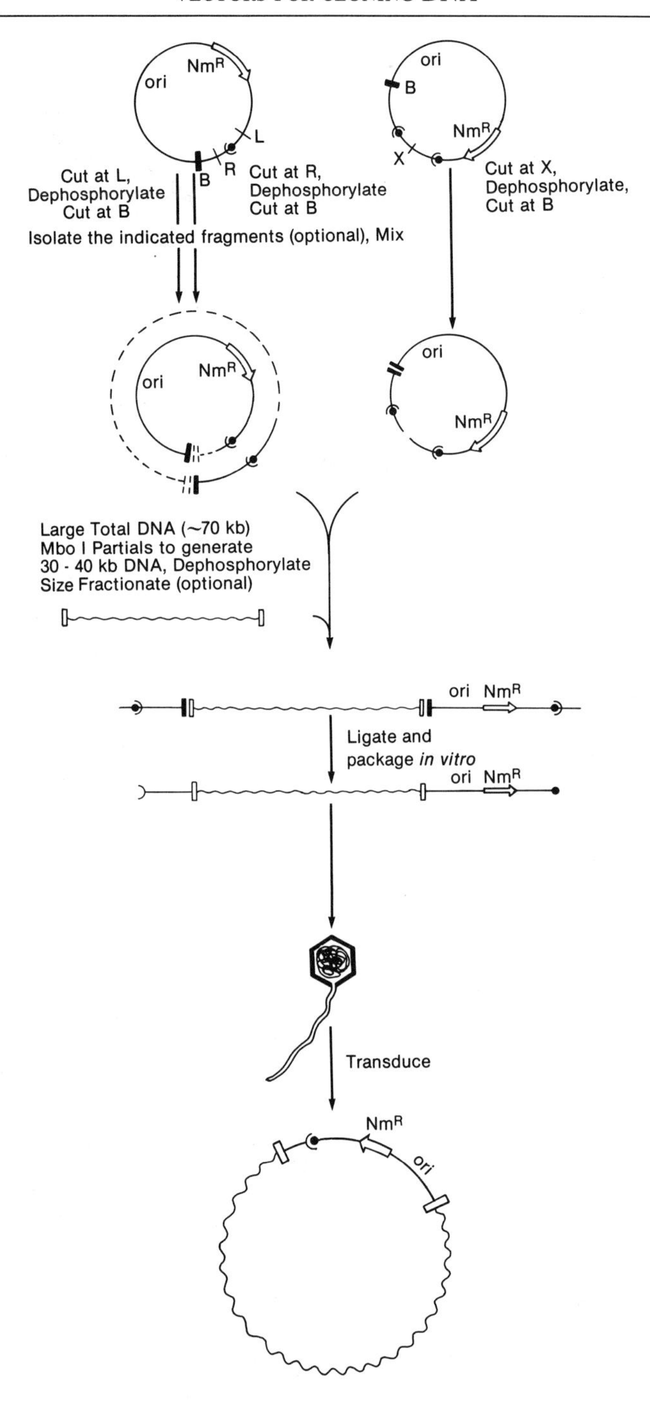
ori
Nm^R
L
R
B
Cut at L,
Dephosphorylate
Cut at B
Cut at R,
Dephosphorylate
Cut at B
ori
B
Nm^R
X
Cut at X,
Dephosphorylate,
Cut at B
Isolate the indicated fragments (optional), Mix
ori
Nm^R
ori
Nm^R
Large Total DNA (~70 kb)
Mbo I Partials to generate
30 - 40 kb DNA, Dephosphorylate
Size Fractionate (optional)
ori Nm^R
Ligate and
package in vitro
ori Nm^R
Transduce
Nm^R
ori

Two vector molecules are needed to generate two usable arms. For this to work there must be a unique site R that lies between cloning site B and the *cos* site (Fig. 6). One batch of vector DNA is cut at a unique site L, to linearize the molecule. These ends are treated in such a way (dephosphorylated, or made blunt) that they do not participate in a subsequent ligation step. These molecules are then cleaved at the unique cloning site B, to generate a large fragment (most of the vector but without the *cos* site) and a small fragment (only the *cos* site). It is possible to use this mixture without further purification. If necessary, one can isolate the small *cos* fragment. Another batch of vector DNA is cut at another unique site R, to linearize the molecule and the ends are modified to prevent participation in a subsequent ligation step and the DNA is cut at the cloning site B. This

---

FIG. 6. Principles of cloning in cosmids. The procedures involved in the generation of ligated λ-packageable DNA substrates vary depending on whether one is working with one *cos* containing or two or more *cos* containing vectors. One-*cos* containing vector: B represents a unique cloning site (preferably a *Bam*HI site), R and L represent two other unique restriction sites that are used in the generation of two arms. It is essential that the λ *cos* site is flanked by two unique restriction sites and that the cloning site is outside of these. In one tube the vector DNA is cut at site L, the ends are dephosphorylated, and then the DNA is cut at site B. The small fragment (continuous line) includes the *cos* site. In another tube the vector DNA is cut at site R, the ends are dephosphorylated, and then the DNA is cut at site B. The large fragment (continuous line) includes the plasmid replication functions, the selectable marker, and the *cos* site. These fragments can be used without further purification; however, purification of these two fragments is desirable and the subsequent reactions with the purified fragments is similar to that seen with the two *cos* vector systems. Two-*cos* containing vector: B represents a unique cloning site (preferably a *Bam*HI site), X represents another unique restriction site that separates the two directly repeated λ *cos* sites. The vector DNA is cut at site X, the ends are dephosphorylated, and then the DNA is cut at site B. This generates two fragments, a small fragment carrying one *cos* site and a large fragment carrying another *cos* site, plasmid replication functions, and the selectable marker. At this point, further manipulations of vector DNA are similar for both one-*cos* and two-*cos* vector systems. Large total DNA isolated from the donor strain (strain from where one wants to isolate the genes of interest) is partially digested with *Mbo*I to generate 30- to 40-kb fragments. It is not necessary to size fractionate these molecules. The ends of these molecules are dephosphorylated to prevent noncontiguous DNA segments from coming together during the subsequent ligation steps. The vector arms are ligated to the insert molecules. Each insert molecule should have a *cos* fragment at both ends. When the two arms together are sufficient to reconstitute a complete functional plasmid, the two *cos* sites are present as direct repeats and the total size of the DNA molecule satisfies the packaging requirements of λ system (37.8–50.9 kb), then these molecules can act as substrates for the λ *in vitro* packaging reaction. λ packaging selects for the appropriate insert sizes. Dephosphorylation of insert molecules and the nonreactive vector end prevents the formation of more complicated structures than that shown in the figure. Packaged cosmids are transduced into *E. coli* strain SF8[59] selecting for the resistance marker present on the plasmid. Strain SF8 is used because it lacks *recBC* enzyme and overproduces *E. coli* ligase.

generates a large fragment (most of the vector and with the *cos* site) and a small fragment. If desired one can isolate the large *cos* fragment. The small *cos* fragment and the large *cos* fragment constitute the two vector arms that participate in the ligation reaction with the insert DNA.

2. *Two-cos system.* Both flanking segments are generated from the same cosmid vector molecule from one set of restriction enzyme cleavages. Vector DNA is cut at the unique site X, which is between the two directly repeated *cos* sites (Fig. 6),[59] the ends are modified to prevent participation in a subsequent ligation step, and the vector is then cut at the unique cloning site B. This generates two arms, each with a *cos* site, one end which is nonligatable and another end which is ligatable.

High-molecular-weight total DNA is partially cut with a restriction enzyme that cuts frequently to generate a set of nearly random fragments, varying in size from 30 to 40 kb. The ends are dephosphorylated to prevent DNA segments from different regions of the chromosome coming together in one recombinant plasmid. It is not necessary to size fractionate the DNA. However, such a size-fractionated DNA should increase the frequency of packageable encounters between the donor DNA and the vector DNA.

Appropriately cut vector and donor DNA's are mixed in different proportions, ligated, and then packaged *in vitro* into λ particles.

The λ particles may be fused directly with *Streptomyces* protoplasts[60] or used to transduce *E. coli*. (*E. coli* strains with different recombinational systems present a choice of the recipient strain to be used.)[60a] It is important to recognize that recovery of recombinants with unaltered donor DNA may be strongly influenced by the particular host chosen for the amplification of the packaged library.[60a–62]

Materials

Chemicals

Spiramycin, neomycin, ampicillin, streptomycin, and spectinomycin were obtained from Sigma Chemical Company. All chemicals except spiramycin were dissolved in water at 100 mg/ml and sterilized by

[59] J. R. Cameron, S. M. Panasenko, I. R. Lehman, and R. W. Davis, *Proc. Natl. Acad. Sci. U.S.A.* **72,** 3416 (1975).

[60] P. Matsushima and R. H. Baltz, personal communication.

[60a] K. F. Wertman, A. R. Wyman, and D. Botstein, *Gene* **49,** 253 (1986).

[61] R. Boissy and C. R. Astell, *Gene* **35,** 179 (1985).

[62] A. R. Wyman, L. B. Wolfe, and D. Botstein, *Proc. Natl. Acad. Sci. U.S.A.* **82,** 2880 (1985).

filtration. Spiramycin was dissolved in absolute ethanol at 100 mg/ml. Thiostrepton was obtained from Squibb and Company, and a 100 mg/ml stock solution was made in dimethyl sulfoxide. Apramycin was obtained from Eli Lilly and Company, and a 100 mg/ml stock solution was made in water and sterilized by filtration

Solutions

TE: 0.01 *M* Tris · HCl, pH 8.0/0.001 *M* EDTA, pH 8.0; Sevag (24 parts chloroform : 1 part isoamyl alcohol)

Enzymes

Restriction endonucleases, polynucleotide kinase, T4 DNA ligase, T4 DNA polymerase, DNase were obtained from a number of different commercial sources—New England BioLabs, Bethesda Research Labs, and Pharmacia. *In vitro* packaging extracts were obtained from International Biotechnologies, Inc., or Stratagene. Bacterial alkaline phosphatase was obtained from International Biotechnologies, Inc. All enzymes were used under conditions specified by the manufacturer

Strains

All the bacterial and the plasmid strains used are listed in Table II[63–67]

Media for Streptomyces

TSB: 30 g trypticase soy broth (BBL)/1 liter distilled water. Autoclave for 30 min at 121°. A complex medium used for general growth

TSA: TSB solidified with 15 g of agar/liter

P medium[68]: Base solution is made by dissolving 103 g of sucrose, 0.25 g of K_2SO_4, and 2.03 g of $MgCl_2 \cdot 6H_2O$ in 700 ml of distilled water. Autoclave for 30 min at 121°. Sterilize separately the following stock solutions: KH_2PO_4 (0.5 g/liter), $CaCl_2 \cdot 2H_2O$ (27.8 g/liter), and 0.25 *M* TES, pH 7.2. At the time of use, add 100 ml each of the KH_2PO_4,

[63] D. Hanahan, *J. Mol. Biol.* **166,** 557 (1983).

[64] C. Yanisch-Perron, J. Vieira, and J. Messing, *Gene* **33,** 103 (1985).

[65] J.-L. Pernodet, J.-M. Simonet, and M. Guerineau, *Mol. Gen. Genet.* **198,** 35 (1984).

[66] D. A. Hopwood, T. Kieser, H. M. Wright, and M. J. Bibb, *J. Gen. Microbiol.* **129,** 2257 (1983).

[67] Stratagene, 3770 Tansy Street, San Diego, CA 92121.

[68] P. Matsushima and R. H. Baltz, *in* "Manual of Industrial Microbiology and Biotechnology" A. L. Demain and N. A. Solomon, eds. Am. Soc. Microbiol., Washington, D.C., pp. 170–183, 1986.

TABLE II
LIST OF STRAINS

Strain	Genotype/phenotype	Reference
Escherichia coil		
SF8	C600 *thr⁻ leu⁻ thi⁻ r⁻ m⁻ recB⁻ recC⁻ lop*-11, *lig⁺*	59
DH1	F⁻ *recA*1 *endA*1 *gyrA*96 *thi*-1 *hsdR*17 *supE*44 *relA*7	63
JM109	*recA*1 *endA*1 *gyrA*96 *thi hsdR*17 *supE*44 *relA*1 (*lac–proAB*)∇/*F' traD*36 *proAB⁺ laci^Q ZM*15∇	64
Streptomyces ambofaciens		
ATCC 15154	Wild type	44
3212	pSAM2	65
Streptomyces griseofuscus		
C581	Wild type	19
Streptomyces lividans		
TK54	*his*-2, *leu*-2, *spc*-1	66
TK64	*pro*-2, *str*-6	66
Plasmids		
pKC31	λdv Nm^R	42, 43
pKC222	Am^R Hm^R Ap^R	40
*cos*4	Ap^R Tc^R $(cos)_3$	38, 39
p*cos*2EMBL	Nm^R $(cos)_2$	36
pFJ103		44
pHJL125	Ap^R	19
pKC462a	Ap^R Nm^R Am^R $(cos)_3$	26
pIJ922	tsr^R	41
pHJL202	Ap^R Nm^R	37
Bluescribe M13⁺	Ap^R	67
pSAM2		65

$CaCl_2$, and TES solutions to the 700-ml base solution. P medium is used in the preparation of *Streptomyces* protoplasts

Modified R2 medium[68]: Base solution is made by dissolving 103 g of sucrose, 0.25 g of K_2SO_4, 10.12 g of $MgCl_2 \cdot 6H_2O$, 10.0 g of glucose, 2.0 g of L-asparagine · $1H_2O$, 0.1 g of casamino acids, and 2 ml of trace elements solution (see below) in 700 ml of distilled water. Adjust the pH of the solution to 7.2 with sodium hydroxide. Autoclave for 25 min at 121°. Autoclave the following solutions separately for 30 min at 121°: $CaCl_2 \cdot 2H_2O$ (22.2 g/liter), KH_2PO_4 (0.5 g/liter), 0.25 *M* TES, pH 7.2. Before pouring plates, add aseptically 100 ml each of the $CaCl_2$, KH_2PO_4, and TES solutions to 700 ml modified R2 agar base solution. Modified R2 medium is used to regenerate transformed *Streptomyces* protoplasts

L-Proline (3 g/liter) can be substituted for asparagine in the above medium.

The above medium is solidified with 2.2% of Difco agar for bottom layer and 0.5% of Difco agar for top layer.

Trace elements solution: 40 mg of $ZnCl_2$, 200 mg of $FeCl_3 \cdot 6H_2O$, 10 mg of $CuCl_2 \cdot 2H_2O$, 10 mg of $MnCl_2 \cdot 4H_2O$, 10 mg of $Na_2B_4O_7 \cdot 10H_2O$, and 10 mg of $(NH_4)_6Mo_{24} \cdot 4H_2O$ in 1 liter of distilled water. Sterilize by filtration

PEG 1000: Make a 50% (w/v) solution in P medium. Filter sterilize

Media for E. coli

TY: Dissolve 10 g of Difco Bacto tryptone, 5 g of sodium chloride, and 5 g of Difco Bacto yeast extract in 1 liter of distilled water and adjust the pH with sodium hydroxide to 7.4 Sterilize the medium by autoclaving

TYMM: Supplement TY broth with 10 m*M* $MgSO_4$ and 0.4% maltose

TY Agar: TY broth solidified with 1.5% Difco Bacto agar

SM: 100 m*M* NaCl/10 m*M* $MgSO_4$/50 m*M* Tris · HCl (pH 7.5) and 0.02% gelatin

TM: 10 m*M* Tris · HCl, pH 7.5, and 10 m*M* $MgSO_4$

Method

Now, we will describe the method used to generate a genomic library of *S. ambofaciens* (producer of spiramycin and resistant to spiramycin) chromosome in the shuttle cosmid vector pKC505. This involves (1) preparation of DNA fragments, (2) preparation of recombinants, (3) analysis of the library, (4) isolation of specific clones, and (5) their analysis. Each of these is discussed in detail below.

1. Preparation of DNA Fragments

This consists of making DNA from donor strain, from vector-containing strain, and their digestion with restriction enzymes. These are discussed in detail below.

Isolation of High-Molecular-Weight Donor DNA

1. Inoculate 50 ml of TSB with 5 ml of fresh overnight culture of *Streptomyces ambofaciens*. TSB may be supplemented with $MgCl_2$

(5 mM) and glycine (0.5%). These additions weaken the cell wall and help the lysozyme action.[69] These additions may also decrease the growth rate. Decrease in growth rate, seen after the addition of glycine, is influenced by the glycine concentration and the *Streptomyces* strain.

2. Grow with vigorous shaking (~250 rpm) at 30–32° for 24 hr or until the culture grows to near stationary phase.

3. Homogenize the culture and pellet the cells by centrifuging in a tabletop centrifuge at top speed (3500 rpm) for 10 min. Discard the supernatant.

4. Resuspend the cells in 20 ml of lysis buffer (15% sucrose, 25 mM Tris · HCl, pH 8.0, 25 mM EDTA) plus lysozyme (5 mg/ml; add lysozyme just before use) and split this into two centrifuge tubes (10 ml/tube). Various additions described below refer to each tube with 10 ml of cells.

5. Incubate for 15 min, 37°. (Avoid cell lysis at this stage. Lysozyme addition is meant to facilitate lysis when SDS is added. Synchronous cell lysis at 70° after the addition of SDS, and not before, should help in the recovery of high-molecular-weight DNA.) Add 0.1 ml of 10 mg/ml proteinase K (prepared fresh in lysis buffer). Mix while adding 1.0 ml of 10% SDS. Immediately incubate at 70° for 15 min. The solution should be clear and very viscous.

6. Cool the solutions on ice. Add 2.5 ml of 5 M potassium acetate. Mix by gently inverting. Incubate on ice for 15 min.

7. Extract with an equal volume of phenol which has been equilibrated with TE. Shake gently for 2–3 min. Centrifuge at 10,000 rpm for 10 min in Sorvall SS34 rotor. The aqueous phase will probably be at the bottom. Occasionally the phases get inverted. If the layers have not separated, centrifuge at higher speed (15,000 rpm) for longer periods (15 min) or at room temperature.

8. Using a pipet with the tip broken off, carefully transfer the bottom layer (aqueous phase) to a fresh tube. Because of the viscosity of the aqueous layer, it has a tendency to slip out of the pipet. The transfer should be done quickly from one tube to another. Save the other layer until you are sure that you have taken the correct layer. If you have taken the correct layer, it is not miscible with Sevag in the next step.

9. Extract (gently) with an equal volume of Sevag (24 parts chloroform to 1 part isoamyl alcohol). Centrifuge at 10,000 rpm for 10 min in Sorvall SS34 rotor. The aqueous phase will be on the top.

10. Transfer the aqueous phase as in step 8 to a fresh tube.

11. Add 2 vol of ethanol at room temperature and mix. The DNA will precipitate out.

[69] M. Okanishi, K. Suzuki, and H. Umezawa, *J. Gen. Microbiol.* **80,** 389 (1974).

12. Use a Pasteur pipet to carefully scoop out the precipitated DNA. Wash the DNA with 70% ethanol (do not use 100% ethanol; do not dry pellet).

13. Dissolve in 5 ml of sterile TE. Add RNase A to a final concentration of 50 μg/ml and RNase T1 to a final concentration of 1 μg/ml. Incubate at 37° for 30 min.

14. Do two phenol and two Sevag extractions as described in steps 7 through 9.

15. To the aqueous phase add 2 vol of ethanol. Mix. Scoop out the DNA as described in step 12. Wash with 70% ethanol and then with 100% ethanol. Dry *in vacuo*.

16. Dissolve in appropriate volume of sterile TE (around 1 ml for a 50-ml culture). It will take several hours (overnight or longer) to dissolve the DNA. The yield of the DNA is usually around 500 μg. Estimate the concentration and purity by measuring optical density at 260 and 280 nm.

17. Check the size of the isolated DNA on a 0.3% agarose gel by conventional electrophoresis[70] or by FIGE.[70a] For cosmid cloning the DNA should be 70 kb or larger.

Some *Streptomyces* strains do not yield satisfactory DNA with the above procedure. It is advisable to vary the glycine concentration (use slightly growth-inhibitory concentration) and the stage of the cells at the time of harvest. We presume that rapid lysis of cells under conditions inhibiting DNase activity (step 5) may alleviate this problem. An alternative procedure is to grind cells in liquid nitrogen in a mortar that is cooled in a dry ice/ethanol mixture. Grinding for 3 to 5 min is usually sufficient to break the cells and these are dissolved in lysis buffer (25 m*M* Tris · HCl, pH 8.0, 25 m*M* EDTA, 1% SDS) that is equilibrated with phenol. Further manipulations are described in steps 7 through 17. High-molecular-weight DNA prepared by either procedure is now ready for partial cleavage.

Partial Cleavage and Dephosphorylation of Donor DNA

1. Dilute *S. ambofaciens* DNA in 1 ml of 100 m*M* NaCl, 10 m*M* Tris · HCl, pH 7.4, 10 m*M* $MgCl_2$, 1 m*M* dithiothreitol to a final concentration of 50 μg/ml.

2. Dispense 200-μl aliquots into three different Eppendorf tubes; 400 μl remains in the original tube.

3. To the original tube add 65 U of *Mbo*I (32.5 U/10 μg/200 μl; the exact amount of enzyme to be added is determined empirically). Quickly

[70] T. Maniatis, E. F. Fritsch, and J. Sambrook, "Molecular Cloning: A Laboratory Manual." Cold Spring Harbor Lab., Cold Spring Harbor, New York, 1982.

[70a] G. F. Carle, M. Frank, and M. V. Olson, *Science* **232,** 65 (1986).

mix, take out 200 μl and add to one of the tubes with 200 μl of DNA. This process is repeated three times serially.

4. Incubate all the tubes at 37° for 15 min. Stop the reaction by heating at 70° for 10 min. Store the DNA at −70° until the analysis of the digested DNA is complete.

5. Analyze 0.5 μg of the DNA from each reaction mixture on a 0.3% agarose gel. Digests yielding smears around 50 kb are considered optimal for the vector pKC505. Maximal insert capacity of pKC505 is 30 kb and maximum number of molecules are present at one-half the molecular weight of the most intense smear seen on the gel.[70] The greatest dilution of *Mbo*I gave the most favorable digestion pattern.

6. Scale up the reaction by digesting 200 μg of chromosomal DNA with 85 U of *Mbo*I in 1000 μl.

7. Incubate at 37° for 3, 6, 10, or 15 min. At each time point, take out 250-μl aliquots and stop the reaction by heating at 70° for 10 min. Store the DNA at −70° until the analysis of the digested DNA is complete.

8. Monitor the digestion as in step 5. DNA digested for 3 min had the suitable distribution of partially digested DNA.

9. Extract the DNA from step 8 once with phenol, once with Sevag, and precipitate with ethanol (one-tenth volume of 3 *M* sodium acetate, 3 vol of ethanol). Collect the precipitate by centrifugation in a microfuge for 15 min.

10. Dissolve the DNA in 100 μl of water. Save ~5 μg of DNA to be used in control reactions. To the rest of the DNA (~45 μg) in 100 μl add 20 μl of 10× phosphatase buffer (0.5 *M* Tris · HCl, pH 8.0, 0.5 *M* NaCl) and 80 μl (24 U/ml) of bacterial alkaline phosphatase.

11. Incubate the DNA at 70° for 1 hr.

12. Add an additional 80 μl of phosphatase and incubate for another hour.

13. Extract the DNA as in step 9 and dissolve in 50 μl TE.

14. To determine if the dephosphorylation is complete use 0.5 μg of DNA in a ligase reaction. Electrophorese ligated and unligated DNA's side by side on a 0.3% agarose gel with suitable size standards. There should be no difference between ligated and unligated DNA's if the dephosphorylation is complete. A more sensitive way to assess the extent of dephosphorylation of the chromosomal DNA is to ligate it with a small plasmid DNA with dephosphorylated *Bam*HI ends and known to be able to ligate to phosphorylated *Mbo*I fragments. It is not easy to monitor the completeness of dephosphorylation of a heterogeneous collection of DNA fragments.

The vector DNA can be isolated either from *E. coli* or from *S. griseofuscus,* a relatively nonrestricting[71] and readily transformable host.[19] It is easy to make large amounts of vector DNA from *E. coli,* and it is the method of choice. However, if the *Streptomyces* strain that is to be used as a host restricts DNA coming from *E. coli,* it is possible to isolate the vector DNA from *Streptomyces* and to fuse packaged λ particles with the *Streptomyces* recipient.[60] This method is not discussed in this chapter as it has not been optimized as yet.

Isolation of Vector DNA from E. coli

1. Grow a 250-ml culture of *E. coli* carrying the vector in TY broth supplemented with 100 μg/ml apramycin overnight at 30° with aeration. (Plasmid pKC505 replicates well at 30° but does not replicate well at 42°. The reason for this is not known. All *E. coli* cultures carrying pKC505 or its derivatives are grown at 30°.)
2. Centrifuge the cells at 8000 rpm for 10 min in a GSA rotor. Decant the supernatant.
3. Resuspend the cells in 7 ml of 25% sucrose, 50 m*M* Tris · HCl, pH 8.0.
4. Add 0.25 ml of 5 mg/ml lysozyme (freshly prepared) in the solution used to resuspend the cells, 0.4 ml of 0.5 *M* EDTA (pH 8), and 0.05 ml of 5 mg/ml RNase A.
5. Incubate for 15 min at 37°.
6. Add 0.75 ml of Triton lytic mix (150 m*M* Tris · HCl, pH 8.0, 3% Triton X-100, 200 m*M* EDTA). Mix. Incubate for 15 min on ice. If the lysis is not good, incubate for about 5 min at 37°.
7. Centrifuge in a Sorvall SS34 rotor at 20,000 rpm for 40 min.
8. Remove the supernatant. Make up a $CsCl_2$ gradient ($\rho = 1.55$). For large tubes add 28.65 g of $CsCl_2$ to 31.2 ml of DNA solution. Mix to dissolve (this will be approximately 39.1 ml) and transfer this to large quick-seal tubes. Fill with 10 mg/ml solution of ethidium bromide (~0.6 ml). Seal the tubes and mix. Centrifuge in VTi 50 rotor at 49,000 rpm for 18 hr. For small tubes add 4.24 g of $CsCl_2$ to 4.14 ml of DNA solution, mix to dissolve (this will be approximately 5.1 ml) and transfer this to small quick-seal tubes. Fill with 10 mg/ml solution of ethidium bromide (~0.1 ml). Seal the tubes and mix. Centrifuge in VTi 80 rotor at 55,000 rpm for 18 hr.
9. Collect the lower band of plasmid DNA and remove the ethidium bromide by extracting four to five times with isoamyl alcohol.

[71] K. L. Cox and R. H. Baltz, *J. Bacteriol.* **159,** 499 (1984).

10. Dialyze against 2 liters of TE, and after 2 hr replace with fresh TE.

11. Do two phenol and two Sevag extractions.

12. Ethanol precipitate (add one-tenth volume of 3 *M* sodium acetate, 3 vol of ethanol. Centrifuge for 10 min at 10,000 rpm in a Sorvall HB4 rotor), wash with 70% and then with 100% ethanol, dry, and then dissolve in about 250 μl of sterile TE.

13. Estimate the concentration and the purity by measuring optical density at 260 and 280 nm.

Isolation of Vector DNA from Streptomyces

1. Grow 250 ml of *Streptomyces* carrying the vector in TSB supplemented with 25 μg/ml of apramycin overnight at 30° with aeration.

2. Pellet the cells by centrifugation at 10,000 rpm for 10 min in a GSA rotor.

3. Wash the cells once in 50 ml of 25 m*M* Tris · HCl, pH 8.0, 25 m*M* EDTA, 15% sucrose solution. Divide the cell suspension into two tubes, centrifuge for 10 min at 10,000 rpm in HB4 rotor.

4. Resuspend each pellet in 4 ml of the same solution containing 2 mg/ml lysozyme.

5. Incubate for 30 min at 37°.

6. Add 3 ml of 0.3 *M* NaOH, 2% SDS solution (preheated to 70°).

7. Incubate for 10 min at 70°.

8. Cool to room temperature, chill on ice for 5 min. Add 4 ml of 3 *M* NaOAc, pH 4.8. Vortex and chill on ice for 15 min.

9. Add 4 ml of acid phenol : Sevag (1 : 1, phenol which is not equilibrated with buffer is called acid phenol), vortex, and centrifuge for 10 min at 10,000 rpm in an HB4 rotor.

10. To the aqueous layer add 1 vol of isopropyl alcohol and precipitate the DNA for 5 min at room temperature.

11. Collect the precipitate by centrifugation for 10 min at 10,000 rpm in an HB4 rotor and dissolve the pellet in 5 ml TE.

12. Extract once with phenol and once with Sevag.

13. Precipitate the DNA at room temperature with 3 vol of ethanol. Collect the precipitate as in steps 10 and 11.

14. Dissolve the pellet in 2.84 ml of TE and add 4.24 g of CsCl, 0.5 ml EDTA (0.25 *M*, pH 8.0), and 0.8 ml of ethidium bromide (5 mg/ml). This makes an approximately 5.1 ml gradient (1.55 g/ml CsCl).

15. Centrifuge in VTi 80 rotor at 65,000 rpm for 5 hr. Centrifugation can be done for 16 hr as well.

16. Collect the bottom band of plasmid DNA. Remove the ethidium bromide by extracting four to five times with an equal volume of isoamyl alcohol.

17. Dialyze against 2 liters of TE, and after 2 hr replace with fresh TE. Ethanol precipitate (add one-tenth volume of 3 *M* sodium acetate, 3 vol of ethanol).

18. Collect the precipitate by centrifugation as in step 11 and dissolve the pellet in 50 μl sterile TE.

19. Estimate the concentration and the purity by measuring optical density at 260 and 280 nm.

Cleavage of Vector DNA and the Preparation of the Two Arms

1. Digest 50 μg of vector DNA (pKC505) with 50 U of *Hpa*I in a 100-μl reaction at 37° for 4 hr. Monitor the reaction by analyzing 0.5 μg of the DNA on a gel. Complete digestion should produce one band migrating at 18.7 kb.

2. After the digestion is complete, extract once with an equal volume of phenol saturated with TE and once with Sevag.

3. Precipitate the aqueous layer by adding 3 vol of ethanol, and centrifuging for 5 min in a microfuge.

4. Redissolve the DNA in 100 μl of water and add 20 μl of 10× buffer (0.5 *M* Tris · HCl, pH 8.0, 0.5 *M* NaCl) and 80 μl of bacterial alkaline phosphatase (24 U/ml).

5. Incubate for 1 hr at 70°.

6. Extract the DNA and precipitate it as in steps 2 and 3. Dissolve the precipitate in 50 μl of sterile 5 m*M* sodium chloride.

7. Use 0.5 μg of the DNA in a ligase reaction to check the extent of dephosphorylation. If the dephosphorylation is not complete (as seen by the appearance of ligation products) then dephosphorylate the DNA some more.

8. Digest the vector DNA with 50 U of *Bam*HI in a 100-μl reaction at 37° for 2 hr. Avoid overdigestion. Monitor the reaction by running 0.5 μg of the DNA. Complete digestion should produce two bands—16.7 and 2 kb.

9. Extract, precipitate, and dissolve the DNA in 50 μl of sterile TE as in steps 2, 3, and 4.

10. Use 0.5 μg of the DNA in a ligase reaction to check the ligatability of the *Bam*HI ends. If the dephosphorylation of *Hpa*I ends is complete and the *Bam*HI ends are ligatable, ligation should produce three bands—33.4, 18.7, and 4 kb.

2. Preparation of Recombinants

Suitably cut vector and donor DNA's are mixed, ligated, and packaged *in vitro* into λ particles. These are transduced into *E. coli* cells. The details of these steps are discussed below.

Ligation of Donor to Vector DNA

In practice, one does not know the exact concentration of donor and vector DNA's. This is because of the involvement of a number of extraction and precipitation steps present in the preparation of donor and vector DNA's and also because the donor DNA is a collection of molecules of heterogeneous lengths. Because of these uncertainties ligations are done at different ratios of donor to vector DNA's, and these are checked by gel analysis and by *in vitro* packaging to estimate the success of ligation step. Where the materials are limiting, these test runs are best made with some other material that comes closest to the experimental material.

1. Ligate 2 μl (approximately 1 μg) of vector, pKC505, DNA that has been *Hpa*I digested, dephosphorylated, and *Bam*HI digested with 4 μl (approximately 1.2 μg) of donor DNA (dephosphorylated *Mbo*I partials) in a 10-μl reaction with 400 U of T4 DNA ligase for 16 hr at 16°.
2. Monitor the ligation by analyzing 5 μl of the ligation reaction mixture on a 0.3% agarose gel. Include an unligated DNA control. If ligations are successful, then one should see a significant shift of the chromosomal smear to a higher molecular weight and the disappearance of the vector bands. Sometimes, the extent of shift seen may not be large, yet it may be satisfactory for generating the library. This can only be ascertained by *in vitro* packaging and by titering the transducing particles.

In Vitro Packaging of Ligated DNA into λ Particles

There are several manufacturers who supply ready-to-use packaging extracts. Basically, these are of two types: one-component system (Packagene) and two-component system (Stratagene). We have used both and find that the two-component Stratagene system gives 5–10× higher packaging efficiency than the single-component Packagene system. It is possible to make packaging extracts in the lab,[72] and a simple one-strain system has been described in this volume.[73] The following is a description of a two-component system protocol that utilizes packaging extracts supplied by Stratagene.

1. Add 2.5 μl of ligation to the tube containing freeze–thaw extract (10 μl).
2. Add 15 μl of the sonic extract to the above tube.
3. Gently mix, centrifuge briefly.
4. Incubate for 2 hr at room temperature (24°).

[72] S. M. Rosenberg, M. M. Stahl, I. Kobayashi, and F. W. Stahl, *Gene* **38,** 165 (1985).
[73] S. M. Rosenberg, this volume [7].

5. Add 0.5 ml of SM and 25 μl of chloroform. Gently mix, clarify by centrifuging for 1 min in a microfuge. Transfer the supernatant to another Eppendorf tube and add 10 μl chloroform. This can be stored for several months at 4°.

Transduction of E. coli

1. Inoculate strain SF8 into 5 ml of TYMM. Incubate overnight at 37° without aeration. Under these conditions, the culture stops growing at about a density of 5×10^7 cells/ml and upon subsequent transfer the culture starts growing without an appreciable delay.
2. Add overnight culture to 50 ml of TYMM. Grow for 3 hr at 37° with aeration.
3. Centrifuge the cells at 6000 rpm in SS34 rotor for 5 min. Resuspend the pellet in 3 ml of TM.
4. Infect 0.2 ml of cells with 10 or 50 μl of *in vitro* packaged phage. Make sure that no chloroform is transferred.
5. Allow the phage to adsorb for 10 min at 37°.
6. Add 1 ml of TY broth and incubate for 2 hr at 30°.
7. Plate aliquots (0.1 ml) on TY plates supplemented with 100 μg/ml apramycin and incubate at 30° overnight.
8. Where 10 μl packaged lysate was used 27 transductants/0.1 ml were obtained and where 50 μl packaged lysate was used 130 transductants/0.1 ml were obtained. This corresponds to about 13,000 transducing particles in the total lysate. This is an evidence of successful cloning experiment. If necessary, one could use a control lysate—where vector DNA ligated in the absence of donor DNA is packaged and used for transduction. Twelve transductants were picked and grown in liquid culture to examine the nature of insert DNA present in them (see next section on analysis of the library for the details).
9. Having shown that the packaged DNA is good for transduction, do a scale-up with the remaining packaged lysate. Add the remaining packaged extract (~500 μl, avoid taking any chloroform that is present at the bottom of the tube) to 1.5 ml of TM in a 50-ml Erlenmeyer flask. Shake at 30° for 15 min to evaporate dissolved chloroform.
10. Prepare SF8 cells as before (steps 1–3) except that the pellet is resuspended in 0.5 ml TM. Add these cells to the phage. Incubate at 37° for 10 min without shaking.
11. Add 10 ml of TY broth. Incubate at 30° for 90 min with shaking.
12. Centrifuge the cells, 6000 rpm for 5 min. Resuspend the cells in 3 ml of TM and plate 0.1 ml/plate on 30 TY plates supplemented with 100 μg/ml of apramycin. Incubate overnight at 30°. Approximately 1000 colonies/plate and a total of ~30,000 colonies were obtained.

It is important to store a master culture, even before evaluating the usefulness of the library. The master culture should have grown the least number of generations. This is to avoid any unnecessary adverse selections which might operate on the cloned inserts. If the initial selections are made on nitrocellulose membranes placed on TY plates with 5% glycerol, the nitrocellulose membranes with the colonies can be stored as masters (for details see Hanahan and Meselson[74]). Alternatively, one can pick a number (determined by the genome size of the donor DNA and the insert size in the recombinant plasmids, 2000–5000 is sufficient for *Streptomyces* libraries) of colonies into microtiter dishes (see Weiss and Milcarek[75]) for storage and analysis. In our experiment we scraped all the transductants into 150 ml of TY broth. Ten milliliters was stored frozen at −70° with 40% glycerol as the primary pool. Twenty milliliters was used to inoculate 1.5 liters of TY broth supplemented with 100 μg/ml of apramycin for a secondary plasmid pool. The remaining cells (~120 ml) were used to make a primary plasmid DNA pool. Note that the cells used in making the primary plasmid DNA pool were grown only once on primary selection plates. One can also store packaged phage or ligated DNA as a master culture.

After storing the master colonies, one must determine how good the library is for the purposes for which it was made. This is described in the next section.

3. Analysis of the Library

An ideal library will consist of all the sequences present in the donor DNA, in the same relative order and in the same relative proportion. This ideal is rarely, if ever, achieved. Any information on the completeness of the library will be of value in designing experiments to isolate specific clones. If a nucleic acid probe to detect the desired clone is available, it can be used to determine if those sequences are represented in the library. In its absence one can use the total number of primary clones obtained[27] as an indication of the completeness of the library.

Initial analysis is done on random transductants obtained from the first transduction experiments. Pick 12 colonies. Grow them in TY broth supplemented with 100 μg/ml of apramycin at 30° and make plasmid DNA as described in the following section on Rapid Isolation of Plasmid DNA from *E. coli,* except that in step 5 incubate at 50° instead of at 70°. Dissolve the DNA from 4 ml of cells in 10 μl TE and use 1 μl for restriction enzyme digests. The plasmid copy number decreases with the increase in

[74] D. Hanahan and M. Meselson, this series, Vol. 100, p. 333.

[75] B. Weiss and C. Milcarek, this series, Vol. 29, p. 180.

plasmid size and plasmid sequences may influence the copy number. Therefore, it is difficult to accurately predict the yields of recombinant DNA's. Digest the isolated recombinant DNA's and the vector pKC505 DNA with *Pst*I. *Pst*I digestion of pKC505 generates two fragments—13.7 and 5 kb: the smaller fragment has the cloning site. *Pst*I digests *S. ambofaciens* DNA to produce a smear of fragments ranging from 30 to 2 kb. *Pst*I digestion of the clones produced a common vector band (13.7 kb) and a number of other fragments (Fig. 7). From these it can be estimated that

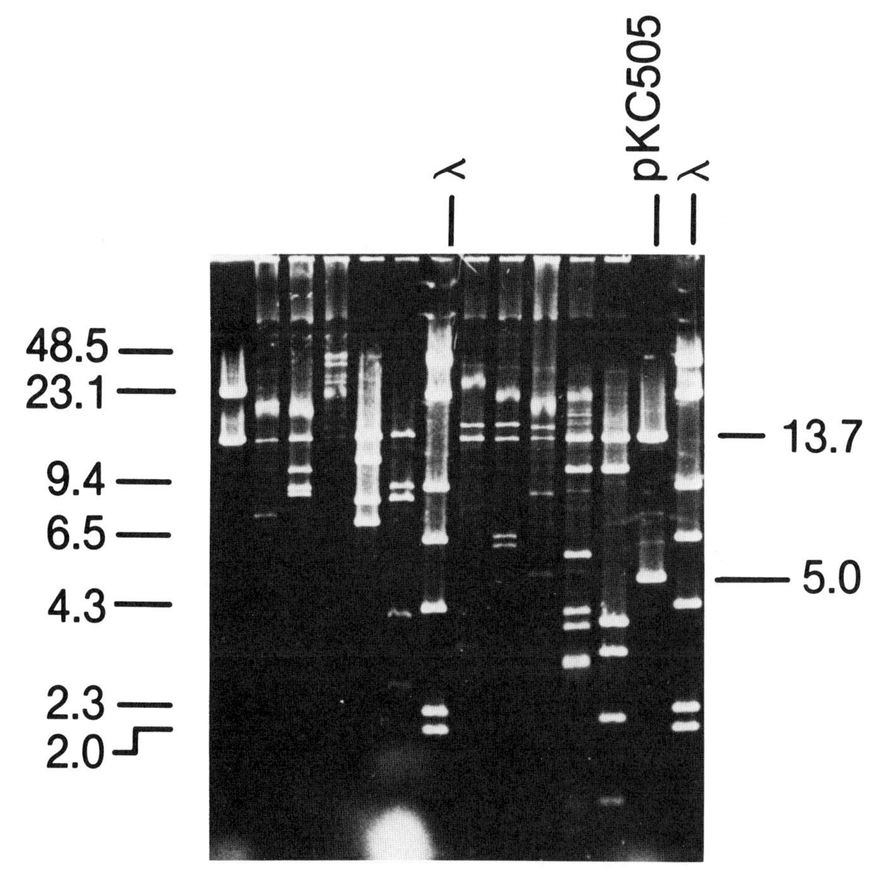

FIG. 7. Restriction enzyme analysis of the plasmid DNA's from 11 randon transductants. The plasmid DNA's were isolated from *E. coli* clones by the rapid procedure, the DNA was digested with *Pst*I, and the fragments were analyzed on 1% agarose gel. As controls, λ DNA cut with *Hin*dIII and plasmid pKC505 cut with *Pst*I were included. In some lanes, *Pst*I digestion did not go to completion. This is particularly true for lane 4. All the experimental lanes have the 13.7-kb vector band and do not have the 5.0-kb vector band that includes the cloning site. All the experimental lanes have a number of different-sized *Pst*I fragments.

the average insert size is 28 kb and that all the 11 clones tested (remaining twelfth clone was not tested) have inserts. DNA from one of the clones was cut poorly with *Pst*I, probably because of some contamination that interfered with *Pst*I digestion.

Rapid Isolation of Plasmid DNA from E. coli

1. Grow a 5 ml overnight culture at 30° in a TY broth supplemented with apramycin at 100 μg/ml.

2. Take 4 ml and pellet the cells in a table top centrifuge. Decant the supernatant. Save the remaining cells at 4° for regrowing the culture, if necessary.

3. Resuspend the cell pellet in 500 μl of 25 m*M* Tris · HCl, pH 8.0, 25 m*M* EDTA. Transfer to a 1.5-ml microfuge tube.

4. Add 250 μl of 0.3 *N* NaOH, 2% SDS. Vortex thoroughly.

5. Incubate at 70° for 10 min.

6. Cool to room temperature. Add 100 μl acid phenol : Sevag (1 : 1). (Do *not* use phenol which has been equilibrated with a buffer.) Vortex well. (Note: Do not add to all tubes and then vortex. Add to each tube and vortex immediately.)

7. Centrifuge for 2 min in a microfuge.

8. The top layer should be fairly clear. If the top layer is milky, or there is no phase separation, add more acid phenol : Sevag, vortex, and spin again. Often this will correct the problem. Remove the top layer and transfer to a fresh 1.5-ml microfuge tube. Add 70 μl of 3 *M* sodium acetate. Fill the tube with 2-propanol and mix well by vortexing. Incubate for 5 min at room temperature.

9. Centrifuge for 5 min in a microfuge. Remove the supernatant.

10. Centrifuge briefly and remove the remaining liquid. It is helpful to orient the tubes in the centrifuge so that all pellets are in the same relative position. Excess liquid can be removed using a tissue paper. Extreme caution must be exercised to avoid touching the area where the pellets are present. Removal of excess fluid is important as salt interferes with the next step.

11. Dissolve the pellet in 500 μl TE. Add 5 μl of 500 m*M* spermine · HCl (5 m*M* final; store spermine stock solution at −20°). Mix, incubate at room temperature for 5 min.

12. Centrifuge for 5 min. Remove the supernatant as in steps 9 and 10.

13. Resuspend the pellet in 300 μl of 0.3 *M* sodium acetate, 0.01 *M* $MgCl_2$. The pellet need not be dissolved.

14. Add 700 μl cold ethanol, vortex, and incubate 5 min at room temperature. Centrifuge for 5 min.

14a. Instead of steps 13 and 14, one can wash the pellet by vortexing in 1 ml of 70% ethanol, 0.3 *M* sodium acetate, 10 m*M* magnesium acetate. Incubate for 5 min at room temperature and centrifuge for 5 min.

15. Remove the supernatant, wash the pellet with 100% ethanol, and dry.

16. Dissolve in TE. For 4 ml of cells use 10 μl of TE. If copy number of the plasmid is similar to pBR322, 1 μl of this should be enough for restriction enzyme digestion and gel electrophoresis. For higher copy number plasmids (the pUC series for example), 0.2 μl may be sufficient. Most restriction enzymes will cut DNA prepared in this fashion. Note that sometimes the final pellet is large and white and sometimes almost invisible. The appearance of this pellet seems to have little (if any) relevance to the yield or quality of DNA. If there are problems with the yield or quality of DNA, prepare a fresh batch of acid phenol : Sevag.

Having shown that all the transductants analyzed have inserts, the next step is to analyze the primary plasmid DNA pool.[76]

Primary Pool Analysis

1. Digest the DNA from the primary pool (library) with *Pst*I and electrophorese this on a gel along with vector alone and donor chromosome alone cut with *Pst*I.

2. Figure 8(a) shows that in both the chromosome and library lane, *Pst*I generates a smear of DNA fragments. This is an indication of the representativeness of the library.

3. Probe the gel with the desired nucleic acid probe (pSAM2) to make sure that the clone library contains the desired sequences. Figure 8(b) shows that in both the chromosome and the library lane that is cut with *Pst*I there is a band at about 15 kb. This is the expected size of the integrated pSAM2 plasmid.[65] This clearly shows that the library includes a clone with an integrated pSAM2 copy.

For functional characterization of the cloned DNA, it must be put in a suitable streptomycete strain. This can be done either with primary plasmid DNA library or with individual clones that have been chosen on the basis of their hybridization with the nucleic acid probes. The procedure involved with either of them is similar, and is discussed below. *Streptomyces griseofuscus* was chosen as a recipient because it is easily transformed[19] and nonrestricting[71] and would allow us to select for spiramycin-resistant colonies.

[76] S. L. Phillips, N. C. Casavant, C. A. Hutchinson III, and M. H. Edgell, *Nucleic Acids Res.* **13,** 2699 (1985).

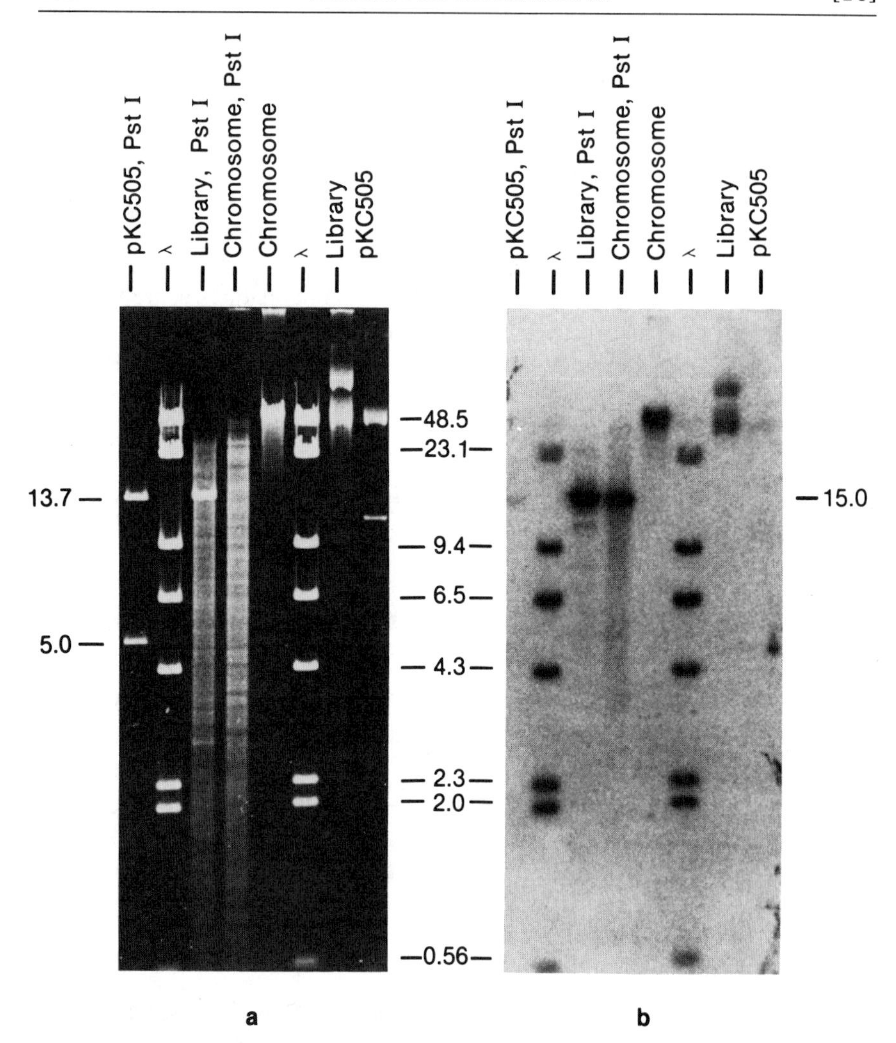

FIG. 8. (a) Restriction enzyme analysis of the primary library DNA. Note the absence of 5.0-kb band and the presence of a smear in *Pst*I-cut library DNA. (b) The same gel probed with nick-translated pSAM2 plasmid. Note the presence of 15-kb band in chromosomal lane and library lane digested with *Pst*I, indicating the presence of a clone with integrated pSAM2.

Transformation of Streptomyces griseofuscus

1. From a fully grown overnight culture of *S. griseofuscus* use 0.5 ml to inoculate 10 ml TSB plus 0.5% glycine (first passage). Incubate at 30° for 24 hr.

2. Homogenize the culture using a tissue grinder. Use 0.5 ml of ho-

mogenate to inoculate 10 ml new TSB with 0.5% glycine culture (second passage). Incubate at 30° for 24 hr.

3. Homogenize the culture. (One can store the cells at this point by adding 5 ml of glycerol, collecting the cells by centrifugation for 10 min at 3500 rpm and freezing the pellets at −70°. To use the frozen cells, thaw them at 37° and resuspend in 10 ml of P medium. Wash the cells three times with P medium before the lysozyme treatment.)

4. Transfer the culture to a 15-ml sterile polystyrene centrifuge tube and centrifuge for 10 min at 3500 rpm to pellet the cells.

5. Wash the cells once with 10 ml of P medium. Pellet the cells as in step 4.

6. Resuspend the cells in 10 ml of P medium with 1 mg/ml lysozyme. Incubate at 30° for 0.5 hr. Monitor protoplast formation by taking small samples and observing under a phase-contrast microscope. When protoplast formation is complete one will see spherical cells.

7. Centrifuge the protoplasts as in step 4. Wash the protoplasts twice in P medium. Resuspend the final pellet in 10 ml of P medium.

7a. Protoplasts can be stored frozen at −70°. Dispense 500 μl of protoplasts per Eppendorf tube. Freeze overnight at −20° and move them next day to −70°. Protoplasts are thawed quickly in the hand and used for transformation.

8. Place 150 μl of protoplasts in a 1.5-ml Eppendorf tube.

9. Add 2 μl of the primary plasmid DNA library. (DNA in a maximum of 10 μl volume can be added to 150 μl of protoplasts.) Gently mix.

10. Immediately add 100 μl of 50% polyethylene glycol (M_r 1000) in P medium.

11. Let sit at room temperature for 2 min.

12. Place 100 μl of transformation mix in 4 ml of modified R2 top agar on dried modified R2 plates.

13. Incubate plates at 30° for 20 hr.

14. Overlay with modified R2 top agar containing enough apramycin to give a final concentration of 25 μg/ml. Incubate the plates at 30°.

15. Two to 3 days after the overlay, transformants should appear.

Streptomyces transformants growing on selective plates must be checked for the presence of appropriate plasmid DNA. This is best done by an analysis of plasmid DNA made from *Streptomyces*. Where this is difficult, the plasmid DNA may be analyzed by DNA blots or it may be analyzed after it has been shuttled into *E. coli*.

Rapid Isolation of Plasmid DNA from S. griseofuscus

1. Grow cells in 25 ml of TSB supplemented with 25 μg/ml of apramycin.

2. Pellet the cells and resuspend them in 5 ml of lysozyme solution (5 mg/ml lysozyme in 0.3 *M* sucrose, 25 m*M* Tris · HCl, pH 8.0, 25 m*M* EDTA).

3. Incubate for 30 min at 37°.

4. Add 2.5 ml of alkaline lysis solution (0.3 *M* sodium hydroxide and 1% SDS). Mix immediately by vigorous vortexing. It is important to mix vigorously. If the solution is viscous, pass it through a plastic disposable syringe without a needle four to six times to reduce the viscosity.

5. Incubate at 50° for 30 min. If the plasmids are 20 kb or smaller, incubation can be carried out at 70° for 20 min. Vortex vigorously. Make sure that there is no leakage and that the tubes do not crack.

6. Add 2 ml of acid phenol : Sevag. Vortex vigorously. Separate the layers by centrifuging for 10 min in a table top centrifuge. The aqueous layer should be clear. If it is not clear, centrifuge it for a longer time at a higher speed and/or add more acid phenol : chloroform and repeat the whole process.

7. Transfer the aqueous layer (~7 ml) into a tube with 0.7 ml of 3 *M* sodium acetate. Add an equal volume of 2-propanol. Mix by vortexing. Incubate for 10 min at room temperature. Centrifuge for 10 min at 10,000 rpm in Sorvall HB4 rotor. Decant, centrifuge for 20 sec to remove liquid adhering to the wall, and remove the last traces of liquid using a tissue paper. Be careful to avoid touching the pellet.

8. Dissolve the pellet in 0.5 ml TE and transfer to an Eppendorf tube with 50 μl of 3 *M* sodium acetate. Extract once with neutral phenol : Sevag, once with Sevag, and then precipitate with an equal volume of 2-propanol. Centrifuge for 2 min and remove all the liquid as in step 7.

9. Redissolve the pellet in 0.5 ml TE and add 5 μl of 0.5 *M* spermine · HCl. Mix, incubate at room temperature for 5 min, and centrifuge for 5 min. Remove all the liquid.

10. Wash the pellet in 1 ml of 70% ethanol, 0.3 *M* sodium acetate, 10 m*M* magnesium acetate. Incubate for 5 min at room temperature and centrifuge for 5 min. Remove all the liquid. Dry the pellet.

11. Redissolve the pellet in 25 μl of TE. Use 1–2 μl/track for restriction enzyme analysis. Note that the copy number of SCP2*-derived plasmids is low.

Very often one finds that there is contaminating chromosomal DNA. This is particularly true for plasmid DNA made from *S. griseofuscus*. It is always helpful to analyze the DNA both before and after digesting with a particular restriction enzyme.

4. *Isolation of Specific Clones*

This can be done either in *E. coli* or in *Streptomyces*. The particular method used will depend on the host organism.

Identification of specific clones in *E. coli* is usually done by colony hybridization[74] using either a short synthetic DNA sequence corresponding to the protein-coding sequence[77] or by using a gene-sized DNA segment presumed to be coming from the areas of interest, either from the same organism[26] or from a related organism. These methods are best applied to the library in *E. coli*. Any screening method that involves the expression of the cloned gene is best done in *Streptomyces*. These methods include the detection of enzymatic activity[78] coded by the desired gene in the clones and the complementation of mutants blocked in the biosynthesis of particular compounds.[79]

Streptomyces ambofaciens is resistant to spiramycin and *S. griseofuscus* is sensitive to spiramycin. We assumed that spiramycin resistance genes from *S. ambofaciens* can be selected in *S. griseofuscus*. To isolate these genes, the *S. ambofaciens* library was transferred to *S. griseofuscus* by selecting for apramycin resistance. All the Am^R transformants ($\sim$10,000) were pooled, grown in TSB supplemented with 25 μg/ml of apramycin, and aliquots were plated on TSA plates containing 25 μg/ml of spiramycin (resistant colonies came at about 10^{-5} frequency). The resistance phenotype was shown to be associated with the recombinant plasmid present in these strains.

5. *Analysis of the Clones*

Mapping of the cloned DNA is a first step in its characterization. This is usually done by digesting the cloned DNA singly or doubly with restriction enzymes that cut the DNA infrequently. When several clones hybridizing to a single DNA probe are isolated, their interrelationship must be verified by DNA blots.[70] Recently, a new blotting method that speeds up the process and eliminates the UV-mediated cross-linking of DNA to nylon membranes has been described.[80]

Before investing time in the analysis of a cosmid clone, it is useful to know that cloned DNA has not undergone any cloning artifact. This can

[77] S. M. Samson, R. Belagaje D. T. Blankenship, J. L. Chapman, D. Perry, P. L. Skatrud, R. M. VanFrank, E. P. Abraham, J. E. Baldwin, S. W. Queener, and T. D. Ingolia, *Nature (London)* **318,** 191 (1985).

[78] G. H. Jones and D. A. Hopwood, *J. Biol. Chem.* **259,** 14151 (1984).

[79] F. Malpartida and D. A. Hopwood, *Nature (London)* **309,** 462 (1984).

[80] K. C. Reed and D. A. Mann, *Nucleic Acids Res.* **13,** 7207 (1985).

be approximated by checking the colinearity of the cloned DNA with that of chromosomal DNA, using one or two restriction enzyme digests. Both cosmid clones and chromosomal DNA should be cut with an appropriate restriction enzyme (one which gives a series of bands in the 0.5- to 10-kb range from the insert DNA). Both should be electrophoresed side by side on a gel and probed with the cosmid DNA. DNA fragments internal to the cloned insert should migrate at identical positions in both the lanes. The two junction fragments from the clone may migrate differently from the corresponding DNA from the chromosomal DNA lane.

Comments

Methods for the handling of *Streptomyces* strains are discussed in greater detail in the John Innes Foundation Manual.[20] Similarly, methods for the handling of *E. coli* strains are discussed in greater detail in the Cold Spring Harbor Laboratory Manual.[70] These two books should be consulted for greater detail of the methods described in this chapter.

When the materials are not in a short supply, it is always helpful to store the materials at all the intermediate stages. When things go wrong at any one stage, one can immediately repeat the experiment starting from the materials stored at the previous stages.

One of the limitations of any cloning experiment is the ability to predict whether the desired clone can be obtained in any one specific experiment. This stems from the complex interactions between cloned DNA, vector DNA, and the genetic background of the recipient used for cloning. When the desired clone is not obtained[60a,62,81] one should vary the method used to prepare the donor DNA, vary the site used in the vector for cloning, vary the vector itself, or vary the recipient (primary and secondary) strain used to amplify the recombinant DNA clones.

When recombinant plasmids are shuttled between diverse genera it is difficult to exclude the existence of variable selective pressures in the two recipients that can affect the cloned DNA. When a clone of a particular phenotype has been isolated in streptomycetes, it is often useful to establish the structural relationship between the cloned DNA and the chromosomal DNA.

Most of the cosmid vectors are based on multicopy plasmid replicons. As the size of the plasmid increases as a consequence of cloning, its copy number may decrease. This could be overcome by replacing plasmid origins with λ replication origins.[46]

[81] L. H. T. Van der Ploeg, D. Valerio, T. De Lange, A. Bernards, P. Borst, and F. G. Grosveld, *Nucleic Acids Res.* **10,** 5905 (1982).

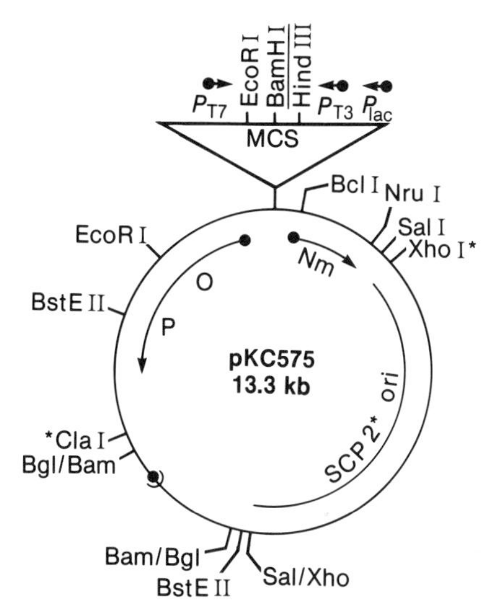

FIG. 9. This plasmid was constructed from pKC521 (7.6 kb in size, for details see legend to Fig. 2), pHJL125,[19] and Bluescribe M13$^+$.[67] Plasmid pKC521 was digested with *Xho*I and ligated to an *Xho*I–*Sal*I fragment (5.1 kb) from pHJL125 that includes SCP2* replication functions. This generated a shuttle cosmid vector pKC529 (12.7 kb). Plasmid pKC529 was digested at the unique *Hin*dIII site that is flanked by the λ replication functions O and P and by the Nm^R gene. Bluescribe M13$^+$ plasmid was digested with *Hae*II (the desired fragment is 0.6 kb in size). These were mixed, the ends were made blunt with T4 DNA polymerase, and then ligated with T4 DNA ligase. The appropriate recombinant plasmid was selected in JM109 as an Nm^R Ap^s blue colony. Plasmid structure was verified using *Bam*HI and *Pst*I restriction enzyme digests. MCS represents multiple cloning sites present in Bluescribe M13$^+$ plasmid, which is same as that present in plasmid pUC18. The unique restriction enzyme sites in plasmid pKC575 include *Bam*HI, *Cla*I, *Xho*I, *Sac*I, *Sph*I, and *Hin*dIII. The large fragment from *Bam*HI and *Cla*I digestion and the large fragment from *Bam*HI and *Xho*I digestion constitute the two arms necessary for *in vitro* packaging. pKC575 can also be used as a shuttle vector by making use of the unique *Hin*dIII, *Bam*HI, and *Xba*I sites present in the MCS fragment. Cloning in these sites will put the cloned fragment under the influence of *lac* promoter or T3 or T7 promoters. Transcripts initiating at the T3 or T7 promoters can be used to generate end-labeled probes.

Mapping large DNA segments is tedious. This has been simplified by the extension[82] (λ *cos* mapping) of Smith and Birnstiel's[83] method of mapping end-labeled DNA molecules. In this method specific end labeling is achieved by hybridizing labeled unique sequences to either left cohesive sequence or to the right cohesive sequence. In principle, cosmids can be packaged *in vivo* into λ heads, allowing the use of λ *cos* mapping proce-

[82] H.-R. Rackwitz, G. Zehetner, A.-M. Frischauf, and H. Lehrach, *Gene* **30,** 195 (1984).
[83] H. O. Smith and M. L. Birnstiel, *Nucleic Acids Res.* **3,** 2387 (1976).

dure. In practice, this is not an efficient process and there is a high frequency of resident prophage being packaged into λ particles. This makes it difficult to use λ *cos* mapping. Use of λ replication origin in the cosmid vector and using a single *cos* strain for *in vivo* packaging overcomes this problem.[46,72] The packaging strain must be heteroimmune with regard to the cloning vector used.

Rapid identification of DNA segments from the ends of the insert will speed up chromosome walking experiments. In the cosmid system used by Stanzak *et al.*[26] the ends are genetically labeled—*Pst*I digestion of the recombinant cosmid will generate one end associated with apramycin resistance gene and another end associated with the 5′-end of ampicillin-resistance gene. In the cosmid system described by Little[84] (loristB), the cloning site is flanked by an SP6 promoter and a T7 promoter. End-specific probes can be generated by transcribing the cloned DNA with SP6 RNA polymerase or with T7 RNA polymerase. We have constructed a cosmid shuttle vector (Fig. 9) where the cloning site is flanked by a T3 promoter and a T7 promoter.

The shuttle vectors discussed in this chapter can shuttle between *Streptomyces* and *E. coli*. The choice of *E. coli* as an alternate host is a historical one, and it is still the most versatile alternative host. *In vitro* packaging systems have been developed for *Bacillus subtilis* phage,[85] and they can also be developed for some of the *Streptomyces* phages.[86] In the future, we can expect the development of a more diverse set of vectors and other alternate hosts to suit different purposes. The further development of *Streptomyces* host–vector systems may ultimately decrease the usefulness of shuttle vectors.

Acknowledgments

The authors are grateful to Drs. H. Lehrach, A. M. Frischauf, T. Kieser, P. F. R. Little, F. W. Stahl, C. L. Hershberger, and D. A. Hopwood for sharing materials, methods, and ideas. The authors thank Drs. R. H. Baltz and D. A. Hopwood for a variety of helpful hints about *Streptomyces*, Drs. R. H. Baltz and J. P. Burnett for support, and Mrs. Cheryl Alexander for typing the manuscript.

[84] S. H. Cross and P. F. R. Little, *Gene* **49,** 9 (1986).

[85] M.-A. Bjornsti, B. E. Reilly, and D. L. Anderson, *Proc. Natl. Acad. Sci. U.S.A.* **78,** 5861 (1981).

[86] T. Morino, H. Takahashi, and H. Sato, *Mol. Gen. Genet.* **198,** 228 (1985).

[11] Host–Vector Systems for Gene Cloning in Cyanobacteria

By C. J. KUHLEMEIER and G. A. VAN ARKEL

Introduction

Cyanobacteria are a diverse group of microorganisms that have two marked things in common: they are gram-negative prokaryotes and they carry out oxygenic, plantlike photosynthesis. For the rest they display considerable variation, especially in structure and development.[1-3] There are unicellular and filamentous forms, fresh and salt water species, and many strains are capable of reducing atmospheric nitrogen and/or chromatic adaptation. It is striking that the GC content of their DNA ranges from 35 to 71%,[4] almost as much as in the whole kingdom of bacteria, emphasizing their genetic heterogeneity. Many of the unique features of cyanobacteria deserve a molecular investigation, but most of the interest in cyanobacterial genetics has arisen from the opportunities for studying photosynthesis and nitrogen metabolism, which have an interest beyond the field of cyanobacteria.

Cyanobacterial and plant photosynthesis are both functionally and structurally similar. In fact, they are so closely related that it has been proposed that plant chloroplasts evolved from ancient cyanobacteria.[5] Not only does the prokaryotic nature of cyanobacteria make them technically more amenable for molecular–genetic analysis, but also the absence of complicated interactions between nucleus and organelles is an added advantage. Heterologous hybridizations with probes from chloroplast genes have been successfully used to isolate the corresponding cyanobacterial genes, e.g., the gene for the 32-kDa herbicide-binding protein[6] and the genes for ribulose-1,5-bisphosphate carboxylase/oxygenase.[7] Mutational analysis of these and other genes is quite feasible in cyanobacteria,

[1] R. Y. Stanier and G. Cohen-Bazire, *Annu. Rev. Microbiol.* **31,** 225 (1977).
[2] N. G. Carr and B. A. Whitton (eds.), "The Biology of Cyanobacteria," Botanical Monogr. Vol. 19. Blackwell, Oxford, 1982.
[3] W. F. Doolittle, *Adv. Microbiol. Physiol.* **20,** 1 (1979).
[4] M. Herdman, M. Janvier, J. B. Waterbury, R. Rippka, R. Y. Stanier, and M. Mandel, *J. Gen. Microbiol.* **111,** 63 (1979).
[5] L. Margulis, "Symbiosis in Cell Evolution." Freeman, San Francisco, 1981.
[6] S. S. Golden and R. Haselkorn, *Science* **229,** 1104 (1985).
[7] B. Y. Reichelt and S. F. Delaney, *DNA* **2,** 121 (1983).

while much harder to perform in higher plants. One aspect of photosynthesis, characteristic to cyanobacteria, is light-harvesting, which is carried out by highly organized protein assemblies, the phycobilisomes. The so-called chromatically adapting species are capable of modulating the composition of the phycobilisomes in response to the wavelength of the incident light.[8,9] Chromatic adaptation has since long attracted the interest as a model system for light regulation of gene expression.

A great number of cyanobacterial strains is capable of reducing atmospheric nitrogen in specialized cells, called heterocysts, which are formed upon starvation for combined nitrogen.[10] The genetics of nitrogen fixation in *Anabaena* PCC 7120 may soon be tackled as gene transfer in this cyanobacterium has recently been demonstrated.[11] Moreover, the differentiation of heterocysts, which involves the deletion of intervening DNA sequences, is an intriguing form of bacterial differentiation.[12] Whereas nitrogen fixation is confined to bacteria, the assimilation of organic nitrogen is a crucial part of plant metabolism and cyanobacteria may serve as an attractive model system in this area as well.[13,14]

Excellent taxonomic work on cyanobacteria has been carried out at the Pasteur Institute[4,15–17] and most reports will now mention the PCC numbers of the strains used. The strains mentioned here without PCC number correspond with the following recommended designations: *Anacystis nidulans* R2 = *Synechococcus* PCC7942; *Agmenellum quadruplicatum* PR-6 = *Synechococcus* PCC7002; *Fremyella diplosiphon* = *Calothrix* PCC7601.

This chapter deals with the developments in the construction of host–vector systems for gene cloning and analysis. If we evaluate the present state of the cyanobacterial systems, it can be said that for *Anacystis nidulans* R2 (unicellular, obligately photoautotrophic, nonnitrogen fixing)

[8] N. Tandeau de Marsac, *Bull. Inst. Pasteur (Paris)* **81,** 201 (1983).

[9] P. B. Conley, P. G. Lemaux, and A. R. Grossman, *Science* **230,** 550 (1985).

[10] W. D. P. Stewart, *Annu. Rev. Microbiol.* **34,** 497 (1980).

[11] C. P. Wolk, A. Vonshak, P. Kehoe, and J. Elhai, *Proc. Natl. Acad. Sci. U.S.A.* **81,** 1561 (1984).

[12] J. W. Golden, S. J. Robinson, and R. Haselkorn, *Nature (London)* **314,** 419 (1985).

[13] C. J. Kuhlemeier, T. Logtenberg, W. Stoorvogel, H. A. A. van Heugten, W. E. Borrias, and G. A. van Arkel, *J. Bacteriol.* **159,** 36 (1984).

[14] C. J. Kuhlemeier, V. J. P. Teeuwsen, M. J. T. Janssen, and G. A. van Arkel, *Gene* **31,** 109 (1984).

[15] R. Rippka, J. Deruelles, J. B. Waterbury, M. Herdman, and R. Y. Stanier, *J. Gen. Microbiol.* **111,** 1 (1979).

[16] M. Herdman, M. Janvier, R. Rippka, and R. Y. Stanier, *J. Gen. Microbiol.* **111,** 73 (1979).

[17] R. Rippka and G. Cohen-Bazire, *Ann. Microbiol.* **134B,** 21 (1983).

a complete system has been developed[18] and the first results have been obtained.[6,13,14,19] In addition, great progress is being made with other organisms, particularly *Synechocystis* PCC6803,[20,21] *Agmenellum quadruplicatum* PR-6[22,23] (both unicellular, facultatively photoheterotrophic, nonnitrogen fixing), *Anabaena* PCC7120[11] (filamentous, obligately photoautotrophic, nitrogen fixing), *Fremyella diplosiphon*[24] (filamentous, facultatively photoheterotrophic, nitrogen fixing, chromatically adapting), and a number of facultatively photoheterotrophic, nitrogen-fixing *Nostoc* strains,[25] of which at least some show chromatic adaptation. Since our experience is almost exclusively with *A. nidulans* R2 we will focus on this organism, treat *Synechocystis* PCC6803 briefly, and give appropriate references to the other systems. After a short discussion of cloning vectors in *A. nidulans* R2, the procedures currently used for DNA isolation, transformation, and mutagenesis in this organism will be described.

Principles

Many reports have been published on gene transfer in cyanobacteria: conjugation, transduction, and transformation.[26] Up to now, however, the only system of practical importance has been the transformation system, originally described by Shestakov and co-workers.[27,28] Wild-type *Anacystis nidulans* R2 was shown to be naturally competent to take up chromosomal DNA isolated from erythromycin- and streptomycin-resistant mutants. Reports concerning transformation of other unicellular strains followed: *Synechococcus* PCC6301[29] and PCC7002,[30] and *Synechocystis*

[18] C. J. Kuhlemeier, A. A. M. Thomas, A. van der Ende, R. W. van Leen, W. E. Borrias, C. A. M. J. J. van den Hondel, and G. A. van Arkel, *Plasmid* **10,** 156 (1983).

[19] V. A. Dzelzkalns, G. C. Owens, and L. Bogorad, *Nucleic Acids Res.* **12,** 8917 (1984).

[20] F. Chauvat, L. de Vries, A. van der Ende, and G. van Arkel, *Photosynth. Prokaryotes, Int. Symp., 5th, 1985* p. 327 (Abstr.).

[21] S. Shestakov, I. Elanskaya, and M. Bibikova, *Photosynth. Prokaryotes, Proc. Int. Symp., 5th, 1985* p. 109 (Abstr.).

[22] J. S. Buzby, R. D. Porter, and S. E. Stevens, *J. Bacteriol.* **154,** 1446 (1983).

[23] J. S. Buzby, R. D. Porter, and S. E. Stevens, *Science* **230,** 805 (1985).

[24] J. Cobley, *Photosynth. Prokaryotes, Proc. Int. Symp., 5th, 1985* p. 105 (Abstr.).

[25] E. Flores and C. P. Wolk, *J. Bacteriol.* **162,** 1339 (1985).

[26] M. Herdman, *in* "The Biology of Cyanobacteria" (N. G. Carr and B. A. Whitton, eds.), p. 263. Blackwell, Oxford, 1982.

[27] S. V. Shestakov and N. T. Khyen, *Mol. Gen. Genet.* **107,** 372 (1970).

[28] G. A. Grigorieva and S. V. Shestakov, *Photosynth. Prokaryotes, Proc. Int. Symp., 2nd, 1976* p. 220.

[29] M. Herdman, *Mol. Gen. Genet.* **120,** 369 (1973).

[30] S. E. Stevens and R. D. Porter, *Proc. Natl. Acad. Sci. U.S.A.* **77,** 6052 (1980).

PCC6308,[31] PCC6714,[32] and PCC6803.[33] For a systematic review of transformation in cyanobacteria the reader is referred to the paper by Porter.[34]

Two strategies have been employed for developing host–vector systems. One uses hybrid plasmids that contain both a cyanobacterial and an *Escherichia coli* origin of replication.[13,14,18,20–24,35–42] The other makes use of the integrational vectors that employ the extremely efficient recombination documented for at least two species, *A. nidulans* R2[35,43–45] and *Synechocystis* PCC6803.[46,47]

Shuttle Vectors

Plasmids, ranging in size from 2.3 to over 100 kb, have been found in many cyanobacteria. They do not seem to code for any of the common plasmid-borne functions, such as antibiotic resistance, and so far they seem to be phenotypically cryptic. *A. nidulans* R2 contains two plasmids, pUH24 and pUH25.[48] The small plasmid, pUH24, has been completely sequenced, and comprises 7845 bp.[49] A restriction map is given in Fig. 1. This plasmid has been used for the construction of most hybrid vectors. It

[31] C. I. Devilly and J. Houghton, *J. Gen. Microbiol.* **98,** 277 (1977).
[32] C. Astier and F. Espardellier, *CR Acad. Sci.* **282,** 795 (1976).
[33] G. Grigorieva and S. Shestakov, *FEMS Microbiol. Lett.* **13,** 367 (1982).
[34] R. D. Porter, *CRC Crit. Rev. Microbiol.* **13,** 111 (1986).
[35] C. J. Kuhlemeier, W. E. Borrias, C. A. M. J. J. van den Hondel, and G. A. van Arkel, *Mol. Gen. Genet.* **184,** 249 (1981).
[36] L. A. Sherman and P. van de Putte, *J. Bacteriol.* **150,** 410 (1982).
[37] D. Friedberg and J. Seijffers, *Gene* **22,** 267 (1983).
[38] S. S. Golden and L. A. Sherman, *J. Bacteriol.* **155,** 966 (1983).
[39] N. Tandeau de Marsac, W. E. Borrias, C. J. Kuhlemeier, A.-M. Castets, G. A. van Arkel, and C. A. M. J. J. van den Hondel, *Gene* **20,** 111 (1982).
[40] S. Gendel, N. Straus, D. Pulleyblank, and J. Williams, *J. Bacteriol.* **156,** 148 (1983).
[41] S. Gendel, N. Straus, D. Pulleyblank, and J. Williams, *FEMS Microbiol. Lett.* **19,** 291 (1983).
[42] K. Shinozaki, N. Tomioka, C. Yamada, and M. Sugiura, *Gene* **19,** 221 (1982).
[43] J. G. K. Williams and A. A. Szalay, *Gene* **24,** 37 (1983).
[44] K. S. Kolowsky, J. G. K. Williams, and A. A. Szalay, *Gene* **27,** 289 (1984).
[45] C. J. Kuhlemeier, E. M. Hardon, G. A. van Arkel, and C. van de Vate, *Plasmid* **14,** 200 (1985).
[46] F. Chauvat, L. de Vries, A. van der Ende, and G. A. van Arkel, *Mol. Gen. Genet.* **204,** 185 (1986).
[47] S. V. Shestakov, I. V. Elanskaya, and M. V. Bibikova, *Dokl.–Akad. Nauk SSR* **282,** 176 (1985). In Russian.
[48] C. A. M. J. J. van den Hondel, S. Verbeek, A. van der Ende, P. J. Weisbeek, W. E. Borrias, and G. A. van Arkel, *Proc. Natl. Acad. Sci. U.S.A.* **77,** 1570 (1980).
[49] P. J. Weisbeek, R. Teertstra, M. van Dijk, G. Bloemheuvel, D. de Boer, J. van der Plas, W. E. Borrias, and G. A. van Arkel, *Photosynth. Prokaryotes, Proc. Int. Symp., 5th, 1985* p. 328 (Abstr.).

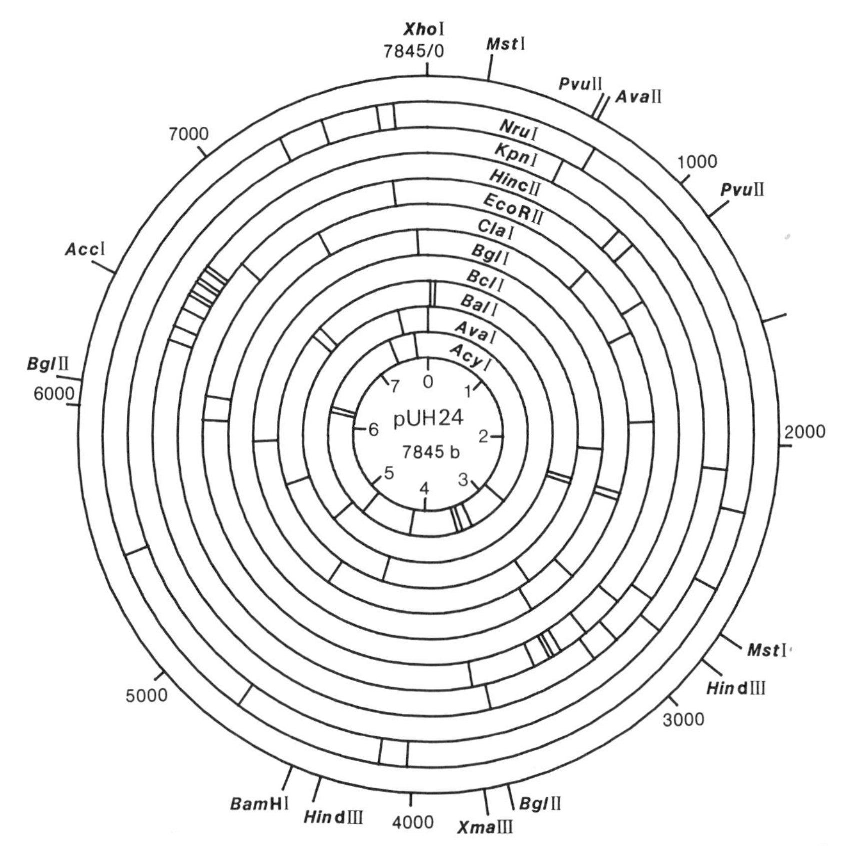

FIG. 1. Restriction map of the endogenous small plasmid pUH24 from *A. nidulans* R2 for a number of restriction endonucleases that recognize specific sequences of five or more nucleotide pairs. The map was derived partly from fragment pattern assays and partly from the complete nucleotide sequence,[49] and is orientated with respect to the single *Xho*I site.

seems to be nonessential to the cell, as it can be cured from the cells without any apparent effect.[18,50] The large plasmid, pUH25, is 50 kb in size and all attempts to cure it have failed. A restriction map of pUH25 has been constructed[51] and it may have two origins of replication.[52]

A transposon-tagged derivative of pUH24 was unable to replicate in *E. coli* and the reverse is also true: plasmids like pBR322, RP4, and RSF1010

[50] F. Chauvat, C. Astier, F. Vedel, and F. Joset-Espardellier, *Mol. Gen. Genet.* **191,** 39 (1983).

[51] D. E. Laudenbach, N. A. Straus, S. Gendel, and J. P. Williams, *Mol. Gen. Genet.* **192,** 402 (1983).

[52] D. E. Laudenbach, N. A. Straus, and J. P. Williams, *Mol. Gen. Genet.* **199,** 300 (1985).

fail to be maintained in *A. nidulans* R2, most likely because they replicate very poorly, if at all, in the heterologous host.[48] The reason for this failure of replication in the heterologous system is not entirely clear. One explanation could be that the RNA polymerases do not recognize foreign promoters, but data on the *in vivo* and *in vitro* specificity of cyanobacterial RNA polymerases are lacking. However, under certain experimental conditions the transformation of *Synechococcus* PCC6301 with pBR322 has been observed recently.[53] An interaction between plasmid and host chromosomal DNA seems to be involved. Further data on that process and on the fate of the plasmids in the transformants have to be awaited.

Many plasmids consisting of a ColE1-like replicon and a cyanobacterial plasmid have been constructed, but they may not all be equally useful for the cloning and analysis of cyanobacterial genes. The main characteristics of a good vector, besides easy manipulation in *E. coli,* are high transformation efficiency and stable maintenance in the cyanobacterium. For *A. nidulans* R2, transformation frequencies vary several orders of magnitude between seemingly similar hybrid plasmids (see Transformation). A complication in the use of pUH24-derived vectors initially was the recombination between the vector and the endogenous pUH24. Use of a pUH24-cured strain, R2-SPc, has alleviated this difficulty[19] and two vectors that have proved their usefulness in practice[13,14,18,39] are shown in Fig. 2. Plasmid pUC303 has two easily selectable antibiotic-resistance markers, transforms *A. nidulans* R2 at high frequency, and is stably maintained in small-plasmid-cured hosts.[18] Plasmid pPUC29 contains the phage λ *cos* site and a gene library constructed in this vector was used successfully for the cloning of nitrate reductase (*nar*) genes by random complementation of *nar* mutants.[14] Interestingly, transformation frequency is not affected by plasmid size, i.e., even 50-kb cosmids can be transformed with the same high efficiency as small plasmids.[13,14]

Recombination

Anacystis nidulans is an obligately photoautotrophic organism, implying that it not only benefits from light but also is continuously exposed to the hazardous effects of solar radiation. Thus it may not be surprising that *A. nidulans* R2 has an extremely efficient system of recombination, as recombination is intimately involved in the repair of radiation damage.[54] Recombinational events, both reciprocal and nonreciprocal in nature,

[53] H. Daniell, G. Sarojini, and B. A. McFadden, *Proc. Natl. Acad. Sci. U.S.A.* **83,** 2546 (1986).

[54] P. C. Hanawalt, P. K. Cooper, A. K. Ganesan, and C. A. Smith, *Annu. Rev. Biochem.* **48,** 783 (1979).

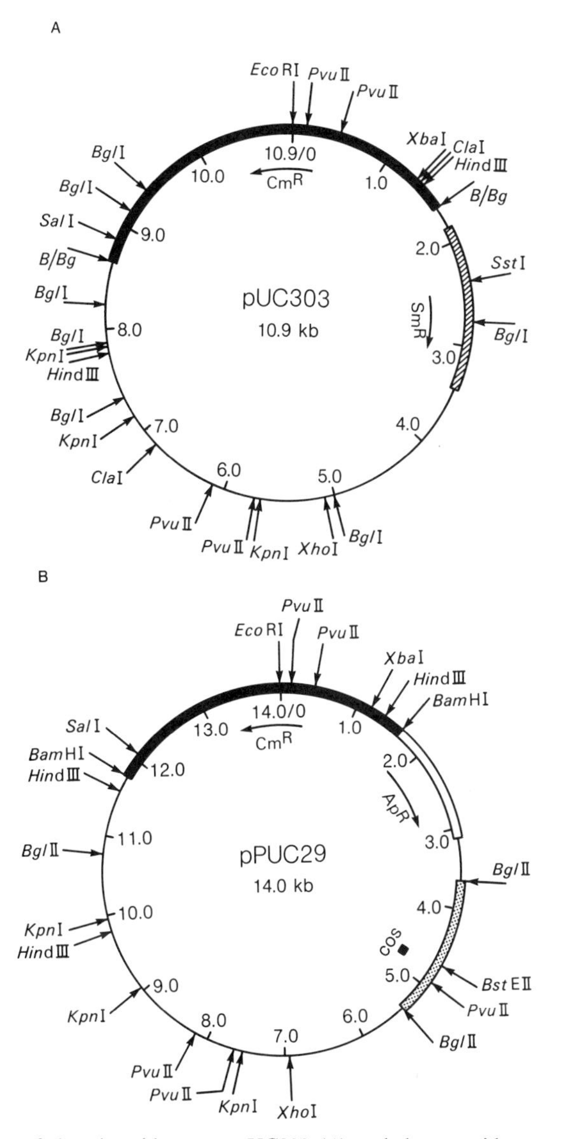

FIG. 2. Maps of the plasmid vector pUC303 (*A*) and the cosmid vector pPUC29 (*B*), oriented with respect to the single *Eco*RI site. Their constructions have been described.[18,39] The various components are shown as follows: parts of pUH24 as thin lines; pACYC184 as filled bars; Sm^R fragment from pRI477S as hatched bar; Ap^R fragment from Tn*901* as open bar; λ *cos* site fragment from pHC79 [B. Hohn and J. Collins, *Gene* **11,** 291 (1980)] as stippled bar. *B*/*Bg* denotes a *Bam*HI/*Bgl*II hybrid site not recognized by either enzyme. pUC303 and pPUC29 have five unique restriction sites, for *Eco*RI, *Xba*I, *Xho*I, and *Sal*I in both plasmids, for *Sst*I in pUC303 and for *Bst*EII in pPUC29.

have been demonstrated in several experiments, although the underlying mechanisms are as yet unclear. When a wild-type DNA fragment cloned in an *E. coli* plasmid was transferred to a mutant strain of R2, wild-type transformants arose at a frequency up to 1%.[13,43] This figure is likely to be a lower limit since uptake of the DNA may not occur in all the cells. It was shown[43] that if a piece of chromosomal DNA containing an insert of foreign DNA was cloned into pBR322 and the resulting plasmid was transferred to R2, three types of recombinants could be obtained depending on the selection applied. In over 90% of the transformants double recombinants were found, which were very stable. The other classes were much less stable.[43,44]

We have extended these observations by merodiploid analysis. Plasmid pUC303 containing a wild-type *met* gene was transferred to a strain of R2 carrying a transposon Tn*901*-inactivated *met* gene and the interactions between the two alleles were studied.[45] The majority of the Sm^R Met^+ transformants retained the parental configuration, but reciprocal exchanges were also found. In Sm^R Met^- transformants the *met*$^+$ allele was lost by nonreciprocal recombination, and both plasmid and chromosome contained the *met*: : Tn*901* allele.

Integrational Vectors

The high recombination efficiency can aptly be applied for practical purposes of gene cloning and analysis. Integrational vectors have been constructed and these ought to be useful, especially for the analysis of mutated genes, which can be introduced at a fixed copy number. In the case of indispensable photosynthetic genes a mutated gene could be introduced, leaving the endogenous gene intact. No data are as yet available on recombinational interactions between a mutated and a wild-type gene when both are located on the chromosome. It should be noted, however, that gene conversion could well take place irrespective of whether the second copy of a gene is chromosomally or plasmid located. Therefore, the same kind of phenomena described above for the wild-type and mutant methionine gene[45] may also occur if integrational vectors are used.

Materials

Strains and Growth Conditions

Anacystis nidulans R2 was originally obtained by K. Floyd (California, 1973) as a fresh water isolate and sent as *Synechococcus* sp. R-2 to S. Shestakov (Moscow State University), in whose laboratory the efficient transformability of the strain was found.[28] It contains two plasmid species, pUH24 of 8 kb, and pUH25 of 50 kb.[48] Cells are grown in BG-11 medium at 28–30° and a light intensity of 4000–5000 lx from white fluorescent tubes. Small, 50-ml liquid cultures are grown in 250-ml Erlenmeyer flasks, closed with a cotton plug. Shaking is not absolutely necessary, but prevents the partial sedimentation of the cells. For large-scale purposes 4-liter cultures are grown in 5-liter Erlenmeyer flasks, slowly bubbled with 5% CO_2 in air, and stirred with a stirring bar. Depending upon the demands, the antibiotic concentrations in liquid medium are as follows: ampicillin, 1 μg/ml, chloramphenicol, 7.5 μg/ml, and streptomycin, 10 μg/ml. For short-term storage (up to 6 weeks), the cultures are kept at room temperature under low light. For long-term storage, cells are pelleted, resuspended in BG-11 with 5% dimethyl sulfoxide (DMSO), and frozen in liquid nitrogen.

Escherichia coli strains HB101[55] and K12-803[18,39] are routinely used for transformation and HB101 for the transduction of cosmids. The standard procedures for growth, handling, and DNA manipulation of *E. coli* are used.[56]

Media, Buffers, and Solutions

Media

BG-11 minimal medium[15] contains (per liter) deionized water, 1.5 g $NaNO_3$, 0.04 g K_2HPO_4, 0.075 g $MgSO_4 \cdot 7H_2O$, 0.036 g $CaCl_2 \cdot 2H_2O$, 0.006 g citric acid, 0.006 g ferric ammonium citrate, 0.001 g EDTA, 0.54 g $Na_2CO_3 \cdot 10H_2O$, and 1 ml of trace element mix A5. This solution contains 2.86 g H_3BO_3, 1.81 g $MnCl_2 \cdot 4H_2O$, 0.222 g $ZnSO \cdot 7H_2O$, 0.39 g $Na_2MoO_4 \cdot 2H_2O$, 0.079 g $CuSO_4 \cdot 5H_2O$, 0.0494 g $Co(NO_3)_2 \cdot 6H_2O$ per liter. Before autoclaving pH is adjusted to 7.5 with HCl

For growth on plates, BG-11 medium is solidified by adding separately autoclaved agar (Difco) to a concentration of 1% (w/v).

Buffers

1. SE buffer: 0.12 *M* NaCl, 0.05 *M* EDTA, pH 8.0
2. TE buffer: 10 m*M* Tris–HCl, 0.1 m*M* EDTA, pH 8.0
3. Lysis buffer: 50 m*M* glucose, 25 m*M* Tris–HCl, 10 m*M* EDTA, pH 8.0

[55] H. W. Boyer and D. Rouilland-Dussoix, *J. Mol. Biol.* **41,** 459 (1969).

[56] T. Maniatis, E. F. Fritsch, and J. Sambrook, "Molecular Cloning: A Laboratory Manual." Cold Spring Harbor Lab., Cold Spring Harbor, New York, 1982.

Extraction Solutions

1. Phenol saturated with 0.1 *M* Tris–HCl, pH 8.0
2. Chloroform/isoamyl alcohol (24 : 1, v/v)

Procedures

Isolation of Plasmid and Chromosomal DNA

Plasmid Minipreps. For the analysis of the plasmid content of *A. nidulans* transformants the following procedures for small-scale DNA isolations is used routinely: Inoculate 50 ml of BG-11 with 0.5 ml of a full-grown culture and incubate for 3–4 days at 28–30° (minimal concentration should be 10^8 cells/ml). Spin down the cells in a Sorvall SS34 rotor for 10 min at 10,000 rpm, wash once with 10 ml SE buffer and once with 10 ml lysis buffer. Resuspend the cells in 0.5 ml lysis buffer, transfer them to an Eppendorf tube, and add 0.25 ml lysozyme (10 mg/ml, Calbiochem) in lysis buffer. After a 1-hr incubation at 37° the cells are lysed by the addition of 0.25 ml freshly prepared 10% SDS (mix very gently). After one more hour at 37°, add 0.25 ml 5 *M* NaCl, mix carefully, and keep the lysed cell overnight at 4°. Remove the cell debris by centrifugation for 30 min at 10,000 *g* at 4°. Transfer the supernatant (~0.75 ml) to a new Eppendorf tube, extract twice with 0.5 ml phenol and once with 0.5 ml chloroform/isoamyl alcohol. The aqueous phase is then transferred to a Beckman SW50 polyallomer tube, 120 μl 3 *M* sodium acetate, pH 5.6, is added, and the total volume made 1.5 ml with TE buffer. Subsequently, fill up the tube by adding 2 vol of cold 96% ethanol, mix, and place at −70° for 30 min or keep overnight at −20°. After this precipitation step, centrifuge the solution for 15 min at 30,000 rpm at 4°, pour off the supernatant, and dry the pellet in a vacuum desiccator for 30 min. Finally, dissolve the DNA in 100 μl TE buffer. Treat with RNase (50 μg/ml) if desired.

Plasmid Maxipreps. For the isolation of plasmid DNA from *A. nidulans* on a preparative scale the following procedure gives good results: Harvest the cells of a full-grown 4-liter culture (2–4 × 10^8 cells/ml) by centrifugation, wash once in SE buffer, and resuspend in lysis buffer to a final volume of 90 ml in a 500-ml Erlenmeyer flask. Add 450 mg lysozyme and incubate for 2 hr at 37° with gentle shaking. Transfer to each of six Beckman R30 tubes 15 ml of the mixture and 3 ml of freshly prepared 10% SDS, mix by gentle rotation, and incubate for 1 hr at 37°. Then add 5 ml of 5 *M* NaCl and after careful mixing keep the tubes for 2 hr in ice or overnight at 4°. Remove the cell debris and most of the chromosomal DNA by centrifugation for 30 min at 25,000 rpm at 4° in an R30 rotor. Collect the cleared lysates of the six tubes, divide among four SS34 tubes, and concentrate by adding 3.6 g polyethylene glycol (PEG 6000) to each tube. Rotate gently overnight at 4° to dissolve the PEG. Centrifuge for 10 min at 5000 rpm, discard the supernatant, and dissolve each precipitate in 4 ml TE buffer at 4°. Transfer the mixtures of the four SS34 tubes to two SW27 tubes (1 × 3.5 inch type) and add TE buffer until the final content of each tube amounts to 26.7 g. Then add 28.7 g CsCl, dissolve carefully, and add 3 ml of an ethidium bromide solution (5 mg/ml). For an additional purification the mixture can be subjected to a short run at 10,000 rpm. Collect the clear supernatants and centrifuge the solution at 45,000 rpm for 20 hr at 15° in a Beckman VTi50 rotor. Screen the gradients with ultraviolet light and collect the plasmid bands. Finally, dilute with at least 2 vol of TE buffer, precipitate the DNA with ethanol, and dissolve in TE buffer. A 10–30% sucrose gradient ultracentrifugation may eventually be done to separate the various plasmid species and to remove the remaining RNA. The yield depends on the plasmid, and is usually 5–50 μg.

Isolation of Chromosomal DNA. For the isolation of chromosomal DNA of *A. nidulans,* cells are grown to midlog phase (~5 × 10^7 cells/ml) in a 4-liter batch, collected by centrifugation, washed once with SE buffer, and resuspended in lysis buffer to a final volume of 150 ml. The suspension is quantitatively transferred to a 1-liter Erlenmeyer flask, 500 mg lysozyme is added, followed by an incubation of 1 hr at 37° under gentle shaking. Subsequently, 50 ml of a freshly prepared 10% SDS solution is added to the shaking suspension and incubated again for 1 hr at 37°. Finally, 22 ml 5 *M* $NaClO_4$ is added while still shaking. The mixture with the lysed cells is then divided over two Sorvall GS3 tubes and extracted at least three times with 100 ml chloroform/isoamyl alcohol. Once the aqueous phase is clear, the DNA is precipitated with 2 vol of cold ethanol (96%) and kept overnight at −20°. (Note: when the ethanol is added without mixing, molecules can be spooled by rotating a glass rod or pipet in the interphase; this DNA is of particularly high molecular weight.) After centrifuging the GS3 tubes for 10 min at 7000 rpm, the pellets are dried in a vacuum dessicator and resuspended in 4 ml TE buffer. This step is followed by a run in a CsCl–EtBr gradient (see the section, Plasmid Maxipreps). The chromosomal DNA band is collected, diluted 3-fold with TE buffer, and precipitated by ethanol treatment. The yield is ~1 mg of DNA/liter of culture. For the isolation of chromosomal DNA from smaller batches (c.f. the standard 50-ml cultures) the procedure can be scaled down proportionally.

Transformation

The following protocol has been successfully used for the transformation of plasmids capable of autonomous replication in *A. nidulans* R2 as well as for plasmids lacking a cyanobacterial origin of replication and thus relying on integrative recombination. The salient feature of transformation in *A. nidulans* R2 is that the cells are naturally competent and that no special treatment is needed. The existence of a phase during the growth cycle in which the cells are especially competent has been reported,[50] but we make no allowance for it. The effect of a dark treatment has been stated to be of great importance[57] and we include this dark step because it never has an adverse effect and sometimes does increase transformation slightly.

Cells of *A. nidulans* R2 are grown to 5–8 × 10^7/ml in 50 ml BG-11 medium, spun down for 10 min at 10,000 rpm in an SS34 rotor at room temperature, washed with BG-11, and concentrated to 1–2 × 10^9 cells/ml in BG-11. To 0.1 ml of cells 0.1 μg DNA is added and incubation is in the light for 1 hr at room temperature under gentle mixing. The transformation mixture is spread on plates containing ~60 ml of BG-11 agar (to prevent desiccation). The plates are inoculated with 10^8 cells for ampicillin selection and with 1–5 × 10^7 cells for streptomycin or chloramphenicol selection and incubated for 16 hr at 30° in the dark followed by 8 hr under standard illumination. Then the agar is partially lifted with a spatula and the antibiotics are added by underlayering 0.5 ml of ampicillin (30–100 μg/ml), streptomycin (1 mg/ml), or chloramphenicol (0.75 mg/ml) to allow slow adaptation of the transformants to increasing concentrations of the drug. These figures represent average values. As the effect of the selection depends on the transforming plasmid and on cell density, it is recommended to test a series of drug concentrations beforehand. The plates are further incubated under the standard light conditions. After 2–4 days the light-green lawn disappears and after 5–6 days resistant colonies are visible and transformants can be isolated. The purification of transformants seems to be a trivial matter. However, after transformation with a recombinant plasmid the resident plasmid is often found to be still present in a plasmid preparation. To overcome this, a transformant colony should be thoroughly resuspended in BG-11 medium and grown for several days under selective conditions. Subsequently, the culture is plated for single colonies and one of these can then be grown on large scale for further analysis.

Similar hybrid plasmids have greatly different transformation efficiencies (e.g., compare pUC303,[18] pSG111,[57] and pPLAN B2,[40] with transformation frequencies of ~10^{-4}, 10^{-6}, and 10^{-5}, respectively). There may be several reasons for this. First, plasmids isolated directly from the cyanobacteria tend to be more efficient than after passage through *E. coli*. This

[57] S. S. Golden and L. A. Sherman, *J. Bacteriol.* **158,** 36 (1984).

may reflect a difference in modification and/or the presence of a restriction enzyme in *A. nidulans* R2.[49,58] Another reason is the selection for antibiotic resistance. In some cases the marker is only weakly expressed and there is only a small window for selection. Especially with ampicillin, which kills the cells more rapidly than other antibiotics, this can be a problem. In such cases, Shestakov's method of applying the antibiotic underneath the agar to allow slow diffusion and a sufficient expression of the resistance gene may give better results than plating the cells on nitrocellulose filters and transferring the filters from nonselective to selective plates. However, even if these factors are taken into account there remains considerable variation. This can only partly be explained by different procedures used in various laboratories, as also in our hands similar vectors behave differently when tested in the same experiment. All newly developed vectors should therefore be carefully tested for their transformation efficiency. The vector pUC303 has a transformation frequency of approximately $10^6/\mu g$ DNA in the linear part of the saturation curve, and at least 10^{-4}/cell when measured with saturating amounts of DNA.[18]

Mutagenesis

Stable mutants of cyanobacteria have been obtained by various mutation induction procedures, with chemical mutagens, with ultraviolet light and X rays, or with transposon insertions. An extensive review on mutation induction, repair, and mutant characterization has been published.[26] We restrict ourselves here to the methods used in our work with *Anacystis nidulans* R2.

Plasmid Curing. Strains lacking the small plasmid can be obtained by inoculating BG-11 medium with cells carrying a pUH24 derivative with an ampicillin resistance marker at a density of 10^3 cells/ml and growing in the presence of 0.003% SDS until stationary phase.[18] Cured cells can be identified by replica plating and then checked for the absence of the small plasmid by gel electrophoresis and of integrated small-plasmid DNA by Southern blotting of total cellular DNA and a pUH24 DNA probe. In contrast to the situation with, e.g., the plasmids pCH1 and pUC1,[48] cells harboring some of the other pUH24 derivatives may become cured also without the SDS treatment. For instance, pUC13 and pUC14[18] are plasmids that are lost during prolonged growth in BG-11 medium without streptomycin selection. It takes ~40 and 70 generations, respectively, to obtain cultures in which 50% of the cells are cured of these small plasmids.

[58] M. L. Gallagher and W. F. Burke, Jr., *FEMS Microbiol. Lett.* **26,** 317 (1985).

Chemical Mutagenesis. Auxotrophic and other mutants are readily generated by *N*-methyl-*N'*-nitro-*N*-nitrosoguanidine (NTG), a potent mutagen for a number of cyanobacteria, although there is a risk that multiple mutations at closely linked sites are induced.[26] The following is a suitable procedure. Grow cells in BG-11 medium to 1×10^8/ml and collect them by centrifugation. Resuspend the pellet in 0.1 *M* sodium citrate, pH 5.5, to a density of 5×10^8 cells/ml. In a Sorvall SS34 tube mix 18 ml of the suspension with 2 ml of a freshly prepared NTG solution (0.5 mg/ml in 0.1 *M* citrate, pH 5.5) to reach a final concentration of 50 μg NTG/ml. Incubate for 15 min at 20° in the light, spin for 5 min at 10,000 rpm, and wash twice with 20 ml BG-11. Spin again and resuspend the cells in 20 ml BG-11 medium (supplemented with the appropriate growth factors[59] if auxotrophic mutants are desired). Transfer 10 ml of the suspension to 50 ml (supplemented) BG-11 in a 250-ml Erlenmeyer flask and grow for 24 hr at 28–30° on a shaker under standard light conditions to allow the expression of the induced mutations. Spin and replace the medium by the same volume of unsupplemented BG-11 to subject the cells to a 24-hr starvation period. Subsequently, transfer 2.5×10^8 cells to 50 ml minimal BG-11 containing 150 μg ampicillin/ml. Incubate on a shaker in the light for 6–24 hr, wash twice and resuspend in 10 ml BG-11, and plate 10^0, 10^{-1}, and 10^{-2} dilutions on (supplemented) BG-11 agar. Finally, check the colonies directly for color mutants or, by inserting a replica-plating step, for auxotrophs. Note: supplemented BG-11 media (liquid as well as solid) should always be tested first for their ability to support growth of the wild-type strain, as sometimes difficulties in this respect arise. If specific auxotrophic mutants are wanted which cannot be obtained by the plain method described above, a more general procedure[60] may be attempted.

Transposon Mutagenesis. In *A. nidulans* R2 transposon mutagenesis has proved to be a useful method.[13,39] In our hands Tn*5* gives less stable mutants than Tn*901*, so the latter transposon is mainly used, even though in general the kanamycin resistance of Tn*5* is a better selective marker than the ampicillin resistance of Tn*901*. A proved procedure for the induction of auxotrophic mutants follows. Cells of *A. nidulans* R2 harboring pCH1 (= pUH24 : : Tn*901*)[48] instead of pUH24 are grown to a density of 8×10^7 cells/ml in BG-11 medium supplemented with the appropriate growth factor and 1 μg/ml of ampicillin. After collecting the cells by centrifugation and extensive washing in BG-11, they are resuspended to a concentration of 2×10^6 cells/ml and incubated for 18–24 hr in BG-11

[59] M. Herdman and N. G. Carr, *J. Gen. Microbiol.* **70,** 213 (1972).

[60] M. Herdman, S. F. Delaney, and N. G. Carr. *Arch. Microbiol.* **124,** 177 (1980).

minimal medium, to deplete the intracellular growth factor pool. As an enrichment step there follows a further incubation for 16 hr in minimal medium with 150 μg/ml ampicillin (cycloserine in the same concentration can effectively replace ampicillin). The cells are then washed in BG-11 and plated on BG-11 agar supplemented with the growth factor. Finally, the colonies are screened for auxotrophy by replica plating. Note: when the antibiotic resistance of a strain is solely mediated by the β-lactamase gene of a Tn*901* inserted in the chromosome, the selective ampicillin concentration must be lowered to 0.3 μg/ml.

The frequencies of mutation induction by transposons are low, and the sophisticated tools used for enrichment in *E. coli* do not exist in *A. nidulans* R2. Therefore, we would like to propose an alternative way of mutagenesis combining the advantages of transposon mutagenesis with the high rate of recombination found in *A. nidulans* R2 (probably some 100-fold higher than transposition). The first step would involve the construction of a gene library of the cyanobacterial genome in an *E. coli* plasmid, e.g., pACYC184. The next step would be transposon mutagenesis in *E. coli*. Finally, the mutated library would be isolated and used to transform *A. nidulans*, selecting for the transposon marker.

Other Host–Vector Systems

In addition to *Anacystis nidulans* R2, host–vector systems have been developed in the facultatively photoheterotrophic unicellular cyanobacteria *Synechocystis* PCC6803 and *Synechococcus* PCC7002. These systems also rely on transformation and are similar in nature to the R2 system. Transformation in *Synechococcus* PCC7002 was complicated by the presence of a restriction enzyme,[22] but recently a strain with decreased restriction was isolated.[23]

For *Synechocystis* PCC6803 a host strain with improved transformation properties was selected and vectors integrating into one of the endogenous plasmids and into the chromosome were constructed. *Synechocystis* PCC6803 contains cryptic plasmids of two sizes, pUG1 (2.27 kb) and pUG2 (5.2 kb).[46] The pUG1 type comprises three different plasmids, which have the same molecular weights but show different restriction fragment patterns. Since the strain was known to be transformable by chromosomal DNA[33] but failed to be transformed by the vectors developed with *A. nidulans* R2, shuttle vectors able to replicate in both *Synechocystis* PCC6803 and *E. coli* were constructed.[46] Into the β-lactamase gene of the *E. coli* plasmid pACYC177 (Ap^R, Km^R), a 1.7-kb *Hpa*I fragment from two of the pUG1-type plasmids was inserted. The resulting hybrid plasmids, pUF3 and pUF12, conferred kanamycin resistance on both hosts, but transformation efficiencies and stabilities in the cyanobacterial host were low. After transformation with pUF3 most of the *Synechocystis* clones harbored in a high copy number a new plasmid of 8.0 kb, pUF311, which originated from recombination of pUF3 with one of the endogenous pUG1-type plasmids. It transforms wild-type *Synechocystis* to a 100-fold higher level of kanamycin resistance than pUF3 but with the same low frequency of 3×10^{-7}/cell. pUF311 can already be used as a shuttle vector, but its utility was improved by inserting the chloramphenicol resistance gene of pACYC184 as a second, in cyanobacteria very efficient, selection marker. The resulting plasmid, pFCLV7 (9.35 kb), replicates well in *E. coli* and *Synechocystis* PCC6803 and expresses chloramphenicol and kanamycin resistance in both hosts. It has single restriction sites for *Cla*I, *Sma*I, and *Xho*I in the Km^R gene, and for *Pvu*II and *Eco*RI in the Cm^R gene, so the method of insertional inactivation can be easily applied. The efficiency of transformation of *Synechocystis* PCC6803 is raised to 2×10^{-5}/cell when strain SUF311, carrying pUF311, is used as the host.[46]

A completely different approach was taken for the introduction of DNA into filamentous *Anabaena* and *Nostoc* strains. It is based on the principle of promoting vector transfer by the conjugational properties and transfer functions provided by helper plasmids.[61] Such a triparental mating system was developed to introduce shuttle vectors devoid of all or most of the recognition sites for the endogenous restriction enzymes in *Anabaena* PCC7120.[11] This method proved effective also in other filamentous cyanobacteria,[24,25] and may have widespread applicability. So far, however, no integration into the chromosome has been reported.

Appendix

All of the wild-type strains used in molecular–genetic studies have been deposited in the Pasteur Culture Collection of Cyanobacteria. They are available at a small charge; contact Dr. R. Rippka, Institut Pasteur, 28 Rue du Dr. Roux, Paris 75015, France. The small-plasmid-cured strain *A. nidulans* R2-SPc, and the vectors pUC303 and pPUC29 can be obtained from Dr. W. E. Borrias, Laboratory of Molecular Cell Biology, University of Utrecht, Padualaan 8, 3584 CH Utrecht, The Netherlands. An elaborate restriction map of pUH24, comprising all the recognition sites

[61] G. Ditta, S. Stanfield, D. Corbin, and D. R. Helinski, *Proc. Natl. Acad. Sci. U.S.A.* **77,** 7347 (1980).

of over 40 restriction enzymes, is available from Dr. P. J. Weisbeek, Utrecht, address as above.

Acknowledgments

We thank Dr. Mies Borrias for helpful comments, Dr. Peter Weisbeek for the restriction map of pUH24, Dr. Ron Porter for making data available prior to publication, and Dr. Pamela Green for critically reading the manuscript.

[12] Genetic Engineering of the Cyanobacterial Chromosome

By SUSAN S. GOLDEN, JUDY BRUSSLAN, and ROBERT HASELKORN

The cyanobacteria are prokaryotic organisms which possess a number of unique biological characteristics. All species carry out oxygenic photosynthesis using a thylakoid apparatus which is very similar to that of plant choroplasts.[1] Some are able to alter the protein composition of their light-gathering antennae to match the quality of incident light.[2] Many species are capable of the fixation of elemental dinitrogen, and some differentiate specialized cell types for this process and for other functions.[3] In addition, cyanobacterial genomes have been shown to encode some genes in multiple copies.[4-6] The study of these and other properties has been facilitated by the development of methods for gene transfer in cyanobacteria and the application of these techniques to the production and analysis of mutants. Several cyanobacterial strains can be transformed by exogenously added DNA[7] or can receive mobilizable plasmid DNA by conjugation from *Escherichia coli*.[8] Much effort has been devoted to the production of autonomously replicating plasmid vectors for these strains; some of these methods and constructions are described elsewhere in this vol-

[1] K. K. Ho and D. W. Krogmann, *in* "The Biology of Cyanobacteria" (N. G. Carr and B. A. Whitton, eds.), p. 191. Univ. of California Press, Berkeley, California, 1982.

[2] G. Cohen-Bazire and D. A. Bryant, *in* "The Biology of Cyanobacteria" (N. G. Carr and B. A. Whitton, eds.), p. 143. Univ. of California Press, Berkeley, California, 1982.

[3] C. P. Wolk, *in* "The Biology of Cyanobacteria" (N. G. Carr and B. A. Whitton, eds.), p. 359. Univ. of California Press, Berkeley, Calfornia, 1982.

[4] S. Robinson, personal communication.

[5] S. E. Curtis and R. Haselkorn, *Plant Mol. Biol.* **3,** 249 (1984).

[6] S. S. Golden, J. Brusslan, and R. Haselkorn, *EMBO J.* **5,** 2789 (1986).

[7] R. D. Porter, *CRC Crit. Rev. Microbiol.* **13,** 111 (1986).

[8] C. P. Wolk, A. Vonshak, P. Kehoe, and J. Elhai, *Proc. Natl. Acad. Sci. U.S.A.* **81,** 1561 (1984).

ume.[9] The strains which can be transformed directly by exogenous DNA are also amenable to manipulations which depend only upon the recombination of cloned homologous or chimeric DNA with the chromosome. Applications of these methods include substitution of a mutated gene for the wild-type allele,[10] inactivation of a gene,[6] duplication of a chromosomal locus,[11] and ectopic mutagenesis.[12] This chapter describes methods for direct genetic engineering of the cyanobacterial chromosome and techniques for the analysis of DNA and RNA from the resulting transformants.

Transformation of Cyanobacteria by Chromosomal Recombination

Several unicellular cyanobacterial strains classified as *Synechococcus* or *Synechocystis*[13] are transformable by exogenously added homologous DNA which carries a selectable marker. A recent review by Porter[7] describes these strains and their transformation properties. Four of the strains which are currently in use as hosts for molecular genetic manipulation are listed in Table I. Suitable donor DNA for transformation of *Anacystis nidulans* R2 and *Agmenellum quadruplicatum* PR-6 includes endogenous plasmids into which antibiotic resistance genes have been introduced by insertion of a transposon[14] or by ligation to an *E. coli* plasmid *in vitro*.[12,15,16] In addition, chromosomal DNA which carries a positively selectable trait can be used to transform cells of any of these strains by recombination within the chromosomal locus homologous to the marked allele. This property can be exploited to move heterologous DNA into the chromosome using flanking homologous DNA to direct it to a particular locus.[12,17]

The homologous DNA fragment used for transformation can either be linear or cloned as part of a covalently closed circular molecule by ligation to an *E. coli* plasmid which does not replicate in the cyanobacterium. The probable outcome of the recombination event leading to integration of the

[9] C. J. Kuhlemeier and G. A. van Arkel, this volume [11].

[10] S. S. Golden and R. Haselkorn, *Science* **229,** 1104 (1985).

[11] S. S. Golden, J. Brusslan, and R. Haselkorn, unpublished observations.

[12] J. S. Buzby, R. D. Porter, and S. E. Stevens, *Science* **230,** 805 (1985).

[13] R. Rippka, J. Deruelles, J. B. Waterbury, M. Herdman, and R. Y. Stanier, *J. Gen. Microbiol.* **111,** 1 (1979).

[14] C. A. M. J. J. van den Hondel, S. Verbeek, A. van der Ende, P. J. Weisbeek, W. E. Borrias, and G. A. van Arkel, *Proc. Natl. Acad. Sci. U.S.A.* **77,** 1570 (1980).

[15] C. J. Kuhlemeier, W. E. Borrias, C. A. M. J. J. van den Hondel, and G. A. van Arkel, *Mol. Gen. Genet.* **184,** 249 (1981).

[16] S. S. Golden and L. A. Sherman, *J. Bacteriol.* **155,** 966 (1983).

[17] J. G. K. Williams and A. A. Szalay, *Gene* **24,** 37 (1983).

TABLE I
TRANSFORMABLE CYANOBACTERIAL STRAINS USED AS HOSTS FOR GENETIC ENGINEERING EXPERIMENTS

Common strain name	Alternate names	PCC number[a]	ATCC number[b]	Transformation references[c]
Anacystis nidulans R2	*Synechococcus* R2 *Synechococcus* 7942	7942	None	*d, e, f, g*
Agmenellum quadruplicatum PR-6	*Synechococcus* 7002	7002	27264	*h, i*
Synechocystis 6803	*Aphanocapsa* 6803	6803	27184	*j*
Aphanocapsa 6714	*Synechocystis* 6714	6714	27178	*k*

[a] Pasteur Culture Collection.
[b] American Type Culture Collection.
[c] References to transformation by homologous recombination with the chromosome.
[d] G. A. Grigorieva and S. V. Shestakov, *Photosynth. Prokaryotes, Proc. Int. Symp., 2nd, 1976* p. 220.
[e] J. G. K. Williams and A. A. Szalay, *Gene* **24,** 37 (1983).
[f] K. S. Kolowsky, J. G. K. Williams, and A. A. Szalay, *Gene* **27,** 289 (1984).
[g] S. S. Golden and L. A. Sherman, *J. Bacteriol.* **158,** 36 (1984).
[h] S. E. Stevens and R. D. Porter, *Proc. Natl. Acad. Sci. U.S.A.* **77,** 6052 (1980).
[i] J. S. Buzby, R. D. Porter, and S. E. Stevens, *Science* **230,** 805 (1985).
[j] G. Grigorieva and S. Shestakov, *FEMS Microbiol. Lett.* **13,** 367 (1982).
[k] C. Astier and F. Espardellier, *C. R. Acad. Sci. Paris, Ser. D* **282,** 795 (1976).

latter form of donor DNA can be controlled by selection. Consider transformation of *A. nidulans* R2 by the chimeric plasmid DNA shown in Fig. 1A and B. Selecting for a marker within the cyanobacterial insert usually results in recombination within the homologous DNA without integration of any heterologous sequences. This appears to occur either by a gene conversion event which is restricted to the homologous region or by reciprocal recombination, as depicted in Fig. 1A. However, if the transformed cells are selected for an antibiotic resistance marker within the *E. coli* vector, the resulting transformants contain the entire plasmid molecule integrated at the chromosomal locus of the cloned cyanobacterial sequence (Fig. 1B). This results in the duplication of the region of cyanobacterial DNA which was present in the plasmid and is the expected product of a single crossover event. The duplication can be maintained by continued selection for the antibiotic resistance conferred by the integrated *E. coli* plasmid. Duplications should prove useful for determining dominance of a mutation in merodiploids and in studying gene dosage effects.

An important application of these techniques is the ability to inactivate a gene in the chromosome by recombination with a cloned, altered allele. This method has been used to inactivate *psbA* genes in *A. nidulans* R2,[6]

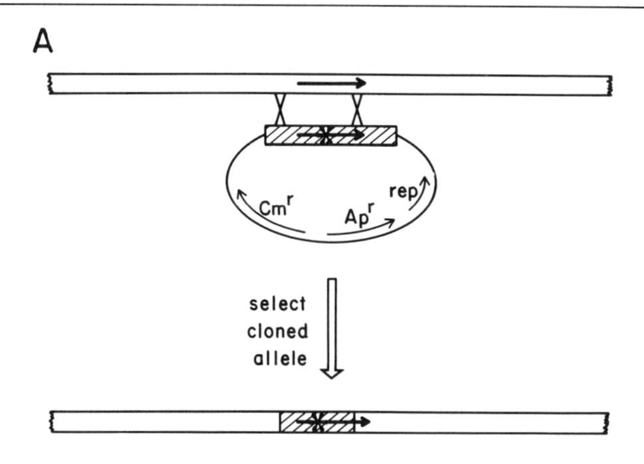
A
Cm^r
Ap^r
rep
select
cloned
allele

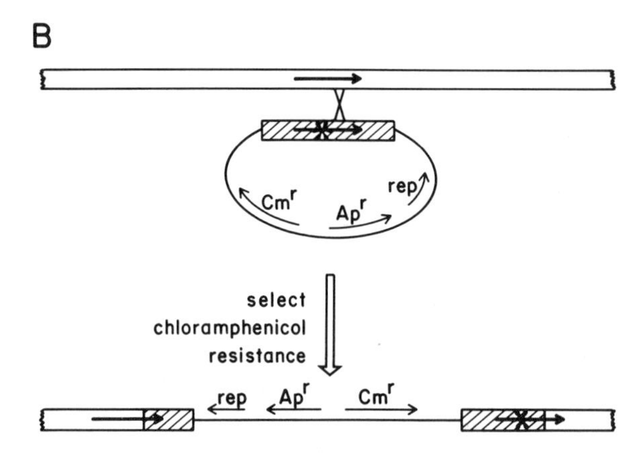
B
Cm^r
Ap^r
rep
select
chloramphenicol
resistance
rep
Ap^r
Cm^r

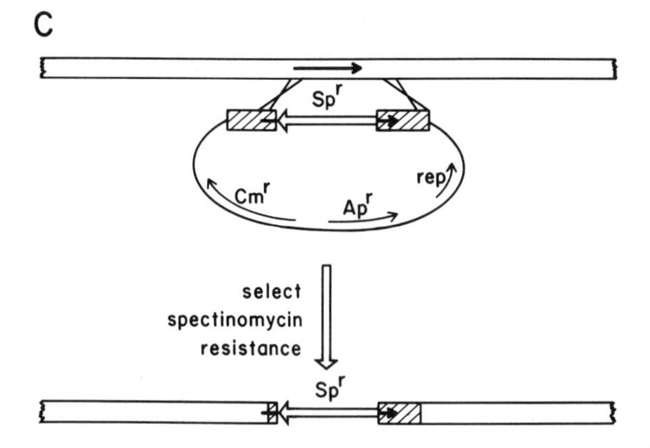
C
Sp^r
Cm^r
Ap^r
rep
select
spectinomycin
resistance
Sp^r

phycobiliprotein genes in *A. quadruplicatum* PR-6,[18] and photosystem II component genes in *Synechocystis* 6803.[19] The cloned gene can be rendered nonfunctional by the insertion of a transposon[20] or a DNA "cassette" which encodes a selectable marker, such as antibiotic resistance.[21,22] The cyanobacterial gene can be deleted completely by removing a restriction fragment containing the coding sequence and re-

[18] D. A. Bryant, R. deLorimier, G. Guglielmi, and S. E. Stevens, manuscript in preparation.
[19] J. G. K. Williams, this series, Vol. 167, in press.
[20] G. B. Ruvkun and F. M. Ausubel, *Nature (London)* **289,** 85 (1981).
[21] S. K. Shapira, J. Chou, F. V. Richaud, and M. J. Casadaban, *Gene* **25,** 71 (1983).
[22] P. Prentki and H. M. Krisch, *Gene* **29,** 303 (1984).

FIG. 1. Schematic representation of recombination events between cloned cyanobacterial DNA fragments and the chromosome. Open bars represent the chromosome and cross-hatched bars are cloned fragments of cyanobacterial DNA, aligned with the chromosome at the homologous locus. The gene of interest is depicted on the chromosome and on the cloned fragment as a heavy horizontal arrow. The site of a selectable mutation on the gene is shown as an asterisk. The *E. coli* plasmid pBR328 is shown as an oval carrying the cloned cyanobacterial fragment and genes, depicted by thin arrows, encoding resistance to chloramphenicol (Cm^r) and ampicillin (Ap^r). The origin of replication in *E. coli* (rep) is also indicated. Each point of recombination is represented by an X. (A) Replacement of a chromosomal gene with a marked, homologous allele. Selection for the marker on the cyanobacterial allele results in gene conversion or reciprocal recombination which replaces the chromosomal allele with the selected sequence. Heterologous DNA is not transferred to the chromosome and is lost from the cell. Recombination may occur at any point on each side of the selectable marker; one end-point is depicted within the coding region of the gene, producing a functional chimeric gene. The restriction pattern of the chromosome is unaltered by recombination. (B) Gene duplication by integration of a chimeric plasmid. Selection for a marker on the nonreplicating heterologous vector yields cells in which the plasmid integrated into the chromosome. Cloned cyanobacterial DNA recombines at its homologous chromosomal locus in a single crossover event. The plasmid is integrated at the site of recombination, causing duplication of the cloned cyanobacterial DNA at the chromosomal locus. Each of the duplicated regions contains some DNA from the chromosome and some from the cloned fragment. The junctions of heterologous plasmid and cyanobacterial DNA are identical to those in the clone. The restriction pattern of chromosomal DNA in the transformant is altered by the integration of heterologous sequences. (C) Gene inactivation by replacement of the wild-type gene with a cloned, inactivated allele. The cloned cyanobacterial gene has been inactivated *in vitro* by the replacement of an internal restriction fragment with a gene cassette encoding spectinomycin resistance (Sp^r, open arrow). The remainder of the cyanobacterial gene and its flanking DNA are still present in the construction. Selection for spectinomycin resistance results in reciprocal recombination or gene conversion which replaces the wild-type gene with the inactivated allele. Transcription of the gene is abortive. The junctions of the spectinomycin resistance cassette and cyanobacterial DNA are identical to those in the cloned construction. The restriction pattern of the chromosome in the transformant is altered by the loss of cyanobacterial sequences and the insertion of heterologous DNA.

placing it *in vitro* with an antibiotic resistance cassette. When any of these constructions is used to transform wild-type cells with selection for the appropriate antibiotic resistance, transformants contain the inactivated allele substituted for the wild-type allele in the chromosome (Fig. 1C). Inactivation of more than one gene in the same strain can be achieved by using altered alleles which carry different selectable markers.[6]

Transformation of cells with alleles of unknown phenotype can be accomplished by selection for a linked marker. Cotransformation of *A. nidulans* R2 with two linked markers was demonstrated by the insertion of the transposon Tn*5*, encoding kanamycin resistance, 1 kb downstream from the coding region of a cloned *psbA* allele[10] which confers resistance to the herbicide Diuron. Transformation of wild-type cells with this construction and selection for kanamycin resistance resulted in cotransformation with the unselected diuron-resistance marker in 24 out of 25 clones.[6] This strategy is suitable for the introduction of alleles which have been altered by site-directed mutagenesis but which may not confer a selectable phenotype.

Analysis of DNA from transformants is necessary to determine the chromosomal structure following recombination. The recombination events illustrated in Fig. 1 are those which are usually seen, but aberrant events have been observed when the cloned region homologous to the chromosome is small. These events include deletion of a chromosomal segment[11] and integration at a related locus in the case of a gene family.[6] Southern hybridization can determine whether the expected event has, in fact, occurred. When the donor DNA is a homologous, selectable allele cloned into a nonreplicating plasmid (Fig. 1A), this analysis should confirm that no heterologous sequences were transferred to the chromosome. The chromosomal restriction pattern is preserved during homologous recombination; restriction site polymorphism should not be detected unless the transforming allele carries a mutation which creates or destroys a restriction site. However, when an event is selected which transfers heterologous sequences to the chromosome (Fig. 1B and C), the chromosomal restriction pattern will be changed at that locus. When the restriction maps of the transforming heterologous DNA and flanking cyanobacterial DNA are known, the sizes of the affected restriction fragments can be predicted based on the type of integration event selected.

It is also important to analyze RNA from transformants which have been engineered by these methods. A homologous recombination event which replaces a wild-type gene with another functional allele should result in the production of a full-length mRNA which initiates and terminates properly. When a gene has been inactivated or deleted as shown in Fig. 1C, no full-length message should be detected by Northern analysis.

The inactivated gene may, however, produce aberrant transcripts, which terminate prematurely or read into or out of the cassette or transposon which disrupts the gene. Fused transcripts can be identified, using Northern analysis, on the basis of their homology to both the cyanobacterial gene and the heterologous sequences.

We have used the following methods for the mutagenesis of *A. nidulans* R2 and for the analysis of DNA and RNA from this strain and the filamentous cyanobacterium *Anabaena* 7120. We provide as examples the results of Southern and Northern analyses for *A. nidulans* R2 strains in which members of the *psbA* gene family (*psbAI, psbAII, psbAIII*) have been inactivated as depicted in Fig. 1C. However, similar experiments have been carried out by others with *A. quadruplicatum* PR-6[18] and *Synechocystis* 6803.[19] These procedures should be applicable to other strains as well.

Materials and Reagents

Cyanobacterial Growth Conditions

We recommend growth of *Anacystis nidulans* R2 (also called *Synechococcus* R2, PCC7942) in liquid BG-11 medium[23] at 30° with continuous illumination of 250–800 foot candles of fluorescent or mixed tungsten–fluorescent light. Prepare solid medium for plating transformed cells as follows: autoclave separately equal volumes of twice-concentrated (2×) BG-11 liquid medium and a solution of agar (Difco, 3% in distilled water), cool to pouring temperature, and mix. Sterilize a 100 m*M* solution of sodium thiosulfate ($Na_2S_2O_3$) by passage through a 0.22-μm Nalgene filter and add 10 ml/liter of BG-11 agar for a final concentration of 1 m*M*. Dispense 40 ml of agar/100-mm Petri dish. The addition of sodium thiosulfate was previously found to encourage growth on solid medium when the distilled water supply contained inhibitory impurities,[24] and we now add it routinely to solid medium. Growth in liquid BG-11 medium has been satisfactory without added sodium thiosulfate.

Equipment

1. Eppendorf microcentrifuge or equivalent
2. Pipetman automatic pipettors (or equivalent) and sterile disposable tips

[23] M. M. Allen, *J. Phycol.* **4,** 1 (1968).
[24] S. S. Golden and L. A. Sherman, *J. Bacteriol.* **158,** 36 (1984).

3. Sorvall superspeed centrifuge with SS34, HB4, and GSA rotors, or equivalent
4. Beckman ultracentrifuge with SW41 rotor, or equivalent

Materials and Reagents for Transformation

1. Sterile 1.5-ml microcentrifuge tubes
2. Sterile 30- to 50-ml screw-capped centrifuge tubes
3. Sterile 10 m*M* NaCl
4. BG-11 liquid and solid growth media as described above

Materials and Reagents for DNA Minipreps

1. 1.5-ml microcentrifuge tubes
2. SE: 120 m*M* NaCl, 50 m*M* EDTA, pH 8.0
3. Lysis buffer: 25% sucrose, 50 m*M* Tris, 100 m*M* EDTA, pH 8.0
4. Lysozyme
5. Sarkosyl: 25% sodium lauryl sarcosine in distilled water
6. Proteinase K: 20 mg/ml in distilled water
7. Phenol: As a solution of phenol plus 0.1% 8-hydroxyquinoline, equilibrated with 100 m*M* Tris, pH 8.0, 0.2% 2-mercaptoethanol
8. Chloroform: As chloroform : isoamyl alcohol (24 : 1)
9. 70% ethanol in water
10. TE: 10 m*M* Tris, 1 m*M* EDTA, ph 8.0

Materials and Reagents for DNA Isolation Using Glass Fines

1. Silica 325 mesh, powdered flint glass from ceramic stone, supplied by Cutter Ceramics, Valley View, Ohio
2. Nitric acid
3. Sodium iodide (NaI) and sodium sulfite (Na_2SO_3) for chaotropic salt solution
4. 70% ethanol in distilled water
5. 250-ml centrifuge bottles

Materials and Reagents for Isolation of Total RNA

1. Sterile 25-ml glass screw-capped centrifuge tubes
2. 50/100 TE: 50 m*M* Tris, 100 m*M* EDTA, pH 8.0
3. TE: 10 m*M* Tris, 1 m*M* EDTA, pH 8.0
4. Breakage buffer: A solution containing 30 ml of 50/100 TE, 2 ml of vanadyl ribonucleoside complex,* 0.15 ml of Triton X-100, 0.6 ml of 25% Sarkosyl solution, 0.6 ml of 20% SDS solution

* Optional, prepared as described by Maniatis *et al.*[25]

[25] T. Maniatis, E. F. Fritsch, and J. Sambrook, "Molecular Cloning: A Laboratory Manual." Cold Spring Harbor Lab., Cold Spring Harbor, New York, 1982.

5. Phenol and chloroform as described above
6. 0.45-mm glass beads: Soaked in chromic acid, washed well with distilled water, and baked at 200° for 4 hr
7. CsCl: Baked at 200° for 4 hr

Methods

Gene Constructions for Transformation

All of the methods shown in Fig. 1 for the recombination of cloned sequences into the cyanobacterial chromosome are independent of autonomous replication of the cloned DNA. We use the *E. coli* plasmids pBR322[26] and pBR328[27] as vectors for propagating cyanobacterial genes in *E. coli* and for reintroducing them into *A. nidulans* R2 as covalently closed circular molecules. These plasmids do not replicate in the cyanobacterium; the heterologous sequences are lost during recombination, unless specifically selected for chromosomal integration as shown in Fig. 1B.

Two methods of gene inactivation have been described in other systems and work well for cloned cyanobacterial genes. These methods require routine molecular biological manipulations which are described elsewhere.[25] One approach is the insertion of a transposon, by transposition from a defective bacteriophage λ, into the cloned gene while it is propagated in *E. coli*.[20] The other requires the *in vitro* insertion of a gene cassette encoding resistance to an antibiotic[21,22]; such cassettes can be removed from a plasmid with one or more restriction enzymes and ligated to compatible sites within the cloned cyanobacterial gene. Cassettes and antibiotics which we have found useful for selection of sequences recombined into the *A. nidulans* R2 chromosome are listed in Table II. Although two of these cassettes encode resistance to two antibiotics, in each case one antibiotic was found to provide cleaner selection than the other in *A. nidulans* R2. Kanamycin was preferable to neomycin for selection of Tn*5*- and pSKS101-derived constructions, and spectinomycin provided better selection than streptomycin for the pHP45Ω cassette.

Transformation of A. nidulans R2

Grow *A. nidulans* R2 cells under standard conditions of continuous illumination. Incubation of cultures in darkness prior to transformation reduces the overall efficiency as much as 40-fold.[24] Any liquid culture of

[26] F. Bolivar, R. L. Rodriguez, P. J. Greene, M. C. Betlach, H. L. Heyneker, H. W. Boyer, J. H. Crosa, and S. Falkow, *Gene* **2,** 95 (1977).

[27] X. Soberon, L. Covarrubias, and F. Bolivar, *Gene* **9,** 287 (1980).

TABLE II
ANTIBIOTICS USED TO SELECT TRANSFORMANTS OF *Anacystis nidulans* R2

Antibiotic	Supplier/trade name	Final concentration (μg/ml)	Source of resistance gene
Chloramphenicol hydrochloride	Sigma	7.5	pBR328[a] pSKS114[b] gene cassette
Spectinomycin	UpJohn/Trobicin[c]	40	pHP45Ω[d,e] gene cassette
Kanamycin sulfate	Sigma	50	Tn*5*[f,g] pSKS101[b,g] gene cassette
Ampicillin[h]	Bristol Laboratories/ Polycillin N[i]	0.5	pBR328[a] pCH1[j]

[a] X. Soberon, L. Covarrubias, and F. Bolivar, *Gene* **9,** 287 (1980).
[b] S. K. Shapira, J. Chou, F. V. Richaud, and M. J. Casadaban, *Gene* **25,** 71 (1983).
[c] Trobicin is prepared for intravenous injection, and contains glucose and salts; 64 μg/ml Trobicin is equivalent to 40 μg/ml spectinomycin.
[d] P. Prentki and H. M. Krisch, *Gene* **29,** 303 (1984).
[e] Also encodes resistance to streptomycin.
[f] G. B. Ruvkun and F. M. Ausubel, *Nature* (*London*) **289,** 85 (1981).
[g] Also encodes resistance to neomycin.
[h] Useful concentration determined for selection of autonomously replicating plasmids; not yet tested for selection of chromosomal integration.
[i] More reliable than ampicillin from other suppliers. Dissolve entire vial in sterile distilled water at 100 mg/ml ampicillin and store 10-μl aliquots at $-20°$. Do not refreeze.
[j] C. A. M. J. J. van den Hondel, S. Verbeek, A. van der Ende, P. J. Weisbeek, W. E. Borrias, and G. A. van Arkel, *Proc. Natl. Acad. Sci. U.S.A.* **77,** 1570 (1980).

cells is appropriate for transformation, regardless of culture age; however, logarithmic phase or early stationary phase cultures have a higher proportion of viable cells in the total population than very old stationary cultures. The R2 strain is readily transformable by homologous DNA; the following procedure can be altered considerably and still yield transformants.

To prepare cells for transformation, pellet them from 30 ml of culture by centrifugation at 5000 rpm in a Sorvall SS34 rotor for 5 min, and resuspend the pellet in 15 ml of 10 m*M* NaCl. Repeat the centrifugation and resuspend the cell pellet in 3 ml of BG-11 liquid. The initial volume of cells used for transformation and the factor by which they are concentrated can be varied; the frequency of transformation increases linearly with increasing cell concentration. We typically concentrate cells to between 5×10^8 and 1×10^9 cells/ml.

Transfer a 300-μl aliquot of cells to a sterile 1.5-ml microcentrifuge tube and add 10 ng to 1 μg of transforming DNA in 1–20 μl of TE. The frequency of transformation may vary as much as 100-fold depending on properties of the transforming donor DNA, including the marker to be selected, the extent of homology with the chromosome, the conformation of the molecule (linear or circular), and the distance between the marker and the end of a linear fragment.[10,24] Wrap the tubes in aluminum foil to exclude light and incubate at 30° for 4–16 hr with gentle agitation to keep the cells in suspension. Although transformation also occurs in the light, we consistently obtain a 2- to 10-fold increase in frequency by carrying out this step in darkness.[24]

Spread 100- to 150-μl aliquots of each transformation reaction on 100-mm plates containing 40 ml of BG-11 agar and incubate under standard illuminated growth conditions for 4–6 hr prior to the addition of the required selective agent. Add these agents by lifting the agar slab with an alcohol-flamed spatula and dispensing 400 μl of a 100×-concentrated stock underneath. Distribute the solution by rotating the spatula under the agar and removing it gently to reseat the agar slab. The incubation period between plating and antibiotic addition and the underlaying technique for adding the antibiotic are both important for obtaining a high yield of transformants.[24] Antibiotics which we have used for selection of transformants are shown in Table II. Transformed colonies appear within 4–7 days of incubation in the light at 30°.

Propagate the transformants by streaking a colony with a sterile toothpick across a sector of a fresh BG-11 agar plate which has been underlayed with selective agent. As many as 15 colonies can be streaked onto a single fresh plate, in about 1-cm^2 areas. After growth for 3 to 4 days, restreak the inoculated area with a sterile toothpick to enlarge the patch and redistribute the cells. After 2 to 3 days of additional growth, the patch should be a thick, uniform mat of cyanobacteria. Cells can be removed from the patch for inoculating liquid medium, in which the culture will establish more quickly and reliably than if inoculated from a single colony. The patch can also be used directly in a DNA miniprep for Southern analysis.

Total DNA Miniprep

To prepare small quantities of total DNA from transformants, scrape the cells from a patch and resuspend them in 500 μl of SE in a 1.5-ml microcentrifuge tube. Pellet the cells in a microcentrifuge and remove the supernatant. Resuspend the pellet in 340 μl of lysis buffer, containing lysozyme freshly added to a final concentration of 2 mg/ml. Incubate the sample at 37° for 60 min, then add 2 μl of a 20 mg/ml stock of proteinase K and 8 μl of 25% Sarkosyl. Incubate at 50° for 30 min, after which the sample should be clear and bright green to yellow–green in color. Add 7 μl of 5 *M* NaCl and extract the sample once with 1 ml of a 1 : 1 mixture of phenol and chloroform prepared as described in the reagents list. Mix very well to form a uniform emulsion for several minutes before separating the phases. Reextract the aqueous phase with 500 μl of chloroform. The sample should be clear and almost colorless. Add 700 μl of chaotropic salt solution and 10 μl of glass fines, prepared as described below. Bind the DNA to the glass fines by incubating at room temperature for 15 min. At this temperature, RNA will not bind to the glass. Remove the chaotropic salt solution by pelleting the glass fines in a microcentrifuge for 1 min and aspirating the liquid. Wash the glass pellet once by adding 1 ml of chaotropic salts, vortexing, and repelleting. Wash the pellet twice with 70% ethanol to remove traces of salt, and repellet. Carefully remove all liquid and allow the pellets to air dry briefly. Resuspend the glass in 20 μl of TE and elute the DNA at 68° for 15 min. Store the sample at 4° until needed, and pellet glass before removing each aliquot of DNA. This procedure produces enough DNA of sufficient purity for two to four restriction digests for Southern analysis.

Some cyanobacterial strains may not lyse well under the conditions described for *A. nidulans* R2. If cells are broken by another method, binding to glass in the presence of the chaotropic salt solution may still be useful to remove tightly bound membrane fragments and other cellular components which can interfere with restriction enzymes.

Choose a restriction enzyme which should distinguish between wild-type and transformant DNA sequences if heterologous DNA was inserted into the chromosome. Separate the products of restriction digestion by electrophoresis in a 0.7% agarose gel prepared in EB gel buffer (80 m*M* Tris, 20 m*M* sodium acetate, 2 m*M* EDTA, pH 8.0) at 2–3 V/cm for 12–16 hr for maximum separation of chromosomal fragments. Transfer to nitrocellulose paper or other DNA transfer membrane by established methods,[25] and probe with a nick-translated restriction fragment or plasmid which contains the cyanobacterial portion of the DNA used to produce the transformant. Figure 2 shows a DNA gel and Southern blot of total DNA from wild-type *A. nidulans* R2 and six transformants. The quality of total DNA prepared by the miniprep method (lanes C–G) is comparable to that from large-scale procedures which involve banding in CsCl density gradients (lanes A–B). The transformed strains (lanes B–G) had the *psbAI* gene inactivated by the insertion of Tn*5* using the scheme outlined in Fig. 1C. The probe used for Southern analysis hybridizes to all three of the *psbA* genes, located on *Bam*HI restriction fragments of 12, 8, and 3.3 kb. The experiment shows that the restriction fragment bearing the *psbAI* gene (3.3 kb) in wild-type cells is replaced by two new restriction fragments in the transformants. This is as expected, because *Bam*HI recognizes a single site within the Tn*5* kanamycin-resistance gene which bisects the region of chromosomal DNA homologous to the probe.

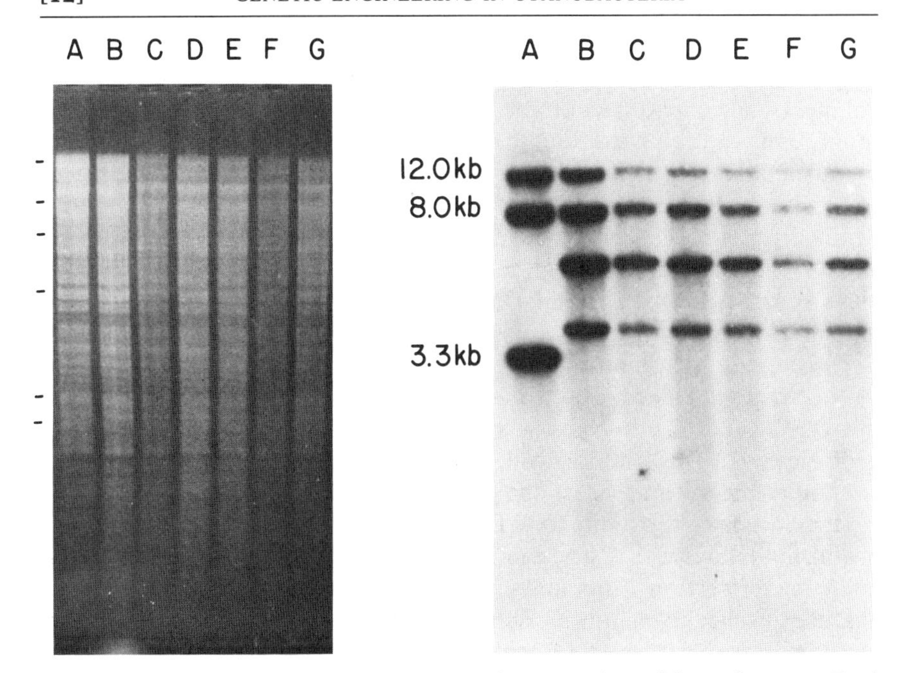

FIG. 2. Southern analysis of miniprep DNA from cyanobacterial transformants. Total DNA was isolated from wild-type *A. nidulans* R2 (lane A) and from six different kanamycin-resistant clones (lanes B–G) which were transformed by a cloned *psbAI* gene into which Tn*5* had been inserted. The DNA in lane A was isolated by subjecting the cell lysate from a 500-ml culture to CsCl/ethidium bromide buoyant density centrifugation. The DNA in lane B was a by-product of the RNA isolation procedure described in this chapter. Lanes C–G contain DNA which was prepared by the total DNA miniprep method. All samples (approximately 250 ng or one-fourth of a miniprep) were digested with the enzyme *Bam*HI and subjected to agarose gel electrophoresis and Southern analysis. The ethidium bromide-stained gel is shown at left. The migration positions are marked for fragments of *Hin*dIII-digested λ DNA (top to bottom, size in kilobases: 23.1, 9.4, 6.6, 4.4, 2.3, and 2.0). The probe, which is homologous to all three *A. nidulans* R2 *psbA* genes, contained all of *psbAI* inserted into pBR328. The autoradiogram is shown at right. The wild-type *psbA* genes are internal to the 12-, 8-, and 3.3-kb *Bam*HI fragments. The 3.3-kb *Bam*HI fragment is replaced in the transformants by two new restriction fragments generated by the single *Bam*HI site in Tn*5* which disrupts *psbAI*.

Preparation of Glass Fines for Binding DNA from Solution

The following instructions describe the preparation of glass fines used in the cyanobacterial total DNA miniprep procedure. This procedure was adapted from the method of Vogelstein and Gillespie.[28]

Suspend 250 ml of 325-mesh silica powder in water for a total volume (glass plus water) of 500 ml and stir for 1 hr. Let the glass settle for 1 hr,

[28] B. Vogelstein and D. Gillespie, *Proc. Natl. Acad. Sci. U.S.A.* **76,** 615 (1979).

then take the *supernatant* and pellet fines in a Sorvall GSA rotor at 5000 rpm for 5 min. Resuspend the pellet in 150 ml of water and add an equal volume of nitric acid. Heat to near boiling in a fume hood and let cool to room temperature. Wash the fines four times with sterile distilled water by suspending fines in water, pelleting as above, and resuspending in fresh sterile distilled water. Store fines as a 50% slurry (v/v) in sterile distilled water at 4°. The procedure yields about 20 g of glass in 40 ml of slurry. This should be enough fines for months or years of use (2 μl of glass slurry will bind 1 μg of DNA) and the preparation can be stored indefinitely.

To prepare the chaotropic salt solution dissolve 90.8 g of NaI and 1.5 g of Na_2SO_3 in distilled water for a total volume of 100 ml. Filter the solution through a 0.22-μm Nalgene sterilizing filter and add an additional 0.5 g of Na_2SO_3. The solution will be saturated, with crystals of Na_2SO_3 settling to the bottom. Store the solution in the dark at 4°.

In addition to use in the total DNA miniprep procedure, glass fines are useful for recovering restriction fragments from agarose gels for use in cyanobacterial transformation. This is accomplished by melting the agarose gel slice in chaotropic salt solution at 50° and binding the DNA to glass fines as described by Vogelstein and Gillespie.[28]

RNA Isolation

The isolation of high-quality RNA from cyanobacteria is necessary for the analysis of transcripts of genes of interest, including those which have been altered by chromosomal engineering. The following procedure, developed by S. Robinson and J. Golden, works well for *A. nidulans* R2 and *Anabaena* 7120, and should be applicable to other strains since it depends on mechanical rather than enzymatic cell breakage.

All solutions and tubes should be sterile and kept on ice, gloves should be worn at all times, and pipettor tips should be factory packed and autoclaved without having been touched. Centrifuge speeds and times are given for a Sorvall HB-4 swinging bucket rotor unless otherwise stated.

Harvest a 500- to 1000-ml cyanobacterial culture by centrifugation at 5000 rpm for 5 min in a Sorvall GSA rotor. Resuspend the cell pellet in 5 ml of 50/100 TE in a sterile 25-ml glass screw-capped centrifuge tube and add 1 ml of chloroform; leave on ice for 5 min and shake occasionally. Centrifuge at 5000 rpm for 5 min and decant the aqueous and chloroform layers, retaining the cell pellet, which will be a tight disk resting on the chloroform. Resuspend the pellet in 5 ml of breakage buffer; add 5 ml of 0.45-mm acid-washed, baked glass beads and 5 ml of a 1 : 1 mixture of phenol and chloroform. Vortex at top speed in three bursts of 3 min each, chilling the tube for a few minutes after each burst. Centrifuge at 10,000 rpm for 10 min to separate the organic and aqueous phases. Transfer the aqueous upper layer to a new tube; increase the volume to 7 ml with 50/100 TE and extract twice more with 1 : 1 phenol : chloroform and twice with chloroform. The glass beads can be recovered for reuse by rinsing with water until all color is gone, followed by acid washing and baking. The final aqueous layer should be clear and may be pink. Add one-tenth volume of 3 *M* sodium acetate and 2.5 vol of ethanol; chill at −20° for at least 2 hr. Collect the nucleic acids by centrifugation at 10,000 rpm for 10 min, dry the pellet, and resuspend in 6 ml of TE. Add 3 g baked CsCl, layer atop 2.5 ml of 5.7 *M* CsCl (in 50/100 TE) in a Beckman SW41 tube, and top off the tube with mineral oil. Spin in a Beckman ultracentrifuge in an SW41 rotor for >12 hr at 35,000 rpm and 20°. The DNA can also be recovered by collecting 1–2 ml of solution near the interface of the RNA solution and the 5.7 *M* CsCl. Chromosomal DNA isolated in this manner is of sufficient average size for Southern analysis. To recover the RNA, remove the liquid to within 1 ml of the bottom of the tube and cut off the tube to make a small cup containing the RNA pellet. Carefully aspirate the remaining liquid and rinse the surface of the pellet with 0.2 ml TE to remove CsCl. To rinse the pellet, add the TE with a pipettor tip, roll the cup to wash the RNA surface, and suck the solution back up into the tip to remove. Dissolve the RNA in 0.4 ml of TE; vortex or pipette up and down if necessary. Add 40 μl of 3 *M* sodium acetate and 2.5 vol of ethanol to precipitate the RNA. Collect the pellet by centrifugation at 10,000 rpm for 10 min, wash the pellet with 70% ethanol, and dry under vacuum. Resuspend the RNA in 0.1 ml TE and determine the concentration by optical density. Dispense into microcentrifuge tubes for long-term storage at −80° or short-term storage at −20°. For very long storage, ethanol precipitate aliquots, remove the ethanol, add 70% ethanol over the pellet, and store at −80°.

This RNA is suitable for Northern analysis and for mapping the 5′ ends of transcripts. The conditions for 5′ end mapping by primer extension with reverse transcriptase[29] and by S1 nuclease protection[30] are described elsewhere. We prepare RNA for Northern analysis using the glyoxal denaturation, gel electrophoresis, and nitrocellulose transfer procedures described by Thomas.[31] Figure 3 shows an example of Northern analysis of RNA from *A. nidulans* R2 wild type and two gene-inactivated transformants. The probe in this experiment was a fragment of DNA

[29] G. A. Kassavetis and E. P. Geiduschek, *EMBO J.* **1,** 107 (1982).

[30] N. E. Tumer, S. J. Robinson, and R. Haselkorn, *Nature* (*London*) **306,** 337 (1983).

[31] P. S. Thomas, this series, Vol. 100, p. 255.

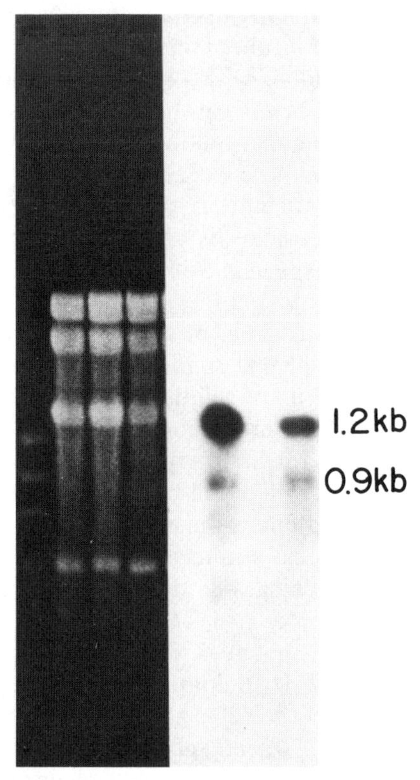

FIG. 3. Northern analysis of RNA from wild-type *A. nidulans* R2 and gene-inactivated transformants. RNA was isolated from wild-type (lane B), *psbAI*-inactivated (lane C), and *psbAII*-inactivated (lane D) strains as described in the Methods section. Samples were denatured with glyoxal and subjected to electrophoresis and Northern analysis as described by Thomas.[31] The acridine orange-stained gel is shown at left. Lane A contains glyoxal-treated *Hae*III fragments of ϕX-174 DNA (top to bottom, in kilobases: 1.35, 1.08, 0.87, 0.6). The RNA blot was hybridized to a radioactive probe which recognizes only full-length transcripts from the *psbAI* gene (1.2 and 0.9 kb). The autoradiogram at right shows that the *psbAI* transcripts are absent from the strain in which this gene was inactivated, but present in the *psbAII*-inactivated strain.

flanking the 3′ end of the *psbAI* gene; it recognizes only complete transcripts of the *psbAI* gene. The probe hybridizes to 1.2- and 0.9-kb RNA species in wild-type cells (lane B) and a *psbAII*-inactivated transformant (lane D), but contains no homology to RNA from the transformant in which the *psbAI* gene was interrupted (lane C).

Comments

The genetic engineering of the cyanobacterial chromosome includes, but is not limited to, the techniques described in this chapter. We have concentrated on methods which employ wholly homologous transforming DNA, or heterologous sequences which are flanked by cyanobacterial DNA on either side. Others have shown the usefulness of different constructions as tools for mutagenesis in cyanobacteria. Buzby *et al.*[12] used random *Sau*3AI fragments of *A. quadruplicatum* PR-6 ligated to Tn*1* as a means of promoting recombination of the selectable ampicillin-resistance gene with the chromosome at multiple sites. From the pool of transformants they obtained a mutant which was a better recipient for subsequent transformation with chimeric plasmid DNA. This mutagenesis procedure (ectopic mutagenesis) used selectable heterologous DNA which was linked to cyanobacterial DNA at one end, rather than flanked on both sides as depicted in Fig. 1C. Dzelzkalns and Bogorad[32] have observed a cyanobacterial transformation event which is independent of either autonomous replication or homologous recombination. They obtained stable, random insertion of *E. coli* plasmids into the genome of *Synechocystis* 6803 following low-level ultraviolet irradiation. Another important development in gene transfer techniques in cyanobacteria is conjugation of plasmid DNA from *E. coli* as described in a recent publication.[8] This promises to extend the methods described here for genetic engineering of chromosomes in the nontransformable filamentous cyanobacteria.

Acknowledgments

We are indebted to Drs. Steven Robinson and Jim Golden, who developed the RNA isolation protocol, Dr. Karen Kolowsky for suggestions which influenced the DNA miniprep procedure, and Dr. Aladar Szalay, who convinced us of the value of direct chromosomal engineering in cyanobacteria.

[32] V. A. Dzelzkalns and L. Bogorad, *J. Bacteriol.* **165,** 964 (1986).

[13] Conjugal Transfer of Plasmids to Cyanobacteria

By TERESA THIEL and C. PETER WOLK

Introduction

Cyanobacteria include some of the most colorful, complex, and interesting microorganisms in nature. Although their ultrastructure and the composition of their walls are typical of those of gram-negative bacteria,[1] their photosynthetic processes are similar to those of eukaryotic plants. The filamentous, heterocyst-forming cyanobacteria, which include the genera *Anabaena* and *Nostoc,* are of particular interest because they are capable of oxygenic photosynthesis, aerobic nitrogen fixation, and cellular differentiation with associated pattern formation. Heterocysts, which differentiate from vegetative cells in response to deprivation of fixed nitrogen, fix gaseous nitrogen[2,3]; akinetes, sporelike derivatives of vegetative cells which can survive adverse environments, germinate to form new filaments. Although many cyanobacteria form spreading colonies, strains of *Anabaena* (nonmotile by the definition of the genus[4]) form discrete colonies. Some can be replica plated.[5] The filamentous nature of the organisms presents difficulties both for quantitation of events in individual cells and for the isolation of strains containing recessive mutations. Bath cavitation of cultures in flasks to produce very short, viable filaments (<2 cells)[5] has proved useful for isolating auxotrophic mutants of *Anabaena variabilis*.[6] Facultatively photoheterotrophic or chemoheterotrophic filamentous strains offer promise for obtaining and characterizing mutants defective in photosynthesis.[7]

A complete understanding of the unique capabilities of cyanobacteria depends upon the development of systems for their genetic analysis. Some unicellular cyanobacteria are transformable,[8–11] but no reproducible

[1] C. P. Wolk, *Bacteriol. Rev.* **37,** 32 (1973).
[2] P. Fay, W. D. P. Stewart, A. E. Walsby, and G. E. Fogg, *Nature* (*London*) **220,** 810 (1968).
[3] R. B. Peterson and C. P. Wolk, *Proc. Natl. Acad. Sci. U.S.A.* **75,** 6271 (1978).
[4] R. Rippka, J. Deruelles, J. B. Waterbury, M. Herdman, and R. Y. Stanier, *J. Gen. Microbiol.* **111,** 1 (1979).
[5] C. P. Wolk and E. Wojciuch, *Arch. Microbiol.* **91,** 91 (1973).
[6] T. C. Currier, J. F. Haury, and C. P. Wolk, *J. Bacteriol.* **129,** 1556 (1977).
[7] P. W. Shaffer, W. Lockau, and C. P. Wolk, *Arch. Microbiol.* **117,** 215 (1978).
[8] S. V. Shestakov and N. T. Khyen, *Mol. Gen. Genet.* **107,** 372 (1970).
[9] C. Astier and F. Espardellier, *C. R. Hebd. Seances Acad. Sci., Ser. D* **282,** 795 (1976).

system for transformation exists for filamentous forms. Although many cyanophages have been isolated,[12–14] transduction has not been reported for any cyanobacterium. Shuttle vectors, plasmids capable of replication in alternate hosts, have been constructed for two unicellular cyanobacteria.[15–18] Such vectors can introduce (1) wild-type homologs of mutant alleles, for complementation analysis; (2) transposons, for mutagenesis[19]; (3) cloned and altered genes, for site-specific mutation[20]; and (4) genes with an easily assayed product, such as β-galactosidase[21] or luciferase,[22] to permit study of transcriptional control. We will describe techniques that have been used to construct shuttle vectors for strains of *Anabaena* and *Nostoc,* and conditions for conjugal transfer of those vectors from *Escherichia coli* to cyanobacterial hosts.

Principle of the Method

Many gram-negative bacteria can receive broad-host-range, P-incompatibility group plasmids by conjugation, and can maintain those plasmids.[23–25] Whereas the capacity to receive a conjugative plasmid probably depends only on the structure of the outer cell membrane, which is quite similar among gram-negative bacteria, maintenance of a plasmid requires both that the plasmid escape restriction by endonucleases in the new host, and that it be able to replicate in that host. If broad-host-range plasmids cannot be maintained in a host, DNA cloning techniques may be used to construct a chimeric plasmid. Such a plasmid contains sequences that are

---

[10] S. E. Stevens and R. D. Porter, *Proc. Natl. Acad. Sci. U.S.A.* **77,** 6052 (1980).
[11] G. Grigorieva and S. V. Shestakov, *FEMS Microbiol. Lett.* **13,** 367 (1982).
[12] L. A. Sherman and R. M. Brown, Jr., *in* "Comprehensive Virology" (H. Frankel-Conrat and R. Wagner, eds.), p. 145. Plenum, New York, 1978.
[13] N.-T. Hu, T. Thiel, T. H. Giddings, Jr., and C. P. Wolk, *Virology* **114,** 236 (1981).
[14] M. I. Mendzhul, N. V. Nesterova, V. A. Goryushin, and T. G. Lysenko, "Tsianofagi-Virusi Tsianobakteri," 146 pp. Naukova Dumka, Kiev, 1985.
[15] C. J. Kuhlemeier, W. E. Borrias, C. A. van den Hondel, and G. A. van Arkel, *Mol. Gen. Genet.* **184,** 249 (1981).
[16] J. S. Buzby, R. D. Porter, and S. E. Stevens, *J. Bacteriol.* **154,** 1446 (1983).
[17] L. A. Sherman and P. van de Putte, *J. Bacteriol.* **150,** 410 (1982).
[18] S. S. Golden and L. A. Sherman, *J. Bacteriol.* **155,** 966 (1983).
[19] G. B. Ruvkun and F. M. Ausubel, *Nature (London)* **289,** 85 (1981).
[20] S. Scherer and R. W. Davis, *Proc. Natl. Acad. Sci. U.S.A.* **76,** 4951 (1979).
[21] B. A. Castilho, P. Olfson, and M. J. Casadaban, *J. Bacteriol.* **158,** 488 (1984).
[22] J. Engebrecht, M. Simon, and M. Silverman, *Science* **227,** 1345 (1985).
[23] G. Ditta, S. Stanfield, D. Corbin, and D. R. Helinski, *Proc. Natl. Acad. Sci. U.S.A.* **77,** 7347 (1980).
[24] Y. Mooraka, N. Takizawa, and T. Harada, *J. Bacteriol.* **145,** 358 (1981).
[25] C. Thomas, *Plasmid* **5,** 10 (1981).

necessary and sufficient for replication in different hosts (e.g., derived from endogenous plasmids of those hosts) and can therefore serve as a shuttle vector, able to replicate in both those hosts.[15] The shuttle vector should carry one or more markers that allow selection for the plasmid in the two hosts of interest, and cloning sites. It may also be desirable for such a vector to possess as few sites as possible that are susceptible to restriction by the endonucleases of the hosts. Derivatives of pBR322 can be mobilized by several conjugative plasmids (including RP-4) to various gram-negative bacteria[26,27] if the vector contains the *bom* (basis of mobility) region of pBR322 and if *trans*-acting factors, required for mobilization, are provided by helper plasmids.[28]

To obtain conjugal transfer of the shuttle vector, one mixes the recipient strain with the donor strain or strains (usually *E. coli*) carrying the shuttle vector, the conjugative plasmid, and the helper plasmid. If the plasmids are in different donor strains, the conjugative plasmid must first transfer itself to the strain carrying the shuttle vector. In practice, the helper plasmid is normally present in the donor strain bearing the shuttle vector. In the presence of the helper plasmid, the conjugative plasmid then mobilizes the shuttle vector from *E. coli* to the recipient. Perhaps all three plasmids are transferred, but only the shuttle vector normally survives restriction and selection. After sufficient time is allowed for transfer of the shuttle vector to the recipient cells and for the expression of an antibiotic resistance encoded by that plasmid, those cells are subjected to conditions that select for the presence of the vector, and are freed of donor cells.

Construction of Shuttle Vectors

Shuttle vectors for certain strains of *Anabaena* and *Nostoc* have been constructed from plasmid pDU1 from *Nostoc* PCC 7524 and plasmid pBR322 from *E. coli*.[29] The original chimeric plasmid, pVW1, had the eight *Ava*II sites and one *Ava*I site of pBR322 and carried genes conferring resistance to ampicillin (Ap) and tetracycline (Tc). Plasmid pVW1 was modified by inserting the chloramphenicol (Cm) acetyltransferase

[26] D. P. Taylor, S. N. Cohen, W. G. Clark, and B. L. Marrs, *J. Bacteriol.* **154,** 580 (1983).

[27] E. van Haute, H. Joos, M. Maes, G. Warren, M. van Montagu, and J. Schell, *EMBO J.* **2,** 411 (1983).

[28] J. Finnegan and D. Sherratt, *Mol. Gen. Genet.* **185,** 344 (1982).

[29] C. P. Wolk, A. Vonshak, P. Kehoe, and J. Elhai, *Proc. Natl. Acad. Sci. U.S.A.* **81,** 1561 (1984).

gene from pBR328,[30] on a *Sau*3AI fragment, into the *Bam*HI site of pVW1 yielding plasmid pVW1C.

The large number of sites for restriction enzymes from *Anabaena* in pVW1C posed a potential problem. Upon transfer to many strains of *Anabaena* or its close relatives,[31] restriction of the plasmid might lower, perhaps greatly, the probability that exconjugants would retain the plasmid and therefore survive selection. Therefore, the *Ava*I and *Ava*II sites were removed, as follows. The plasmid was cut with *Ava*II, terminal phosphates were removed to prevent religation of *Ava*II sites, and the fragments were further cut with *Sau*96I. Because of the treatment with phosphatase, of the resulting 15 fragments only 4 contained 2 potentially ligatable ends. Because *Sau*96I restricts DNA at incompletely specified sequences, GGNCC, different fragments contained different terminal sequences. In consequence, any particular terminus could be ligated only to particular other termini. Four different recombinant plasmids, denoted pRL1, pRL2, pRL3, and pRL4, contained no sites for *Anabaena* endonucleases *Ava*I, *Ava*II, *Ava*III, or *Avr*II. The efficiency of conjugal transfer of pRL1 to *Anabaena* M-131 was much greater than observed for pVW1C. Plasmid pRL1 lacks not only the restriction sites mentioned; it also lacks large regions of pBR322. It is therefore difficult to attribute the increased efficiency to any single difference between pRL1 and pVW1C. Matings with shuttle vectors which differ only in the number of *Ava*II sites have indicated that each such site has relatively little effect on mating efficiency. However, we have consistently observed a correlation between the presence of large numbers of sites for *Ava*I and *Ava*II in chimeric plasmids intended for transfer to *Anabaena,* and low frequency of transfer per recipient cell. There is also evidence that modification and restriction account for large differences in the efficiency of plating of cyanophages on different cyanobacterial hosts.[13,32] Because such differences do not always correlate with the detectable presence or absence of known restriction endonucleases,[31] those cyanobacterial strains may have other, hitherto undetected restriction enzymes. The fact that pDU1-based plasmids transfer poorly or not at all to some strains of *Anabaena* and *Nostoc* may be due to restriction of the plasmids by host-specific endonucleases. Alternatively, pDU1 may fail to replicate, or the antibiotic resistance determinants may not be expressed, in those strains.

[30] P. Prentki, F. Karch, S. Iida, and J. Meyer, *Gene* **14,** 289 (1981).

[31] M. G. C. Duyvesteyn, J. Korsuize, A. de Waard, A. Vonshak, and C. P. Wolk, *Arch. Microbiol.* **134,** 276 (1983).

[32] T. C. Currier and C. P. Wolk, *J. Bacteriol.* **139,** 88 (1979).

In order to increase the probability that an expressible antibiotic resistance determinant would be available, the following derivatives of pRL1 were constructed. Plasmid pRL5 contains the streptomycin (Sm) resistance determinant from plasmid R300B,[33] pRL6 and pRL7 contain the kanamycin (Km)/neomycin (Nm) resistance gene from Tn*5*,[34] and plasmids pRL10 (replacing pRL8) and pRL11 contain the erythromycin (Em) resistance gene from pE194,[35] in opposite orientations. (Plasmid pRL8, which contains an *Ava*II site as a result of a cloning accident, has been retired.) Maps of pRL1, pRL5, pRL6, and pRL8 have been published.[29] Cm, Sm, Nm/Km, and Em have all been successfully used as selective agents for conjugation into strains of *Anabaena* or *Nostoc* as has Hg^r (resistance to mercuric ion) for the unicellular cyanobacterium, *Aphanocapsa*.[36] Resistance to Ap has been shown to be expressed strongly by exconjugant *Anabaena* sp. PCC 7120 that has received the β-lactamase gene of pBR322 on a shuttle vector.[37]

Derivatives of some of these plasmids have been constructed that lack the pDU1 region, and therefore cannot replicate in cyanobacteria. These vectors should be useful for introducing transposons and for promoting integration of a conjugally transferred plasmid into the cyanobacterial chromosome. The original plasmids have also been modified by reduction in size, and by inclusion (1) of polylinker regions to facilitate cloning, (2) of the λ *cos* site to enable packaging in λ, and (3) of features permitting positive selection for insertion into cloning sites.[37]

Mating Procedures

We present here a detailed description of the mating procedure, and a discussion of the effect, on the efficiency of conjugation, of varying certain of the experimental parameters.

In all matings, *E. coli* donors are mixed with the cyanobacterial recipients on filters on an agar medium. Normally, conditions are such that both the donor and recipient can grow for at least a few hours. The filter is then transferred to a selective medium and incubated under conditions that are optimal for growth of exconjugant cyanobacteria, until colonies appear. Control matings contain one or neither of the two strains of *E. coli* em-

[33] P. T. Barth, L. Tobin, and G. S. Sharpe, *in* "Molecular Biology, Pathogenicity, and Ecology of Bacterial Plasmids" (S. B. Levy, R. C. Clowes, and E. L. Koenig, eds.), p. 439. Plenum, New York, 1981.

[34] E. Beck, G. Ludwig, E. A. Auerswald, B. Reiss, and H. Schaller, *Gene* **19,** 327 (1982).

[35] S. Horinouchi and B. Weisblum, *J. Bacteriol.* **150,** 804 (1982).

[36] G. S. Bullerjahn and L. A. Sherman, *Abstr. Am. Soc. Microbiol.* **I64,** 157 (1985).

[37] J. Elhai and C. P. Wolk, unpublished observations.

ployed in these triparental[23] matings. The use of filters, a convenience introduced by Williams and Szalay,[38] is presumably not essential, because antibiotic can be added directly to the agar.[39] We use 25-mm filters (T.T.: MSI #E04WP02500, 0.45 μm, mixed cellulose ester) or 82-mm filters (C.P.W.: Schleicher and Schuell BA85; Nuclepore "Membra-Fil" membrane filters, 0.45 μm, 90 mm[29]) although, as will be pointed out below, many variations are possible.

E. coli HB101, containing both the shuttle vector and the helper plasmid, and *E. coli* J53, containing the conjugative plasmid RP-4, are grown overnight at 37° in L broth with Km at 50 μg ml^{-1}, Ap at 50 μg ml^{-1}, and Cm at 25 μg ml^{-1} as appropriate. Cultures are diluted 1 : 40 into L broth containing antibiotic, grown at 37° with shaking to optical density $(OD)_{600}$ of 0.45 (approximately 2.5 hr), and 0.75 ml centrifuged, washed once in L broth, and resuspended in 60 μl of L broth. For mixtures of *E. coli*, 0.75-ml portions of each strain are washed as described, mixed, recentrifuged, and resuspended in 60 μl of L broth. Such suspensions are left to mate at room temperature while the cyanobacteria are prepared.

Strains of *Anabaena* are grown in 50-ml volumes of cyanobacterial nutrient solution in 125-ml flasks, shaken at 100 rpm at 30–32° under cool white fluorescent lamps at 60–70 μE m^{-2} sec^{-1} (T.T.: 3500 erg cm^{-2} sec^{-1} [29,40]). We use, as nutrient solution, either BG-11[41] for liquid and solid media (T.T.) or AA/8 with 5 m*M* nitrate for liquid and AA with nitrate for solid medium.[29] Cultures about 1 week old are adjusted to an OD_{700} of 0.7 (T.T.); or to 5 μg chlorophyll ml^{-1} [40]; or not adjusted[29] and concentrated 20-fold by centrifugation. A concentrated suspension is then diluted 10-fold (2×) and 100-fold (0.2×) with nutrient solution.

Portions of the suspensions of bacteria and of cyanobacteria that are to be mated are placed on sterile filters atop 1-day-old Petri plates and allowed to dry. The Petri plates contain cyanobacterial nutrient medium supplemented with 5% (v/v) L broth and solidified with 1% Difco agar, purified in our laboratories.[42] For matings of pRL10 and pRL11 the medium also contains Em at 0.1 μg ml^{-1}. Ten-microliter portions of the cyanobacterial concentrate and of each dilution are spotted either alone or on top of 5-μl spots of the suspension of *E. coli* (T.T.) or, alternatively, 2-μl portions of the suspension of *E. coli* are spotted on top of streaks of

[38] J. G. K. Williams and A. A. Szalay, *Gene* **24,** 37 (1983).

[39] C. A. M. J. J. van den Hondel, S. Verbeek, A. van der Ende, P. J. Weisbeek, W. E. Borrias, and G. A. van Arkel, *Proc. Natl. Acad. Sci. U.S.A.* **77,** 1570 (1980).

[40] E. Flores and C. P. Wolk, *J. Bacteriol.* **162,** 1339 (1985).

[41] R. Y. Stanier, R. Kunisawa, M. Mandel, and G. Cohen-Bazier, *Bacteriol. Rev.* **35,** 171 (1971).

[42] A. C. Braun and H. N. Wood, *Proc. Natl. Acad. Sci. U.S.A.* **48,** 1776 (1962).

20-μl portions of the cyanobacterial suspensions (C.P.W.). To prevent undesirable spreading of the second organism, the first organism should be permitted to dry onto the filter before the second is added. Moreover, when one microorganism is spotted atop another, a fresh micropipet tip must be used for each addition of the top organism: the volumes used do not drop from the pipet tip, and any contact between the droplet on the pipet tip and the filter results in cross-contamination. The plates are incubated under the conditions used for growth of the cyanobacteria. After 24 hr, the filters are transferred to Petri dishes of BG-11 agar or AA with nitrate agar, containing antibiotic. For transfer of pRL6 (pDS4101 serving as helper plasmid) and pRL11 (usually using pGJ28 as helper plasmid) into either *Anabaena* sp. U. Tokyo M-131 or *Anabaena* sp. PCC 7120, we normally include Nm at 25 μg ml^{-1}, or Em at 5 μg ml^{-1}. Petri plates are incubated under cyanobacterial growth conditions until colonies appear (3–10 days).

Variant Protocols

Age and Amount of Donor. One of us (T.T.) has sought to determine the effect, on the efficiency of mating, of varying a number of the experimental parameters. In order to determine whether exponentially growing donor cells are important for high efficiency of transfer, the standard mating was compared with matings in which exponentially growing donor cells were replaced with stationary phase cells (overnight cultures) of the *E. coli* strains used for mating. The latter were washed with L broth and then adjusted to approximately the same optical density as the concentrated exponential phase cells of the standard mating. In addition, 5-μl portions of overnight cultures (without washing or concentration) of each *E. coli* donor were mixed, placed on filters on L agar plates without antibiotic, and incubated at 37° for 4 hr. The filters were then transferred to plates of BG-11 agar, supplemented with 5% (v/v) L broth, and portions of suspensions of *Anabaena* spotted on the *E. coli* as in the standard protocol. In later matings, overnight cultures of *E. coli* were used directly for matings (5 μl of each donor) without washing or adjustment of cell density (the optical density of a 1 : 5 dilution was 0.7–1.0).

For the matings of pRL6 and pRL11 with *Anabaena* strain M-131 there was little or no difference in the number of exconjugants regardless of the age or method of growth of the donor. For matings to *Anabaena* strain 7120 the highest yields of exconjugants were obtained with exponential phase donor cells, but the other methods of preparing the donor cells resulted in a decrease in yield of less than a factor of two compared to exponential phase cells. Except for matings in which the frequency of exconjugants is very low, it appears to make little difference whether exponential or stationary phase bacterial cells are used.

The amount of donor was varied by concentrating exponential phase cells at an OD_{600} of 0.45 by factors of up to 100. Matings used 5 μl of each donor. In matings of pRL6 or pRL11 with *Anabaena* strains M-131 and 7120 differences seen with different amounts of donor were slight (less than 2-fold) and inconsistent.

Conjugative Plasmids. Would P-incompatibility group plasmids other than RP-4, e.g., R702 (Km^r) and R751 (trimethoprim resistant), or plasmids of other incompatibility groups, e.g., W-incompatibility group plasmids S-a (Sm^r) and R7K (Sm^r), also support transfer? When grown and used for mating experiments as described for RP-4, plasmid R702 mediated transfer of pRL6 and pRL11 to both strains of *Anabaena* as well as did RP-4, and the efficiency of mobilization by R751 was nearly (within a factor of two) as great. Plasmids S-a and R7K did not promote transfer or did so very weakly.

Age of Recipient. Cultures of *Anabaena* subcultured weekly (1.0 to 50 ml of BG-11 liquid) were standardly used after 7–10 days (OD_{700} of about 0.7). To determine the effect of age on the efficiency of conjugation, both 4-day (OD_{700} ~0.3)- and 4-week (OD_{700} ~1.4)-old cultures were used as recipients after adjusting the cell density to correspond to the normal 20×, 2×, and 0.2× concentrations of recipient cells. There was very slight or no lowering of conjugation efficiency with old cells compared to young or standard age cells for all mating systems. The age of the recipient therefore appears not to be critical.

Filters. Most matings are performed on MSI brand mixed cellulose ester filters which contain a detergent. These filters are wrapped in aluminum foil in stacks of 25 filters and autoclaved at 20 psi for 20 min on fast exhaust. We also washed these filters in distilled water and autoclaved them in water to remove detergent, and used similar filters that are detergent free (MSI #E04TF02500, 0.45 μm). In addition we used MSI brand nylon filters (#N02SP02500, 0.22 μm). This particular set of experiments was done using only conjugal transfer of pRL6 to *Anabaena* strain M-131. There were no significant differences in the efficiency of conjugation with any of these filters; the ones used in the standard mating protocol are the least expensive. Moreover, the size of the filter used is a matter of convenience and economy.

Timing of Transfer of Filters. In order to determine whether our standard 24-hr period of time for plasmid transfer and for expression of antibiotic resistance prior to selection was optimal, filters were set atop antibiotic-containing medium immediately, or 24 or 48 hr after the inception of mating. Although there was little difference between the efficiency of conjugation for filters transferred at 24 or 48 hr, controls grew more if filters were not transferred until 48 hr. Exposure to antibiotic from the start of mating decreased the efficiency of conjugation at least 100-fold.

For matings of pRL11, conferring inducible resistance to Em, a subinhibitory concentration of Em was included in the medium during the first 24 hr of incubation. In one set of experiments with both strains of *Anabaena* this inducing amount of Em was omitted from some matings. No consequent reduction in the efficiency of conjugation was observed.

Concentration of Antibiotics. The concentration of antibiotic used for selection must be determined for each new cyanobacterium, and may vary for different media. Cyanobacteria are often more sensitive to antibiotics in liquid than on solid media; hence, the selective dose of antibiotic must often be reduced several-fold in order to ensure growth of exconjugants in liquid media. For example, whereas Nm is required at 10–25 μg ml^{-1} for selection of pRL6-containing exconjugates on BG-11 agar medium, 3–10 μg ml^{-1} suffices to inhibit completely the growth of wild-type cells in liquid medium.

Medium for Mating. In order to determine whether the addition of 5% L broth enhanced efficiency, matings were done according to the standard mating protocol except that one set of filters for each system was incubated on BG-11 agar medium without added L broth. In addition, one set of filters was changed from medium with 5% L broth to medium without L broth after 4 hr.[29] All media contained either no antibiotic, or a subinhibitory concentration of Em. Filters were transferred to selective media after 24 hr. The standard mating, with L broth present for 24 hr, gave the best results for all matings; however, the number of exconjugants was reduced by less than a factor of two, and in some experiments not reduced at all, for either no L broth or L broth present for only 4 hr. Because it is simple to add L broth to the medium, we have continued to do so, despite its limited effect.

In Summary. Conjugation works well with either exponential or stationary phase donor cells, and, above the lowest amount we used (about 10^7 cells of each donor strain on each spot), the amount of donor is not important. The age of the recipient is also not critical; however, none of the cultures that we used was senescent. In general, about 1.5×10^5 cells of the recipient (which corresponds to 10 μl of the $2\times$ concentration in the experiments described above) suffice for visualization of conjugation, but for systems of low or unknown efficiency up to 2×10^6 cells may be included in a mating mixture. For greater numbers of cyanobacterial cells within the same area, the protective effect of a higher concentration of cells could necessitate the use of higher concentrations of antibiotics, which might kill exconjugants. Our experience with brands of filters is limited, but we have had no failures that we can attribute to any particular brand or type of filter. Similarly it is likely that any medium that supports growth of the recipient strain is suitable for this procedure. The minimum lethal concentration of each antibiotic must be determined for each strain under its normal conditions of growth both in liquid and on solid medium. Although the addition of L broth to support growth of the donor during mating has only marginally beneficial effects, it or any substances required for growth of the donor are probably worth including. Matings should proceed for 24 hr prior to transfer of filters to selective media.

Efficiency of Transfer

The efficiency of transfer depends on both the shuttle vector and the recipient strain. The ratio of exconjugant colonies to total cyanobacterial cells in a spot is generally 10^{-2}–10^{-4}. However, if cyanobacterial filaments (each of which may give rise to only a single colony) are broken by bath cavitation to an average of 1 to 1.5 cells per fragment, frequencies of $\geq$0.03 have been measured for transfer of pRL1[29] and pRL11 to *Anabaena* strain M-131. *Nostoc* sp. strain ATCC 27896 gave an efficiency of 1 to 3 $\times$ 10^{-3} in matings of very short filaments with *E. coli* bearing pRL6 or pRL8.[40] Transfer of pRL6 to filaments of *Anabaena* strains 7120 and M-131 occurs at about 10^{-3}/cell,[29] i.e., much less than one exconjugant per filament, but has not been quantitated with very short filaments.

Purification of Exconjugants and Recovery of Plasmid

Colonies of cyanobacteria growing on antibiotic-containing medium are streaked on the same selective medium used in mating. In general, only colonies from spots containing both donor strains survive restreaking.[43] Axenic colonies can normally be obtained on the first restreaking, but further streakings are sometimes required. We have tried to counterselect the *E. coli,* other than by use of a medium such as BG-11 or AA that does not support its growth. Specifically, we have added colicin E1, arsenate (most cyanobacteria are resistant to 100–200 m*M*), trimethoprim

[43] One of us (T.T.) has on several occasions isolated antibiotic-resistant cyanobacterial colonies from control filters that contain *E. coli* with a shuttle vector but that lack a conjugative plasmid. These colonies have been shown by plasmid extraction and by transformation of *E. coli* to contain the shuttle vector, and may therefore have resulted from transformation. However, conditions that give rise to such colonies have not been identified.

(most cyanobacteria are resistant to 50–100 μg/ml), or streptomycin (using Sm^s *E. coli* and Sm^r cyanobacteria received courtesy of N. Wood, Roosevelt University) during or after antibiotic selection for the plasmid. Although these agents have great potential to eliminate the donor cells, a fully satisfactory protocol has not yet been developed.

Colonies tested as axenic are grown in liquid media, containing antibiotic, until they are a moderate to dense green. A portion of each liquid culture is subcultured to L broth to test whether *E. coli* is indeed absent. The presence of the shuttle vector in cyanobacterial cells can be verified initially by visualization of a plasmid of appropriate molecular weight on agarose gels, and subsequently by isolation and characterization of the plasmid from the cyanobacterium. More frequently, DNA isolated from the cyanobacterium is used to transform *E. coli,* and plasmid isolated from *E. coli* is then more fully characterized.[29,40] Sometimes, little or no new plasmid is observed in extracts of cyanobacterial cells, but plasmid-containing transformants of *E. coli* are obtained using these extracts. In such instances, it is desirable to subculture the cyanobacteria extensively, and then to reisolate the plasmid. Such transformants might have arisen from nonviable, plasmid-bearing cells of *E. coli* present in the original suspension. Two methods for isolation of plasmids from cyanobacteria have been most successful in our laboratories: the method of Simon[44] has been most effective for isolation of plasmids from *Anabaena* strains 7120 and M-131 and from several other filamentous strains; the method of Kuhlemeier *et al.*[15] has been effective for unicellular strains and for a few filamentous strains. Processing of 100 to 250 ml of a dark green culture normally provides sufficient amounts of plasmid to permit easy visualization.

Cyanobacterial Strains to Which Conjugation Has Been Demonstrated

Plasmids have been transferred by conjugation to other strains of *Anabaena* and *Nostoc* in addition to those strains mentioned above[29,40,45]; to *Fremyella,*[46] which is also a filamentous cyanobacterium; and to the unicellular cyanobacteria, *Anacystis* (*Synechococcus*)[29] and *Aphanocapsa* (*Synechocystis*).[36] In principle, the technique can be used to transfer plasmids to a wide range of cyanobacteria and, in fact, to a wide range of gram-negative bacteria.

[44] R. D. Simon, *J. Bacteriol.* **136,** 414 (1978).

[45] C. P. Wolk, E. Flores, G. Schmetterer, A. Herrero, and J. Elhai, *in* "Nitrogen Fixation Research Progress" (N. J. Evans, P. J. Bottomley, and W. E. Newton, eds.), p. 491. Nijhoff, Dordrecht, The Netherlands, 1985.

[46] J. Cobley, personal communication.

Acknowledgements

I (T.T.) thank Nancy Cross for help in experiments in which the protocol for mating was varied. Research described from the laboratory of T.T. was supported by N.S.F. Grant DCB-8508542 and in the laboratory of C.P.W. by the U.S. Dept. of Energy under Contract DE-AC02-76ER01338 and by N.S.F. Grant PCM-8402500.

[14] Site-Directed Chromosomal Rearrangements in Yeast

By RICHARD T. SUROSKY and BIK-KWOON TYE

Transformation of yeast with nonreplicating plasmids yields transformants in which the plasmid DNA is stably integrated into the genome.[1] The integration occurs at a region of homology between the plasmid and the yeast genome. It has been observed that linear DNA fragments transform at higher frequencies than their circular counterparts,[2] presumably because free-ended DNA is capable of strand invasion and therefore is recombinogenic. These properties of yeast transformation have allowed the development of several techniques for altering DNA structure *in vivo* such as gene interruption[3] and gene disruption.[4] Gene disruption, sometimes known as gene replacement, allows the substitution of the gene of interest by an altered gene. In this procedure, cells are transformed using a DNA fragment containing the 3′ and 5′ ends of the gene of interest, but with the internal portion of the gene replaced by a selectable marker. In many of the resulting transformants, the wild-type copy of the gene on the chromosome is replaced by the disrupted gene. In this chapter, we describe an extension of this technique which allows the replacement of large regions of the genome with small DNA fragments. Using this technique, we have generated a series of large chromosomal deletions,[5,6] as well as a circular derivative of a chromosome.

[1] A. Hinnen, J. B. Hicks, and G. R. Fink, *Proc. Natl. Acad. Sci. U.S.A.* **75,** 1929 (1978).
[2] T. L. Orr-Weaver, J. W. Szostak, and R. J. Rothstein, *Proc. Natl. Acad. Sci. U.S.A.* **78,** 6354 (1981).
[3] D. Shortle, J. E. Haber, and D. Botstein, *Science* **217,** 371 (1982).
[4] R. J. Rothstein, this series, Vol. 101, p. 228.
[5] R. T. Surosky and B.-K. Tye, *Proc. Natl. Acad. Sci. U.S.A.* **82,** 2106 (1985).
[6] R. T. Surosky, C. S. Newlon, and B.-K. Tye, *Proc. Natl. Acad. Sci. U.S.A.* **83,** 414 (1986).

General Outline of Method

Diploid strains are used in the construction of large chromosomal deletions because extensive deletions are lethal in haploid cells. We begin with a diploid that is homozygous for a mutation in the *Z* gene (Fig. 1). The diploid is also heterozygous for a second gene, *B*/*b*, located in the region that is to be deleted. The phenotype of the diploid is Z^-B^+. The diploid is then transformed to Z^+ using a DNA fragment containing the wild-type *Z* gene with sequences from the *A* region of the chromosome at one end and *C* region at the other end. Transformants containing the deletion are identified by screening for a B^- phenotype, presumably resulting from a deletion between *A* and *C* on the chromosome containing the *B* gene. These Z^+B^- transformants are analyzed further to determine if the correct deletions have been made. Three criteria are used to identify the desired deletions: (1) Other markers in the deleted region of the chromosome are lost; (2) diploids sporulate to yield only two viable spores; (3) the restriction map determined by genomic blot analysis agrees with the predicted restriction map of the desired deletion. We chose to make all of

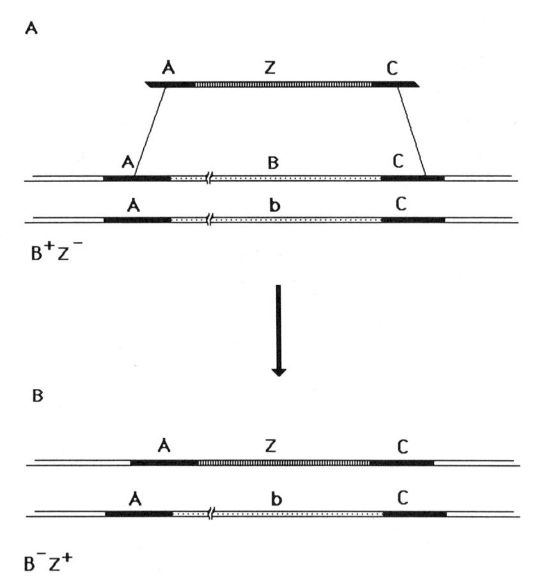

FIG. 1. Construction of large chromosomal deletions. (A) A diploid cell with the phenotype Z^- and B^+ (*B*/*b*) is transformed using a linear DNA fragment containing the selectable marker *Z*, DNA homologous to one region of the chromosome at one end, *A*, and DNA homologous to another region of the chromosome at the other end, *C*. (B) The fragment replaces all sequences between regions *A* and *C* including the *B* gene, producing a transformant that is Z^+ B^-.

our deletions on chromosome III (Fig. 2) for several reasons. This chromosome is one of the smallest in the yeast genome,[7] and there are multiple genetic markers that can be used in the screening for deletions. More importantly, many regions of the chromosome have been cloned (C. Newlon, personal communication) so that deletions from different regions can be made.

In the plasmids used for the construction of the chromosomal deletions, both the position and the orientation of the yeast sequences are important for generating the proper deletion. When cut with the appropriate restriction enzymes, the plasmids yield a fragment with a selectable marker flanked by yeast sequences at both ends which will serve as the endpoints of the deletion. The relative orientation of these sequences at the ends of the fragment determines the type of deletions to be made. If the end sequences are in the direct orientation, i.e., the same as their orientation in the chromosome, then internal deletions will be generated (Fig. 2). If the end sequences are in the inverted orientation, then distal deletions will be generated, resulting in a circular chromosome (Fig. 3). We successfully constructed deletions using fragments with as little as 0.7 kb of homologous sequence at one end. It is possible that even shorter homologous sequences would suffice.

Materials and Methods

Strains and Media

Saccharomyces cerevisiae strain 8534-8C (*MATα ura3 his4 leu2*) was obtained from G. Fink and strain GM119 (*MATα ura3 leu2*) was obtained from G. Maine. Strains 320-13B (*MATα ura3 trp1 met3 MAL2*) and 320-16C (*MATα ura*3 *leu*2 *trp*1 *met*3 *MAL*2) were constructed for this study. The diploid 168C was constructed by crossing 320-16C with 8534-8C. The *Escherichia coli* strain HB101 (*thr leuB hsr hsm recA*) was used as the host for plasmid amplification. YEPD, synthetic complete medium, and MAL indicator medium have been described.[8]

DNA Preparation and Transformation

Purified plasmid DNA was prepared by the method of Birnboim and Doly[9] and rapid DNA minipreps were prepared by the boiling technique of

[7] G. F. Carle and M. Olson, *Proc. Natl. Acad. Sci. U.S.A.* **82,** 3756 (1985).

[8] F. Sherman, G. R. Fink, and D. W. Lawrence, "Methods in Yeast Genetics." Cold Spring Harbor Lab., Cold Spring Harbor, New York, 1974.

[9] H. C. Birnboim and J. Doly, *Nucleic Acids Res.* **7,** 1513 (1979).

Holmes and Quigley.[10] Total yeast DNA was prepared as described[11] and circular yeast DNA was prepared by the method of Devenish and Newlon.[12] Transformation of *E. coli*[13] and yeast[1] was carried out as described.

Other Procedures

The preparation of nick-translated ^{32}P-labeled DNA probes,[14] transfer of DNA to nitrocellulose,[15] and hybridization conditions[16] have been described. Meiotic tetrad analyses were performed according to published procedures.[8]

Results

Deletions on One Arm of the Chromosome

Plasmid DNA was digested with the appropriate restriction enzyme(s) to completion and used directly to transform yeast cells. The diploids used in the construction of the deletions on the left arm of the chromosome were made by crossing strain 8534-8C with strain 320-13B. For the deletions on the right arm, we used diploids made by crossing strain 8534-8C with strain 320-16C or strain GM119. It is important to use haploids with the same *ura3* or *leu2* allele, otherwise *URA3* or *LEU2* prototrophs may result from recombination between the two mutant alleles. In all of these deletions, one of the endpoints was the repetitive telomeric *Y′* sequence which is present on most yeast telomeres.[17,18] In a typical transformation, 20 μg of plasmid and 2×10^9 cells were used and greater than 10,000 transformants were obtained.

To identify transformants containing a deletion chromosome, we first screened the transformants for the desired phenotype and then examined those with the proper phenotype using genetic and hybridization analysis. Approximately 500 transformants were picked onto 2 sets of selective plates and then replica plated onto the appropriate tester plates. When screening for deletions on the left arm of the chromosome, we looked for

[10] D. S. Holmes and M. Quigley, *Anal. Biochem.* **114,** 193 (1981).
[11] D. R. Cryer, R. Ecclesshall, and J. Marmur, *Methods Cell Biol.* **12,** 39 (1975).
[12] R. J. Devenish and C. S. Newlon, *Gene* **18,** 277 (1982).
[13] D. A. Morrison, *J. Bacteriol.* **132,** 349 (1977).
[14] P. W. J. Rigby, M. Dieckmann, C. Rhodes, and P. Berg. *J. Mol. Biol.* **113,** 237 (1977).
[15] E. Southern, *J. Mol. Biol.* **98,** 503 (1975).
[16] C. S. M. Chan and B.-K. Tye, *Proc. Natl. Acad. Sci. U.S.A.* **77,** 6329 (1980).
[17] C. S. M. Chan and B.-K. Tye, *J. Mol. Biol.* **168,** 505 (1983).
[18] C. S. M. Chan and B.-K. Tye, *Cell* **33,** 563 (1983).

loss of *HIS4* using plates with complete media without histidine (Cm-his). When screening for deletions on the right arm, we used either MAL plates to indicate the loss of *MAL2* or mating lawns to indicate the loss of *MATa*. We obtained positives in this screen at a frequency of 1–5%. We determined that the changes in phenotypes were not caused by chromosome loss by examining the markers on the other arm of the chromosome. We verified that His$^-$ diploids were still nonmaters and thus retained both the *MATα* and *MATa* alleles, and Mal$^-$/Matα diploids were still His$^+$. The events that produced the other transformants were not determined, but they may have resulted from gene conversion of *ura3* or *leu2* using the wild-type gene on the DNA fragment.

The diploids that displayed the proper phenotype were then sporulated and the spore viability in these tetrads was determined. In each case, mostly two, but no more than two, viable spores were obtained, suggesting that an extensive deletion was made and the deletion was haploid lethal.

Physical Analysis of Deletion Chromosomes

If the correct deletions have been made, the restriction maps of these derivative chromosomes at the site of the deletion should conform to that of the transforming DNA fragment. Furthermore, the regions that were adjacent to the deleted sequences should now be joined. For example, the diploid strain 168C was transformed with the plasmid pPLY-1, which was cleaved with the restriction enzyme *Sst*I to yield a DNA fragment with homology to the *PGK1* region at one end and the telomeric *Y′* sequence at the other (Fig. 2a). Genomic DNA was prepared from two transformants, DPT6 and DPT11, with the desired phenotype and the parent strain 168C. The structure of the region surrounding the integrated *LEU2* gene in each of the deletion chromosomes was determined by digesting the DNA with different restriction enzymes and probing the DNA blots with the 1.0-kb *Sal*I/*Eco*RI fragment from the *LEU2* region (Fig 2a).

When the DPT6 strain was examined in conjunction with its parent strain, the probe always hybridized to a single fragment from the normal *LEU2* region in the 168C strain and to an additional fragment from the deleted region in the DPT6 strain (Fig. 2b). From the sizes of the hybridizing fragments in the deletion strain, we can deduce the restriction map of the region surrounding the deletion (Fig. 2a). The map of the *Y′* region adjacent to the *LEU2* gene on the deletion chromosome is identical to that of the 6.7-kb *Y′* sequence on the transforming plasmid. On the other side of the 6.7-kb *Y′* repeat is another 5.5 kb of *Y′* sequence. At the end of these tandemly repeated *Y′* sequences is an apparent cluster of restriction

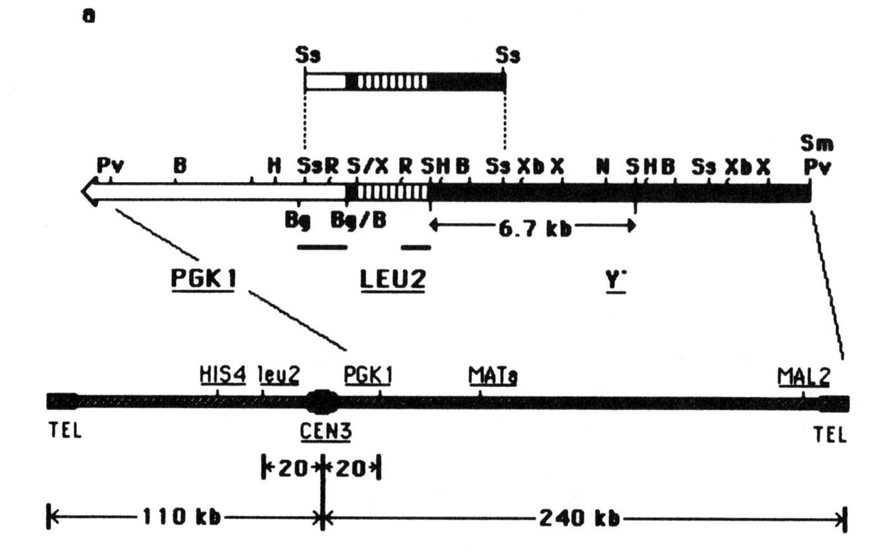

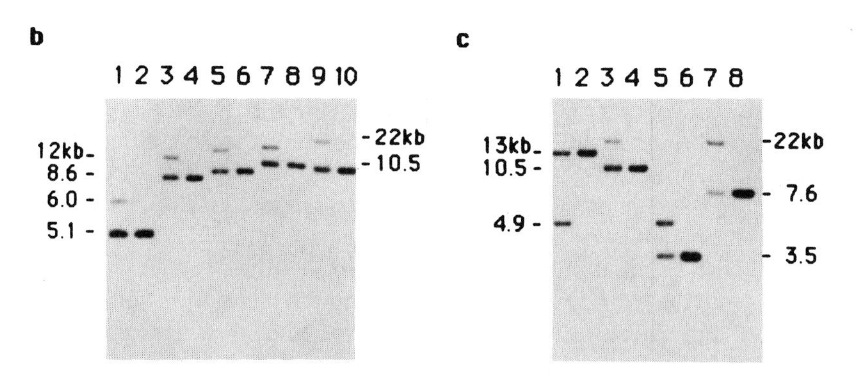

FIG. 2. Analysis of the deletion on chromosome III in the DPT6 strain. (a) Restriction maps of the rearranged region in chromosome III of the transformant DPT6 and the transforming DNA fragment (shown as bar above the rearranged region). Restriction enzyme sites shown are as follows: B, *Bam*HI; Bg, *Bgl*II; H, *Hin*dIII; N, *Nco*I; Pv, *Pvu*II; R, *Eco*RI; S, *Sal*I; Sm, *Sma*I; Ss, *Sst*I; X, *Xho*I; Xb, *Xba*I. Bars shown below the restriction map represent probes used for hybridization in Fig. 2b and c. Genetic markers on chromosome III are also shown to indicate their approximate locations on the chromosome. ■, pBR322 DNA. (b) Two micrograms of genomic DNA from each of the strains DPT6 and 168C were digested with different enzymes, separated on an agarose gel, transferred to a nitrocellulose filter, and hybridized with a ^{32}P-labeled probe. The hybridization probe used was the 1.0-kb *Sal*I/*Eco*RI fragment from the *LEU2* region. The DNA and the restriction enzymes used are as follows: lane 1, DPT6, *Sst*I; lane 2, 168C, *Sst*I; lane 3, DPT6, *Pvu*II/*Xba*I; lane 7, DPT6, *Pvu*II/*Nco*I; lane 8, 168C, *Pvu*II/*Nco*I; lane 9, DPT6, *Pvu*II; lane 10, 168C, *Pvu*II. (c) DNA was prepared as described in (b). Lanes 1–4, the 1.0-kb *Sal*I/*Eco*RI fragment from the *LEU2* region was used as hybridization probe; lanes 5–8, the 1.6-kb *Bgl*II fragment from the *PGK1* region was used as hybridization probe. Lanes 1 and 5, DPT6 DNA digested with *Hin*dIII; lanes 2 and 6, 168C DNA digested with *Hin*dIII; lanes 3 and 7, DPT6 DNA digested with *Pvu*II; lanes 4 and 8, 168C DNA digested with *Pvu*II.

enzyme sites (*Pvu*II, *Bss*HI, and *Sma*I) which are not found in any of the cloned *Y'* sequences. The position of these sites distal to the *Y'* sequence is diagnostic of the termini of most yeast chromosomes.[5,6,17,18] These tandem *Y'* repeats were also observed on the right arm of the deletion chromosome in strain DPT11. Thus, the *LEU2–Y'* junction is 12 kb from the telomere in the deletion chromosome of the DPT strains.

To verify that the *LEU2* region is adjacent to the *PGK1* region, we examined the DPT6 diploid for junction fragments that hybridized to probes from the *LEU2* and *PGK1* regions. DNA from diploid DPT6 digested with the enzyme *Hin*dIII and probed with the 1.0-kb *Sal*I/*Eco*RI fragment from the *LEU2* region or the 1.6-kb *Bgl*II fragment from the *PGK1* region produced a band of 4.9 kb (Figs. 2b and c). These probes hybridized to the same 22-kb fragment in *Pvu*II digests of DPT6 DNA. Similar results were obtained from the DPT11 strain. In these strains, we have deleted approximately 220 kb of DNA (Fig. 2a) between the telomere and the *PGK1* region on the right arm of chromosome III.

Deletions on Both Arms of the Chromosome

In the construction of double deletions, we have isolated diploids in which both the left and right arm deletions were generated on the same chromosome as well as diploids in which a left or a right arm deletion has been generated on each of the homologs. To make these secondary deletions, we began with diploids that contained a left arm deletion. These diploids were Leu$^-$ as a result of the removal of the *LEU2* allele by the deletion on the left arm. The same plasmids used to make the primary deletions on the right arm in previous constructions were used to construct these deletions. By screening for loss of one of the mating type alleles, transformants which contained a deletion on either copy of chromosome III could be identified.

The frequencies that these secondary deletions were obtained was comparable to that of the primary deletions. The frequency of secondary deletions generated on the same or a different chromosome was also similar. It is important to examine the spore viabilities in tetrads derived from the diploids with putative secondary deletions. Diploids which contain a chromosome which has a deletion on both the left and right arm should produce tetrads with no more than two viable spores. Some tetrads with one or no viable spores may be obtained as a result of recombination between the normal and deletion chromosome. In tetrads derived from diploids with a deletion on each of the homologs, nearly half should contain no viable spore. In one case, a diploid which contained deletions on both arms of chromosome III had undergone a mitotic recombination with its homolog on the right arm. Progenies derived from this diploid, which we believed to contain both the primary and secondary deletions on the same chromosome, actually transferred one of its deletions to the homolog. This was only determined after genetic analysis of these transformants.

The Deletion Derivatives of Chromosome III

Using this procedure, we constructed a series of deletion derivatives of chromosome III (Table I). The size of the original chromosome is 350 kb[7] and the derivatives range in size from 270 to 29 kb. The 29-kb chromo-

TABLE I
DELETION DERIVATIVES OF CHROMOSOME III

Strain	Structure	Size	Deleted Sequence
Wild-type	H L 3 P ————*————	350	
DLT	L 3 P ——*————	270	*LEU2* → *Y′*
DC_LT	3 P *————	250	*CEN3* → *Y′* (left arm)
DPT	H L 3 P ————*——	150[a]	*PGK1* → *Y′*
DC_RT	H L 3 ————*—	120	*CEN3* → *Y′* (right arm)
DLMT	L 3 P Cr ——*———	120	*LEU2* → *Y′*, *CRY1* → *Y′*
DC_LMT	3 P Cr *———	100	*CEN3* → *Y′* (left arm), *CRY1* → *Y′*
DHPT	H L 3 P ————*——	90	*HIS4* → *Y′*, *PGK1* → *Y′*
DHC_RT	H L 3 ————*	70	*HIS4* → *Y′*, *CEN3* → *Y′* (right arm)
DLPT	L 3 P ——*——	65	*LEU2* → *Y′*, *PGK1* → *Y′*
DC_LPT	3 P *——	47[a] 40	*CEN3* → *Y′* (left arm), *PGK1* → *Y′*
DLC_RT	L 3 ——*	40	*LEU1* → *Y′*, *CEN3* → *Y′* (right arm)
DC_LC_RT	3 —*	29[a]	*CEN3* → *Y′* (left arm), *CEN3* → *Y′* (right arm)

[a] Contains an extra copy of *Y′* at the right telomere.

some contains only the centromere, the telomeres, *URA3, LEU2,* and pBR322 DNA. The structure of these derivatives varies from telocentric to nearly metacentric. All of these derivatives were made by using one or two of six different plasmids. On the left arm, we deleted sequences between the left telomere and the *HIS4, LEU2,* or *CEN3* region. On the right arm, we deleted sequences between the right telomere and the *PGK1, CRY1,* or *CEN3* region. By combining different left and right arm deletions, we constructed a series of chromosomes of different sizes and structures.

On some deletion chromosomes, an extra copy of *Y'* observed at the right telomere. However, none of the deletion chromosomes contained a similar tandem repeat of the *Y'* sequence at the left telomere. This suggests that there may be tandem copies of the *Y'* sequence at the right telomere of the normal chromosome III and that recombination between the DNA fragment and the chromosome can take place at either one of the repeats. The structure of one such deletion chromosome containing a tandem *Y'* repeat is shown in Fig. 2.

Construction of Ring Chromosomes

We have also constructed a ring derivative of chromosome III using this same procedure. A diploid made by crossing strain 8534-8C with strain GM119 was transformed using a DNA fragment containing the selectable marker *URA3* with sequences from the *LEU2* region at one end and the *PGK1* region at the other end (Fig. 3). The orientation of the *LEU2* and *PGK1* sequences determines that a ring chromosome rather than an internal deletion will be generated. Transformants were screened for loss of the *HIS4* gene which is distal to *LEU2*, and the *MATa* gene which is distal to *PGK1*. Three percent of the transformants screened displayed the proper phenotype, a frequency similar to that obtained in the generation of internal deletions. DNA from the transformants displaying the proper phenotype was prepared by an alkaline extraction method[12] which enriches for circular DNA. After fractionation of the DNA on an agarose gel and hybridization with the appropriate probes, the ring chromosome could be identified by its slower migration as compared to linear DNA.[12,19]

Concluding Remarks

Although all of the examples of chromosomal deletions illustrated here were constructed on chromosome III, in principle, unless barred from

[19] R. T. Surosky and B.-K. Tye, *Genetics* **110,** 397 (1985).

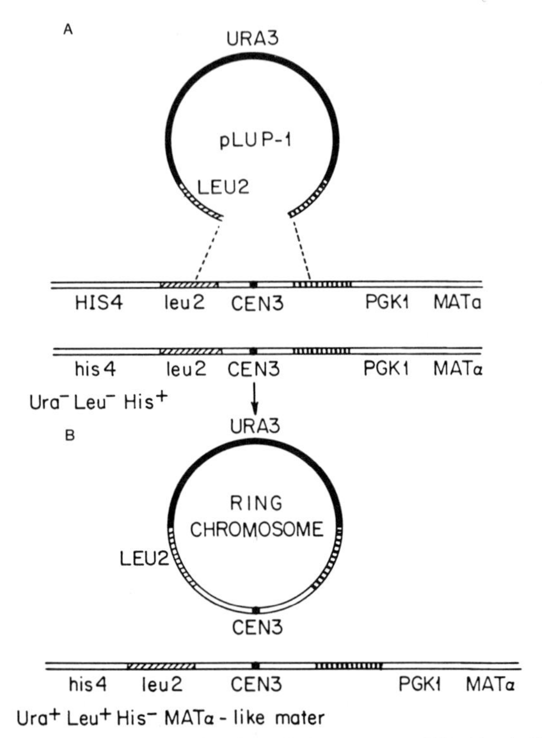

FIG. 3. Construction of a ring derivative of chromosome III. (A) A diploid strain was transformed with a DNA fragment with *URA3* and sequences from the *LEU2* region at one end and the *PGK1* region at the other end. (B) Formation of a ring chromosome may result in loss of sequences distal to *LEU2* and *PGK1*, producing a His⁻ diploid that mates as a *MATα* cell.

lethal dosage effects, similar deletions could be constructed on other chromosomes. The only requirements for the construction of these deletions are that the sequences at the endpoints of the deletion are available, and that their orientations on the chromosome are known. There should be a suitable genetic marker in the region between the two deletion endpoints. The frequency at which the deletions are obtained appeared independent of either the size of the deletion or the size of the remaining chromosome, so this method could be used for constructing longer deletions on the larger yeast chromosomes.

The ease with which one can alter the structure of yeast chromosomes is very useful in the study of the replication, recombination, and segregation of chromosomes during mitosis and/or meiosis. We used the deletion chromosomes to study the effects of chromosome size and structure on mitotic stability.[5,6] Not only could we examine the stabilities of chromosomes ranging in size from 270 to 29 kb, but we could also compare the stabilities of telocentric, acrocentric, and metacentric chromosomes of similar size. The technique described here may be used to construct other types of rearrangements. Chromosomal translocations may be generated by transforming cells with a DNA fragment that contains sequences from one chromosome at one end of the fragment and sequences from another chromosome at the other end. These translocations must be made such that acentric or dicentric chromosomes are not generated. It also may be possible to fuse two chromosomes together, provided that the centromere of one of the chromosomes to be fused can be deleted simultaneously.

By changing the orientation of the homologous sequences on the DNA fragment, we constructed a ring derivative of chromosome III. Studies on artificial circular chromosomes have suggested that factors affecting their mitotic stability may be different from those affecting linear chromosomes.[20] Using this method, the mitotic behavior of various circular derivatives of a natural chromosome could be investigated. The ability to readily circularize regions of a chromosome may aid in the isolation of genes or other sequences which cannot be isolated by simple selection. If the location of the sequence has been defined genetically, then the region can be circularized. The ring chromosome could be purified[17] and then subcloned for analysis.

Acknowledgment

This work was supported by Grant MV-233 from the American Cancer Society.

[20] P. Heiter, C. Mann, M. Snyder, and R. W. Davis, *Cell* **40**, 381 (1985).

[15] Improved Vectors for Plant Transformation: Expression Cassette Vectors and New Selectable Markers

By S. G. Rogers, H. J. Klee, R. B. Horsch, and R. T. Fraley

Introduction

Much of the recent progress in plant molecular biology, particularly in the area of gene function, is due to the availability of facile systems for plant transformation. Improvements in the basic vectors and selectable markers for identification of transformants are appearing at an increasing pace. These are leading to increased efficiency and ease of producing transgenic plants as well as to an extension of the range of plant species which may be transformed. This chapter will review the vectors and methods based on the *Agrobacterium tumefaciens* Ti plasmid and will describe, in detail, the construction and use of several new cassette plasmids for expression of gene-coding sequences in plants.

Agrobacterium-Derived Plant Transformation Systems

By far, the *Agrobacterium tumefaciens* Ti plasmid-derived vectors are the easiest and most utilized of the various schemes for introduction of DNAs into plants. In nature, *A. tumefaciens* infects most dicotyledonous and some monocotyledonous plants[1] by entry through wound sites. The bacteria bind to cells in the wound and are stimulated by phenolic compounds released from these cells[2] to transfer a portion of its endogenous, 200-kilobase (kb) Ti (tumor inducing) plasmid into the plant cell. The transferred portion of the Ti plasmid (T-DNA) becomes covalently integrated into the plant genome where it directs the biosynthesis of phytohormones from enzymes which it encodes. These phytohormones stimulate the undifferentiated growth of the infected and surrounding tissue, leading to the formation of a tumor or gall.

Two regions of the Ti plasmid are essential for the mobilization and integration of the transferred DNA into the plant cell. The *vir* region encodes a group of bacterial genes whose products are required for the excision of the T-DNA from the Ti plasmid, and for its transfer and integration into the plant genome.[3,4] The expression of these genes has been shown to be induced by compounds released from damaged plant cells at the wound site. The second regions essential for transfer are the T-DNA border sequences; these are almost perfect, direct repeats of 25 bp that flank the T-DNA. The border sequences apparently function as sites both for nuclease cleavage[5] and for integration of the T-DNA.

The Ti plasmid-based vectors all adapt these two essential features for the introduction of new DNAs into plant cells. Since no portion of the T-DNA other than the borders is required for transfer of the DNA into plant cells, vectors could be developed that retain the border sequences but selectable markers that function in plants replace the normal comple-

[1] M. De Cleene and J. De Ley, *Bot. Rev.* **42,** 389 (1976).

[2] S. Stachel, E. Messens, M. Van Montagu, and P. Zambryski, *Nature (London)* **318,** 624 (1985).

[3] G. Ooms, P. Klapwijk, J. Poulis, and R. Schilperoort, *J. Bacteriol.* **144,** 82 (1980).

[4] H. Klee, F. White, V. Iyer, M. Gordon, and E. Nester, *J. Bacteriol.* **153,** 878 (1983).

[5] Z. Koukolikova-Nicola, R. Shillito, B. Hohn, K. Wang, M. Van Montagu, and P. Zambryski, *Nature (London)* **313,** 191 (1985).

ment of T-DNA genes. The T-DNA functions that are removed encode phytohormone biosynthetic genes that specify the synthesis of cytokinin and auxin, which prevent the morphogenesis of transformed tissue and regeneration of plants. The altered Ti plasmids that have had the phytohormone genes deleted or replaced are referred to as "disarmed." An example is the disarmed pTiB6S3-SE[6] plasmid which is an octopine-type Ti plasmid in which most of the T_L T-DNA, including the phytohormone genes and octopine synthase gene and entire T_R DNA, have been replaced with a bacterial kanamycin resistance marker. Only the left border and approximately 2 kb of the T_L DNA remains. The T-DNA segment that remains is referred to as the left inside homology (LIH) and is essential for formation of the cointegrate T-DNAs. Another approach to disarm the Ti plasmid was taken by Zambryski and co-workers.[7] This group created pGV3850 in which the entire T-DNA, save the border sequences, was replaced with pBR322. In this case, the pBR322 sequences provide homology for cointegrate formation as described below.

Cointegrating Intermediate Vectors

The great size of the disarmed Ti plasmid and lack of unique restriction endonuclease sites prohibit direct cloning into the T-DNA. Instead, intermediate vectors such as pMON200 (Fig. 1) were developed for introducing genes into the Ti plasmid. This method was adapted from one published by Ruvkun and Ausubel[8] for mutagenesis of the large plasmids of *Rhizobium*. The intermediate vectors are smaller plasmids, usually derivatives of pBR322, that are easily manipulated in *Escherichia coli* and then transferred to *A. tumefaciens* using standard bacterial genetic techniques. Intermediate plasmids that are derivatives of pBR322 cannot replicate in *A. tumefaciens,* therefore these vectors carry DNA segments homologous to the disarmed T-DNA to permit recombination to form a cointegrate T-DNA structure.[9,10]

The pMON200 vector carries a 1.6-kb fragment of the disarmed T-DNA called the left inside homology (LIH) fragment which provides a region of homology for recombination. Once recombination occurs, the cointegrated pMON200 plasmid is replicated by the Ti plasmid origin of

[6] R. Fraley, S. Rogers, D. Eichholtz, J. Flick, C. Fink, N. Hoffmann, and P. Sanders, *Bio/Technology* **3,** 629 (1985).

[7] P. Zambryski, H. Joos, C. Genetello, J. Leemans, M. Van Montagu, and J. Schell, *EMBO J.* **2,** 2143 (1983).

[8] G. Ruvkun and F. Ausubel, *Nature* (*London*) **289,** 85 (1981).

[9] L. Comai, C. Schilling-Cordaro, A. Mergia, and C. Houck, *Plasmid* **10,** 21 (1983).

[10] L. Van Haute, H. Joos, M. Maes, G. Warren, M. Van Montagu, and J. Schell, *EMBO J.* **2,** 411 (1983).

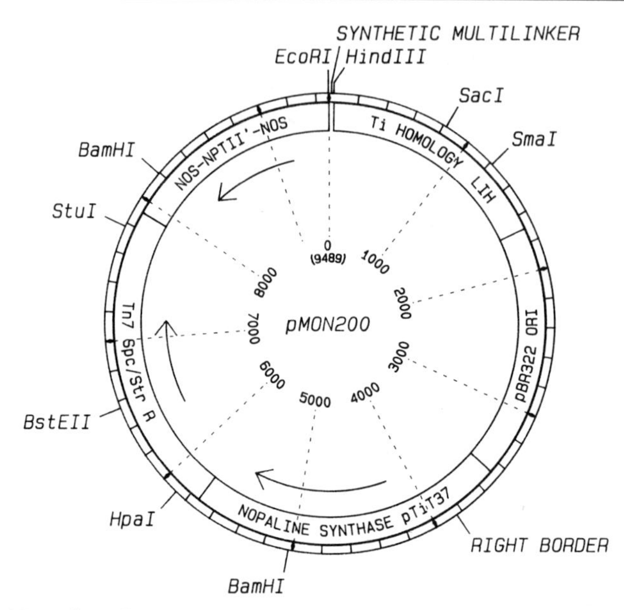

FIG. 1. Map of pMON200: A cointegrating intermediate transformation vector. The pMON200 plasmid has been described in Fraley *et al.*[6] and in the Appendix to this chapter. The plasmid is made up of a 1.6-kb LIH (left inside homology) segment derived from the octopine-type pTiA6 plasmid which provides a region of homology for recombination with a resident Ti plasmid in *A. tumefaciens,* a 1.6-kb segment carrying the pBR322 origin of replication, a 2.4-kb segment of the nopaline-type pTiT37 plasmid that carries the right border of the nopaline T-DNA and intact nopaline synthase (NOS) gene, a 2.2-kb segment of Tn7 carrying the spectinomycin/ streptomycin resistance determinant, a 1.6-kb segment encoding a chimeric NOS-NPTII'-NOS gene that provides selectable kanamycin resistance in transformed plant cells, and a synthetic multilinker containing unique restriction sites for insertion of other DNA segments. This and the following maps were generated using the UWGCG program PLASMID [J. Devereux, P. Haeberli, and O. Smithies, *Nucleic Acids Res.* **12,** 387 (1984)].

replication. Such intermediate plasmid vectors are called cointegrating or cis vectors. Use of this system has been fully described in a previous volume of this series.[11]

Binary Vectors

Hoekema *et al.*[12] and de Framond *et al.*[13] demonstrated that the T-DNA did not have to be physically joined to the Ti plasmid for transfer

[11] S. Rogers, R. Horsch, and R. Fraley, this series, Vol. 118, p. 627.

[12] A. Hoekema, P. Hirsch, P. Hooykaas, and R. Schilperoort, *Nature (London)* **303,** 179 (1983).

[13] A. de Framond, K. Barton, and M.-D. Chilton, *Bio/Technology* **1,** 262 (1983).

into plant cells to occur. This finding led to the development of binary or trans vectors. In these systems the intermediate plasmid has an origin of replication that functions in both *E. coli* and *A. tumefaciens*. When a binary intermediate plasmid is transferred into *A. tumefaciens,* it is able to replicate independently of the Ti plasmid. Binary vectors must carry the border sequences for the *vir* region products to transfer the DNA of interest into plant cells. For small binary vectors, such as pMON505 described below, a single border is sufficient.[14] At this writing, several groups have published the construction and use of binary vector systems for plant transformation.[15–18]

The binary vectors offer the advantage of higher frequency transfer from *E. coli* to *A. tumefaciens* since cointegration is no longer required for their maintenance. The frequency of transfer for the binary vectors is routinely 10% since this percentage of the exconjugant *A. tumefaciens* contains the binary vector. This can be as high as 100% and may be four to six orders of magnitude higher than the rate of cointegrate formation using the pMON200 plasmid.

Design of Expression Cassette Vectors

The expression cassettes have been constructed with the promoter region, transcription initiation site, and a portion of the 5′ nontranslated leader of the promoter of interest joined to a synthetic multilinker and followed by a polyadenylation signal carried in the 3′ nontranslated portion of a plant gene. This design permits the efficient initiation of transcription at the promoter's natural mRNA start site and maintains any features of the 5′ end of the messenger RNA required for efficient ribosome interactions. No additional, upstream AUG translational initiator signals are present in the 5′ leader sequence to ensure that translation will begin on the AUG of the inserted coding sequence and therefore to maximize translation of the coding sequence.[19,20]

The promoter–leader segment is followed by a synthetic multilinker with unique restriction endonuclease cleavage sites for the insertion of coding sequences derived from prokaryotic or eukaryotic genes, including genomic sequences with introns or cDNAs. Although cDNA clones

[14] R. Horsch and H. Klee, *Proc. Natl. Acad. Sci. U.S.A.* **83,** 4428 (1986).

[15] M. Bevan, *Nucleic Acids Res.* **12,** 8711 (1984).

[16] G. An, B. Watson, S. Stachel, M. Gordon, and E. Nester, *EMBO J.* **4,** 277 (1984).

[17] H. Klee, M. Yanofsky, and E. Nester, *Bio/Technology* **3,** 637 (1985).

[18] P. van den Elzen, K. Lee, J. Townsend, and J. Bedbrook, *Plant Mol. Biol.* **5,** 149 (1985).

[19] M. Kozak, *Cell* **15,** 1109 (1978).

[20] S. Rogers, R. Fraley, R. Horsch, A. Levine, J. Flick, L. Brand, C. Fink, T. Mozer, K. O'Connell and P. Sanders, *Plant Mol. Biol. Rep.* **3,** 111 (1985).

often contain the signal and site of polyadenylation, including a portion of the poly(A) tail, it is not clear if these signals or transcribed runs of poly(A) are efficient in providing correct message stability and transport from the nucleus.[21] Therefore, an additional, functional polyadenylation signal from the Ti plasmid T-DNA nopaline synthase gene is present downstream of the multilinker.

Two different expression cassettes have been developed using the two cauliflower mosaic virus promoters: the 19 S (P66 or gene VI) promoter and the 35 S (full-length transcript) promoter. Both the 19 and 35 S promoters are constitutively expressed in transformed plants.[22,23] The relative strengths of these promoters have been determined in our laboratory by comparison to the nopaline synthase promoter. The 19 S promoter is at most 5-fold stronger than the nopaline synthase promoter. In contrast, the 35 S promoter is approximately 50 times stronger than the nopaline synthase promoter and is the strongest constitutive promoter identified at this time for expression of foreign coding sequences in transformed plants.[24] It has the additional advantage of functioning in both dicots and monocots.[25,26] The availability of both of these cassettes allows construction of chimeric genes that differ in expression levels over a 10-fold range. Details of the construction of the cassettes and the vectors in which they are carried follow.

pMON237: The 19 S-NOS Cassette Vector

The CaMV 19 S promoter fragment was isolated from plasmid pOS-1, a derivative of pBR322 carrying the entire genome of CM4-184 as a *Sal*I insert.[27] This plasmid was kindly provided by R. J. Shepherd. The CM4-184 strain is a naturally occurring deletion mutant of strain CM1841. The nucleotide sequences of the CM1841[28] and Cabb-S[29] strains of CaMV have been published as well as some partial sequences for a different CM4-184

[21] H. Okayama and P. Berg, *Mol. Cell. Biol.* **2,** 161 (1982).

[22] J. Paszkowski, R. Shillito, M. Saul, V. Mandak, T. Hohn, B. Hohn, and I. Potrykus, *EMBO J.* **3,** 2717 (1984).

[23] J. Odell, F. Nagy, and N.-H. Chua, *Nature (London)* **313,** 810 (1985).

[24] S. Rogers, K. O'Connell, R. Horsch, and R. Fraley, *in* "Biotechnology in Plant Science" (M. Zaitlin, P. Day, and A. Hollander, eds.), p. 219. Academic Press, Orlando, Florida, 1986.

[25] M. Fromm, L. Taylor, and V. Walbot, *Proc. Natl. Acad. Sci. U.S.A.* **82,** 5824 (1985).

[26] M. Fromm, L. Taylor, and V. Walbot, *Nature (London)* **319,** 791 (1986).

[27] A. Howarth, R. Gardner, J. Messing, and R. Shepherd, *Virology* **112,** 678 (1981).

[28] R. Gardner, A. Howarth, P. Hahn, M. Brown-Luedi, R. Shepherd, and J. Messing, *Nucleic Acids Res.* **9,** 2871 (1981).

[29] A. Franck, H. Guilley, G. Jonard, K. Richards, and L. Hirth, *Cell* **21,** 285 (1980).

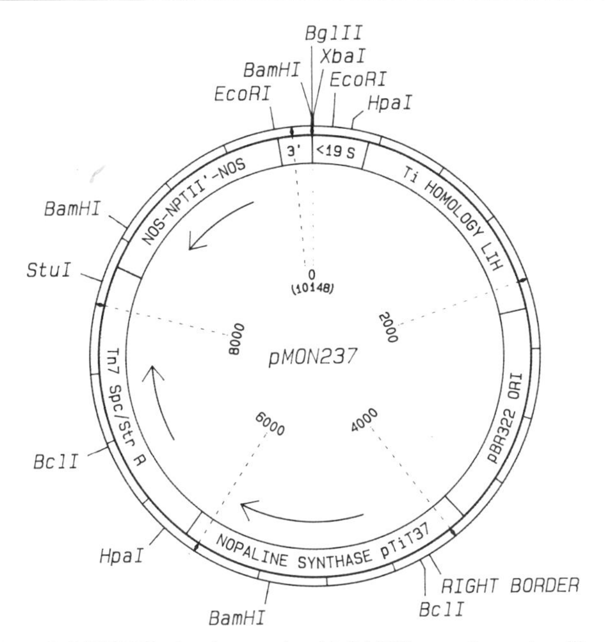

FIG. 2. Map of pMON237: A cointegrating 19 S-NOS cassette vector. The construction of this expression vector is described in the text. <19 S, The CaMV 19 S promoter segment; 3′, the NOS 3′ nontranslated segment. The additional segments are described in the legend to Fig. 1. The complete sequence of the cassette appears in Fig. 3.

clone.[30] The references to nucleotide numbers (*n*) in the following discussion are those for the sequence of CM1841.[28] The nucleotide sequences of both 19 and 35 S promoter regions of these three isolates are essentially identical. A 476-bp fragment extending from the *Hin*dIII site at bp 5372 to the *Hin*dIII site at bp 5848 was cloned into M13 mp8 for site-directed mutagenesis[31] to introduce an *Xba*I site immediately 5′ of the first ATG translational initiation signal in the 19 S transcript.[30] The resulting 400-bp *Hin*dIII–*Xba*I fragment was isolated and joined to a synthetic linker containing *Bgl*II and *Bam*HI sites and then cloned adjacent to the nopaline synthase 3′ nontranslated region. This cassette was inserted into pMON200 (Fraley *et al.*[6]) between the *Eco*RI and *Hin*dIII sites to give pMON237 (Fig. 2). The complete sequence of the 19 S promoter-NOS cassette is given in Fig. 3.

Plasmid pMON237 is a cointegrating type intermediate vector with unique *Xba*I and *Bgl*II sites for the insertion of coding sequences carrying

[30] R. Dudley, J. Odell, and S. Howell, *Virology* **117,** 19 (1982).

[31] M. Zoller and M. Smith, *Nucleic Acids Res.* **10,** 6487 (1982).

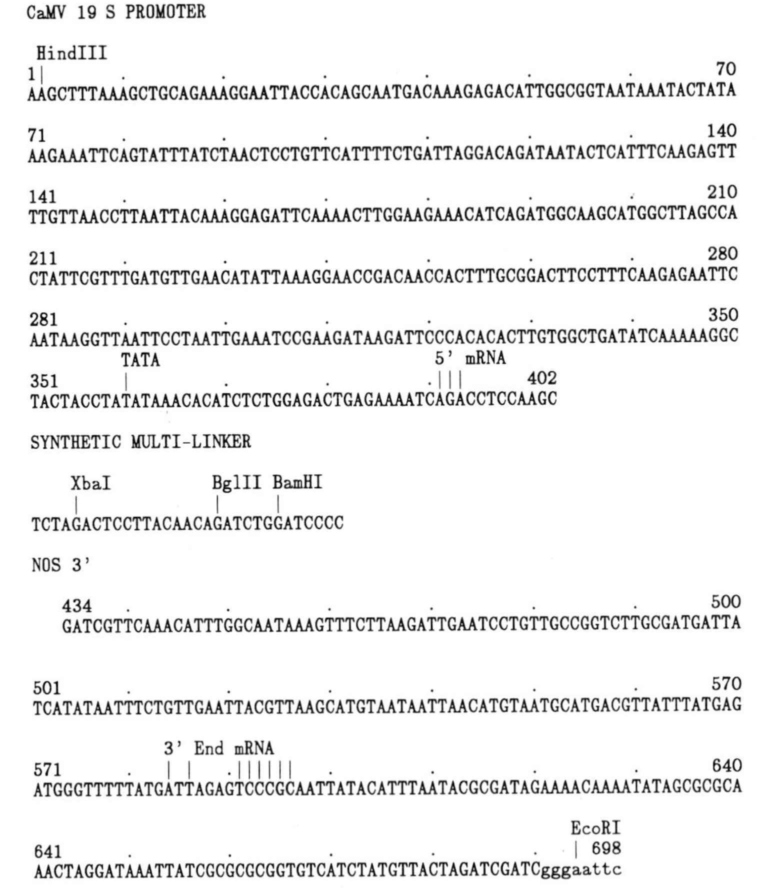

FIG. 3. Nucleotide sequence of the CaMV 19 S promoter–NOS 3′ cassette. The 5′ end of the natural 19 S mRNA is from Dudley *et al.*[30] The sequence of the NOS 3′ segment and the 3′ end of the NOS mRNA were determined by Bevan *et al.* [M. Bevan, W. M. Barnes, and M.-D. Chilton, *Nucleic Acids Res.* **11,** 370 (1983)].

their own translational initiation signals immediately adjacent to the 19 S transcript leader sequence. The pMON237 plasmid retains all of the properties of pMON200, including spectinomycin resistance for selection in *E. coli* and *A. tumefaciens,* as well as a chimeric kanamycin gene (NOS-NPTII′-NOS) for selection of transformed plant tissue and the nopaline synthase gene for ready scoring of transformants and inheritance in progeny. The pMON237 plasmid is used exactly as is pMON200.[6,11]

pMON316 and pMON530: The 35 S-NOS Vectors

The 35 S promoter was isolated from the pOS-1 clone of CM4-184 as an *Alu*I ($n = 7143$)–*Eco*RI* ($n = 7517$) fragment which was inserted first into pBR322 cleaved with *Bam*HI, treated with the Klenow fragment of DNA polymerase I, and then cleaved with *Eco*RI. The promoter fragment was then excised from pBR322 with *Bam*HI and *Eco*RI, treated with Klenow polymerase, and inserted into the *Sma*I site of M13 mp8 so that the *Eco*RI site of the mp8 multilinker was at the 5′ end of the promoter fragment. The nucleotide numbers refer to the sequence of CM1841.[28] Site-directed mutagenesis was then used to introduce a G at nucleotide 7464 to create a *Bgl*II site. The 35 S promoter fragment was then excised from the M13 as a 330-bp *Eco*RI–*Bgl*II fragment which contains the 35 S promoter, transcription initiation site, and 30 nucleotides of the 5′ nontranslated leader but contains none of the CaMV translational initiators nor the 35 S transcript polyadenylation signal that is located 180 nucleotides downstream from the start of transcription.[32,33] The 35 S promoter fragment was joined to a synthetic multilinker and the NOS 3′ nontranslated region and inserted into pMON200 to give pMON316 (Fig. 4). Plasmid 316 contains unique cleavage sites for *Bgl*II, *Cla*I, *Kpn*I, *Xho*I, and *Eco*RI located between the 5′ leader and the NOS polyadenylation signals. The 316 plasmid retains all of the properties of pMON200. The complete sequence of the 35 S promoter, multilinker, and NOS 3′ segment is given in Fig. 5. This sequence begins with the *Xmn*I site created by Klenow polymerase treatment to remove the *Eco*RI site located at the 5′ end of the 35 S promoter segment.

The 35 S-NOS cassette was also inserted into our binary vector, pMON505. Plasmid pMON505 is a derivative of pMON200 in which the Ti plasmid homology region, LIH, has been replaced by a 3.8-kb *Hin*dIII to *Sma*I segment of the mini RK2 plasmid, pTJS75.[34] This segment contains the RK2 origin of replication, *ori*V, and the origin of transfer, *ori*T, for conjugation into *Agrobacterium* using the triparental mating procedure. The detailed construction of pMON505 has been published elsewhere.[14]

Plasmid pMON505 (Fig. 6) retains the important features of pMON200 including the synthetic multilinker for insertion of desired DNA fragments, the chimeric NOS-NPTII′-NOS gene for kanamycin resistance in plant cells, the spectinomycin/streptomycin resistance determinant for

[32] S. Covey, G. Lomonosoff, and R. Hull, *Nucleic Acids Res.* **9,** 6735 (1981).
[33] H. Guilley, R. Dudley, G. Jonard, E. Balaz, and K. Richards, *Cell* **30,** 763 (1982).
[34] T. Schmidhauser and D. Helinski, *J. Bacteriol.* **164,** 446 (1985).

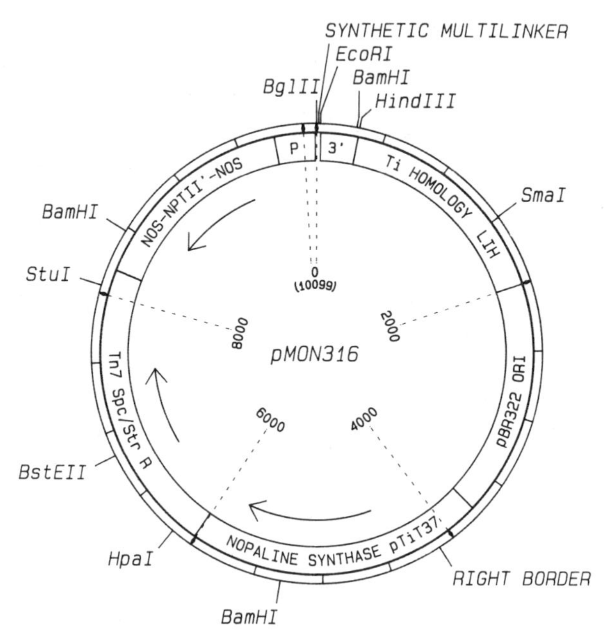

FIG. 4. Map of pMON316: A cointegrating 35 S-NOS cassette vector. The construction of this expression vector is described in the text. P, The CaMV 35 S promoter segment; 3′, the NOS 3′ nontranslated sequences. The additional segments are described in the legend to Fig. 1. The complete sequence of the cassette appears in Fig. 5.

selection in *E. coli* and *A. tumefaciens*, an intact nopaline synthase gene for facile scoring of transformants and inheritance in progeny, and a pBR322 origin of replication for ease in making large amounts of the vector in *E. coli*. The 505 plasmid contains a single T-DNA border derived from the right end of the pTiT37 nopaline-type T-DNA. Experiments in our laboratory[14] have shown that this single border sequence is both necessary and sufficient for high-frequency transfer of the pMON505 into plant cells and its stable integration into the plant genome. Southern analyses have shown that the 505 plasmid and any DNA that it carries are integrated into the plant genome. The entire plasmid acts as a T-DNA and is inserted into the plant genome. One end of the integrated DNA is located between the right border sequence and the nopaline synthase gene and the other end is between the border sequence and the pBR322 sequences.

A pMON505 derivative carrying the 35 S-NOS cassette was created by transferring the 2.3-kb *Stu*I–*Hin*dIII fragment of pMON316 into pMON526. Plasmid pMON526 is a simple derivative of pMON505 in which the *Sma*I site was removed by digestion with *Xma*I, treatment with

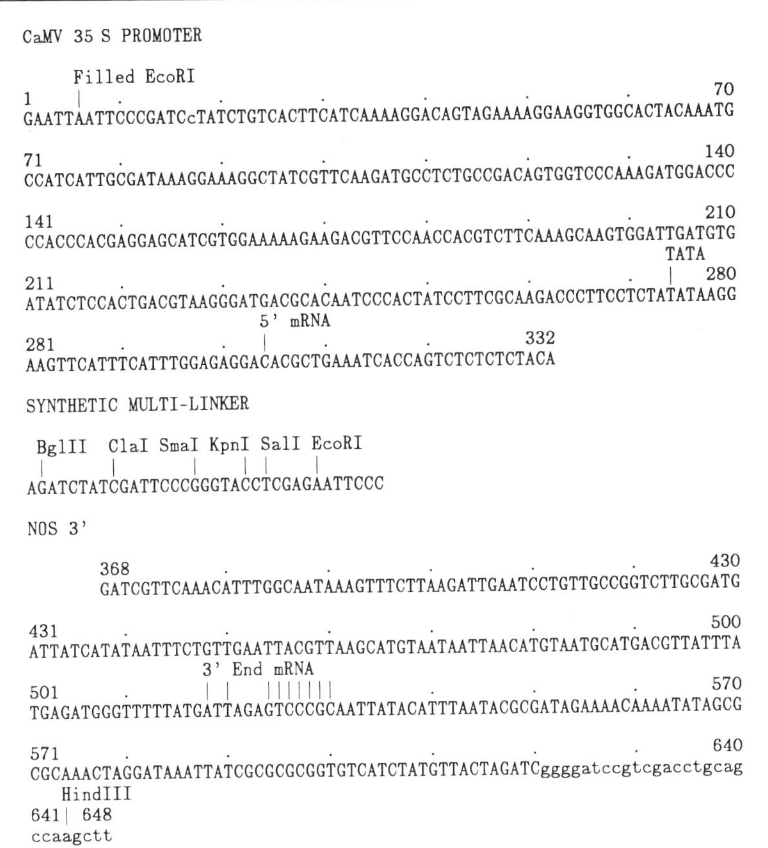

FIG. 5. Nucleotide sequence of the CaMV 35 S promoter-NOS 3′ cassette. The 5′ end of the 35 S mRNA is from Odell *et al.*[23] The sequence of the NOS 3′ segment and the 3′ end of the NOS mRNA were determined by Bevan *et al.* [M. Bevan, W. M. Barnes, and M.-D. Chilton, *Nucleic Acids Res.* **11,** 370 (1983)].

Klenow polymerase, and ligation. The resultant plasmid, pMON530 (Fig. 7), retains the properties of pMON505 and the 35 S-NOS expression cassette now contains a unique cleavage site for *Sma*I between the promoter and polyadenylation signals. Tables listing the major restriction endonuclease cleavage sites useful for cloning and analysis of clones and transformants are included in the Appendix to this chapter. The nucleotide sequences of most of the segments that comprise the pMON200 and pMON505 vectors have been determined and published by other workers. As an aid to mapping inserts and analysis of transformants, the appropriate references and directions for the assembly of these sequences to give pMON200 and pMON505 are provided in the Appendix.

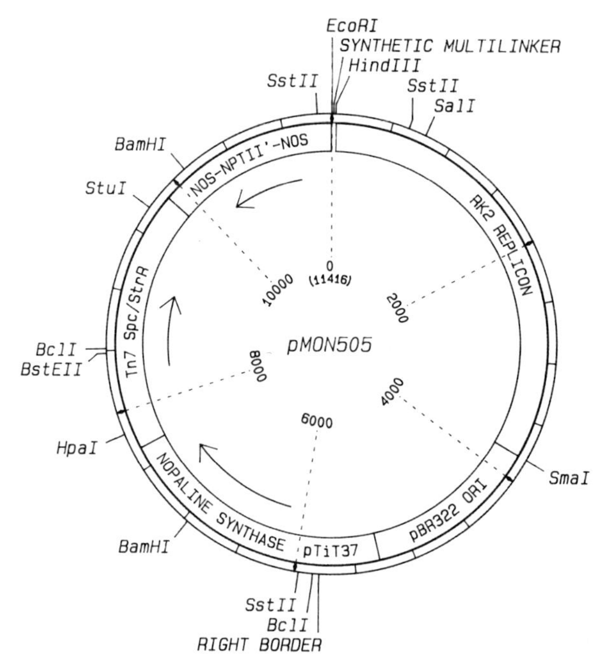

FIG. 6. Map of pMON505: A binary transformation vector. Details of the construction of pMON505 have been described in Horsch and Klee.[14] The 3.8-kb RK2 replicon fragment permits independent replication in *A. tumefaciens*. The additional segments are described in the legend to Fig. 1. More complete information on pMON505 appears in the Appendix.

Development of New Selectable Markers

Chimeric genes for the expression of the Tn*5* neomycin phosphotransferase II (NPTII) coding sequence have become the standard markers for plant transformation vectors. The kanamycin resistance provides for direct selection of transformants in a wide variety of plant species, including tobacco, petunia, tomato, lettuce, canola, wheat, and maize.

There are species such as *Arabidopsis thaliana* and certain legumes where the selectability of kanamycin resistance is marginal. New dominant selectable markers are required for these species. Additional selectable markers would also permit retransformation of previously transformed tissues or plants to introduce additional copies of the same gene of interest, or several different genes for enzymes of a biosynthetic pathway, or several different members of the same gene family for the study of their relative expression levels. For genetic studies, additional selectable mark-

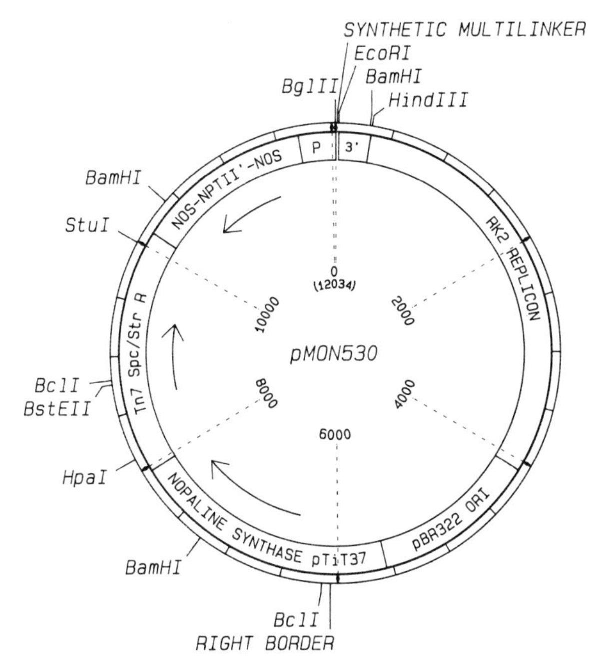

FIG. 7. Map of pMON530: A binary 35 S-NOS cassette vector. The construction of this expression vector is described in the text. P, The CaMV 35 S promoter segment; 3′, the NOS 3′ nontranslated sequences. The additional segments are described in the legend to Fig. 1. The complete sequence of the cassette appears in Fig. 5.

ers provide the means to follow the segregation of several introduced genes in the progeny of transformants. All of these reasons led us to develop other dominant selectable markers for selection of transformed plants.

Plant tissues are sensitive to many antibiotics that are toxic to bacteria and antimetabolites that are toxic to mammalian cells. This property provides a wide range of potential selective agents for the development of new markers for plants. In addition, many bacterial genes that encode enzymes for the detoxification of antibiotics and eukaryotic genes for target proteins that are insensitive to antimetabolites have been identified. The expression cassette vectors described above and current methods of DNA modification make the construction of chimeric genes and their testing as selectable markers a straightforward process. Specific use of these vectors and methods for the development of two marker genes is described below.

pMON410: pMON530 Carrying a Coding Sequence for Hygromycin Resistance

Hygromycin B is an aminocyclitol antibiotic that inhibits protein synthesis in prokaryotic and eukaryotic cells, including those of plants.[35] Two groups[36,37] have isolated and described a gene from a bacterial R-factor that encodes a hygromycin phosphotransferase (HPH)[38] which inactivates the antibiotic. Chimeric genes that act as dominant selectable markers for yeast[36,39] and mammalian cells[40] have been constructed. These results led us to construct and test a HPH marker for transformation of plants.

Gritz and Davies[36] created a set of fragments carrying the HPH gene by S1 digestion and addition of *Bam*HI linkers. J. Davies provided us with two of their plasmids, pLG89 and 83, which carry synthetic *Bam*HI linkers near the 5′ and 3′ ends of the HPH coding sequence, respectively. A full-length HPH coding sequence was assembled by joining the 0.26-kb *Bam*HI–*Eco*RI fragment from pLG89 to a 0.9-kb *Eco*RI–*Bam*HI fragment from pLG83 and by inserting the 1.2-kb fragment into the *Bam*HI site of M13 mp8. A spurious ATG translational initiator signal, located 5 bp 5′ to the HPH initiator, was removed by site-directed mutagenesis. The sequence of the reassembled, altered HPH coding region is shown in Fig. 8. The resulting 1.2-kb *Bam*HI fragment was then transferred to the *Bgl*II site of pMON316 to create pMON408, and into the *Bgl*II site of pMON530 to create pMON410 (Fig. 9).

Two other groups have assembled chimeric HPH genes that provide hygromycin resistance in plant tissues using either the octopine promoter[41] or the nopaline synthase promoter.[42] Neither of these groups has removed the spurious ATG initiator codon found in the bacterial leader to increase translational efficiency. Deletion of the extra ATG can be expected to increase the levels of HPH protein by 5-fold based on a similar alteration of the neomycin phosphotransferase II leader to remove a spurious ATG codon.[20] This prediction has been confirmed in our laboratory

[35] A. Gonzalez, A. Jimenez, D. Vasquez, J. Davies, and D. Schindler, *Biochim. Biophys. Acta* **521,** 459 (1978).

[36] L. Gritz and J. Davies, *Gene* **25,** 179 (1983).

[37] K. Kaster, S. Burgett, R. Rao, and T. Ingolia, *Nucleic Acids Res.* **11,** 6895 (1983).

[38] R. Rao, N. Allen, J. Hobbs, W. Alborn, H. Kirst, and J. Paschal, *Antimicrob. Agents Chemother.* **24,** 689 (1983).

[39] K. Kaster, S. Burgett, and T. Ingolia, *Curr. Genet.* **8,** 353 (1984).

[40] R. Santerre, N. Allen, J. Hobbs, R. Rao, and R. Schmidt, *Gene* **30,** 147 (1984).

[41] C. Waldron, E. Murphy, J. Roberts, G. Gustafson, S. Armour, and S. Malcolm, *Plant Mol. Biol.* **5,** 103 (1985).

[42] P. van den Elzen, J. Townsend, K. Lee, and J. Bedbrook, *Plant Mol. Biol.* **5,** 299 (1985).

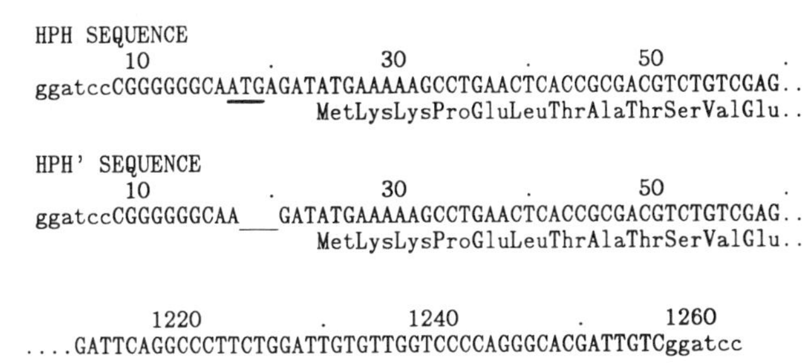

FIG. 8. Partial sequence of the wild-type (HPH) and modified (HPH′) hygromycin phosphotransferase coding sequences. The underlined extra ATG signal has been deleted in the HPH′ sequence. The nucleotides in lower case are the *Bam*HI linkers added to the coding sequence clone. The nucleotide numbers are from Gritz and Davies.[36]

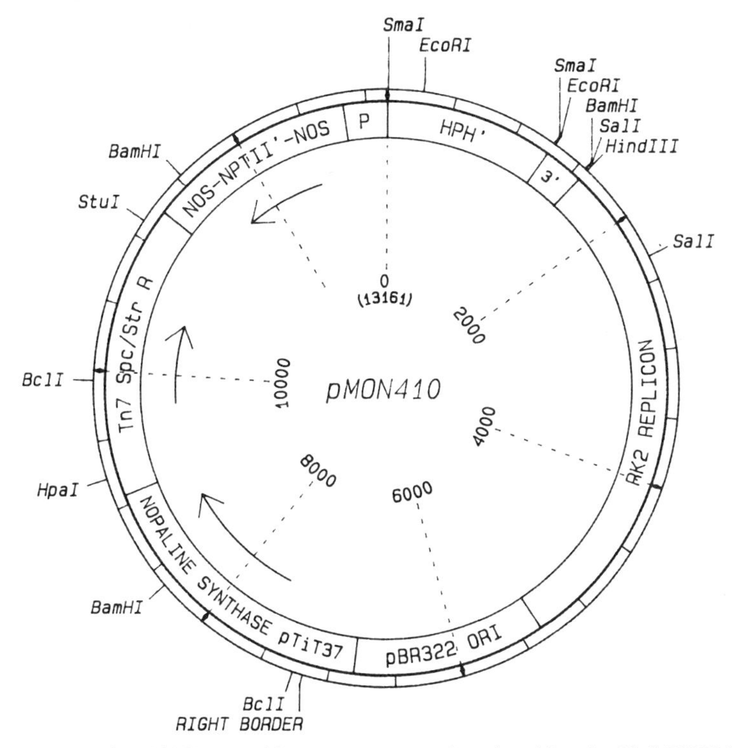

FIG. 9. Map of pMON410: A binary vector carrying the chimeric 35 S-HPH′-NOS selectable marker gene. The insertion of the HPH′ coding sequence into pMON530 is described in the text.

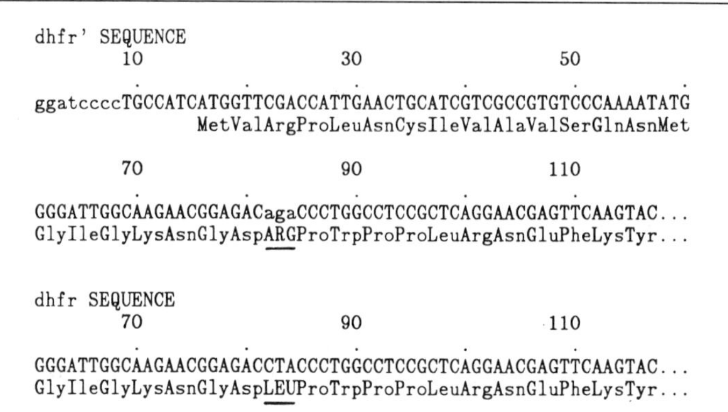

FIG. 10. Partial sequence of the modified and wild-type mouse dihydrofolate reductase (dhfr') coding regions.[44] The positions of the Leu to Arg substitution are underscored.

by comparison of the transformation efficiency and growth characteristics of transformed plant tissue carrying either the single or two ATG chimeric genes.

pMON321: pMON237 Carrying a Coding Sequence for Methotrexate Resistance

Methotrexate (mtx) is an antimetabolite that inhibits eukaryotic dihydrofolate reductase (dhfr), ultimately preventing biosynthesis of glycine, thymine, and purines. Plant cells and tissues are quite sensitive to mtx. Certain mammalian cell lines selected for resistance to mtx show amplification of the *dhfr* gene[43] while others produce an altered *dhfr* enzyme that does not bind mtx. Simonsen and Levinson[44] cloned the cDNA for an altered mouse *dhfr* from cell line 3T6-R400. The altered *dhfr* produced by this cell line shows a 270-fold decrease in mtx binding compared to the wild-type enzyme. These authors demonstrated that the resistance was due to a single base change that caused an arginine to be substituted for a leucine at position 22 in the protein. We tested the ability of this altered *dhfr* coding sequence to confer mtx resistance on plant cells and tissues in the following manner.

The wild-type mouse *dhfr* coding sequence was obtained on a 1.1-kb *Fnu*4HI fragment from pDHFR-11, a cDNA clone in pBR322[45] provided by R. Schimke. The *Fnu*4HI fragment was treated with Klenow poly-

[43] F. Alt, R. Kellems, J. Bertino, and R. Schimke, *J. Biol. Chem.* **253,** 1357 (1976).

[44] C. Simonsen and A. Levinson, *Proc. Natl. Acad. Sci. U.S.A.* **80,** 2495 (1983).

[45] J. Nunberg, R. Kaufman, A. Chang, S. Cohen, and R. Schimke, *Cell* **19,** 355 (1980).

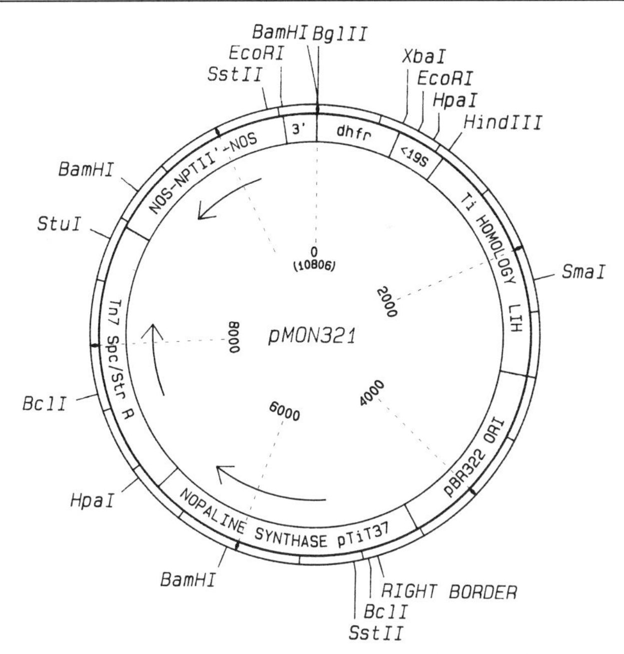

FIG. 11. Map of pMON321: A cointegrating vector carrying the chimeric 19 S-dhfr′-NOS selectable marker gene. The insertion of the *dhfr*′ coding sequence into pMON237 is described in the text.

merase and inserted into the *Sma*I site of M13 mp8 for site-directed mutagenesis. Mutagenesis was performed to replace the wild-type 5′-CTA sequence with 5′-AGA resulting in an arginine substitution for leucine at amino acid 22. A partial sequence of the *Bam*HI–*Bgl*II fragment showing the alteration that results in an mtx-resistant *dhfr* enzyme is given in Fig. 10. A 660-bp *Bam*HI to *Bgl*II fragment was then isolated from the mp8 and inserted into the *Bgl*II site of pMON237 to give pMON321 (Fig. 11). A plasmid in which the modified *dhfr* coding sequence is expressed from the CaMV 35 S promoter-NOS 3′ cassette was also made. In this plasmid, pMON806, the NOS-NPTII′-NOS selectable marker gene of pMON505, was replaced with the methotrexate-resistant *dhfr* gene. Plasmid pMON806 will be described in a subsequent publication.[46]

Introduction of the pMON Plasmids into *A. tumefaciens*

These pMON plasmids were transferred into *A. tumefaciens* using a simplified triparental mating procedure. Aliquots (0.1 ml) of fresh over-

[46] H. Klee, manuscript in preparation.

night cultures of *E. coli* cells carrying the pMON plasmids were spotted with 0.1 ml of *E. coli* cells carrying the pRK2013 helper plasmid[47] and 0.1 ml of *Agrobacterium tumefaciens* strain 3111 carrying the disarmed octopine-type pTiB6S3-SE plasmid[6] or strain A208 carrying the disarmed nopaline-type pTiT37-SE plasmid[48] on a fresh Luria agar plate and the cells are mixed with a sterile loop. As a control the pRK2013 cells are spotted on a plate with the 3111-SE or A208-SE cells and mixed. After overnight incubation at 28°, aliquots of these mating mixes are transferred to selection plates. For the binary vectors (pMON505 and derivatives) this is simply done by streaking a loopful of the cells on selective plates for single colonies. For the cointegrating vectors (pMON200 and derivatives), the entire mating mix is resuspended in 3 ml of 10 m*M* $MgSO_4$ with a glass rod, transferred to a culture tube, washed with an additional 3 ml of 10 m*M* $MgSO_4$, and then a 0.2-ml aliquot is spread on a selection plate. Selection plates contain 25 μg/ml chloramphenicol, 50 μg/ml kanamycin, and 100 μg/ml spectinomycin. After 2–3 days at 28° single colonies are visible. These are picked into LB containing the same antibiotics at the same concentration as used in the selection plates and grown at 28°. We use the 3111-SE or A208-SE *Agrobacterium* strains for both the cointegrating and binary vectors. We have found that these disarmed strains are more efficient for transfer of our binary vectors in petunia and tobacco transformation than is the commonly used LBA4404 disarmed, virulence plasmid strain.[12]

Use of the Hygromycin and Methotrexate Markers for Transformation

A. tumefaciens 3111-SE or A208-SE cells containing a pMON vector can efficiently transform plant cells using the leaf disk technique.[11,49] Petunia or tobacco leaf discs are cultured for 2 days on medium containing 4.3 g/liter MS salts (Gibco, Grand Island, N.Y.), 30 g/liter sucrose, B_5 vitamins, 1 mg/liter BA, 0.1 mg/liter NAA, and 0.8% agar before infection. Following brief immersion in a liquid culture of the bacteria, the leaf disks are blotted dry and cocultured for 2 days on the same medium before being transferred to medium supplemented with 500 μg/ml carbenicillin to kill residual bacteria, and the appropriate selective agent to inhibit growth of nontransformed plant cells.

The efficiency of selection for antibiotic resistance is influenced by

[47] G. Ditta, S. Stanfield, D. Corbin, and D. Helinski, *Proc. Natl. Acad. Sci. U.S.A.* **77,** 7347 (1980).

[48] S. Rogers and D. Fischhoff, manuscript in preparation.

[49] R. Horsch, J. Fry, N. Hoffmann, M. Wallroth, D. Eichholtz, S. Rogers, and R. Fraley, *Science* **227,** 1229 (1985).

two separate factors: the degree of resistance conferred by the gene construct and the effects of the dying cells that comprise the great mass of tissue from which the resistant cells must grow. The degree of resistance is itself determined by the inherent toxicity of the selective agent compared to the level and specificity of the resistance conferred by the gene construct. Selectability is a combination of this degree of resistance and the toxicity of materials released from the surrounding, dying cells. In extreme cases, the resistant cells may all be poisoned by their dying neighbors despite high-level resistance to the selective agent.[50]

While transformed cells containing pMON410 are resistant to up to 400 μg/ml of hygromycin B, the selectability is poor: many fewer colonies grow from leaf disks on 40 μg/ml hygromycin than on 300 μg/ml kanamycin. The rapid bleaching and browning of the disks on hygromycin may indicate a hypertoxic effect on wild-type cells compared to the effect of kanamycin.

Excellent resistance to high levels (30 μM) of methotrexate is conferred by pMON321, but initially the selectability was poor. Delaying selection for 2 to 3 days after completion of transformation greatly improved the recovery of resistant colonies from leaf disks. This improved selectability could have been due to allowing growth of single transformed cells into colonies of several cells before selection or due to accumulation of more of the resistant *dhfr* protein before selection. The latter was demonstrated to be the case by using the stronger CaMV 35 S to drive expression in plasmid pMON806, resulting in efficient selection immediately after transformation. Thus the cells did not have time to divide but only to accumulate sufficient quantities of the resistant enzyme before selection.

In routine practice, we use 50 μg/ml hygromycin B (CalBehring) or 1 μM methotrexate (Sigma, St. Louis, MO) to select transformed cells and shoots with tobacco and petunia. Hygromycin-selected shoots are transferred to medium without added phytohormones for rooting. In contrast, methotrexate-resistant shoots are transferred to solid medium without added phytohormones containing 1 μM mtx for rooting so that selection may be maintained throughout the regeneration procedure. This represents an improvement over the previously reported chimeric mtx-resistant bacterial *dhfr* marker gene[51] which could only be selected at low levels of mtx as callus and not during morphogenesis.

Expression of hygromycin or mtx resistance may also be assayed by

[50] R. Horsch and G. Jones, *Mutat. Res.* **72,** 91 (1980).

[51] M. De Block, L. Herrera-Estrella, M. Van Montagu, J. Schell, and P. Zambryski, *EMBO J.* **3,** 1681 (1984).

the leaf callus assay by culturing sterile leaves on medium with phytohormones containing 50 μg/ml hygromycin or 1 μM methotrexate.

Hygromycin resistance is the best marker we have found for use in transformation of *Arabidopsis thaliana,* where 20 to 50 μg/ml is optimal for efficient selection.[52] The kanamycin resistance conferred by pMON200 does not function well in *Arabidopsis:* the degree of resistance is low and overlaps with spontaneous resistance that readily develops in control tissues. Here too the hygromycin selection is removed during the regeneration step.

Other Uses of the Expression Cassette Vectors

The expression cassette vectors described here have been used to express coding sequences for a wide range of bacterial, mammalian, and plant genes. These include a plant coding sequence for 5-enolpyruvylshikimate-3-phosphate (EPSP) synthase, the target of glyphosate, the active ingredient of Roundup® herbicide.[53] Transgenic plants containing the chimeric EPSP synthase gene show greatly enhanced tolerance to Roundup®.

A mammalian hormone cDNA for the α-subunit of human chorionic gonadotropin has been inserted into the pMON316 vector. Transformed plant tissues produce the hormone as determined by radioimmunoassay and protein antibody blots.[54]

Recently, the tobacco mosaic virus (TMV) coat protein has been expressed from a cDNA inserted into pMON316. Tobacco and tomato plants expressing this normally cytoplasmic mRNA from a nuclear gene show resistance to superinfecting TMV.[55]

The availability of these simple-to-use expression cassette vectors will permit continued rapid progress in the introduction and testing of many additional prokaryotic and eukaryotic coding sequences in plants.

Appendix

The information given below is to assist researchers in analyzing recombinants of pMON200 or pMON505 and derivatives carrying a new

[52] A. Lloyd, A. Barnason, S. Rogers, M. Byrne, R. Fraley, and R. Horsch, *Science* **234,** 464 (1986).

[53] D. Shah, R. Horsch, H. Klee, G. Kishore, J. Winter, N. Tumer, C. Hironaka, P. Sanders, C. Gasser, S. Aykent, N. Siegel, S. Rogers, and R. Fraley, *Science* **233,** 478 (1986).

[54] S. Rogers *et al.,* manuscript in preparation.

[55] P. Abel, R. Nelson, B. De, N. Hoffmann, S. Rogers, R. Fraley, and R. Beachy, *Science* **232,** 738 (1986).

insert and for mapping the inserts of these vectors in transformed plant tissue and plants. For those who wish to assemble the pMON200 and pMON505 plasmid sequences in their own computer systems, we have provided directions, sequences, coordinates, and references for the various segments that comprise the vectors. It must be noted that neither of the sequences are complete. Regions for which sequence is lacking are represented by inserts of spacer null nucleotides of the appropriate size based on extensive restriction endonuclease mapping.

Table I gives the endpoints of the various segments comprising pMON200 as well as the coordinates of the cleavage sites for the 6-bp and larger recognition site restriction endonucleases. A list of endonucleases that do not cleave the vector are also included.

Assembly of pMON200 begins with the *Eco*RI site of the synthetic multilinker which has the sequence

5′-GAATTCATCGATATCTAGATCTCGAGCTCGCGAAAGCTT

The multilinker ends with a *Hin*dIII site. The next fragment is the LIH or left inside homology segment derived from pTiA6. This sequence is the reverse of bp 1618 to 3395 of the pTi15955 octopine-type T-DNA.[56] This segment begins with a *Hin*dIII site and ends with a *Bgl*II site that was made flush ended by treatment with Klenow polymerase and the four nucleotide triphosphates. This fragment was ligated to a fragment of pBR322[57] from the *Pvu*II site (bp 2069) to the *Pvu*I (bp 3740). This *Pvu*I site was joined to a *Pvu*I site located in the pTiT37 Ti plasmid approximately 150 bp from the end of the published sequence. At this point in the assembled sequence, 150 N's were inserted. Then bp 1 to bp 2102 of the nopaline T-DNA right border flanking sequence, the right border, and entire nopaline synthase gene ending at a *Cla*I site, were added.[58] This segment was joined to the Tn*7* dihydrofolate reductase sequence[59] beginning at the *Cla*I site at bp 560 and ending at the end of the published sequence at bp 883. The next segment consists of the other unknown segment. Here 250 N's were inserted. The next segment of sequence comes from the Tn*7* spectinomycin/streptomycin resistance determinant[60] starting at bp 1 and ending after the second T of the *Eco*RI site at bp 1614. This *Eco*RI site was treated with Klenow polymerase and the four nucleo-

[56] R. F. Barker, K. B. Idler, D. V. Thompson, and J. D. Kemp, *Plant Mol. Biol.* **2,** 335 (1983).

[57] J. G. Sutcliffe, *Proc. Natl. Acad. Sci. U.S.A.* **75,** 3737 (1978).

[58] A. Depicker, S. Staechel, P. Dhaese, P. Zambryski, and H. M. Goodman, *J. Mol. Appl. Genet.* **1,** 561 (1982).

[59] M. Fling and C. Richards, *Nucleic Acids Res.* **11,** 5147 (1983).

[60] M. Fling, J. Kopf, and C. Richards, *Nucleic Acids Res.* **13,** 7095 (1985)

TABLE I
MAJOR REGIONS AND RESTRICTION ENDONUCLEASE CLEAVAGE SITES OF pMON200

Segment name	Coordinates
Synthetic mutilinker	1–36
pTiA6 fragment LIH	36–1700
pBR322 origin fragment	1700–3476
Right border	3842
pTiT37 fragment	3485–5730
Nopaline synthase coding region	4217–5455
Tn*7 Spc/StrR* fragment	5731–7911
Spc/StrR coding region	6390–7144
NOS-NPTII′-NOS kanamycin^{R} fragment	7934–9489
NPTII′ coding region	8366–9159

Endonuclease[a]	Cleavage site	Endonuclease[a]	Cleavage site
*Acc*I	739, 1990	*Nae*I	7040, 7551, 8251, 8534
*Afl*II	5509, 8154	*Nar*I	9034
*Asu*II	8352	*Nco*I	504, 4607, 8600
*Ava*I	21, 1214, 3745	*Nde*I	1252, 2041, 5978
*Bal*I	7694, 8954	*Nhe*I	523, 4115, 9284
*Bam*HI	5028, 8191	*Nru*I	30, 4768
*Bbe*I	9037	*Pst*I	3356, 8987, 9190
*Bcl*Id	3917, 6587	*Pvu*I	3481, 6770, 7934, 8187, 9175
*Bgl*I	3231	*Pvu*II	367, 594, 4637, 7897, 8930
*Bgl*II	17	*Rsr*II	8517
*Bst*EII	6549	*Sac*I	28, 822
*Bvu*I	28, 822, 8675	*Sac*II	4033, 9372
*Cla*I	8	*Sma*I	1216
*Cla*Id	510, 4389, 5731, 7934	*Sna*BI	1023
*Cvn*I	4100, 9300	*Spe*I	7809
*Dra*I	629, 702, 749, 2977, 2996, 3814, 3891	*Sph*I	191, 3771, 5212, 5316, 8635
*Dra*III	6526, 7143	*Stu*I	7776
*Eco*RI	1	*Tth*I	1965, 8917
*Eco*RV	12, 723	*Xba*Id	14
*Fsp*I	3333, 8934	*Xho*I	21
*Hinc*II	4547, 4715, 5883	*Xma*III	9126
*Hind*III	34	*Xmn*I	617, 6034, 7133, 7924
*Hpa*I	5883		
*Mst*I	3333, 8934		

Enzymes that do not cut

*Aat*II *Apa*I *Avr*II *Bst*XI *Dra*II *Kpn*I
*Mlu*I *Not*I *Ppu*MI *Sal*I *Sca*I *Sfi*I

[a] d, The cleavage site is protected by *dam* methylation.

tide triphosphates and joined to a similarly treated *Eco*RI site at the end of a synthetic linker joined to the 3′ end of the NOS 3′ nontranslated region. The resultant sequence is

5′-GAATTAATTCCCGATCGATC

The ATCGAT is the *Cla*I site located at bp 2102 of the nopaline synthase sequence.[58] The NOS 3′ sequence ends at the *Sau*3A site at bp 1847 of the nopaline synthase sequence and adjoins the following linker sequence

5′-GGGGATCCGGGGG

The last three G's of this sequence are from one-half of the *Sma*I site located at bp 1118 of the Tn*5* sequence.[61] The Tn*5* neomycin phosphotransferase II (NPTII) segment extends from the *Sma*I site at bp 1118 to the *Sau*3A site at bp 140. This *Sau*3A site is immediately adjacent to the following linker

5′-GTCTAGGATCTGCAG

The T of the *Pst*I site (CTGCAG) is the 3′ end of the nopaline synthase promoter segment which begins at bp 584 and ends with the *Bcl*I site at bp 284 of the nopaline synthase sequence.[58] The *Bcl*I site was cleaved with *Sau*3A and joined to a linker to give the following sequence which includes the half *Bcl*I site and the *Eco*RI site which is the origin of the pMON200 plasmid

5′-TGATCCGGGGAATTC

When assembled following the above instructions, the total size of the complete pMON200 plasmid will be 9489 bp.

The coordinates of the various segments comprising pMON505 and selected restriction endonuclease cleavage sites are given in Table II. Directions for assembly of the pMON505 plasmid are as follows. Starting with pMON200 the LIH region beginning after the unique *Hin*dIII site is replaced with the following segments. First, 74 N's are inserted followed by bp 617 to bp 1 of the published *ori*V sequence.[62] Next is a run of 54 N's and a *Sal*I site (5′-GTCGAC) and then 694 N's. This is followed by the bp 400 to bp 1618 sequence of the *trfA** gene.[63] Next are 155 N's followed by an *Nco*I site (CCATGG) followed by 500 N's and a second *Nco*I site (CCATGG). This is followed by another run of 435 N's and then the *Sma*I site that is the end of the original *Hin*dIII to *Sma*I fragment from

[61] E. Beck, G. Ludwig, E.-A. Auerswald, B. Reiss, and H. Schaller, *Gene* **19,** 327 (1982).

[62] D. Stalker, C. Thomas, and D. Helinski, *Mol. Gen. Genet.* **181,** 8 (1981).

[63] C. Smith and C. Thomas, *J. Mol. Biol.* **175,** 251 (1984).

TABLE II
MAJOR REGIONS AND RESTRICTION ENDONUCLEASE CLEAVAGE SITES OF pMON505

Segment name	Coordinates
Synthetic multilinker	1–36
RK2 origin	36–3801
pBR322 origin fragment	3802–5273
Right border	5778
pTiT37 fragment	5274–7665
Nopaline synthase coding region	6153–7391
Tn*7 Spc/StrR* fragment	7666–9838
Spc/StrR coding region	8317–9071
NOS-NPTII′-NOS kanamycin^{R} fragment	9839–11412
NPTII′ coding region	10293–11086

Endonuclease[a]	Cleavage site	Endonuclease[a]	Cleavage site
*Acc*I	785	*Nco*I	2860, 3360, 6543, 10527
*Afl*II	7445, 10081		
*Asu*II	10279	*Nde*I	2423, 3838, 7914
*Ava*I	21, 3800, 5681	*Nhe*I	6051, 11211
*Bal*I	491, 9621, 10881	*Not*I	634
*Bam*HI	6964, 10118	*Nsi*I	7146, 7254, 7536, 9998
*Bbe*I	10964		
*Bcl*Id	5853, 8514	*Ppu*MI	1835, 2192, 2365
*Bgl*II	17	*Pst*I	5153, 10914, 11117
*Bsp*MI	9843	*Pvu*II	6573, 9824, 10857
*Bst*EI	8476	*Rsr*II	10444
*Cla*I	8	*Sal*I	784
*Cla*Id	6325, 7667, 9861	*Sac*I	28
*Cvn*I	6036, 11227	*Sst*II	640, 5969, 11299
*Dra*I	212, 4774, 4793, 5750, 5827	*Sfi*I	1866
		*Sma*I	3802
*Eco*RI	1	*Spe*I	9736
*Eco*RV	12	*Sph*I	5707, 7148, 7252, 10562
*Fsp*I	5130, 10861		
*Hin*cII	537, 572, 786, 2051, 6483, 6651, 7819	*Ssp*I	204, 6060, 6590, 7263, 11206
		*Stu*I	9703
*Hin*dIII	34	*Tth*I	10844
*Hpa*I	7819	*Xba*Id	14
*Mst*I	5130, 10861	*Xho*I	21
*Nar*I	10961	*Xma*II	11053
		*Xmn*I	7970, 9060, 9851

Enzymes that do not cut
*Aat*II *Aap*I *Avr*II *Bst*XI *Eco*K *Kpn*I *Mlu*I *Sca*I *Sna*BI

[a] d, The cleavage site is protected by *dam* methylation.

pTJS75[34] used to construct pMON505. The next segment is a small 43-bp *Sma*I to *Nde*I fragment derived from the pMON200 LIH segment and has the following sequence

5′-CCCGGGATGG CGCTAAGAAG CTATTGCCGC CGATCTTCAT ATG

This is joined to the *Nde*I (bp 2297) to *Pvu*I (bp 3740) fragment of pBR322.[57] This is followed by the pMON200 sequences described above. The restriction endonuclease sites shown on the maps have been verified by restriction endonuclease cleavage. However, since the sequences are incomplete, certain of the sites listed in the tables must be regarded as preliminary.

[16] Vectors for Cloning in Plant Cells

By R. Deblaere, A. Reynaerts, H. Höfte, J.-P. Hernalsteens, J. Leemans, and M. Van Montagu

Introduction

Transfer of foreign DNA into plant cells is most often performed using plasmid vector systems derived from the tumor-inducing (Ti) plasmids of *Agrobacterium tumefaciens*. A portion of these plasmids, the T-DNA, is transferred into the plant chromosomes during the naturally occurring process of crown gall tumor formation (for a review, see Gheysen *et al.*[1]). *Agrobacterium* has a very wide host range and, recently, transformation of Liliaceae and Amaryllidaceae by *Agrobacterium* has been reported.[2,3] This contradicted the classical observations, which indicated that the applicability of the *Agrobacterium* system would be limited to dicotyledonous plants. However, *Agrobacterium*-mediated transformation of Gramineae, including the important cereal crops, has not yet been achieved.

Other emerging techniques to introduce foreign DNA into plants are the uptake of naked DNA by protoplasts after polyethylene glycol treat-

[1] G. Gheysen, P. Dhaese, M. Van Montagu, and J. Schell, *Adv. Plant Gene Res.* **2,** 11 (1985).

[2] J.-P. Hernalsteens, L. Thia-Toong, J. Schell, and M. Van Montagu, *EMBO J.* **3,** 3039 (1984).

[3] G. M. S. Hooykaas-Van Slogteren, P. J. J. Hooykaas, and R. A. Schilperoort, *Nature (London)* **311,** 763 (1984).

ment,[4,5] electroporation,[6] or a combination of both.[7] Although this method can be applied to both monocot and dicot species, the need to use protoplasts very often limits the possibilities to regenerate normal plants after transformation.

At present, a wide variety of *Agrobacterium* vector systems has been developed. They allow easy cloning of foreign genes and they are transferred in a single step from *Escherichia coli* to *Agrobacterium*. Most important, transformation can be obtained after infection of wounded plant parts, such as stems,[8] leaf segments,[9] tuber slices in potato (L. Willmitzer, personal communication), root slices in *Daucus*,[10] and probably an even wider range of tissue types, which will hopefully include embryonic or meristematic tissues.

Two observations allowed the design of the presently used *Agrobacterium* vectors. First, the T-DNA genes responsible for phytohormone-independent tumorous growth of crown gall cells are not required for T-DNA transfer and integration.[10] Second, the T-DNA borders, including the directly repeated sequences of 25 bp, were shown to be the only essential cis-acting sequences involved in T-DNA transfer.[11]

Since successful transformation usually depends on the use of proper selectable marker genes, chimeric genes, consisting of a plant promoter sequence and a bacterial antibiotic resistance gene, have been constructed. These genes confer an easily selectable phenotype to plant cells[12–15] (Table I).

[4] F. A. Krens, L. Molendijk, G. J. Wullems, and R. A. Schilperoort, *Nature (London)* **296,** 72 (1982).

[5] R. Hain, P. Stabel, A. P. Czernilofsky, H. H. Steinbiss, L. Herrera-Estrella, and J. Schell, *Mol. Gen. Genet.* **199,** 161 (1985).

[6] M. Fromm, L. P. Taylor, and V. Walbot, *Proc. Natl. Acad. Sci. U.S.A.* **82,** 5824 (1985).

[7] I. Potrykus, R. D. Shillito, M. W. Saul, and J. Paszkowski, *Plant Mol. Biol. Rep.* **3,** 117 (1985).

[8] P. Zambryski, H. Joos, C. Genetello, J. Leemans, M. Van Montagu, and J. Schell, *EMBO J.* **2,** 2143 (1983).

[9] R. B. Horsch, J. E. Fry, N. L. Hoffmann, D. Eichholtz, S. G. Rogers, and R. T. Fraley, *Science* **227,** 1229 (1985).

[10] J. Leemans, R. Deblaere, L. Willmitzer, H. De Greve, J.-P. Hernalsteens, M. Van Montagu, and J. Schell, *EMBO J.* **1,** 147 (1982).

[11] K. Wang, L. Herrera-Estrella, M. Van Montagu, and P. Zambryski, *Cell* **38,** 455 (1984).

[12] L. Herrera-Estrella, M. De Block, E. Messens, J.-P. Hernalsteens, M. Van Montagu, and J. Schell, *EMBO J.* **2,** 987 (1983).

[13] M. W. Bevan, R. B. Flavell, and M.-D. Chilton, *Nature (London)* **304,** 184 (1983).

[14] R. T. Fraley, S. G. Rogers, R. B. Horsch, P. R. Sanders, J. S. Flick, S. P. Adams, M. L. Bittner, L. A. Brand, C. L. Fink, J. S. Fry, G. R. Galluppi, S. B. Goldberg, N. L. Hoffmann, and S. C. Woo, *Proc. Natl. Acad. Sci. U.S.A.* **80,** 4803 (1983).

[15] C. Waldron, E. B. Murphy, J. L. Roberts, G. D. Gustafson, S. L. Armour, and S. K. Malcolm, *Plant Mol. Biol.* **5,** 103 (1985).

TABLE I
DOMINANT SELECTABLE MARKER GENES FOR TOBACCO CELLS

Resistance gene	Origin	Resistance
Neomycin phosphotransferase	Tn*5*	Kanamycin Geneticin (G418)
Hygromycin phosphotransferase	pJR225	Hygromycin B
Mtx^R dihydrofolate reductase	R67	Methotrexate
Chloramphenicol acetyltransferase	Tn9	Chloramphenicol

This chapter presents recent octopine Ti plasmid-derived vectors and their use to transfer and express foreign genes in plants. All T-DNA genes have been removed from the T-DNA vector, but the border fragments are still present. Plant cells transformed with this avirulent T-DNA retain their intrinsic capacity to redifferentiate into complete plants. Moreover, the T-DNA vectors contain a dominant selectable marker gene, mainly a chimeric neomycin phosphotransferase gene that provides the transformed plant cell with a selectable kanamycin resistance trait.

Methods

The Agrobacterium Acceptor Strain

All vectors which will be presented are to be used in *Agrobacterium* strain C58C1Rif^R (pGV2260).[16] pGV2260 is a derivative of the octopine Ti plasmid pTiB6S3 in which the entire T-region (both T_L and T_R) including the 25-bp border sequences has been deleted and substituted by pBR322 sequences, including the carbenicillin resistance gene. pGV2260 contains the intact *vir* region which encodes Ti plasmid functions necessary for T-DNA transfer and/or integration. These functions can act *in trans* to the T-DNA.[17]

Cointegration Vectors

A first type of T-DNA vector is based on the common cloning vehicle pBR322. They replicate in *E. coli* and are directly mobilized into *Agrobacterium* strain C58C1Rif^R (pGV2260). pBR322 itself contains a *bom*

[16] R. Deblaere, B. Bytebier, H. De Greve, F. Deboeck, J. Schell, M. Van Montagu, and J. Leemans, *Nucleic Acids Res.* **13,** 4777 (1985).

[17] A. Hoekema, P. R. Hirsch, P. J. J. Hooykaas, and R. A. Schilperoort, *Nature (London)* **303,** 179 (1983).

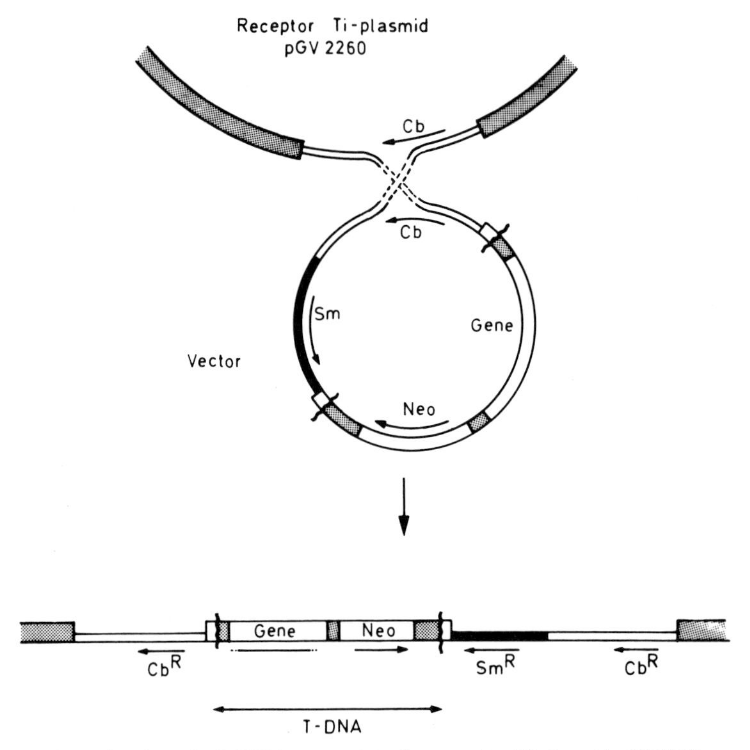

FIG. 1. Introduction of a T-DNA vector into the receptor Ti plasmid pGV2260 occurring via a homologous recombination between the pBR322 sequences present on both plasmids. In the resulting cointegrate, the T-DNA is flanked by a direct repeat of pBR322 sequences. T-DNA border sequences are indicated as wavy lines. Abbreviations: Cb^R, carbenicillin resistance; Sm^R, streptomycin resistance; neo, chimeric *npt*-II gene.

site, and mobilization as well as transfer functions can be complemented *in trans*.[18,19] Since the vector cannot replicate in *Agrobacterium,* it can be maintained only if a recombination occurs between the homologous pBR322 sequences present in the vector and in the acceptor Ti plasmid pGV2260.

This results in the formation of a cointegrate in which the T-DNA is flanked by directly repeated pBR322 sequences (Fig. 1). pGV1500 is the prototype cointegration vector (Fig. 2). Its replicon consists of pBR322 sequences and of a streptomycin resistance gene (Sm^R). This gene pro-

[18] G. Ditta, S. Stanfield, D. Corbin, and D. R. Helinski, *Proc. Natl. Acad. Sci. U.S.A.* **77,** 7347 (1980).

[19] E. Van Haute, H. Joos, M. Maes, G. Warren, M. Van Montagu, and J. Schell, *EMBO J.* **2,** 411 (1983).

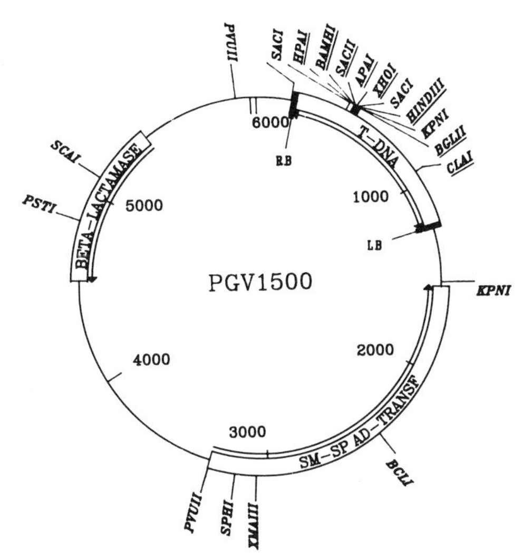

FIG. 2. Cointegration vector pGV1500, derived from pGV825[16] and containing a multilinker sequence between the T-DNA borders. Unique restriction sites for cloning of foreign DNA within the T-DNA are underlined. Abbreviations: LB, left border; RB, right border; SM-SP AD-TRANSF, streptomycin/spectinomycin adenyltransferase.

vides an excellent marker in *Agrobacterium,* allowing the selection and maintenance of the cointegrate structure. Moreover, the vector contains a 1550-bp T-DNA fragment including the left and right 25-bp terminal repeat sequences of the octopine T_L-DNA and a synthetic multilinker containing several unique restriction sites for the introduction of foreign DNA. The steps to introduce a foreign gene into a Ti plasmid vector are as follows:

1. Insert the gene of interest into one of the unique restriction sites in pGV1500, using classical recombinant DNA techniques.[20]
2. Select *E. coli* clones carrying the recombinant vector on LB plates, supplemented with 20 μg/ml streptomycin and 50 μg/ml spectinomycin.
3. Start three bacterial cultures for mobilization, using the triparental mating procedure[18]: *E. coli* carrying the T-DNA vector, the mobilization helper strain HB101(pRK2013), and the acceptor C58C1 Rif^R (pGV2260).
4. Plate 0.1 ml of a 1 : 1 : 1 mixture of these three overnight cultures on an LB plate and incubate overnight at 28°.
5. Collect the mobilization mixture in 2 ml 10 m*M* $MgSO_4$ and plate serial dilutions (in 10 m*M* $MgSO_4$) on LB plates supplemented with 100

[20] T. Maniatis, E. F. Fritsch, and J. Sambrook, "Molecular Cloning: A Laboratory Manual." Cold Spring Harbor Lab., Cold Spring Harbor, New York, 1982.

μg/ml rifampicin (Rif), 300 μg/ml streptomycin (Sm), and 100 μg/ml spectinomycin (Sp) (LB $Rif_{100}Sm_{300}Sp_{100}$).

6. Incubate the selection plates for 2 days at 28°; Sm^R/Sp^R transconjugants are obtained at an average frequency of 10^{-5} per *Agrobacterium* recipient.

7. Streak transconjugants on LB $Rif_{100}Sm_{300}Sp_{100}$ to obtain purified colonies.

8. Start overnight cultures (28°) of two isolates in 3 ml LB medium for the preparation of total bacterial DNA.[21]

9. Verify the cointegrate structure using classical Southern blot hybridization techniques.[20]

Binary Vectors

An approach to avoid the cointegration step is the use of binary vectors. In this system the T-DNA is carried by a vector molecule capable of autonomous replication in *Agrobacterium.* Ti plasmid functions required for T-DNA transfer are complemented *in trans* by the *vir* genes, present on a Ti plasmid derivative.[17,22] *Agrobacterium* strains are equally easy to construct using cointegration or binary vectors, but the use of binary vectors omits the need to verify the recombinant Ti plasmid by Southern blot and hybridization analysis. Because of its higher transfer frequency from *E. coli* to *Agrobacterium,* it is hoped that complete libraries of plant genomic DNA, constructed in cosmid derivatives of binary vectors, can be transferred en masse into plant cells. This could allow the complementation of biochemical mutants or selection for the expression of dominant genes. In the resulting plant cell lines the gene of interest would be easily recognized due to the cotransferred T-DNA sequences.

The binary vectors presently described are based on broad-host-range plasmids derived from the P-type plasmid RK2.[23-25] However, a major consideration in the use of this replicon is its considerable instability under nonselective conditions.[26] Therefore, we looked for other plasmids which might lead to more stable *Agrobacterium* replicons.

[21] P. Dhaese, H. De Greve, H. Decraemer, J. Schell, and M. Van Montagu, *Nucleic Acids Res.* **7,** 1837 (1979).

[22] A. J. de Framond, K. A. Barton, and M.-D. Chilton, *Bio/Technology* **1,** 262 (1983).

[23] M. Bevan, *Nucleic Acids Res.* **12,** 8177 (1984).

[24] G. An, B. D. Watson, S. Stachel, M. P. Gordon, and E. W. Nester, *EMBO J.* **4,** 277 (1985).

[25] H. J. Klee, M. F. Yanofsky, and E. W. Nester, *Bio/Technology* **3,** 637 (1985).

[26] G. Ditta, T. Schmidhauser, E. Yakobson, P. Lu, X-.W. Liang, D. R. Finlay, D. Guiney, and D. R. Helinski, *Plasmid* **13,** 149 (1985).

pVS1 is a 30-kb *Pseudomonas aeruginosa* plasmid which confers resistance to sulfonamide and mercury ions[27] and replicates stably in *Agrobacterium* C58 (unpublished results). A binary vector, pGV941, was constructed based on the stability and replication functions of pVS1 (Fig. 3), which are carried by a 3.8-kb *Bam*HI/*Sac*II fragment.[28] This vector or its derivatives are also mobilized into C58C1Rif^R (pGV2260), but do not integrate into the acceptor Ti plasmid. The pVS1-derived plasmid proved completely stable (>99% after 30 generations) in nonselectively grown *Agrobacterium* cultures as well as under conditions of a classic tobacco protoplast cocultivation experiment.

These binary vectors are used in the following way:

1. Clone the foreign DNA into the unique *Bam*HI or *Hpa*I site, using classic recombinant DNA techniques.[20]
2. Select Sm^R *E. coli* clones containing the recombinant vector.
3. Start three bacterial cultures in 3 ml liquid LB medium for a triparental mating: *E. coli* strain carrying a binary vector, helper strain HB101(pRK2013), and acceptor strain C58C1Rif^R (pGV2260).
4. Plate 0.1 ml of a 1 : 1 : 1 mixture of these strains on an LB plate and incubate overnight at 28°.
5. Collect the conjugation mixture and plate serial dilutions on LB plates containing rifampicin (100 μg/ml), streptomycin (300 μg/ml), and spectinomycin (100 μg/ml); incubate for 2 days at 28°: Sm^R/Sp^R transconjugants are obtained at an average frequency of 10^{-2} per *Agrobacterium* recipient.
6. Colony purify transconjugants on LB $Rif_{100}Sm_{300}Sp_{100}$.

Expression of Chimeric Genes in Plant Cells

In many cases, the gene one desires to express in plant cells will not carry the sequences required for proper transcription. A set of plant-specific promoter sequences has been described and has been used to express chimeric genes (Table II). Figure 4 presents two T-DNA expression vectors derived from pGV1500. pGSH160 and pGSJ280 each contain an *npt*-II gene to confer kanamycin resistance to the transformed plant cells. In pGSH160 the resistance gene is under control of the T_R-DNA gene *1′*, in pGSJ280 under that of the nopaline synthase promoter. pGSH160 contains the T_R-DNA gene *2′* promoter fragment followed by the T_L-DNA gene *7* polyadenylation signal. pGSJ280 contains a promoter

[27] V. A. Stanisich, P. M. Bennett, and M. H. Richmond, *J. Bacteriol.* **129,** 1227 (1977).
[28] Y. Itoh, J. M. Watson, D. Haas, and T. Leisinger, *Plasmid* **11,** 206 (1984).

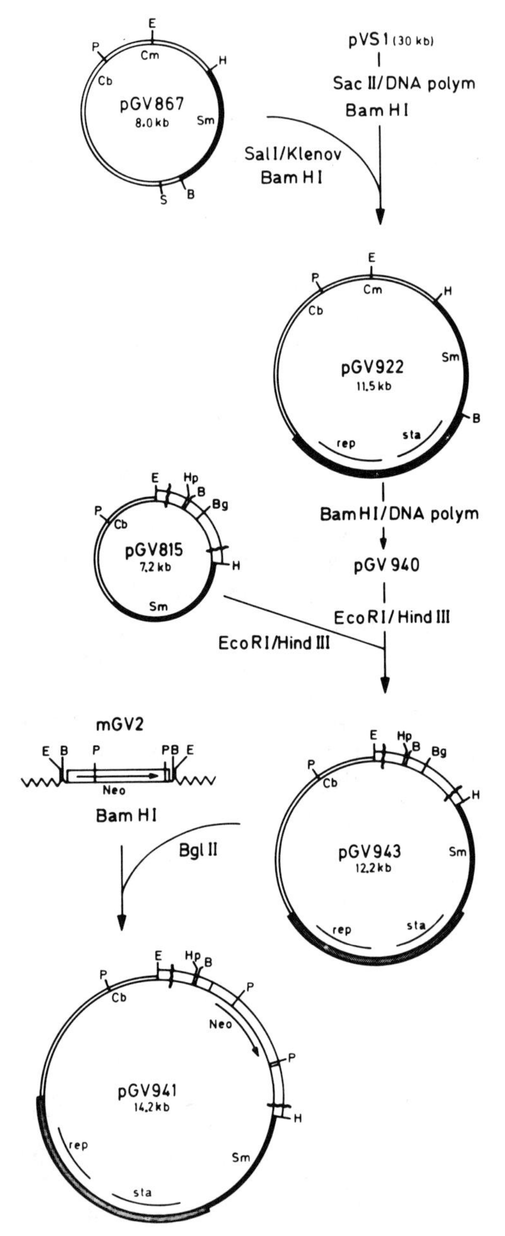

FIG. 3. Construction of the binary vector pGV941 based on the replication (rep) and stability (sta) functions of pVS1.[28] pGV867 is a pBR325 derivative, which contains a 2.3-kb *Hind*III/*Bam*HI fragment encoding streptomycin resistance (Sm^R). pGV815 is a T-DNA cointegration vector.[16] mGV2 is a M13mp7 clone containing the chimeric *npt*-II gene (*neo*) isolated from pLGVneo1103 [H. Lörz, B. Baker, and J. Schell, *Mol. Gen. Genet.* **199,** 178 (1985)]. T-DNA border sequences are indicated as wavy lines.

TABLE II
PLANT-SPECIFIC PROMOTERS FOR THE EXPRESSION OF CHIMERIC GENES IN PLANTS

Promoter	Origin	Reference
Constitutive		
Pnos	Nopaline synthase gene of pTiC58	12
P1′2′	Genes *1′* and *2′* of the T_R-DNA of pTiB6S3	*a*
Pocs	Octopine synthase gene of pTiB6S3	*b*
P19 S	Cauliflower mosaic virus 19 S RNA	*c*
P35 S	Cauliflower mosaic virus 35 S RNA	29
Light regulated		
Pssu	Gene encoding pea small subunit of ribulose-1,5-bisphosphate carboxylase	30
Pcab	Gene encoding pea light-harvesting chlorophyll *a*/*b*-binding protein	31
Pchs	Chalcone synthase gene of *Antirrhinum majus*	*d*

[a] J. Velten and J. Schell, *Nucleic Acids Res.* **13,** 6981 (1985).
[b] N. Murai, D. W. Sutton, M. G. Murray, J. L. Slightom, D. J. Merlo, N. A. Reichert, C. Sengupta-Gopalan, C. A. Stock, R. F. Barker, J. D. Kemp, and T. C. Hall, *Science* **222,** 476 (1983).
[c] M. G. Koziel, T. L. Adams, M. A. Hazlet, D. Damm, J. Miller, D. Dahlbeck, S. Jayne, and B. J. Staskawicz, *J. Mol. Appl. Genet.* **2,** 549 (1984).
[d] H. Kaulen, J. Schell, and F. Kreuzaler, *EMBO J.* **5,** 1 (1986).

fragment derived from cauliflower mosaic virus DNA from which the 35 S RNA is transcribed.[29] Both vectors contain a *Cla*I and *Bam*HI site downstream of the promoter for the introduction of foreign coding sequences (Fig. 4). Coding sequences lacking a start codon are inserted into the *Bam*HI site and are then fused to the ATG start codon, if the reading frames match. Coding sequences including their initiation codon are cloned into the *Cla*I site. This site is nonmethylated, in contrast to the second *Cla*I site near the left T-DNA border and can thus be used as a unique site.

The promoters used here are from constitutively expressed genes. Recently, promoter sequences allowing light-regulated and chloroplast-dependent expression of chimeric genes have been isolated from the pea gene for the small subunit of ribulose-1,5-bisphosphate carboxylase (*rbcS*) and from the gene for the light-harvesting chlorophyll *a*/*b*-binding protein. These promoters are contained within a DNA segment of 900 and 400 bp,

[29] J. T. Odell, F. Nagy, and N.-H. Chua, *Nature* (*London*) **313,** 810 (1985).

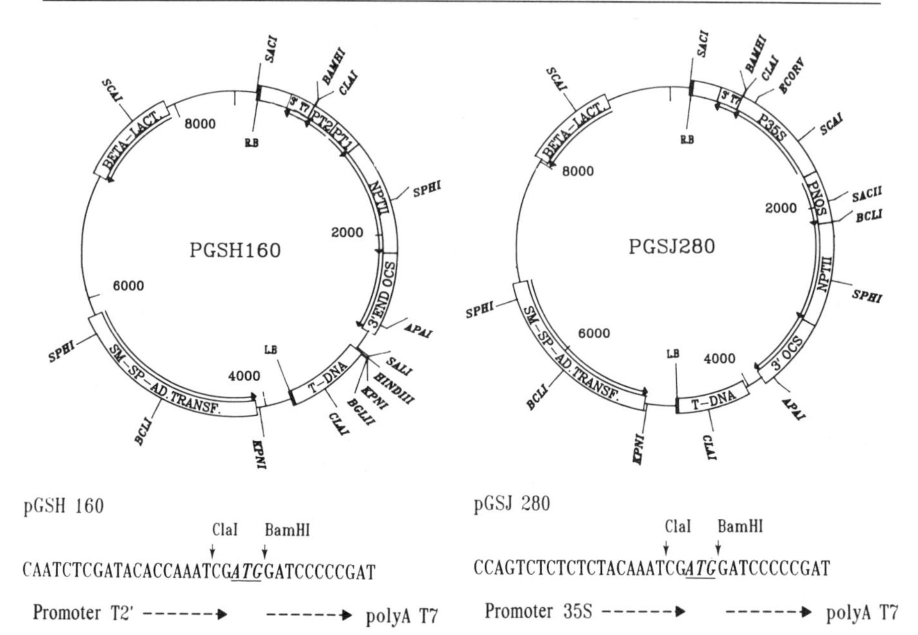

FIG. 4. Two expression vectors, pGSH160 and pGSJ280, in which foreign DNA is expressed under control of the T_R-DNA gene *2'* (PT2) and CaMV 35 S (P35 S) promoter, respectively. The DNA sequence at the junction of the promoter and the polyadenylation fragment (from T_L-DNA gene *7*) is depicted below. The unique *Bam*HI and *Cla*I sites as well as the ATG start codon are shown.

respectively, and have been used to express chimeric genes.[30,31] It can be expected that signals which regulate the expression of other plant genes will also be located at the 5' end of such genes, and that analogous constructions can be made to allow expression of foreign genes under various regulatory regimes. Possible candidate promoters include those from seed-specific storage protein genes in bean,[32] soybean,[33] or maize,[34] from root-specific and tuber-specific genes from potato (L. Willmitzer, personal communication). Soybean heat shock genes have been cloned and thermoinduced expression of these genes in transformed sunflower tissue has been demonstrated.[35]

[30] L. Herrera-Estrella, G. Van den Broeck, R. Maenhaut, M. Van Montagu, J. Schell, M. Timko, and A. Cashmore, *Nature* (*London*) **310,** 115 (1984).

[31] J. Simpson, M. P. Timko, A. R. Cashmore, J. Schell, M. Van Montagu, and L. Herrera-Estrella, *EMBO J.* **4,** 2723 (1985).

[32] J. L. Slightom, S. M. Sun, and T. C. Hall, *Proc. Natl. Acad. Sci. U.S.A.* **80,** 1897 (1983).

[33] R. B. Goldberg, G. Hoschek, and L. O. Vodkin, *Cell* **33,** 465 (1983).

[34] P. Langridge and G. Feix, *Cell* **34,** 1015 (1983).

[35] F. Schöffl and G. Baumann, *EMBO J.* **4,** 1119 (1985).

Plant Transformation

Cocultivation of plant protoplasts with *Agrobacterium*[36] has been used to obtain transgenic plants.[12,37,38] However, this method is slow and its use is restricted to a limited number of plant species in which plant regeneration from protoplasts is possible. Therefore, it was important to develop techniques to transform somatic plant tissues and to regenerate transformed plants from them. Indeed, for a variety of plant species, complete plants can be regenerated from tissue cultures, although protoplast cultivation techniques are not available.

Currently we use the following method which reproducibly yields large numbers of transformed tobacco plants within 6–8 weeks. A similar method has been described to transform *Petunia hybrida*.[9]

Transformation of Tobacco Leaf Fragments

1. Excise leaves from 4-week-old *Nicotiana tabacum* cv. "Petit Havana" SR1 plants. Tobacco plants have been propagated *in vitro* from shoots, on half-strength Murashige and Skoog (MS) medium,[39] solidified with 0.7% agar.
2. Cut leaf segments of ±0.5 cm^2 and transfer approximately 15 segments into a Petri dish containing 10 ml of liquid complete MS medium. Take care to float the leaf segments upside down.
3. Add 10 μl of a bacterial culture/1 ml tissue culture medium. Agrobacteria are grown overnight in minimal medium,[40] supplemented with 300 μg/ml spectinomycin (Sigma) and/or 1000 μg/ml streptomycin (Sigma).
4. Incubate plates for 48 hr at 25° in the plant culture room.
5. Wash the infected fragments twice with liquid complete MS medium (2× initial volume).
6. Transfer the leaf segments to a plate with the shoot-inducing medium putting the lower epidermis toward the agar. For tobacco shoot regeneration, the complete MS medium is supplemented with 0.8% agar, 1.0 mg/liter benzylaminopurine (Sigma), and 0.1 mg/liter naphthalene ace-

[36] L. Márton, G. J. Wullems, L. Molendijk, and R. A. Schilperoort, *Nature (London)* **277**, 129 (1979).

[37] M. De Block, L. Herrera-Estrella, M. Van Montagu, J. Schell, and P. Zambryski, *EMBO J.* **3**, 1681 (1984).

[38] R. B. Horsch, R. T. Fraley, S. G. Rogers, P. R. Sanders, A. Lloyd, and N. Hoffman, *Science* **223**, 496 (1984).

[39] T. Murashige and F. Skoog, *Physiol. Plant.* **15**, 473 (1962).

[40] J. H. Miller, "Experiments in Molecular Genetics." Cold Spring Harbor Lab., Cold Spring Harbor, New York, 1972.

tic acid (Sigma). In addition, 500 μg/ml cefotaxime (Calbiochem) is added to the medium to eliminate the *Agrobacterium* cells as well as kanamycin (100 μg/ml) (Sigma) to select the growth of transformed tobacco shoots.

7. Incubate plates at 25° and 16 hr light (±2000 lx)/8 hr dark cycle.
8. Small shoots will appear after 3–4 weeks.
9. Transfer the shoot-forming leaf disks to fresh medium and grow for another 3 weeks.
10. Transfer 5 shoots (±0.5 cm large) into a 500-ml glass container, containing hormone-free agar-solidified MS medium, supplemented with 500 μg/ml cefotaxime. The shoots are subsequently transferred to half-strength solid MS medium in which the concentration of cefotaxime is gradually decreased to zero during the next two transfers in order to obtain axenic plantlets.
11. When the plantlets are 9–10 cm in height, transfer them to soil. Remove medium from the roots and transfer the plantlets into Jiffy-7 peat disks. Place the pots in a growth chamber in which humidity can be controlled. An important factor in the survival of the new plantlets is their protection from desiccation during the first weeks. Plants can be hardened off by gradually reducing the humidity in the growth container: this can be done by gradually opening the lid over several days.

Analysis of Gene Expression in Transformed Plants

All the T-DNA vectors described in this chapter contain a chimeric neomycin phosphotransferase gene to select kanamycin resistance in plant cells. The expression of this dominant selectable marker gene can be monitored in the transformed plant cells in two ways: (1) by the ability of transformed plant tissue to form callus on kanamycin-containing medium; (2) by direct assay for neomycin phosphotransferase II (NPT-II) activity in a crude plant extract.

Callus Induction Assay. The phenotypic expression of the chimeric *npt*-II gene in plant cells can be assayed indirectly by placing leaf segments (±0.5 cm^2) on a callus-inducing medium [complete MS medium supplemented with 0.8% agar, 1.0 mg/liter 2,4-dichlorophenoxyacetic acid (Sigma), and 0.1 mg/liter 6-benzylaminopurine], containing 50 μg/ml kanamycin. Leaf tissue from kanamycin-resistant plants will produce callus after 2–3 weeks whereas callus induction from wild-type plants is completely inhibited at this concentration of kanamycin. In some cases, callus induction is observed on medium containing up to 500 μg/ml kanamycin. Considerable variation in the level of kanamycin resistance is observed in transformed plants, regenerated from a single leaf fragment. Molecular analysis showed that a copy number effect cannot explain these differences (unpublished results). A possible explanation for this variability is the position effect due to insertion of the T-DNA in different sites of the genome. Similar variability has been observed by Jones *et al.*[41] using octopine synthase as a reporter gene.

Neomycin Phosphotransferase Assay. This assay is based on the electrophoretic separation of the NPT-II protein from plant endogenous phosphorylating enzymes during a nondenaturing gel electrophoresis and the *in situ* phosphorylation of kanamycin in the presence of radioactively labeled adenosine triphosphate (ATP).

1. Extract 100 mg plant tissue in 100 μl buffer (1% 2-mercaptoethanol, 50 m*M* Tris–HCl, pH 6.8, and 0.13 mg/ml leupeptin) on ice and transfer into a microcentrifuge tube.
2. Centrifuge in an Eppendorf centrifuge at 4° for 2 min.
3. Add 15 μl of loading buffer (50% glycerol, 0.2% SDS, 10% 2-mercaptoethanol, 0.005% bromphenol blue) to 75 μl of plant extract.
4. Separate on a 10% nondenaturing polyacrylamide gel.[42] The enzyme will migrate near the bromphenol blue front.
5. Wash the gel twice for 10 min in distilled water and equilibrate for 30 min in 2× reaction buffer (100 m*M* Tris–HCl, pH 7.5, 50 m*M* $MgCl_2$, 400 m*M* NH_4Cl, 1 m*M* DTT).
6. Transfer the gel onto a glass plate and overlay with a 1% agarose gel containing 30 μg/ml kanamycin sulfate, and 100 μCi [γ-^{32}P]ATP in 1× reaction buffer.
7. Incubate for 30 min at room temperature.
8. Cover the gel sandwich with a sheet of Whatman P81 phosphocellulose paper, two sheets of Whatman 3MM paper, and a stack of blotting paper.
9. Incubate for 2 hr to allow binding of phosphorylated kanamycin to the P81 paper.
10. Wash the P81 paper twice in hot water (80°) for 5 min, twice in cold water (4°) for 5 min, and in 50 m*M* sodium phosphate buffer (pH 7.0) for 5 hr (or overnight).
11. Dry the P81 paper, wrap in a Saran Wrap, and expose to an X-ray film using an intensifying screen at −70°.

T-DNA Organization in Transformed Plants

The following microscale procedure[43] is used to isolate tobacco plant DNA for subsequent physical analysis. Only small amounts of plant mate-

[41] J. D. G. Jones, P. Dunsmuir, and J. Bedbrook, *EMBO J.* **4,** 2411 (1985).
[42] U. K. Laemmli, *Nature (London)* **227,** 680 (1970).
[43] S. L. Dellaporta, J. Wood, and J. B. Hicks, *Plant Mol. Biol. Rep.* **1,** 19 (1983).

rial are needed and DNA can be prepared simultaneously from a large number of plant samples. Purification via CsCl gradients is not required.

DNA Microscale Preparation

1. Freeze 0.7–1 g plant material (callus or leaves) in liquid nitrogen. Grind in a mortar to obtain a fine powder. Transfer into a 50-ml polypropylene centrifuge tube.
2. Add 15 ml extraction buffer (100 m*M* Tris–HCl, pH 8; 50 m*M* EDTA, pH 8; 500 m*M* NaCl; 10 m*M* 2-mercaptoethanol).
3. Add 10 ml 20% SDS; mix vigorously and incubate for 10 min at 65°.
4. Add 5 ml 5 *M* potassium acetate; mix vigorously and incubate for 20 min on ice.
5. Centrifuge for 20 min at 15,000 rpm in a Sorvall SS34 rotor.
6. Pass the supernatant through a Miracloth filter (Calbiochem) and collect it in a new 50-ml centrifuge tube; add 10 ml 2-propanol; mix and incubate for 30 min at −20°.
7. Pellet DNA (15,000 rpm for 15 min) and remove the supernatant.
8. Resuspend the DNA pellet in 0.7 ml TE (50 m*M* Tris–HCl, pH 8; 20 m*M* EDTA, pH 8); transfer into microcentrifuge tube and spin for 10 sec to remove nonsoluble debris.
9. Transfer the supernatant to a new microcentrifuge tube; add 20 μl of a 10 mg/ml RNase solution and incubate for 10 min at 37°.
10. Add 100 μl Tris-equilibrated phenol; mix carefully and spin down for 10 min.
11. Transfer the supernatant to a new tube; add 200 μl chloroform–isoamyl alcohol (24 : 1); mix and spin down for 5 min.
12. Transfer the aqueous phase to a new tube and repeat chloroform extraction.
13. Add 70 μl 3 *M* sodium acetate; mix and add 500 μl 2-propanol.
14. Mix gently until a DNA web is formed; centrifuge DNA down for 30 sec; wash the pellet in 80% ethanol and dry the pellet.
15. Resuspend the pellet in 100 μl TE (10 m*M* Tris–HCl, pH 8; 1 m*M* EDTA, pH 8). This DNA is suitable for subsequent restriction enzyme digestion and Southern blot hybridization analysis.

Molecular Analysis of the T-DNA Structure

Ten micrograms of total DNA is digested to completion with the appropriate restriction endonuclease(s) and loaded on a 1% agarose gel. After gel electrophoresis, the DNA is transferred to a nylon filter (Hybond N, Amersham).[44] We prefer nylon filters over nitrocellulose filters, since they can be reprobed several times. DNA–DNA hybridizations using a radioactively labeled probe are performed according to the manufacturers' instructions.

Molecular analysis of the T-DNA structure in plants transformed with either cointegration or binary vectors showed that (1) the T-DNA border sequences present on the vector are recognized as integration signals to delineate the T-DNA inserts in the majority of transformed plants analyzed; (2) the copy number of T-DNA inserts does not differ from those obtained with the wild-type Ti plasmid pTiB6S3 (one to four copies per genome) (Ref. 16; R. Deblaere, unpublished results).

Inheritance of Transferred Genes

The segregation of the genes introduced in plants by the T-DNA vectors can be studied by classic genetic techniques. Chimeric antibiotics resistance genes are the most convenient markers for this aim. They often allow a direct selection for resistant seedlings after sowing of the seeds on antibiotic-containing medium.[37]

Usually surface-sterilized seeds (e.g., by incubation for 10–30 min in 10% commercial bleach with a trace of detergent and subsequent rinsing in sterile H_2O) are used for this analysis. However, sterile seeds can also be recovered directly from unopened ripe capsules after a rapid swabbing of the surface with ethanol. This procedure allows a considerable gain of time when large numbers of samples must be analyzed and also avoids damage to the sensitive seeds of some species due to the bleach treatment.

It is important to carefully adjust the concentration of antibiotics by preliminary experiments. This concentration depends on the culture conditions. *Nicotiana tabacum* seedlings are, for example, completely inhibited by 25 mg/ml of kanamycin when growing on half-strength MS mineral medium[39] while 50 mg/liter kanamycin is required if this medium is supplemented with 15 g/liter sucrose as a carbon source (F. Budar, unpublished results).

The procedure is also species dependent. *Nicotiana plumbaginifolia* seeds must, for instance, be treated for 2–3 days with a 0.5–1 m*M* gibberellic acid solution or incubated at different day and night temperatures to obtain efficient and homogeneous germination. Moreover, in this species we were unable to obtain reliable segregation data by sowing the

[44] E. M. Southern, *J. Mol. Biol.* **98,** 503 (1975).

seeds directly on kanamycin-containing medium. Therefore, leaf fragments from the seedlings are incubated on callus induction medium (complete MS medium with 0.5 mg/liter 2,4-D and 0.5 mg/liter BAP) with 25 mg/liter kanamycin to test their resistance.

An alternative strategy to follow the segregation of the T-DNA would be to test for the expression of an opine synthase gene[45] or another gene linked to the T-DNA.

Most of the plants obtained by T-DNA-mediated transformation have a low number of T-DNA insertion loci. A large number of plants transformed by T-DNA vectors harboring chimeric nopaline synthase-neomycin phosphotransferase genes were recently studied. About 80% of the plants segregate a single kanamycin resistance locus while two independent loci are present in 12% of the plants.[46] One or more closely linked T-DNA copies can be present at each locus.

Acknowledgments

The authors wish to thank M. De Cock, K. Spruyt, and A. Verstraete for preparing the manuscript and for drawings and photographs. R.D. is a Research Assistant and J.P.H. is a Research Associate of the National Fund for Scientific Research (Belgium).

[45] L. Otten, H. De Greve, J.-P. Hernalsteens, M. Van Montagu, O. Schieder, J. Straub, and J. Schell, *Mol. Gen. Genet.* **183,** 209 (1981).

[46] F. Budar, L. Thia-Toong, J.-P. Hernalsteens, and M. Van Montagu, *Genetics* **114,** 303 (1986).

[17] Binary Ti Vectors for Plant Transformation and Promoter Analysis

By GYNHEUNG AN

Introduction

Agrobacterium tumefaciens is able to transfer a specific segment of tumor-inducing (Ti) plasmid, the T-DNA, to wounded plant cells where it is integrated into the plant nuclear genome.[1-3] In recent years several

[1] M.-D. Chilton, M. H. Drummond, D. J. Merlo, D. Sciaky, A. L. Montoya, M. P. Gordon, and E. W. Nester, *Cell* **11,** 263 (1980).

[2] M.-D. Chilton, R. K. Saiki, N. Yadav, M. P. Gordon, and R. Quetier, *Proc. Natl. Acad. Sci. U.S.A.* **77,** 4060 (1980).

[3] L. Willmitzer, M. De Beuckeleer, M. Lemmers, M. Van Montagu, and J. Schell, *Nature (London)* **287,** 359 (1980).

studies have demonstrated the use of the naturally occurring Ti plasmids as transformation vectors for higher plants.[4–7] In this technique, exogeneous DNA is either inserted into the T-DNA region of the Ti plasmid by homologous recombination using an intermediate vector system,[8–10] or directly into one of the recently designed binary vectors.[11,12]

Because binary Ti vectors need only the cis-acting elements required for efficient DNA transfer, they are much smaller than natural Ti plasmids. The cis elements are the T-DNA borders,[13,14] a selectable marker[8–11] expressible in plants, and a wide-host-range replicon that replicates in both *A. tumefaciens* and *Escherichia coli*.[15] Other necessary functions are provided *in trans* by a helper Ti plasmid[16,17] and the chromosome[18,19] of *A. tumefaciens*. Foreign DNA can be inserted into the binary vector using *E. coli* as a host and the recombinant molecule can be transferred to *A. tumefaciens* carrying a helper Ti plasmid.

Various plant species have been transformed by the cocultivation method in which plant cells are cocultured with *A. tumefaciens* containing the binary vector and a helper Ti plasmid.[11,12,20,21] Transformed cells selected with an appropriate marker present on the binary vector have been regenerated to whole plants.[21]

[4] K. A. Barton and M.-D. Chilton, this series, Vol. 101 [33].

[5] A. Caplan, L. Herrera-Estrella, D. Inze, E. Van Haute, M. Van Montagu, J. Schell, and P. Zambryski, *Science* **222,** 815 (1983).

[6] A. Hoekema, P. R. Hirsch, P. J. J. Hooykaas, and R. A. Schilperoort, *Nature (London)* **303,** 179 (1983).

[7] L. W. Ream and M. P. Gordon, *Science* **218,** 854 (1982).

[8] M. W. Bevan, R. B. Fravell, and M.-D. Chilton, *Nature (London)* **304,** 184 (1983).

[9] R. T. Fraley, S. G. Rogers, R. B. Horsch, P. R. Sanders, J. S. Flick, S. P. Adams, M. L. Bittner, L. A. Brand, C. L. Fink, J. S. Fry, G. R. Galluppi, S. B. Goldberg, N. L. Hoffmann, and S. C. Woo, *Proc. Natl. Acad. Sci. U.S.A.* **80,** 4803 (1983).

[10] L. Herrera-Estrella, M. Deblock, E. Messens, J. P. Hernalsteens, M. Van Montagu, and J. Schell, *EMBO J.* **2,** 987 (1983).

[11] G. An, B. D. Watson, S. Stachel, M. P. Gordon, and E. W. Nester, *EMBO J.* **4,** 277 (1985).

[12] M. W. Bevan, *Nucleic Acids Res.* **12,** 8711 (1984).

[13] C. H. Shaw, M. P. Watson, G. H. Carter, and C. H. Shaw, *Nucleic Acids Res.* **12,** 6031 (1984).

[14] K. Wang, L. Herrera-Estrella, M. Van Montagu, and P. Zambryski, *Cell* **38,** 455 (1984).

[15] G. Ditta, S. Stanfield, D. Corbin, and D. R. Helinski, *Proc. Natl. Acad. Sci. U.S.A.* **77,** 7347 (1980).

[16] V. N. Iyer, H. J. Klee, and E. W. Nester, *Mol. Gen. Genet.* **188,** 418 (1982).

[17] R. C. Lundquist, T. J. Close, and C. I. Kado, *Mol. Gen. Genet.* **193,** 1 (1984).

[18] D. J. Garfinkel and E. W. Nester, *J. Bacteriol.* **144,** 732 (1980).

[19] C. J. Douglas, R. J. Staneloni, R. A. Rubin, and E. W. Nester, *J. Bacteriol.* **161,** 850 (1985).

[20] G. An, *Plant Physiol.* **79,** 568 (1985).

[21] G. An, B. D. Watson, and C. C. Chiang, *Plant Physiol.* **81,** 301 (1986).

Vectors

Binary Vector

A segment of DNA can be transferred from a Ti plasmid into the plant genome only when the plasmid is present in *Agrobacterium*. *E. coli*, which is most commonly used for *in vitro* genetic manipulation, cannot directly serve as a donor for plant transformation. The binary vector system is therefore a convenient way to transform higher plants since foreign DNA can first be inserted into the vector and manipulated using *E. coli* as a host. The constructed molecules are then transferred into *Agrobacterium* where they are maintained as autonomous replicons. Finally the desired segment of DNA can be incorporated into a plant genome.

An example binary Ti vector is shown in Fig. 1. This molecule, pGA482,[22] is 13.2 kbp long and has the following important features for simple and efficient plant transformation.

1. DNA fragment (about 700 bp) containing the nopaline T-DNA right border and a fragment (about 600 bp) containing the nopaline T-DNA left border[23] are included. Although one 25-bp direct repeat is sufficient for T-DNA transfer,[11,14] the DNA sequence surrounding these direct repeats appear to increase the efficiency of the transformation and are therefore included in the vector.

2. A chimeric fusion of the nopaline synthase promoter and the neomycin phosphotransferase gene[11] provides the selectable marker that confers kanamycin resistance to the transformed plant cells.

3. The vector also contains the colE1 origin of replication and the bacteriophage λ *cos* site in order to facilitate the cloning of large DNA fragments and the retrieval of the transferred DNA from the plant genome.

4. There are nine unique sites within the T-DNA borders for cloning foreign DNA. These are *Hin*dIII, *Xba*I, *Sac*I, *Hpa*I, *Kpn*I, *Cla*I, *Bgl*II, *Sca*I, and *Eco*RI from right to left border. The first seven sites are clustered as multiple cloning sites.

5. pGA482 includes pTJS75,[24] a derivative of the RK2 plasmid,[15] that carries a wide-host-range replicon and the tetracycline-resistance gene. This allows for the stable maintenance of pGA482 in both *E. coli* and *Agrobacterium*. The bacterial cells carrying the binary vector can be

[22] G. An, *Plant Physiol.* **81,** 86 (1986).

[23] N. S. Yadav, J. Vanderleyden, D. R. Bennett, W. M. Barnes, and M.-D. Chilton, *Proc. Natl. Acad. Sci. U.S.A.* **79,** 6322 (1982).

[24] T. J. Schmidhauser and D. R. Helinski, *J. Bacteriol.* **164,** 446 (1985).

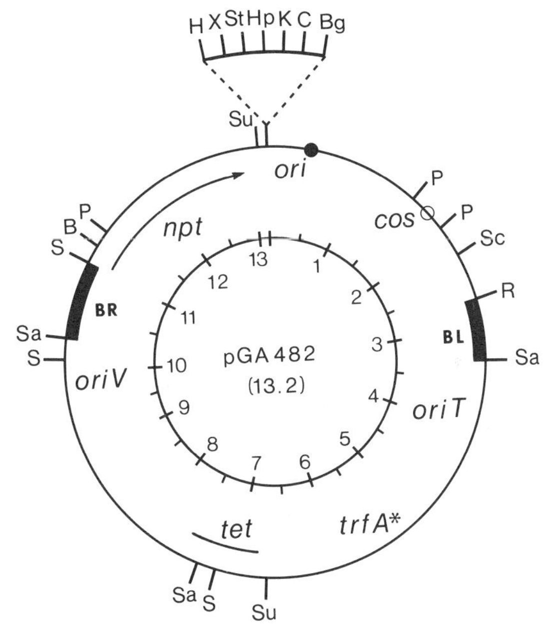

FIG. 1. Map of the pGA482 plasmid. Restriction endonuclease sites are indicated on the outer circle. The size of 13.2 kbp is indicated by the coordinates shown on the inner circle. B, *Bam*HI; Bg, *Bgl*II; BL, the T-DNA left border; BR, the T-DNA right border; C, *Cla*I; *cos*, bacteriophage λ *cos* site; H, *Hin*dIII; Hp, *Hpa*I; K, *Kpn*I; *npt*, neomycin phosphotransferase gene; ori, colE1 origin of replication; oriT, RK2 origin of conjugal transfer; oriV, RK2 origin of replication; P, *Pst*I; R, *Eco*RI; S, *Sst*II; Sa, *Sal*I; Sc, *Sca*I; St, *Sst*I; Su, *Stu*I; *tet*, tetracycline resistance gene of RK2 plasmid; *trfA**, a segment code for a replication protein; X, *Xba*I.

maintained with 3 μg/ml of tetracycline and 10 μg/ml of kanamycin in their culture medium.

Promoter Expression Vector

Much of differential gene expression is regulated at the level of transcription initiation.[25] In order to facilitate the study of the regulatory elements involved in transcription initiation, promoter expression vectors were developed based on the binary Ti vector system.[22] pGA492 (Fig. 2) was constructed from pGA482 by replacing the 2.7-kbp *Bgl*II–*Eco*RI fragment containing the colE1 origin of replication and the *cos* site with a 1.5-kbp DNA fragment containing the chloramphenicol acetyltransferase

[25] J. R. Navins, *Annu. Rev. Biochem.* **52,** 441 (1983).

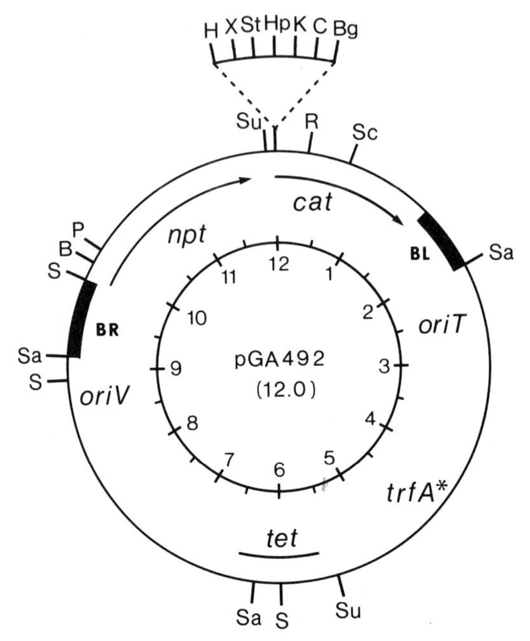

FIG. 2. Map of the pGA492 plasmid. Restriction endonuclease sites are indicated on the outer circle. The size of 12.0 kbp is indicated by the coordinates shown on the inner circle. *cat*, chloramphenicol acetyltransferase gene. Other symbols, see Fig. 1.

(*cat*) gene and a plant terminator sequence derived from one of the octopine T-DNA transcripts. A plant promoter fragment can be inserted into any of the multiple cloning sites located immediately upstream of the promoterless *cat* gene and then tested for its ability to express the *cat* gene in plants.

There are two ways to fuse a promoter to the *cat* gene. In the first, the genes are fused such that the start codon of the natural *cat* gene is the first AUG at the 5′ end of the recombinant primary transcript. It has been observed that translation initiation of most eukaryotic mRNA occurs predominantly at the first AUG triplet from the 5′ end of the message.[26] Translation initiation from internal AUG codons is found to be much less efficient in animals[27] and plants.[28] Therefore, the inserted promoter fragment should not introduce any extra start codons. The second way to fuse the genes is to make an in-reading-frame translational fusion so that the

[26] M. Kozak, *Cell* **15,** 1109 (1978).

[27] R. C. Mulligan and P. Berg, *Mol. Cell. Biol.* **1,** 449 (1981).

[28] G. An, P. R. Ebert, B.-Y. Yi, and C.-H. Choi, *Mol. Gen. Genet.* **203,** 245 (1986).

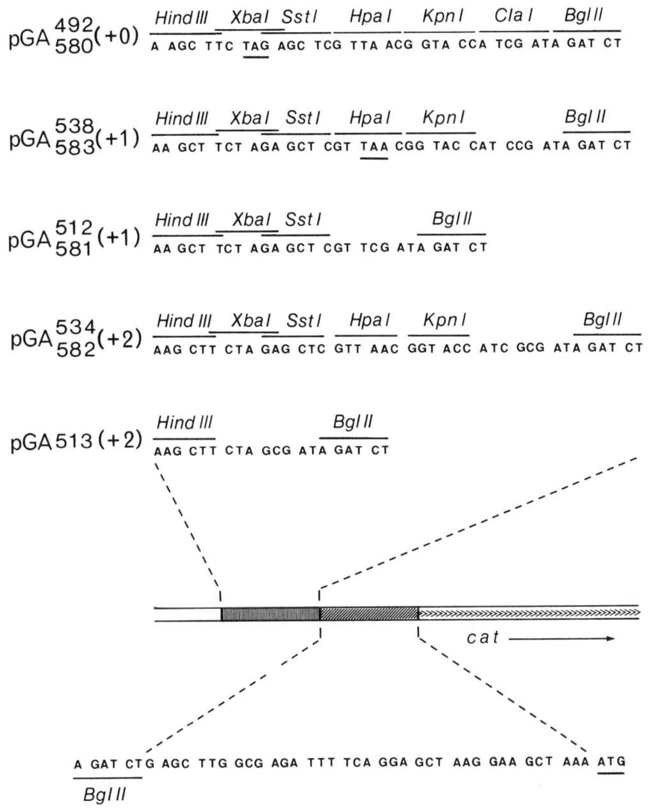

pGA492/580 (+0)
HindIII XbaI SstI HpaI KpnI ClaI BglII
A AGC TTC TAG AGC TCG TTA ACG GTA CCA TCG ATA GAT CT

pGA538/583 (+1)
HindIII XbaI SstI HpaI KpnI BglII
AA GCT TCT AGA GCT CGT TAA CGG TAC CAT CCG ATA GAT CT

pGA512/581 (+1)
HindIII XbaI SstI BglII
AA GCT TCT AGA GCT CGT TCG ATA GAT CT

pGA534/582 (+2)
HindIII XbaI SstI HpaI KpnI BglII
AAG CTT CTA GAG CTC GTT AAC GGT ACC ATC GCG ATA GAT CT

pGA513 (+2)
HindIII BglII
AAG CTT CTA GCG ATA GAT CT

cat →

A GAT CTG AGC TTG GCG AGA TTT TCA GGA GCT AAG GAA GCT AAA ATG
BglII

FIG. 3. The DNA sequence of the multiple cloning sites in the binary vectors. The initiation codon and in-frame termination codons to the *cat* open reading frame are underlined.

start codon belonging to the foreign promoter fragment becomes the first codon of a reading frame that encodes a hybrid *cat* fusion protein. It has been observed that chimeric *cat* fusion proteins carring a foreign peptide at the amino terminus retain their enzymatic activity.[29,30] In order to accommodate a gene in all three reading frames, vectors were developed that have a + 1 (pGA538 and pGA512) and + 2 (pGA534 and pGA513) shift in reading frame relative to pGA492 (Fig. 3). These promoter expression vectors that create *cat* fusion proteins not only make it easier to

[29] D. S. Goldfarb, R. L. Rodriguez, and R. H. Doi, *Proc. Natl. Acad. Sci. U.S.A.* **79,** 5886 (1982).
[30] G. An, *Mol. Gen. Genet.* **207,** 210 (1987).

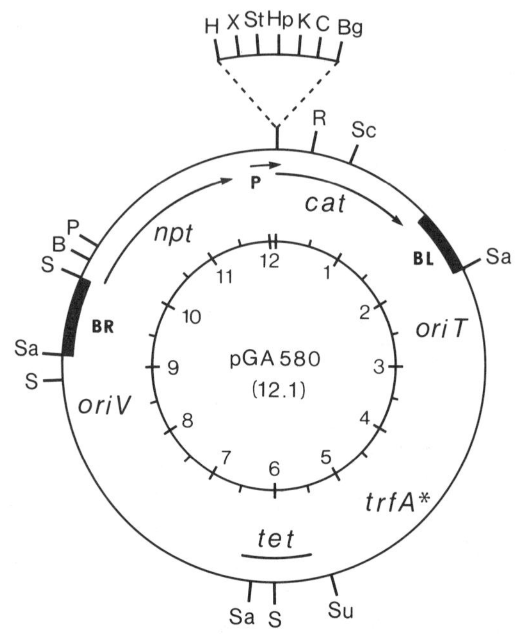

FIG. 4. Map of the pGA580 plasmid. Restriction endonuclease sites are indicated on the outer circle. The size of 12.1 kbp is indicated by the coordinates shown on the inner circle. P, The *lac* promoter. Other symbols, see Figs. 1 and 2.

produce a functional fusion, but also allow one to study posttranscriptional control mechanisms, such as mRNA stability, protein transportation, and protein modification.

In order to facilitate detection of DNA insertion, the promoter expression vectors were further modified. The *lac* promoter carried on the 181-bp *Pvu*II–*Hin*dIII fragment of pUC19[31] was inserted upstream of the *cat* coding region by replacing the 86-bp *Stu*I–*Hin*dIII fragment located between the kanamycin resistance gene and the multiple cloning sites (Fig. 4). The *cat* gene in these modified vectors (pGA580–pGA583) is under the control of the *lac* promoter and therefore confers chloramphenicol resistance to *E. coli* at 20 μg/ml. Insertion of a DNA fragment into the multiple cloning sites attenuates the transcription initiated from the *lac* promoter and reduces the level of chloramphenicol resistance. The degree of reduction in the resistance depends on the nature of the inserted fragment, but screening for inserts is usually straightforward. These vectors are not only useful for cloning and testing a promoter fragment but also may be used to clone any DNA fragment to be transferred into plants.

[31] J. Vieira and J. Messing, this volume [1].

Media

MGL medium (per liter): 2.5 g yeast extract, 5 g tryptone, 5 g NaCl, 5 g mannitol, 1.16 g monosodium glutamute, 0.25 g KH_2PO_4, 0.1 g $MgSO_4 \cdot 7H_2O$, 1 mg biotin, pH 7.0

LB medium (per liter): 5 g yeast extract, 10 g tryptone, 10 g NaCl, pH 7.0

AB medium

20× AB salt (per liter): 20 g NH_4Cl, 6 g $MgSO_4 \cdot 7H_2O$, 3 g KCl, 0.2 g $CaCl_2$, 50 mg $FeSO_4 \cdot 7H_2O$

20× AB buffer (per liter): 60 g K_2HPO_4, 23 g $NaH_2PO_4 \cdot H_2O$

Glucose solution (per liter): 5 g glucose

Autoclave all three solutions separately and then mix 50 ml 20× AB salt, 50 ml 20× AB buffer, and 900 ml glucose solution to make 1 liter

YEP medium (per liter): 10 g yeast extract, 10 g peptone, 5 g NaCl, pH 7.0

DNA Transfer from *E. coli* to *Agrobacterium*

Once a foreign DNA is inserted into the Ti vector the recombinant molecule is transferred into *Agrobacterium* containing an appropriate helper Ti plasmid. The host *Agrobacterium* strain and the helper Ti plasmid it carries should be carefully selected depending upon the plant species to be transformed. This is because the host specificity of *Agrobacterium* and Ti plasmids varies significantly.[32,33] Two different methods are widely used to move the binary vector to *Agrobacterium*.

Triparental Mating

The binary vectors can be transferred from *E. coli* to *Agrobacterium* by conjugation.[15] Since the vector does not carry the mobilization function necessary for the conjugal transfer, this function should be provided by a mobilization plasmid carried in a helper strain. The advantage of triparental mating is a much higher frequency of transformation compared to the direct DNA transfer method which will be described in the next section.

Day 1. Grow *Agrobacterium* (acceptor) on an MGL agar plate at 28° and the *E. coli* (helper) strain containing the mobilization plasmid pRK2073[34] on an LB agar plate at 37°.

[32] A. R. Anderson and L. Moore, *Phytopathology* **69,** 320 (1979).

[33] M. DeCleene and J. Deley, *Bot. Rev.* **42,** 389 (1976).

[34] M. Better and D. Helinski, *J. Bacteriol.* **155,** 311 (1983).

Day 2. Grow the *E. coli* (donor) strain containing the vector on LB agar medium containing 3 μg/ml tetracycline and 10 μg/ml kanamycin at 37°.

Day 3. Mix an equal amount (about 10^8 cells) of all three bacteria and incubate on MGL agar medium without antibiotics overnight at 28°.

Day 4. Streak out the bacterial mixture on AB agar medium containing 5 μg/ml tetracycline and 25 μg/ml kanamycin for 3 to 4 days at 28° until the transconjugants appear. Higher drug concentrations inhibit the transformation frequency.

It may be necessary to streak out once again the transformants on the same AB medium to isolate single colonies.

Direct DNA Transfer

Although the previous method provides high-efficiency transformation, the procedure requires at least 7 to 10 days. The direct DNA uptake method is less efficient but much faster than the triparental mating method.

1. Grow *Agrobacterium* in 50 ml YEP medium (220 rpm at 28°) until OD_{600} reaches about 0.5.
2. Centrifuge the cells at 5000 rpm for 5 min and resuspend the cells in 10 ml 0.15 *M* NaCl.
3. Centrifuge the cells as above and resuspend the pellet in 1 ml of ice-cold 20 m*M* $CaCl_2$.
4. Add 1 μg DNA to 0.2 ml bacterial cells in a 1.5-ml microcentrifuge tube and incubate it on ice for 30 min.
5. Freeze the mixture in liquid nitrogen for 1 min.
6. Thaw the cells in a 37° water bath.
7. Add 1 ml of YEP medium and incubate the cells at 28° for 2 to 4 hr with gentle shaking.
8. Centrifuge the cells in a microcentrifuge for 1 min and resuspend the cells in 0.1 ml YEP medium.
9. Plate the cells on YEP or MGL agar medium containing 3 μg/ml tetracycline and 25 μg/ml kanamycin. Higher drug concentrations inhibit the transformation frequency. Optimum drug concentrations may be different depending on *Agrobacterium* strain.

Transformed colonies will appear in 2 to 3 days.

The structure of the plasmid transferred to *Agrobacterium* can be analyzed by alkaline lysis method[35] or by mating back to *E. coli* as de-

[35] H. C. Birnboim and J. Doly, *Nucleic Acids Res.* **7,** 1513 (1979).

scribed above. *E. coli* recipient cells should be hsdR$^-$ in order to accept the plasmids replicated in *Agrobacterium*. A stationary phase *Agrobacterium* culture grown in either MGL or AB medium containing 5 μg/ml of tetracycline and 25 μg/ml of kanamycin is used for plant transformation by the cocultivation method as described below.

Plant Transformation

The technique currently used for *in vitro* transformation of plants by the Ti vector system is based on the cocultivation method[36] in which plant cells are cocultured with *Agrobacterium* for about 2 days and transformed plant cells are plated on an appropriate selective medium. The material can be either protoplast,[9–11,36] callus,[20] or organ,[21,37] depending upon the plant species. Cocultivation with organs such as leaf pieces is the simplest method. Protoplasts may be the best material for clonal transformation. Since callus cultures transform efficiently and grow rapidly,[20] callus may be the best material for rapidly assaying the expression of foreign genes.

Protoplast Transformation

Tobacco and petunia protoplasts have been transformed with Ti vectors.[9–11,36] This method is slower than transformation of callus or organ. However, clonal transformation is enhanced by the protoplast cocultivation method.

1. Isolate protoplasts and incubate them in a protoplast culture medium under dim light until 1 day prior to the first cell division. It is necessary to determine the time of the first cell division in a previous experiment.
2. Add about 100 *Agrobacterium* cells/protoplast into the protoplast culture to give about 10^7 bacteria/ml.
3. Incubate the cell mixture in a Petri plate without agitation. The first plant cell division must occur during the cocultivation period to obtain a good transformation frequency.
4. After cocultivating for 2 days wash off the bacterial cells using low-speed centrifugations and incubate the plant cells in a protoplast regeneration medium containing 500 μg/ml of carbenicillin. If the Ti vector also carries the β-lactamase gene, 250 μg/ml of cefotaxime (Calbiochem) must

[36] L. Marton, G. S. Wullems, L. Molendijk, and R. A. Schilperoort, *Nature (London)* **277,** 129 (1979).
[37] R. B. Horsch, J. E. Fry, N. L. Hoffmann, D. Eichholtz, S. G. Rogers, and R. T. Fraley, *Science* **227,** 1229 (1985).

be included into the culture medium to suppress growth of residual bacteria.

5. Grow the plant cells to the 10-cell stage and gradually reduce the osmoticum of the culture to half that of the original protoplast culture medium.

6. Plate the minicalli on either callus or shoot induction medium[38] with 200 μg/ml of kanamycin and 500 μg/ml of carbenicillin. If necessary include 250 μg/ml of cefotaxime in the culture medium to prevent growth of carbenicillin resistant *Agrobacterium*.

Callus Transformation

Cultured tobacco calli can be transformed by the Ti vector system.[20] The transformation efficiency is strictly dependent on the physiological stage of the cultured cells. Only actively growing cells are transformed by the *Agrobacterium* cocultivation method.

1. Put 4 ml of 3- to 4-day-old, exponentially growing, suspension-cultured tobacco cells into a 10-cm Petri dish.
2. Add about 10^8 *Agrobacterium* cells to the plant culture and incubate the mixture at 28° for 2 days.
3. Wash the plant cells two to three times with the plant culture medium to remove bacterial cells.
4. Plate the plant cells on a callus culture medium containing 200 μg/ml of kanamycin, 500 μg/ml of carbenicillin, and, if necessary, 250 μg/ml of cefotaxime.

Transformed tobacco calli are visible in 3 weeks.

Organ Transformation

Plant organs have also been transformed by cocultivation.[21,37] Since plant regeneration is faster and easier using organ pieces, this method is widely used for the transformation of various plant species. Leaf slices are most commonly used for the transformation although other organs such as stem or cotyledon are also used. Selection of the specific organ depends on the plant species.[21] For example, both leaf and stem of potato are equally transformed, whereas hypocotyls and cotyledons of tomato are more highly transformable than other tomato organs. Kanamycin resistant *Arabidopsis thaliana* transformants were obtained only from stems, primarily at the axil.[21]

[38] D. A. Evans, W. R. Sharp, P. V. Ammirato, and Y. Yamata, "Handbook of Plant Cell Culture," Vol. 1. Macmillan, New York, 1983.

1. Plant material can be grown in either a greenhouse or sterile bottles. Select healthy plants for the experiment. If greenhouse grown, sterilize the materials by treatment with 10% bleach for 10 min followed by extensive washes with sterile water.

2. Cut leaves or cotyledons to about 0.5-cm^2 sections and stems or hypocotyls to about 1 cm long.

3. Wound the plant sections several times with forceps.

4. Place the plant cells in 10-cm Petri dishes and add 4 ml of callus induction medium and about 10^8 *Agrobacterium* cells (grown overnight in MGL medium).

5. After cocultivation for 2 days at 28°, wash out the bacterial cells and grow the plant slices on callus or shoot induction medium containing 200 μg/ml of kanamycin, 250 μg/ml carbenicillin, and, if necessary, 250 μg/ml cefotaxime. Concentration of kanamycin can be increased to 500 μg/ml if the plant slices are highly resistant to kanamycin. High concentration of carbenicillin may inhibit shoot or root formation of some plant species. In this case, replace carbenicillin with cefotaxime in the culture medium. It will be necessary to transfer the plant samples to a fresh medium every 2 weeks since cefotaxime is unstable under light.

6. Incubate the cocultivated plant tissues in the dark for callus induction. Light reduces transformation efficiency of some plant species.[21] However, transformed shoots must be generated under light, since shoot induction is often inhibited in dark. Therefore, plants exhibiting light-sensitive transformation, such as *Arabidopsis thaliana,* must be transformed in dark. Shoots may then be induced under the light from the transformed calli.

Analysis of Transformants

Transformed plant calli can be selected on 100 to 500 μg/ml of kanamycin when the binary vectors described in this chapter are used. Untransformed calli do not normally grow on the kanamycin medium. Authentic transformants can be quickly identified by assaying for neomycin phosphotransferase[39] or chloramphenicol acetyltransferase[40] activity in the plant materials. Less than 0.1 g of tissue is required for these assays.

Extraction of Plant Tissue

1. Place about 0.1 g of a plant tissue in a microcentrifuge tube. All the following procedures are carried out on ice.

[39] M. J. Hass and J. E. Dowding, this series, Vol. 43 [48].

[40] C. M. Gorman, L. F. Moffat, and B. H. Howard, *Mol. Cell. Biol.* **2,** 1044 (1982).

2. Add 1 ml/g of a plant extraction buffer (0.5 *M* sucrose, 0.1% ascorbic acid, 0.1% cysteine–HCl, 0.1 *M* Tris–HCl, pH 7.5).

3. Lyse the cells by homogenization with a pestle (Kontes Scientific Glassware) which fits into the microcentrifuge tube. Fine calli can be more efficiently lysed by sonication with a microtip probe.

4. Centrifuge the lysate for 5 min in a microcentrifuge and use the supernatant for the enzyme assay.

5. Measure the protein concentration of the extracts and dilute to the same concentration (usually 1 mg/ml).

Neomycin Phosphotransferase Assay

1. The reaction mixture contains the following:

10 μl of the plant extract
10 μl of the assay buffer (40 m*M* $MgCl_2$, 400 m*M* NH_4Cl, 2 m*M* dithiothreitol, 65 m*M* Tris–HCl, pH 7.5)
2 μl of 1 mg/ml kanamycin sulfate
10 μl of ATP solution (50 μ*M* ATP and 0.1 μCi [^{32}P]ATP, 3000 Ci/mmol)

2. Incubate the mixture at 37° for 20 min.

3. Spot the solution onto 1 cm^2 of Whatmann P-81 phosphocellulose paper and dry briefly under a heat lamp.

4. Wash the paper four times in 80° water.

5. Count in a scintillation counter. Data is expressed as cpm/10 μl of a sample.

Chloramphenicol Acetyltransferase Assay

1. The reaction mixture contains the following:

10 to 20 μl of the plant extract
100 μl of 0.25 *M* Tris–HCl, pH 7.8
10 μl of 4 m*M* acetyl coenzyme A
50 nCi of [^{14}C]chloramphenicol (57 mCi/mmol, New England Nuclear)

2. Incubate the mixture at 37° for 20 min. The incubation time and amount of protein extract can be changed depending upon the enzyme activity.

3. Stop the reaction by adding 1 ml of ethyl acetate to the reaction and mixing it well.

4. Centrifuge the mixture for 1 min in a microcentrifuge.

5. Transfer the upper (ethyl acetate) phase into a new microcentrifuge tube.

6. Evaporate the solvent in a SpeedVac concentrator with heat for 30 min.

7. Dissolve the pellet in 30 μl of ethyl acetate.

8. Apply 15 μl of the sample onto a silica gel thin-layer chromagram sheet (13179 silica gel without fluorescent indicator, Eastman Kodak).

9. Run the thin-layer chromagram until the solvent front migrates 5 to 7 cm from the origin in a solvent containing 95 ml chloroform and 5 ml methanol. Presaturate the air in the developing tank with the solvent by using a paper wick. The thin-layer sheet should be free standing to obtain an even migration.

10. Expose an X-ray film to the chromagram sheet overnight and develop the film.

11. Cut out and count the spots of chloramphenicol and acetylated products. Data are expressed as units per milligram total protein. One unit of the enzyme catalyzes acetylation of 1 nmol of chloramphenicol/min at 37°. The assay is linear up to 20% conversion of chloramphenicol.

Acknowledgments

This work was supported in part by a grant from the National Scientific Foundation (DCB-8417721) and the United States Department of Agriculture (85-CRCR-1-1746).

[18] Detection of Monocot Transformation via *Agrobacterium tumefaciens*

By PAUL J. J. HOOYKAAS and ROB A. SCHILPEROORT

General Introduction

The soil bacterium *Agrobacterium tumefaciens* induces the crown gall disease in dicotyledonous plants.[1] Crown galls are tumors that are formed on wounded plants after infection with the bacterium. The molecular mechanism underlying the induction of this neoplastic disease has partially been elucidated (for recent reviews, see Refs. 2 and 3). After attachment to the plant cell wall *Agrobacterium* introduces a segment of its

[1] E. F. Smith and C. O. Townsend, *Science* **25,** 671 (1907).

[2] P. J. J. Hooykaas and R. A. Schilperoort, *Adv. Genet.* **22,** 209 (1984).

[3] E. W. Nester, M. P. Gordon, R. M. Amasino, and M. F. Yanofsky, *Annu. Rev. Plant Physiol.* **35,** 387 (1984).

DNA, called the transferred DNA (T-DNA), into the plant cell, where it becomes integrated in the nuclear DNA. In the bacterium the T-region forms a minor part of a large (more than 200 kbp) plasmid called the Ti (tumor-inducing) plasmid. The T-DNA contains a number of genes that are expressed in the transformed plant cells. Two of these genes (T-DNA gene 1[4] and T-DNA gene 2[5–7]) are *onc* genes which code for enzymes that cooperate in the synthesis of the phytohormone indole acetic acid (an auxin), while T-DNA gene 4[8–10] is an *onc* gene determining the production of an enzyme that mediates the formation of the phytohormone isopentenyladenosine monophosphate (a cytokinin). The production of both an auxin and a cytokinin in the transformed plant cells explains why these cells grow without limit both *in vivo* and *in vitro* even in the absence of phytohormones. There are a number of different types of Ti plasmids. They are classified according to the opines that are produced in the tumors they induce, such as octopine-type, nopaline-type, and leucinopine-type Ti plasmids. Opines are formed via enzymes that are encoded by the T-DNA. Since they have never been detected in normal plant cells, they can be considered as reliable indicators for transformation of plant cells via *Agrobacterium*.

When a foreign piece of DNA is linked to the T-region of the Ti plasmid, this DNA is cotransferred to the plant cell together with the T-region. Therefore, Ti plasmids can be used as vectors for the genetic manipulation of higher plants. They would be of limited practical use, however, if only cells with a tumorous nature without regenerative capacity are obtained. Fortunately, it has been found that the elimination of the genes that determine phytohormone production and hence tumorous growth does not affect the transfer of the T-DNA to the plant cells. Cells transformed by such disarmed T-DNAs do have the capacity to regenerate into normally developing plants. This finding formed the rationale

[4] M. F. Thomashow, S. Hugly, W. G. Buchholz, and L. S. Thomashow, *Science* **231,** 616 (1986).

[5] L. S. Thomashow, S. Reeves, and M. F. Thomashow, *Proc. Natl. Acad. Sci. U.S.A.* **81,** 5071 (1984).

[6] D. Inzé, A. Follin, M. Van Lijsebettens, C. Simoens, C. Genetello, M. Van Montagu, and J. Schell, *Mol. Gen. Genet.* **194,** 265 (1984).

[7] G. Schröder, S. Waffenschmidt, E. W. Weiler, and J. Schröder, *Eur. J. Biochem.* **138,** 387 (1984).

[8] D. E. Akiyoshi, H. Klee, R. M. Amasino, E. W. Nester, and M. P. Gordon, *Proc. Natl. Acad. Sci. U.S.A.* **81,** 5994 (1984).

[9] G. F. Barry, S. G. Rogers, R. T. Fraley, and L. Brand, *Proc. Natl. Acad. Sci. U.S.A.* **81,** 4776 (1984).

[10] I. Buchmann, F.-J. Marner, G. S. Schröder, S. Waffenschmidt, and J. Schröder, *EMBO J.* **4,** 853 (1985).

TABLE I
STRAINS OF *Agrobacterium*[6]

Strain	Ti plasmid	Ri plasmid	Opine type	Virulence
LBA 288	–	–	—	–
LBA 1010	+	–	Octopine	+
LBA 1023	+	+	Octopine	+
LBA 1516	+	–	Octopine	– *virB* mutation
LBA 2318	+	–	Nopaline	+
LBA 2347	+	+	Nopaline	+

behind the development of a number of Ti plasmid-based vector systems for plants.[11]

Detection of Transformation in the Absence of Tumor Formation—Principle of the Method

Agrobacterium forms tumors on a wide range of dicotyledonous plants, but not on monocotyledonous plants.[12] We noted that, at most, some small swellings may occasionally arise around the infected sites on monocots.[13] Tumor formation is a response of the transformed plant cells to the phytohormones produced. At the time we argued that it might be the case that monocots are transformed by *Agrobacterium,* but that these plants do not react with tumor formation. Therefore, we tested whether opines were produced in the plant tissues surrounding the wound sites that had been infected with *Agrobacterium.* In this way we found that species from the monocot families Liliaceae (*Chlorophytum, Hyacinthus*) and Amaryllidaceae (*Narcissus*) are transformed by *Agrobacterium.*[13] Transformation has been demonstrated also for *Asparagus.*[14]

Materials and Reagents

Bacterial strains are described in Table I. They are cultured at 29° on LB or TY plates solidified with 1.8% Difco Bacto agar. Octopine and nopaline are from Sigma, phenanthrene quinone from Merck. Electrophoresis is done on Macherey-Nagel MN214 paper.

[11] T. L. Thomas and T. C. Hall, *BioEssays* **3,** 149 (1985).

[12] M. De Cleene and J. De Ley, *Bot. Rev.* **42,** 389 (1976).

[13] G. M. S. Hooykaas-Van Slogteren, P. J. J. Hooykaas, and R. A. Schilperoort, *Nature (London)* **311,** 763 (1984).

[14] J. P. Hernalsteens, L. Thia-Toong, J. Schell, and M. Van Montagu, *EMBO J.* **3,** 3039 (1984).

LB: 1% Difco Bacto tryptone, 0.5% Difco yeast extract, 0.8% NaCl
TY: 0.5% Difco Bacto tryptone, 0.3% Difco yeast extract
Extraction buffer: 0.1 *M* Tris–HCl, 0.5 *M* sucrose, 0.1% ascorbic acid, 0.1% cysteine–HCl, pH 8.0
Incubation mixture for octopine synthase: 30 m*M* L-arginine–HCl, 75 m*M* sodium pyruvate, 20 m*M* NADH in 0.2 *M* sodium phosphate buffer (pH 7.0)
Incubation mixture for nopaline synthase: 60 m*M* L-arginine–HCl, 60 m*M* sodium α-ketoglutarate, 16 m*M* NADH in 0.2 *M* sodium phosphate buffer (pH 6.8)
Electrophoresis buffer: 5% (v/v) formic acid, 15% (v/v) acetic acid, 80% (v/v) H_2O
Staining reagent: A freshly made 1 : 1 (v/v) mixture of 0.02% phenanthrene quinone in ethanol and 10% NaOH in 60% ethanol. The components can be kept separately if stored at −20°.

Method for the Detection of Plant Transformation by *Agrobacterium*

Plants that are to be tested for their susceptibility to transformation by *Agrobacterium* can be treated as follows. The plants are wounded with a cocktail picker at the leaves and/or flower stems (preferably at the youngest, still growing parts) and are then infected with a culture of *Agrobacterium* taken from a colony on a fresh plate. The strains of *Agrobacterium* should include a nopaline strain (LBA 2318), an octopine strain (LBA 1010), a strain lacking a Ti plasmid (LBA 288), and a strain carrying a Ti plasmid with an intact octopine synthase gene, but with a *virB* mutation so that it is defective in T-region transfer (LBA 1516).[13,15] The *virB* mutant is required as a control to prove that *Agrobacterium* itself does not synthesize opines when the bacterium is present in the wound. After 3–6 weeks the plant tissues surrounding the wound sites are excised and analyzed for the presence of opines (Fig. 1). Materials from wound sites that has not been infected with *Agrobacterium* is treated similarly as a control. If the plant species tested is sensitive to *Agrobacterium*, the following results should be obtained. Octopine should be detected after infection with the octopine strain, nopaline after infection with the nopaline strain, but no opines should be detected after wounding without infection, or after infection with the avirulent strains lacking Ti or with the *virB* mutant Ti plasmid. In this way we have shown that certain species of monocots can indeed be transformed by *Agrobacterium*, although no tumor formation occurs.[13] The reason for this absence of tumor formation might be that the

[15] P. J. J. Hooykaas, M. Hofker, H. den Dulk-Ras, and R. A. Schilperoort, *Plasmid* **11**, 195 (1984).

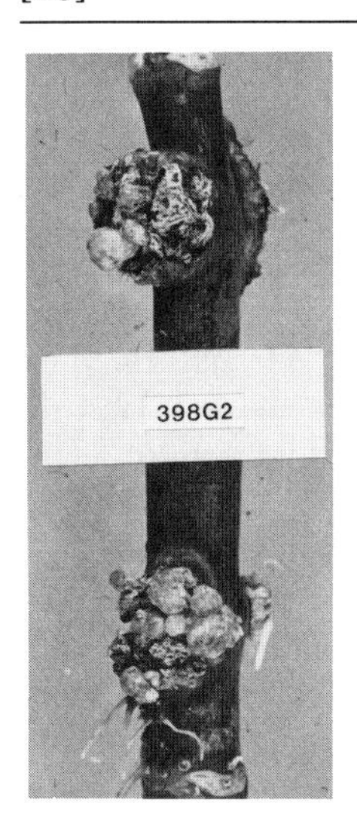

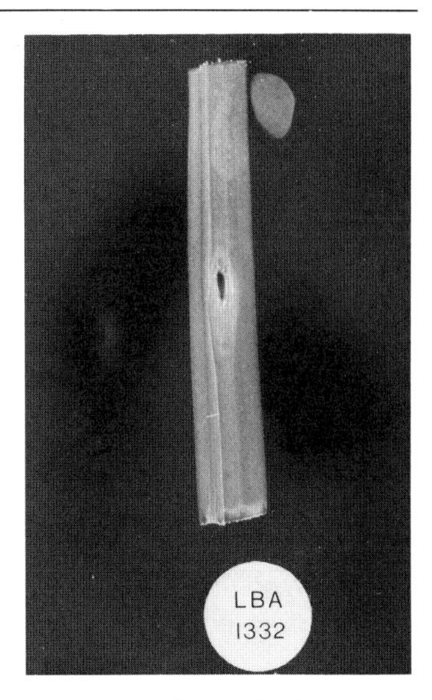

FIG. 1. Infection of a dicot (*Kalanchoe* is shown) with *A. tumefaciens* leads to tumor formation (left), whereas inoculation with *A. rhizogenes* results in rooting (middle). The infection of monocots (*Narcissus* is shown) with *A. tumefaciens* or *A. rhizogenes* neither gives tumor nor root induction; at the most a small swelling is formed surrounding the infected wound sites (right).

onc genes in the T-DNA are not properly expressed in monocots, or more likely that the monocots simply respond in a different way to the phytohormones produced than dicots. That the Ti *onc* genes can be expressed in the monocot species tested was established via experiments in which it was shown that the auxin genes but not the cytokinin gene are required for the formation of the swellings (our unpublished results). It is known already that monocot tissues in *in vitro* culture require different regimes of phytohormones for growth than tissues from dicots. Nevertheless for certain plant species it might still be impossible to obtain positive results after infections with *Agrobacterium*. There can be several reasons for this, since the transfer of T-DNA from *Agrobacterium* to plant cells is a process that is quite complex. First of all plants may produce compounds that are toxic for *Agrobacterium*. The use of resistant mutants might solve this problem. Then the inability of *Agrobacterium* to attach to the cell walls of certain plant species might restrict its host range. It is interesting to note here that the bacterium is able to attach to the cell walls of bamboo,[16] a species belonging to the plant family of Gramineae for which

[16] C. Douglas, W. Halperin, M. Gordon, and E. Nester, *J. Bacteriol.* **161,** 764 (1985).

transformation by *Agrobacterium* has not been demonstrated yet. Genes in *Agrobacterium* that are involved in the transfer of the T-region to the plant cells are called *vir* genes and are located in a 40-kbp region of the Ti plasmid called the virulence (*Vir*) region. The genes in the *Vir* region are only expressed after induction by particular compounds that are present in exudates of wounded plant tissues.[17,18] It can be expected that plant species which do not excrete these compounds cannot be transformed by *Agrobacterium*. The addition of exudate from plant species which do excrete the inducing compounds to plant species which do not do so might lead to transformation by *Agrobacterium*. Alternatively, *Agrobacterium* strains which have a mutation rendering the *vir* genes constitutive might be used for the transformation of such plants. Both in the Ti plasmid and in the chromosome there are a number of genes of unknown function which determine the eventual host range of the host bacterium for tumor induction.[15] Different strains of *Agrobacterium* may have a different host range for tumor induction and this may in fact reflect the presence or absence of certain of these host-range-determining genes.[19] Moreover, different strains can have a different efficiency with which (particular) plants are transformed.[20] It will therefore be obvious that it is advisable to use a variety of (wild-type) strains when screening plant species or varieties for sensitivity to transformation by *Agrobacterium*. In this respect it is worth noting that the frequency with which we found transformation of *Narcissus* and *Chlorophytum* was higher after infection with strains carrying, besides a Ti plasmid, a Ri plasmid (strains LBA 1023 and LBA 2347) than after infection with strains harboring only a Ti plasmid (unpublished results). The Ri (root inducing) plasmid originates from *Agrobacterium rhizogenes*. This bacterium is capable of transferring a piece of DNA from the Ri plasmid to plant cells at the infection sites. A last point that must be stressed is that the method of scoring transformation influences the eventual results. As stated in the beginning of this chapter, screening for tumor formation is not a method of choice, since many plant species may not react with tumor formation to the phytohormones produced. The detection of transformation by assaying for opines or preferably for opine synthase activity (see below) is therefore preferred. Still-transformed plant cells may not produce (enough) opine synthase activity to be detect-

[17] R. J. H. Okker, H. Spaink, J. Hille, T. A. N. van Brussel, B. Lugtenberg, and R. A. Schilperoort, *Nature (London)* **312,** 564 (1984).

[18] S. E. Stachel, E. Messens, M. Van Montagu, and P. Zambryski, *Nature (London)* **318,** 624 (1985).

[19] A. R. Anderson and L. W. Moore, *Phytopathology* **69,** 320 (1979).

[20] E. E. Hood, G. Jen, L. Kayes, J. Kramer, R. T. Fraley, and M.-D. Chilton, *Bio/Technology* **2,** 702 (1984).

able, either because the gene for this enzyme is not transcribed at a sufficient level, or because the protein is degraded rapidly so that no opines are formed. To increase transcription the opine synthase coding region may be linked to a strong promoter such as the cauliflower mosaic virus 35 S promoter or to a strong promoter of an endogenous gene of the host plant to be tested. It is worth noting here that in electroporation experiments expression of a chloramphenicol acetyltransferase gene was detectable in corn protoplasts only when the coding sequence was linked to the 35 S promoter but not when fused to the promoter of the nopaline synthase gene.[21]

Assays for Octopine and Nopaline Synthase Activities

Octopine synthase (previously called lysopine dehydrogenase) is present in plants cells transformed by *Agrobacterium* strains with an octopine Ti plasmid, while the enzyme nopaline synthase (previously called nopaline dehydrogenase) is present in tissues transformed by strains with a nopaline Ti plasmid. Octopine synthase activity leads to the production of a number of amino acid derivatives such as octopine, lysopine, octopinic acid, and histopine, which are N^2-(D-1-carboxyethyl) derivatives of arginine, lysine, ornithine, and histidine, respectively. Nopaline synthase mediates the formation of nopaline and ornaline, which are N^2-(1,3-dicarboxypropyl) derivatives of arginine and ornithine, respectively. Octopine and nopaline can easily be detected after paper electrophoresis of tumor extracts by staining with phenanthrene quinone reagent, which is specific for guanidine compounds. However, especially in the case of octopine, the levels present in transformed plant tissues are frequently too low to be detected in this way, and therefore it is usually advisable to analyze plant tissues for the presence of opine synthase activity according to Otten and Schilperoort[22] instead (Fig. 2). The principle of the assay for opine synthase activity is to isolate a crude extract from transformed plant tissue, to incubate this with the precursors of an opine, and then to analyze the formation of this opine via paper electrophoresis. A small piece of plant tissue (about 30 mm) is sufficient for an opine synthase test. The plant material is ground with a small rod in an equal volume of extraction buffer (0.1 *M* Tris–HCl, 0.5 *M* sucrose, 0.1% ascorbic acid, 0.1% cysteine–HCl, pH 8.0). After centrifugation for 2 min in an Eppendorf centrifuge, the supernatant, which contains the opine synthase activity, is removed and added to an identical volume of an

[21] M. Fromm, L. P. Taylor, and V. Walbot, *Proc. Natl. Acad. Sci. U.S.A.* **82,** 5824 (1985).
[22] L. A. B. M. Otten and R. A. Schilperoort, *Biochim. Biophys. Acta* **527,** 497 (1978).

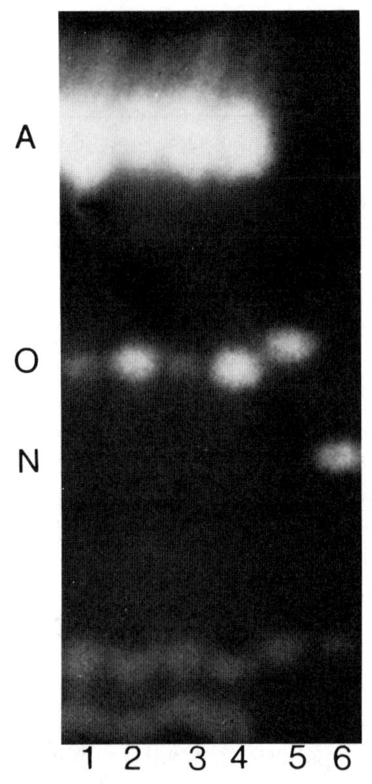

FIG. 2. Octopine synthase assay according to Otten and Schilperoort.[22] Lanes show (1) extract from infected tissue (*Narcissus*) treated as described in the text at $t = 0$; (2) the same after a 1-hr incubation (octopine has been formed); (3) a similar extract at $t = 0$; (4) extract shown in (3) after a 1-hr incubation; (5) octopine control; (6) nopaline control. A, Position of arginine; N, position of nopaline; O, position of octopine.

incubation mix with the precursors of an opine. In the case of an assay for octopine synthase the incubation mix consists of 30 mM arginine–HCl, 75 mM sodium pyruvate, 20 mM NADH in 0.2 M sodium phosphate buffer (pH 7.0). In the case of nopaline synthase the incubation mix is 60 mM arginine–HCl, 60 mM sodium α-ketoglutarate, 16 mM NADH in 0.2 M sodium phosphate buffer (pH 6.8). Incubations are done at room temperature (about 20–25°). Samples are taken from the assay mixtures at the start of the incubation and after 1 hr of incubation, and are analyzed for the presence of octopine or nopaline. In order to do this the samples are spotted onto the anodal side of an MN 214 paper sheet of 20 × 25 cm. The paper is dried and then used for high-voltage electrophoresis (20 V/cm) in 5% (v/v) formic acid, 15% (v/v) acetic acid, 80% (v/v) H_2O, pH 1.8.

Methyl green, which runs slightly slower than arginine, can be used as a visible marker during electrophoresis, and samples of stock solutions of arginine, nopaline, and octopine are used as markers to help the interpretation of the electropherogram. After electrophoresis (about 1 hr) the paper is dried and then stained with phenanthrene quinone reagent (a freshly prepared mix of equal volumes of 0.02% phenanthrene quinone in ethanol and 10% NaOH in 60% ethanol). After staining the paper is dried with cold air. Areas containing guanidine compounds such as arginine, octopine and nopaline can be visualized as yellow spots by irradiation with long-wave ultraviolet (360 nm).

[19] Direct Gene Transfer to Protoplasts of Dicotyledonous and Monocotyledonous Plants by a Number of Methods, Including Electroporation

By RAYMOND D. SHILLITO and INGO POTRYKUS

Introduction

Recent progress in cell and molecular biology has made available a great variety of new techniques for modifying the information content of plant cells. Moreover, in the last few years it has become possible to genetically engineer the plant cells by introduction of defined DNA sequences. This has been achieved mainly by exploiting the natural gene transfer system of the pathogen *Agrobacterium tumefaciens* (for a review, see Ref. 1). This is a convenient and efficient technique, but it can be used only with plants which fall inside its host range. This includes most herbaceous dicots, but only a limited number of monocots,[2] and, as yet, not the graminaceous monocots which make up the bulk of our crop species.

The introduction of genes by DNA-mediated transformation is a well-established procedure for bacterial, fungal, and animal systems and has proved to be a very powerful technique in the analysis of gene function. It has recently been shown that DNA can be introduced into plant protoplasts and integrated into the chromosomal DNA without intervention of

[1] G. Gheysen, P. Dhaese, M. Van Montagu, and J. Schell, *in* "Genetic Flux in Plants" (B. Hohn and E. S. Dennis, eds.), p. 12. Springer-Verlag, Vienna, Austria, 1985.

[2] M. De Cleene and J. De Ley, *Bot. Rev.* **42,** (1976).

a bacterial intermediary (direct gene transfer).[3–7] It is also possible, using this technique, to transform graminaceous cells.[8,9] A number of other techniques have also been developed recently, employing a viral vector,[10] liposomes,[11] bacterial spheroplasts,[12] and microinjection[13] to deliver the DNA. Thus a number of routes are available for introduction of defined pieces of DNA into plant cells, from which one can choose, on the basis of the culture system available, a suitable procedure for the particular species or cell type under study. Transformation of plant cells has the added advantage that in many cases the totipotency of the transformed cells allows regeneration of large numbers of whole plants and genetic and molecular analysis of progeny, which is not available in such an easily accessible manner with transformed animal cells.

Methods for the direct introduction of genes into plant protoplasts are presented, along with methods for the subsequent selection of transformed colonies, regeneration of the genetically altered fertile plants, and characterization of the introduced DNA by molecular, biological, and genetic techniques. The methods are drawn from two complementary fields: recombinant DNA and plant tissue culture. In the interests of continuity, methods from both these fields will be presented mixed with one another in the chronological order in which they are needed in a transformation experiment.

The frequency previously obtained by these methods was low (10^{-5}–10^{-4} of recoverable colonies).[5,6] Such low frequencies seriously hampered the application of the technique of direct gene transfer as a general method for introducing genes into plant cells. However, recent advances

[3] M. R. Davey, E. C. Cocking, J. Freeman, N. Pearce, and I. Tudor, *Plant. Sci. Lett.* **18,** 307 (1980).

[4] J. Draper, M. R. Davey, J. P. Freeman, E. C. Cocking, and B. G. Cox, *Plant Cell Physiol.* **23,** 451 (1982).

[5] F. A. Krens, L. Molendijk, G. J. Wullems, and R. A. Schilperoort, *Nature* (*London*) **296,** 72 (1982).

[6] J. Paszkowski, R. D. Shillito, M. Saul, V. Mandak, T. Hohn, B. Hohn, and I. Potrykus, *EMBO J.* **3,** 2717 (1985).

[7] R. Hain, P. Stabel, A. P. Czernilofsky, H.-H. Steinbiss, L. Herrera-Estrella, and J. Schell, *Mol. Gen. Genet.* **199,** 161 (1985).

[8] I. Potrykus, M. Saul, J. Petruska, J. Paszkowski, and R. D. Shillito, *Mol. Gen. Genet.* **199,** 183 (1985).

[9] H. Loerz, B. Baker, and J. Schell, *Mol. Gen. Genet.* **199,** 178 (1985).

[10] N. Brisson, J. Paszkowski, J. R. Penswick, B. Gronenborn, I. Potrykus, and T. Hohn, *Nature* (*London*) **310,** 511 (1984).

[11] A. Deshayes, L. Herrera-Estrella, and M. Caboche, *EMBO. J.* **4,** 2731 (1985).

[12] R. Hain, H.-H. Steinbiss, and J. Schell, *Plant Cell Rep.* **3,** 60 (1984).

[13] T. Reich, V. N. Iyer, and B. Miki, *Proc. Congr. Plant Mol. Biol., 1st, 1985* p. 28 (Abstr.).

have increased frequencies into the percentage range.[14] Protocols for such efficient transformation, and improvements of the original protocols, are given in this publication.

Abbreviations

MES: 2(*N*-morpholino)ethanesulfonic acid
NAA: Naphthaleneacetic acid
2,4-D: 2,4-Dichlorophenoxyacetic acid
BAP: 6-Benzylaminopurine
SDS: Sodium dodecyl sulfate
TE: 10 m*M* Tris–HCl, 5 m*M* EDTA, pH 7.5
dsDNA: Double-stranded DNA

Media

Bacterial media, as specified in Ref. 15
Plant culture media are shown in Table I

Materials

E. coli strain DH1[15]: Sources for the plant material are given with each individual protocol
Table-top centrifuge: Universal 2, (Hettich Zentrifugen, Tuttlingen, West Germany)
Osmometer: Roebling Micro-Osmometer (Infochroma AG, Zug, Switzerland)
Rocking table: Heidolph Reax 3 rocking table (Salvis AG, Reussbuehl, Switzerland)
Stainless steel sieves: Saulas and Company (Paris, France)
The 10-cm-diameter containers used for the "bead-type" culture are obtained from Semadani AG (Ostermundigen, Switzerland)
Petri dishes: These are obtained from a range of suppliers
Electroporator: "DIA-LOG" Elektroporator: DIA-LOG GmbH., (Dusseldorf, West Germany)
Resistance meter: AVO B183 LCR meter (Thorn-EMI Ltd., Dover, England)
SeaPlaque agarose: Marine Colloids, FMC Corporation (Rockland, Maine)

[14] R. D. Shillito, M. W. Saul, J. Paszkowski, M. Mueller, and I. Potrykus, *Bio/Technology* **3,** 1099 (1985).

[15] T. Maniatis, E. F. Fritsch, and J. Sambrook, "Molecular Cloning: A Laboratory Manual." Cold Spring Harbor Lab., Cold Spring Harbor, New York, 1982.

TABLE I
COMPOSITION OF THE MEDIA USED[a]

A. Inorganic salts used in tissue culture media

Medium[b]:	T[1]	MS[2]LS[3]	K3[4]	H/J[5]/K0[6]	CC[7]	MDS[8]	RPZ[9]	NT[10]
Macroelements[c]: Final concentration (mg/ml)								
KCl								
KNO_3	950	1900	2500	1900	212	100	273	950
KH_2PO_4	68	170		170	136	85	271	680
$NaNO_3$								
NH_4NO_3	720	1650	250	600	640			825
$NaH_2PO_4 \cdot H_2O$			150					
$CaCl_2 \cdot 2H_2O$	220	440	900	600	588	220		438
$MgSO_4 \cdot 7H_2O$	185	370	250	300	247	185	57	123
$(NH_4)_2SO_4$			134				233	
$Ca(NO_3)_2 \cdot 4H_2O$							416	
$Mg(NO_3)_2 \cdot 6H_2O$							392	
Microelements[c]: Final concentration (mg/liter)								
Na_2EDTA	74.6	74.6	74.6	74.6	37.3	37.3	74.6	74.6
$FeSO_4 \cdot 7H_2O$					27.8			
$FeCl_3 \cdot 6H_2O$	27.0	27.0	27.0	27.0		13.5	27.0	27.0
H_3BO_3	10.0	6.2	3.0	3.0	3.1	0.06	0.12	6.2
KI		0.83	0.75	0.75	0.83	0.15	0.166	0.83
$MnCl_4 \cdot 4H_2O$								
$MnSO_4 \cdot H_2O$	17.25	16.9	10.0	10.0	11.15	2.0	3.38	16.9
$ZnSO_4 \cdot 7H_2O$	10.0	8.6	2.0	2.0	5.76	0.4	1.72	8.6
$CuSO_4 \cdot 5H_2O$	0.025	0.025	0.025	0.025	0.025	0.005	0.005	0.025
$Na_2MoO_4 \cdot 2H_2O$	0.25	0.25	0.25	0.25	0.24	0.05	0.05	0.25
H_2Mo_4								
$CoCl_2 \cdot 6H_2O$			0.025	0.025		0.005		
$CoSO_4 \cdot 7H_2O$		0.03			0.028		0.006	0.03

B. Vitamins and other organics used in culture media: Concentration (mg/liter)

Medium:	T	LS	K3	H/J[def]	K0[de]	CC	MDS[h]	RPZ	NT
myo-Inositol	100	100	100	100	100	90	100	100	100
Vitamins[g]									
Biotin	0.05			0.01	0.01			0.05	
Pyridoxine–HCl	0.50		0.10	1.00	1.00	1.0	0.5	0.5	
Thiamin–HCl	0.50	0.04	1.00	10.00	10.00	8.5	5.0	0.5	1
Nicotinamide				1.00	1.00				
Nicotinic acid	5.00		0.10			6.0	0.5	5.0	
Folic acid	0.50			0.40	0.40			0.5	
D-Calcium pantothenate				1.00	1.00				
p-Aminobenzoic acid				0.02	0.02				
Choline chloride				1.00	1.00				
Riboflavin				0.20	0.20				
Ascorbic acid				2.00	2.00				
Vitamin A				0.01					
Vitamin D_3				0.01	0.01				
Vitamin B_{12}				0.02	0.02				
Glycine	2.0			0.10		2.0		2.0	
Coconut water (%v/v)						10.0			

C. Carbohydrates and phytohormones

Medium:	T	LS	K3A	K3C	K3E	H	J	K0	CC	MDS	RPZ	NT
Sugars and sugar alcohols (g/liter)												
Sucrose	10.0	30.0	102.96	102.96	36.0	0.25	0.25		20.0	30.0	0.5	
Glucose						68.40	21.0	68.4				10
Mannitol						0.25	0.25	0.25	36.43	50.0	18.0	
Sorbitol						0.25	0.25	0.25				

(continued)

TABLE I (*continued*)

Cellobiose						0.25	0.25	0.25				
Fructose						0.25	0.25	0.25				
Mannose						0.25	0.25	0.25				
Rhamnose						0.25	0.25	0.25				
Ribose						0.25	0.25	0.25				
Xylose			0.25	0.25	0.25	0.25	0.25	0.25				
Hormones (mg/liter)												
2,4-D		0.05	0.10	0.05	0.05	0.10		2.0	2.0			
p-CPA												
NAA		2.00	1.00	2.00	2.00	1.00	1.0			0.05		
BAP		0.10	0.20	0.10	0.10	0.20	0.25	0.5		0.1		
Kinetin		0.10		0.10	0.10							
Zeatin											0.25	
Final pH	5.5	5.8	5.8	5.8	5.8	5.8	5.8	6.0	5.8	5.7	5.6	5.8

[a] Where the inorganic component is common, i.e., media H/J/K0, these have been given together.

[b] Key to the reference source: [1] J. P. Nitsch and C. Nitsch, *Science* **163,** 85 (1969); [2] T. Murashige and F. Skoog, *Physiol. Plant.* **15,** 473 (1962); [3] E. M. Linsmaier and F. Skoog, *Physiol. Plant.* **18,** 100 (1965); [4] J. I. Nagy and P. Maliga, *Z. Pflanzenphysiol.* **78,** 453 (1976); [5] K. N. Kao and M. R. Michayluk, *Planta* **126,** 105 (1975); [6] H. Koblitz and D. Koblitz, *Plant Cell Rep.* **1,** 147 (1982); [7] I. Potrykus, C. T. Harms, and H. Loerz, *Theor. Appl. Genet.* **54,** 209 (1979); [8] I. Negrutiu, R. Dirks, and M. Jacobs, *Theor. Appl. Genet.* **66,** 341 (1983); [9] P. Installé, I. Negrutiu, and M. Jacobs, *J. Plant Physiol.* **119,** 443 (1985); [10] T. Nagata and I. Takebe, *Planta* **99,** 12 (1971).

[c] Macroelements are usually made up as a 10× concentrated stock solution and microelements as a 1000× concentrated stock solution.

[d] Citric, fumaric, and malic acid (each 40 mg/liter final concentration) and sodium pyruvate (20 mg/liter) are prepared as a 100× concentrated stock solution, adjusted to pH 6.5 with NH_4OH, and added to these media.

[e] Adenine (0.1 mg/liter), and guanine, thymine, uracil, hypoxanthine, and cytosine (0.03 mg/liter) are prepared as a 1000× concentrated stock solution, adjusted to pH 6.5 as above, and added to these media.

[f] The following amino acids are added to this medium using a 10× stock solution (pH 6.5 with NH_4OH) to yield the given final concentrations: Glutamine (5.6 mg/liter), alanine, glutamic acid (0.6 mg/liter), cysteine (0.2 mg/liter), asparagine, aspartic acid, cystine, histidine, isoleucine, leucine, lysine, methionine, phenylalanine, proline, serine, threonine, tryptophan, tyrosine, and valine (0.1 mg/liter).

[g] Vitamin stock solution is normally prepared 100× concentrated.

[h] Ammonium succinate (770 mg/liter) is added to this medium as a source of reduced nitrogen.

Cleaned agar: This is prepared by washing with water, acetone, and ethanol in succession[16,17]

Tween 80: ICI (Runcorn, England) or Merck-Schuchardt (Munich, West Germany)

Greenzit^R: Ciba-Geigy AG (Basel, Switzerland)

Restriction enzymes (ligase, etc.) can be obtained from a number of commercial sources

Cellulase "Onozuka" R10 and macerozyme R10: Yakult Pharmaceutical Industries Company, Ltd. (Nishinomiya, Japan)

Driselase: Fluka AG (Chemische Fabrik, Buchs, Switzerland)

Hemicellulase: Sigma Chemical Company (St. Louis, MO)

Pectinol: Roehm GmbH (Chemische Fabrik, Darmstadt, FRG)

Antibiotics: Kanamycin sulfate: Serva, (Heidelberg, FRG); ampicillin: "Penbritin" Beecham SA (Bern, Switzerland); G418: Gibco

Polyethylene glycol (PEG): Merck PEG 6000 and PEG 4000

All other organic and inorganic substances are of the highest purity available from the usual commercial sources.

Protocols for the Preparation, Transformation, and Culture of Protoplasts and Regeneration of Plants

We describe here protocols which are in everyday use in our laboratory for four plant species.

A number of factors are common to these protocols: (1) Centrifugations are carried out at 60 *g* except where otherwise stated. (2) Washing solutions (osmoticum) in all protocols are buffered with 0.5% (w/v) MES and adjusted to pH with KOH except where otherwise stated. (3) Counting of protoplasts is carried out by placing a drop of a 1 : 10 dilution of the suspension in the wash solution (where protoplasts will sediment) or in 0.17 *M* calcium chloride in a haemocytometer, counting, and estimating the density in the original suspension.

1. Source of the Hybrid-Selectable Gene

Several hybrid marker genes for use in plant cell transformation, using *A. tumefaciens*-mediated transformation, have been described in the last

[16] A. C. Braun and H. N. Wood, *Proc. Natl. Acad. Sci. U.S.A.* **48,** 1776 (1962).

[17] R. D. Shillito, J. Paszkowski, and I. Potrykus, *Plant Cell Rep.* **2,** 244 (1983).

2 years.[18–21] The elements necessary in such a construction can be summarized as follows:

1. Plant gene expression signals, i.e., promoter and terminator regions for an RNA, which are best derived from a constitutively and highly expressed plant or plant viral gene
2. A protein-coding region joined precisely to the above expression signals which when expressed will give an active product which allows easy selection at the plant cell level: e.g., detoxification of antibiotics lethal for plant cells
3. For DNA-mediated transformation, a region on the bacterial vector plasmid which allows recombination into the plant genome without disruption of the selectable gene

A plasmid fulfilling the above requirements has been constructed (pABD1). Details of the construction are given elsewhere.[6] The expression signals used are derived from gene VI of the plant dsDNA virus cauliflower mosaic virus (CaMV).[22] The selectable marker gene joined to these sequences is aminoglycoside phosphotransferase type II [APH(3′)II][23] and the bacterial plasmid containing this construction is pUC8.[24] Before using the construction in direct DNA transformation experiments it was tested for biological activity and for the ability to be selected by introduction into *Nicotiana tabacum* cells via the *A. tumefaciens* method. Any suitable construction which satisfies the above criteria can be used for direct gene transfer.

Preparation of the DNA for Protoplast Transformation

Purification. The plasmid pABD1 is grown in *Escherichia coli* strain DH1 in the presence of 50 μg/ml ampicillin and isolated by a cleared lysate method.[25] After lysis in Triton X-100 containing lytic mix, supercoiled DNA is purified by a single round of CsCl/ethidium bromide gradient centrifugation. Ethidium bromide is removed by repeated extraction

[18] M. W. Bevan, R. B. Flavell, and M. D. Chilton, *Nature (London)* **304,** 184 (1983).
[19] R. T. Fraley, S. G. Rogers, R. B. Horsch, P. R. Sanders, J. S. Flick, S. P. Adams, M. L. Bittner, L. A. Brand, C. L. Fink, J. S. Fry, G. R. Gallupi, S. B. Goldberg, N. L. Hoffman, and S. C. Woo, *Proc. Natl. Acad. Sci. U.S.A.* **80,** 4803 (1983).
[20] L. Herrera-Estrella, M. De Block, E. Messens, J.-P. Hernalstens, M. Van Montagu, and J. Schell, *EMBO J.* **2,** 987 (1983).
[21] L. Herrera-Estrella, A. Depicker, M. Van Montagu, and J. Schell, *Nature (London)* **303,** 209 (1983).
[22] H. Guilley, R. G. Dudley, G. Jonard, E. Balazs, and K. Richards, *Cell* **30,** 763 (1982).
[23] S. J. Rothstein and W. S. Reznikoff, *Cell* **23,** 191 (1981).
[24] J. Messing and J. Vieira, *Gene* **19,** 269 (1982).
[25] Y. M. Kuperstock and D. Helsinki, *Biochem. Biophys. Res. Commun.* **54,** 1451 (1973).

with CsCl-saturated 2-propanol solution. The DNA is then precipitated with ethanol (1 vol DNA solution plus 2 vol TE plus 6 vol 96% ethanol) at −20° overnight. The precipitate is collected by centrifugation at 5000 *g* for 10 min, washed in 70% ethanol, dried briefly in an air stream, and redissolved in TE buffer. After spectrophotometric determination of the DNA concentration this is adjusted to 1 mg/ml. DNA for transformation is linearized by digestion with *Sma*I or *Bgl*II overnight and precipitated by addition of one-tenth volume of 3 *M* potassium acetate followed by 3 vol of ethanol. The precipitate is collected by centrifugation as above, washed in 70% ethanol and 100% (v/v) ethanol in succession, dried in an air stream, and redissolved in sterile double-distilled water at 0.4 mg/ml. All manipulations with the DNA after this sterilization step are carried out under aseptic conditions in a laminar flow cabinet.

Physical form of the transforming DNA: All of our early transformation experiments, which have already been well analyzed, were carried out with supercoiled plasmid DNA. We can conclude at this time that both linear and supercoiled molecules can be successfully taken up into plant protoplasts and integrated into the plant genome. However, linear molecules are clearly superior in the efficiency of transformation, amounting to a factor of 3–10 depending on the precise conditions used.

Carrier DNA: Experiments are generally carried out using high-molecular-weight calf thymus DNA (Sigma) as carrier, as described by Krens *et al.*[5] for experiments involving transformation of protoplasts with isolated Ti plasmid. Calf thymus DNA is dissolved in water, precipitated in 70% and washed with 70 and 100% (v/v) ethanol, dried, and redissolved at 2 mg/ml in sterile double-distilled water. The carrier DNA is mixed at a ratio of five times the amount of pABD1 DNA (equal volumes of the two DNA solutions as given). Trials with carrier DNA of other types have shown that salmon sperm DNA and *N. tabacum* DNA give comparable results but also that transformation is possible at reduced efficiency without any carrier DNA.

2. *Preparation, Transformation, and Culture of Protoplasts from a Sterile Shoot Culture of* N. tabacum, *and Regeneration of Plants*

The example given is for protoplasts from shoot cultures of the widely used genotype of *N. tabacum* cv. "Petit Havana," SR1.[26] This material is grown as sterile axenic shoot cultures.

The protocol for protoplast isolation is modified from that of Nagy and Maliga.[27] The culture method uses complex media based on that of Kao

[26] P. Maliga, A. Breznovitz, and L. Marton, *Nature (London) New Biol.* **244,** 29 (1973).
[27] J. I. Nagy and P. Maliga, *Z. Pflanzenphysiol.* **78,** 453 (1976).

and Michayluk[28] (Table I) and the agarose "bead-type" system[17] to obtain high division frequencies and rates of conversion of protoplasts to colonies. Transformation is carried out using one of the four methods given.

Colonies are transferred to agar-solidified medium for one subculture and then placed on regeneration medium to promote the formation of shoots. Regenerated shoots are cultured on the original shoot culture medium.

Source of Material. Shoot cultures are established from seeds which are sterilized using mercury chloride (see protocol 2) or sodium hypochlorite [5 min, 1.4% (w/v) containing 0.05% (w/v) Tween 80]. The plants arising are serially subcultured every 6 weeks as cuttings on T medium[29] (Table I) solidified with 0.8% (w/v) Difco Bacto agar and grown at 26° with 16 hr of light (1000–2000 lx) per day in a growth chamber.

Preparation of Protoplasts. Just fully expanded leaves of 6-week-old shoot cultures are removed under sterile conditions and wetted thoroughly with enzyme solution. The leaves are then cut into 1- to 2-cm squares and floated on enzyme solution [1.2% (w/v) cellulase "Onozuka" R10, 0.4% (w/v) macerozyme R10 in K3A medium with 0.4 *M* sucrose] in Petri dishes (~1 g leaves in 12 ml enzyme solution in a 9-cm-diameter Petri dish). These are sealed and incubated overnight at 26° in the dark.

The digest is gently agitated and then left for a further half-hour to complete digestion. The solution is filtered through a 100-μm stainless steel mesh sieve and washed through with one-half volume of 0.6 *M* sucrose (MES, pH 5.6), distributed into capped centrifuge tubes and centrifuged for 10 min.

The protoplasts collect at the upper surface of the medium. The medium is then removed from under the protoplasts. A simple method of doing this uses a sterilized cannula (A. R. Howell, Ltd., London, England) attached to a 20-ml disposable plastic syringe. This must be done slowly so as to avoid disturbing the layer of protoplasts excessively. Alternatively, the protoplasts can be collected using a pipet (with a medium orifice).

The protoplasts are resuspended in K3A medium (Table I) containing 0.4 *M* sucrose. Washing of the protoplasts is carried out by repeated (3×) flotation and replacing of the medium in this way. A sample is taken for counting before the last centrifugation, and the protoplasts resuspended for the last time in the appropriate medium for the transformation protocol to be used.

[28] K. N. Kao and M. R. Michayluk, *Planta* **126,** 105 (1975).

[29] J. P. Nitsch and C. Nitsch, *Science* **163,** 85 (1969).

Transformation

Method 1. "F medium" method: This method is a modification of the original method described by Paszkowski *et al.*[6,30] for transformation of protoplasts of *N. tabacum,* which was in turn a modification of the method of Krens *et al.*[5] We have added a heat shock step and changed the order of addition of DNA and PEG.

After counting, the protoplasts are adjusted to 2×10^6/ml in K3A medium and aliquots of 1 ml are distributed to 10 ml sterile plastic tubes. Heat shock is then administered by immersing the tubes for 5 min in a water bath at 45°, followed by cooling to room temperature on ice. Then DNA solution is added to the samples (10 μg of pABD1 plus 50 μg of calf thymus DNA in 50 μl sterile distilled water), followed by gentle mixing. Finally, 0.5 ml of PEG solution [40% (w/v) PEG 6000 in F medium (Table I)] is added with shaking.

The protoplasts are incubated with DNA and PEG for 30 min at room temperature with occasional gentle mixing. Then five aliquots of 2 ml of F medium are added at intervals of 5 min. We have noted that the pH of F medium drops to 4.3–4.6 after autoclaving. Since this is likely to be harmful to many protoplast systems we recommend adjustment of the pH after autoclaving with KOH to 5.8.

Following transformation, the protoplasts are sedimented by centrifugation for 5 min, resuspended in 2 ml of K3A culture medium, and transferred in 1-ml aliquots to 9-cm Petri dishes. To each dish is added 10 ml of a 1 : 1 mixture of K3A and H media (Table I) containing 0.6% (w/v) SeaPlaque agarose, and the protoplasts dispersed by gentle swirling. This protocol gives transformation efficiencies in the range of 10^{-4} to 10^{-3}.

Method 2. Electroporation of protoplasts[14]: Electroporation is a process in which cells or protoplasts are treated with high-voltage electric fields for short periods in order to induce the formation of pores across the membrane.[31,32] In this way it is possible to induce uptake of DNA into animal cells or plant protoplasts, leading to transient expression[33] or to stable transformation.[14,32]

Samples of protoplasts are pulsed with high-voltage pulses in the chamber of a "DIA-LOG" Elektroporator. This chamber is cylindrical in form with a distance of 1 cm between parallel steel electrodes and has a

[30] J. Paszkowski and M. W. Saul, this series, Vol. 118, p. 668.

[31] R. Benz, F. Beckers, and U. Zimmermann, *J. Membr. Biol.* **48,** 181 (1979).

[32] E. Neumann, M. Schaeffer-Ridder, Y. Wang, and P. H. Hofschneider, *EMBO J.* **1,** 841 (1982).

[33] M. Fromm, L. P. Taylor, and V. Walbot, *Proc. Natl. Acad. Sci. U.S.A.* **82,** 5824 (1985).

pulsed volume of 0.32 ml.[32] The pulse is applied by discharge of a capacitor across the cell. The decay constant of the pulse (time taken to decay to $1/e$ of the initial voltage) is in the order of 10 μsec using a capacitor of 10 nF and a chamber resistance of 1 kΩ. The resistance across the chamber is measured using an alternating current multimeter operating at 1 kHz.

The protocol given is for leaf mesophyll protoplasts of *N. tabacum*. These have an average diameter of 42 μm. The field strength required is inversely proportional to the diameter of the protoplasts being treated, and may vary a little from species to species in the field strength required for a given size of protoplast. In addition, protoplasts originating from suspension cultures generally require a slightly higher field strength than leaf mesophyll protoplasts.

Protoplasts are resuspended, following the last flotation step, in 0.4 *M* mannitol containing 1 g/liter MES (pH 5.6 with KOH), and containing 6 m*M* magnesium chloride to stabilize the protoplasts, at a population density of 1.6×10^6/ml. An aliquot of 0.37 ml is transferred to the chamber of the electroporator and the resistance measured.

In order to bring the resistance of the mannitol solution in which the protoplasts are suspended to the correct value, it is necessary to add ionic salts. Magnesium chloride is used to adjust the resistance to a value of 1–1.1 kΩ, by adding 1–3% (v/v) of a 0.3 *M* solution to the protoplast suspension.

Heat shock is carried out before distributing the preparation into tubes, by treating the protoplasts for 5 min at 45°, followed by cooling to room temperature on ice.

Aliquots of 0.25 ml of protoplast suspension are placed in 5-ml-capacity polycarbonate tubes and DNA [4 μg pABDI linearized with *Sma*I and 20 μg of calf thymus DNA (Sigma) in 20 μl water/aliquot] and one-half volume of the PEG solution [24% (w/v) PEG 6000: Merck] added. This PEG is prepared in mannitol with sufficient magnesium chloride added (~30 m*M*) to bring the resistance when measured in the chamber of the electroporator into the region of 1.2 kΩ.

Ten minutes after addition of the DNA and PEG, samples are transferred to the chamber of the electroporator and pulsed three times at 10-sec intervals with pulses of an initial field strength of 1.4–1.5 kV/cm. They are then returned to a 6-cm-diameter Petri dish and held at 20° for 10 min before addition of 3 ml of a 1 : 1 mixture of K3A and H media (Table I) containing 0.6% (w/v) SeaPlaque agarose. This is the optimized protocol. It gives the highest frequencies we have as yet obtained—in the region of 1 to 3% for all colonies recoverable without selection. It is not necessary when using this method to cool the protoplasts on ice, due to the low heating effects when using such short pulses, in contrast to when using the method of Langridge *et al.*[34,35]

Method 3. Quick method for transformation without electroporation: Protoplasts are treated in an identical manner to that described for method 2 above. However, there is no need to adjust the resistance of the mannitol with magnesium sulfate, and 40% PEG is added to the protoplasts in place of the 24% (v/v) PEG. The protoplasts are transferred to the Petri dishes 10 min after addition of the PEG, and agarose added 10 min later. This protocol gives efficiencies in the range of 10^{-4} to 10^{-3}.

Method 4. Cotransformation: This is carried out using method 2 given above. The gene to be cotransformed into the protoplasts is added instead of the carrier DNA, i.e., at 50 μg/ml (20 μg/sample). This may be linearized before use, as this appears to give a higher rate of cotransformation. In experiments using a zein gene as the nonselected gene, 88% of the kanamycin-resistant colonies recovered contained sequences hybridizing to the zein gene sequences, with 27% containing a full copy of the zein gene clone used.[36]

Protoplast Culture, Selection of Transformed Lines, and Regeneration of Plants. Selection in the agarose bead-type culture system[17,37] has been found to be superior to selection in other culture systems tested. In this way a nearly constant selection pressure is maintained during the first 4 weeks of culture, thus suppressing any possibility of background colonies arising due to reduction in the selection pressure by the decay of the drug.

The dishes containing the protoplasts are sealed with Parafilm and incubated at 24° for 1 day in the dark followed by 6 days in continuous dim light (500 lx, cool fluorescent Sylvania "daylight" tubes). The agarose containing the protoplasts is then cut into quadrants and cultured in a 1 : 1 mixture of K3A and H media in the bead-type culture system[17] using one container with 30 ml medium for each 6-cm Petri dish from methods 2 and 3, and three containers with 30 ml of medium/10-cm dish from method 1. Kanamycin sulfate (50 μg/ml) is added to this and to all subsequent media for selection of transformants. Aliquots of the agarose containing the protoplasts can be cultured as a nonselected control in medium lacking kanamycin. After 1 week one-half of the medium is replaced with a 1 : 1

[34] W. H. R. Langridge, B. J. Li, and A. A. Szalay, *Plant Cell Rep.* **4,** 355 (1985).

[35] W. H. R. Langridge, B. J. Li, and A. A. Szalay, this volume [20].

[36] R. Scocher, R. D. Shillito, M. W. Saul, J. Paszkowski, and I. Potrykus, *Bio/Technology* **4,** 1093 (1985).

[37] I. Potrykus and R. D. Shillito, this series, Vol. 118, p. 549.

mixture of KC3 and J media (Table I), and thereafter one-half the medium is replaced weekly with a 1 : 1 mixture of K3E and J media (Table I).

Resistant clones are first seen 3–4 weeks after the start of the experiment, and after a total time of 5–6 weeks (when 2–3 mm in diameter) they are transferred to LS medium[38] (Table I) solidified with 0.8% (w/v) cleaned agar.

Regeneration of Plants. After 3–5 weeks, depending on the size of the original colony, these should reach 1 cm in diameter. Each colony is then split into four parts and two placed on fresh LS medium as above, and two on LS medium with 0.2 mg/liter BAP as the only phytohormone (regeneration medium). These latter dishes are incubated in the dark for 1 week and thereafter in the light (3000–5000 lx).

Shoots arising from the callus on regeneration medium are cut off when 1–2 cm long and placed on LS medium as above, but without hormones, where they produce shoots. When the shoots reach 3–5 cm in length they are transferred to T medium and treated as shoot cultures (see above). They can be transferred to soil once they have an established root system: the agar is gently washed away and the plantlets potted up. They require a humid atmosphere for the first week and can then be hardened off and grown under normal greenhouse conditions.

Genetic Analysis. When regeneration of fertile plants from tissue culture is possible then the introduced trait can be followed in its transmission to progeny. The earliest opportunity to observe this is by culture of the male gametes via another culture.[39,40] Haploid plantlets developed from microspores can be tested under selective conditions by transfer to kanamycin-containing media (300 mg/liter kanamycin sulfate) when they are at the "seedling" stage. For a single dominant gene one expects approximately 50% of the plants to be resistant due to the segregation of the trait during meiosis.

For genetic analysis, plants of *N. tabacum* are grown in the greenhouse to flowering. They are then selfed and crossed in both directions with wild-type SR1 plants. Seeds are collected from individual capsules, and stored in a dry place at room temperature. Nonsterilized seed is sown on half-strength NN69 medium[29] (Table I) containing 300 μg/ml kanamycin sulfate and solidified with 0.8% cleaned agar.[41] Seeds germinate at a

[38] E. M. Linsmaier and F. Skoog, *Physiol. Plant.* **18,** 100 (1965).

[39] N. Sunderland and J. M. Dunwell, *in* "Plant Cell and Tissue Culture" (H. E. Street, ed.), p. 233. Univ. of California Press, Berkeley, 1977.

[40] E. Heberle-Bors, *Theor. Appl. Genet.* **71,** 361 (1986).

[41] I. Potrykus, J. Paszkowski, M. Saul, J. Petruska, and R. D. Shillito, *Mol. Gen. Genet.* **199,** 169 (1985).

high frequency and, after 1 week, seedlings can be scored for resistance to kanamycin. Green seedlings are resistant, white sensitive. The counts are compared to the expected segregation ratios for one or more independent loci for the resistance character using the chi-square test.

A high proportion of the regenerated plants are fertile and pass the introduced genes to their progeny in a normal dominant Mendelian fashion, but a small proportion of the plants are disturbed in their fertility.

3. *Preparation, Transformation, and Culture of Leaf Mesophyll Protoplasts from Greenhouse-Grown Plants (Petunia hybrida)*

The method described for protoplast isolation is based on that of Durand *et al.*[42] and modified in our laboratory for use with the "cyanidin type,"[43] Mitchell,[44] and other petunias. A high division frequency and rate of conversion to colonies of protoplasts from a number of *Petunia* species is achieved by this method.

Source of Material. Plants of *P. hybrida* and other *Petunia* species are clonally propagated via cuttings, grown in clay pots in a controlled environment [12 hr light (5000 lx), 27/20° day/night, 50/70% relative humidity], and watered morning and evening with commercial fertilizer [0.01% (v/v) Greenzit[R]].

Preparation of Protoplasts. Young leaves at approximately two-thirds to three-quarters of their final size from 4- to 6-week-old plants are washed with tap water and sterilized by immersion for 8 min in 0.01% (w/v) $HgCl_2$ solution containing 0.05% (w/v) Tween 80, and then washed carefully with five changes of sterile distilled water (each change 5 min).

Leaf halves without midribs are wet with osmoticum P1 [0.375 *M* mannitol, 0.05 *M* $CaCl_2$, 0.2% (w/v) MES, pH 5.7] and arranged in a stack of six on the lid of a 9-cm Petri dish ready for cutting. They are cut diagonally into clean sections 0.5 mm wide, transferred into a small screw-top flask containing 10 ml of enzyme solution [2% (w/v) cellulase "Onozuka" R10, 1% (w/v) hemicellulase, and 1% (w/v) pectinol in 0.3 *M* mannitol, 0.04 *M* $CaCl_2$, 0.2% (w/v) MES, pH 5.7], and vacuum infiltrated until the leaf tissue is translucent. The vacuum infiltration is carried out by placing the screw-top flask containing the leaf pieces and enzyme in a larger chamber (a small desiccator can be used), and applying a vacuum (~700 mm Hg) while gently shaking the material. The air present in the intracellular spaces in the leaf pieces will expand and bubble out. The

[42] J. Durand, I. Potrykus, and G. Donn, *Z. Pflanzenphysiol.* **69,** 26 (1973).
[43] D. Hess, *Planta* **59,** 567 (1963).
[44] A. Z. Mitchell, B.A. thesis. Harvard Univ., Cambridge, Massachusetts.

pressure is then gently allowed back through a sterile filter. It may be necessary to repeat the infiltration one or two times to completely remove the air from the leaf spaces.

The leaf slices are placed in fresh enzyme solution in a Petri dish (0.5 g/10 ml in a 9-cm Petri dish), which is sealed with Parafilm and incubated at 28° for ~3 hr. The incubation mixture is checked periodically under the inverted microscope for the release of protoplasts. The time required may vary, especially with greenhouse-grown material.

The digest is gently agitated, filtered through a 100-μm mesh stainless steel sieve, and transferred in 5-ml aliquots into 10- to 15-ml centrifuge tubes. Osmoticum P1 is added (5 ml) to each tube and, after gentle mixing, these are centrifuged for 5 min to sediment the protoplasts.

The supernatant is carefully pipetted off, and the sediment is gently shaken to free the protoplasts before resuspension in 10 ml of osmoticum P1. Washing by sedimentation is repeated two times. If necessary, the suspension is overlaid on 0.6 *M* sucrose to remove debris and the protoplasts collecting at the interface recovered and resuspended in osmoticum P1. A sample is taken and diluted in osmoticum for counting and the protoplasts sedimented once more and resuspended in medium (K0,[45] Table I) at 1×10^6/ml.

Transformation

Method 1. "F medium" method: Transformation is carried out as described for *N. tabacum* protoplasts above (protocol 2; method 1). After addition of the F medium, the protoplasts are resuspended in K0 culture medium at 2×10^5/ml. To 5 ml of this culture medium in a 9-cm-diameter Petri dish is added 5 ml of liquefied K0 medium containing 1.2% (w/v) SeaPlaque agarose. The protoplasts are dispersed by gentle swirling, and the agarose allowed to solidify. The dishes are sealed with Parafilm and cultured as described below.

Method 2. Quick method: The protoplasts are suspended following the last purification step in osmoticum P1 at 10^6/ml. Aliquots of 1 ml are placed in 10-ml polycarbonate centrifuge tubes, the protoplasts subjected to a heat shock as described above, and DNA (10 μg pABD1 plus 50 μg calf thymus DNA in 50 μl water) added. After 1 min, 0.5 ml of the PEG solution [40% (w/v) PEG 6000 in osmoticum P1) is added, followed 10 min later by 10 ml osmoticum P1. The protoplasts are collected by centrifugation, and plated in culture medium as described in method 1 above at 1×10^5/ml.

[45] H. Koblitz and D. Koblitz, *Plant Cell Rep.* **1,** 147 (1982).

Method 3. Electroporation: The protoplasts are resuspended after purification at a density of 1.2×10^6/ml in 0.4 *M* mannitol containing 6 m*M* $MgCl_2$ as for electroporation of *N. tabacum,* and treated as described for *N. tabacum* above (protocol 2, method 2). A pulse voltage of 1.2 kV is delivered from a capacitor of 10 nF. They are then embedded, in the same way as used for *N. tabacum,* in K0 medium solidified with 0.6% (w/v) SeaPlaque agarose at 1×10^5/ml.

Culture and Selection of Transformants. Agarose-solidified cultures are incubated for 6 days in the dark at 26°. For plates containing 10 ml of protoplasts, half the protoplast-containing agarose gel is transferred to each of two 10-cm-diameter containers each containing 40 ml of liquid K0 medium containing 0.2 mg/liter, 2,4-D, 0.5 mg/liter BAP, and 2% (v/v) coconut milk. These are incubated at 26° on a gyratory shaker (60 rpm, 1.2 cm throw) in the light (500–1000 lx). The liquid medium is replaced weekly, each time reducing the glucose concentration in the original medium by one-quarter so as to reduce the osmotic pressure.

After 5–6 weeks the colonies are 1–2 mm in size and can be cultured further.

Regeneration of Plants. Colonies are transferred directly to the regeneration medium (NT,[46] Table I) containing 16 mg/liter zeatin and solidified with 1.0% (w/v) SeaPlaque agarose (16 colonies/6-cm-diameter Petri dish). These are cultured in the conditions used for the growth of plants.

The dark green calli which arise after 2–4 weeks are transferred to the same medium containing only 2 mg/liter zeatin. Those showing shooting morphology are placed on hormone-free medium to allow outgrowth of these and subcultured further as cuttings. In general, three such passages are necessary before shoots showing a "normal" morphology are obtained and can develop a strong enough root system to be transferred to soil in the greenhouse.

4. *Preparation, Transformation, and Culture of Protoplasts from a Nonmorphogenic Suspension Culture of the Graminaceous Monocot Species Lolium multiflorum*

Protoplasts from leaf or other whole plant tissues of grasses do not in general divide in culture although there have been exceptions to this rule.[47] However, there are a number of suspension cultures of graminaceous species available, and these have been used to produce protoplasts

[46] T. Nagata and I. Takebe, *Planta* **99,** 12 (1971).

[47] I. Potrykus, in "Advances in Protoplast Research" (L. Ferenczy and G. L. Farkas, eds.), p. 243. Hungarian Academy of Sciences, Budapest, 1980.

which divide. There have been reports of division and colony formation of protoplasts from morphogenic suspension cultures of these species,[48] but these have not yet proved to be repeatable.

We describe a protocol developed in our laboratory for the isolation, transformation, and culture of protoplasts from *Lolium multiflorum* (Italian ryegrass) suspension culture cells.

Source of Material. The cell line was originally established by P. J. Dale, who has also used it for protoplast culture.[49]

Suspension cultures are maintained by weekly serial transfer (1:7 dilution) in CC medium[50] (Table I) without mannitol and with 2 mg/liter 2,4-D on a gyratory shaker (110 rpm, 2 cm throw) in low light levels (500 lx).

Preparation of Protoplasts. Cultures are used for protoplasting, 4, 5, or 6 days after subculture. Cells (10 ml) are sedimented by centrifugation (5 min) and resuspended in the same volume of enzyme solution [4% (w/v) Driselase in 0.38 *M* mannitol, 8 m*M* $CaCl_2$, MES, pH 5.6].

The solution is poured into a 9-cm Petri dish and this is sealed and placed on a rocking table for 1 hr at 20° before being incubated overnight (15 hr) without agitation at the same temperature. The preparation is then placed on the rocking table for an hour followed by another hour without agitation.

The protoplasts are filtered through a 100-μm mesh stainless steel sieve, an equal volume of 0.2 *M* $CaCl_2$ (MES, pH 5.8) added, and the suspension distributed into two centrifuge tubes. After centrifugation to sediment the protoplasts (10 min) they are taken up in 3 ml osmoticum L1 (0.25 *M* mannitol, 0.1 *M* $CaCl_2$, MES, pH 5.8) and overlayered on a 5-ml sucrose cushion (0.6 *M* sucrose, MES, pH 5.8).

Protoplasts collecting at the interface after centrifugation are carefully removed and washed twice with osmoticum L1, counted, and resuspended in CC medium (Table I) at a density of 2×10^6/ml.

Transformation

Method 1. "F medium" method: One-milliliter aliquots of the protoplasts, in CC medium, are heat shocked and treated with DNA as described for *N. tabacum* in protocol 2, method 1 above. Following the addition of the "F" medium, they are sedimented by centrifugation (5 min), and resuspended in 2 ml of CC culture medium.

[48] V. Vasil and I. K. Vasil, *Theor. App. Genet.* **56,** 97 (1980).
[49] M. G. K. Jones and P. J. Dale, *Z. Pflanzenphysiol.* **105,** 267 (1982).
[50] I. Potrykus, C. T. Harms, and H. Loerz, *Theor. Appl. Genet.* **54,** 209 (1979).

Culture and Selection of Transformants. The protoplasts are cultured in 3.5-cm-diameter Petri dishes (2 ml/dish) at 26° in the dark. Fresh CC medium (1 ml) with 0.2 *M* mannitol is added after 7 days to dilute the culture and reduce the osmotic pressure. Two weeks after the DNA treatment, the cultures are sedimented and the cells taken up in the 3 ml of CC medium with 0.2 *M* mannitol containing 25 mg/liter G418. After a further 2 weeks these cultures are diluted 5-fold with CC medium without mannitol, and containing G418 as before. After a total of 6 weeks, calli arising from the cultures are transferred to CC medium as used for the suspension cultures solidified with 0.8% (w/v) cleaned agar and grown at 24° in the dark or in the light (2000 lx).

5. *Preparation, Transformation, and Culture of Protoplasts from a Sterile Shoot Culture of N. plumbaginifolia, and Regeneration of Plants*

Source of Material. This material is grown as sterile axenic shoot cultures. They are grown as described for *N. tabacum* above except that the medium used is Murashige and Skoog's medium[51] with 10 g/liter sucrose, as described by Negrutiu.[52]

Preparation of Protoplasts. For isolation of protoplasts, sterile leaves from shoot cultures are sliced carefully into 2-mm-wide strips using a sharp razor blade and these incubated overnight at 26° in enzyme solution [0.5% (w/v) Driselase in 0.5 *M* sucrose, 0.005 *M* $CaCl_2$, pH 5.5, with KOH]. The released protoplasts are filtered through 100- and 66-μm stainless steel sieves, and centrifuged for 5 min. The protoplasts at the surface are collected and washed two times with W5 salt solution (154 m*M* NaCl, 125 m*M* $CaCl_2$, 5 m*M* KCl, 5 m*M* glucose, pH 5.5, with KOH[53]). They are resuspended in W5 solution at a final density of $1–1.6 \times 10^6$ and used immediately for transformation. Alternatively, they can be stored in the W5 solution for up to 6 hr at 6–8° before use.

Transformation

Method 1. PEG method: Aliquots (1 ml) of protoplasts are placed in centrifuge tubes, and a heat shock applied as described for *N. tabacum*. A mixture of linearized plasmid DNA (10 μg/ml) and carrier DNA (50 μg/ml calf thymus DNA) is added, followed by 1.0–1.5 ml PEG solution [45% (w/v) PEG 4000, 2% (w/v) $Ca(NO_3)_2 \cdot 4H_2O$, 0.4 *M* mannitol, pH adjusted

[51] T. Murashige and F. Skoog, *Physiol. Plant.* **15,** 473 (1962).

[52] I. Negrutiu, *Z. Pflanzenphysiol.* **104,** 431 (1981).

[53] L. Menczel, G. Galiba, F. Nagy, and P. Maliga, *Genetics* **100,** 487 (1982).

to 9 with KOH repeatedly over a period of 4 hr and autoclaved; stored at −20°, final pH 7.5–8.5] with gentle shaking to give a final concentration of 22–27% PEG.

After incubation with the PEG for 20–30 min, W5 solution is added 3× in 1-ml aliquots at 2- to 5-min intervals, and the protoplasts centrifuged for 5 min. The pellet is resuspended in 10 ml culture medium (K3 medium with 0.4 *M* glucose[27]) and cultured in a 10-cm-diameter dish. Alternatively, the protoplasts can be embedded in the same culture medium solidified with 0.6% (w/v) SeaPlaque agarose as described above for *N. tabacum*.

Method 2. Electroporation: The protoplasts can be transformed by electroporation as described for *N. tabacum* above. Following the enzyme incubation, the released protoplasts are cleaned and transformed exactly as described for *N. tabacum* in protocol 2.

Culture and Selection of Transformants. At the 2- to 4-cell stage, the developing cultures are diluted 6- to 8-fold (to ~2000 surviving colonies/ml) in low-hormone medium MDS[54] with 25 mg/liter kanamycin sulfate. In the case of agarose-embedded cells, these are suspended at a similar dilution factor as used for liquid medium in liquid MDS medium with (30 ml) or without (10 ml) shaking. The dishes are incubated at 1000–1500 lx at 26° for 3–5 weeks, when colonies should become visible.

Transformation frequencies of 10^{-4} to 10^{-3} should be obtained. It has not yet been possible to attain higher frequencies with this species despite many experiments to this end.

Calli are transferred to MDS medium solidified with 0.8% (w/v) cleaned agar containing 50 mg/liter kanamycin. After a further 2–4 weeks they can be transferred to RPZ medium[55] for regeneration, and shoots arising are cultured as shoot cultures as described for the source material, and transferred to the greenhouse.

Evaluation of Results

Colonies which are selected as being resistant should be further analyzed for proof of transformation. Although with the system described we have never observed resistant colonies from control cultures with *N. tabacum* or *P. hybrida,* variations in the many factors present, particularly when using other protoplast systems, may lead to misinterpretation of apparently resistant colonies as transformants.

[54] I. Negrutiu, R. Dirks, and M. Jacobs, *Theor. Appl. Genet.* **66,** 341 (1983).
[55] P. Installé, I. Negrutiu, and M. Jacobs, *J. Plant Physiol.* **119,** 443 (1985).

We feel that the minimal criteria for confirmation of a transformation event should be the following:

1. A phenotypic change to resistance, or growth under other selective conditions, in a selection scheme which is proved to be "clean"
2. The presence for the transforming DNA in the selected lines in form expected for transformed DNA (integrated in the genome or autonomously replicating)
3. Expression of the foreign DNA at the RNA/protein level

In addition, if the plant cell tissue culture system being used is capable of regenerating plants, then genetic data should be obtained.

Phenotypic Change

The assumption of direct selection on kanamycin (or G418) is that only transformed cell lines will be phenotypically resistant to the drug. Ideally the level of selection should permit recovery of only transformed clones. Therefore resistant clones should only appear after transformation with the correct vector and not in any control treatments. Selection conditions should be adjusted to produce this situation. The resistant phenotype should be rechecked at later stages in culture by comparison with wild type, for instance at the callus level.

In the case of *N. tabacum,* shoots can be regenerated from transformed callus under selective (100 mg/liter kanamycin sulfate) or nonselective conditions. These shoots can be rooted in medium containing 150 mg/liter kanamycin sulfate. Wild-type SR1 shoots regenerated in the absence of kanamycin never form roots, bleach, and die when cultured in the presence of kanamycin. In order to confirm the resistant phenotype of the cells of such regenerated transformed plants or to show the possible loss of the introduced trait during plant development, mesophyll protoplasts from these plants can be isolated and their resistance checked by culture in kanamycin-containing media.

Molecular Analysis of the DNA of Transformed Clones and Regenerated Plants

Analysis of DNA. DNA of transformed cell lines, regenerated plants, and their progeny should be analyzed using standard Southern blot techniques.[56] We will omit the exact procedures used since they are now standard in molecular biology and the choice of restriction enzymes, etc.,

[56] E. M. Southern, *J. Mol. Biol.* **98,** 503 (1975).

will depend on the transforming DNA being analyzed. It should, however, be possible to prove unequivocally the presence of the transforming DNA in high-molecular-weight nuclear DNA of transformed lines in an integrated form in the absence of any hybridization to DNA from control untransformed lines. We shall concentrate here only on a method for the efficient isolation of DNA from small amounts of callus or plant tissue. The relatively slow growth rate of plant cell tissue cultures (~1 week doubling time for callus on solid media) means that the adaption of DNA extraction procedures to small amounts of tissue allows a significant shortening in the length of an experiment. We have adapted a method of Thanh Huynh (personal communication, Dept. of Biochemistry, Stanford University), which allows the extraction of pure DNA from ~0.5 g of tissue:

Samples of 0.5 g of callus or leaf tissue are homogenized in a Dounce homogenizer in 3 ml of a buffer containing 15% sucrose, 50 m*M* EDTA, 0.25 *M* NaCl, 50 m*M* Tris–HCl, pH 8.0. Centrifugation of the homogenate for 5 min at 1000 *g* results in a crude nuclear pellet which is resuspended in 2 ml of a buffer containing 15% sucrose, 50 m*M* EDTA, 50 m*M* Tris–HCl, pH 8.0. SDS is added to a final concentration 0.2% (w/v). Samples are heated for 10 min at 70°. After cooling to room temperature, potassium acetate is added to a final concentration of 0.5 *M*. After incubation for 1 hr at 0° the precipitate formed is sedimented for 15 min in an Eppendorf centrifuge at 4°. The DNA in the supernatant is precipitated with 2.5 vol of ethanol at room temperature and redissolved in 10 m*M* Tris–HCl, pH 7.5, 5 m*M* EDTA. The DNA samples are then run in a cesium chloride/ethidium bromide gradient in the vertical rotor (Beckman VTi 65) for 17 hr at 48,000 rpm. The DNA is removed from the gradient with a wide-bore hypodermic syringe needle and the ethidium bromide extracted as for plasmid DNA (see above). The DNA obtained is of high molecular weight and susceptible to various restriction enzyme.

For Southern analysis 5–10 μg DNA is electrophoresed in a 1% agarose gel, transferred to a nitrocellulose membrane, and hybridized with nick-translated DNA[57] (5–10 $\times$ 10^8 cpm/μg). Filters are washed with 2$\times$ SSC at 65°, 3 $\times$ for 1 hr, and subsequently exposed to X-ray film with intensifying screens for 24–48 hr.

Activity Assay for the Product of the Transforming Gene

The assay for activity of the transformed gene will of course depend on the expected product. In the case described here, a method developed

[57] W. J. Rigby, M. Dieckmann, C. Rhodes, and P. Berg, *J. Mol. Biol.* **113,** 237 (1977).

by Reiss *et al.*[58] allowed us to detect activity of the APH(3')II gene product in transformed lines.

Callus or leaf pieces (100–200 mg) are crushed in an Eppendorf centrifuge tube with 20 μl extraction buffer [100% (v/v) glycerol, 0.1% (w/v) SDS, 5% mercaptoethanol, 0.005% (w/v) bromphenol blue, 0.06 M Tris, pH 6.8]. Extracts are centrifuged for 5 min at 12,000 g. Proteins in 35 μl of the supernatant are separated on a 10% nondenaturing polyacrylamide gel. The gel is incubated with kanamycin and ^{32}P-labeled ATP and then blotted onto Whatman P81 phosphocellulose paper. The paper is washed five times with deionized water, wrapped in plastic film, and exposed to X-ray film with intensifying screens for 24–48 hr. Kanamycin binds to this paper but ATP does not, therefore radioactive bands on the paper reveal bands on the gel with an activity which transfers radiolabeled phosphate from the [^{32}P]ATP to the kanamycin, i.e., aminoglycoside phosphotransferases. Other transformant nonspecific bands are seen at the top of the gel. These are protein phosphorylases, and the bands can be removed from the P81 paper by treatment with protease if required.[59]

General Comments

Comparison with Other Gene Transfer Systems

A. tumefaciens-mediated gene transfer was, for a while, the only possible way to introduce foreign genes into plants. Although it is a well-established method, there has been little attempt until recently to study the fate of genes introduced in this way. There is also increasing evidence that the integration patterns found when using this method can be complex and show rearrangements of the original T-DNA. One of the great drawbacks of the method is the limited host range, which excludes, for instance, the graminaceous monocots. However, this will remain the method of choice for introduction of foreign DNA in most cases.

DNA viruses have been suggested as possible gene vectors. To date there is one example of this, in which cauliflower mosaic virus carrying a modified bacterial methotrexate-resistance gene was used to infect a plant. The foreign gene was thus transported into and systemically spread in the plant.[10,60] The advantages of this system are the ease of infection, systemic spread within the plant, and multiple copies of the gene being present per cell. The disadvantages are the narrow host range, lack of

[58] B. Reiss, R. Sprengel, M. Willi, and H. Schaller, *Gene* **30,** 217 (1984).

[59] P. H. Schreier, E. A. Seftor, J. Schell, and H. J. Bonnert, *EMBO J.* **4,** 25 (1985).

[60] N. Brisson and T. Hohn, this series, Vol. 118, p. 659.

transfer to sexual offspring and the limited space available for passenger DNA.

Liposome fusion has been shown to be a method for transformation of plant cells.[11] As with direct gene transfer, this requires protoplasts. It will be of interest to see whether the pattern of integration of DNA by this method is different than that seen when using other methods.

Microinjection is an efficient means of transforming plant protoplasts.[13] However, the number of transformants that one can obtain is limited by the number which can be injected by the operator. The main use of this technique may be that it could be possible eventually to inject cells, thus circumventing the need for protoplasts which is inherent in other direct gene transfer techniques such as described above.

Conclusions

The procedures described above require a protoplast system which allows regeneration of callus from protoplasts and a suitable selectable marker gene. If these requirements are satisfied the methods for DNA delivery and selection of transformants are flexible. The protocols described here are certainly applicable to other systems and are reproducible in our hands.

Acknowledgments

The authors thank all the members of our laboratory, particularly M. W. Saul, J. Paszkowski, I. Negrutiu, and S. KrugerLebus for help in supplying information and protocols for this publication.

[20] Uptake of DNA and RNA into Cells Mediated by Electroporation

By W. H. R. Langridge, B. J. Li, and A. A. Szalay

Introduction

To study the regulation of eukaryotic gene expression, a reliable and efficient method for introduction of altered genes into cells is required. Zimmerman[1] described an electric shock method (electroporation) for the

[1] U. Zimmermann, *in* "Target Drugs" (E. Goldberg, ed.), p. 153. Wiley, New York, 1983.

introduction of drugs into mammalian red blood cells. The transformation of *Bacillus cereus* spheroplasts with plasmid DNA was stimulated by high-voltage electrical pulses.[2] High-voltage pulses have been used to mediate uptake of DNA encoding herpes virus thymidine kinase (TK) into mouse cells.[3,4] A substantial increase in frequency of DNA transformation was obtained when mouse lymphocytes were treated with colchicine prior to electroporation.[5] Recently, experiments were described in which *Agrobacterium* Ti plasmid DNA,[6] a chimeric antibiotic resistance marker gene,[7] plant genomic DNA,[8] and tobacco mosaic virus RNA[9] were successfully introduced into plant protoplasts by electroporation.

In the following sections we outline simple methods which allow the uptake of plasmid DNA and RNA into plant protoplasts. The DNA uptake is mediated by a series of high-voltage, direct current electrical pulses.

Electroporation Theory

When a cell membrane is rapidly polarized (less than 10 to 100 μsec) at high voltage, a physical breakdown of the membrane occurs which is associated with a reversible increase in membrane conductivity and permeability.[9,10] The effects caused by the electrical breakdown in the membrane are reversible; after a short time, the original membrane resistance and impermeability are restored. If the field strength, defined as volts/centimeter distance between the electrodes, exceeds the voltage required to reach the point of physical membrane breakdown by a factor of 2 to 6, or if the exposure time of the membrane to the electric field is excessive, e.g., a pulse duration of milliseconds to seconds, the electrical breakdown phenomenon becomes irreversible and the cells do not survive the treatment.

Available membrane breakdown data are consistent with the view that

[2] N. Shivarova, W. Forster, H. E. Jacob, and R. Grigorava, *Z. Allg. Mikrobiol.* **23,** 595 (1983).

[3] T. K. Wong and E. Neumann, *Biochem. Biophys. Res. Commun.* **107,** 584 (1982).

[4] E. Neumann, M. Schafer-Ridder, Y. Young, and P. H. Hofschneider, *EMBO J.* **1,** 841 (1982).

[5] H. Potter, L. Weir, and P. Leder, *Proc. Natl. Acad. Sci. U.S.A.* **81,** 7161 (1984).

[6] W. H. R. Langridge, B. J. Li, and A. A. Szalay, *Plant Cell Rep.* **4,** 355 (1985).

[7] M. Fromm, L. P. Taylor, and V. Walbot, *Proc. Natl. Acad. Sci. U.S.A.* **82,** 5824 (1985).

[8] I. Potrykus, R. D. Shillito, M. W. Saul, and J. Paszkowski, *Plant Mol. Biol. Rep.* **3,** 117 (1985).

[9] M. Nishiguchi, W. H. R. Langridge, M. Zaitlin, and A. A. Szalay, *Plant Cell Rep.*, in press (1986).

[10] V. Zimmermann, *Biochim. Biophys. Acta* **694,** 227 (1982).

the membrane is compressed locally due to the high field strength.[11] The local electromechanical compression of the membrane leads to pore formation. The creation of pores in the membrane permits the exchange of intracellular and extracellular components; however, the size of the molecules passing through the membrane depends on the field strength and electrical pulse duration.[12] With pulses of high voltage, or with increased pulse duration, the permeability of the membrane may increase to the point where plasmid DNA molecules are capable of passage. When the pulse length exceeds 1 to 5 μsec at high field strengths, irreversible mechanical breakdown is observed in the membranes of mammalian cells.[1] The electric field, high current densities inside the cell, and the associated osmotic changes which occur inside the cell can create adverse side effects in the organelles and cytoplasm once breakdown has occurred. Since membrane breakdown occurs within 10 nsec,[13] the exposure time of the cells to the electrical field pulse is one of the most critical factors.

Materials and Reagents

In order to pass an electrical current through cells or protoplasts, an electroporation chamber is required. The chamber can be of simple construction,[5] consisting of a flat-sided, open-topped plastic (0.5 ml vol) cuvette, in which two electrodes of platinum or stainless steel are placed along opposite walls (Fig. 1A). Other chamber configurations available commercially consist of a plastic container into which a plastic cone holding two helically wound platinum wire electrodes can be inserted (Fig. 1C) or a plastic slide on which two platinum wire electrodes are mounted in a parallel configuration (Fig. 1B).

The distance between the electrodes in the electroporation chamber determines the voltage required to obtain pore formation in the cell membrane according to the formula $E = V/d$, where E is the electric field strength, V is the voltage applied to the preparation, and d is the distance in centimeters between the electrodes in the electroporation chamber.

Electrical Field Requirements for Cell Membrane Poration

When an electrical field is applied to a suspension of animal cells or plant protoplasts, pores are formed in the cell membranes when the membrane potential exceeds 200–250 mV.[14] During electroporation experiments with carrot and tobacco protoplasts, we found that the electrical potential required for successful poration was dependent not only on the

[11] V. Zimmermann and J. Wienken, *J. Membr. Biol.* **67,** 165 (1982).

[12] H. Coster, E. Steudle, and V. Zimmermann, *Plant Physiol.* **58,** 636 (1976).

[13] R. Benz and V. Zimmermann, *Biochim. Biophys. Acta* **597,** 637 (1980).

[14] J. Tessié and T. Y. Tsong, *Biochemistry* **20,** 1548 (1981).

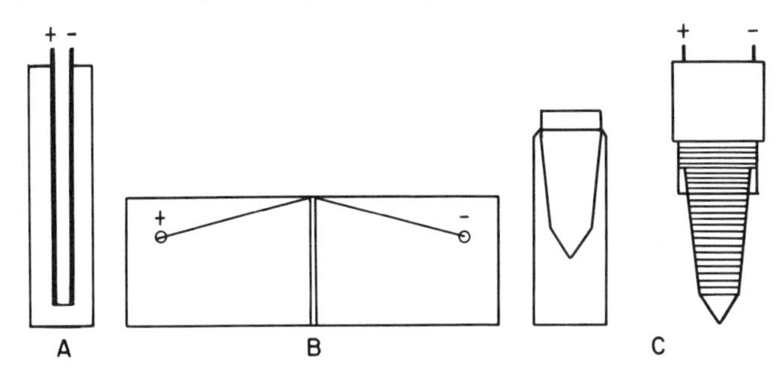

FIG. 1. Three types of electroporation chambers. (A) A simple chamber consisting of a 1.5-ml volume plastic cuvette into which has been inserted two steel or platinum flat plate electrodes. The electrodes are cemented to opposite walls of the cuvette to fix their position. (B and C) Commercially available electroporation chambers (GCA Corporation, Chicago, IL). Chamber B consists of a plastic plate the size of a microscope slide, to which two parallel platinum wire electrodes are attached. The distance between the wire electrodes is fixed at either 100 or 200 μm. Approximately 10–15 μl of sample material can be layered between the electrodes of this chamber. Chamber C is a two-piece chamber; the receptacle, which receives the sample, is made out of autoclavable plastic. After the sample is introduced into the chamber the cone, which contains platinum wire electrodes wrapped in a helical configuration around the cone, is inserted into the receptacle. The distance between opposite electrodes on the cone is 200 μm, which is also the distance between the electrodes and the inner wall of the chamber. The maximum sample capacity of this chamber is approximately 400 μl.

plant species but also on the origin of the protoplasts, e.g., routinely protoplasts obtained from plant leaf tissue were porated at lower voltages than protoplasts isolated from suspension culture cells. In addition to the above-mentioned factors, successful electroporation of plant protoplasts depends on the components of the electroporation medium. Electroporation media may be either nonionic or of high electrolyte content.[3,5–7,15,15a] Vital dye uptake experiments and TMV-RNA uptake experiments conducted in our laboratory indicate that a lower field strength is required to successfully electroporate carrot and tobacco protoplasts when they are suspended in a medium without electrolytes. A summary of the electroporation conditions found to be successful for nucleic acid uptake into both plant and animal cells is presented in Table I.

Types of Electric Pulse Generators

Two types of electrical pulses are presently used for creating pores in the membranes of plant and animal cells. One type of electroporation instrument generates a square wave field[5,8,15b] with a pulse decay duration

[15] D. Falkner, E. Neumann, and H. G. Zachau, *Hoppe Seyler's Z. Physiol. Chem.* **365,** 1331 (1984).

[15a] M. E. Fromm, L. P. Taylor, and V. Walbot, *Nature (London)* **319,** 791 (1986).

[15b] D. Zerbib, F. Amalric, and J. Teissié, *Biochem. Biophys. Res. Commun.* **129,** 611 (1985).

TABLE I
SUMMARY OF ELECTROPORATION CONDITIONS FOR NUCLEIC ACID UPTAKE INTO PLANT AND ANIMAL CELLS

Investigator	Cells	Nucleic acid[a]	Field strength (kV/cm)	Pulse number	Pulse length	Temperature (°C)	Transformants/10^6 cells/μg nucleic acid	Cells/ml	DNA (μg/ml)
Wong and Neuman[3]	Mouse	Tk-gene	5–10	3	3.0 μsec	20	67	1.2×10^5	10
Neumann *et al.*[4]	Mouse	Tk-gene	8	3	3.0 μsec	20	100	5.0×10^6	20
Potter *et al.*[5]	Mouse	Tk-gene	4–8	1	—	—	200–300	1.2×10^7	20
Zerbib *et al.*[15]	Mouse	Tk-gene	6	3	11.0 μsec	20	150–200	2.0×10^6	8
Langridge *et al.*[6]	Carrot	Ti-plasmid	2	6	40.0 μsec	4	1.5×10^3	3.5×10^6	10
Nishiguchi *et al.*[9]	Tobacco	TMV-RNA	5	9	90.0 μsec	4	7.0×10^5	1.5×10^5	10
Shillito *et al.*[24]	Tobacco	NPTII	1–1.5	3	3.5 msec	20	90	1.0×10^6	10
Falkner *et al.*	Lymphoid	K-gene	7.5	3	4.6 μsec	20	15	1.3×10^7	10
Fromm *et al.*[7]	Carrot, tobacco	CAT	1.5	3	50.0 msec	20	75	2.0×10^6	8
Fromm *et al.*	Corn	NPTII	0.5	1	2.4 msec	20	80	4.0×10^6	50

[a] TK, Thymidine kinase; Ti, tumor-inducing plasmid from *Agrobacterium tumefaciens;* TMV, tobacco mosaic virus; NPTII, neomycin phosphotransferase; K, immunoglobulin K gene; CAT, chloramphenicol acetyltransferase.

in the microsecond range. A second type of instrument generates an exponential wave field with a pulse decay time in the millisecond range. Zerbib *et al.*[15] compared the efficiency of both types of pulse generators by comparing transformation frequencies obtained after introduction of the herpes virus thymidine kinase (TK) gene containing vector into murine TK^- cells. Using the exponential decay system, the authors obtained a maximum yield of 1.5 transformants/μg plasmid DNA at a field strength of 6 kV/cm. When a square wave generator was used as the source of electrical pulses, a maximum of 0.75 TK^+ transformants/μg plasmid DNA was calculated at a field strength of only 1.5 kV/cm. Thus, both types of pulse generators can be used for introduction of foreign DNA into animal cells.

Both exponential decay and square wave pulse generators have been successfully used for the introduction of DNA and RNA molecules into plant protoplasts.[6,9] Equipment which generate exponential pulses depends on the discharge of a single capacitor or a battery of capacitors previously charged by a common laboratory electrophoresis power supply.[5,7] Thus, capacitor discharge systems are relatively inexpensive to build and utilize already available laboratory equipment. However, the high field strength required to create pores in the plasma membrane and the relatively long capacitor discharge time (milliseconds) may result in excessive heat generation over the duration of the pulse, which can significantly decrease cell viability.[7]

For investigators who do not wish to construct their own capacitor discharge electroporation device, one may be obtained from Interactive Radiation, Inc., Northvale, New Jersey. Two companies currently marketing square wave electroporation devices are the GCA/Precision Scientific Group, Chicago, Illinois, and Compudigital Industries, Davis, California. The Compudigital square wave pulse generator is used in conjunction with a laboratory power supply and is therefore considerably less expensive than the GCA instrument, which is equipped for cell fusion as well as electroporation. The commercial electroporation devices described currently range in price from approximately $1000 to $14,000.

Method

Recent results from our laboratory indicate that stable integration and expression of plasmid-encoded genes[6] and tobacco mosaic virus RNA[9] can be detected in plant protoplasts following electroporation. The following method is described for electric pulse-mediated introduction of radioactively labeled plasmid DNA and tobacco mosaic virus RNA into carrot protoplasts.

To study effects of field strength (volts/centimeter) and number of electrical pulses on the uptake of DNA into plant cells, and to determine the optimum conditions required for electrotransformation, carrot protoplasts are prepared from suspension culture and electroporated in a buffer containing a radioactively labeled plasmid DNA. After electroporation, the amount of plasmid DNA introduced into the carrot protoplasts is determined by measurement of the amount of DNase-resistant radioactivity associated with the protoplasts.

Experimental Protocol

Carrot WOO1C Protoplast Preparation

Enzyme buffer (100 ml)
- Distilled H_2O, 50.0 ml
- 3.0 mM MES, 58.50 mg
- 0.7 mM NaH_2PO_4, 10.00 mg

Adjust the pH to 6.5, then add the following:

- 0.07 M $CaCl_2 \cdot 2H_2O$, 100.00 mg
- 0.35 M mannitol, 6.37 g
- 0.35 M sorbitol, 6.37 g
- H_2O, up to 100 ml

1. To 100 ml of enzyme buffer, add the following:

   Cellulase (cellulysin, 2.0%), 2.0 g
   Pectinase (macerozyme, 1.0%), 1.0 g

2. The enzyme mixture is centrifuged at 7000 rpm for 5 min and vacuum filtered through a 0.22-μm pore size Nalgene filter.
3. Mix 5–10 ml of carrot WOO1C suspension culture cells with an equal volume of the enzyme mixture in four sterile 10-cm-diameter Petri plates.
4. The plates are sealed with Parafilm and incubated at 26° for 12–16 hr (or 3 hr at 32°) on a rotary shaker at 40 rpm.
5. At the end of the incubation period, the cell suspension is pipetted up and down through a 10-ml pipet several times to separate cell clumps, the plates are resealed, and incubation continued for 1 hr.
6. The protoplast suspension is filtered through a layer of sterile absorbent cotton supported on a metal screen and centrifuged at 100 g for 5 min at room temperature in a table-top centrifuge.
7. The protoplast pellet is resuspended in 10.0 ml of wash medium and

centrifuged at 100 *g* for 5 min at room temperature. The wash procedure is repeated two times.

Wash medium
- 0.37 *M* glucose, 66.6 g/liter
- 1.50 m*M* $CaCl_2 \cdot H_2O$, 200.0 mg/liter

8. The protoplast pellet is resuspended in sterile electroporation buffer[16] modified as previously described,[6] to a final concentration of 1×10^6 protoplasts/ml.

Electroporation buffer (pH 7.5)
- $MgCl_2$, 200.00 m*M*
- NaCl, 140.00 m*M*
- KCl, 5.00 m*M*
- Na_2HPO_2, 0.75 m*M*
- Glucose, 5.00 m*M*
- Sucrose, 0.50 *M*

9. Plasmid DNA in the size range of 10 to 20 kb is labeled by nick translation[17] to a specific activity of 1×10^8 dpm/μg DNA and mixed with the protoplasts at an approximate ratio of 1×10^2 to 1×10^3 DNA molecules per protoplast.

10. Aliquots (200 μl) of the protoplast–DNA mixture are transferred to an ice-cold electroporation chamber and subjected to square wave electrical pulses of selected field strength, number, and duration.

11. The electroporated protoplasts are held at 0° for 10 min, transferred to a 1.5-ml Eppendorf tube at room temperature, and incubated at 30° for 30 min with DNase I (Sigma), concentration 0.4 μg DNase/ml protoplast suspension.

12. Each protoplast sample is washed twice with 1.5 ml K3 medium[18] containing 0.5 *M* sorbitol in place of sucrose.

13. The samples are transferred to glass fiber filters (Whatman GF/C), and the high-molecular-weight labeled DNA precipitated by washing the filters repeatedly with 10 ml of 10% TCA. The amount of radioactivity in each fraction is determined by liquid scintillation spectrometry.

14. The number of plasmid molecules taken up per protoplast under the selected electroporation conditions is calculated based on the specific activity of the ^{32}P-labeled plasmid DNA molecules. The effects of field

[16] F. A. Krens, L. Molendijk, G. J. Wullems, and R. A. Schilperoort, *Nature (London)* **296,** 72 (1982).

[17] P. W. J. Rigby, M. Dieckmann, C. Rhodes, and P. Berg, *J. Mol. Biol.* **113,** 237 (1977).

[18] K. N. Kao and M. R. Michayluk, *Planta* **126,** 105 (1975).

strength and number of electrical pulses on the uptake of plasmid pUCD9k3 DNA in a typical experiment are presented in Table II.

K3 Carrot Tissue Culture Medium[18]

Solutions	Concentration	Method of sterilization
	(g/liter)	
I. $NaH_2PO_4 \cdot 1H_2O$	1.5	Filter (0.22-μm-pore diameter)
$CaCl_2 \cdot 2H_2O$	9.0	
KNO_3	25.0	
NH_4NO_3	2.5	
$(NH_4)_2SO_4$	1.3	
	(g/100 ml)	
II. $FeSO_4 \cdot 7H_2O$	0.557	Filter, store at 4°
Na_2EDTA	0.745	
III. $CuSO_4 \cdot 7H_2O$	0.00125	Autoclave
$CoCl_2 \cdot 6H_2O$	0.00125	
IV. KI	0.0075	Autoclave
$MnSO_4 \cdot 7H_2O$	0.100	
$ZnSO_4 \cdot 7H_2O$	0.020	
H_3BO_3	0.030	
$Na_2MoO_4 \cdot 2H_2O$	0.0025	
$MgSO_4 \cdot 7H_2O$	2.50	
V. Thiamin–HCl	0.100	Filter, store at 4°
Nicotinic acid	0.010	
Pyridoxine–HCl	0.010	
myo-Inositol	1.000	
Xylose	2.5	

For 1.0 liter of K3 medium, mix stock solutions I through V, as indicated below, adjust medium to pH 5.8, add phytohormones, filter sterilize, and add to previously autoclaved water and sucrose:

Solution	Concentration
I	100.0 ml/liter
II	5.0 ml/liter
III	2.0 ml/liter
IV	10.0 ml/liter
V	10.0 ml/liter
NAA	1.0 mg/liter
KT	0.2 mg/liter
0.4 *M* sucrose	137.0 g/liter or 72.0 g/liter glucose (0.4 *M*)
H_2O	873.0 ml

TABLE II
EFFECT OF ELECTRICAL FIELD STRENGTH AND NUMBER OF PULSES ON THE UPTAKE OF pUCD9k3 DNA BY CARROT PROTOPLASTS[a]

Field strength (kV/cm)	Number of pulses[b] (6 μsec)	Number of pUCD9k3 DNA molecules incorporated/protoplast
A. 0	0	1.6
0.5	6	9.5
1.0	6	3.6
1.5	6	2.2
2.0	6	2.3
4.0	6	5.4
B. 0	0	1.6
2	6	2.3
2	12	2.6
2	30	4.5
2	60	3.2

[a] Each data point represents the average of 10 electroporation events.
[b] Time between individual pulses = 0.2 sec.

Uptake and Expression of TMV-RNA in Carrot Protoplasts

Up to 80% of *Nicotiana tabacum* cv. xanthi protoplasts can be transfected with TMV-RNA by electroporation in an appropriate medium.[8] We describe here a procedure for detecting the electrotransfection of carrot protoplasts with TMV-RNA. Until recently,[9] carrot cells were considered to be resistant to TMV infection.

Experimental Protocol

1. Carrot protoplasts are prepared exactly as described in steps 1–8 of the general protocol for DNA electroporation experiments.

2. For TMV-RNA electroporation experiments the carrot protoplasts are resuspended at a concentration of 1×10^7 protoplasts/ml in a nonionic electroporation buffer, and mixed with TMV-RNA to a final concentration of 10 μg TMV-RNA/ml.

EP buffer (pH 6.5)
0.5 *M* mannitol
10.0 m*M* MES
5.0 m*M* $CaCl_2$

3. Aliquots (15 μl) of the protoplasts preparation are transferred to an ice-cold electroporation chamber and subjected to square wave electrical pulses of selected voltage and duration. The optimum electroporation conditions for TMV-RNA transfection of carrot protoplasts are eight pulses of 90-μsec duration at a field strength of 5 kV/cm. Appropriate control samples are included, e.g., carrot protoplasts electroporated in the absence of TMV-RNA and protoplasts incubated with TMV-RNA without electroporation.

4. After electroporation, the protoplasts are transferred to an ice-cold 1.5-ml volume Eppendorf tube and incubated at 0° for 10 min. The protoplast preparation is centrifuged at 50 *g*, for 5 min, and the pellet washed once with 1.5 ml of EP buffer.

5. After centrifugation, the protoplast pellet is resuspended in 1.5 ml of K3 medium containing 0.4 *M* sucrose and transferred to a sterile 6-cm plastic culture dish or multiwell culture plate. The dish is sealed with Parafilm and incubated at 26° subdued light conditions for 24 hr.

Immunological Detection of TMV-RNA Transfected Carrot Protoplasts[19]

1. After the 24-hr incubation, the electrotransfected and control protoplast cultures with TMV-RNA without electroporation are transferred to Eppendorf tubes, mixed 1 : 1 with phosphate-buffered saline (PBS), and pelleted at 100 *g* for 4.0 min at room temperature.

PBS	g/liter
NaCl	8.5
KH_2PO_4	0.455
$Na_2HPO_4 \cdot 7H_2O$	1.85

2. The supernatant is removed, the protoplast pellet resuspended in 1.5 ml ice-cold 95% ethanol, and the preparation incubated for 15 min at room temperature.

3. The fixed protoplasts are centrifuged as previously described, resuspended in 1.5 ml of PBS, and incubated for 10 min at room temperature.

4. The protoplast preparation is centrifuged for 4 min at 50 *g* and the pellet resuspended in 15 μl of TMV coat protein antiserum diluted 1 : 10 with PBS. The preparation is incubated for 30 min at room temperature,

[19] M. Sulzinski and M. Zaitlin, *Virology* **121,** 12 (1982).

diluted with PBS to 1.5 ml, and pelleted at 100 *g* for 5 min at room temperature.

5. The protoplasts are resuspended in 15 μl of FITC-conjugated goat anti-rabbit IgG, diluted 1 : 20 with PBS, and incubated for 30 min at room temperature in a dark, humid box.

6. The preparation is diluted up to 1.5 ml of PBS, centrifuged as before, and the protoplast pellet resuspended in 30 μl of glycerol–PBS (9 : 1).

7. The stained protoplast preparations (10 μl) are transferred to glass slides and observed at a magnification of 100–400× under a fluorescence microscope equipped with appropriate FITC excitation (450–490 nm) and barrier (520 nm) filters.

Methods to Increase Specificity and Sensitivity of the Immunological Reaction

Method A

To reduce nonspecific binding of the FITC-conjugated goat anti-rabbit IgG to the carrot protoplasts, the protoplasts are preadsorbed with the FITC-conjugated immunoglobulin according to the following procedure:

1. *Daucus carota* WOO1C suspension culture cells (2.0 ml) are centrifuged at 100 *g* for 5 min. The cell pellet is resuspended in 95% ethanol at room temperature for 2 min.

2. The cells are centrifuged at 100 *g* for 15 min at room temperature. The ethanol is removed and the fixed cells are resuspended in 0–6.0 ml PBS.

3. The cells are transferred into two sterile centrifuge tubes, 4.0 ml of cells in one tube and 2.0 ml in the other. The cells are pelleted by centrifugation at 100 *g* for 5 min.

4. The supernatant is removed from the cells and the FITC-conjugated goat anti-rabbit IgG is added to the tube containing 4.0 ml fixed carrot cells and the preparation incubated for 30 min at 37°.

5. The mixture is centrifuged and the preadsorbed antiserum supernatant is transferred to the second tube and incubated for 15–20 min at 37°.

6. The cells are pelleted by centrifugation and the preadsorbed FITC-conjugated goat ant-rabbit IgG is transferred into sterile Eppendorf tubes (50 μl/tube) and stored at −20°.

7. Immediately prior to immunoassay, the FITC-conjugated IgG is diluted 1 : 5, 1 : 10, and 1 : 20 with PBS to determine the optimal dilution for fluorescent staining, e.g., high-intensity staining of transfected protoplasts with low background fluorescence.

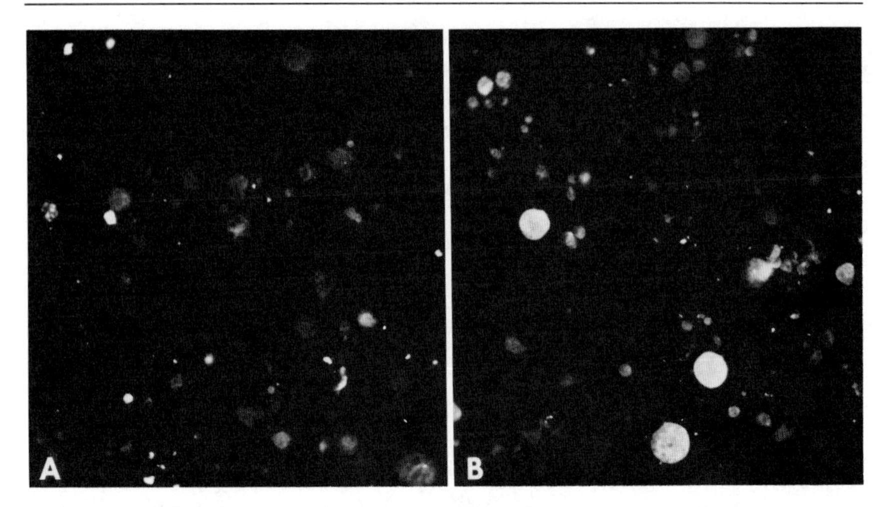

FIG. 2. FITC-antibody staining of carrot WOO1C protoplasts transfected with TMV-RNA. (A) Carrot protoplasts electroporated in the absence of TMV-RNA and photographed in the fluorescence microscope 24 hr after electroporation; background fluorescence is detected in the protoplasts. (B) Carrot protoplasts electroporated in the presence of 10 μg/ml TMV-RNA and prepared for fluorescence microscopy as described above. In this photograph, three cells clearly display the distinct fluorescence characteristic of TMV infection. $\times 100$.

Method B

In the event of persistent high background fluorescence, preadsorption of the TMV antiserum is accomplished by mixing the viral antiserum with *D. carota* WOO1C suspension culture cells as described for the FITC-conjugated goat anti-rabbit IgG (Method A, steps 1–6). After the final centrifugation (Step A, 6), the preadsorbed antiserum is diluted 1 : 2 with PBS immediately prior to use.

Following the procedures outlined above, approximately 5–10% of the carrot protoplasts are found to contain TMV coat protein (Fig. 2).

Comments

Factors Which Influence the Frequency of Electrotransformation

In addition to using protoplasts in good physiological condition for optimum levels of transformation, the components of the electroporation

[20] I. Potrykus, M. W. Saul, J. Petruska, J. Paszkowski, and R. D. Shillito, *Mol. Gen. Genet.* **199,** 183 (1985).

medium can affect the frequency of transformants obtained after electroporation.[20] DNA uptake in mammalian cells is thought to be preceded by DNA binding to the cell membrane.[13] As previously described for yeast and mammalian cells,[21,22] polycations such as poly-L-ornithine and polyvinyl alcohol may improve the binding of DNA to the protoplast membrane. Formation of a calcium phosphate-plasmid DNA coprecipitate as described for the transformation of animal cells[23] may also increase the frequency of transformants via binding of the Ca^{2+} DNA complex to the protoplast membrane prior to DNA uptake, or by protection of the transforming DNA from nuclease degradation.[23] Thus, an electroporation medium which enhances binding of the foreign DNA to the protoplast membrane should be used. Polyethylene glycol (PEG), known to facilitate membrane fusion, was omitted from our electroporation medium.[6] However, Potrykus *et al.*[8] have reported increased levels of kanamycin-resistant tobacco transformants when PEG was included in the medium up to 8.0% (w/v).

The results of our electroporation experiments indicate that $MgCl_2$ is essential for electric pulse-mediated transformation of carrot protoplasts. When $MgCl_2$ was omitted from the electroporation medium or when Ca^{2+} was substituted for $MgCl_2$, almost no teratoma-like somatic embryos were obtained after electroporation with pTiC58 DNA.[6] Electric pulse-mediate DNA uptake experiments conducted by Neumann *et al.*[4] indicate that $MgCl_2$ may inhibit stable transformation of mouse L cells by plasmid DNA molecules containing the HSV TK DNA. Thus, the role of Mg in DNA uptake into plant protoplasts is uncertain.

The highest efficiency of transfection was obtained when tobacco protoplasts were electroporated with TMV-RNA in 0.5 *M* mannitol.[9] Thus, nonionic electroporation buffers can be used to obtain high-frequency transformation of plant protoplasts.

Physical factors, such as increases in electroporation field strength and number of electrical pulses, substantially increase the amount of exogenous DNA taken up by carrot protoplasts (Table II). The temperature of the electroporation medium may be a critical factor for electrotransformation of plant cells. Potrykus *et al.*[8] found that incubation of protoplasts for 5 min at 45° prior to electroporation gave an increased efficiency of transformation of tobacco protoplasts with a chimeric antibiotic resistance marker gene. Successful electrotransformation experiments conducted in our laboratory were performed at 4°.[6] In addition, the

[21] A. Hinnen, J. B. Hicks, and G. R. Fink, *Proc. Natl. Acad. Sci. U.S.A.* **75,** 1929 (1978).
[22] S. Kawai and M. Nishizawa, *Mol. Cell. Biol.* **4,** 1172 (1984).
[23] F. L. Graham and A. J. Van der Eb, *Virology* **52,** 456 (1973).

electroporated protoplasts were held on ice for 10 min after electroporation to prolong the time during which pores generated in the membrane remained open.[6]

Results from immunoglobulin gene electrotransformation experiments in murine cells indicate that increased numbers of transformants may be obtained if the nuclear envelope is removed by colchicine treatment prior to electrotransformation.[5] Significant increases in animal cell transformation are obtained if the cells are electroporated with linear rather than circular DNA.[5] The presence of calf thymus carrier DNA up to 50 μg/ml in the electroporation buffer substantially increased the number of transformants obtained after electroporation of tobacco protoplasts in the presence of a chimeric antibiotic resistance marker gene.[24]

In summary, the successful introduction of foreign nucleic acids into plant cells mediated by high-voltage electrical pulses depends on empirical determination of several critical factors, which include the size and origin of the cells (plant or tissue culture) to be electroporated, the type of electrical field used for electroporation, ionic strength of the electroporation medium, field strength, number and duration of the electrical pulses, the temperature at which electroporation is conducted, the molecular weight and single- or double-stranded nature of the nucleic acid to be electroporated, and the amount of carrier DNA used during electroporation. The labeled DNA uptake experiments described provide a guideline for establishment of optimal conditions for stable electrotransformation of animal cells and plant protoplasts. The RNA electroporation experiments described provide optimal conditions for RNA uptake and transient expression from DNA transcripts in the host cell. Preliminary experiments in our laboratory to introduce labeled plasmid DNA into chloroplasts indicate that, in addition to mediating DNA uptake into protoplasts, electroporation may also be used to introduce nucleic acids directly into isolated subcellular organelles. Thus, electroporation may now be applied for the genetic modification of genes involved in photosynthesis which are transcribed in the chloroplasts of economically important crop plants.

Acknowledgment

The authors wish to express their appreciation to Ms. Deborah Bridwell for secretarial assistance. This work was supported by NSF Grant #PCM-84-10753.

[24] R. D. Shillito, M. W. Saul, J. Paszkowski, M. Muller, and I. Potrykus, *Bio/Technology* **3**, 1099 (1985).

[21] Electroporation of DNA and RNA into Plant Protoplasts

By MICHAEL FROMM, JUDY CALLIS, LOVERINE P. TAYLOR, and VIRGINIA WALBOT

Direct DNA transfer techniques have been developed for plant transformation in response to the host-range limitations of *Agrobacterium tumefaciens*-mediated DNA transfer. One of the most useful and general of these techniques is electroporation. Electroporation refers to the process of applying a high-intensity electric field to reversibly permeabilize bilipid membranes.[1] Electroporation of cell membranes has proved useful for cell fusions as well as gene transfer into plant[2,3] and animal[4] cells. The technique may be applicable to all cell types since all biomembranes can be reversibly permeabilized by high-intensity electric fields.[1] In this chapter we describe the procedures for electrically transferring DNA and RNA into monocot and dicot plant protoplasts. The electrical fundamentals and equipment design are presented first, followed by the detailed protocols and parameters that affect electroporation-mediated gene transfer. Many of the features described for plant cells are relevant to electroporation of animal cells.

Electrical Theory and Apparatus

Most electroporation machines use capacitors (see below) to store electrical energy for later delivery to the electroporation cuvette. This is true even when a power supply is used directly to deliver the electric pulse.[2,4] The internal capacitors and resistors, as well as complex control circuits, determine the nature of the electric pulse delivered directly from a power supply. In order to have an electric pulse with easily controlled parameters, we have used a simple capacitor charge/discharge circuit instead of a power supply to generate the electric pulses used for electroporation. After a brief description of the electrical properties of capacitor circuits, we describe the type of circuit used in the electroporation device and the purpose of the components. The voltages and currents handled by the electroporation device are dangerous and must be handled with pre-

[1] U. Zimmermann and J. Vienken, *J. Membr. Biol.* **67,** 165 (1982).
[2] M. Fromm, L. P. Taylor, and V. Walbot, *Proc. Natl. Acad. Sci. U.S.A.* **82,** 5824 (1985).
[3] R. D. Shillito, M. W. Saul, J. Paskowski, M. Muller, and I. Potrykus, *Bio/Technology* **3,** 1099 (1985).
[4] H. Potter, L. Weir, and P. Leder, *Proc. Natl. Acad. Sci. U.S.A.* **81,** 7161 (1984).

cautions appropriate for high-voltage electrical equipment. We strongly recommend that the electroporation device be constructed by someone familiar with electrical equipment construction or purchased commercially. As a guideline to price, the cost of components is $100 to $200 and the assembled apparatus will probably be about twice that amount. Commercial machines cost considerably more.

Capacitors and RC Circuits

Capacitors are electrical devices for storing electrical charge. Capacitors differ from batteries in that capacitors store charge through the use of electric fields while batteries store charge through the use of chemical reactions. The three important parameters of a capacitor are (1) the capacitance, typically in microfarads (μF), (2) the maximum safe working voltage of the capacitor, and (3) the polarity of the capacitor (many nonelectrolytic capacitors do not have a polarity).

The capacitance, C, of a capacitor is defined by the amount of electrical charge q that it contains at voltage V, according to the equation $q = CV$. The energy stored in a capacitor at a voltage V is $\frac{1}{2}CV^2$. Large-capacitance capacitors hold more charge and hence more energy than smaller capacitors when both are charged to the same voltage. The capacitance of identical-size capacitors is additive when placed in parallel and is divided by the number placed in series. The insulating material of a capacitor determines the maximum voltage it can be charged with. Voltages greater than the voltage rating of the capacitor destroy the insulating material of the capacitor. The large capacitors used in electroporation devices often employ an electrolytic insulating material which has a voltage polarity-dependent breakdown voltage. Therefore electrolytic capacitors are marked to ensure that they are charged with the correct polarity. The + pole of the capacitor should be connected to the + side of the circuit. The + side of a circuit is defined as the direction toward which negatively charged particles such as electrons move; DNA migrates toward the + side of an electrophoresis device.

The electric pulse used in electroporation is generated when the charged capacitor is suddenly discharged through the electroporation cuvette containing the cells and DNA. The ionic composition of the solution in the electroporation cuvette determines the electrical resistance, R, of this capacitor discharge circuit. A simple capacitor discharge circuit is called an RC circuit and the voltage of the applied pulse decreases exponentially with time according to the equation $v(t) = V_0 e^{-t/RC}$. V_0 is the voltage of the capacitor at time 0, R is the resistance of the circuit, and C is the capacitance of the capacitor. A graph of this equation is shown in

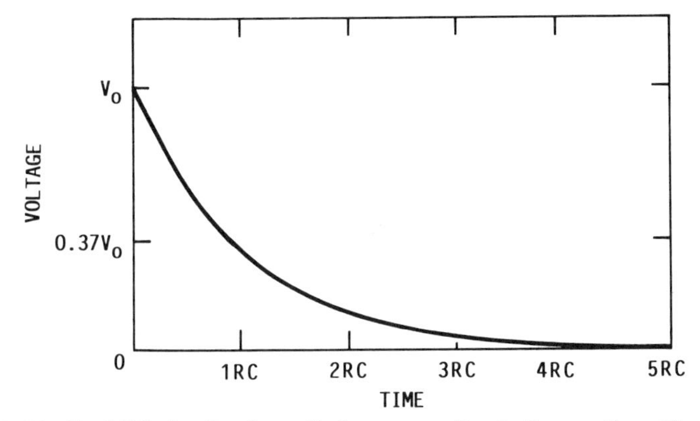

FIG. 1. Idealized RC circuit voltage discharge according to the equation $v(t) = V_0e^{-t/RC}$. V_0 is the initial capacitor voltage at $t = 0$, R is the resistance of the circuit, and C is the capacitance of the capacitor. At $t = RC$ the voltage is 0.37 V_0 and at $t = 5RC$ the voltage is less than 1% of V_0.

Fig. 1. We can see from this equation that when $t = RC$ the initial voltage has decreased by a factor of e^{-1}. RC is called the RC time constant of a capacitor discharge circuit. As either R or C increases, such as reducing the amount of salt in the electroporation buffer or increasing the size of the capacitor, the time constant increases, meaning that the electric pulse lasts longer. Note that by $t = 5RC$ the voltage has decreased to V_0e^{-5} or 1% of the original value (Fig. 1). This can be considered the end of the pulse for practical purposes.

A schematic diagram of an electroporation apparatus is shown in Fig. 2. It uses a power supply to charge the capacitor to the desired voltage and then discharges the capacitor through the electroporation cuvette. The capacitance of the capacitor used and the electrical resistance of the electroporation buffer form an RC circuit which determines the electric pulse length, which has a shape like that shown in Fig. 1.

When switch S_1 is closed (with switch S_2 open, Fig. 2) the power supply charges the capacitor through resistor R_1 to the power supply voltage setting. (The voltage meter reading of the power supply will drop as the capacitor is charged and then return to the original setting when the capacitor is charged.) Switch S_1 is then opened to disconnect the power supply from the capacitor. The capacitor stores the charge until switch S_2 is closed (Fig. 2). S_2 connects the charged capacitor to the electroporation cuvette which has a resistance determined by the buffer composition, typically about 15 Ω. For a 500-μF capacitor discharging through a 15-Ω resistance the RC time constant (the units of R are ohms and C are farads

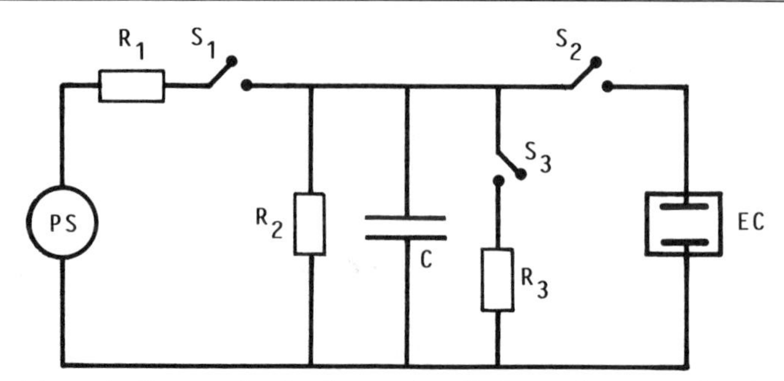

FIG. 2. Schematic diagram of a simple capacitor discharge electroporation circuit. Refer to the text for a description of the circuit operation and purpose of the components. PS, Power supply; R, resistors; S, switches; C, capacitor; EC, electroporation cuvette.

to give an *RC* value in seconds) is 7.5 msec and the pulse length is 37.5 msec (5*RC*).

Electroporation Apparatus Components

A case surrounding the electrical components of the electroporation device is essential to prevent accidental contact with any of the components.

Resistors. R_1 limits the initial current surge from the power supply to the capacitor and forms an RC charge circuit with time constant R_1C.

R_2 is a safety bleed resistor that slowly discharges the capacitor and prevents the capacitor from staying charged after the experiment is finished. This prevents accidental discharge at some later time when the machine is "off."

R_3 provides a path to quickly discharge the capacitor using switch S_3. For instance, if the capacitor is charged and you realize you forgot to attach the leads to the next cuvette, it would be wise to discharge the capacitor through R_3 before switching the leads. R_3 can also be used for trial discharges that are monitored by an oscilloscope.

Switches. The actual switches controlling the charging and discharging of the capacitor must have high voltage (greater than 500 V) and current (greater than 50 A) ratings. In practice small switches on the exterior of the case handle the low voltages and currents that control the large switches inside the case. The internal switches can be either electromechanical relays or solid state switches such as power transistors or silicon-controlled rectifiers.

Capacitors. A useful arrangement is to have several identical capacitors that can be switched from parallel to series arrangements by an external switch. A good working range would be 250, 500, and 1000 μF (the voltage rating of capacitors in series is additive, allowing higher voltages to be used). Capacitors should have at least 450 V dc voltage ratings and will probably be electrolytic and therefore they will have defined polarities that must be followed. A high-voltage diode on the power supply input will prevent damage from accidental reversal of the polarity.

Capacitor Charge Indicator. Either a voltage meter or light-emitting diode (LED) will indicate when the capacitor is charged or discharged. This is useful both as a safety feature and as an indicator that electroporation has occurred (if not monitored by oscilloscope).

Oscilloscope Attachment. An oscilloscope is optional. The capacitor discharges should be identical and the capacitor charge indicator is adequate for most work. But in the event of problems or modifications an oscilloscope is very helpful. A voltage divider and oscilloscope hook-up can easily be added to the electroporation circuit. Millisecond pulses are easily detected on nonstorage oscilloscopes but a storage oscilloscope allows for taking pictures and detailed work.

Power Supply. Since the capacitor circuit determines the pulse length, the sole purpose of the power supply is to charge the capacitor to the desired voltage. Some power supplies have safety circuits that will not function with low-resistance circuits and might require a large R_1 and even adjustment of the R_2 value (Fig. 2). We routinely use Heathkit model IP-17 or an Isco model 494 power supply, but any power supply should work.

Cuvettes. The important parameter with respect to electroporation chambers is the electric field intensity which is the applied voltage divided by the distance in centimeters between the electroporation electrodes. Parallel plate electrodes generate the most uniform electric field but almost any design of electroporation cuvette holding an adequate volume should function. We use 1-ml disposable spectrophotometric cuvettes (Starstadt) as described by Potter *et al.*[4] Either aluminum foil (heavy duty) or platinum foil (Fisher, 0.025 mm) can be used as electrodes in the cuvette. The platinum is nonreactive but expensive. We use aluminum foil for routine work. Either aluminum or platinum foil is epoxied (5-min epoxy glue) to the 1-cm sides of the cuvette such that the electrodes are separated by 0.4 cm. We usually have our cuvettes sterilized with ethylene oxide gas (a service available at most hospitals). The electroporation device leads are connected to the cuvette electrodes with an alligator clip.

Alternatively, reusable, movable stainless steel or platinum electrodes can be used as described by Potter *et al.*[4]

Electroporation Protocol Outline

1. Plant protoplasts are prepared from either plant tissue or suspension culture cells. They are washed once in isolation buffer without enzymes and once in electroporation buffer.

2. The isolated protoplasts are incubated with either DNA or RNA in electroporation buffer and subjected to a high-voltage electric pulse to permeabilize the cell membrane.

3. After 10 min at 0° and 10 min at 22° to reseal cell membranes, the electroporated protoplasts are diluted with protoplast growth media and incubated at 26° in the dark for the desired length of time to allow for expression of the introduced nucleic acid, typically 6 hr for RNA and 48 hr for DNA.

4. The protoplasts are analyzed for expression of the introduced nucleic acid or stable transformants are selected using the expression of the introduced gene as a marker.

Electroporation Parameters

The degree of membrane permeabilization achieved depends primarily on the electric field intensity and duration of the electric pulse perceived by the cells. Higher levels of membrane permeabilization increase the amount of gene transfer but decrease cell viability. The trade-off between cell viability and gene transfer efficiency is roughly as follows: low gene transfer efficiency with no loss of viability; medium gene transfer efficiency with reduced cell division; high gene transfer efficiency with metabolically active but nondividing cells. Too much permeabilization leads to rapid loss of cell metabolic activity and is not useful. Either the electric field intensity or the electric pulse length, or both, can be adjusted for the desired degree of membrane permeabilization. At any fixed voltage an increase in pulse length increases the degree of membrane permeabilization. At any fixed pulse length an increase in the voltage increases the degree of membrane permeabilization. Because of this dependence on two adjustable variables a very wide range of voltages and pulse length combinations can be used. In our hands electric pulses with 5 to 15 msec RC time constants (250- to 1000-μF capacitors with 15-Ω resistance) and electric field intensities of 500 to 1000 V/cm have proved the most efficient for nucleic acid transfer by electroporation.

We have examined square wave electric pulses, which have a constant voltage for the pulse duration, and found that they are also effective for electroporation. For comparison, when the same degree of gene transfer is obtained at a fixed electric field intensity, the square wave pulses will have a much shorter pulse duration than exponentially decreasing pulses. We have used capacitor discharge pulses for most of our electroporations.

The electroporation solution composition is also an important yet flexible parameter affecting electroporation efficiency. Minimally the solution is composed of a nonionic osmotic agent, typically mannitol or sorbitol, a pH buffer, and some salt because protoplast viability decreases when incubated in nonionic media. Any solution containing these components can be used as the electroporation solution but the solution composition will have an effect on the gene transfer efficiency and protoplast viability. One important aspect is that the ionic composition of the solution determines its resistance and hence the RC time constant of the electric pulse. This change can be compensated for by altering the voltage, capacitor size, or by using a resistor in parallel to keep the composite resistance of the circuit the same. The ionic composition of the electroporation solution may also affect a number of microscopic events such as local electric field intensity, salt shielding of charges, and the size of the pores in the permeabilized cell. These and other processes affect the trade-off between gene transfer and loss of cell viability. It has been our experience that while not essential, some salt is beneficial, generally 90 to 140 m*M* NaCl. We have also found 4 m*M* $CaCl_2$ to increase protoplast survival and hence gene transfer.[2] Considerable modification of the ionic composition of the electroporation solution is possible, primarily for the purpose of increasing protoplast viability without decreasing gene transfer.

The pH of the electroporation solution is important as well. Gene transfer efficiency increases with pH over the range from pH 5 to pH 9. The difference between pH 7.2 and 9.0 is only 2-fold and pH 7.2 seemed appropriate since it is approximately the intracellular pH.

Electroporation can be performed at either 0° or room temperature. If the power ($\frac{1}{2}CV^2$) is greater than 20 W then electroporation should probably be performed at 0° to prevent heating. We routinely perform our electroporations with the cuvette at 0° in an ice/water bath.

Increasing levels of gene expression are obtained with increasing DNA concentrations in the electroporation solution, up to about 1 mg/ml. Typically the DNA concentration is 10 to 100 μg/ml. There appears to be little toxic effect from DNA concentrations less than 500 μg/ml. The ability to increase the number of genes transferred at increasing DNA concentrations without any decrease in cell viability is extremely useful. Ecker and Davis[5] have found that increasing amounts of sonicated salmon

[5] J. R. Ecker and R. W. Davis, *Proc. Natl. Acad. Sci. U.S.A.* **83,** 5372 (1986).

sperm DNA increase the signal observed from a constant amount of expressing plasmid DNA. This "carrier effect" is not as effective as increasing the amount of expressing plasmid DNA but can help conserve DNA. Both linear and supercoiled DNA appear to be transferred and expressed at similar levels. We routinely use supercoiled plasmid DNA in 10 m*M* Tris, pH 8, and 1 m*M* EDTA for electroporation. The DNA concentration is adjusted such that less than 0.05 ml is added to each electroporation sample. Plasmid DNA is isolated from *E. coli* by the alkali–SDS lysis[6] protocol and purified by two equilibrium centrifugations in ethidium bromide, cesium chloride gradients.

Protoplast Isolation and Electroporation

We describe the protocol we use to isolate protoplasts from Black Mexican Sweet (BMS) maize suspension culture cells. However, any protocol that yields viable protoplasts should be compatible with gene transfer by electroporation. The best criterion for protoplast viability is cell wall resynthesis and subsequent cell division. Cytoplasmic streaming and fluorescein diacetate[7] are useful as quick indicators of viable cells.

Reagents

BMS media: Murashige and Skoog salts (Gibco), 20 g/liter sucrose, 2 mg/liter (2,4-dichlorophenoxy)acetic acid, 200 mg/liter inositol, 130 mg/liter asparagine, 1.3 mg/liter niacin, 0.25 mg/liter thiamin, 0.25 mg/liter pyridoxine, 0.25 mg/liter calcium pantothenate, final solution adjusted to pH 5.8

Protoplast isolation solution: 7.35 g/liter $CaCl_2 \cdot 2H_2O$, 1 g/liter anhydrous sodium acetate, 45 g/liter mannitol, adjust solution to pH 5.8

Protoplast isolation enzyme solution: To 100 ml of the protoplast isolation solution add 1 g cellulase (Cellulysin, Calbiochem), 0.5 g hemicellulase (Rhozyme, Genencor), 20 mg pectinase (Pectolyase Y23, Seishin Pharmaceutical Co.), 0.5 g bovine serum albumin (Sigma), and 0.05 ml 2-mercaptoethanol. Dissolve the proteins and 2-mercaptoethanol, centrifuge at 10,000 *g* for 10 min to remove insoluble components, and filter sterilize by passage through a 0.45-μm filter (Nalgene). We pressure filter instead of vacuum filter to minimize foaming

Electroporation solution: 20 mg/liter KH_2PO_4, 115 mg/liter $NaHPO_4$,

[6] T. Maniatis, E. F. Fritsch, and J. Sambrook, "Molecular Cloning: A Laboratory Manual." Cold Spring Harbor Lab., Cold Spring Harbor, New York, 1982.

[7] J. M. Widholm, *Stain Technol.* **47,** 189 (1972).

7.5 g/liter NaCl, 36.4 g/liter mannitol. Usually the solution does not require adjustment to have a pH of 7.2. Autoclave the solution and add $CaCl_2$ to 4 m*M* after cooling. Autoclaving in the presence of $CaCl_2$ leads to a calcium phosphate precipitate

Protoplast growth media: To BMS media add 25 ml/liter coconut water (Gibco) and 55 g/liter mannitol. Final volume is 800 ml (but made up as if for 1000 ml) as conditioned BMS media (see below) is added to 20% just prior to culturing the protoplasts

Conditioned BMS media: The supernatant, after centrifugation of the cells in BMS media, is saved and filtered through a 60-μm nylon mesh screen to remove residual cells.

BMS cells (ATCC #54022) are maintained as a finely divided suspension culture in BMS media (reagents). Forty-milliliter cultures in 125-ml Erlenmeyer flasks are shaken at 150 rpm at 26° either in the dark or with a 12-hr light/dark cycle. The culture is diluted with an equal volume of fresh medium every 3 days. Protoplasts are isolated from cells 1 to 2 days after adding fresh medium. It is important that the cells be actively growing at the time of protoplast isolation since low levels of transcriptional activity exists in stationary phase cells. The cells are pelleted at 200 *g* in a swinging bucket table top centrifuge. Approximately 6 ml of packed cell volume is obtained from a 40-ml culture. The supernatant is removed and saved for use as conditioned medium in culturing the protoplasts. Forty milliliters of the enzyme mix (reagents) is added to the cells and 11.5-ml aliquots are placed in 10-cm Petri dishes. The Parafilmed dishes are placed on a rotary platform set at 40 rpm and at 26°. Digestion of the cells proceeds for 2–2.5 hr after which time 50 to 80% of the cells have become protoplasts. The protoplasts are then filtered through a 60-μm mesh nylon screen and collected by centrifugation at 200 *g*. The protoplast pellet is washed once in 60 ml of the same solution without enzymes followed by a second wash in 40 ml of electroporation buffer. The washed protoplast pellet is resuspended in electroporation solution (reagents) at a protoplast density of 4×10^6/ml and is ready for electroporation. Any protoplast density can be used with the amount of gene expression proportional to the number of protoplasts present and assayed.

Electroporation of Protoplasts

Adequate mixing of the DNA with the protoplasts is important. The DNA (generally less than 0.05 ml to avoid diluting the electroporation solution too much) is thoroughly mixed with electroporation solution (without protoplasts present as yet) to give a final volume of 0.5 ml and this is added to 0.5 ml of electroporation solution containing approximately 2 million protoplasts. Both solutions are prechilled prior to mixing and electroporation. Preincubation of prechilled DNA and protoplasts together at 0° for 5 min or 30 sec results in the same level of gene transfer, indicating that interactions prior to the electric pulse do not seem important. The mixed 1-ml solution containing DNA and protoplasts is transferred into a electroporation cuvette sitting in an ice/water bath and an electric pulse of the desired voltage and RC time constant is delivered to the cuvette. We generally try to electroporate the protoplasts before they have time to settle to the bottom of the cuvette.

Postelectroporation incubations are to allow time for the DNA to diffuse into the permeabilized protoplasts and for the protoplast membrane to reseal. The protoplasts are more fragile than normal immediately after electroporation. We allow them to remain at 0° for 10 min and then pipette them into a 10-cm Petri dish and allow them to sit at room temperature for another 10 min before adding 8 ml of protoplast growth medium. Washing the protoplasts after electroporation is optional and will probably result in some losses but might be desirable if you need to remove DNA and electroporation buffer.

Protoplasts are incubated at 26° in the dark for the desired interval. We have found that transient gene expression from DNA templates is maximal between 24 and 48 hr after electroporation and can be detected as early as 3 hr and as late as 2 weeks after electroporation. Gene expression in selected transformants carrying the introduced DNA in the host chromosomes appears to be stable over many months.

Electroporation of RNA

The electroporation-mediated transfer of RNA shows the same parameters as those described for DNA transfer by electroporation. We have observed RNA transfer by electroporation in both plant and animal cells (Callis *et al.*[8]) and it would appear to be a very useful technique for studying plant and animal RNA viruses. One to 5 μg/ml of 5'-capped, 3'-polyadenylated CAT mRNA gives signals comparable to 10 to 20 μg/ml of pCaMVCAT DNA in plants and of pSV2CAT in animal cells. Due to the susceptibility of RNA to RNases we always prechill all solutions and mix the RNA solution with the solution containing the protoplasts immediately before electroporation (elapsed time between mixing and electric pulse delivery is about 20 sec). We have used *in vitro* transcription with SP6 RNA polymerase to generate RNA for transfer. We have found that both a 5'-cap structure and a 3'-polyadenylated tail are required for maximal RNA translation and/or stability *in vivo*. Since *in vitro* preparation of

[8] J. Callis, M. Fromm, and V. Walbot, *Nucleic Acids Res.*, in press.

RNA for electroporation is limited to special interests, the details are described elsewhere (Callis *et al.*[8] and references therein).

Assay of Chloramphenicol Acetyltransferase Activity

The transient expression of introduced genes is monitored by measuring the production of protein or RNA. Enzymatic activity is a convenient and sensitive measurement. We routinely measure the level of gene expression by the amount of CAT enzymatic activity present in extracts of transfected protoplasts (see Gorman *et al.*[9] for the initial characterization of CAT expression in eukaryotic cells; see Shaw[10] for descriptions of bacterial CAT proteins and their properties). Other assayable gene products such as β-galactosidase, neomycin phosphotransferase II, and hygromycin phosphotransferase are also easily measured using the appropriate assay.

The basis of the CAT assay is the shift in chromatographic mobility when the [^{14}C]chloramphenicol substrate reacts with acetyl-coenzyme A in the presence of the enzyme to yield the acetylated forms of [^{14}C]chloramphenicol. Enzymes capable of performing this reaction are present in some plants, therefore it is important to check whether extracts from nontransformed cells have high background levels. We have found this to be a problem in several *Brassica* species, but backgrounds are low in maize, tobacco, and carrot cells. A second potential problem with the CAT assay is the presence of materials in some plant species that inhibit CAT activity. This inhibiting material can often be inactivated by a brief heat treatment. The CAT enzyme is resistant to 60° but loses some activity at 70°.[10] We treat for 10 min at 60°, but we recommend doing reconstruction experiments by mixing CAT protein (either commercially available or from bacterial extracts containing CAT activity) with extracts from nontransformed cells and comparing activities from reactions either without a heat treatment step or with a 10-min heat treatment (try both 60 and 65° treatments).

Reagents

Tris base, 0.25 *M*, pH 7.9

4 m*M* Acetyl-coenzyme A: 1.7 mg/0.5 ml of 0.25 *M* Tris, pH 7.9. This solution is made fresh each time. The lithium salt of acetyl-coenzyme A appears less hygroscopic than the sodium salt. Store the bottle desiccated at −20°

[9] C. M. Gorman, L. F. Moffat, and B. H. Howard, *Mol. Cell. Biol.* **2,** 1044 (1982).

[10] W. V. Shaw, this series, Vol. 43, p. 737.

[^{14}C]Chloramphenicol: 40–60 mCi/mmol (New England Nuclear) stored in ethanol at $-20°$ to retard degradation products that increase reaction backgrounds
Silica gel thin-layer chromatography plates (20 × 20 cm)
Ethyl acetate
Chloroform : methanol, 95 : 5

Procedure

After incubation of the protoplasts (typically 2 million/sample) for the desired length of time, typically 24 to 48 hr after electroporation, the protoplasts and media are poured from the Petri dish into a 15-ml conical screw-cap tube. A rubber policeman can be used to facilitate this. The protoplasts are collected by centrifugation at 200 *g* in a swinging bucket table top centrifuge. The supernatant is removed and 0.4 ml of 0.25 *M* Tris, pH 7.9, is added to the protoplast pellet, typically 0.1 to 0.2 ml. The resuspended protoplasts are transferred to 1.5-ml microfuge tubes and disrupted by sonication for 20 sec using a microtip probe with the sonicator power setting at 75 W. It is important not to have the extract foam during sonication. Inspection under a microscope should reveal broken cells. Cellular debris is removed by centrifugation for 5 min in a microfuge at 4°. The extract supernatant (0.18 ml) is placed in a new 1.5-ml microfuge tube and heated for 10 min at 65°. We recommend checking whether this step is working properly (see above). Denatured proteins should be evident after the heat treatment. These denatured proteins will form an interphase between the aqueous and organic layers at the ethyl acetate extraction step. The tubes are cooled to room temperature and 22.5 μl of a cocktail containing 20 μl of 4 m*M* acetyl-coenzyme A and 0.5 μCi [^{14}C]chloramphenicol is added to each tube. The tubes are incubated in a water bath at 37° for 0.5 to 3 hr, depending on the amount of CAT activity present. The reaction is stopped by removing the tubes from the water bath and adding 0.6 ml of ethyl acetate per tube. The tubes are shaken vigorously for 1 min and then spun in a microfuge for 2 min to separate the phases. The upper ethyl acetate phase now contains the acetylated and nonacetylated forms of [^{14}C]chloramphenicol and is transferred to a fresh microfuge tube. The lower aqueous phase is discarded. The ethyl acetate is evaporated either by 30 min of vacuum centrifugation or by overnight evaporation in a hood. Ethyl acetate (10 μl) is added to dissolve the slight residue in the tube and spotted onto a silica gel chromatography plate by repeated applications of 1-μl aliquots. The samples are chromatographed by a 95 : 5 chloroform : methanol solvent until the solvent has migrated approximately 13 cm (this takes about 40 min). The chromatographed silica gel is then autoradiographed with the film adjacent to the silica gel (no Saran Wrap). The spots can be cut out and counted or compared to spots containing known amounts of [^{14}C]chloramphenicol. (Chromatograph various dilutions of unreacted label for standards. These do not need to be on the same silica gel but should be placed under the same film as the experiment.)

Gene Expression in Electroporated Protoplasts

Figure 3 shows an autoradiograph of a CAT assay of carrot protoplasts electroporated with CAT plasmids. The electric pulse was 220 V/cm and with an RC time constant of 16 msec and gene expression was assayed 48 hr after electroporation. We choose this example of early work[2] because it contains a number of important controls that should be performed when initially characterizing a system. Lane g shows a positive control enzymatic reaction with an extract from an *Escherichia coli* strain carrying the CAT gene of pBR325 and shows the migration positions of the acetylated forms of [^{14}C]chloramphenicol (CAM) and serves as a control for each substrate cocktail. Lane a shows the migration of unreacted label; this particular batch of [^{14}C]chloramphenicol has an impurity that moves just ahead of the chloramphenicol. This impurity does not interfere with the assay and is not present in many lots of [^{14}C]chloramphenicol. The unreacted control lane also serves to check the level of impurities that comigrate with the acetylated forms of chloramphenicol. These often increase with storage of [^{14}C]chloramphenicol, especially at 4° in water.

Electroporation of carrot protoplasts without CAT DNA present does not produce any CAT activity (Fig. 3, lane c), indicating that carrot protoplasts contain little endogenous CAT activity and that none is induced by an electric pulse. Incubation of pNOSCAT plasmid DNA with protoplasts without an electric pulse does not produce CAT activity (lane d), indicating an electric pulse is required to introduce DNA into protoplasts. Electroporation of protoplasts in the presence of pNOSCAT or pCaMVCAT DNA, genes proved to express in carrot cells by *Agrobacterium*-mediated DNA transfer, produces CAT activity (lanes e and f). Results similar to these have been obtained with tobacco and maize protoplasts.[2]

Direct enzymatic assay of the pNOSCAT DNA stock solution shows that no contaminating CAT activity is present in the DNA solution (Fig. 3, lane b). Another control we always perform is to spot 25-μl aliquots of the protoplasts and medium on nutrient broth plates at the time we harvest the protoplasts for analysis. This simple assay checks for any microbial contamination that might contain CAT activity or have inhibited protoplast metabolic activity.

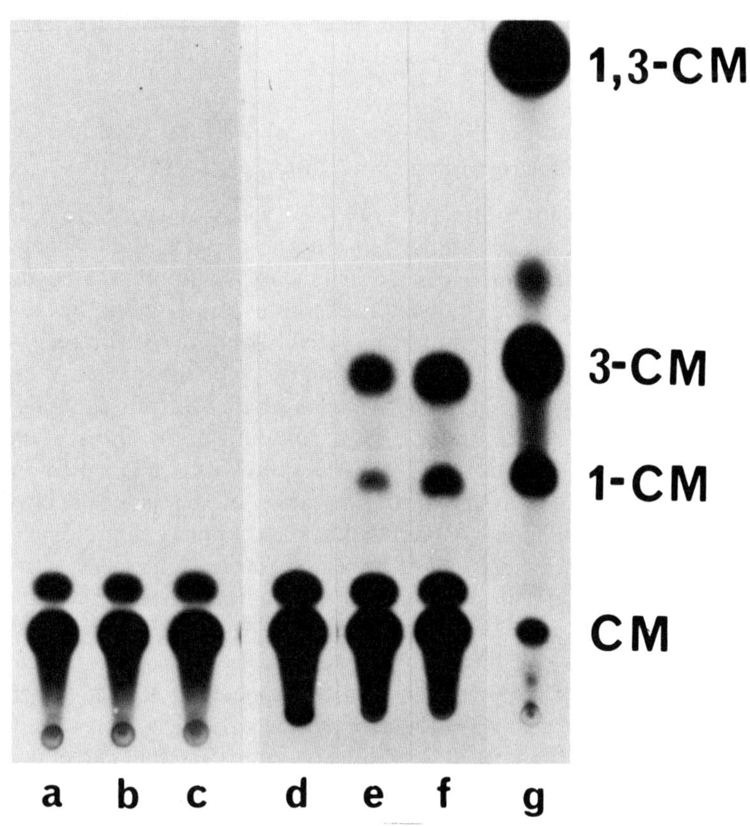

FIG. 3. CAT gene expression after DNA transfer by electroporation into carrot protoplasts. The CAT activities of carrot protoplasts electroporated with a 220 V/cm electric pulse with a 16-msec RC time constant was measured 48 hr after electroporation as described in the text. An autoradiograph of the chromatographed enzymatic products is shown. The positions of unreacted [^{14}C]chloramphenicol (CAM) as well as 1-acetylchloramphenicol, 3-acetylchloramphenicol, and 1,3-acetylchloramphenicol are indicated. Lanes: a, unreacted chloramphenicol; b, 10 μg of pNOSCAT DNA was added directly to the enzyme assay; c, an electric pulse was delivered to protoplasts without plasmid DNA present; d, protoplasts with 10 μg/ml of pNOSCAT plasmid DNA present but without an electric pulse delivered; e, an electric pulse was applied to protoplasts with 10 μg/ml pNOSCAT DNA present; f, an electric pulse was applied to protoplasts with 10 μg/ml of pCaMVCAT DNA present; g, enzymatic products from an extract of an *E. coli* strain containing pBR325, which carries a CAT gene. Reprinted from Fromm *et al.*[2]

Discussion

We have found electroporation to be a convenient, reproducible, and efficient method for introducing nucleic acids into plant protoplasts. By assaying for β-galactosidase expression *in situ* after electroporation of tobacco protoplasts with a β-galactosidase vector (unpublished data), we estimate that the number of protoplasts taking up and transiently expressing genes transferred by electroporation can be greater than 10% of the total protoplasts. In general the level of transient expression in the protoplast population is comparable to the levels from an equivalent number of cells containing one to several copies of the same gene integrated into the cell's chromosomes. Thus, electroporation-mediated transient assays allow rapid characterization of many aspects of transcription, RNA processing and translation, and protein stability and activity. However, it is important to verify that the protoplasts transcribe the particular gene of interest since protoplasts may transcribe a different set of genes than the cells they were derived from.

When difficulties arise in detecting gene expression with a transient or stable transformation system, whether by electroporation or any other transfer method, the problem can be in either the transfer technique or the lack of expression of the transforming DNA. Deciphering which category the problem fits into is halfway to the solution.

We recommend the following steps to solve electroporation/expression problems or as an exercise to gain familiarity with electroporation.

1. Use a cuvette with electroporation buffer as a test of the electroporation machine and use an oscilloscope to monitor the electric pulses delivered to the cuvette at several different voltage and pulse length settings.
2. Perform reconstruction assays of mixing the desired gene product, such as CAT protein, with extracts from nontransformed cells to determine that the assay is working and not inhibited by materials in the plant extract.
3. Set up a test system using well-behaved protoplasts known to express a gene such as pNOSCAT or pCAMVCAT in carrot or tobacco protoplasts.[2] Successful results show that your protoplast isolation, electroporation, and incubation techniques are functional and the assay is sensitive enough to detect gene expression.
4. Perform an electroporation on your experimental protoplasts using your gene and a proved gene, if available, as a positive control. Vary the electric field intensity from 250 V/cm to about 1250 V/cm using pulse lengths with RC time constants in the 5- to 15-msec range. The highest

voltage, longest pulse of your series should noticably affect protoplast viability.

5. In the event of difficulties and if a positive control gene is not available, the transfer of a mRNA of the gene of interest can help localize the problem. If successful, this experiment provides data on electroporation transfer, mRNA translation, and protein detection sensitivity in the cells of interest. We initially could not detect any gene expression from electroporated DNA in maize protoplasts and the transfer of 5′-capped, 3′-polyadenylated CAT mRNA demonstrated that electroporation of nucleic acids was occurring in maize protoplasts. This result indicated that the problem was probably due to poor quality protoplasts and weak promoters. However, there is an additional uncertainty not answered by a successful RNA transfer experiment. The mechanism by which DNA, after transfer through the cell membrane, is transported to the cell nucleus is unknown and transfer of RNA to the cytoplasm does not prove DNA transfer to the nucleus is occurring. Keeping this uncertainty in mind, electroporation-mediated gene transfer has been successful in the cells of enough plant and animal species to indicate that gene transfer to the nucleus will occur in most if not all higher eukaryotic cells.

6. Transient gene expression appears to be a useful indicator of gene expression in transformed callus. This allows rapid characterization of promoters and marker gene expression prior to undertaking transformation experiments. Our characterization of the cauliflower mosaic virus 35 S promoter as a functional promoter in maize transient assays lead to its use in obtaining transformed maize callus resistant to kanamycin.[11]

[11] M. Fromm, L. P. Taylor, and V. Walbot, *Nature (London)* **319,** 791 (1986).

[22] Cloning Vectors of Mitochondrial Origin for Eukaryotes

By CHRISTINE LANG-HINRICHS and ULF STAHL

Introduction

Although molecular cloning in prokaryotes has achieved great success, cloning in eukaryotes is limited to only a few species. One of the main reasons for this difference is attributable to the lack of appropriate DNA species for the construction of eukaryotic cloning vehicles. Plasmids which are commonly used for this purpose are widely distributed among bacteria but seem to be the exception among eukaryotes. One of the few examples of eukaryotic cloning vehicles based on plasmid DNA is the 2-μm DNA of the baker's yeast *Saccharomyces cerevisiae*, which has been successfully used to establish an efficient, autonomously replicating cloning vector.[1] Comparable vector/host systems based on endogenous plasmids have also been developed in two other yeasts, *Schizosaccharomyces pombe* and *Kluyveromyces lactis*.[2,3] If the organism to be manipulated does not harbor a suitable plasmid, it is possible to construct a *nonreplicative* or *integrative* cloning vector, containing a marker gene for selection. Because it lacks an origin of replication, such a vector is not able to replicate autonomously and must be integrated into the genome of the host to be inherited. Nonreplicative vectors show a decreased transformation rate and copy number per haploid genome compared to a replicative vector and are difficult to reisolate. However, their stability during cell divisions of the host is very high. Examples of this approach include *S. cerevisiae* and filamentous fungi such as *Aspergillus* and *Neurospora*,[4–8] and also virus-based vectors for mammalian cells.

To establish a *replicative* vector, on the other hand, one can use chromosomal DNA for the induction of replication. The location and isolation of an appropriate DNA sequence with the ability to replicate autonomously (autonomously replicating sequence, or ars) requires an efficient test system. At present *S. cerevisiae* is usually employed as host to identify ars's of homologous and heterologous origin.

Besides using chromosomal DNA as a source of ars's in organisms without plasmids, replicons may be obtained from organelles, especially mitochondrial (mt) DNA sequences. mtDNA is ubiquitous in eukaryotes, but it is less complex than chromosomal DNA and thus easier to analyze. For example, phenotypically identified sequences can be simply localized on the physical or genetic map. The principal availability of mt ars's for the construction of vectors was demonstrated in *S. cerevisiae*, where

[1] J. D. Beggs, *Nature (London)* **275,** 104 (1978).

[2] P. Fournier, C. Gaillardin, L. de Louvencourt, H. Heslot, B. F. Lang, and F. Kaudewitz, *Curr. Genet.* **6,** 31 (1982).

[3] L. de Louvencourt, H. Fukuhara, H. Heslot, and M. Wesolowski, *J. Bacteriol.* **154,** 737 (1983).

[4] A. Hinnen, J. B. Hicks, and G. R. Fink, *Proc. Natl. Acad. Sci. U.S.A.* **75,** 1929 (1978).

[5] K. Wenars, T. Goosen, L. M. J. Wennekes, J. Visser, C. J. Bos, H. W. J. van den Broek, R. F. M. van Gorcom, C. A. M. J. J. van den Hondel, and P. H. Pouwels, *Curr. Genet.* **9,** 361 (1985).

[6] S. S. Dhawale and G. A. Marzluf, *Curr. Genet.* **10,** 205 (1985).

[7] B. L. Miller, K. Y. Miller, and W. E. Timberlake, *Mol. Cell. Biol.* **5,** 1714 (1985).

[8] J. M. Kelly and M. J. Hynes, *EMBO J.* **4,** 475 (1985).

random or defined mtDNA fragments exhibiting replication activities were detected.[9,10] The strategy described in the following sections will bring about a more general view of the usefulness of "mitochondrial vectors." According to the examples given (*S. cerevisiae, Schizosaccharomyces pombe, Penicillium chrysogenum*), replicating cloning vectors based on the host's mitochondrial genome can be constructed for different eukaryotes and may open up an alternative approach for the development of replicating vectors for any eukaryotic host.

Principle of the Method

The mtDNA to be used for the construction of a cloning vehicle is isolated and, if possible, physically mapped. The restriction fragments of the mtDNA are *randomly* joined to a common *E. coli* plasmid and cloned in this host for amplification. Each of the hybrid molecules formed carrying a different fragment of mtDNA is then combined with a selection marker functional in the eukaryotic host. One can use either a resistance gene (e.g., resistance for gentamycin or copper) or a gene complementing a nutritional deficiency (auxotrophy). The replication ability of the whole molecule is subsequently checked by transfer into the host (which donated the mtDNA) and analysis of the transformants for "plasmid" content.

In contrast to identifying replicons among random pieces of chromosomal DNA, the identification of mtDNA requires no special selection scheme, since the mtDNA fragments may be checked individually by size, using electrophoresis. Thus the probability of finding a replicating sequence seems to be much higher than among random chromosomal fragments.

In addition to the approach of identifying replicating sequences directly in the homologous host, it may sometimes be advantageous or less laborious to preselect replicons in an intermediate or heterologous host like *S. cerevisiae* for which well-tested vector/host systems are available. In this connection it should be emphasized that screening for ars regions following the outlined strategy may simply lead to *S. cerevisiae*-specific replicons which perhaps are not functional in the donor of the mtDNA.

[9] H. Blanc and B. Dujon, *in* "Mitochondrial Genes" (P. Slonimski, P. Borst, and G. Attardi, eds.), p. 279. Cold Spring Harbor Lab., Cold Spring Harbor, New York, 1982.

[10] B. C. Hyman, J. Harris Cramer, and R. H. Rownd, *Proc. Natl. Acad. Sci. U.S.A.* **79,** 1578 (1982).

Materials and Reagents

Isolation of DNA

Plasmid DNA was isolated from *Escherichia coli* by alkaline lysis[11] and purified by CsCl EtBr gradient centrifugation.

Cloning Procedures

Methods for *in vitro* recombination used for the construction of DNA vectors were adapted from Maniatis *et al.*[12]

Enzymes

Restriction endonucleases, T4 DNA ligase, and alkaline phosphatase were purchased from Boehringer–Mannheim, S1 nuclease was from Pharmacia, and protoplasting enzyme mix Novozym 234 was from Novo Enzymes (Copenhagen, Denmark).

Strains

E. coli K-12, strain SF8 (*recBC lop*11 *tonA*1 *thr*1 *leuB*6 *thi*1 *lacY*1 *supE*44 hsm^- hsr^-), kindly furnished by R. Eichenlaub (Bielefeld), was used for all plasmid amplifications and for transformation with whole-cell DNA of yeast and fungi transformants.

Saccharomyces cerevisiae strain AH 22[4] was used as a yeast host (a *leu*2-3 *leu*2-112 *his*4-519 *can*1 cir^+). *S. cerevisiae* a-1/1R/Z1 (from G. Bernardi/Paris, France) is a supersuppressive *petite* whose mitochondrial genome consists of 416-bp repeat units carrying a mitochondrial origin of replication (ori1).[13]

Schizosaccharomyces pombe strain *leu*1-32 h^- (from K. Wolf/Munich, West Germany) can be transformed to leucine prototrophy with the *leu*2 gene of *S. cerevisiae* and was used as a host strain. *S. pombe* mtDNA was prepared from strain *ade*7-50 h^- [14] (from K. Wolf).

Penicillium chrysogenum was supplied by Biochemie Kundl/Austria.

[11] H. C. Birnboim and J. Doly, *Nucleic Acids Res.* **7,** 1513 (1979).

[12] T. Maniatis, E. F. Fritsch, and J. Sambrook, "Molecular Cloning: A Laboratory Manual." Cold Spring Harbor Lab., Cold Spring Harbor, New York, 1982.

[13] G. Bernardi, G. Baldacci, G. Faugeron-Fonty, C. Gaillard, R. Goursot, A. Huyard, M. Mangin, R. Marotta, and M. de Zamaroczy, *in* "The Organization and Expression of the Mitochondrial Genome" (A. M. Kroon and C. Saccone, eds.), p. 21. Elsevier, Amsterdam, 1980.

[14] P. Q. Anziano, P. S. Perlman, B. F. Lang, and K. Wolf, *Curr. Genet.* **7,** 273 (1983).

Media

Complete media used were LB (1% Bacto tryptone, 0.5% yeast extract, 0.5% NaCl, pH 7.2) for *E. coli;* YE (0.5% yeast extract, 2% glucose, pH 6.3) for *S. cerevisiae* and *S. pombe;* Czapek-Dox (Difco) for *P. chrysogenum*. Defined medium for *E. coli* was M9[15] plus amino acids as required. Selective medium was YNB (Difco yeast, nitrogen base, with 2% glucose) for *S. cerevisiae* AH22 transformation, and PMM[16] for *S. pombe* transformation.

For *P. chrysogenum* transformation selective medium was supplied with G418 (30 μg/ml).

Procedures

The following section will present a general strategy for constructing mitochondrial vectors for three different organisms (*Saccharomyces cerevisiae, Schizosaccharomyces pombe,* and *Penicillium chrysogenum*). Since the overall scheme apart from minor adaptions is applicable to each of these hosts, a subdivision of the whole procedure according to the successive steps required is adequate. In addition, specific modifications of the general scheme for each organism will be discussed.

Isolation of mtDNA

mtDNA of yeast and fungi is most readily prepared from a crude mitochondria preparation. It is obtained from mechanical disruption of the cells or hyphae followed by an enrichment of the lysate for mitochondria by differential centrifugation. Alternatively, the centrifugational enrichment can be substituted by a chromatographic separation using hydroxylapatite, which not only purifies DNA (from protein and carbohydrates), but also fractionates nucleic acids (separation of RNA and DNA).[17,18] This enables the isolation of pure cell DNA which is subsequently separated into nuclear and mitochondrial components by several cycles of CsCl density gradient centrifugation.

The isolation of mtDNA from crude mitochondria was performed for the yeasts *S. cerevisiae* and *S. pombe* according to the following protocol:

1. Cells are grown for 2 days (1-liter culture) and harvested by centrifugation (10,000 *g*, 10 min), yielding 10–20 g wet weight.

[15] E. H. Anderson, *Proc. Natl. Acad. Sci. U.S.A.* **32,** 120 (1946).
[16] P. Thuriaux, W. D. Heyer, and A. Strauss, *Curr. Genet.* **6,** 13 (1982).
[17] M. Shoyab and A. Sen, this series, Vol. 68, p. 199.
[18] A. Colman, M. J. Byers, S. B. Primrose, and A. Lyons, *Eur. J. Biochem.* **91,** 303 (1978).

2. Cells are suspended in 30 ml of 0.05 *M* Tris–HCl (pH 7.4), 0.01 *M* EDTA, and 0.5 *M* sucrose (TES buffer), cooled in ice, and slowly poured into a cooled mortar filled with liquid nitrogen. The frozen cells are then ground to a sandlike powder; the liquid nitrogen is refilled at intervals to ensure that cells are always covered by liquid.

3. Ground cells are allowed to thaw slowly by stirring them in 1 liter of TES buffer (see above) at 4° for 1 hr.

4. Crude mitochondria are separated from cell debris by centrifugation at 1500 *g* for 10 min (4°); the supernatant (containing mitochondria) is kept on ice, while the cell pellet is resuspended in TES buffer and homogenized using a glass hand homogenizer. Centrifugation and homogenization are repeated twice. Mitochondria are subsequently pelleted from the supernatants by centrifugation at 10,500 *g* for 15 min (4°).

5. Mitochondria are lysed by addition of 14 ml of lysis buffer (0.5 *M* NaCl, 0.05 *M* Tris–HCl, pH 8.0, 0.05 *M* EDTA, 2% SDS). Cesium chloride (14.8 g) is added, and the lysate is incubated at 60° for 30 min to achieve denaturation of proteins.

6. The lysate is cooled and centrifuged at 27,000 *g* for 15 min at 0° to separate denatured proteins and remaining cell debris from the clear CsCl nucleic acid solution.

7. The cleared lysate is then mixed either with ethidium bromide or DAPI (4,6-diamidino-2-phenylindole) at 200 μg/ml, adjusted to a refractive index of 1.390, and centrifuged for 21 hr in a Sorvall TV 865 vertical rotor at 45,000 rpm at 15°.

8. The nucleic acid banding pattern can be monitored under long-wave UV light. mtDNA has a lower buoyant density (due to a lower GC content) than nuclear DNA and thus bands above it.

Isolation of whole-cell DNA and purification by chromatography as outlined above is applicable not only for unicellular but also for hyphal organisms. As shown in the following procedure, it was successfully applied, with some modification, to fungal mtDNA (*Penicillium chrysogenum*):

Steps 1 and 2 are as described above except that cultures were grown in up to 500-ml vol and growth was allowed for 4 days; mycelium is harvested by filtration and soaked with glycerol prior to grinding (10–20 g wet weight).

3. To achieve lysis of cells, ground mycelium is suspended in lysis buffer at 1 ml/g wet weight (4% SDS in 30 m*M* Tris, 5 m*M* EDTA, 50 m*M* NaCl, pH 8.0 buffer) and rapidly thawed by incubation at 60°. The thawed mycelium is thoroughly mixed and incubation at 60° is continued for 1 hr.

4. After pelleting of cell debris (10,000 *g*, 10 min) the supernatant is made 1 *M* in NaCl and kept in an ice bath for at least 4 hr.

5. Precipitated material is removed by centrifugation (16,000 *g*, 20 min). The supernatant is mixed with 40% (w/v) PEG 6000 (in water) to a final concentration of 10% and left on ice for at least 4 to 6 hr.

6. Precipitating material is pelleted (1500 *g*, 20 min) and resuspended in 240 m*M* phosphate buffer, pH 6.8, 8 *M* urea, 0.8% SDS.

7. A hydroxylapatite column is packed according to the instructions of the supplier (Bio-Rad, BioGel HTP DNA grade) and washed with "washing buffer" (8 *M* urea, 240 m*M* phosphate buffer, pH 6.8). The nucleic acid solution is pipetted onto the column and washed through with washing buffer. Urea is eluted with "low buffer" (10 m*M* phosphate buffer, pH 6.8) and the nucleic acids subsequently are eluted with "elution buffer" (300 m*M* phosphate buffer, pH 6.0).

8. The nucleic acid fraction is subjected to CsCl density gradient centrifugation (refractive index: 1.3985; TV 865 rotor, 45,000 rpm, 21 hr, 15°). Depending on the desired purity of the mtDNA, the density gradient centrifugation may have to be repeated once or twice.

Using either of these procedures, mtDNA, essentially free from other cellular DNA species, has been isolated.

Cloning of mtDNA Fragments

To obtain a (more or less) complete mtDNA bank (gene bank) of a prospective host organism, different restriction endonucleases are used to digest mtDNA. For cloning mtDNA fragments an enzyme is chosen which yields fragments of readily clonable size. Fragment sizes of 5 to 10 kb are most useful since they are small enough to be cloned in common *E. coli* vectors, yet large enough to cover the whole mitochondrial genome with up to 20 cloning events (20 hybrid plasmids are still a reasonable number to be tested individually in a transformation assay). If the origin of replication has been predetermined (e.g., by electron microscopy studies or deletion analysis) and therefore its position on the mtDNA is known, only the restriction fragment comprising the appropriate region has to be cloned. Both of these approaches have been used for cloning origins of replication.

In *S. cerevisiae,* for example, the mitochondrial origin of replication was previously defined as a common part of the mitochondrial genomes of supersuppressive rho$^-$ mutants[13] and was later shown to provide start signals for DNA replication.[19] The ori1 *in vivo* origin of replication of the

[19] G. Baldacci, B. Chérif-Zahar, and G. Bernardi, *EMBO J.* **3,** 2115 (1984).

TABLE I
COMPILATION OF HYBRID VECTORS OBTAINED FROM EITHER *Schizosaccharomyces pombe* OR *Penicillium chrysogenum* mtDNA

Organism/donor of mtDNA	Vector	Bacterial part	mtDNA fragment	Reference
S. pombe	pPM201	pHP34	*Bgl*II-4 (3.8 kb)	This study
	pPM217	pHP34	*Bgl*II-2 (4.3 kb)	This study
	pFM111	pACYC184	*Eco*RI-1 (11.7 kb)	Lang and Wolf[22]
	pFM141	pACYC184	*Eco*RI-4 (1.7 kb)	Lang and Wolf[22]
P. chrysogenum	pSP530	pDam1	*Eco*RI-4	This study;
	pSP533	pDam1	*Eco*RI-5	Stahl *et al.*[22a]
	pSP1015	pDam1	*Eco*RI-2	This study
	pScP1	pDam1	*Hin*dIII-1	
	pScP8	pDam1	*Hin*dIII-8	
	pScP9	pDam1	*Hin*dIII-9	

S. cerevisiae mitochondrial genome (400-bp *Sau*3A fragment) could be isolated, inserted into a nonreplicative yeast vector (pDam1[20]), and subsequently tested for replication activity in the yeast host.

Since the positions of the mitochondrial replicons are not known in either *S. pombe* or *P. chrysogenum*, random fragments were cloned comprising the whole mitochondrial genome or most of the mitochondrial genome, respectively. In Fig. 1 the allocation of the cloned mtDNA fragments to the genetic map of *S. pombe*[14,21] and the physical map of *P. chrysogenum*[22] is given. The characteristics of the resulting hybrid plasmids are compiled in Table I.

Cloning of a Selection Marker

In addition to an ars, a functional hybrid vector requires at least one marker gene for selection in the eukaryotic host. Appropriate auxotrophic

[20] D. Beach, M. Piper, and S. Shall, *Nature (London)* **284,** 185 (1980).

[21] B. F. Lang, F. Ahne, S. Distler, H. Trinkl, F. Kaudewitz, and K. Wolf, *in* "Mitochondria" (R. J. Schweyen, K. Wolf, and F. Kaudewitz, eds.), p. 313. de Gruyter, Berlin, 1983.

[22] B. F. Lang and K. Wolf, *Mol. Gen. Genet.* **196,** 465 (1984).

[22a] U. Stahl, E. Leitner, and K. Esser, *Eur. J. Appl. Microbiol. Biotechnol.* **26,** 237 (1987).

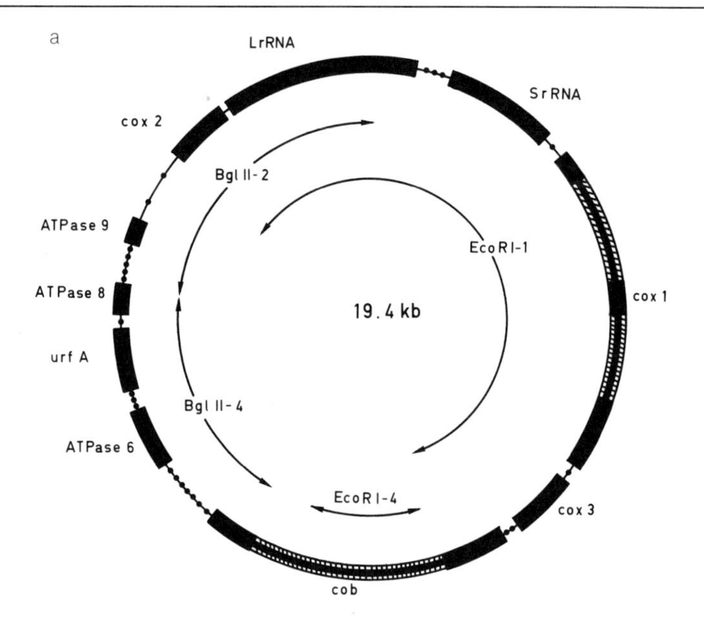

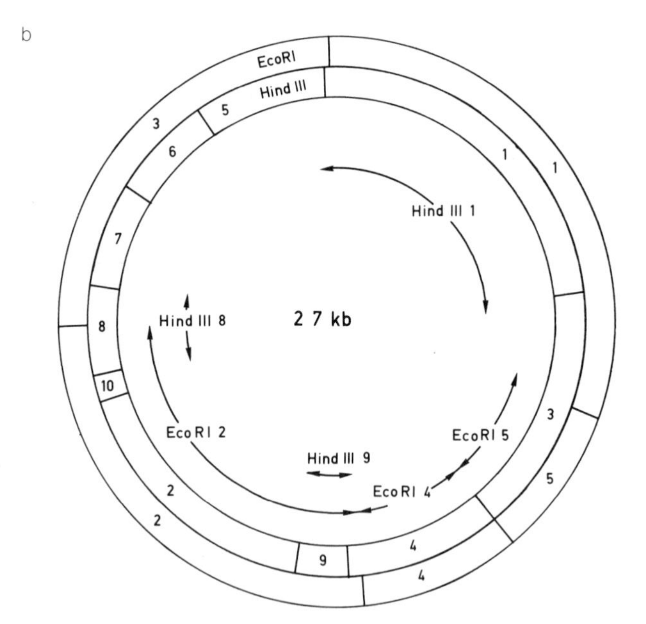

FIG. 1. Allocation of cloned mtDNA fragments to the physical (and genetic) map of the donor organisms. Restriction and gene map of *Schizosaccharomyces pombe* (a) is according to Refs. 14 and 21; *Penicillium chrysogenum* (b) mtDNA was mapped previously.[22a]

mutants of the yeasts *S. cerevisiae* and *S. pombe* can be transformed to prototrophy (e.g., by the *leu2* gene of *S. cerevisiae*). Since this gene is readily available on several plasmid vectors (e.g., pDam1), it was integrated into the described mtDNA hybrid plasmids and used for all transformation experiments involving these two yeast hosts. For the fungus *P. chrysogenum* such auxotrophic mutants and appropriate complementing alleles are not yet available. In this system the bacterial gene for kanamycin resistance, which induces G418 resistance in eukaryotes, was used for selection. Sensitivity toward the aminoglycoside antibiotic G418 is a common feature of diverse eukaryotes and transformation to resistance by using one of the bacterial kanamycin resistance genes has been shown to be possible in several hosts.[23]

Test for Replication Activity of Mitochondrial Hybrid Vectors

Hybrid vectors possessing both an appropriate marker gene and an mtDNA fragment to be tested for its replication activity are next transferred to a eukaryotic host. For this purpose either the mtDNA donor (i.e., the "homologous" host; this approach was used for mitochondrial vectors of *S. cerevisiae* and of *S. pombe*) or a heterologous host suitable as primary tester for replication activity can be used. *S. cerevisiae* is commonly used as a heterologous test organism. By this step preselection of replicating sequences prior to transformation of the homologous host can be achieved. This approach was used to establish mitochondrial vectors for *P. chrysogenum*. As compiled in Table I, hybrid vectors were tested for their capacity to induce autonomous replication in the heterologous host *S. cerevisiae* prior to their transfer to *P. chrysogenum*. Four out of six mitochondrial vectors proved to be autonomously replicating in the intermediate host, namely those containing mtDNA fragments *Eco*RI-4, *Eco*RI-5, *Hin*dIII-1, and *Hin*dIII-9. Two of them (*Eco*RI-4 and *Eco*RI-5) were used for subsequent *Penicillium* transformations (pSP530 and pSP533).

The critical step at this point is the availability of an efficient transformation procedure which is well established for yeast,[4,24] but generally not for hyphal fungi. The procedure we applied for the transformation of *P. chrysogenum* is detailed below:

1. Germination of conidiospores (about 10^8) is allowed for 24 hr at 25° in Czapek-Dox medium.

[23] K. Esser, U. Kück, C. Lang-Hinrichs, P. Lemke, H. D. Osiewacz, U. Stahl, and P. Tudzynski, "Plasmids of Eukaryotes." Springer-Verlag, Berlin, 1986.

[24] H. Ito, Y. Fukuda, K. Murata, and A. Kimura, *J. Bacteriol.* **153,** 163 (1983).

2. Germinated spores (mycelium) are collected by centrifugation (16,000 *g*, 15 min), resuspended in 20 ml 0.15 *M* Soerensen phosphate buffer, pH 5.8, containing 1.1 *M* sorbitol and 290 mg/ml sodium thioglycolate, and incubated for 1 hr at 25° with shaking.

3. For protoplast preparation the mycelium again is collected by centrifugation (16,000 *g*, 10 min) and resuspended in protoplasting buffer (0.15 *M* Soerensen phosphate buffer, pH 5.8, 1.1 *M* sorbitol) with 10 mg/ml Novozym 234. The suspension is incubated at 25° for 30 to 45 min. At this point about 70% of the mycelium had become osmotically fragile as determined microscopically.

4. Protoplasts are collected by centrifugation (16,000 *g*, 10 min), washed once with 15 ml of protoplasting buffer (see above), and passed through defatted cotton wool to eliminate nonprotoplasted mycelial fragments.

5. For transformation, 10^6 protoplasts are suspended in 0.5 ml protoplasting buffer containing 5 μg of vector DNA.

6. The transformation mix is kept in an ice bath for 30 min and subsequently warmed to 45° for 1 min.

7. Protoplasts are allowed to cool to room temperature and 9 ml 20% PEG 4000 (containing 1.48 mg/ml $CaCl_2$) is added. The preparation is carefully mixed and incubated at 25° for 15 min.

8. Protoplasts are collected by centrifugation (6000 *g*, 10 min), gently resuspended in 1 ml protoplasting buffer, and plated on selective media at 2×10^5 protoplasts/plate.

9. Transformed colonies appear after 7 days at 25°.

Transformation frequencies obtained with the different hosts are compiled in Table II. The data show that only in *S. cerevisiae* is a significant difference in transformation rate conceivable for integrative (pDam1) and replicative vector (pScL3). In both *S. pombe* and *P. chrysogenum* the transformation frequencies are not directly correlated to the replication activity of the transferred vectors, since negative controls, i.e., plasmids without mtDNA inserts, pHP341 and pSP325, respectively, transform at the same frequency as mtDNA containing vectors. A discrimination between replicative and integrative vectors is only possible on the molecular level which is provided by an analysis of the cellular DNA of transformed colonies.

Molecular Analysis of Transformants

In order to ascertain that phenotypically selected transformed colonies have arisen as a result of genetic transformation, molecular analysis is performed. Such analysis normally involves preparation of whole-cell

TABLE II
TRANSFORMATION FREQUENCIES IN *Saccharomyces cerevisiae*, *Schizosaccharomyces pombe*, AND *Penicillium chrysogenum* USING MITOCHONDRIAL VECTORS[a]

Host	Vector DNA	Selection	Transformants/ μg DNA	Transformants/10^5 regenerated protoplasts
S. cerevisiae AH22	pDam1 (control)	Leucine prototrophy	1	
	pScL3	Leucine prototrophy	$0.5–1 \times 10^3$	
S. pombe	pHP341 (control)	Leucine prototrophy	1.2×10^3	
	pPM201	Leucine prototrophy	1.4×10^3	
	pPM217	Leucine prototrophy	1.6×10^3	
	pFM111	Leucine prototrophy	0.8×10^3	
	pFM141	Leucine prototrophy	1.0×10^3	
P. chrysogenum	pSP525 (control)	G418 resistance		4
	pSP530	G418 resistance		5
	pSP533	G418 resistance		3

[a] Plasmids without an mtDNA fragment were employed as controls.

DNA and both biochemical and biological assays for the transforming vectors.

Preparation of Whole-Cell DNA of Transformants

The amount of DNA needed is provided by small-scale preparations of cellular DNA. For *S. cerevisiae* a previously described method[25] is used with minor modifications. The recovery of plasmid DNA from the yeast *S. pombe* and from *P. chrysogenum* proved to be more efficient after purification of whole-cell DNA by CsCl density gradient centrifugation. The procedure used for these two organisms involves the following steps:

1. Transformed colonies are grown under selective conditions in 100- to 200-ml cultures, harvested, and mechanically ruptured as described

[25] K. Struhl, D. T. Stinchcomb, S. Scherer, and R. W. Davis, *Proc. Natl. Acad. Sci. U.S.A.* **76,** 1035 (1979).

above or (in the case of *S. pombe*) protoplasted by addition of 2.5 ml protoplasting buffer (1.2 *M* sorbitol, 10 m*M* EDTA, 100 m*M* sodium citrate, pH 5.8) and protoplasting enzyme (5 mg Zymolase 20 T, Miles Lab.) and incubated at 37° for 1–2 hr with shaking.

2. Ground mycelium or spheroplasted cells are lysed completely by addition of an equal amount of lysis buffer (0.5 *M* NaCl, 0.05 *M* Tris–HCl, pH 8.0, 0.05 *M* EDTA, 2% SDS).

3. The suspension is carefully mixed; 7.4 g of cesium chloride is added directly to the lysis mix which is subsequently incubated at 60° for 30–45 min to achieve denaturation of proteins.

4. The suspension is subjected to centrifugation (18,000 *g*, 15 min, 0°) to separate the protein layer from the nucleic acids solution.

5. The cleared nucleic acid CsCl fraction is collected, supplied with ethidium bromide, adjusted to a refractive index of 1.390, and centrifuged as described above.

Molecular Analysis

Hybridization analysis of cellular DNA provides information on the presence of the transferred vector and its intracellular conformation (integrated or extrachromosomal). When hybridization is used in combination with restriction analysis, structural rearrangements of the vector DNA which occurred in the host can be determined.

A second approach is based on the recovery of free plasmid molecules by transfer of the cellular DNA into an appropriate *E. coli* host and selection for the bacterial part of the hybrid vectors. This method is more sensitive than hybridization assays and is also suitable for quantifying the amount of plasmid DNA per haploid genome (or per microgram cellular DNA), i.e., for estimating the copy number. Furthermore, plasmids recovered from *E. coli* can easily be analyzed with respect to their structural integrity and rearrangement of vector sequences, which seems to occur rather frequently in eukaryotes.

Both of these approaches were employed to analyze mitochondrial vectors; data obtained by the *E. coli* transformation method are compiled in Table III. It must be stressed that plasmids are classified as "autonomously replicating" only when their structural integrity is preserved in the eukaryotic host and they are readily recovered.[26] Together with the hybridization data these results show that in *S. pombe* and *P. chrysogenum* the capacity of autonomous replication of plasmid vectors cannot be inferred from transformation frequency; that in all organisms tested

[26] K. Maundrell, A. P. H. Wright, M. Piper, and S. Shall, *Nucleic Acids Res.* **13,** 3711 (1985).

TABLE III
ASSAY FOR FREE PLASMID MOLECULES IN WHOLE-CELL DNA OF *Saccharomyces cerevisiae*, *Schizosaccharomyces pombe*, AND *Penicillium chrysogenum* TRANSFORMANTS USING THE *E. coli* RETRANSFORMATION METHOD[a]

Transformant	Plasmid recovery by *E. coli* transformation	*E. coli* transformants/μg DNA
S. cerevisiae AH 22/pDam1 (control)	−	0
S. cerevisiae AH 22/pScL3	+	200
S. pombe/pHP341 (control)	−	0
S. pombe/pPM201-4	+	13
S. pombe/pPM217-4	+	4
S. pombe/pFM111-3	−	0
S. pombe/pFM141-1	−	0
P. chrysogenum/pSP525 (control)	−	0
P. chrysogenum/pSP530	+	2
P. chrysogenum/pSP533	+	2

[a] Transformation frequencies are normalized to 1×10^7 transformants using purified plasmid DNA.

distinct fragments of the mitochondrial genome carry determinants for autonomous replication (*S. cerevisiae:* ori1; *S. pombe:* fragments *Bgl*II-2 and *Bgl*II-4; *P. chrysogenum:* fragments *Eco*RI-4 and *Eco*RI-5) and are thus able to provide autonomous replication of hybrid vectors in their homologous host; and that preselection of ars's in *S. cerevisiae* as an intermediate host may be an adequate possibility to reduce the total number of hybrid vectors to be tested in the final host, as shown for *P. chrysogenum* vectors.

Maintenance of Mitochondrial Vectors in the Host

Maintenance of vectors in a host cell is mainly characterized by the mitotic stability of the vector. Mitotic stability seems to be influenced by replication efficiency (efficiency of the replication origin) and by the copy number per cell or efficiency of the distribution to daughter cells. The most common methods for vector characterization are measuring the mitotic stability of the vector-bound selectable phenotype and estimating the

TABLE IV
MITOTIC STABILITY OF MITOCHONDRIAL VECTORS OF *Saccharomyces cerevisiae* AND *Schizosaccharomyces pombe*

Strain	Percentage leu$^+$ cells after 10–15 generations under: Selective growth conditions	Nonselective growth conditions
S. cerevisiae AH22/pJDB248 (control)	89	40
S. cerevisiae AH22/pScL3	43	10
S. pombe/pPM201-4	28	0
S. pombe/pPM217-4	40	0

copy number. The assay detailed below is only useful for single-cell organisms like yeast; in hyphal fungi its application to conidiospores should lead to comparable results.

Assay for Mitotic Stability

1. Transformants are grown under selective and under nonselective conditions in liquid culture. At stationary phase cells are diluted and plated on both selective and nonselective media.
2. The growth cycle is repeated at least twice to be able to follow vector maintenance for 10–20 generations.
3. Stability is calculated for both selectively and nonselectively grown cultures as the percentage of cells carrying the vector-associated phenotypic marker.

Stability values obtained for the mitochondrial vectors are listed in Table IV; for *S. cerevisiae* data of conventional vectors are given for comparison.

Comments

The method described in this chapter represents a novel approach to the construction of autonomously replicating vectors for yeast and filamentous fungi. The use of the host's mtDNA as the origin of replication in vector molecules has now been shown to yield replicating plasmids (vectors) in *S. cerevisiae, S. pombe,* and *P. chrysogenum*. In *S. cerevisiae* a predefined origin of replication (ori1) could be integrated in hybrid vectors; in *S. pombe* distinct fragments of mtDNA (*Bgl*II-2 and *Bgl*II-4) were shown to possess replicating activities. In these two examples transformation tests were performed in the homologous host directly; the approach with *P. chrysogenum* took advantage of *S. cerevisiae* as intermediate host to preselect replicating plasmids which subsequently were analyzed in *Penicillum,* the final host. In all of these cases, mtDNA could provide replication functions in the homologous host.

The prerequisites of establishing mitochondrial vectors are an efficient procedure for the isolation of organelle DNA and protocol and selection scheme to obtain transformants. The nonrandom cloning of restriction fragments from mtDNA makes the screening for replication plasmids much easier than the isolation of ars's from chromosomal DNA.

Plasmids[23] which have been detected in a variety of organisms may be used for the construction of replicating vectors similar to mitochondrial genomic DNA,[27,28] as has been shown for the ascomycetes *Podospora anserina* and *Neurospora crassa.*

When using the procedure detailed above, attention should be paid to the following points:

1. The use of *S. cerevisiae* as an intermediate host may not unequivocally result in replicating vectors for the homologous host. It may just lead to the selection of *S. cerevisiae*-specific ars's, since it seems that mutual recognition of replication origins of different eukaryotes does not generally[26,29] exist.
2. The behavior of replicating plasmids in hyphal fungi will be different from that in unicellular organisms, especially with regard to transmission and mitotic stability. Because of considerable mixing of the cytoplasm, compartments or hyphal regions which have lost their plasmids will show no selective disadvantage as do single cells.
3. Structural rearrangements of vectors, which have often been observed in diverse eukaryotic hosts, may be a great impediment for subsequent vector application in gene bank constructions or gene expression studies.

Using the selection and screening scheme described, new cloning vehicles for any host should, in principle, be obtainable. Further studies on replicating vectors in different eukaryotes should provide more information on the possibility of interchanging replication units among different species and the replication and transmission characteristics of replicating

[27] U. Stahl, P. Tudzynski, and K. Esser, *Proc. Natl. Acad. Sci. U.S.A.* **79,** 3641 (1982).
[28] L. L. Stohl and A. M. Lambowitz, *Proc. Natl. Acad. Sci. U.S.A.* **80,** 1058 (1983).
[29] C. Lang-Hinrichs, *Bibl. Mycol., Vol. 102.* J. Cramer, Berlin, 1986.

vectors in hyphal fungi. Mitochondrial vectors may be convenient tools in this respect and might also prove useful for molecular genetic studies in eukaryotic hosts.

Acknowledgment

The experimental work was supported in part by the Deutsche Forschungsgemeinschaft/Bonn.

Section II

Vectors for Expression of Cloned Genes

[23] Short Homopeptide Leader Sequences Enhanced Production of Human Proinsulin in *Escherichia coli*[1]

By WING L. SUNG, FEI-L. YAO, and SARAN A. NARANG

Small foreign proteins which can be produced in bacteria via recombinant DNA are often degraded rapidly by the proteolytic enzyme system of the host. A widely used approach is to produce the target protein as fusion product with a host protein,[1a–4] which often constitutes an undesirably large portion of the fused polypeptide. This lowers the yield and complicates the purification of the desired protein.

To overcome these problems we have designed a small protective "cap" for the efficient expression of the desired polypeptide.[5] A short duplex oligonucleotide, which encodes a homooligopeptide of hydrophilic and "protease-resistant" amino acid residues, was fused to the proinsulin gene at the amino terminus. The fused gene was found to be expressed efficiently in *Escherichia coli*. Structure of the cell wall was distorted because of accumulation of the new polypeptide which resisted degradation. After removal of the homopeptide leader sequence by cyanogen bromide, the fused polypeptide was converted to proinsulin.

Materials

Bacterial Strain. *E. coli* K-12 strain JM103 was used in all cloning and expression experiments.

Proinsulin Gene-Containing Plasmid pNSY. A synthetic proinsulin gene[6] was inserted between the *Eco*RI and *Bam*HI sites in the multiple cloning region of plasmid pUC8 (Bethesda Research Laboratories). This

[1] NRCC publication number 27919.

[1a] K. Talmadge and W. Gilbert, *Proc. Natl. Acad. Sci. U.S.A.* **79,** 1830 (1982).

[2] L.-H. Guo, P. P. Stepien, J. Y. Tso, R. Brousseau, S. Narang, D. Y. Thomas, and R. Wu, *Gene* **29,** 251 (1984).

[3] D. V. Goeddel, D. G. Kleid, F. Bolivar, H. L. Heyneker, D. G. Yansura, R. Crea, T. Hirose, A. Kraszewski, K. Itakura, and A. D. Riggs, *Proc. Natl. Acad. Sci. U.S.A.* **76,** 106 (1979).

[4] S.-I. Sumi, A. Hasegawa, S. Yagi, K. Miyoshi, A. Kanezawa, S. Nakagawa, and M. Suzuki, *J. Biotechnol.* **2,** 59 (1985).

[5] W. L. Sung, F.-L. Yao, D. M. Zahab, and S. A. Narang, *Proc. Natl. Acad. Sci. U.S.A.* **83,** 561 (1986).

[6] R. Brousseau, R. Scarpulla, W. Sung, H. M. Hsiung, S. A. Narang, and R. Wu, *Gene* **17,** 279 (1982).

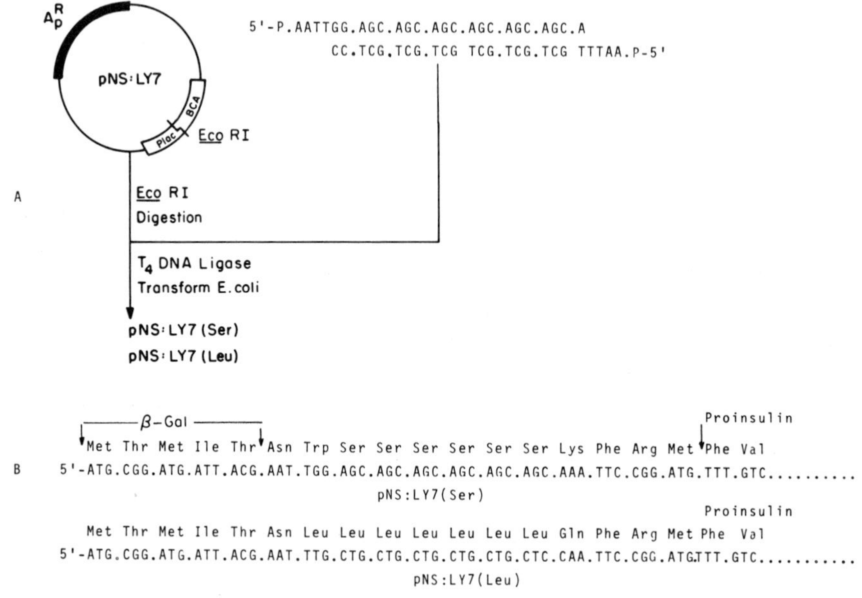

FIG. 1. (A) Construction of plasmids pNSY-Ser-1 and pNSY-Leu-1 with respective leader peptides $(Ser)_6$ and $(Leu)_7$. Ap^R, Ampicillin-resistance gene; plac, *lac* promotor; BCA, proinsulin gene. (B) Nucleotide and encoded amino acid sequences of the 5′ ends of β-galactosidase–oligonucleotide–proinsulin fusion gene. Sites of cleavage by cyanogen bromide are indicated by arrows.

plasmid has a *lac* promotor and eight amino acid codons of the β-galactosidase gene upstream of the *Eco*RI site.

Oligonucleotides Encoding the Homooligopeptide Leader Sequences. (See Fig. 1.) Synthetic DNA encoding a homopolymer of six or seven amino acids was designed with *Eco*RI ends for cloning. Thirteen pairs of 25-mer oligonucleotides were prepared to encode all 20 amino acids for testing the effectiveness of these leader sequences (Table I).

Methods

Insertion of Synthetic Oligonucleotide Leader Sequences. (See Fig. 1.) Two complementary oligonucleotides (0.3 pmol)

$^{5'}$HO · AAT TGG AGC AGC AGC AGC AGC AGCA
CC TCG TCG TCG TCG TCG TCGT TTAA · OH$^{5'}$

were phosphorylated separately. The reaction mixture contained 1 μl (10 U) of T4 DNA kinase, 1 μl of 10× kinase buffer (0.7 *M* Tris–HCl, pH 7.5, 0.1 *M* $MgCl_2$, 0.1 *M* KCl, 0.05 *M* DTT), 1 μl of 1 m*M* ATP, and 6 μl H_2O;

TABLE I
SYNTHESIS OF HUMAN PROINSULIN IN *E. coli*

Peptide leader[a]	Oligonucleotides[b] (5′ → 3′)	Proinsulin in percentage of bacterial protein[c,d]	
		Induced	Noninduced
$(Asn)_6$	$AATTCC(AAC)_6A$	26 (16)	2.1
$(Gln)_7$	$AATTCA(CAA)_6C$	23 (20)	5.5
$(Thr)_6$	$AATTCC(ACC)_6A$	22 (20)	3.0
$(Ser)_6$	$AATTGG(AGC)_6A$	12 (10)	4.3
$(Ala)_6$	$AATTTA(GCA)_5GCGG$	9.8	
$(His)_6$	$AATTCC(CAC)_6A$	8.0	5.8
$(Cys)_7$	$AAT(TGT)_7G$	6.0 (6)	2.5
$(Trp)_6$	$AATTTG(TGG)_6G$	0.6	
$(Pro)_6$	$AATTGG(CCA)_6G$	0.6	
$(Ile)_6$	$AATTCC(ATC)_6A$	0.3	
$(Tyr)_6$	$AATTCG(TAT)_6A$	0.2	
$(Asp)_6$	$AATTTC(GAT)_5GACC$	0.2	
$(Glu)_7$	$AATTTC(GAA)_6G$	0.2	
$(Arg)_6$	$AATTTA(CGA)_5CGCG$	0.08	
$(Lys)_7$	$AATTGG(AAG)_6A$	0.05	

[a] Homopeptide leaders are presented, according to their effectiveness in enhancing proinsulin production.
[b] Oligonucleotide directly encoding the homopeptide leader. However, duplexes of oligonucleotides were used in all insertions to yield two different progenies with different leaders.
[c] Calculated by RIA and method of Lowry, values in parentheses were estimated by way of densitometric scanning of gel.
[d] Other homopeptide leaders, such as $(Leu)_7$, $(Val)_6$, $(Gly)_6$, $(Met)_6$, and $(Phe)_7$, which yielded proinsulin less than 0.05% of total protein, are not presented.

incubation was at 37° for 1 hr. Solutions of the two fragments were then combined and heated at 70° for 10 min.[5] After cooling slowly to room temperature, the solution was mixed with 100 ng (0.03 pmol) of *Eco*RI-digested and calf intestinal alkaline phosphatase-treated plasmid pNSY, 1 μl (3 U) of T4 DNA ligase, 1 μl of 4 m*M* ATP, 1 μl of 10× kinase buffer, and 6 μl of H_2O. After 18 hr at 12°, the ligation mixture was used to transform competent cells of JM103 and plated on ampicillin-containing YT plates. Colony hybridization with one of ^{32}P-labeled 25-mers selected positive progenies. DNA sequencing by Sanger's dideoxynucleotide chain-termination method confirmed the insert sequence upstream of the proinsulin gene.

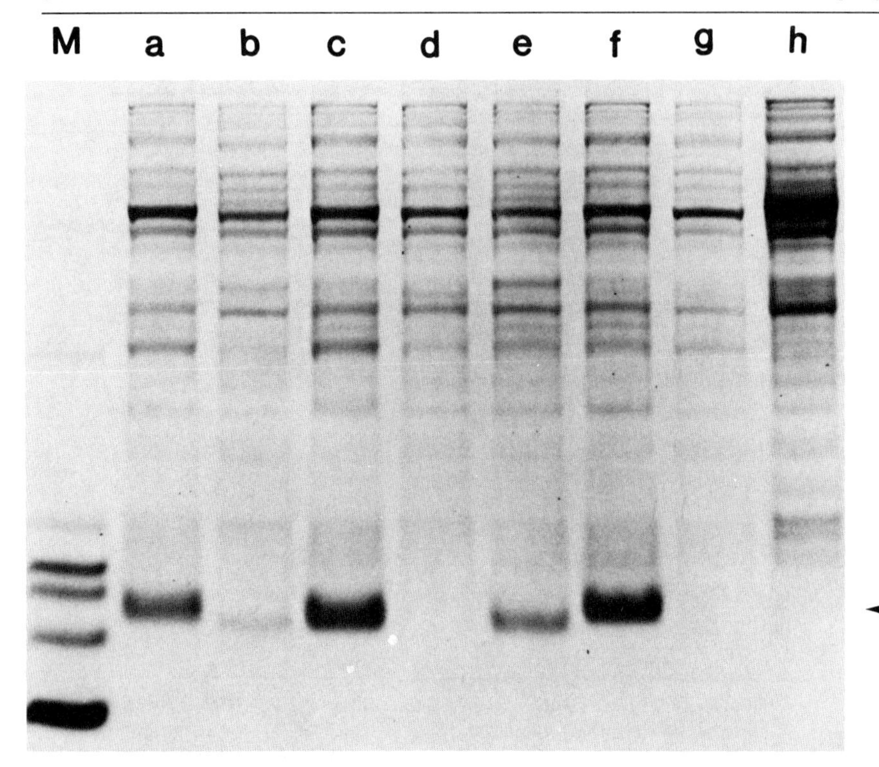

FIG. 2. Proinsulin polypeptide in different bacterial lysates, separated by $NaDodSO_4$/15% polyacrylamide gel electrophoresis, and stained with Coomassie brilliant blue. Lanes a–g: total protein, equivalent to about 100 μl of original bacterial culture, of cells containing plasmids with leader peptides $(Asn)_6$ (lane a), $(Cys)_7$ (lane b), $(Gln)_7$ (lane c), $(Leu)_7$ (lane d), $(Ser)_6$ (lane e), $(Thr)_6$ (lane f), and $(Val)_6$ (lane g). Lane h: non-plasmid-containing JM103. Lane M: protein standards: from top to bottom M_r 43,700, 25,700, 18,400, 14,300, 12,300, 6200, and 3000. Arrow at right indicates the band of the proinsulin-fused protein (M_r 11,700).

After insertion of 13 25-mer duplexes, 26 progenies, encoding homo-oligopeptides of all 20 amino acids including certain degenerate codons, were prepared.

Expression of the Fused Protein-Containing Proinsulin. A bacterial culture was grown overnight at 37° in 2YT medium containing ampicillin. Then 2YT medium (8 ml) with ampicillin was inoculated with 0.08 ml of the overnight culture. After 2 hr at 37°, isopropyl β-D-thiogalactoside was added to give a final concentration of 0.7 m*M*. Cells were harvested after 24 hr by centrifugation at 1600 *g* for 10 min. They were then suspended in 6 *M* guanidine hydrochloride (pH 7.0) or 1% $NaDodSO_4$ and sonicated. After centrifugation at 1600 *g* for 20 min, the lysate was analyzed by radioimmunoassay to detect the C-peptide of proinsulin (Table I) and $NaDodSO_4$/15% PAGE (Fig. 2). Seven homooligopeptide leader sequences alanine, asparagine, cysteine, glutamine, histidine, serine, and threonine were shown to effectively enhance the production of proinsulin, which amounted to 26% of the bacterial protein. Accumulation of such large amounts of proinsulin can also be observed directly by phase-contrast and electron microscopy.

Cyanogen Bromide Cleavage of the Fused Protein to Give Intact Proinsulin. Fused protein isolated by preparative $NaDodSO_4$/15% PAGE was treated with a 50-fold molar excess of cyanogen bromide in 70% formic acid for 24 hr at room temperature.[2] The cleavage product was identical to authentic human proinsulin as analyzed by $NaDodSO_4$/15% PAGE.

Comments. A definite relationship between the efficiency of the homooligopeptide leader sequences and the nature of their side chains is shown in Table I. Whereas oligopeptide leader sequences with side chains such as amide (Asn and Gln) and hydroxy (Thr and Ser) groups are very effective in proinsulin production, the other leader sequences constituted by basic (Lys and Arg), acidic (Asp and Glu), and hydrophobic (Phe, Val, etc.) groups are the least efficient.

The 7 homopolypeptide leader sequences shown at the top of Table I, which are at least 10 times more efficient than the others, are hydrophilic in nature and are generally considered to be less vulnerable to proteases.

Codon selection for a specific homopeptide leader sequence also has a direct effect on the production of proinsulin.[5] Usage of codons with TC or CT sequences dramatically reduced the yield of proinsulin, presumably due to interference between the respective UC and CU region of mRNA and the ribosomal binding site.[7]

Further work is needed to establish the mechanism(s) by which the homooligomeric leader sequences affect the accumulation of proinsulin.

[7] M. N. Hall, J. Gabay, M. Debarbouille, and M. Schwartz, *Nature* (*London*) **295,** 616 (1982).

[24] Expression of Bovine Growth Hormone Derivatives in *Escherichia coli* and the Use of the Derivatives to Produce Natural Sequence Growth Hormone by Cathepsin C Cleavage

By HANSEN M. HSIUNG and WARREN C. MACKELLAR

The semisynthetic bGH[1] gene has been efficiently expressed in *Escherichia coli* by Seeburg *et al.*[2] and by George *et al.*[3] However, efficient expression cannot be achieved by simply inserting a strong promoter (*trp* or *lpp*) and a "good" ribosome-binding site (*trp* or *lpp*) upstream from the bGH gene and a good transcription terminator (*lpp*) downstream from the gene in a multicopy plasmid.[4] One approach to increase the level of bGH expression uses a "two-cistron" construction.[4,5] An artificial cistron with an efficient ribosomal binding site and short coding sequences (first cistron) is inserted before the bGH gene (second cistron). This approach dramatically increases the expression of bGH. This observation prompted us to further study the effects of changing the 5′-coding sequence on the levels of bGH expression.

The 5′-coding region of the bGH gene in expression plasmids was changed by replacing the natural sequence with various synthetic gene fragments (Fig. 1). The levels of expression varied greatly as a result of one to several nucleotide changes in the 5′-coding region. Our results also showed that replacing the 5′-terminal codons with A–T-rich codons enhanced the expression level. The codon after the initiation codon was especially important in modulating expression.

The amino terminus of authentic bGH is heterogeneous,[6] having either alanine or phenylalanine as the amino terminal amino acid. The phenylalanyl form of bGH, when expressed in *E. coli*, generally contains an

[1] Abbreviations: bGH, bovine growth hormone; Km^r, kanamycin resistance; bp, base pair; kb, kilobase pair; HPLC, high-pressure liquid chromatography; SDS–PAGE, sodium dodecyl sulfate–polyacrylamide gel electrophoresis; TEAB, triethylammonium bicarbonate; SD, Shine–Dalgarno.

[2] P. H. Seeburg, S. Sias, J. Adelman, H. A. de Boer, J. Hayflick, P. Jhurani, D. V. Goeddel, and H. L. Heyneker, *DNA* **2,** 37 (1983).

[3] H. J. George, J. J. L'Italien, W. P. Pilacinski, D. L. Glassman, and R. A. Krzyzek, *DNA* **4,** 273 (1985).

[4] B. E. Schoner, H. M. Hsiung, R. M. Belagaje, N. G. Mayne, and R. G. Schoner, *Proc. Natl. Acad. Sci. U.S.A.* **81,** 5403 (1984).

[5] B. E. Schoner, R. M. Belagaje, and R. G. Schoner, this volume [25].

[6] C. H. Li and L. Ash, *J. Biol. Chem.* **203,** 419 (1953).

A. pCZ152

```
         Taq I                                                        Hgi A I
              Met Ala Phe Pro Ala Met Ser Leu Ser Gly Leu Phe Ala Asn Ala Val
5'- CGACCATGGCTTTTCCGGCTATGTCTCTGTCCGGCCTGTTTGCCAACGCTGTGCT  -3'
3'-    TGGTACCGAAAAGGCCGATACAGAGACAGGCCGGACAAACGGTTGCGAC      -5'
```

B. pCZ161

```
         Xba I                                                                  Hgi A I
                         Met Phe Pro Ala Met Ser Leu Ser Gly Leu Phe Ala Asn Ala Val
5'- CTAGAGGGTAT TAATAATGTTTCCAGCTATGTCTCTATC TGGTCTGTTTGCCAACGCTGTGCT  -3'
3'-     TCCCATA ATTATTACAAAGGTCGATACAGAGATAGACCAGACAAACGGTTGCGAC      -5'
```

FIG. 1. (A) Synthetic bGH gene fragments used to construct pCZ152. Similar fragments were used to construct other *trp* bGH expression plasmids. (B) Synthetic bGH gene fragment used to construct pCZ161. Similar fragments were used to construct other *lpp* expression plasmids.

amino terminal methionine,[2,4] however, the alanyl form expressed in *E. coli* appears to have the initiator methionine removed.[3]

In order to produce the phenylalanyl form of bGH without an amino-terminal methionine, we developed an efficient enzymatic cleavage method using cathepsin C (diaminopeptidyl peptidase 1, Boehringer–Mannheim).

Cathepsin C is a diaminopeptidase which removes two amino acids, as a unit, from the amino terminal end of the molecule.[7] The enzyme is unreactive with a substrate when the amino terminal amino acid is blocked.[8] Cathepsin C will not cleave on either side of the imino nitrogen of proline,[9] and the enzyme is inactive if lysine or arginine is the amino terminal amino acid of the substrate.[10] The enzyme requires halide ions and a sulfhydryl compound for its activity.[9]

Materials and Methods

Bacterial Strains and Plasmids

E. coli K-12 RV308 [*su*⁻, Δ*lac*X74, *gal* IS II::OP308, *str*A][11] was the host strain for all of the recombinant plasmids. The promoter *trp*PO region was obtained from pHI7.[12] The plasmids used for the expression

[7] R. M. Metrione, A. G. Neves, and J. S. Frutor, *Biochemistry* **5,** 1597 (1966).

[8] J. K. McDonald and C. Schwabe, *in* "Proteinases in Mammalian Cells and Tissues" (A. J. Barrett, ed.), p. 311. Elsevier/North-Holland, New York, 1977.

[9] J. K. McDonald, P. X. Callahan, S. Ellis, and R. E. Smith, *in* "Tissue Proteinases" (A. J. Barrett and J. T. Dingles, eds.), p. 69. North-Holland, Amsterdam, 1971.

[10] J. K. McDonald, P. K. Callahan, and S. Ellis, this series, Vol. 25, p. 272.

[11] R. Maurer, B. J. Meyer, and M. Ptashne, *J. Mol. Biol.* **139,** 147 (1980).

[12] P. R. Rosteck and C. L. Hershberger, *Gene* **24,** 29 (1983).

studies of bGH were constructed from pIMIA,[13] which has a thermo-inducible runaway replicon originally derived from pKN402.[14] The plasmid pIMIA has a copy number of 10–15 per cell below 30° and 1000–2000 per cell at 37°. It also contains a kanamycin resistance (Km^r) marker and the *E. coli* lipoprotein (*lpp*) promoter and ribosome-binding site. The bGH cDNA clone was obtained from W. L. Miller and J. D. Baxter.[15]

Chemical Synthesis and Purification of Oligonucleotides

The chemical synthesis of oligonucleotides, 10 to 20 nucleotides long, was performed by the manual modified phosphotriester synthesis[16] or by automated phosphite triester synthesis[17] using an Applied Biosystem 380A synthesizer. The oligonucleotides were purified by 7 *M* urea–20% polyacrylamide gel electrophoresis. The desired bands were visualized by UV shadowing using a long-wavelength UV lamp (318 nm). The oligonucleotide products were then excised from the gel and were eluted by a "crush and soak" procedure[18] with 1 *M* TEAB buffer, pH 8.0. The eluted oligonucleotide samples were then dried and redissolved in 1 ml of H_2O. The samples were exchanged into a volatile buffer (1 *M* TEAB, pH 8.0) on a DE-52 (Whatman) ion exchange column before drying. The dried oligonucleotide were dissolved in 10 m*M* Tris–HCl, 1 m*M* EDTA, pH 7.5 (TE) buffer and a small sample was phosphorylated with [γ-^{32}P]ATP and T4 polynucleotide kinase to determine the purity and size of oligonucleotides on denaturing polyacrylamide gels containing 7 *M* urea.

Enzymatic Ligations

The standard ligation for preparing the bGH gene fragments (Fig. 1) was performed by mixing the phosphorylated oligonucleotides with the unphosphorylated oligonucleotide comprising the 5′-cohesive end in a ligation buffer (50 m*M* Tris–HCl, 10 m*M* $MgCl_2$, 10 m*M* DTT, and 0.5 m*M* ATP) with T4 DNA ligase (20–100 U/nmol). Final ligation products were analyzed and purified by 15% polyacrylamide gel electrophoresis.

The ligated DNA product was detected by autoradiography and eluted from the gel by "crush and soak" procedure[18] using 1 *M* TEAB, pH 8.0.

[13] Y. Masui, J. Coleman, and M. Inouye, *in* "Experimental Manipulation of Gene Expression" (M. Inouye, ed.), p. 15. Academic Press, New York, 1983.

[14] B. E. Uhlin, S. Molin, P. Gustafsson, and K. Nordstrom, *Gene* **6,** 91 (1979).

[15] W. L. Miller, J. A. Martial, and J. D. Baxter, *J. Biol. Chem.* **255,** 7521 (1980).

[16] H. Hsiung, S. Inouye, J. West, B. Sturm, and M. Inouye, *Nucleic Acids Res.* **11,** 3227 (1983).

[17] M. H. Caruthers, *Science* **230,** 281 (1985).

[18] A. M. Maxam and W. Gilbert, this series, Vol. 65, p. 499.

The eluted DNA was desalted on a Sephadex G50 (superfine) column and the peak fractions were pooled and lyophilized. The lyophilized sample was redissolved in TE buffer. An aliquot of the sample (10–20 pmol) was phosphorylated on the 5′-cohesive end with [γ^{32}P]ATP and T4 polynucleotide kinase. The phosphorylated synthetic bGH gene fragment was then ready for vector construction.

Constructions of bGH Expression Plasmids

All restriction endonucleases were used in the conditions recommended by the manufacturer (New England BioLabs or Boehringer–Mannheim). The plasmid, pIMIA,[13] was used as a starting plasmid for all expression vector constructions. The special features of pIMIA, making it an attractive starting material, are described in another chapter[5] in this volume.

To construct a bGH expression plasmid (Fig. 2A) with the *E. coli lpp* promoter and SD sequence, the *Xba*I/*Bam*HI-digested pIMIA (9.4 kb) was ligated to a synthetic 63-bp *Xba*I/*Hgi*AI bGH gene fragment (Fig. 1B) and a 600-bp *Hgi*AI/*Bam*HI bGH cDNA fragment (a *Bam*HI site had been inserted after the stop codon of bGH cDNA by using a *Bam*HI synthetic adaptor). To construct a bGH expression plasmid with the *trp* promoter and SD sequence (Fig. 2B), the 55-bp synthetic gene fragment (Fig. 1A) was ligated to three other fragments: (1) 9.2 kb, *Eco*RI/*Bam*HI pIMIA fragment, (2) 600 bp, *Hgi*AI/*Bam*HI bGH cDNA fragment, and (3) 200 bp, the *Eco*RI/*Cla*I-digested pHI7[12] fragment which contained the *trp* promoter and SD sequence.

The ligation mixtures were used to transform competent RV308 cells. Transformants were selected on TY broth (Difco) agar plates containing 50 μg/ml kanamycin. Plasmid DNA from the transformants was extracted and subjected to restriction analysis. The appropriate restriction fragments (*Xba*I/*Bam*HI or *Hpa*I/*Bam*HI) containing the synthetic bGH gene fragment were then sequenced by Maxam and Gilbert method.[18]

Analysis of bGH Expression

SDS–PAGE[19] was used to analyze the bGH expression in logarithmic phase RV308 cells. The *E. coli* cells harboring the bGH plasmid were grown in TY medium for 16 hr at 25°. A 20-μl culture was then inoculated into 2 ml of fresh medium and the plasmids were amplified by growth at 37° for 6 hr prior to harvest. Cell pellets from 1-ml cultures were suspended in 200 μl of sample buffer (0.125 *M* Tris–HCl, pH 6.8, 2% SDS,

[19] U. K. Laemmli, *Nature (London)* **227**, 680 (1970).

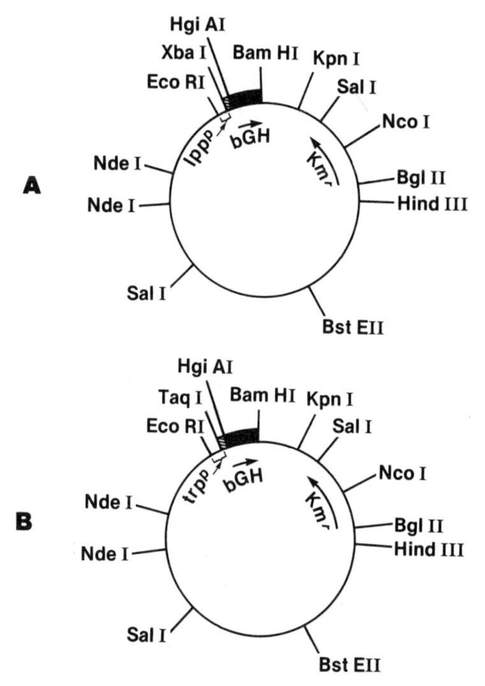

FIG. 2. Structure of bGH expression plasmids. (A) The *lpp* expression plasmids (10 kb) are derived from pIMIA. The *lpp* promoter is located in the region between *Eco*RI and *Xba*I. The shaded block (*Xba*I/*Hgi*AI) is a synthetic bGH gene fragment and the solid block (*Hgi*AI/*Bam*HI) is the remaining portion of the bGH gene derived from cDNA. (B) The *trp* expression plasmids are also derivatives of pIMIA but contain the *trp* promoter located in the region between *Eco*RI and *Taq*I, from pHI7. The shaded block (*Taq*I/*Hgi*AI) is a synthetic bGH gene fragment and the solid block (*Hgi*AI/*Bam*HI) is the remaining portion of the bGH gene derived from cDNA. The location and direction of transcription of the bGH and the kanamycin phosphotransferase gene are indicated by the arrows.

30% glycerol, 2% 2-mercaptoethanol, 6 *M* urea, and 0.05% bromphenol blue). The samples were heated at 90° for 5 min before loading onto an SDS–polyacrylamide gel described in the legend of Fig. 4. The gel was stained with Coomassie blue (Bio-Rad) and analyzed with a Shimadzu 910 gel scanner on line with a Hewlett-Packard 2100 computer that integrated the areas under the peaks.

Met-Asp-bGH or Met-Val-bGH Preparation for Cathepsin C Cleavage

The bacteria were disrupted mechanically with a Gaulin homogenizer.[20] The material from the Gaulin was centrifuged at low speed and the

[20] C. R. Engler and C. W. Robinson, *Biotechnol. Bioeng.* **23,** 765 (1981).

pellet containing the Met-Asp-bGH derivative was washed with water, pH 9.0, and 1 *M* urea, respectively. The pellet after the washes was frozen at −20° and labeled as bGH granules.

BGH granules were dissolved in a 0.2% SDS solution to make a final protein concentration of 1 mg/ml. The protein concentration was determined spectrophotometrically using the BCA protein assay system.[21] The pH was raised to 12.5 with NaOH and cysteine–HCl was added to make a final concentration of 2.5 m*M*. This sample was stirred overnight at room temperature.

The solubilized granule solution was mixed with an equal volume of 2 *M* ammonium sulfate and incubated at room temperature for 5 min. This solution was centrifuged at 12,000 *g* for 20 min and the supernatant was removed and discarded. The pellet was dissolved in 50 ml of 10 m*M* sodium phosphate buffer, pH 6.0. This material was dialyzed for 4 hr against 5 liters of 10 m*M* sodium phosphate buffer, pH 6.0, to remove excess ammonium sulfate.

The cleavage reaction was performed by adding an equal volume of the buffer (20 m*M* sodium phosphate, 4 m*M* sodium chloride, and 10 m*M* 2-mercaptoethanol, pH 6.0) to a volume of the dialyzed Met-Asp-bGH sample. Cathepsin C was then added (enzyme to substrate molar ratio, 1 : 2000) and the reaction mixture was incubated at 37°. Aliquots were taken at various time points and the enzymatic reaction was quenched by the addition of SDS (final concentration 0.2%). The conversion of Met-Val-bGH to bGH was performed in a similar manner as that of Met-Asp-bGH.

Two analytical procedures were used to monitor the conversion of the precursor molecule to the authentic sequence molecule. (1) Reversed-phase HPLC using a Zorbax C-8, 150-Å pore size column: This HPLC was run isocratically using 50 m*M* ammonium phosphate buffer, pH 7.4, containing 30% *n*-propanol at a temperature of 45°; (2) SDS–PAGE[19] utilizing a 15% resolving gel: see the legend of Fig. 6 for details.

Results

Constructions of bGH Expression Plasmids

The various bGH expression vectors differ, primarily, with respect to the promoters (*E. coli lpp* or *trp*) and the DNA sequence at or near the 5′-end of the bGH coding region. The variations in the expression plasmids

[21] BCA Protein Assay, Pierce catalog 1985/1986, p. 354. Pierce Chemical Co., Rockford, Illinois, 1985/1986.

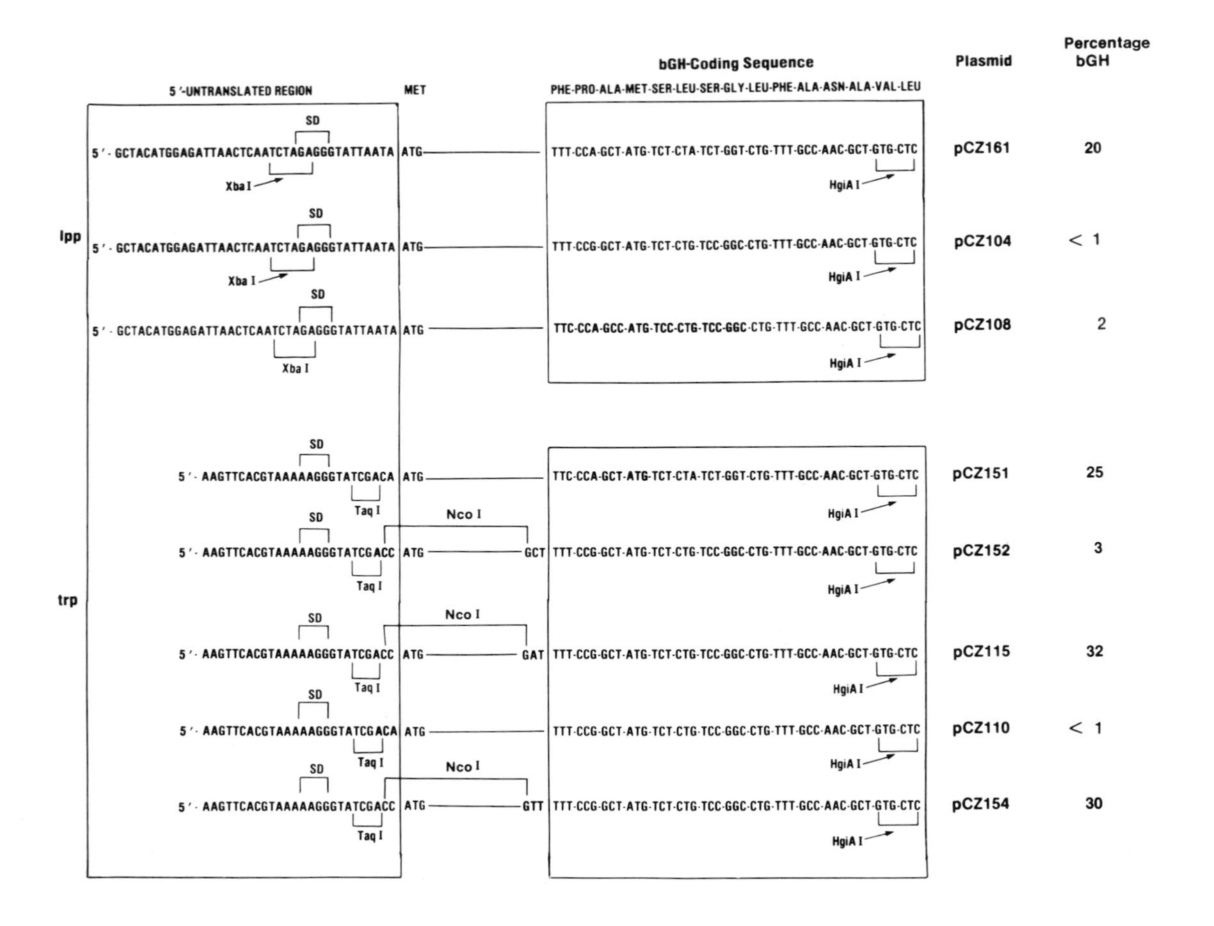

5′-UNTRANSLATED REGION
MET
bGH-Coding Sequence
PHE-PRO-ALA-MET-SER-LEU-SER-GLY-LEU-PHE-ALA-ASN-ALA-VAL-LEU
Plasmid
Percentage bGH
lpp
SD
5′- GCTACATGGAGATTAACTCAATCTAGAGGGTATTAATA ATG
TTT-CCA-GCT-ATG-TCT-CTA-TCT-GGT-CTG-TTT-GCC-AAC-GCT-GTG-CTC
Xba I
HgiA I
pCZ161
20
SD
5′- GCTACATGGAGATTAACTCAATCTAGAGGGTATTAATA ATG
TTT-CCG-GCT-ATG-TCT-CTG-TCC-GGC-CTG-TTT-GCC-AAC-GCT-GTG-CTC
Xba I
HgiA I
pCZ104
< 1
SD
5′- GCTACATGGAGATTAACTCAATCTAGAGGGTATTAATA ATG
TTC-CCA-GCC-ATG-TCC-CTG-TCC-GGC-CTG-TTT-GCC-AAC-GCT-GTG-CTC
Xba I
HgiA I
pCZ108
2
trp
SD
5′- AAGTTCACGTAAAAAGGGTATCGACA ATG
TTC-CCA-GCT-ATG-TCT-CTA-TCT-GGT-CTG-TTT-GCC-AAC-GCT-GTG-CTC
Taq I
HgiA I
pCZ151
25
SD
Nco I
5′- AAGTTCACGTAAAAAGGGTATCGACC ATG
GCT
TTT-CCG-GCT-ATG-TCT-CTG-TCC-GGC-CTG-TTT-GCC-AAC-GCT-GTG-CTC
Taq I
HgiA I
pCZ152
3
SD
Nco I
5′- AAGTTCACGTAAAAAGGGTATCGACC ATG
GAT
TTT-CCG-GCT-ATG-TCT-CTG-TCC-GGC-CTG-TTT-GCC-AAC-GCT-GTG-CTC
Taq I
HgiA I
pCZ115
32
SD
5′- AAGTTCACGTAAAAAGGGTATCGACA ATG
TTT-CCG-GCT-ATG-TCT-CTG-TCC-GGC-CTG-TTT-GCC-AAC-GCT-GTG-CTC
Taq I
HgiA I
pCZ110
< 1
SD
Nco I
5′- AAGTTCACGTAAAAAGGGTATCGACC ATG
GTT
TTT-CCG-GCT-ATG-TCT-CTG-TCC-GGC-CTG-TTT-GCC-AAC-GCT-GTG-CTC
Taq I
HgiA I
pCZ154
30

are highlighted in Fig. 3. The first three plasmids (pCZ161, 104, 108) use the *E. coli lpp* promoter and ribosome-binding site, and the remaining five plasmids (pCZ151, 152, 115, 110, 154) use the *trp* promoter and ribosome-binding site to express bGH.

Expression of bGH Derivatives

SDS–PAGE was used to analyze bGH synthesis by cells harboring the expression vectors.

The results (Fig. 4) showed that Met-bGH expression was low (2% of total cell protein) in the cells harboring pCZ108, in which the natural bGH codons were used (Fig. 4, lane 1). As an effort to increase the expression level, the first eight codons of the bGH gene were replaced with codons used in highly expressed *E. coli* genes (called *E. coli* codons). The cells harboring the resulting plasmid (pCZ104) accumulated little bGH (<1%, Fig. 4, lane 3). Furthermore, the *lpp* promoter and ribosome-binding site sequences were replaced with the *trp* sequences and the cells harboring the resulting plasmid (pCZ110, Fig. 4, lane 2) still accumulated little bGH (<1%).

Since both the *E. coli* codons and the mammalian codons used for constructing the bGH expression plasmids had a high G–C content and neither set of codons allowed high levels of expression, we suspected the mRNA transcribed from the genes might form secondary structures that inhibited translation. The first eight codons of the bGH gene were, therefore, changed to A–T-rich codons. The resulting plasmids (pCZ151 and 161, Fig. 3) increased the level of expression from less than 1 to 25% (Fig. 4, lane 7) and to 20% (Fig. 4, lane 5), depending on whether the *trp* or *lpp* promoter was used.

When one additional codon (GAT) was added to the 5′-end of the bGH gene in the plasmid pCZ110 (which contained G–C-rich *E. coli* codons and did not facilitate high-level expression), the cells harboring the resulting plasmid (pCZ115) yielded a high expression level (32%, Fig. 4, lane 4).

The most striking result was obtained from the study of the plasmid pCZ152 which differed from pCZ115 by only one nucleotide (GAT to GCT). Although the cells harboring pCZ115 yielded 32% expression of

FIG. 3. The 5′-end sequence of cloned bGH genes in the expression plasmids. The first three lines represent the DNA sequence corresponding to the 5′-end of the *lpp* mRNA extending to the *Hgi*AI site within the bGH coding region of the *lpp* expression plasmids (see Fig. 2A). The next five lines represent the sequence of the 5′-end of the *trp* mRNA in the *trp* expression plasmids (Fig. 2B). The sequences beyond the *Hgi*AI site are identical in all plasmids.

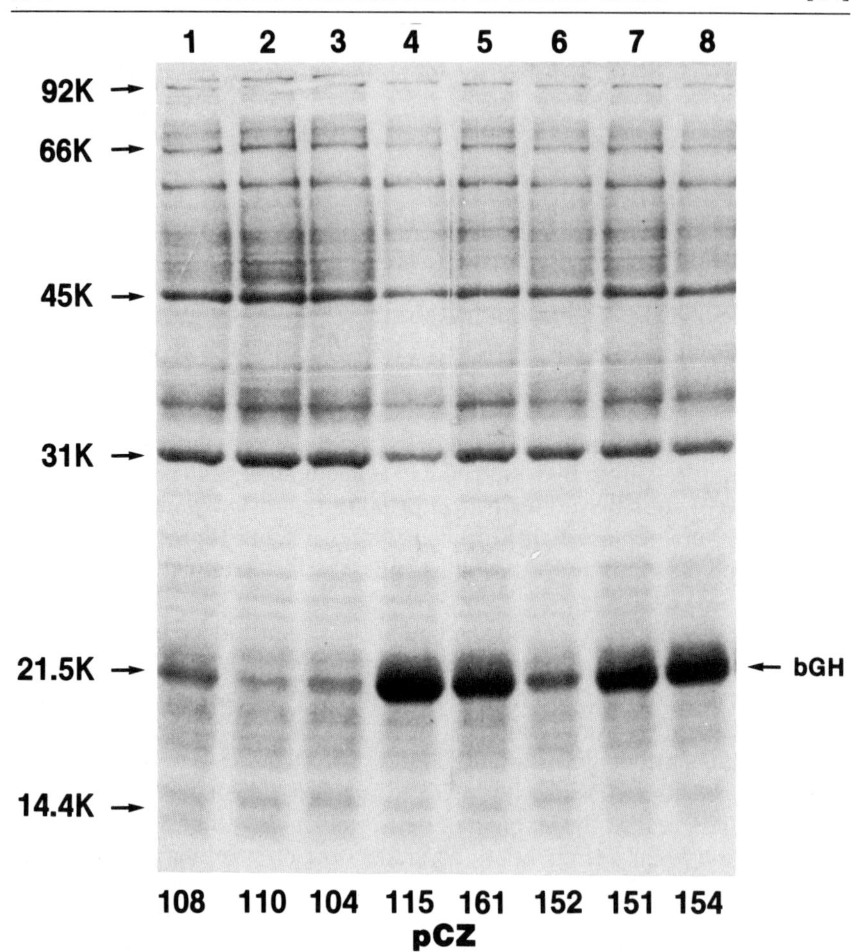

FIG. 4. SDS–polyacrylamide gel analysis of bGH synthesis. Lysates were prepared from the cells harboring the bGH plasmids. The samples (20 μl) were loaded on a 12.5% polyacrylamide gel (1.5 × 18 × 18 mm) and run at 50 mA for 3 hr. Lanes: 1, pCZ108; 2, pCZ110; 3, pCZ104; 4, pCZ115; 5, pCZ161; 6, pCZ152; 7, pCZ151; 8, pCZ154. The position of bGH standard was marked with an arrow on the right side of the picture and the positions of molecular weight standards were marked on the left side.

bGH (Fig. 4, lane 4), those harboring pCZ152 yielded only 3% expression (Fig. 4, lane 6). The nucleotide C in the GCT codon seemed to be responsible for the poor expression of bGH since another plasmid (pCZ154), differing from pCZ152 by one nucleotide (GTT vs GCT), restored the high level of expression (30%, Fig. 4, lane 8) of bGH.

Production of Natural Sequence Growth Hormone Using Cathepsin C

Reversed-phase HPLC and SDS–PAGE were used to monitor the conversion of Met-Asp-bGH or Met-Val-bGH to natural sequence bGH. HPLC analysis (Fig. 5) showed that Met-Asp-bGH, having a retention time centered at 945 sec, was converted to bGH (retention time, 1880 sec) when treated with cathepsin C over a period of 2 hr (Fig. 5). Time-coursc analysis of the conversion of Met-Asp-bGH to bGH was performed using an enzyme/substrate molar ratio of 1 : 2000. The analysis showed that the conversion was 50% complete within the first 15 min of the reaction and was 100% complete within 2 hr (Fig. 5). The conversion was also analyzed by 15% SDS–PAGE. The Met-Asp-bGH sample (Fig. 6, lane 4) ran slightly slower than the pituitary-derived bGH (Fig. 6, lane 3). After cathepsin C cleavage of Met-Asp-bGH, the resulting protein (Fig. 6, lane 5) comigrated with pituitary bGH. The Met-Val-bGH was converted to natural sequence bGH by cathepsin C in a similar manner to that used for the Met-Asp-bGH conversion. The conversion was analyzed by both HPLC (data not shown) and by 15% SDS–PAGE (Fig. 6; lanes 1 and 2). The results showed that Met-Val-bGH (Fig. 6, lane 2) was converted efficiently to natural sequence bGH (Fig. 6, lane 1).

The cathepsin C-converted bGH sample was then subjected to Edman amino acid sequence analysis. The analysis showed that the amino terminal sequence of the sample was Phe-Pro-Ala (data not shown),

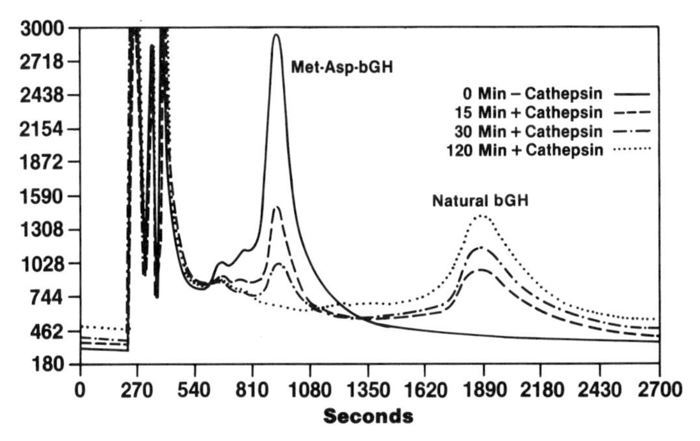

FIG. 5. Time course analysis for the conversion of Met-Asp-bGH to bGH. Aliquots of the conversion reaction, as described in Methods, were taken at 15, 30, and 120 min. The sample was analyzed by reversed-phase HPLC.

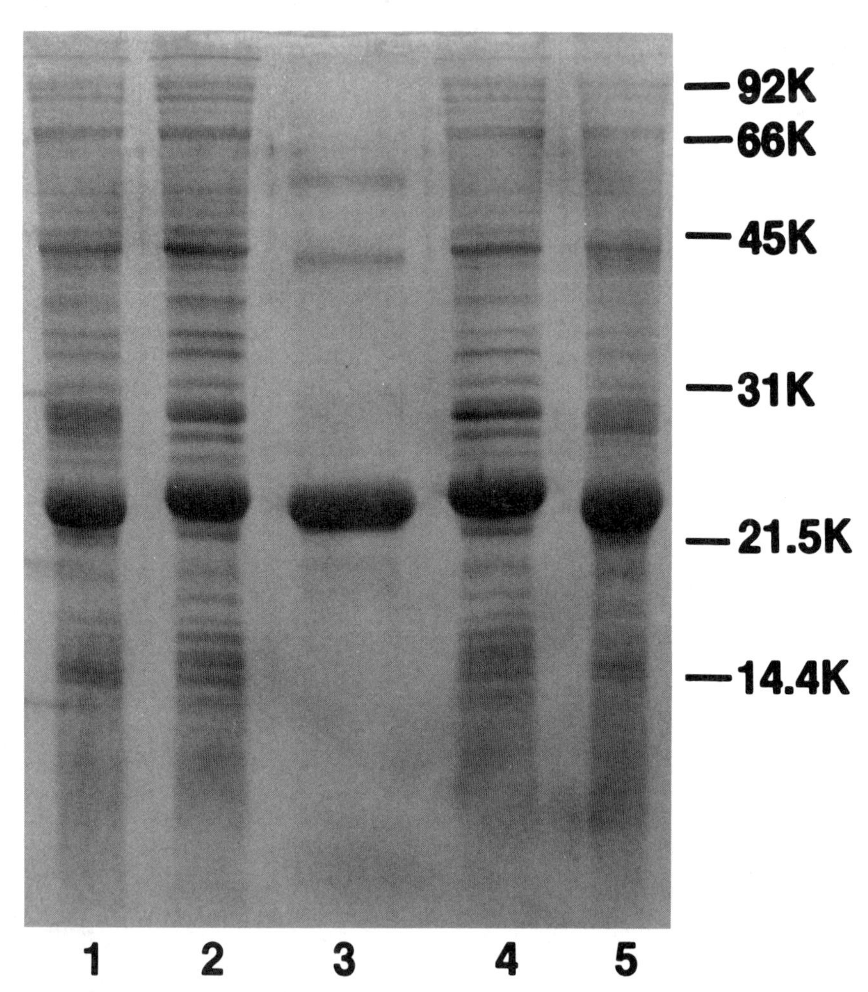

FIG. 6. SDS–PAGE analysis of the conversion of Met-Asp-bGH and Met-Val-bGH to bGH. Met-Asp-bGH and Met-Val-bGH were incubated in the presence of cathepsin C as we described in Methods. After a 2-hr incubation at 37°, an equal volume of 2× sample buffer (20% glycerol, 10% 2-mercaptoethanol, 4% SDS, and 0.125 *M* Tris–HCl, pH 6.8) was added to each sample. The samples were loaded onto a 15% polyacrylamide gel and SDS–PAGE was run at a constant voltage (150 V) for 1.5 hr. The gel was stained with Coomassie blue. Lanes: 1, converted Met-Val-bGH; 2, unconverted Met-Val-bGH; 3, pituitary bGH; 4, unconverted Met-Asp-bGH; 5, converted Met-Asp-bGH.

thereby confirming the correct processing of Met-Asp-bGH to bGH by cathepsin C.

Conclusion

The DNA sequence required for high-level expression of bGH derivatives in *Escherichia coli* was investigated using temperature-sensitive copy control plasmids as cloning vectors. Using the synthetic DNA fragments to alter the 5′-terminal coding region of the bGH gene, the expression level was dramatically altered. The *E. coli* cells harboring certain bGH expression plasmids yielded bGH greater than 30% of the total cell protein while the cells harboring the plasmid with original cDNA Met-bGH gene accumulated little bGH (2%).

The results showed that minor changes (from one to a few nucleotides) in the 5′-terminal coding region of the bGH gene had a profound effect on bGH expression. In several instances the level of bGH expression was increased dramatically by changing the 5′-terminal bGH codons to A–T-rich codons. The changes on the first codon after the initiation codon were especially important in modulating levels of expression.

A method was also developed using cathepsin C to cleave a dipeptide from the amino termini of two highly expressed bGH derivatives (Met-Asp-bGH and Met-Val-bGH); this cleavage yielded a natural sequence bGH.

Acknowledgment

The authors are indebted to Mr. Dennis P. Smith for the technical assistance and to Ms. Sherry J. Pike for typing the manuscript. The authors also want to express their appreciation to Drs. J. Paul Burnett and Edward L. Smithwick for their suggestions, support and encouragement throughout this research.

[25] Expression of Eukaryotic Genes in *Escherichia coli* with a Synthetic Two-Cistron System

By BRIGITTE E. SCHONER, RAMA M. BELAGAJE, and RONALD G. SCHONER

The overproduction of many prokaryotic and eukaryotic proteins is now possible through the use of recombinant DNA technology. In recent years, this technology has greatly advanced our knowledge of medically and agriculturally important proteins, and several of them are already available for clinical research and therapy.

The design of the first generation expression vectors for cloned eukaryotic genes was based on the assumption that efficient synthesis of the mRNA from strong bacterial promoters would lead to efficient synthesis of the gene product. During the past few years, however, it has become clear that overproduction of the mRNA does not guarantee overproduction of the protein. One serious problem is degradation of the foreign gene product by *Escherichia coli* proteases. This problem is most severe with proteins of low molecular weight (usually less than 10,000) and was overcome by creating gene fusions[1,2] and polyproteins.[3]

A second, perhaps more elusive cause of low protein yield, is inefficient translation of some mRNAs containing eukaryotic sequences. Initially, it was thought that eukaryotic sequences would be efficiently translated from bacterial ribosome-binding sites that are normally associated with abundant bacterial proteins. However, this is not always the case.[4-6] Extensive studies in several laboratories revealed that efficient translation requires that the distance between the Shine–Dalgarno (SD) sequence[7] and the translational start codon, and the nucleotide composition in this "spacer" region, be optimized for each gene, usually by empirical methods.[8-10] Through this work it was also recognized that the sequence at the 5′ end of the eukaryotic gene strongly influences translational efficiency. Further support for this idea came from several successful attempts to improve the translational efficiency by altering the codon selection at the 5′ end of genes without altering the amino acid sequence.[11,12] In addition,

[1] D. V. Goeddel, D. G. Kleid, F. Bolivar, H. L. Heyneker, D. G. Yansura, R. Crea, T. Hirose, A. Kraszewski, K. Itakura, and A. D. Riggs, *Proc. Natl. Acad. Sci. U.S.A.* **76,** 106 (1979).

[2] K. Itakura, T. Hirose, R. Crea, A. D. Riggs, H. L. Heyneker, F. Bolivar, and H. W. Boyer, *Science* **198,** 1056 (1977).

[3] S.-H. Shen, *Proc. Natl. Acad. Sci. U.S.A.* **81,** 4627 (1984).

[4] B. E. Schoner, H. M. Hsiung, R. M. Belangaje, N. G. Mayne, and R. G. Schoner, *Proc. Natl. Acad. Sci. U.S.A.* **81,** 5403 (1984).

[5] K. Nagai and H. C. Thogersen, *Nature (London)* **309,** 810 (1984).

[6] R. Varadarajan, A. Szabo, and S. G. Boxer, *Proc. Natl. Acad. Sci. U.S.A.* **82,** 5681 (1985).

[7] J. Shine and L. Dalgarno, *Nature (London)* **254,** 34 (1975).

[8] M. G. Shepard, E. Yelverton, and D. V. Goeddel, *DNA* **1,** 125 (1982).

[9] H. A. de Boer, A. Hui, L. J. Comstock, E. Wong, and M. Vasser, *DNA* **2,** 231 (1983).

[10] A. Hui, J. Hayflick, K. Dinkelspiel, and H. A. de Boer, *EMBO J.* **3,** 623 (1984).

[11] P. H. Seeburg, S. Sias, J. Adelman, H. A. de Boer, J. Hayflick, P. Jhurani, D. V. Goeddel, and H. L. Heyneker, *DNA* **2,** 37 (1983).

[12] H. J. George, J. J. L'Italien, W. P. Pilacinski, D. L. Glassman, and R. A. Krzyzek, *DNA* **4,** 273 (1985).

certain derivatives of eukaryotic proteins with minor modifications at their amino termini showed dramatic differences in expression levels.[4] These observations did not predict, however, which bases are important nor did they reveal the mechanism by which these bases exert their effect.

One possibility is the formation of local secondary structure in the mRNA involving the 5′ untranslated region and the 5′ end of the coding region. These structures can decrease translation initiation dramatically when the SD sequence and/or the initiation codon are sequestered.[13,14] In some mRNAs, secondary (and perhaps tertiary) structures form because of interactions between distal regions in the mRNA. Currently available computer programs are not adequate to predict such complex sequence interactions, particularly in the absence of corroborating genetic analyses.[15] Therefore, the development of universal expression vectors for these genes has not been particularly successful.

In this chapter, we describe a new strategy to improve the translational efficiency of hybrid mRNAs. In the currently available expression systems, the ribosome-binding site usually encompasses the 5′ end of the gene,[16–19] which severely limits the number of base changes that can be introduced to optimize translation initiation. To eliminate this problem, we have placed a short coding sequence (the first cistron) in front of the gene to be expressed (the second cistron), such that the 5′ end of the gene is no longer part of the ribosome-binding site.[4,20] We refer to this arrangement as a two-cistron expression system. The role of the first cistron in this system is to provide an optimal sequence for translation initiation. Since the polypeptide encoded by the first cistron has no apparent function, its sequence can be manipulated and optimized for mRNA translation without the need to preserve a particular amino acid sequence. Further, since the first cistron separates the second cistron from the 5′ untranslated region, potential local interactions between these sequences are minimized. Initially, we designed a "generic" first cistron sequence for the expression of bovine growth hormone (bGH). Based on sequence changes that we subsequently introduced into this first cistron, we have

[13] G. Simons, E. Remaut, B. Allet, R. Devos, and W. Fiers, *Gene* **28,** 55 (1984).

[14] P. Stassens, E. Remaut, and W. Fiers, *Gene* **36,** 211 (1985).

[15] M. Zuker and P. Stiegler, *Nucleic Acids Res.* **9,** 133 (1981).

[16] L. Gold, D. Pribnow, T. Schneider, S. Shinedling, B. S. Singer, and G. Stormo, *Annu. Rev. Microbiol.* **35,** 365 (1981).

[17] J. A. Steitz, *Nature (London)* **224,** 957 (1969).

[18] J. Hindley and D. H. Staples, *Nature (London)* **224,** 964 (1969).

[19] S. L. Gupta, J. Chen, L. Schaefer, P. Lengyel, and S. M. Weissman, *Biochem. Biophys. Res. Commun.* **39,** 883 (1970).

[20] B. E. Schoner, R. M. Belagaje, and R. G. Schoner, *Proc. Natl. Acad. Sci. U.S.A.* **83,** 8506 (1986).

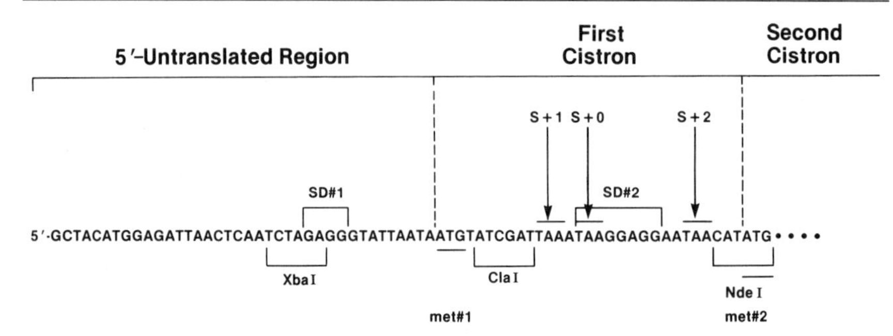

FIG. 1. DNA sequence of a functional first cistron. The sequence 5′ to the ATG codon for Met #1 is identical to the 5′ untranslated region (including the SD sequence) of the *E. coli lpp* mRNA; the sequence 3′ to the ATG #1 represents the synthetic first cistron. Restriction sites and SD sequences are indicated by brackets, stop codons are indicated by arrows.

established some general rules and have shown that these rules also apply to the expression of other eukaryotic genes that were substituted for bGH.[20]

Design of a Two-Cistron Expression System

The basic components of a two-cistron expression system are (1) a multicopy plasmid containing a strong promoter and a ribosome-binding site, (2) a chemically synthesized first-cistron sequence, and (3) the coding sequence of the protein to be expressed at a high level. In this section, we discuss the main considerations in designing a first cistron and present an example (Fig. 1). In the next section, we describe how this cistron can be cloned to create a functional two-cistron expression system.

Because the purpose of the first cistron is to improve the translational efficiency of any gene, one important consideration in designing the first cistron is that the sequence be "portable" such that it can be readily inserted into a variety of existing expression plasmids. For this purpose, we placed an *Xba*I restriction site at the 5′ end of the portable region. The site is already present in a number of useful expression vectors constructed by Masui *et al.*[21] At the 3′ end of the cistron, we placed an *Nde*I site (CATATG). The ATG within the *Nde*I recognition sequence can serve as a translational initiation codon for any gene properly inserted at the *Nde*I site. The 5′ untranslated region (Fig. 1) is identical to the native *lpp* 5′ untranslated region and was used in our constructions because it

[21] Y. Masui, J. Coleman, and M. Inouye, *in* "Experimental Manipulation of Gene Expression" (M. Inouye, ed.), p. 15. Academic Press, New York, 1983.

was reported to be free of secondary structure and because of the convenient location of the *Xba*I restriction site just upstream from the SD sequence.[22]

The first cistron is 31 bases in length and extends from the ATG for Met #1 to the ATG for Met #2. Because we anticipated that the cistron would be translated, we kept it relatively short to avoid unnecessary expenditure of energy and amino acid precursors for the peptide synthesis. Except for the assignment of ATG #1 and ATG #2, and the inclusion of an SD sequence and a stop codon (as discussed below), the cistron sequence is potentially flexible. Since we could not predict exactly what specific bases would be optimal for translation initiation, we enriched the sequence for AT bases (23 out of 31 positions) to discourage the formation of secondary structures that might involve the SD sequence and/or the ATG #1 start codon. The addition of a *Cla*I restriction site facilitated the introduction of subsequent sequence changes that became necessary. (The T between the ATG and the *Cla*I site was included to break up a *dam* methylation site that would have prevented digestion with *Cla*I.)

Although the sequence shown in Fig. 1 contains three nonsense codons (TAA), only one, (S + 2), is in frame with the ATG #1. When the reading frame was shifted by base insertions or deletions, translation initiated at the ATG #1 is terminated at the S + 1 or S + 0 stop codons (Fig. 2).[20] We found that when this occurs, the protein encoded by the second cistron is expressed at a very low level, presumably because the ribosomes do not continue translation into the second cistron. Thus, we defined a "functional" first cistron as one that allows for read-through translation and for high-level expression of the protein encoded by the second cistron. For high-level expression of this protein, the stop codon that is in frame with the ATG #1 must lie 3′ to the SD #2 and within a few bases (and 5′ to) the restart codon. In addition, we have shown by deletion that the SD #2, spaced properly with respect to the translational start codon for the second cistron, is also required (Fig. 2).[20]

Methods and Reagents

Bacterial Strains and Plasmids

E. coli K-12 RV308[23] [su^-, Δ*lac*X74, *gal*ISII::OP308, *str*A] was the host strain for all two-cistron expression plasmids. Some of the intermedi-

[22] K. Nakamura, R. M. Pirtle, I. L. Pirtle, K. Takeishi, and M. Inouye, *J. Biol. Chem.* **255,** 210 (1980).

[23] R. Maurer, B. J. Meyer, and M. Ptashne, *J. Mol. Biol.* **139,** 147 (1980).

Sequence	Plasmid	% bGH
TCTAGAGGGTATTA —————— ∧ —————— CATATG	pCZ 140	< 0.4
TCTAGAGGGTATTAATA [ATG] TATCGATTTAAA [TAA] GGAGGAA TAA CATATG	pCZ 143	< 0.4
TCTAGAGGGTATTAATA [ATG] TATC (GC ↓) GATTTAAA TAA GGAGGAA [TAA] CATATG	pCZ 144	24
TCTAGAGGGTATTAATA [ATG] TATCGAT ∧ TAAA TAA GGAGGAA [TAA] CATATG	pCZ 145	24
TCTAGAGGGTATTAATA [ATG] TATCGATTTAAA ∧ A ——— ∧ (GT) [TAA] CATATG	pCZ 148	< 0.4

FIG. 2. DNA sequences of two-cistron expression systems for Met-bGH. The sequences shown extend from the *Xba*I site in the 5′ untranslated region to the *Nde*I site at the beginning of the bGH gene. The sequence changes are relative to the sequence in pCZ143. The arrows indicate positions at which bases were inserted into the sequence, the carets (∧) indicate deletions. The predicted reading frame of the first cistron in each plasmid begins at the boxed in ATG codon and terminates at the boxed in TAA codon. The amounts of bGH measured in cells containing these plasmids are expressed as percentage of total cell protein.

ate plasmids in the constructions were propagated in *E. coli* K-12 RRI [*hsd*S20(r^-_B,m^-_B), *ara*-14, *leu*B6, *pro*A2, *lac*Y1, *gal*K2, *rps*L20, *xyl*-5, *mtl*-1, *sup*E44] ATCC #31343. Plasmid pIMAI was obtained from M. Inouye,[21] the bGH-cDNA clone was obtained from W. Miller and J. D. Baxter,[24] and the human protein C-cDNA clone was isolated by Beckmann *et al.*[25]

Enzymes

Conditions for the use of restriction endonucleases and T4 DNA ligase were those recommended by the manufacturers (New England Biolabs or Boehringer–Mannheim).

[24] W. L. Miller, J. A. Martial, and J. D. Baxter, *J. Biol. Chem.* **255,** 7521 (1980).

[25] R. J. Beckmann, R. J. Schmidt, R. F. Santerre, J. Plutzky, G. R. Crabtree, and G. L. Long, *Nucleic Acids Res.* **13,** 5233 (1985).

Gel Purification of DNA Restriction Fragments

Plasmid DNA was digested with appropriate restriction endonucleases, and the resulting fragments were separated on horizontal 0.8–1.2% (w/v) agarose gels. After electrophoresis in Tris–phosphate buffer (100 m*M* Tris–phosphate, 10 m*M* EDTA), the gel was stained with ethidium bromide (0.5 μg/ml) and the desired fragment was located using a 360-nm ultraviolet lamp. The desired fragment was excised and the DNA recovered by electroelution onto a strip of DEAE membrane (Schleicher and Schull, NA45, prepared according to the manufacturer's instructions). The membrane strip was rinsed several times in buffer containing 150 m*M* NaCl, 10 m*M* Tris–HCl, pH 7.8, prior to elution of the DNA in 300 μl of buffer containing 1 *M* NaCl, 10 m*M* Tris–HCl, pH 7.8. The DNA was precipitated by the addition of 3 vol (900 μl) of 95% ethanol and collected by centrifugation.

Chemical Synthesis of DNA Linkers

All deoxyribonucleotides were synthesized with a DNA synthesizer (Applied Biosystems, model 380A) according to the procedure recommended by the manufacturer. The oligonucleotides were purified by gel electrophoresis or by HPLC on a Whatman Partisil-10 SAX column using a 60% formamide (pH 6.5) potassium phosphate buffer gradient (0.001–0.4 *M*).[26] Several fractions were collected and desalted on a DE-52 ion exchange column before lyophilization. The lyophilized fractions were dissolved in 10 m*M* Tris–HCl, pH 7.5, and a small sample was phosphorylated with [^{32}P]ATP and polynucleotide kinase to determine the purity and size of the oligonucleotides by electrophoresis in denaturing polyacrylamide gels containing 7 *M* urea and autography.[27] In general, the desired oligonucleotides were found in the final fractions. Longer DNA fragments were prepared by joining oligonucleotides (synthesized as described above) enzymatically using T4 DNA ligase.

Cloning Synthetic Linkers

Linkers with Blunt Ends. If the fragment to which the linker was ligated had 5′ extensions, these were "filled in" prior to the addition of the linker. To fill in 5′ extensions, about 1 μg (or less) of plasmid DNA, digested with an appropriate restriction enzyme, was treated with 1 U of

[26] M. J. Gait and R. C. Sheppard, *Nucleic Acids Res.* **8,** 1081 (1980).

[27] E. L. Brown, R. Belagaje, M. J. Ryan, and H. G. Khorana, this series, Vol. 68, p. 109.

DNA polymerase I (the Klenow fragment) for 30 min at room temperature in 10–20 μl of buffer containing 50 m*M* Tris–HCl, pH 7.2, 10 m*M* $MgSO_4$, 1 m*M* dithiothreitol, and the four deoxyribonucleotides (each 1 m*M*) dissolved in 10 m*M* Tris–HCl, pH 8.0. The linkers (in TE, 10 m*M* Tris–HCl, pH 7.8, 1 m*M* EDTA) were heated to 95° and cooled to room temperature. They were ligated (using T4 DNA ligase) with the Klenow-treated fragment in at least 100 *M* excess at 20° for several hours. Because the linkers are not phosphorylated, they cannot self-ligate, which ensures that only one copy of the linker is cloned.

Linkers with Compatible Ends. Chemically synthesized linkers were annealed by heating them in equal molar amounts to 95° and slowly cooling them to 15° in a large beaker of water. They were then ligated in a 100 *M* excess with appropriate, gel-purified restriction fragments, using T4 DNA ligase.

Transformation and DNA Isolation

Plasmid DNA was transformed into RV308 by the $CaCl_2$ procedure,[28] and the cells were plated on TY media [1% (w/v) tryptone, 0.5% (w/v) NaCl, 0.5% (w/v) yeast extract, adjusted to pH 7.2 with NaOH] with 100 μg/ml kanamycin. The plates were incubated at 25° for 2 days. Recombinants that grew were streaked for single colonies on the same medium. For plasmid isolation, a single colony was inoculated into 1.5 ml TY broth (with 100 μg/ml kanamycin), grown overnight at 25°, and in the morning diluted (1 : 20) into fresh broth. The cultures were incubated at 37° for 1–3 hr and harvested. Plasmid DNA was extracted by the method of Birnboim and Doly[29] and examined by restriction analysis.[30] In all plasmids where base changes were made, the region containing the change was sequenced by the method of Maxam and Gilbert.[31] Throughout these procedures it is very important to maintain the cultures at the specified temperature because we found that plasmids containing the "runaway" replicon begin to amplify at temperatures above 25°.

Protein Production

For protein production, overnight cultures grown in TY broth containing 100 μg/ml of kanamycin at 25° were diluted 1 : 100 into the same medium and growth was continued at 25° for 2–3 hr. The temperature was

[28] P. C. Wensink, D. J. Finnegan, J. E. Donalson, and D. S. Hogness, *Cell* **3,** 315 (1974).
[29] H. C. Birnboim and J. Doly, *Nucleic Acids Res.* **7,** 1513 (1979).
[30] K. J. Danna, this series, Vol. 65, p. 449.
[31] A. M. Maxam and W. Gilbert, this series, Vol. 65, p. 499.

shifted to 37° and, after 6 hr, 1-ml cultures were pelleted for polyacrylamide/SDS gel analysis and scanning. The cells were also examined under the phase-contrast microscope (970× magnification) for the presence of protein granules. It has been our experience that these granules form in cells overproducing bGH and other eukaryotic proteins.[32] Since these granules are clearly visible with the phase-contrast microscope, we were able to screen quickly a large number of cultures following transformation to distinguish between producers and nonproducers.

Polyacrylamide Gel Electrophoresis

Polyacrylamide gels (12.5%, w/v) were used to analyze cells for protein production.[33] Cells from 1 ml of thermally induced cultures were pelleted by centrifugation for 1 min in an Eppendorf centrifuge. The liquid was carefully decanted, and the cell pellet was dissolved in 300 μl of modified Laemmli sample buffer containing the following: 125 m*M* Tris–HCl, pH 6.8; 2% (w/v) SDS; 30% (v/v) glycerol; 1 *M* 2-mercaptoethanol, and 6 *M* urea. The urea and the high concentration of 2-mercaptoethanol were used to solubilize the protein granules that form when bGH and other proteins are overproduced in *E. coli*. The samples were boiled for 3 min and usually 20 μl was loaded onto a discontinuous gel (18 cm × 18 cm × 1.5 mm) described by Laemmli.[33] Electrophoresis was for 3 hr at 50 mA. After electrophoresis, the gels were stained with Coomassie blue and destained in 10% (v/v) acetic acid. The gels were dried at 80° under vacuum and scanned with a Shimadzu scanner, model 930.

Plasmid Constructions

In this section, we describe two approaches for constructing two-cistron expression plasmids. In the first case, we assume that the gene of interest is already cloned into an expression vector, but no protein production is observed. In the second case, we assume that the cDNA clone for the gene of interest is available and that the DNA sequence of the insert is known, but no prior knowledge exists of what is required to express the gene at a high level. To illustrate the method, we describe the individual steps in the construction of two-cistron expression vectors for bovine growth hormone (Fig. 3) and human protein C (HPC; Fig. 4).

The starting plasmid for both approaches was pIMAI.[21] This plasmid was derived from pKN402[34] and contains a thermoinducible runaway

[32] R. G. Schoner, L. F. Ellis, and B. E. Schoner, *Bio/Technology* **3,** 151 (1985).

[33] U. K. Laemmli, *Nature* (*London*) **227,** 680 (1970).

[34] B. E. Uhlin, V. Schweickart, and A. J. Clark, *Gene* **22,** 255 (1983).

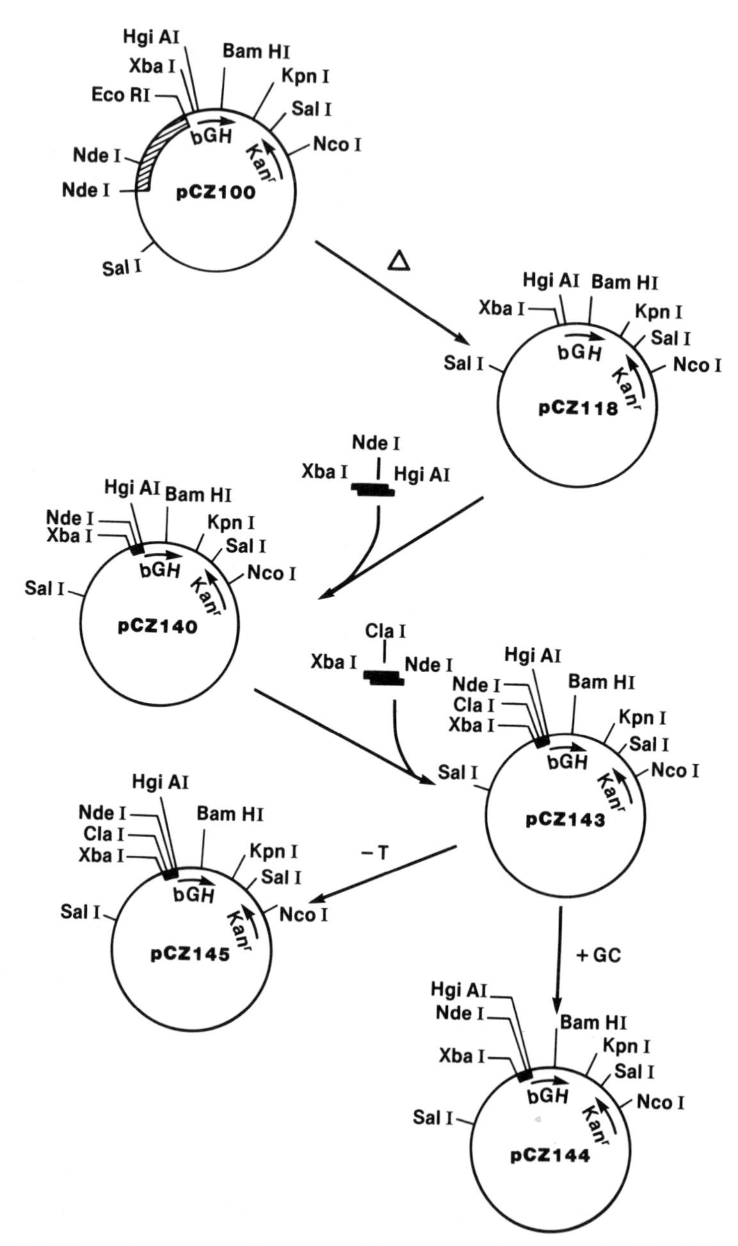

FIG. 3. Scheme for constructing a two-cistron expression vector for Met-bGH. The arrows show the location and the direction of transcription of the bGH gene and the kanamycin phosphotransferase (Kan^r) gene. The plasmids are not drawn to scale.

replicon. At temperatures below 30°, the plasmid copy number is maintained at about 10 copies per cell; a shift to 37° causes uncontrolled plasmid replication and leads to the accumulation of about 2000 copies per cell. pIMAI contains the constitutive *E. coli lpp* promoter; therefore, the expression of genes that depend on the *lpp* promoter can be controlled only indirectly by thermally regulating plasmid copy number. This type of regulation was adequate for producing bGH and HPC and certain other eukaryotic and prokaryotic proteins as long as the low-temperature growth was properly maintained such that mutant plasmids do not accumulate in the cultures. When we replaced the *lpp* promoter with the regulated *trp* promoter for the expression of bGH, we did not observe any significant differences in plasmid stability or the level of bGH expression.[4]

The pIMAI vector contains several convenient features including the presence of a unique *Xba*I site located between the *lpp* promoter and the *lpp* SD sequence and a polylinker for *Eco*RI, *Hin*dIII, and *Bam*HI at the 5′ end of the lipoprotein gene.[21] The region beyond the polylinker codes for the 3′ end of the *lpp* mRNA and has extended regions of self-complementarity that can potentially form stable hairpin structures and thus impart stability to hybrid mRNAs.[35] Transcription from the *lpp* promoter is terminated by a *rho*-independent transcription termination signal.[35] The selectable marker on pIMAI is the kanamycin resistance gene from Tn*5*.[21]

The bGH gene (with its native codons) does not express at high levels in conventional vectors that have been successfully used to overproduce other eukaryotic proteins.[4,12] Therefore, we chose this gene to determine if the use of a two-cistron system would improve its expression. To convert the bGH expression plasmid, pCZ100, into a two-cistron expression plasmid (Fig. 3) the 5′ end of the bGH coding region was first reconstructed prior to the addition of the first cistron sequence. We started with pCZ100, which was derived from pIMAI and contains the coding sequence for bGH. The two *Nde*I sites in pCZ100 were eliminated (to obtain pCZ118) by digesting pCZ100 with *Nde*I and *Eco*RI, treating the fragments with DNA polymerase I (Klenow fragment), and reclosing the plasmid with T4 DNA ligase. An *Hgi*AI site near the 5′ end of the bGH-coding sequence was used to reconstruct the 5′ end of the gene. This was accomplished by chemically synthesizing a 66-bp *Xba*I to *Hgi*AI fragment (that contains an *Nde*I site) and cloning it into pCZ118. Since *Hgi*AI cuts only once in the bGH gene, but several times elsewhere in the plasmid, the synthetic fragment was cloned in a three-piece ligation reaction containing an *Xba*I to *Bam*HI vector fragment, an *Hgi*AI to *Bam*HI bGH fragment, and the synthetic 66-bp fragment. The desired recombinant was

[35] R. M. Pirtle, I. L. Pirtle, and M. Inouye, *J. Biol. Chem.* **255,** 199 (1980).

named pCZ140. To construct pCZ143, pCZ140 DNA was digested with *Xba*I and *Nde*I and the larger fragment was gel purified and ligated with a chemically synthesized *Xba*I to *Nde*I fragment containing the first cistron sequence. In this sequence, the S + 0 stop codon is in frame with the initiator ATG for the first cistron. Hence no read-through translation into the bGH gene occurs and very low amounts of bGH are produced. We subsequently altered the first cistron sequence by either inserting two bases (GC) at the *Cla*I site (in pCZ144) or by deleting one base (T) at the *Aha*III site (in pCZ145) to shift the reading frame of the first cistron to the S + 2 stop codon. These changes increased the expression level of bGH from less than 0.4% of total cell protein (with pCZ143) to over 20% of total cell protein (with pCZ144 and pCZ145).

Another approach for constructing two-cistron expression plasmids is illustrated in Fig. 4. Using HPC as an example, we have outlined steps to clone a gene directly into a two-cistron expression vector without examining first if high-level expression requires the use of a cistron. We chose this approach because we had shown previously that placement of a functional first cistron in front of genes that can be expressed without a first cistron does not lower the expression level.[20]

In the scheme outlined below, the starting plasmids were (1) pCZ11, a derivative of pIMAI that contains the first cistron sequence and appropriate restriction sites for inserting HPC, and (2) the cDNA clone for HPC. Prior to cloning the synthetic first cistron sequence into pIMAI to create pCZ11, we needed to eliminate the two existing *Nde*I sites from pIMAI (for details, see construction of pCZ118) and introduce a new *Nde*I site into the polylinker region. This was accomplished by digesting pIMAI with *Bam*HI, filling in the 5′ extensions with DNA polymerase I (Klenow fragment), and religating the plasmid in the presence of *Nde*I linkers. The resulting intermediate plasmid (with the new *Nde*I site) was digested with *Xba*I and *Nde*I and the synthetic cistron fragment was inserted. This plasmid, pCZ11, can be used for cloning HPC and any other gene that is bounded by *Nde*I and *Bam*HI restriction sites.

The addition of an *Nde*I site to the 5′ end of the HPC gene was accomplished through the use of a synthetic DNA linker. As shown in Fig. 4, the HPC gene contains two internal *Bam*HI sites, one of which is located in the leader peptide region and the other near the 3′ end of the gene. Since we wanted to express a derivative of mature protein C that includes 10 amino acids from the leader peptide, a 20-bp linker was synthesized which is bounded by an *Nde*I site and a *Bam*HI site and codes for the translational start (ATG) and six amino acids of the leader peptide. The ATG located within the *Nde*I restriction site is aligned with the downstream reading frame of the HPC gene. This synthetic linker was

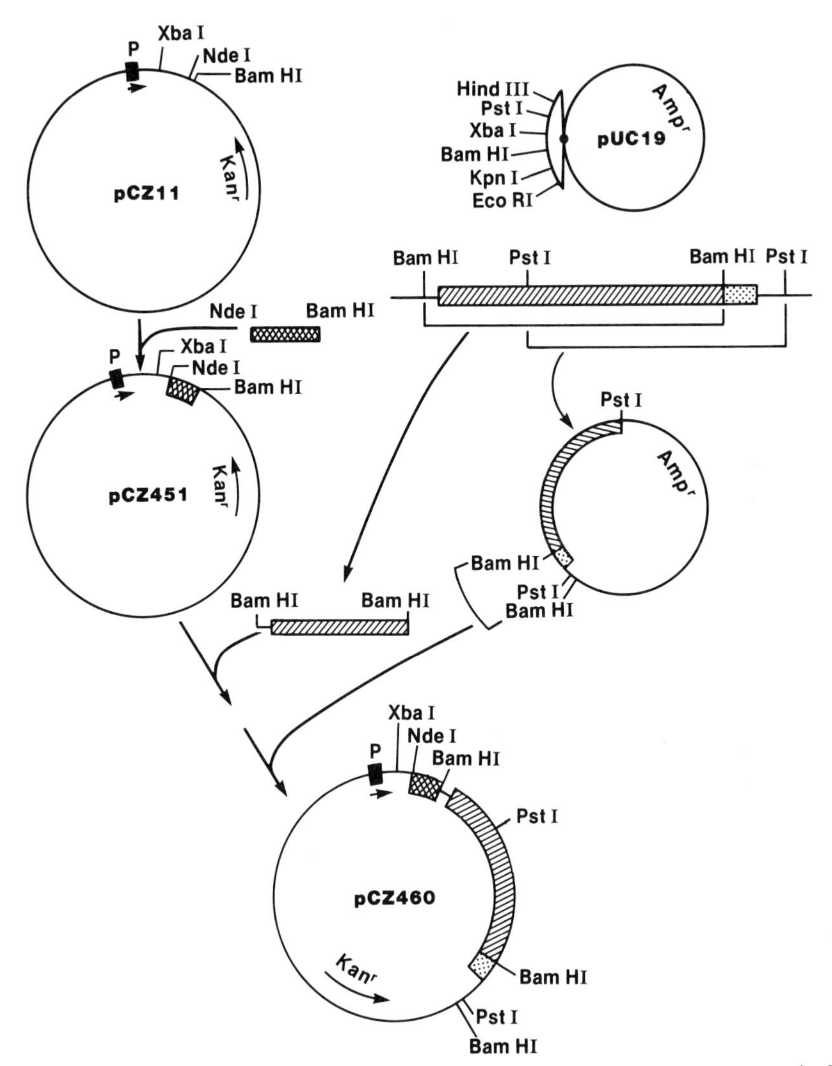

FIG. 4. Scheme for constructing a two-cistron expression vector for human protein C (HPC). The cross-hatched segment represents the 5′ end of the HPC gene, the striped segment represents the central portion of the gene, and the stippled segment represents the 3′ end of the gene. The arrow shows the location and the direction of transcription of the kanamycin phosphotransferase (Kan^r) gene. The plasmids are not drawn to scale.

cloned into *Nde*I- and *Bam*HI-digested pCZ11, yielding pCZ451. To add the remaining coding sequence for HPC to pCZ451, we needed to place a *Bam*HI site at the 3′ side of the HPC-coding region. This was accomplished by first cloning a *Pst*I fragment (containing the 3′ end of the HPC gene) into the *Pst*I site of the polylinker region in pUC19. When this *Pst*I fragment is inserted in the orientation shown in Fig. 4, the *Bam*HI site in the polylinker of pUC19 lies proximal to the *Bam*HI site near the 3′ end of HPC. Thus, an 80-bp fragment could be isolated that contains the 3′ end of the HPC gene as well as its translational stop codon (TAG). This small 80-bp *Bam*HI fragment and the large *Bam*HI fragment (containing most the HPC-coding sequence) were inserted in two successive steps into pCZ451, yielding plasmid pCZ460. This construction was somewhat complicated by the inconvenient distribution of restriction sites (including the two *Bam*HI sites) in the HPC cDNA clone. To simplify constructions of this type, it is usually possible to find a unique restriction site in the 3′ untranslated region of a cDNA clone. This site can then be used to transfer the entire gene from the cDNA clone into pCZ11, which only requires the reconstruction of the 5′ end of the gene and the conversion of the *Bam*HI site in pCZ11 to the unique restriction site in the 3′ untranslated region of the gene.

While we have outlined above two specific approaches for constructing two-cistron expression plasmids, only slight modifications are required to adapt this scheme to other eukaryotic and prokaryotic genes. Any one of several expression vectors (described elsewhere in this volume) can serve as starting plasmids.[36,37] At a minimum, the vector must contain a selectable marker, a promoter, and preferably two unique cloning sites, just downstream from the promoter. In general, plasmids that allow for controlled gene expression either through regulated transcription and/or plasmid copy number are most appropriate.

Summary and Comments

Certain hybrid mRNAs containing eukaryotic sequences are not efficiently translated in *E. coli,* even though they possess a "good" ribosome-binding site.[4–6] This has considerably limited the development of universal vectors for the high-level expression of eukaryotic genes in *E. coli*. Studies in several laboratories revealed that the sequence context of the ribosome-binding site (which extends into both the 5′ untranslated region and the 5′ end of the coding region) is important for efficient

[36] E. Remaut, A. Marmenout, G. Simons, and W. Fiers, this volume [26].
[37] G. Milman, this volume [30].

translation.[4,10,11] As different genes are cloned into expression vectors, the sequence context of the ribosome-binding site necessarily changes and in some cases lowers the efficiency of ribosome binding and translation initiation. Hence, the observed yields of desired proteins are substantially lowered.

To preserve the sequence context around the ribosome-binding site, we have designed the two-cistron expression system described in this chapter. In this system, the first cistron contains all the information necessary for efficient translation initiation and the second cistron contains the coding information for the gene of interest. For maximal expression of the second cistron, the stop codon that terminates translation of the first cistron must be located 3′ to the SD sequence for the second cistron and within a few bases and 5′ to the restart codon for the second cistron. An SD sequence for the second cistron must be included and must be properly spaced with respect to the ATG start codon for the second cistron. In principle, any sequence that allows for efficient ribosome binding and translation initiation should be suitable as a first cistron. Such sequences can be synthesized chemically and inserted into appropriate expression vectors. This approach provides the greatest amount of flexibility regarding the choice of a cistron sequence, its length, and the distribution of suitable restriction sites.

Alternatively, two-cistron expression systems can be derived directly from "one" cistron expression systems that encode hybrid (or fusion) proteins. Hybrid proteins are created by inserting protein-coding sequences into expression vectors, at restriction sites that are located downstream from translational start codons for the synthesis of these proteins.[38] Although usually small, the number of additional amino acids attached to the amino termini of these hybrid proteins is dictated by the number of codons that exist between the translational start and the restriction site. Larger hybrid proteins are typically formed when the eukaryotic coding sequences are fused in phase to highly expressed *E. coli* genes such as those encoded by *trpE*,[39,40] *lacZ*,[1,2] or λCII.[5,6] These larger hybrid proteins are often biologically inactive and the native protein must be released either by cyanogen bromide treatment or by enzymatic cleavage at specific amino acid residues. In some cases, especially when the eukaryotic protein is relatively large, it may be desirable to express the

[38] R. Crowl, C. Seamans, P. Lomedico, and S. McAndrew, *Gene* **38,** 31 (1985).

[39] A. R. Davis, D. P. Nayak, M. Ueda, A. L. Hiti, D. Dowbenko, and D. G. Kleid, *Proc. Natl. Acad. Sci. U.S.A.* **78,** 5376 (1981).

[40] D. G. Kleid, D. Yansura, B. Small, D. Dowbenko, D. M. Moore, M. J. Grubman, P. D. McKercher, D. O. Morgan, B. H. Robertson, and H. L. Bachrach, *Science* **214,** 1125 (1981).

"unfused" protein directly in a two-cistron expression vector. This simply entails the conversion of the mRNA encoding the fusion protein to a two-cistron mRNA. In general, such conversions require only a few base changes that create a translational stop codon at the end of the *E. coli* sequence, a translational start codon for the eukaryotic gene, and an SD sequence near the end of the *E. coli* sequence and 5′ to the beginning of the eukaryotic gene. These changes can be introduced by DNA linker replacement or by *in vitro* mutagenesis.[41,42]

[41] D. Shortle and D. Botstein, this series, Vol. 100, p. 457.
[42] M. J. Zoller and M. Smith, this series, Vol. 100, p. 468.

[26] Expression of Heterologous Unfused Protein in *Escherichia coli*

By ERIK REMAUT, ANNE MARMENOUT, GUUS SIMONS, and WALTER FIERS

Introduction

One of the major achievements of recombinant DNA technology is the high-level expression of adventitious genes in *Escherichia coli*. In particular, a steadily increasing number of eukaryotic proteins that are difficult to obtain from their natural sources are now being produced in massive amounts by "engineered" bacteria. The cloning of a eukaryotic gene onto a bacterial plasmid and its introduction into *E. coli* does not, in general, lead to an efficient synthesis of the corresponding protein, because the eukaryotic DNA lacks the specific signals necessary for it to be recognized by the host's transcription–translation machinery. To remedy this problem, expression vectors have been developed that incorporate the essential control elements to ensure efficient transcription and translation of essentially any coding region in *E. coli*. The salient features of such an expression vector are (1) the presence of a strong and preferably regulatable promoter and (2) the presence of an efficient ribosome-binding site and initiation codon which are easily accessible for insertion of adventitious coding sequences. The most widely used strategy to clone eukaryotic genes is reverse transcription of (semi-) purified mRNA. The cDNA clones obtained in this way contain sequences upstream from the coding region that are not suitable for expression in bacterial systems. In cases where the mRNA codes for a secreted protein the coding region of the mature protein is preceded by the signal peptide which is removed during the process of secretion. In order to obtain efficient expression of the mature protein in *E. coli,* the cDNA needs to be tailored in such a way that thc codon for the first amino acid of the mature protein can be precisely fused to the initiation codon of the bacterial expression vector.

This chapter describes the use of plasmid expression vectors based on the inducible leftward promoter of coliphage λ. These vectors have been engineered to incorporate bacterial ribosome binding sites designed for easy insertion of adventitious coding sequences. Examples of tailoring cloned cDNA to obtain precise fusion of coding sequences to the prokaryotic translation signals are discussed.

Principle of the Method

There are two aspects to be considered in the overall design of an expression system for unfused proteins: (1) construction of an expression vector suitable for easy insertion of adventitious coding sequences, and (2) methods to tailor coding sequences such that these can be precisely joined to the expression signals of the vector.

1. Design of an Expression Vector

The expression vectors used in this work are derivatives of pBR322[1] containing a 247-bp DNA fragment carrying the leftward operators and leftward promoter (P_L) of phage λ.[2] This strong promoter is tightly regulated by a repressor protein, product of the phage gene *cI*. Mutants are available that synthesize a thermolabile repressor.[3] This property allows the control of the activity of the P_L promoter by a simple temperature shift. The P_L expression plasmids[2] are propagated at 28° in *E. coli* strains which synthesize a temperature-sensitive repressor. While repression of the P_L promoter is virtually complete at 28°, full activity is obtained by raising the temperature to 42°. The ability to regulate expression is an important feature of an expression system since continuous high-level expression of an adventitious protein may be lethal to the cell.[4] The *cI* repressor product can be supplied to the system in a variety of ways.

[1] F. Bolivar, R. L. Rodriguez, P. Y. Greene, M. C. Betlach, H. L. Heynecker, H. W. Boyer, Y. H. Crosa, and S. Falkow, *Gene* **2,** 95 (1977).

[2] E. Remaut, P. Stanssens, and W. Fiers, *Gene* **15,** 81 (1981).

[3] M. Lieb, *J. Mol. Biol.* **16,** 149 (1966).

[4] E. Remaut, P. Stanssens, and W. Fiers, *Nucleic Acids Res.* **11,** 4677 (1983).

Strains of *E. coli* are available which synthesize a thermolabile repressor from a defective prophage, e.g., strain K12ΔH1Δ*trp*.[5] Alternatively, the mutant *cI* gene can be present on a compatible plasmid such as pRK248cIts,[5] pcI857,[6] or on an F′ episome.[7] The dual plasmid system allows the use of essentially any *E. coli* strain as a host for P_L expression vectors. This versatility is of special interest, as it may be that proper choice of the host results in elevated levels of accumulation of a given protein. While the presence of a strong regulatable promoter on an expression plasmid generally ensures efficient transcription of cloned DNA sequences, efficient translation into protein depends on the presence on the cloned sequence of a ribosome-binding site that is recognized by the *E. coli* host. As outlined in the Introduction, eukaryotic genes are unlikely to display this feature. It is therefore necessary to develop expression vectors which incorporate a strong ribosome-binding site of bacterial origin. Moreover, the nucleotide sequence at the initiation codon has to be engineered in such a way that it becomes easily accessible for precise fusion of a eukaryotic coding region to the ATG codon. This is most frequently accomplished by "creating" a unique restriction site which partly overlaps with the initiation codon. Ways to achieve this end, some of which are detailed in the Methods and Results section, are outlined below (see Table I).

a. Removal of 5′ Protruding Ends with Single-Strand-Specific Nucleases. The initiation codon ATG is engineered in such a way that it is part of a restriction site recognized by a restriction enzyme generating 5′ protruding ends. Following cleavage with the restriction enzyme and removal of the 5′ protruding end with a single-strand-specific nuclease, the ATG codon can be exposed as a blunted end (see Table IA). Single-strand-specific nucleases currently available are *Aspergillus oryzae* S1 nuclease[8] and mung bean nuclease.[9] It should be noted that these nucleases are often not very reliable in their action, i.e., base-paired nucleotides adjacent to the single strand may be removed during the reaction.

b. Filling in of 5′ Protruding Ends. The enzyme *Nco*I cleaves the sequence 5′-CCATGG-3′, between the two C residues, generating 5′ protruding ends. These ends can be filled in using the 5′ to 3′ polymerizing reaction of DNA polymerases, thus generating a blunted ATG codon (see Table IB). The large fragment of *E. coli* DNA polymerase I (Klenow

[5] H.-U. Bernard, E. Remaut, M. V. Herschfield, H. K. Das, D. R. Helinski, C. Yanovsky, and N. Franklin, *Gene* **5,** 59 (1979).

[6] E. Remaut, H. Tsao, and W. Fiers, *Gene* **22,** 103 (1983).

[7] M. Mieschendal and B. Müller-Hill, *J. Bacteriol.* **164,** 1366 (1985).

[8] V. M. Vogt, *Eur. J. Biochem.* **33,** 192 (1973).

[9] W. D. Kroeker, D. Kowalski, and M. Laskowski, *Biochemistry* **15,** 4463 (1976).

TABLE I

METHODS TO EXPOSE THE ATG CODON OF AN EXPRESSION VECTOR AS A BLUNTED END[a]

	Specific example	Alternatives
A. Removal of 5′-protruding ends with single-strand-specific nucleases	5′-ATG↓AATTC- 3′-TACTTAA↑G- ↓ *Eco*RI 5′-ATG 3′-TACTTAA ↓ S1 5′-ATG 3′-TAC	5′-ATG↓TGCAC *Apa*LI 5′-ATG↓GTACC *Asp*718 5′-ATG↓GATCC *Bam*HI 5′-ATG↓CGCGC *Bss*HII 5′-ATG↓CTAGC *Nhe*I 5′-ATG↓TCGAC *Sal*I
B. Filling in of 5′- protruding ends	5′-C↓CATGG- 3′-GGTAC↑C- ↓ *Nco*I 5′-C 3′-GGTAC ↓ Klenow *Pol*I or T4 Pol 5′-CCATG 3′-GGTAC	Not known to date
C. Resection of 3′-protruding ends	5′-ATGGTAC↓C 3′-C↑CATGG ↓ *Kpn*I 5′-ATGGTAC 3′-TAC dGTP ↓ Klenow *Pol*I or T4 Pol 5′-ATG 3′-TAC	5′-ATGGGCC↓C *Apa*I 5′-ATGACGT↓C *Aat*II 5′-ATGGCGC↓C *Bbe*I 5′-ATGAGCT↓C *Sac*I 5′-ATGCATG↓C *Sph*I

[a] Examples of sequences in which an ATG codon partly overlaps with a restriction site are illustrated. Methods to expose the ATG codon as a blunted end are worked out for a specific example. Alternative possibilities are indicated with only one strand of the duplex DNA being displayed. Only hexameric recognition sequences were considered. The small arrows indicate cleavage points of the restriction enzymes.

fragment)[10] or T4 DNA polymerase are frequently used for this purpose. Both enzymes lack the 5′ to 3′ exonuclease activity. In the presence of deoxyribonucleoside triphosphates, the 3′ to 5′ exonuclease activity of these enzymes is inhibited in favor of the polymerizing reaction. The fidelity of this reaction is very high.

c. Resection of 3′ Protruding Ends. A number of restriction enzymes recognize a palindromic sequence starting with a 5′ G residue and cleave that sequence leaving 3′ protruding ends. These 3′ ends can be resected by the 3′ to 5′ exonuclease activity of, e.g., Klenow fragment or T4 DNA polymerase. When the reaction is performed in the presence of dGTP, 3′ to 5′ resection into the duplex DNA is inhibited, because the much more active 5′ to 3′ polymerizing activity takes over by exchanging the G residue opposite the C. Restriction sites of the type mentioned here can be engineered to partly overlap with an ATG codon so that the enzymatic reactions will result in blunt-end exposure of the ATG codon (see Table IC). As in the previous example the reaction can be easily controlled.

2. Tailoring of Coding Regions for Precise Fusion to the Initiation Signals of an Expression Vector

Only in rare cases will a cloned cDNA possess a restriction site positioned thus that the coding region can be precisely fused to the ATG codon of an expression vector by simple cleavage and joining reactions. Far more commonly one has to rely on a restriction site present some distance downstream from the desired fusion point. In this approach the gene is cleaved at the internal restriction site and synthetic DNA fragments are used to establish the link between the ATG codon of an expression vector and the restriction site of the gene. The nucleotide sequence of the DNA linkers is designed to restore the missing coding information between the desired point of fusion and the restriction site used (see Methods and Results section for specific examples). A prerequisite of this general approach is, of course, that the nucleotide sequence of the gene has been determined.

Materials and Reagents

Enzymes and Chemicals

1. Restriction enzymes: New England BioLabs, Beverly, Massachusetts, or Boehringer–Mannheim, West Germany

[10] H. Jacobsen, H. Klenow, and K. Overgaard-Hansen, *Eur. J. Biochem.* **45,** 623 (1974).

2. T4 DNA ligase: Purified from the overproducing strain C600 (pcI857) (pPLc28lig8)[6]
3. DNA polymerase I (Klenow fragment): Boehringer–Mannheim, West Germany
4. T4 DNA polymerase: New England BioLabs, Beverly, Massachusetts
5. S1 nuclease: Boehringer–Mannheim, West Germany
6. T4 polynucleotide kinase: New England Nuclear Corp., Boston, Massachusetts
7. Serva Blue R: Serva, Heidelberg, West Germany
8. LB medium: 1% Bacto tryptone, 0.5% Bacto yeast extract, 0.5% NaCl (Difco Corp., Detroit, MI)
9. Synthetic DNA linker fragments: Oligonucleotide linker fragments were synthesized following the phosphoramidite method.[11] The fragments were purified by HPLC chromatography
10. Gene Screen: New England Nuclear Corp., Boston, Massachusetts
11. EN^3HANCE: New England Nuclear Corp., Boston, Massachusetts

Bacterial Strains

All bacterial strains used are derivatives of *E. coli* K-12.

K12ΔH1Δ*trp*[5]: Sm^R, *lacZ*am, Δ*bio-uvrB*, Δ*trpE*A2, (λ*Nam*7*Nam*-53*cI*857ΔH1)

MC1061[pcI857][12,13]: Sm^R, *araD*139, Δ(*ara leu*)7697, Δ*lacX*74, *galU*, *galK*, $r_K^- m_K^+$, [pcI857]

C600[pcI857][13,14]: *thr*-1, *leuB*6, *thi*-1, *sup*E44, *lacY*1, *tonA*21, [pcI857]

GM119(λ)[15]: *dam*-3, *dcm*-6, *metB*1, *galK*2, *galT*22, *lacY*1, *tsx*-78, *supE*44(λwt)

Plasmids

pcI857: A derivative of pACYC177 that contains the *cI*857 allele. This plasmid specifies resistance to kanamycin (50 μg/ml) and is compatible with ColE1-derived replicons[6]

pPLc236*trp*: A ColE1-type plasmid derived from pBR322; specifies resistance to carbenicillin (100 μg/ml); carries the P_L promoter and a

[11] A. Chollet, E. Ayala, and E. Kawashima, *Helv. Chim. Acta* **67,** 1356 (1984).

[12] M. J. Casadaban and S. N. Cohen, *J. Mol. Biol.* **138,** 179 (1980).

[13] G. Simons, E. Remaut, B. Allet, R. Devos, and W. Fiers, *Gene* **28,** 55 (1984).

[14] R. K. Appleyard, *Genetics* **39,** 440 (1954).

[15] This strain was kindly provided by M. G. Marinus.

ribosome-binding site derived from the tryptophan attenuator region[13]

pPLc245: As above, except that the ribosome-binding site is derived from the replicase gene of the RNA phage MS2[4]

pPLcmu299: As above, except that the ribosome-binding site is derived from the *ner* gene of phage Mu[16]

pAT153: A general ColE1-type cloning vector, specifying resistance to carbenicillin and tetracycline[17]

Plasmid constructions engineered to give optimal expression of a specific cloned gene are detailed in the Methods and Results section.

Methods and Results

General DNA Methodology

Procedures for preparation of plasmid DNA, restriction and ligation of DNA fragments, as well as conditions for bacterial transformation have been described.[18]

S1 nuclease was used to remove 5′ protruding ends in a restriction buffer consisting of 25 m*M* sodium acetate, pH 4.4, 250 m*M* NaCl, 4.5 m*M* $ZnSO_4$. About 5 μg of CsCl-purified plasmid DNA was incubated in 100 μl of buffer with 100 U of enzyme at 18° for 30 min.

E. coli DNA polymerase I, Klenow fragment was used to fill in 5′ sticky ends. A typical reaction mixture contained 1 pmol of DNA termini and 1 U of enzyme in 30 μl of a buffer consisting of 25 m*M* Tris · HCl, pH 7.4, 10 m*M* $MgCl_2$, 10 m*M* dithiothreitol, and 0.1 m*M* of all four deoxyribonucleoside triphosphates. The reaction was performed at 18° for 45 min.

Resection of 3′ protruding ends was carried out with T4 DNA polymerase. The reaction was performed at 15° for 3 hr in a 30-μl reaction mixture containing 3 pmol of DNA termini in 65 m*M* Tris · HCl, pH 7.9, 20 m*M* KCl, 10 m*M* $MgCl_2$, 5 m*M* dithiothreitol, and 0.1 m*M* of any one of the four deoxynucleoside triphosphates.

Enzymatic reactions were stopped by phenol extraction. The aqueous layer was extracted twice with 5 vol of ethyl ether. The DNA was precipitated from the aqueous phase by addition of 2 vol of ethanol, 0.1 vol of 3 *M* potassium acetate, pH 4.5, and standing at −20° for at least 2 hr. The

[16] G. Buell, J. Delamarter, G. Simons, E. Remaut, and W. Fiers, *Basic Life Sci.* **30,** 949 (1985).

[17] A. J. Twigg and D. Sheratt, *Nature (London)* **283,** 216 (1980).

[18] E. Remaut, P. Stanssens, G. Simons, and W. Fiers, this series, Vol. 119, p. 366.

precipitated DNA was pelleted in an Eppendorf centrifuge (5 min at maximum speed). The pellet was washed with 70% ethanol and after careful draining resuspended in the desired buffer.

Specific DNA fragments were purified by electrophoresis in 0.8–2% agarose gels and recovered by the squeeze–freeze method[19] or in more recent experiments by filtration through Gene Screen membranes.[20] The latter procedure combines high yields of recovery and excellent quality of the DNA with respect to secondary restriction and ligation.

Construction of Expression Plasmids

1. Human Immune Interferon. cDNA clones containing the coding sequence for human immune interferon (IFN-γ) were obtained by reverse transcription of an enriched mRNA fraction prepared from splenocyte cultures stimulated with staphylococcal enterotoxin A.[21,22] The IFN-γ cDNA was originally obtained via oligo-dG tailing and annealing with a filled in *Bam*HI site which had been tailed with oligo-dC. In this way, the *Bam*HI site is restored and the fragment can be cleaved out with *Bam*HI.[22] The nucleotide sequence was determined using the chemical degradation method of Maxam and Gilbert.[23] As IFN-γ is a secreted protein, the coding sequence for mature protein is preceded by a signal peptide which is cleaved off during the process of secretion. At the time these studies were performed the exact N-terminus of the mature protein had not been determined. By analogy with the known N-terminal amino acid sequence of human IFN-α, the cleavage point was postulated to occur before a cysteine residue at position 21 in the IFN-γ coding sequence.[24,25] The constructions detailed below aimed at precise fusion of this cysteine codon to the ATG codon of an expression vector.[26]

[19] R. W. J. Thuring, J. P. H. Sanders, and P. Borst, *Anal. Biochem.* **66,** 213 (1975).

[20] J. Zhu, W. Kempenaers, D. Van der Straeten, R. Contreras, and W. Fiers, *Biotechnology* **3,** 1014 (1985).

[21] R. Devos, H. Cheroutre, Y. Taya, and W. Fiers, *J. Interferon Res.* **2,** 409 (1982).

[22] R. Devos, H. Cheroutre, Y. Taya, W. Degrave, H. Van Heuverswyn, and W. Fiers, *Nucleic Acids Res.* **10,** 2487 (1982).

[23] A. M. Maxam and W. Gilbert, *Proc. Natl. Acad. Sci. U.S.A.* **74,** 560 (1977).

[24] N. Mantei, M. Schwarzstein, M. Streuli, S. Panem, S. Nagata, and C. Weissman, *Gene* **10,** 1 (1980).

[25] P. W. Gray, D. W. Leung, D. Pennica, E. Yelverton, R. Najarian, C. C. Simonsen, R. Derynck, P. J. Sherwood, D. M. Wallace, S. L. Berger, A. D. Levinson, and P. V. Goedell, *Nature (London)* **295,** 503 (1982).

[26] It has since been established that mature IFN-γ starts with a GIN residue at position 24 of the coding region [E. Rinderknecht, B. H. O'Connor, and H. Rodriguez, *J. Biol. Chem.* **259,** 6790 (1984)].

The coding sequence for mature IFN-γ contains a unique *Ava*II site located at a position corresponding to the fifth amino acid of the mature protein. This site was used as a starting point for plasmid constructions. The *Ava*II site overlaps with an *Eco*RII site and is consequently refractory to cleavage because of C-methylation. To alleviate this problem, plasmids were propagated in a *dam* strain [GM119(λ)] which is defective in C-methylation. *Ava*II cleavage of IFN-γ DNA removes the first four amino acid residues of the mature protein. The missing coding information was restored using synthetic linker fragments as schematically shown in Fig. 1. As a result of the enzymatic reactions the TGT codon, corresponding to the presumptive N-terminal cysteine residue of the mature protein, was exposed as a blunt end. This end was then ligated to the ATG codon of an expression vector. In this approach we used a vector which carries the ATG codon as part of an *Eco*RI site. Following *Eco*RI cleavage, this site was blunted with S1 nuclease and ligated to the treated IFN-γ sequence. Full details of the constructions are given in the legend to Fig. 1. In the example discussed the vector contains the P_L promoter and a manipulated ribosome-binding site derived from the *E. coli* tryptophan attenuator region.[13] In a similar way IFN-γ expression plasmids were constructed which contain the ribosome-binding site of the phage MS2 replicase gene.[4,13] The construction of further derivatives of the expression modules which differ in the vector part (e.g., presence of a transcription terminator) will not be detailed here. The characteristics of these vectors are schematically shown in Table II. Full details on their construction can be found in Simons *et al.*[13]

2. *Human Tumor Necrosis Factor*. The construction and identification of cDNA clones containing the coding sequence for human tumor necrosis factor (TNF) have been described.[27]

The TNF cDNA sequence was engineered for expression of mature protein according to the procedure detailed in Fig. 2. A unique *Ava*I site is present in the coding region of TNF and overlaps with the eighth amino acid residue of the mature protein. Following cleavage with *Ava*I, the sequence coding for the first seven amino acids is removed and the triplet of the eighth amino acid is exposed as a 5′ protruding *Ava*I end. Two synthetic DNA linker fragments were synthesized which, after annealing, form a double strand that codes for the first seven amino acids and has one 5′ protruding *Ava*I end, and another 5′ protruding *Nco*I end. The *Ava*I site was used to link the fragment to the body of TNF coding region

[27] A. Marmenout, L. Fransen, J. Tavernier, J. Van der Heyden, R. Tizard, E. Kawashima, A. Shaw, M.-J. Johnson, D. Semon, R. Müller, M.-R. Ruysschaert, A. Van Vliet, and W. Fiers, *Eur. J. Biochem.* **152,** 515 (1985).

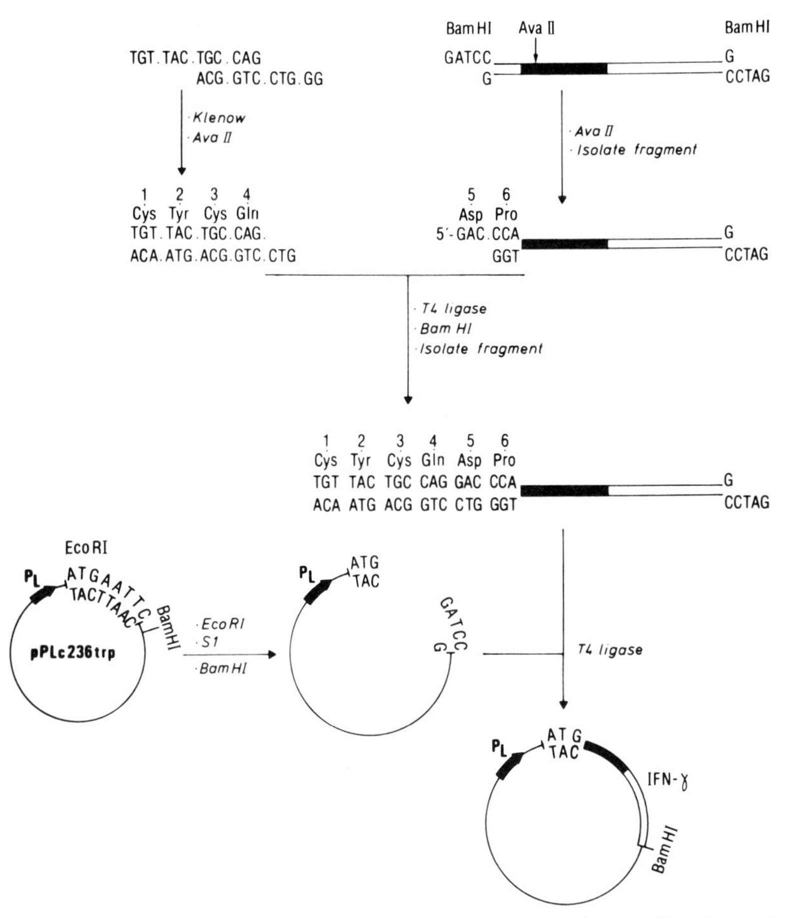

FIG. 1. Construction of a vector that expresses mature human immune interferon. As a consequence of the cDNA cloning procedure,[22] the IFN-γ fragment can be obtained as a *Bam*HI fragment. This fragment contains an *Ava*II site located at the fifth amino acid residue of mature IFN-γ. Following cleavage with *Ava*II, a 1100-bp fragment was isolated by gel electrophoresis and elution. Two synthetic linker molecules were phosphorylated using T4 polynucleotide kinase, annealed to each other, and filled in with Klenow fragment of DNA polymerase I. Following cleavage with *Ava*II, a duplex is obtained containing (1) a blunt end, (2) codons for the first four amino acids of mature IFN-γ, and (3) a sticky *Ava*II end. Following ligation with the isolated IFN-γ fragment, the mixture was digested with *Bam*HI to resolve dimers originating from ligated *Bam*-HI ends. The fragment was purified by gel electrophoresis and ligated to a vector molecule having an ATG codon as a blunt end and a sticky *Bam*HI end downstream. The plasmid used, pPLc236trp, contains a unique *Eco*RI site partially overlapping with the ATG codon. To expose the ATG codon and the *Bam*HI site, the plasmid was sequentially cleaved with *Eco*RI, treated with S1 nuclease, and cleaved with *Bam*HI.

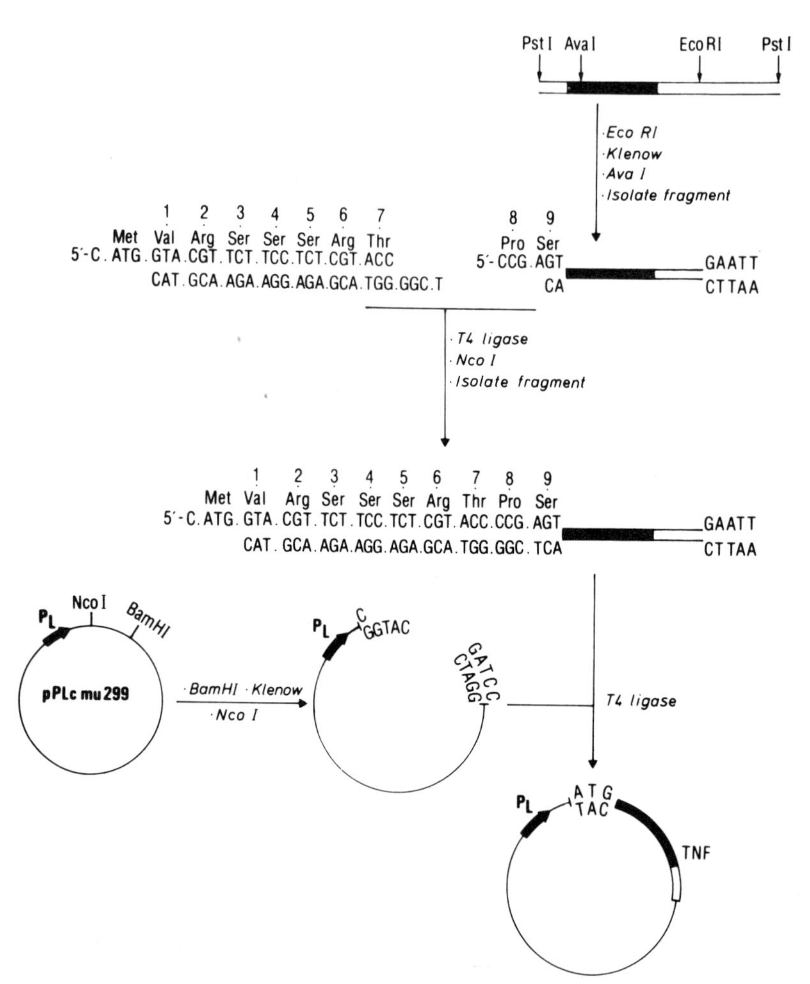

FIG. 2. Construction of a vector that expresses mature human tumor necrosis factor. A *Pst*I fragment obtained from the TNF cDNA clone[27] was sequentially digested with *Eco*RI, treated with Klenow fragment of DNA polymerase I, and cleaved with *Ava*I. The resulting 669-bp fragment was purified by gel electrophoresis and ligated to a double-stranded synthetic DNA fragment containing (1) a sticky *Nco*I end, (2) codons corresponding to the first seven amino acids of mature TNF, and (3) a sticky *Ava*I end overlapping with the eight amino acid residue. Following ligation, the reaction mixture was cut with *Nco*I to resolve dimers and the fragment was purified on an agarose gel. This fragment was ligated into the expression vector pPLcmu299. Plasmid DNA was first blunted at the *Bam*HI site by cleavage with *Bam*HI and treatment with Klenow fragment of DNA polymerase I, and subsequently opened at the unique *Nco*I site. The isolated TNF fragment having an *Nco*I end and a blunted *Eco*RI end was ligated between the *Nco*I end and the blunted *Bam*HI end of the vector.

TABLE II
LEVELS OF ACCUMULATED PROTEIN

Expressed gene	Ribosome-binding site[a]	Transcription termination[b]	Percentage of total protein[c]
IFN-γ	MS2	—	3.5
		T4	10
	trp	—	15
		T4	24
		fd	24
TNF	*ner*	—	9
	MS2	—	0.5
	trp	—	0.1
IL-2	*ner*	—	5
	trp	—	10

[a] Origin of ribosome binding site. MS2, Derived from the replicase gene of the RNA phage MS2 [E. Remaut, P. Stanssens, and W. Fiers, *Nucleic Acids Res.* **11,** 4677 (1983)]; *trp*, derived from the *E. coli* tryptophan attenuator [G. Simons, E. Remaut, B. Allet, R. Devos, and W. Fiers, *Gene* **28,** 55 (1984)]; *ner*, derived from the *ner* gene of phage Mu [G. Buell, J. Delamarter, G. Simons, E. Remaut, and W. Fiers, *Basic Life Sci.* **30,** 949 (1985)].

[b] Origin of the terminator cloned downstream from the expressed gene. T_4, Derived from gene 32 of phage T4 [H. M. Krisch and B. Allet, *Proc. Natl. Acad. Sci. U.S.A.* **79,** 4937 (1982)]; fd, the central terminator of phage fd.

[c] This value was determined as described in the text.

while the *Nco*I site was used to join the restored coding region to an expression vector. For this purpose we used a vector in which the ATG codon overlaps with an *Nco*I site. Plasmid pPLcmu299 contains the ribosome-binding site of the *ner* gene of phage Mu.[17] Similar constructions were obtained in pPLctrp321 which carries likewise a *Nco*I site, in this case part of the ribosome-binding site derived from the *E. coli* tryptophan attenuator region.

3. Human Interleukin 2. cDNA clones coding for human interleukin 2 (IL-2) were obtained by reverse transcription of an enriched mRNA fraction prepared from splenocyte cultures[28] using the method of oligo-dG tailed cDNA annealed into a filled in and oligo-dC tailed *Bam*HI site as described above.[22]

[28] R. Devos, G. Plaetinck, H. Cheroutre, G. Simons, W. Degrave, J. Tavernier, E. Remaut, and W. Fiers, *Nucleic Acids Res.* **11,** 4307 (1983).

The several construction steps performed to obtain expression of mature IL-2 are outlined in Fig. 3. In this case all steps were carried out using cutting and joining reactions without the need for synthetic DNA linker fragments. Mature IL-2 starts with the sequence Ala-Pro. A unique *Hgi*AI site is present at this region. Following *Hgi*AI cleavage and resec-

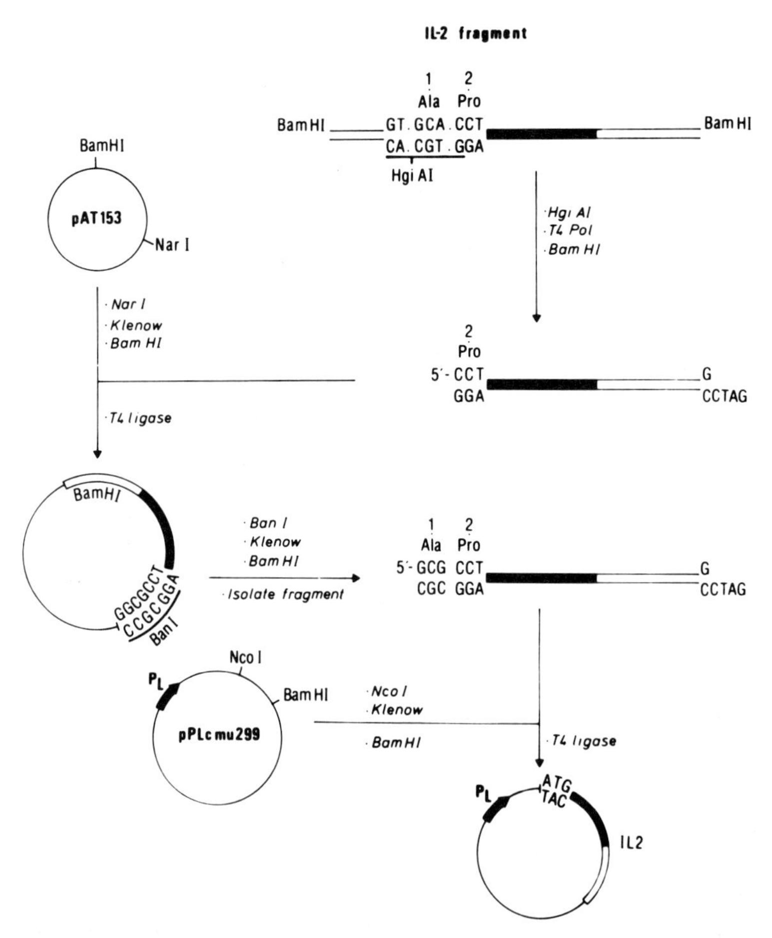

FIG. 3. Construction of a vector that expresses mature human interleukin 2. A pAT153 derivative containing the IL-2 cDNA as a *Bam*HI fragment was sequentially cleaved with *Hgi*AI, treated with T4 DNA polymerase in the presence of dGTP, and cleaved with *Bam*HI. The resulting 700-bp fragment was ligated between the *Bam*HI site and a filled in *Nar*I site of pAT153. The reconstructed *Nar*I site was cleaved by the enzyme *Ban*I. The 5′ protruding ends were filled in with Klenow polymerase, and following *Bam*HI restriction a 700-bp fragment was isolated. This fragment was cloned between a filled in *Nco*I site and a *Bam*HI site of pPLcmu299.

tion of the 3′ protruding ends with T4 DNA polymerase, the CCT codon of the proline residue is exposed as a blunt end. To add an alanine codon, the CCT codon was blunt-end ligated to a filled in *Nar*I site of an acceptor plasmid. The resulting sequence GGCGCC constitutes a *Ban*I site. The latter enzyme cleaves between the G residues, leaving 5′ protruding ends. After filling in, a GCG codon (Ala) is added to the proline codon. The GCG codon is then ligated to a filled in *Nco*I site of the expression vector pPLcmu299.

4. Detection of Expressed Protein. Bacteria harboring one of the expression plasmids discussed above were inoculated at a density of 2×10^6/ml in LB medium containing 5 μCi/ml U-^{14}C-labeled protein hydrolysate. The cultures were incubated at 28° with vigorous agitation to a density of 2×10^8/ml. Half of the culture was then shifted to 42° and incubation was continued for up to 6 hr. At various time points after induction, aliquots were collected by centrifugation. The pellet was dissolved in sample buffer and boiled for 5 min before electrophoresis in SDS–polyacrylamide gels (15%). The gels were fixed in 10% TCA and stained with 0.05% Serva Blue R in 30% methanol and 7% acetic acid. Autoradiographs were obtained after treatment with EN^{3}HANCE and exposure of the dried gel to X-ray film at −70°. To determine the percentage of accumulated protein, the relevant band was excised from the dried gel and its radioactivity compared to the total radioactivity present in the same track. Under the conditions used, the cellular proteins have been uniformly labeled so that the incorporated radioactivity is an accurate measure of the amount of protein synthesized.

Examples of expressed protein using different expression plasmids and different host cells are shown in Fig. 4. A survey of the efficiency of expression obtained with different plasmids is given in Table II.

Comments

In the present communication we have emphasized the mechanics of precisely joining a coding region to the exposed ATG codon of an expression vector. Of the methods discussed, the one using single-strand-specific nucleases is the less reliable one. Indeed, it has been observed that removal of the single strand is not always accurate, so that the ATG codon may be damaged in the process. The methods of filling in 5′ protruding ends or resecting 3′ protruding ends (see Nishi *et al.*[29]), on the other hand, have proved to be very reliable. When choosing an expression vector for optimal expression of an unfused protein, other consider-

[29] T. Nishi, M. Sato, A. Saito, S. Itoh, C. Takaoka, and T. Taniguchi, *DNA* **2**, 265 (1983).

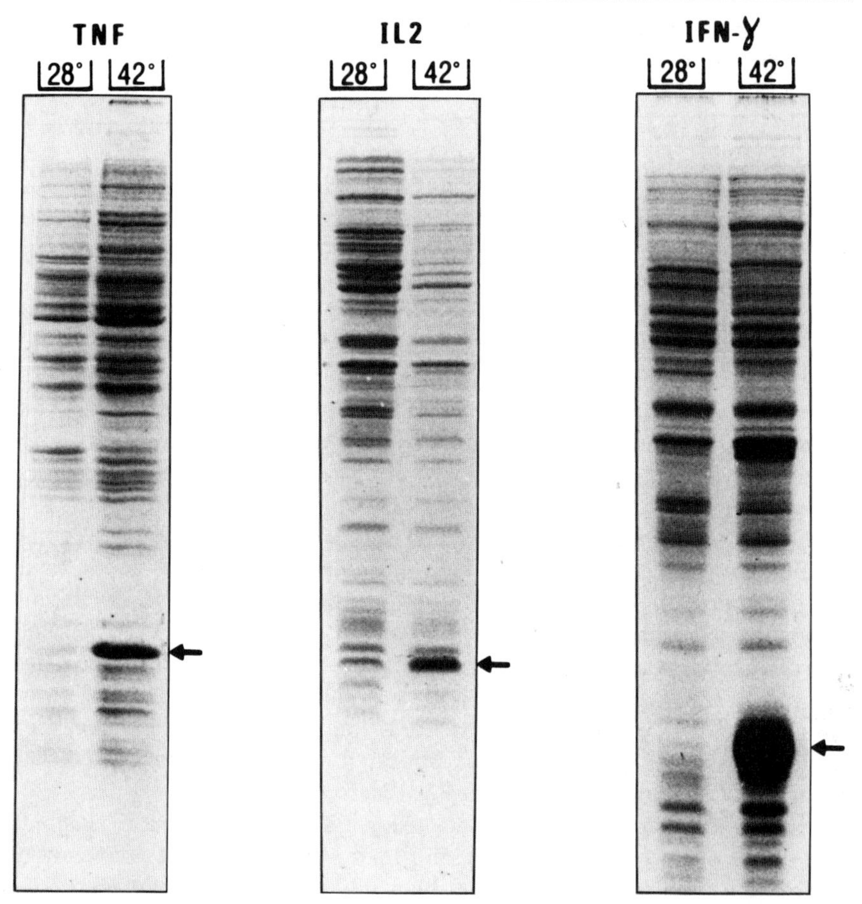

FIG. 4. Stained protein profiles obtained following SDS–PAGE of extracts from induced (42°) and uninduced (28°) cells containing an expression plasmid for IFN-γ, TNF, and IL-2, respectively. The induction period was 4 hr.

ations regarding the type of ribosome-binding site used, as well as other features of the host–plasmid sytem, become important.

The nature of the ribosome-binding site may greatly influence expression levels. For evident reasons, expression vectors were provided with ribosome-binding sites known to be very effective in their natural context. It soon became clear, however, that the efficiency of a ribosome-binding site is not an absolute parameter but depends to a large extent on the nature of the downstream coding region. It is believed that the secondary structure of the mRNA in the vicinity of the ATG codon plays a major role in determining efficiency of expression.[30,31] It should be noted, however, that current models to predict secondary structure are frequently inadequate to foresee the effect on expression levels.[32] It is therefore worthwhile to try out a number of ribosome-binding sites in combination with the gene to be expressed. Table II lists some examples of the large differences that can be observed.

Another parameter (probably of less consequence) in optimizing expression is the presence of a strong transcription terminator downstream from the cloned gene. In the case of clones expressing immune interferon, this feature increased the yield of the product by a factor of 1.5–3 (Table II). This is not primarily related to an increased efficiency of synthesis, but is rather due to a prolonged maintenance of the level of synthesis.[33]

The bacterial strain used is another, poorly documented, element affecting accumulation of the expressed protein. For instance, human fibroblast interferon (not discussed in this chapter) was found to accumulate to levels of 0.4 and 2%, respectively, in *E. coli* strains MC1061[pcI857] and K12ΔH1Δ*trp*.[4] Human interleukin 2, on the other hand, was recovered in higher yields from the former strain than from the latter.[34] The phenomenon has not been studied in any detail. Variations between strains in mRNA and/or protein stability might conceivably contribute to the observed differences.

Acknowledgments

This research was supported by Biogen, SA, Geneva, and by a grant from the Gekoncerteerde Onderzoeksakties of the Belgian Ministry of Science.

[30] D. Iserentant and W. Fiers, *Gene* **9,** 1 (1980).
[31] D. Gheysen, D. Iserentant, C. Derom, and W. Fiers, *Gene* **17,** 55 (1982).
[32] P. Stanssens, E. Remaut, and W. Fiers, *Gene* **36,** 211 (1985).
[33] G. Simons, unpublished results (1984).
[34] R. Leemans, unpublished results (1985).

[27] Directing Ribosomes to a Single mRNA Species: A Method to Study Ribosomal RNA Mutations and Their Effects on Translation of a Single Messenger in *Escherichia coli*

By ANNA HUI, PARKASH JHURANI, and HERMAN A. DE BOER

Introduction

Several purines located about eight nucleotides upstream from the initiation codon are found in most prokaryotic messenger RNAs. This purine tract, which is known as the Shine–Dalgarno (SD) sequence, has the consensus sequence 5′-GGAGG and is complementary to a sequence near the 3′ terminus of 16 S-rRNA.[1] This anti-Shine–Dalgarno (ASD) sequence, 5′-CCUCC, is part of the highly conserved 3′ end of the 16 S-rRNA found in prokaryotes.[2] Synthetic oligonucleotide probes can be bound specifically to the ASD region of 16 S-rRNA in intact 30 S ribosomal subunits, implying that this sequence is not complexed with proteins or RNA.[3,4] Shine and Dalgarno[1] proposed that the SD sequence in mRNA interacts with the ASD region of 16 S-rRNA and that this mRNA–rRNA base pairing plays an important role in the protein initiation process. A large body of biochemical and genetic evidence supports this view (reviewed in Ref. 5). The base-paired complexes can be physically isolated[6] and mutations introduced in the mRNA which alter the complementarity have profound effects on mRNA translatability.[7–10] However, the mRNA–rRNA duplex formation does not appear to be an absolute requirement for translation initiation as there are a few natural mRNAs

[1] J. Shine and L. Dalgarno, *Proc. Natl. Acad. Sci. U.S.A.* **71,** 1342 (1974).

[2] R. van Charldorp and P. H. Van Knippenberg, *Nucleic Acids Res.* **10,** 1149 (1982).

[3] C. Backendorf, C. J. C. Ravensbergen, J. van der Plas, J. H. van Boom, G. Veeneman, and J. Van Duin, *Nucleic Acids Res.* **9,** 1425 (1981).

[4] J. Van Duin, C. J. C. Ravensbergen, and J. Doornebos, *Nucleic Acids Res.* **12,** 5079 (1984).

[5] M. Kozak, *Microbiol. Rev.* **47,** 1 (1983).

[6] J. A. Steitz and K. Jakes, *Proc. Natl. Acad. Sci. U.S.A.* **72,** 4734 (1975).

[7] J. J. Dunn, E. Buzash-Pollert, and F. W. Studier, *Proc. Natl. Acad. Sci. U.S.A.* **75,** 2741 (1978).

[8] M. Schwartz, M. Roa, and M. Debarbouille, *Proc. Natl. Acad. Sci. U.S.A.* **78,** 2937 (1981).

[9] B. S. Singer, L. Gold, T. Shinedling, M. Colkitt, H. Hunter, D. Pribnow, and M. Nelson, *J. Mol. Biol.* **149,** 405 (1981).

[10] C. Chapon, *EMBO J.* **1,** 369 (1982).

lacking an SD sequence.[11,12] These mRNAs are poorly translated, indicating that the base pairing enhances the efficiency of initiation rather than being required for the mechanism itself.

Although several studies have described the effect of different SD sequences and their neighboring nucleotides on mRNA translatability,[13-15] little is known about the actual role of the ASD sequence in translation efficiency and mRNA specificity.

So far, very few mutants in 16 S-rRNA have been reported and none of them maps in the 3′ region. It is technically difficult to select for ribosomal-RNA mutants since there are seven nearly identical copies of rRNA operons in *Escherichia coli*. Recessive mutations in any one operon will be masked by the remaining functional operons. Moreover, many deletions or substitutions introduced in 16 S-rRNA (via recombinant DNA techniques) appear to be lethal to the cell.[16] Although it is not clear why this is the case, it is likely that subunits with certain mutations in the 16 S-rRNA are still capable of binding to mRNA, while being impaired in other aspects of the protein biosynthetic cycle. The protein synthesis process is likely to be crippled by such dominating mutated ribosomes.

To study the role of the ASD sequence in translation and ribosome specificity, we developed a system overcoming these physiological barriers. For this purpose, the ASD sequence of the 16 S-rRNA itself was altered. In addition, a spectinomycin resistance mutation was introduced in a plasmid-borne 16 S-rRNA operon with a mutated ASD sequence. Thus the performance of these mutated ribosomes could be evaluated after inactivating the spectinomycin-sensitive ribosomes derived from chromosomal *rrn* operons with spectinomycin. It is thought that spectinomycin inhibits protein synthesis by inhibiting ribosomal movement along the messenger.

The assembly of this novel expression system was done as follows: after mutating the ASD sequence of the 16 S-rRNA gene of a plasmid-borne *rrnB* operon, a gene encoding a specialized messenger with an SD sequence complementary to the altered ASD sequence was introduced into the same plasmid. The transcription of the specialized messenger was designed to be constitutive, whereas the transcription of the mutated 16 S-

[11] M. Ptashne, K. Backman, M. Humayun, A. Jeffrey, R. Maurer, B. Meyer, and R. T. Sauer, *Science* **194,** 156 (1976).

[12] V. Pirrotta, *Nucleic Acids Res.* **6,** 1495 (1979).

[13] H. A. de Boer, A. Hui, L. J. Comstock, E. Wong, and M. Vasser, *DNA* **2,** 231 (1983).

[14] A. Hui, J. Hayflick, K. Dinkelspiel, and H. A. de Boer, *EMBO J.* **3,** 623 (1984).

[15] H. M. Shepard, E. Yelverton, and D. V. Goeddel, *DNA* **1,** 125 (1982).

[16] A. Hui and H. A. de Boer, unpublished observations.

rRNA is under the regulation of the inducible λP_L promoter. The transcription of the mutated 16 S-rRNA was not detectable under uninduced conditions. However, when the temperature was increased to 42°, the mutated *rrnB* operon is transcribed and its products are assembled into functional 30 S subunits. Upon addition of spectinomycin, wild-type ribosomes were inactivated, leaving only the specialized ribosomes which preferentially translate the single specialized mRNA species.

Construction of the Specialized Ribosome System

The *rrnB* operon on plasmid KK3535 was used throughout this work.[17] It contains the entire *rrnB* operon, including its tandem promoter P1 and P2, and its transcriptional terminators. The construction of the specialized system involved the following steps (see Fig. 1):

1. Removal of both *rrnB* promoters P1 and P2
2. Replacement of the natural ASD region at the 3′ end of *rrnB* (CCUCCAAU-3′) with the mutated ASD sequence: GGAGGAAU-3′ (system IX) or with CACACAAU-3′ (system X)
3. Introduction of the spectinomycin resistance mutation into these mutated 16 S-rRNA genes
4. Replacement of the natural *trp* promoter/operator SD region preceding the coding region of the human growth hormone (hGH) gene on pHGH207-1*[18] to create a constitutive promoter and an altered SD sequence that is complementary to the mutated ASD sequence. In summary:

System	SD sequence	ASD sequence
Wild type (VIII)	5′-GGAGG	3′-CCUCC
System IX	5′-CCUCC	3′-GGAGG
System X	5′-GUGUG	3′-CACAC

The nucleotide changes in system IX were chosen to study the effect of complete reversal of the SD and ASD sequence. Since we could not predict whether such a dramatic change might still allow for ribosome function, we designed simultaneously system X, which is a compromise between the wild-type system and system IX

[17] J. Brosius, T. J. Dull, D. D. Sleeter, and H. F. Noller, *J. Mol. Biol.* **14,** 107 (1981).

[18] H. A. de Boer, L. J. Comstock, and M. Vasser, *Proc. Natl. Acad. Sci. U.S.A.* **80,** 21 (1983).

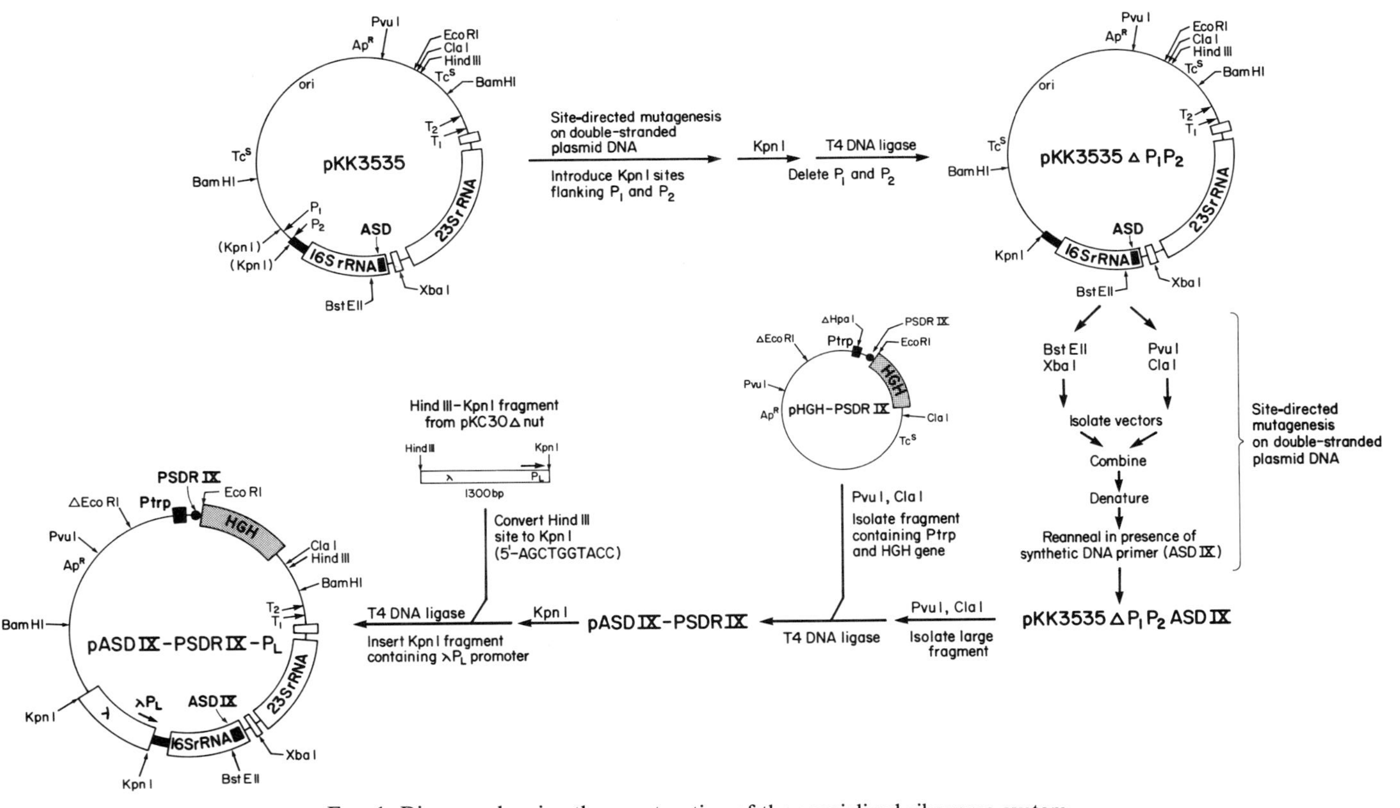

FIG. 1. Diagram showing the construction of the specialized ribosome system.

5. Insertion of the "specialized" hGH gene into the plasmids bearing the mutated ASD sequences
6. Insertion of the P_L promoter before the modified *rrnB* operon

In the following we will describe in detail steps 1–6.

Step 1: Removal of the Tandem rrnB Promoters P1 and P2

In our attempts to mutagenize the ASD region directly on an intact *rrnB* operon, we always obtained mutants with extensive deletions or rearrangements in the 16 S-rRNA gene. Similar observations were reported by others while studying mutants elsewhere in the 16 S-rRNA gene.[19] Apparently, certain mutated 16 S-rRNA species can have a lethal effect on the cells. Therefore the endogenous *rrn* promoters were removed first before any mutations were introduced.

The excision of P1 and P2 must be precise since the precursor region of the 16 S-rRNA contains essential information with regard to the processing of the rRNA precursor molecules.[20] It also provides the antitermination properties required for successful transcription through the *rrn* operon.[21] Using site-directed mutagenesis on double-stranded plasmid DNA,[22] we introduced two unique *Kpn*I sites, flanking P1 and P2, into the plasmid (Fig. 1). The downsteam *Kpn*I site was introduced at the transcriptional start site, leaving the *rrn* precursor region intact. Thus P1 and P2 could be excised precisely, yielding a functional but silent *rrnB* operon which is now amenable for mutagenesis without causing any lethal effects on the cells.

There are several advantages to using double-stranded plasmid DNA for site-directed mutagenesis over methods requiring single-stranded DNA subcloned in phage M13: (1) Some DNA fragments, especially large ones, cannot be stably maintained in an M13 vector; (2) it is sometimes not practical to subclone a proper size fragment into M13 due to lack of convenient restriction sites; and (3) the expression of the mutated gene in a plasmid vector can be examined immediately without any further manipulation, whereas the mutated DNA in M13 must be transferred back into an expression vector. Often the reconstruction of very large plasmids such as the ones described here are complicated due to the lack of unique restriction sites to be used for reinsertion of the mutated gene. For the

[19] M. J. R. Stark, R. J. Gregory, R. L. Gourse, D. L. Thurlow, C. Zwieb, R. A. Zimmerman, and A. E. Dahlberg, *J. Mol. Biol.* **178,** 303 (1984).
[20] R. A. Young and J. A. Steitz, *Proc. Natl. Acad. Sci. U.S.A.* **75,** 3593 (1978).
[21] W. E. Holben and E. A. Morgan, *Proc. Natl. Acad. Sci. U.S.A.* **81,** 6789 (1984).
[22] Y. Morinaga, T. Franceschini, S. Inouye, and M. Inouye, *Bio/Technology* **2,** 637 (1984).

latter reason, we chose to do all the mutagenesis on a plasmid vector. However, the low yield of mutants is a disadvantage of this procedure.[22]

Figure 2 illustrates the steps involved in this site-directed mutagenesis procedure. The mutagenesis procedure was carried out according to Morinaga *et al.*[22] with slight modifications. Two sets of vectors were made. In one set the region to be mutagenized was removed. In the other set a different region of the plasmid was deleted in order to reduce the background of wild-type plasmids. Both vectors and the mutagenesis primers were mixed, heat denatured, and cooled slowly to allow formation of heteroduplexes with two single-stranded regions. Mutagenesis primers hybridize to the single-stranded region of the heteroduplex. Subsequently, the Klenow fragment of DNA polymerase I, T4 DNA ligase, and deoxynucleotides were used to fill the single-stranded gaps between the primers and the vector.

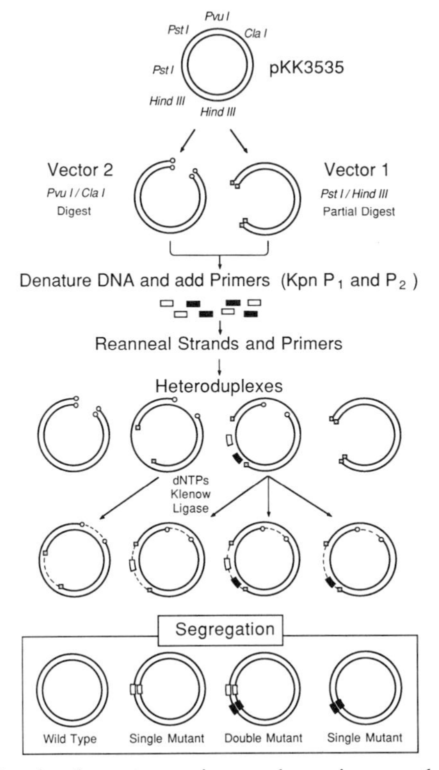

FIG. 2. Diagram showing the mutagenesis procedure using a synthetic oligonucleotide on heteroduplexed plasmids.

Detailed Mutagenesis Procedure of Double-Stranded Plasmid DNA

1. Generation of vector 1 (partially cut *Pst*I/*Hind*III large fragment of pKK3535): 5 μg of pKK3535 was incubated with 10 U of *Pst*I and 10 U of *Hind*III in a 100-μl reaction mixture at room temperature. Twenty microliters of the sample was removed at 2, 5, 10, 15, and 30 min and immediately frozen in a dry ice/ethanol bath. All the samples were pooled and extracted with phenol/chloroform and precipitated with ethanol. The DNA pellet was washed and dried and resuspended at 100 ng/μl in TE buffer (10 m*M* Tris–HCl, pH 8.0/0.1 m*M* EDTA).

2. Generation of vector 2 (*Pvu*I/*Cla*I large fragment of pKK3535): 5 μg of pKK3535 was digested with 10 U of *Pvu*I and 10 U of *Cla*I in a reaction mixture of 100 μl for 2 hr at 37°. The sample was treated as described above. The final pellet was resuspended in 200 μl of 0.1 *M* Tris–HCl, pH 8.0, and 2 μl (10 U) bacterial alkaline phosphatase was added. The mixture was incubated at 68° for 1 hr, then extracted with phenol/chloroform, ethanol precipitated, dried, and resuspended at about 100 ng/μl. The *Pvu*I/*Cla*I digestion cuts into the ampicillin resistance (Amp) gene and the phosphatase treatment prevents self-ligation of the vector. Thus the background of wild-type transformants could be greatly reduced.

3. Phosphorylation of the synthetic oligonucleotide primers 1 and 2: The following two primers of 26 nucleotides in length were used to introduce *Kpn*I sites into the heteroduplexes of pK3535.

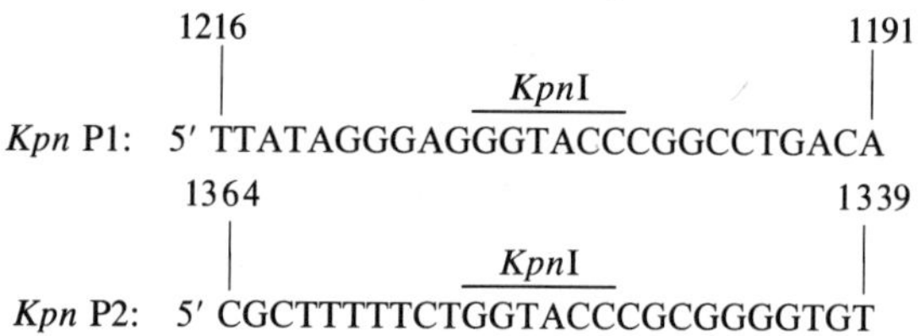

The numbers indicate the corresponding complementary positions in *rrnB* of pKK3535.[17] In both primers, the *Kpn*I site (the mismatched bases) is flanked by 10 bases that are complementary to the sense strand of pKK3535. The primers can only hybridize to one strand to form stable complexes (see Fig. 2). The primers were kinased as follows: 0.2 nmol of the oligonucleotide was mixed with 2 μl 10× kinase buffer (0.5 *M* Tris–HCl, pH 7.6/0.1 *M* $MgCl_2$/50 m*M* DTT), 2 μl 10 m*M* ATP, 1 μl T4 polynucleotide kinase (2 U, BioLabs), and H_2O (final volume, 20 μl). The reaction was incubated at 37° for 1 hr. The kinase was inactivated by heat treatment (68° for 5 min).

4. Generation of heteroduplexes: 1 μl of each opened vector (1 and 2, see Fig. 2), 2.5 μl each of the kinased primers (25 pmol), 9.6 μl 0.1× TE buffer (1 m*M* Tris–HCl, pH 7.5/0.1 m*M* EDTA), and 2 μl 10× poly-

merase–ligase buffer (1 *M* NaCl/65 m*M* Tris–HCl, pH 7.5/80 m*M* $MgCl_2$/10 m*M* DTT) were mixed in an Eppendorf tube. The tube was incubated in a 100° temp-block for 3 min to denature the DNA fragments. The temp-block with the sample tube was removed from the heater and allowed to cool slowly to room temperature. The tube was then transferred to a 4° bath for 10 min and subsequently 4 μl of a dNTP mix (2.5 m*M* each), 2 μl 10 m*M* ATP, 1 μl T4 DNA ligase (10 U/μl, BioLabs), 1 μl Klenow fragment of DNA polymerase I (Boehringer–Mannheim, 5 U/μl), and H_2O were added to a final volume of 20 μl. The reaction mixture was incubated at 12° for a minimum period of 5 hr.

5. Transformation of *E. coli* 294: 10 μl ligation mixture was added to 200 μl thawed competent cells of *E. coli* 294 (F^- *supE*44, *endA*1, *thi*-1, *hsdR*4). The cells were incubated at 4° for 30 min, heat shocked at 42° for 1 min, then plated on L broth agar plates supplemented with 50 μg/ml of ampicillin. Plates were incubated at 37° overnight.

6. Identification of the mutants: Colony hybridization: To identify mutants that carry both primer-induced mutations, it was necessary to probe the colonies separately with both primers (*Kpn* P1 and *Kpn* P2). Duplicate nitrocellulose filters were made for the screening process as follows:

a. 100 μl LB-ampicillin (50 μg/ml) medium was pipetted into the wells of a 96-well microtiter plate. Individual transformants were picked with sterile toothpicks and the toothpick was dipped into a microtiter well and discarded. One of the wells was left empty to orient the plate. The microtiter plate was then incubated at 37° for 1–2 hr.
b. A nitrocellulose filter was laid gently on top of LB-Amp agar in a large Petri dish. The colonies from the microtiter plate were stamped on top of the filter. The cells were grown on the filters overnight at 37°. Two filters were made per microtiter plate. The microtiter plates were sealed with Parafilm and stored at 4° for future use.
c. The nitrocellulose filters were lifted off the LB agar plates and placed on paper towels soaked with 0.5 *M* NaOH/1.5 *M* NaCl for 3 min.
d. The filters were transferred to paper towels soaked with neutralization solution (3 *M* NaCl/0.5 *M* Tris–HCl, pH 7.5) for 15 min.
e. The filters were soaked in 2× SSC (30 m*M* sodium citrate/0.3 *M* NaCl) for 10 min.
f. The filters were dried on 3MM Whatman paper for 1 hr at 37° or at room temperature.
g. The filters were baked in an 80° oven under vacuum for 1 hr.

h. The filters were prehybridized in hybridization buffer for 1–2 hr at room temperature with slow shaking on a rotary platform.

Hybridization buffer: 0.1 m*M* ATP/1 m*M* NaH_2PO_4/1 m*M* sodium pyrophosphate/0.9 *M* NaCl/1× Denhardt's solution/0.5% NP40/0.2 mg/ml *E. coli* tRNA/90 m*M* Tris–HCl, pH 7.5/6.0 m*M* EDTA

i. The oligonucleotide primers (used for mutagenesis) were phosphorylated using [γ-^{32}P]ATP as follows: 25 pmol of the oligonucleotide was mixed with 2 μl 10× kinase buffer, 6 μl [γ-^{32}P]ATP (60 μCi; Amersham), and 1 μl polynucleotide kinase in a 20-μl reaction mixture. The kinase reaction was done at 37° for 30 min.

j. A 10-μl aliquot of the kinase reaction mixture was added to the prehybridized filters and the hybridization was carried out at room temperature overnight.

k. The filters were washed once with 2× SSC at room temperature and twice with 0.3× SSC/0.1% SDS at 50°. The washing was carried out in a heated incubator on a slowly shaking platform.

l. The filters were air dried, sandwiched between Saran Wrap, and exposed to an X-ray film in a film cassette holder with an intensifier screen. The exposure was carried out in −80° for an appropriate length of time depending on factors such as the specific radioactivity of the probes.

m. Candidates hybridizing to both probes were selected for further analysis. The yield of colonies carrying one or more mutations was about 3%. The yield of colonies with a double mutation was approximately 0.2%.

7. Cell purification: Candidates which hybridized positively to both probes, indicating the presence of both *Kpn* mutations, were picked. However, in addition to the mutated pKK3535, the cell also harbors the wild-type pKK3535 (see Fig. 2). Therefore, further plasmid segregation was necessary to obtain cells with a single homogeneous plasmid population. Cell purification was carried out as follows:

a. A candidate carrying a double mutation was identified on the microtiter master plate (see previous section). This colony was used to prepare an overnight culture for miniscreen DNA isolation.

b. Miniscreen DNA was isolated by the alkaline SDS lysis method.[23]

c. Miniscreen DNA (5–10 ng) was used to transform *E. coli* 294.

[23] T. Maniatis, E. F. Fritsch, and J. Sambrook, "Molecular Cloning: A Laboratory Manual." Cold Spring Harbor Lab., Cold Spring Harbor, New York, 1982.

d. Positive colonies that contain mutated pKK3535 were identified by the same colony hybridization method as described in the previous section (probing with one primer is sufficient in this case).

e. Miniscreen DNA was isolated from a positive colony and digested with *Kpn*I. The restriction pattern was analyzed on a 6% polyacrylamide gel. The presence of a 250-bp *Kpn*I fragment indicated the introduction of two *Kpn*I sites into pKK3535. This plasmid is called pKK3535Kpn. Segregation of the plasmid is an absolutely necessary step to obtain a homogeneous plasmid population as we have seen that after segregation about 50% of the colonies are wild type.

8. Deletion of the promoters P1 and P2:

a. The plasmid pKK3535Kpn was cut with *Kpn*I (see Fig. 1), purified as described previously, and resuspended in TE buffer at 100 ng/μl.

b. One microliter of cut vector was mixed with 2 μl 10× ligase buffer (0.5 *M* Tris–HCl, pH 8.0/0.1 *M* $MgCl_2$/20 m*M* DTT), 2 μl 10 m*M* ATP, and 1 μl ligase (final volume, 20 μl). The ligation reaction was carried out at 14° for a minimum of 1 hr.

c. A 10-μl aliquot of the ligation mix was transformed into competent *E. coli* 294 and the cells were spread on LB Amp plates.

d. Transformants were grown and DNA was isolated. The DNA was digested with *Kpn*I and analyzed on a 6% polyacrylamide gel. The absence of a 250-bp *Kpn*I fragment indicates that the promoters were excised. Thus pKK3535ΔP1P2 was obtained (see Fig. 1).

Step 2: Mutagenesis of the ASD Region of the 16 S rRNA Gene

Mutagenesis was carried out on pKK3535ΔP1P2 by the site-directed mutagenesis method as described in the previous sections. The following primers were used:

ASDIX–5′TCTTTAAGGTAACCTCCTGATCCAACCGC
ASDX –5′TCTTTAAGGTAAGTGTGTGATCCAACCGC

To generate heteroduplexes for mutagenesis, the vectors were obtained by digesting (1) pKK3535ΔP1P2 with *Bst*EII and *Xba*I and (2) with *Pvu*I and *Cla*I (see Fig. 1). Transformants were isolated, screened, and segregated as described above. The sequence of positive clones was confirmed by DNA sequence analysis of supercoiled DNA.[24] The resultant plasmids are pKK3535ΔP1P2ASDIX and pKK3535ΔP1P2ASDX (their ASD sequences are 5′-GGAGG and 5′-CACAC, respectively).

[24] E. Y. Chen and P. H. Seeburg, *DNA* **4,** 165 (1985).

Step 3: Introduction of the Spectinomycin Mutation in pKK3535ΔP1P2-ASDIX and X

It has been shown that a single base change of a C to a T at position 1192 of 16 S-rRNA renders ribosomes spectinomycin resistance.[25] We introduced this mutation into pKK3535ΔP1P2-ASDIX and X by site-directed mutagenesis on double-stranded plasmid DNA using the primer with the sequence 5′-ATGATGACTTGACATCATCCCCACCTT. Two plasmid samples were cut with (1) *Sma*I and (2) *Cla*I and *Pvu*I. The Spc^r mutation was confirmed by DNA sequencing of supercoiled plasmid DNA.

Step 4: Replacement of the trp Operator and SD Sequence of pHGH207-1 with Synthetic Oligonucleotide Fragments*

pHGH207-1* contains the entire coding sequence of the hGH gene (Ref. 26 and Fig. 3). The transcription of the hGH gene occurs from the *trp* promoter. The 3′-end of the hGH gene was joined to the *Hin*dIII site of the tetracycline resistance gene and the *Bam*HI site within the *tet* gene was converted to a *Cla*I site. Thus pHGH207-1* is tetracycline sensitive due to the disruption within the *tet* gene. Since we wanted to use the inducible P_L promoter to direct transcription of the mutated 16 S-rRNA gene, we constructed a constitutive (*trp*) promoter to direct transcription of the specialized (hGH) gene. In addition, we wished to alter the SD sequence of hGH gene such that it complements the mutated ASD sequence of 16 S-rRNA. To achieve this, we replaced the natural *trp* promoter/operator and adjacent SD region with a synthetic DNA fragment. Two complementary fragments were synthesized, as shown for each system.

```
                                             SD
HGH VIII (WT)  5′ TAATGTGTGGAAGCTTTGGAGGTCTAG
                  ATTACACACCTTCGAAACCTCCAGATCTTAA
                                             SD
HGH IX         5′ TAATGTGTGGAAGCTTTCCTCCTCTAG
                  ATTACACACCTTCGAAAGGAGGAGATCTTAA
                                             SD
HGH X          5′ TAATGTGTGGAAGCTTTGTGTGTCTAG
                  ATTACACACCTTCGAAACACACAGATCTTAA
```

The synthetic DNA fragments used here encode part of the *trp* promoter/operator and the SD sequences. The blunt-ended left hand side is joined

[25] C. D. Sigmund, M. Ettayebi, and E. A. Morgan, *Nucleic Acids Res.* **12,** 4653 (1984).
[26] H. A. de Boer, P. Ng, and A. Hui, *UCLA Symp. Mol. Cell. Biol.* **30,** 419 (1986).

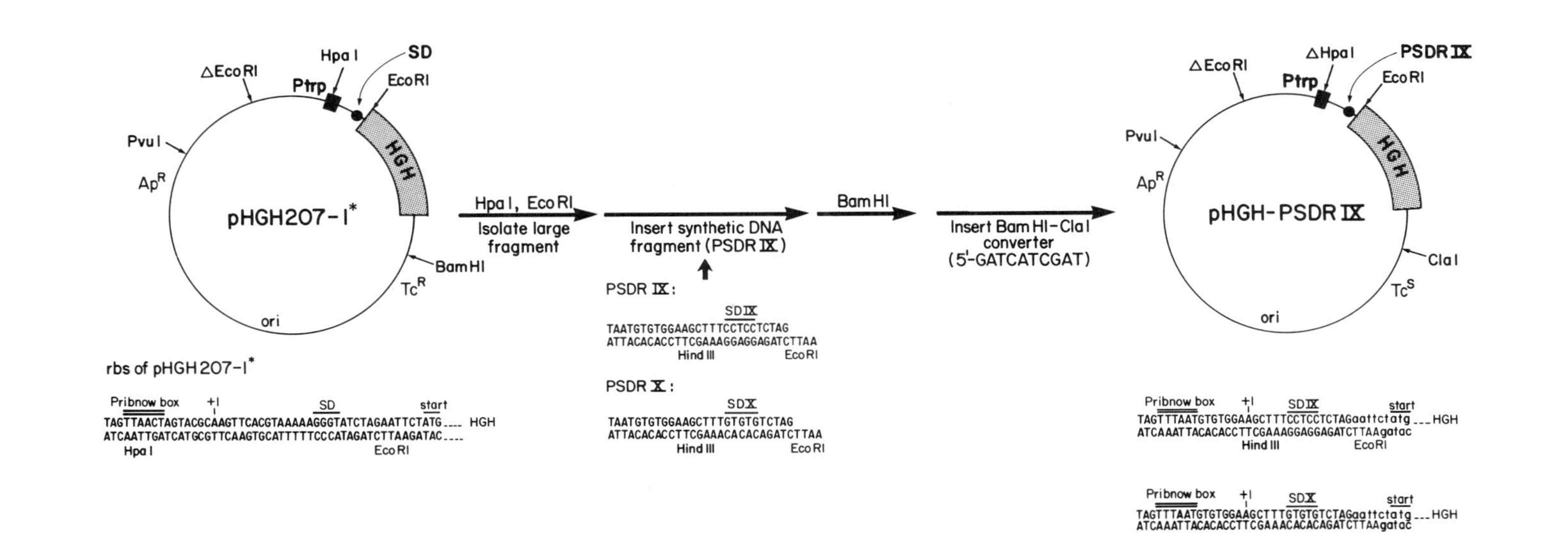

FIG. 3. Construction of the plasmid containing a constitutively transcribed hGH gene with a specialized ribosome-binding site.

to the *Hpa*I site which cuts within the *trp* Pribnow box.[27] The altered operator sequence results in a constitutive *trp* promoter (see Fig. 3). This was accomplished as follows:

1. pHGH207-1* was cut with *Hpa*I and *Eco*RI, thus removing part of the Pribnow box/*trp* repressor binding site and SD sequence. The cut vector was isolated and resuspended at 100 ng/μl in TE buffer.
2. The complementary synthetic fragments shown above were phosphorylated and mixed with 100 ng of cut vector (the molar ratio fragment/vector should be approximately 5 : 1) and treated with ligase.
3. The ligation mixture was used to transform *E. coli* 294.
4. Transformants were screened for inserts and the sequences were confirmed by supercoiled DNA sequencing. The resultant plasmids were pHGH207-1*PSDR VIII, IX, and X.

Step 5: Insertion of the Specialized hGH Gene into the Mutated Plasmids Containing an rrn Operon with a Mutated ASD Sequence

pHGH-PSDR (VIII, IX, or X) (Fig. 3) was cut with *Pvu*I and *Cla*I. The 1.8-kb fragment containing part of the ampicillin gene and the entire transcriptional unit of the hGH gene was isolated on a 6% polyacrylamide gel. The appropriate fragments carrying the specialized SD sequence and the hGH gene were inserted into the *Pvu*I/*Cla*I vectors with corresponding ASD sequences (Fig. 1). Transformants were isolated and confirmed by restriction analysis. The resultant plasmid contains the silent, mutated *rrnB* operon and the hGH gene whose SD sequence is complementary to the ASD sequence of the plasmid-borne 16 S-rRNA [the resulting plasmid is pASD-PSDR (Fig. 1)].

Step 6: Insertion of the P_L Promoter into ΔP1P2-ASD-HGH

1. Isolation of P_L fragment: The P_L promoter in pNO2678[28] is flanked by two restriction sites: a *Hin*dIII site at the 5′ end and a *Kpn*I site at the 3′ end near the transcription initiation site. For our construction, we converted the *Hin*dIII site to a *Kpn*I site. To do this, we used a synthetic DNA fragment for site conversion. A synthetic dodecamer was made with the sequence 5′-AGCTCGGTACCG. These oligonucleotides are self-complementary:

```
       KpnI
5' AGCTCGGTACCG
       GCCATGGCTCGA 5'
```

[27] C. Yanofsky, T. Platt, I. P. Crawford, B. P. Nichols, G. E. Christie, H. Horowitz, M. Van Cleemput, and A. M. Wu, *Nucleic Acids Res.* **9**, 6647 (1981).

[28] R. L. Gourse, Y. Takebe, R. A. Sharrock, and M. Nomura, *Proc. Natl. Acad. Sci. U.S.A.* **82**, 1069 (1985).

This duplex was inserted into the *Hin*dIII-cut vector, thus generating a new *Kpn*I site while destroying the *Hin*dIII site (Fig. 1). The procedure was as follows:

a. One microgram pNO2678 DNA was cut with *Hin*dIII, then treated with bacterial alkaline phosphatase as described before.
b. The synthetic fragments were phosphorylated with polynucleotide kinase.
c. The synthetic fragments were ligated with the cut vector at a molar ratio of 5 : 1 and the plasmid was used to transform *E. coli* K5716 [Δ(*lac pro*)*supE*, F'*traD*36, *proAB*, *lacI*^{Q}ZΔM15], a λ^{+} lysogen derived from JM101.
d. Transformants were isolated and screened for the presence of a 1.3-kb *Kpn*I fragment. Multiple linker insertion would not be a problem in this case since upon cutting of the plasmid with *Kpn*I, only one linker would remain at the converted site.

2. Construction of the P_L ASD-HGH plasmids: The P_L-containing *Kpn*I fragment was isolated by gel electrophoresis and ligated into a *Kpn*I-cut pASD-PSDR vector. DNA was transformed into *E. coli* K5716. Cells were plated on LB Amp plates and incubated at 37°. Transformants were screened for the presence of a properly oriented P_L promoter preceding the *rrnB* operon.

In summary, we described the construction of three plasmids containing the hGH gene preceded by an altered SD sequence under the direction of a constitutive *trp*-promoter. This same plasmid contains an *rrnB* operon with a mutated ASD sequence (with or without a spectinomycin mutation) under the control of the P_L promoter. In each of the three cases (wild type or system VIII, system IX, and system X) the ASD sequence in the *rrnB* operon is complementary to the SD sequence of the hGH gene of the same plasmid (Fig. 1).

Properties of the Specialized Ribosome System

1. *Probe Specificity and Induction of Mutated rRNA Species*

In *E. coli* all chromosomal rRNA operons are essentially identical. The system described here allows for a direct measurement of the rRNA derived from a single specialized operon using oligonucleotide probes specific for the 3′ end of a specialized or wild-type *rrn* operon. This is illustrated in Fig. 4. Total RNA was isolated from cells harboring the wild-type system (VIII) or systems IX or X. The RNA was separated on a 1% agarose–6% formaldehyde gel and subjected to a Northern analysis.[23] The

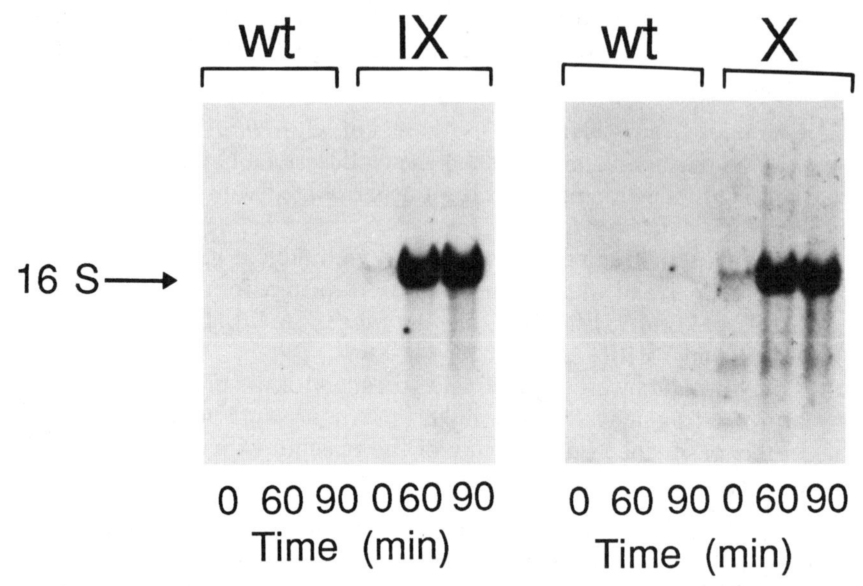

FIG. 4. Northern analysis of the induction of an rRNA operon bearing a mutated ASD region. In the left panel the probe specific for ASD IX was used and in the right panel the probe for ASD X was used. The transcription of the mutated *rrn*B operon was induced at t = 0 min.

The sequence of the ASDIX probe is as follows:
5′TCTTTAAGGTAACCTCCTGATCCAACCGC
The sequence of the ASDX probe is as follows:
5′TCTTTAAGGTAAGTGTGTGATCCAACCGC

left panel shows system VIII (wt) and IX probed with an oligonucleotide that is fully complementary to ASDIX. The nitrocellulose blot was washed under stringent conditions (0.1× SSC/0.1% SDS at 50°). Clearly this probe does not bind to wild-type rRNA but binds specifically to the mutated ASDIX 3′ end under these conditions. A strong induction of the new rRNA species with ASDIX is evident. A very low level of the mutated form is present before induction. Similar results are obtained with system X (right panel).

2. *Induction of hGH Synthesis*

To evaluate the performance of the specialized ribosome system, we followed hGH synthesis during the course of P_L induction. Plasmids were transformed into cells containing a defective λ prophage with a temperature-sensitive repressor [*E. coli* K5637 (*cI*857 Δ*Bam cro*[27] *Oam*[29])]. Overnight cultures grown at 30° in M9 medium supplemented with ampicillin were used to inoculate fresh M9 medium to an OD_{550} of 0.05. Cultures were grown at 30° for 2 hr, shifted to a 42° water bath for 15 min, and incubated further at 30°. One-milliliter samples were taken at the indicated times, the cells were spun down, and the pellet was frozen immediately in a dry ice/EtOH bath. The cell pellets were resuspended in 200 μl lysis buffer (30 m*M* Tris–HCl, pH 7.5, and 2% SDS). Each sample was sonicated for 15 sec at 4°. Serial dilutions of samples were made in horse serum (Hybritech) and the hGH levels were determined using a standard radioimmune assay (HGH-PRIST-KIT, Pharmacia).

Results. A low level of hGH was found in cells containing the hGH gene with the mutated SD sequence but in the absence of specialized ribosomes (Fig. 5). This low hGH level must be due to wild-type ribosomes translating the specialized mRNA with low efficiency. However, when a copy of the expressed, complementary, mutated 16 S-rRNA gene is introduced, a gradual increase of the hGH level is observed after the temperature induction of the mutated *rrn* operon. Within 3 hr of induction, the

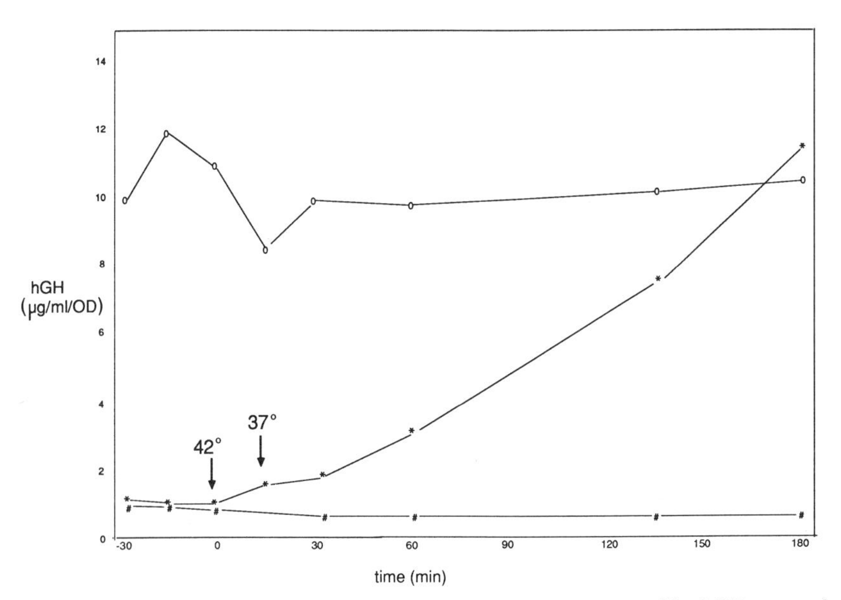

FIG. 5. Induction of the specialized ribosome synthesis as measured by hGH accumulation. O—O, Constitutive control system VIII; #—#, control system with a constitutively expressed specialized hGH mRNA (SD X) but in the absence of complementary specialized ribosomes; *—*, hGH accumulation in the complete specialized ribosome system X. In the experiment all three cultures were grown at 30°, shifted to 42° for 15 min, and incubated further at 37°.

hGH level in the specialized system exceeds that observed in the constitutive control system (VIII). The hGH level after 3 hr is, under these conditions, approximately 10% of total protein. These data show that complementarity between SD and ASD sequences increases the translational efficiency considerably even though unnatural combinations are used.

3. *Specificity of the Specialized System*

In the presence of specialized ribosomes, translation of the specialized hGH messengers occurs more efficiently than in the presence of solely wild-type ribosomes. The key question to be asked is how efficient and how specific the specialized ribosomes are in translating the specialized mRNA. In order to assess the efficiency, the number of specialized ribosomes at any time after P_L induction needs to be measured. We will discuss this issue later. An impression of the specificity of the specialized ribosomes can be obtained only after inactivation of the normal ribosomes. To investigate this *in vivo,* we used P_L-induced spectinomycin-resistant specialized ribosomes in a background of spectinomycin-sensitive wild-type ribosomes. In such a system, spectinomycin could be used to inactivate the wild-type ribosomes in the cells. The specialized ribosomes carrying the spectinomycin-resistance mutation are still functional in the presence of this antibiotic. The cells were pulse labeled with [^{35}S]methionine in the presence of spectinomycin and the distribution of label in cellular proteins was analyzed by SDS–PAGE (Fig. 6). The cells were labeled as follows:

Labeling Cells in Vivo

1. Cells were grown overnight at 30° in M9 (minus casamino acids) medium supplemented with ampicillin (50 μg/ml) and 19 amino acids (20 μg/ml each) except methionine.

2. Fresh M9 ampicillin medium was inoculated to OD_{550} = 0.05 and the cultures were grown at 30° for 2 hr.

3. The cultures were shifted to 42° for 15 min and then to 37° for 2 hr; 500 μg of spectinomycin (Gibco) was added/ml culture. Ten minutes later, [^{35}S]methionine was added to a final concentration of 26 μM (770 μCi/μmol). The cultures were grown for another 90 min and 1 OD_{550} Eq of cells was removed and spun down. The pellets were frozen immediately in a dry ice/ethanol bath.

Protein Profile Analysis. SDS–PAGE was performed according to the procedure of Laemmli.[29]

[29] U. K. Laemmli, *Nature (London)* **227,** 681 (1970).

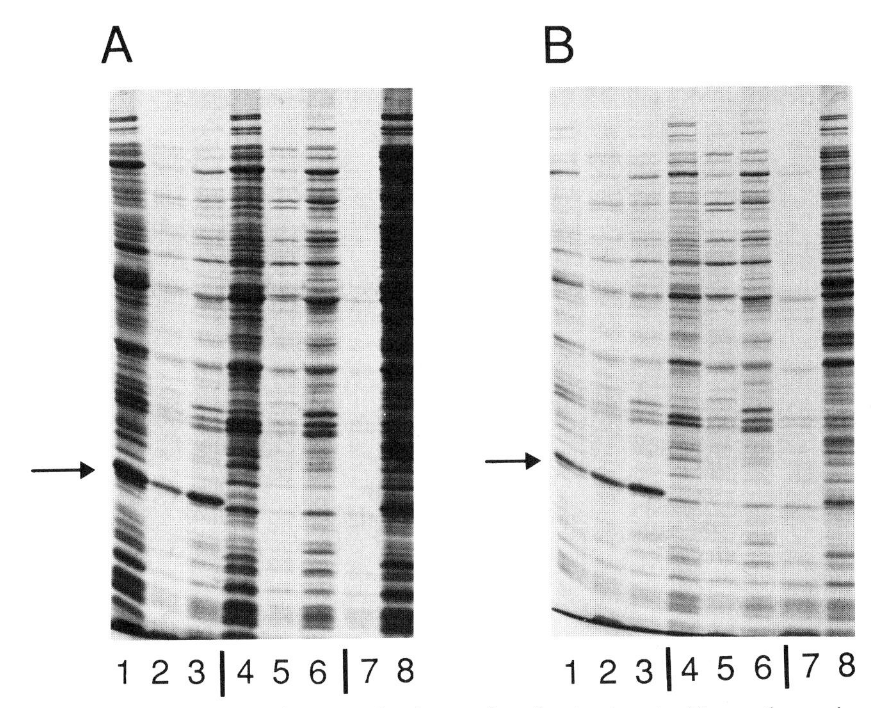

FIG. 6. [^{35}S]Methionine incorporation in proteins after treatment with spectinomycin. The band showing HGH is indicated with an arrow. Lane 1 reflects the Spc^r wild-type system. Lanes 2 and 3 show the complete specialized ribosome system in the presence of spectinomycin. Lanes 4, 5, and 6 are controls showing ^{35}S incorporation caused by specialized (lanes 5 and 6) ribosomes in the absence of an hGH gene. Lane 7 shows background incorporation in the absence of the plasmid-borne spectinomycin resistance mutation and lane 8 shows the same as in lane 7 but in the absence of the antibiotic. Left panel: The same amount of cells was loaded in each lane (0.15 OD/lane). Right panel: The same number of counts was loaded per lane (50,000 cpm/lane). Schematically:

Lane	1	2	3	4	5	6	7	8
ASD:	wt	IX	X	wt	IX	X	wt	wt
SD:	wt	IX	X	-----no hGH gene-----			wt	wt
Spc locus:	R	R	R	R	R	R	S	S
Spectinomycin added:	Yes	Yes	Yes	Yes	Yes	Yes	Yes	No

1. The cell pellets were dissolved in 100 μl of gel sample buffer, boiled at 100° for 5 min, and spun for 5 min in an Eppendorf centrifuge to remove cell debris.

2. Fifteen microliters of sample was loaded per lane on a 12.5% SDS–polyacrylamide gel. The gel was run at 35 mA until the dye front was approximately 1 cm away from the bottom.

3. The gel was dried under vacuum at 60° and exposed overnight at −80° in a film holder with intensifier screens.

Results. The protein profile is shown in Fig. 6. In the cells carrying only chromosomal and plasmid-borne Spc^s *rrn* operons, no [^{35}S]methionine incorporation is observed in the presence of spectinomycin (lane 7). Lane 1 shows that wild-type ribosomes derived from the plasmid-borne Spc^r *rrB* operon are able to translate (probably) all cellular proteins. Lanes 2 and 3 show that specialized ribosomes derived from the mutated Spc^r *rrnB* operon (system IX and X, respectively) translate the hGH messenger preferentially. The overall efficiency of system X appears to be higher than that of system IX. Although hGH is the predominant protein species made by specialized ribosomes, other mRNAs are also translated, albeit with a lower efficiency. Close inspection of this background protein profile shows that the relative intensities of several protein bands differ from those produced by wild-type ribosomes. This shows that the specificity of the ribosomes with respect to wild-type mRNAs depends on the nature of the ASD sequence.

Conclusion. With the specialized ribosome system, we have demonstrated the following:

1. The use of site-directed mutagenesis to introduce novel restriction sites and base substitutions into large double-stranded plasmids

2. The use of synthetic DNA fragments to replace short regions of DNA and their use as restriction site convertor

3. The inducible λP_L promoter to control the transcription of a potentially lethal mutated *rrnB* operon

4. The role of the 3′ end of 16 S-rRNA in determining the specificity of ribosomes

5. The use of the spectinomycin resistance mutation in 16 S-rRNA to assess the specificity of the specialized ribosomes

Applications of the Specialized Ribosome System

1. The Specialized Ribosome System as a Novel System to Express Heterologous Proteins. The current systems used to express a heterologous gene rely on translation of a foreign mRNA by ribosomes that translate all endogenous mRNAs as well. Unfavorable competition between the foreign and all endogenous mRNAs may occur especially when the foreign mRNA is not recognized efficiently by endogenous ribosomes. In the specialized ribosome system described here, ribosomes are translating a specialized mRNA in a noncompetitive fashion.

It is known that more than 50% of all rRNA originates from a single *rrn* operon when present on a high-copy-number plasmid under conditions similar to those described here.[28] Although we have not yet quantified the number of active specialized ribosomes present in our system, there is no reason to assume that the number of specialized ribosomes should be different. Since all of the specialized ribosomes are "programmed" for the translation of a single mRNA species, the level of foreign gene expression that can potentially be obtained after further improvements of the current system is likely to be higher than that using conventional systems.

2. Regulation of rRNA Synthesis. E. coli contains seven rRNA operons whose expression is tightly regulated. The work from Nomura's laboratory[28,30,31] indicates that free ribosomes (or rRNA) directly or indirectly control rRNA synthesis via a negative feedback mechanism. Testing of such a model may depend on the ability to modulate the level of free ribosomes in the cell and measuring its effect on rRNA synthesis. In normal cells, the ratio of free ribosomes to polysomes is constant during exponential growth and cannot be modulated experimentally in a meaningful way. The system described here may be useful in this regard. P_L-induced synthesis of specialized ribosomes in the absence of a specialized mRNA will allow for controlled accumulation of free ribosomes which are unable to assemble efficiently into polysomes. In this way, the free ribosome pool size could be modulated and could lead to a novel approach in the study of rRNA synthesis regulation.

3. Analysis of Mutations in the 16 S-rRNA. Traditionally, elucidation of gene function relies on analysis of mutations obtained with genetic or recombinant DNA techniques. So far, very few mutations in the rRNA gene have been reported. Obtaining mutations in the rRNA gene and their analysis is complicated by at least two factors. Seven genes encode (essentially) indistinguishable products, making mutant selection difficult. Many deletions or substitutions introduced in the 16 S-rRNA gene via recombinant DNA techniques are lethal. Although it is not clear why this is the case, it is likely that base substitutions and deletions that, e.g., still permit subunits to bind to mRNA but are impaired in other aspects of the protein biosynthetic cycle, are highly toxic to the cell. Mutations that are

[30] M. Nomura, *Proc. FEBS Congr., Part B, 16th,* p. 239 (1985).

[31] M. Nomura, R. Gourse, and G. Baughman, *Annu. Rev. Biochem.* **53,** 75 (1984).

not dominantly lethal are likely to be those that are blocked in binding to natural messengers, i.e., the kind of mutations that we described in this chapter. However, it is nearly impossible to select for such mutations in one of the chromosomal *rrn* genes using genetic techniques. Using the P_L-controlled plasmid-borne specialized ribosome system, we have also introduced mutations elsewhere in the 16 S-rRNA gene without seeing a dominant lethal effect on the cell (D. Eaton and H. A. de Boer, unpublished results). With the construction of the specialized ribosome system, we have reduced a genetically complex system to a relatively simple one; mutations in a single *rrn* operon (bearing the altered ASD sequence) can now be studied by analyzing the effects on translation of a single mRNA species.

[28] New Expression Vectors for Identifying and Testing Signal Structures for Initiation and Termination of Transcription

By KLAUS SCHNEIDER and CHRISTOPH F. BECK

Analysis of DNA sequences with regulatory function(s) has been greatly facilitated by the *in vitro* manipulation of DNA. A given regulatory sequence is cloned in front of an appropriate indicator gene, the product of which can be determined easily and quantitatively. When the indicator gene is preceded by a ribosome-binding site and a translation initiation codon, the level of expression of the gene should reflect the strength of the promoter inserted before the gene. Such promoter–probe vectors with the galactokinase (*galK*) gene as indicator have found widespread application.[1] The vectors described here contain pairs of divergently oriented indicator genes, either *lacZ* (β-galactosidase)/*galK* or *lacZ*/*phoA* (alkaline phosphatase), separated by different polylinkers. The combinations are carried on plasmids of different host ranges and copy number to further increase their versatility. The resulting plasmids are of great potential use for isolating linked or overlapping divergent promoter sequences as well as for the isolation of unidirectional promoter sequences.

[1] K. McKenney, H. Shimatake, D. Court, U. Schmeissner, C. Brady, and M. Rosenberg, *in* "Gene Amplification and Analysis. Vol. II: Structural Analysis of Nucleic Acids" (J. B. Chirikjian and T. S. Papas, eds.), p. 383. Elsevier/North-Holland, New York, 1981.

Material, Reagents, and Assays

Strains

The *Escherichia coli* K-12 test strains used are CB454 (Δ*lacZ*, lacY^+, *galK*, *rpsL*, *thi*, *recA*56) and CB806 (Δ*lacZ*, *lac*Y^+, *galK*, *phoA*8, *rpsL*, *thi*, *recA*56).

Growth Media

Minimal media are AB medium[2] [component A (per liter): 2 g of $(NH_4)_2SO_4$, 6 g of Na_2HPO_4, 3 g of KH_2PO_4, 3 g of NaCl, 11 mg of Na_2SO_4; component B (per liter): 200 mg of $MgCl_2$, 10 mg of $CaCl_2$, 0.5 mg of $FeCl_3 \cdot 7H_2O$. A sterile solution of component A (10-fold concentrated stock solution) is added to the medium containing component B (prepared as 40-fold concentrated stock solution)] with 64 m*M* phosphate and Tris medium [per liter: 4.7 g of NaCl, 0.5 mg of $FeCl_3 \cdot 7H_2O$, 1.5 g of KCl, 14.5 g of Tris (base), 2.65 g of $(NH_4)_2SO_4$, 0.25 g of $MgSO_4 \cdot 7H_2O$, and sodium phosphate to 5 m*M*; the pH is adjusted to 7.5 with concentrated HCl]. Glycerol (0.2%) is routinely used as a carbon source. L-broth contains (per liter) 10 g of tryptone, 5 g of yeast extract, 5 g of NaCl, 4 g of glucose. Media are solidified by addition of 20 g of agar per liter. McConkey-lactose and McConkey-galactose indicator plates are prepared as suggested by the manufacturer (Difco). Cloning of a promoter in front of the *phoA* indicator gene can be tested directly on plates after transformation of a $phoA^-$ host strain. The test plates contain Tris medium with 0.2% glycerol and 0.05% α-glycerophosphate but no inorganic phosphate. Flooding of the plates with a solution of 10 m*M* *p*-nitrophenyl phosphate (PNPP) in 1 *M* Tris–HCl buffer, pH 8, reveals alkaline phosphatase-positive colonies by their bright yellow color.

Buffers

NET buffer: 0.15 *M* NaCl, 0.1 m*M* EDTA, 20 m*M* Tris–HCl, pH 8

High-salt NET buffer: 1 *M* NaCl, 0.1 m*M* EDTA, 20 m*M* Tris–HCl, pH 8

Klenow buffer (10-fold strength): 0.33 *M* Tris–acetate, pH 7.9, 0.66 *M* potassium acetate, 0.1 *M* magnesium acetate, 5 m*M* dithiothreitol, 1 mg/ml of bovine serum albumin. This buffer is stored at $-20°$

DNA ligation buffer (5-fold strength): 0.25 *M* Tris–HCl, pH 7.8, 50

[2] D. J. Clark and O. Maaløe, *J. Mol. Biol.* **23,** 99 (1967).

m*M* $MgCl_2$, 100 m*M* dithiothreitol, 0.3 mg/ml bovine serum albumin, 3.5 m*M* spermidine, 5 m*M* ATP. This buffer is stored at −20°

Z-buffer (per liter): 16.1 g of $Na_2HPO_4 \cdot 7H_2O$, 5.5 g of $NaH_2PO_4 \cdot H_2O$, 0.75 g of KCl, 0.246 g of $MgSO_4 \cdot 7H_2O$, 2.7 ml of 2-mercaptoethanol; pH adjusted to 7

GalK reaction mix: 0.25 *M* Tris–HCl, pH 8, 4 m*M* $MgCl_2$, 4 m*M* ATP, 8 m*M* NaF, 3.5 μ*M* [^{14}C]galactose (specific activity, 5 mCi/mmol)

Gal buffer: 50 m*M* Tris–HCl, pH 8, 1 m*M* dithiothreitol, 1 m*M* EDTA

TE buffer: 10 m*M* Tris–HCl, pH 8, 1 m*M* EDTA

E buffer: 40 m*M* Tris (base), 20 m*M* sodium acetate, 2 m*M* EDTA. The pH is adjusted to 8.3 with glacial acetic acid

Agarose Gel Electrophoresis

DNA fragments are separated on a horizontal flat bed gel containing 0.7 to 1.5% of agarose (BRL). The agarose is melted in E buffer and ethidium bromide to a final concentration of 2 μg/ml is added just before pouring of the gel. E buffer is used as electrophoresis buffer. Electrophoresis is carried out at 100–140 mA.

Plasmid DNA Isolation

Plasmid DNA is isolated from overnight cultures in L-broth containing a selective antibiotic by the alkaline lysis procedure.[3]

DNA Restriction Fragment Isolation

The DNA fragment to be isolated is localized by illumination with long-wavelength UV light (366 nm) on the agarose gel. With a razor blade an incision is made below the fragment band and a small piece of DEAE membrane (Schleicher and Schüll DEAE membrane NA 45) is inserted. (The DEAE membrane is prepared by washing it consecutively with 10 m*M* EDTA, pH 7.6, for 10 min, with 0.5 *M* NaOH for 5 min, and finally with H_2O. It is stored at 4° in distilled H_2O.) Electrophoresis is continued until the fragment is completely bound to the DEAE membrane. The DEAE membrane is rinsed with NET buffer and the residual liquid removed by placing it on Whatman 3MM paper. The membrane is then placed in an Eppendorf tube containing 300 μl of high-salt NET buffer, and incubated at 68° for 1 hr with occasional shaking. After removal of the membrane the DNA is concentrated by ethanol precipitation.

[3] H. C. Birnboim, this series, Vol. 100, p. 243.

Filling in of Single-Stranded 3′ Recessed Ends

The reaction mixture comprises 17 μl of DNA solution, 3.4 μl of Klenow buffer (10-fold strength), 1 μl of dNTP mix (2 m*M* each of dTTP, dGTP, dCTP, dATP), and 12.1 μl H_2O. The reaction is started by the addition of 2.5 U of the Klenow fragment of *E. coli* DNA polymerase I (0.5 μl), incubated at room temperature for 10 min, and terminated by phenol extraction.

Restriction and Ligation of DNA

Restriction digests are performed in the buffers described by the enzyme suppliers. DNA ligations are performed in a 20-μl reaction volume containing 1 U of T4 DNA ligase (BRL), 4 μl of ligation buffer, and 0.5 to 1 μg of DNA. Reactions are incubated at 15° overnight. After ligation, the DNA is used directly for transformation.

β-Galactosidase Assay

Plates. These contain AB minimal medium, 0.2% glycerol, 2% agar, and 40 mg/liter of 5-bromo-4-chloro-3-indolyl-β-D-galactoside (X-Gal).

Liquid Cultures. Cells are grown in minimal medium. Samples (0.1 to 0.3 ml) are added to 0.9 and 0.7 ml, respectively, of ice-cold Z-buffer[4] containing 100 μg/ml of chloramphenicol. One drop of toluene is added, cells are vortexed for 30 sec, and the toluene evaporated in a shaking water bath at 37°. The reaction is started by the addition of 0.2 ml of a 4 mg/ml solution of *o*-nitrophenyl-β-D-galactoside (ONPG) to the cells preincubated at 30°. The reaction is stopped when a yellow color is visible—or after 1 hr—by the addition of 0.5 ml of 1 *M* Na_2CO_3. The OD_{410} of the supernatant is measured after centrifugation. Units are 1000 × OD_{410}/time (min) × volume assayed (in ml) × OD_{578} of the culture.[4]

Galactokinase Assay

One-milliliter samples from cultures grown in minimal medium are added to 1 ml 0.9% NaCl on ice. The cells are pelleted by centrifugation, resuspended in 1 ml of Gal buffer, and treated with toluene as described for the β-galactosidase assay. To 60 μl of treated cells or dilutions thereof, 40 μl of GalK reaction mix is added. The reactions and preequilibrations are at 37°. Duplicate 30-μl samples are transferred to DEAE membrane

[4] J. H. Miller (ed.), "Experiments in Molecular Genetics." Cold Spring Harbor Lab., Cold Spring Harbor, New York, 1972.

squares (Schleicher and Schüll DEAE membrane NA 45), stopping the reaction. The squares are immersed in 80% ethanol, washed three times with water, and then dried. The bound radioactivity is determined in a scintillation spectrometer. Units are micromoles of galactose phosphate formed per minute, volume of cells in the assay, and the OD_{578} of the culture.

Alkaline Phosphatase Assay

From cultures grown in Tris medium, 0.2-ml samples are removed and added to 0.8 ml of 50 m*M* Tris–HCl, pH 8, containing 100 μg/ml of chloramphenicol, on ice. The cells are washed once with 1 ml of 50 m*M* Tris–HCl, pH 8, and resuspended in 1 ml of 1 *M* Tris–HCl, pH 8. After toluene treatment (see β-galactosidase assay) the reaction is started by the addition of 10 μl of 100 m*M* *p*-nitrophenyl phosphate (dissolved in H_2O and stored at $-20°$). The reaction is stopped when yellow color is visible—or after 1 hr—by the addition of 90 μl of 1 *M* potassium phosphate, pH 8. Cell debris is removed by centrifugation and the OD_{410} is measured. Units are OD_{410}/time (min) × volume of cells assayed (in ml) × OD_{578} of the culture.

Procedures and Results

Promoter–Probe Vectors

Plasmid pCB302, shown in Fig. 1, has been developed for the cloning of unidirectionally or bidirectionally arranged promoters. This plasmid contains the indicator genes *galK* and *lacZ* and a transcriptional termination signal (t_0 of phage λ) behind *galK*. Gene *galK* and its 5′ flanking sequences up to the *Sma*I site was derived from pK0-1.[1] At the 5′ end of *lacZ*, the codons for the first four amino acids and the translational start signal are from the *E. coli lpp* gene.[5] pCB302 a and b differ with respect to the orientation of the polylinker sequence (Fig. 1). Vector pCB302 b, in addition, harbors a sequence with transcription termination properties located between the *Xba*I and *Sma*I restriction sites (see later).

For the construction of a promoter–probe vector with *phoA* as an indicator gene (pCB267), all *phoA* sequences upstream of the Shine–Dalgarno sequence were deleted.[6] The DNA sequence between *lacZ* and *phoA* of pCB267 is given in Fig. 1. This vector provides indicator genes whose products can be quantitatively determined by simple, reproduc-

[5] K. Nakamura and M. Inouye, *EMBO J.* **1,** 771 (1982).
[6] K. Schneider and C. F. Beck, *Gene* **42,** 37 (1986).

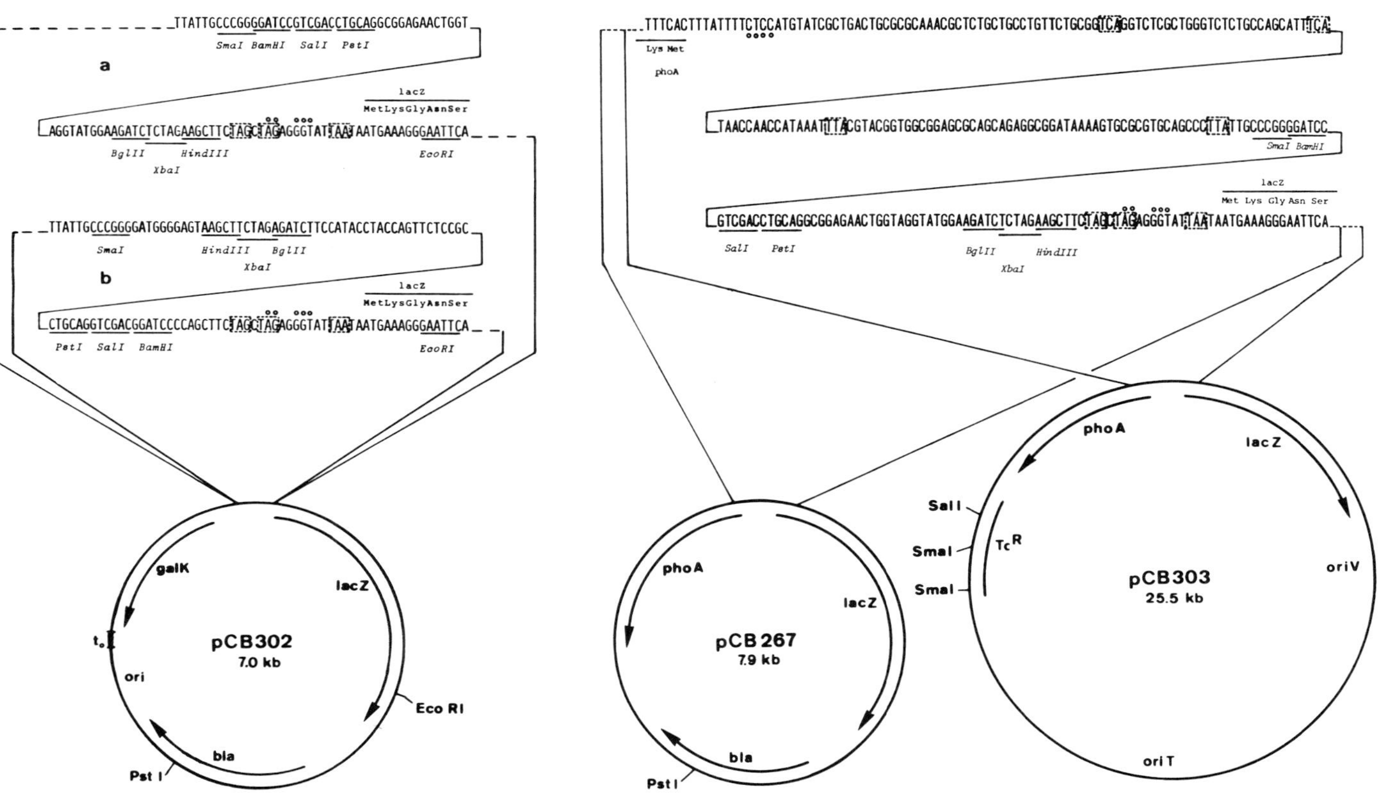

FIG. 1. Vectors with divergently arranged indicator genes. For pCB302 (a) and pCB302 (b), the nucleotide sequences from the *Sma*I site, originally present on the pK0-1 vector, to the beginning of *lacZ* are given. For pCB267 and pCB303 the nucleotide sequence between structural genes *phoA* and *lacZ* is given. Translational stop codons are boxed. The ribosome-binding sites are indicated by small circles.

ible, and inexpensive assays. This vector cannot be used for the cloning of strong promoters in front of *phoA* because overexpression of this gene slows the growth of *E. coli*.

The replication functions and the selectable gene *bla* of vectors pCB302 and pCB267 are from pBR322-derived vector pK0-1.[1] These vectors exhibit high copy number and a restricted host range. Since plasmid copy number will influence the amount of enzyme produced, it is important to use exponentially growing cultures with approximately the same cellular mass per milliliter for determinations of promoter strength. Cloning of very strong promoters may, however, affect plasmid copy number, and enzyme activities must be corrected for copy number.[7] High-copy-number plasmids facilitate the isolation of plasmid DNA and also increase the sensitivity of the test system because of gene dosage.

Promoter–probe vector pCB303 (Fig. 1) has a broad host range (P1 incompatibility group) and an intermediate copy number of about 10 per cell.[8] It was constructed by insertion of indicator genes *lacZ*/*phoA* of pCB267 into the filled-in *Bgl*II site of RK2-derived pRK290.[8] This vector is suitable for regulatory studies in cases where the intracellular concentration of protein(s) regulating the test genes may be limiting. Provision of conjugation functions *in trans* by a second plasmid, e.g., pRK2013,[8] allows transfer of pCB303 to most gram-negative bacteria by triparental mating.

Salient features of the promoter–probe vectors described are (1) polylinker sequences with multiple restriction sites which allow for convenient cloning of DNA fragments carrying unidirectional or divergently arranged promoters, (2) translational stop codons present in all three reading frames upstream of the initiation codons (Fig. 1), (3) terminators for transcription behind *galK* (t_0 from λ) and *phoA*,[6] (4) a very low residual level of expression of the promoterless indicator genes (Table I), (5) the choice of vectors with different copy number and host range, and (6) the detectability of promoter activity directly on plates after transformation.

Screening for Promoters from Uncharacterized DNAs

The DNA is partially or completely digested with restriction endonucleases recognizing 4-base sequences (e.g., *Sau*3A). The fragments are ligated into compatible restriction sites (e.g., *Bam*HI or *Bgl*II) on the desired vector. Alternatively, the polylinkers of the vectors allow for a

[7] C. W. Adams and G. W. Hatfield, *J. Biol. Chem.* **259,** 7399 (1984).

[8] G. Ditta, S. Stanfield, D. Corbin, and D. R. Helinski, *Proc. Natl. Acad. Sci. U.S.A.* **77,** 7347 (1980).

TABLE I

LEVELS OF INDICATOR GENE EXPRESSION OF VARIOUS PROMOTER–PROBE VECTORS

Host strains and relevant genotype	Plasmid	Promoters and their orientations	Specific activities[a]		
			β-Galactosidase	Galactokinase	Alkaline phosphatase
CB454 (*lacZ*$^-$, *galK*$^-$)	None	—	<1	ND	0.004
CB806 (*lacZ*$^-$, *galK*$^-$, *phoA*$^-$)	None	—	<1	ND	0.001
CB454	pCB302a	None	<1	<0.1	ND
CB454	pCB302b	None	<1	<0.1	ND
CB806	pCB267	None	<1	ND	0.007
CB806	pCB303	None	<1	ND	0.005
CB454	pCB237	$tetP_A$–*galK*, $tetP_R$–*lacZ*	316	216	ND
CB454	pCB304	$tetP_R$–*galK*, $tetP_A$–*lacZ*	163	445	ND
CB454	pCB238	$tetP_R$–*galK*, $tetP_A$–*lacZ*	238	44	ND

[a] ND, Not done.

directed insertion of appropriate DNA fragments (e.g., as *Bam*HI–*Xba*I fragments). After transformation of the test vector into a host strain, the presence of promoters transcribing in one or both directions can be determined directly on plates: AB minimal medium plates with galactose as a carbon source and containing X-Gal (as an indicator for β-galactosidase) are used for pCB302 derivatives. Tris medium plates with lactose as a carbon source are used for pCB267 and pCB303 derivatives; the colonies can be screened for alkaline phosphatase activity *in situ* (see Material, Reagents, and Assays).

Insertion of Promoters in pCB302

A DNA fragment with the divergently arranged promoters $tetP_A$ and $tetP_R$, located on plasmid pBR322 in front of the *tet* gene,[9,10] was inserted into test vector pCB302 in two orientations. In pCB237, *galK* and *lacZ* are transcribed by promoters $tetP_A$ and $tetP_R$, respectively. In pCB238 *galK* is transcribed from promoter $tetP_R$. pCB304 is a derivative of pCB238 in which the 19-bp DNA fragment between the *Sma*I and *Xba*I restriction sites has been removed from the polylinker region. The promoter activities are given in Table I. A comparison of the strength of promoter $tetP_R$ and promoter $tetP_A$ (pCB237 vs pCB304) reveals a ratio of approximately 2, irrespective of the orientation of inserted DNA. This indicates that the enzyme activities measured indeed reflect the frequency of transcriptional starts at the promoters. Possible complications arising from DNA sequences located between a promoter and the indicator gene are illustrated by a comparison of the enzyme levels encoded by pCB304 and pCB238 (Table I). These vectors are identical except for the 19-bp DNA fragment between the *Sma*I and *Xba*I restriction sites of the polylinker located between promoter $tetP_R$ and *galK*. The presence of these 19 bp appears to reduce transcription of *galK* drastically.

Comments

The promoter–probe vectors described may be used for cloning, detection, and regulatory analyses of single as well as divergently arranged promoters. Relative strength of promoters can be determined using these vectors, though caution in the interpretation of the data is advised since DNA sequences in the 5′ untranslated regions between a promoter and an

[9] D. Stüber and H. Bujard, *Proc. Natl. Acad. Sci. U.S.A.* **78,** 167 (1981).

[10] G. Klock, B. Unger, C. Gatz, W. Hillen, J. Altenbuchner, K. Schmid, and R. Schmitt, *J. Bacteriol.* **161,** 326 (1985).

indicator gene may strongly influence the expression of the indicator gene.

Acknowledgments

We wish to thank S. Kiedrowski for constructing pCB303. This work was supported by a grant from the Deutsche Forschungsgemeinschaft.

[29] Synthesis and Sequence-Specific Proteolysis of Hybrid Proteins Produced in *Escherichia coli*

By KIYOSHI NAGAI and HANS CHRISTIAN THØGERSEN

In order to express foreign genes in *Escherichia coli,* a strong promoter and a ribosome-binding site from a highly expressed *E. coli* gene are placed in front of the desired coding sequence.[1-3] The structure of the DNA double helix is determined locally, and therefore any strong promoter retains its activity regardless of the gene cloned downstream. However, formation of a complex between the messenger RNA (mRNA) and the ribosome depends not only on the Shine–Dalgarno sequence (ribosome-binding site)[4,5] but also on the three-dimensional structure of the folded mRNA on either side of the ribosome-binding site. If a foreign coding sequence is joined to the ribosome-binding sequence from a highly expressed *E. coli* gene, then the folding of the mRNA may no longer be the same as in natural *E. coli* mRNA and this usually results in poor translation of the message.[1]

To overcome this problem, the foreign coding sequence can be joined in phase to a short coding sequence of a highly expressed *E. coli* gene so that a hybrid protein is produced.[6-9] The short segment of *E. coli* gene

[1] T. J. R. Harris, *in* "Genetic Engineering" (R. Williamson, ed.), Vol. 4, p. 127. Academic Press, London, 1983.

[2] E. Remaut, P. Stanssen, and W. Fiers, *Gene* **15,** 81 (1981).

[3] A. Shatzman, Y.-S. Ho, and M. Rosenberg, *in* "Experimental Manipulation of Gene Expression" (M. Inouye, ed.), p. 1. Academic Press, New York, 1983.

[4] J. Shine and L. Dalgarno, *Nature (London)* **254,** 34 (1975).

[5] J. A. Steitz, *Biol. Regul. Dev.* **1,** 349 (1979).

[6] M. Courtney, A. Buchwalder, L.-H. Tessier, M. Jaye, A. Benavente, A. Ballard, V. Kohli, R. Lathe, P. Tolstoshev, and J.-P. Lecocq, *Proc. Natl. Acad. Sci. U.S.A.* **81,** 669 (1984).

[7] J. A. Lauteuberger, D. Court, and T. S. Papas, *Gene* **23,** 75 (1983).

[8] K. Nagai, and H. C. Thøgersen, *Nature (London)* **309,** 810 (1984).

[9] K. Nagai, M. F. Perutz, and C. Poyart, *Proc. Natl. Acad. Sci. U.S.A.* **82,** 7252 (1985).

directs folding of the mRNA in the vicinity of the ribosome-binding site and thereby ensures high translational efficiency. If the fusion protein thus produced can be cleaved specifically at the junction of the two sequences by an appropriate enzymatic or chemical method, the authentic protein will then be liberated.

We have constructed a fusion protein expression vector pLcII(nic$^-$).[9] This plasmid has multiple cloning sites downstream of a short coding sequence from the λ cII gene,[10] which is under the control of the strong leftward (P_L)[10,11] promoter of λ phage. Therefore, pLcII(nic$^-$) directs the synthesis of fusion proteins consisting of the 31 amino terminal residues of the λ cII protein and the polypeptide sequence encoded by the cloned gene. The following proteins have been produced in this way at the level of 5–10% of total *E. coli* cellular protein: human β-globin,[8,9] human α-globin,[12] human myoglobin,[13] chicken β-actin,[14] chicken myosin light chain,[15] chicken troponin C,[16] human tropomyosin,[17] human c-myc oncogene product,[18] *Xenopus* histones H2A and H2B,[19] tobacco mosaic virus coat protein,[20] yeast MATα1,[20a] yeast SWI-5,[20b] and *Xenopus* TFIIIA.[20c]

In a second stage, we have inserted a sequence encoding the tetrapeptide Ile-Glu-Gly-Arg at the junction in the fusion proteins.[8,9] This tetrapeptide sequence precedes the two-factor X_a cleavage sites in prothrombin and is considered to be the recognition sequence for blood coagulation factor X_a (protease).[21–23] The CIIFX fusion proteins (hereafter a CII fusion

[10] H. Shimatake and M. Rosenberg, *Nature (London)* **292,** 128 (1981).
[11] H.-U. Bernard and D. R. Helinski, this series, Vol. 68, p. 482.
[12] K. Nagai and J. Pagnier, unpublished results.
[13] R. Varadarajan, A. Szabo, and S. G. Boxer, *Proc. Natl. Acad. Sci. U.S.A.* **82,** 5681 (1985).
[14] R. Karlsson and K. Nagai, unpublished results.
[15] F. Reinach, K. Nagai, and J. Kendrick-Jones, *Nature (London)* **322,** 80 (1986).
[16] R. Karlsson, F. Reinach, J. Kendrich-Jones, and K. Nagai, unpublished results.
[17] C. Houlker and A. R. MacLeod, unpublished results.
[18] L. Bulawela, P. van Straaten, and T. H. Rabitts, unpublished results.
[19] M. Waye and T. J. Richmond, unpublished results.
[20] D. R. Turner, unpublished results.
[20a] S. Tan and T. J. Richmond, unpublished results.
[20b] K. Nagai and D. Rhodes, unpublished results.
[20c] P. K. Hansen and H. C. Thøgersen, unpublished results.
[20d] K. P. Nambiar, J. Stackhouse, S. R. Presnell, and S. A. Brenner, *Eur. J. Biochem.* **163,** 67 (1987).
[21] S. Magnusson, T. E. Petersen, L. Sottrup-Jensen, and H. Claeys, *in* "Protease and Biological Control" (E. Reich, D. B. Rifkin and E. Shaw, eds.), p. 123. Cold Spring Harbor Lab., Cold Spring Harbor, New York, 1975.
[22] D. A. Walz, D. Hewett-Emmett, and W. H. Seegers, *Proc. Natl. Acad. Sci. U.S.A.* **74,** 1969 (1977).
[23] R. J. Butkowski, J. Elion, M. R. Downing, and K. G. Mann, *J. Biol. Chem.* **252,** 4942 (1977).

protein with the factor X_a cleavage site is referred to as CIIFX) of α-globin,[24] β-globin,[8,9] actin,[14] myosin light chain,[15] troponin C,[16] tobacco mosaic virus coat protein,[20] yeast MATα1,[20a] TFIIIA,[20c] and pancreatic ribonuclease A[20d] were cleaved at the peptide bond following the tetrapeptide by blood coagulation factor X_a and authentic proteins were liberated.

Some peptide hormones have been cleaved from genetically engineered fusion proteins by cyanogen bromide, trypsin, clostripain, etc.[1,25] But such methods are only applicable under very limited circumstances since large proteins usually contain several cleavage sites: Met for cyanogen bromide, Arg or Lys for trypsin, and Arg for clotripain. In contrast to these methods, factor X_a cleavage is specific to the Ile-Glu-Gly-Arg tetrapeptide sequence, which occurs very rarely in protein sequences.

This chapter describes a general method of producing eukaryotic proteins in *E. coli* using the fusion protein expression vector pLcII(nic$^-$) and subsequent purification and cleavage methods using blood coagulation factor X_a.

A Fusion Protein Expression Vector: pLcIIFXβ-Globin(nic$^-$)

pLcIIFXβ-globin(nic$^-$) is a pUC9[26]-based protein expression vector which directs synthesis of a hybrid protein consisting of the 31 amino terminal residues of the λ cII protein and human β-globin with the tetrapeptide Ile-Glu-Gly-Arg in between (Fig. 1).[9] The hybrid gene is transcribed from the P_L promoter. The host strain must provide a thermosensitive (cI857) repressor so that protein synthesis is completely repressed at 30° but can be induced at 42°. Constitutive expression of foreign proteins is often lethal to the cell and can be avoided by use of this stringent control system.[27] There is also a transcriptional terminator (t_{R1}) between the P_L promoter and the gene, and therefore the λ N-protein must be provided by the host strain.[28] RNA polymerase can then overcome the terminator by interacting with the N-protein at the nut R and nut L sites.

As shown in Fig. 2, the CIIFXβ-globin fusion protein represents 5–10% of total cellular protein after heat induction.[8,9] The protein was partially purified and treated with blood coagulation factor X_a. As shown in Fig. 3, CIIFXβ-globin is cleaved exclusively at the peptide bond following the Ile-Glu-Gly-Arg tetrapeptide and authentic β-globin is liberated.[8,9] Cleavage did not take place elsewhere in the fusion protein despite the

[24] J. Pagnier and K. Nagai, unpublished results.

[25] K. Itakura, T. Hirose, R. Crea, A. D. Riggs, H. L. Heyneker, F. Bolivar, and H. W. Boyer, *Science* **198,** 1056 (1977).

[26] J. Vieira and J. Messing, *Gene* **19,** 259 (1982).

[27] M. Rosenberg, Y.-S. Ho, and A. Shatzman, this series, Vol. 101, p. 123.

[28] D. F. Ward and M. E. Gottesman, *Science* **216,** 946 (1982).

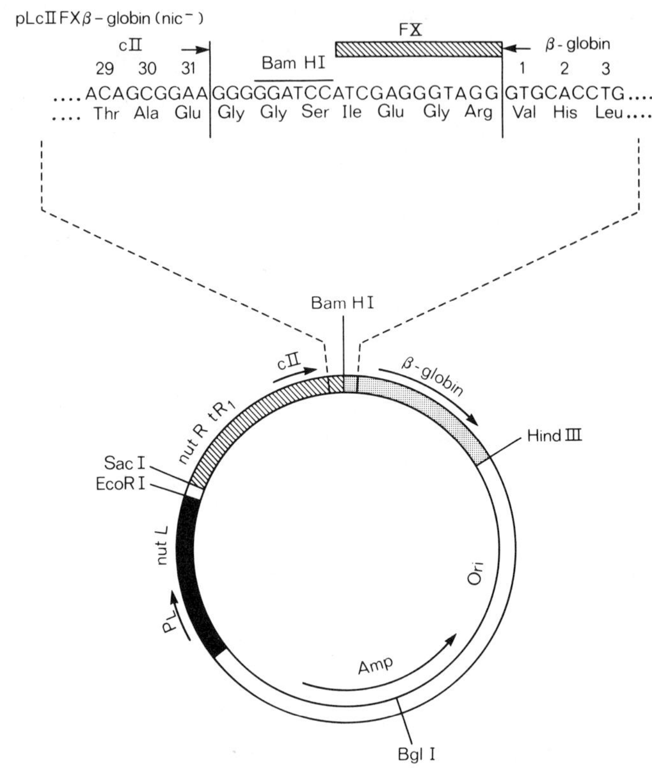

FIG. 1. Cleavable fusion protein expression vector, pLcIIFXβ-globin(nic⁻). The plasmid directs synthesis of a hybrid protein consisting of the 31 amino terminal residues of the λ cII protein, the factor X_a recognition sequence (FX), and human β-globin polypeptide. The *Hin*dIII–*Bgl*I fragment of pLcIIFXβ-globin has been replaced with the *Hin*dIII–*Bgl*I fragment of pUC9 to remove the nic site.

presence of 6 other Arg and 14 Lys residues. It is clear from Fig. 3 that many *E. coli* proteins present in this preparation remained uncleaved by factor X_a treatment, showing that factor X_a is indeed highly specific for its recognition sequence.

Use of pLcII(nic⁻) for Expression of Eukaryotic Proteins and Subsequent Cleavage of Fusion Proteins by Factor X_a

Joining Protein Coding Sequence to the Factor X_a Recognition Sequence

We have prepared a derivative of M13 phage vector mp11 which contains a sequence encoding the Ile-Glu-Gly-Arg tetrapeptide. Oligonu-

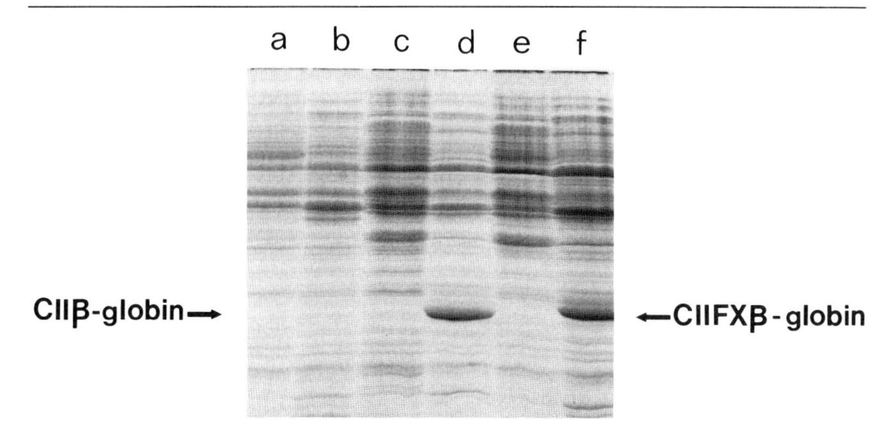

FIG. 2. Expression of β-globin fusion protein in *E. coli*. Total cellular protein was analyzed on an 18% polyacrylamide–SDS gel and visualized with Coomassie blue. (a) pLcII at 30°; (b) pLcII at 42°; (c) pLcIIβ at 30°; (d) pLcIIβ at 42°; (e) pLcIIFXβ at 30°; (f) pLcIIFXβ at 42°. pLcIIβ is identical to pLcIIFXβ except for the absence of the factor X_a recognition sequence. Reprinted by permission from *Nature*, Vol. 309, p. 812.[8] Copyright © 1984 Macmillan Journals Limited.

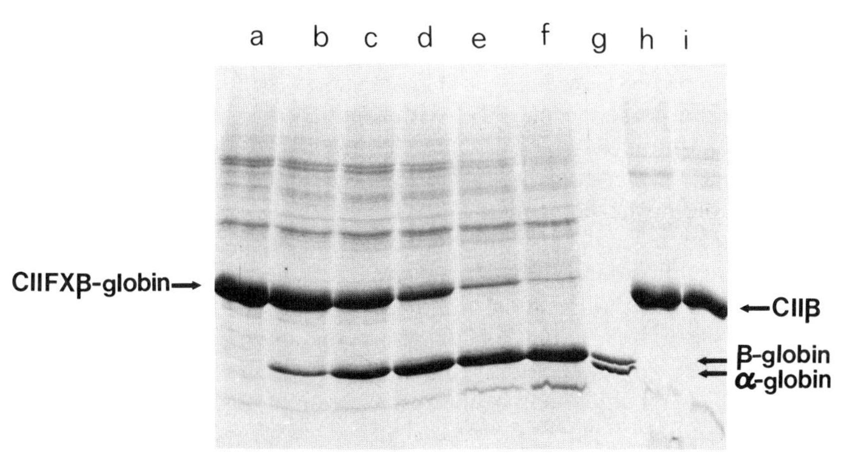

FIG. 3. Digestion of CIIFXβ-globin protein with bovine blood coagulation factor X_a. The CIIFXβ-globin fusion protein was partially purified and digested with factor X_a at an enzyme-to-substrate molar ratio of 1 : 100 at 25°. The samples were analyzed on an 18% polyacrylamide–SDS gel. (a) Undigested CIIFXβ-globin; (b) 5 min; (c) 15 min; (d) 30 min; (e) 60 min; (f) 120 min after addition of factor X_a; (g) human adult hemoglobin comprising the α (faster band) and β (slower band) chains. The CIIβ-globin fusion protein which lacks the cleavage site for factor X_a was treated identically. (h) After 120 min of treatment; (i) untreated CIIβ-globin. Reprinted by permission from *Nature*, Vol. 309, p. 812.[8] Copyright © 1984 Macmillan Journals Limited.

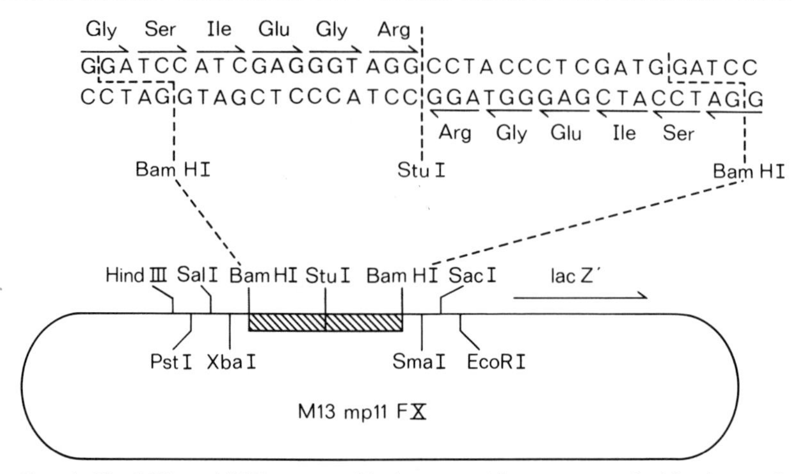

FIG. 4. The M13 mp11FX vector with the recognition sequence for blood coagulation factor X_a. The vector can be linearized with *Stu*I and any coding sequence with blunt ends can be cloned into the *Stu*I site. Reprinted by permission from *Nature*, Vol. 309, p. 811.[8] Copyright © 1984 Macmillan Journals Limited.

cleotides encoding the tetrapeptide sequence were cloned into the *Bam*HI site of M13 mp11 to form M13 mp11FX (Fig. 4).[8] M13 mp11FX gives blue plaques in the presence of IPTG (isopropyl-β-D-thiogalactopyranoside) and BCIG (5-bromo-4-chloro-3-indolyl-β-D-galoctoside, Sigma) so that blue and white selection can be used for cloning. As shown in Fig. 5, this vector can be linearized at the *Stu*I (AGGCCT) site after the Arg codon. Some eukaryotic genes have suitable restriction sites around the initiation codon, and such genes can be cloned into the *Stu*I site after an appropriate end repair reaction. Human α-globin,[24] β-globin,[8] and chicken β-actin[14] genes have an *Nco*I (CCATGG) site at the initiation codon, and they could therefore be cloned into the *Stu*I site after treatment with the single-stranded DNA specific mung bean nuclease which removes both 5′ and 3′ protruding ends.

Methods[8,12,14]

80-μl solution of plasmid DNA (40 μg) cleaved at the *Nco*I site and the site downstream from the coding sequence. DNA is dissolved in 10 m*M* Tris–Cl, pH 8.0, 0.1 m*M* EDTA

5× mung bean nuclease buffer: 150 m*M* sodium acetate buffer, pH 4.6; 250 m*M* NaCl; 5 m*M* $ZnCl_2$; 25% glycerol

2 μl mung bean nuclease (Pharmacia; 100 U/μl)

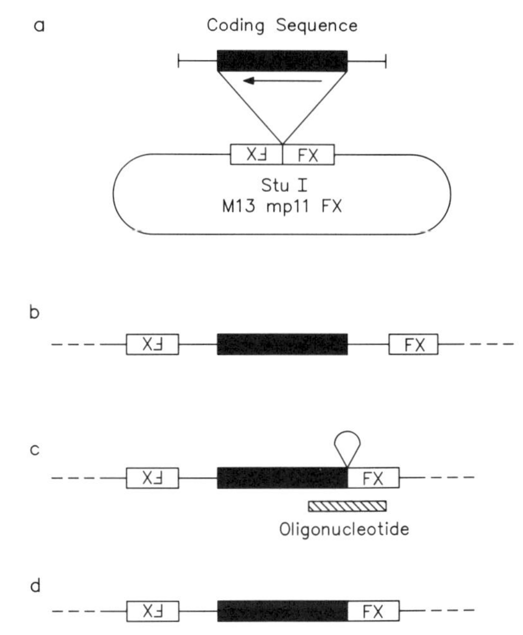

FIG. 5. Joining the first amino acid codon to the factor X_a recognition sequence by deletion mutagenesis.

An Eppendorf tube containing the DNA solution in mung bean nuclease buffer is placed on ice. Two microliters of mung bean nuclease (100 U/ μl, Pharmacia) is added and the solution mixed quickly using a vortex mixer. After 10 min, the reaction is stopped by addition of half a volume of chloroform–isoamyl alcohol mixture (24 : 1, v/v) and half a volume of phenol saturated with 0.1 *M* Tris–HCl, pH 8.0. The DNA solution is extracted with phenol–chloroform–isoamyl alcohol twice and the aqueous phase is extracted with ether to remove phenol. DNA is precipitated with one-tenth volume of 3.5 *M* NaCl and 2.5 vol of ethanol.[29]

The mixture of DNA fragments is cloned into the *Stu*I site of M13 mp11FX. The DNA sequence is determined for several white plaques by the dideoxy chain termination method.[30,31] A clone in which the codon for the first amino acid residue is correctly joined to the Arg codon of the

[29] T. Maniatis, E. F. Fritsch, and J. Sambrook, "Molecular Cloning: A Laboratory Manual." Cold Spring Harbor Lab., Cold Spring Harbor, New York, 1982.

[30] F. Sanger, S. Nicklen, and A. R. Coulson, *Proc. Natl. Acad. Sci. U.S.A.* **74,** 5463 (1977).

[31] M. D. Biggin, T. J. Gibson, and G. F. Hong, *Proc. Natl. Acad. Sci. U.S.A.* **80,** 3963 (1983).

factor X_a recognition sequence is chosen and its replicative form of DNA is prepared. The protein-coding sequence, together with the factor X_a recognition sequence, can then be cut out with *Bam*HI, and this fragment cloned into the *Bam*HI site of pLcII(nic$^-$).

Joining the First Amino Acid Codon to the Factor X_a Recognition Sequence by Deletion Mutagenesis

Most genes do not have convenient restriction sites around the initiation codon. In such cases, a restriction fragment containing the entire coding sequence can be cloned into the *Stu*I site of M13 mp11FX (Fig. 5a). Then, the clone with the coding sequence in opposite orientation to the *lacZ'* gene is chosen by DNA sequence determination (Fig. 5b). The codon for the first amino acid can be brought next to the factor X_a recognition sequence by deletion mutagenesis using an oligonucleotide (Fig. 5c and d). A procedure for deletion mutagenesis is described in detail by Carter.[32] The frequency of mutants is, however, often quite low for deletion mutagenesis and many spurious deletion mutants are produced. Waye *et al.*[33] have therefore modified M13 mp11FX and constructed a new M13 vector mp11RX with four *Eco*K restriction sites between the cloning site and the factor X_a recognition sequence. The gene to be expressed can be inserted into one of the cloning sites and the coding sequence can be joined to the factor X_a recognition sequence by deletion mutagenesis transfecting into an *Eco*K$^+$ host. Any clone which has failed to undergo correct deletion is not viable in the *Eco*K$^+$ strain because it has not lost the *Eco*K sites present in the DNA sequence to be deleted. By using *Eco*K selection, Waye *et al.* have improved the frequency of deletion mutagenesis considerably.[32,33] Thus, any coding sequence can be readily joined to the factor X_a recognition sequence by deletion mutagenesis.

Cloning into the pLcII(nic$^-$) Fusion Protein Expression Vector

Once the coding sequence is joined to the factor X_a recognition sequence in M13 mp11FX or in M13 mp11RX,[33] the gene can be cut out together with the factor X_a recognition sequence with *Bam*HI and can then be cloned into the *Bam*HI site of the fusion protein expression vector pLcII(nic$^-$).[9] As shown in Fig. 6, pLcII(nic$^-$) has multiple cloning sites after the short coding sequence of λ cII gene. The gene can be cloned in

[32] P. Carter, this series, Vol. 154 [20].

[33] M. M. Y. Waye, M. E. Verhoeyen, P. T. Jones, and G. Winter, *Nucleic Acids Res.* **13,** 8561 (1985).

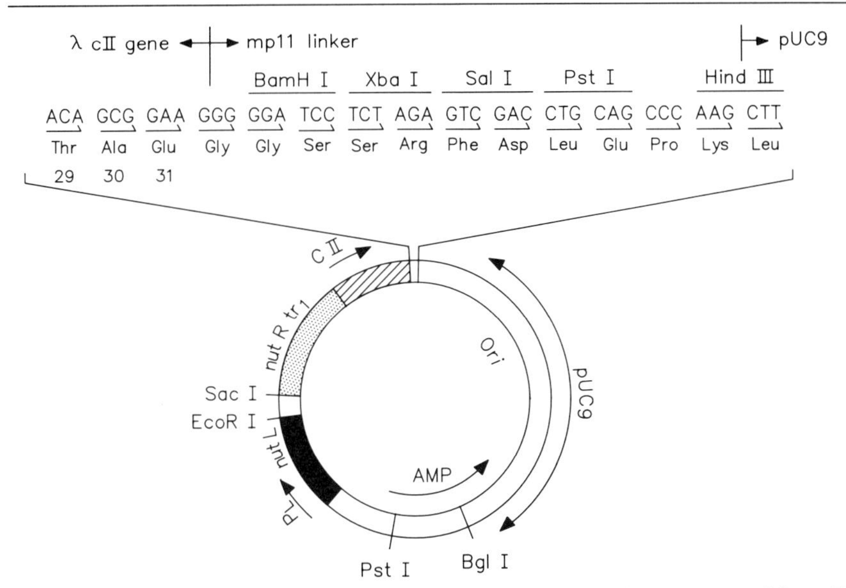

FIG. 6. The fusion protein expression vector, pLcII(nic⁻). The short segment of the λ cII gene is followed by multiple cloning sites.

any of these restriction sites [except the *Pst*I site, which is not unique in pLcII(nic⁻)]. The pLcII vector described by Nagai and Thøgersen[8] included the nic site (the origin for conjugal transfer) but Nagai *et al.*[9] have replaced the *Bgl*I–*Hin*dIII fragment of the original vector with the *Bgl*I–*Hin*dIII fragment of pUC9. Therefore, the plasmid is no longer transmissible so that the experiments do not require any biological containment for most eukaryotic genes. The *Bgl*I–*Eco*RI fragment including the P_L promoter is derived from pLc245.[2]

In order to control expression of the foreign gene, a defective phage λ lysogenic strain with the temperature-sensitive repressor (cI857) gene and an active N gene should be used as a host. QY13 (F^- lac_{am} trp_{am} BB′ bio-256 N^+ cI857 ΔH Sm^r recA), constructed by Dr. Sydney Brenner, is the best strain in our experience.

Methods. An overnight culture of QY13 is diluted 100-fold in 50 ml of 2× TY medium (16 g Bacto tryptone, 10 g yeast extract, 5 g NaCl/1 liter and pH is adjusted to 7.4 with NaOH) and cells are grown until the optical density at 600 nm reaches 1. The cells are spun down and suspended in 50 ml of cold 100 m*M* $CaCl_2$. After 40 min of incubation on ice, the cells are spun down again and resuspended in 1 ml of cold 100 m*M* $CaCl_2$ solution. A 150-μl aliquot of competent cell solution is added to 10 μl of ligation mixture containing 200 ng of vector DNA. After 40 min on ice, the cells are heat shocked for 5 min at 34° and incubated on ice for a further 15 min.[2] One hundred and fifty microliters of 2× TY medium is then added to the cells and they are incubated without shaking for 60 min at 30°. One hundred microliters is plated on a TYE plate (15 g agar, 10 g Bacto tryptone, 10 g yeast extract, 8 g NaCl/1 liter) containing ampicillin (100 μg/ml). After a 36-hr incubation at 30°, 100–200 colonies should be visible per plate.

Small-Scale Protein Induction

Methods. Several colonies are restreaked on fresh TYE plates with ampicillin (100 μg/ml) and overnight cultures are prepared from a single colony on each plate in 2× TY with 50 μg/ml ampicillin. Plasmid DNA is prepared from a 10-ml overnight culture by the alkaline lysis method,[29] and digestion with several restriction enzymes should be carried out to confirm that the construction is correct. An overnight culture is diluted 100-fold in 2× TY (50 μg/ml ampicillin) and cells are grown until OD_{600} = 1. The culture tubes are then immersed in a 43° waterbath for 15 min (shaking is not necessary) and culture is continued in a 37° shaker for 2–4 hr. The cells are spun down and resuspended in 500 μl water and the protein is extracted with 500 μl of phenol solution saturated with 0.1 *M* Tris–HCl, pH 8.0. Protein is precipitated from the phenol phase with an equal volume of ethanol. The precipitate is dissolved in 500 μl of SDS sample buffer and 10–20 μl of the sample is loaded onto a polyacrylamide–SDS gel.[34] Protein extracted from identically treated cells harboring pLcII should be run as a control.[8] A sample of protein from uninduced cells is not an appropriate control because some "heat shock" proteins are produced during the induction process.

Isotope Labeling of Newly Synthesized Proteins after Heat Induction

After heat induction, protein synthesis is predominantly directed to the gene placed under the control of the λ P_L promoter.[2] Therefore, even if the product is not identifiable on a Coomassie blue-stained gel, it can sometimes be visualized by isotopic labeling after heat induction.

As shown by Remaut *et al.*,[2] the synthesis of different proteins persists for different time periods after induction, depending on the level of expression and the toxicity of the product. It is therefore important to radiolabel newly synthesized protein at various times after heat induction, in order to determine the optimum for the protein of interest.

[34] U. K. Laemmli, *Nature (London)* **227,** 680 (1970).

Method.[35] The culture is heat induced as in the previous section. One milliliter of the culture is removed at various times after heat induction and transferred to a small glass culture tube containing 50 μCi [^{35}S]methionine. The culture is continued for 5 min and the cells are harvested and treated as in the previous section. The polyacrylamide gel is processed for fluorography.[36]

If the Fusion Protein Is Not Produced in Significant Quantities

In most cases, pLcII(nic$^-$) directs synthesis of fusion proteins at a level of 5–10% of total cellular protein. If a protein of the expected molecular weight is not produced in a significant quantity, check the following steps: (1) transcription of the gene; (2) stability of mRNA; (3) translation of mRNA; (4) stability of protein product.

Transcriptional activity of *E. coli* promoters depends little on the DNA sequence cloned downstream and it is unlikely that the gene is not transcribed. But if there is a terminator-like sequence within the cloned gene, transcription might be terminated and therefore the expected protein product will not be produced. An *in vitro* transcription experiment can be carried out as described by Queen and Rosenberg[37] to check if the entire coding sequence is transcribed. If the gene is transcribed, poor production of the fusion protein is due either to poor translation of mRNA or to degradation of mRNA or polypeptide.

Mellando and Murray[38] fused histone H3 gene of the yeast *Saccharomyces cerevisiae* to the β-galactosidase gene and placed the hybrid gene under the control of the *lac* promoter. They could not detect the gene product *in vivo* but, using an *in vitro* transcription–translation system, they could show that the gene is properly transcribed and translated. They concluded that the absence of a protein product *in vivo* is attributable to the proteolytic degradation of the newly synthesized polypeptide chain. If the protein product is soluble, proteolysis by endogenous proteases can be a serious problem.[39] A cell-free transcription–translation system free from such enzymes can be prepared as described by Zubay[40,41] and a kit (Amersham) is also available commercially. The β-lactamase

[35] F. Calabi, unpublished results.

[36] R. A. Laskey and A. D. Mills, *Eur. J. Biochem.* **56,** 335 (1975).

[37] C. Queen and M. Rosenberg, *Cell* **25,** 241 (1981).

[38] R. P. Mellando and K. Murray, *J. Mol. Biol.* **168,** 489 (1983).

[39] A. L. Goldberg, K. H. S. Swamy, C. H. Chung, and F. S. Laumore, this series, Vol. 80, p. 680.

[40] G. Zubay, this series, Vol. 65, p. 856.

[41] H.-Z. Chen and G. Zubay, this series, Vol. 101, p. 674.

gene on pLcII(nic$^-$) serves as a good reference for the level of expression. Note that the *in vitro* transcription–translation system does not provide the λ N gene product.[28] Therefore, transcription from the λ P_L promoter is partially terminated at the t_{R1} and the level of expression of the *cII* fusion protein would be underestimated. If the fusion protein is produced well *in vitro* but not *in vivo,* degradation of either mRNA[42] or polypeptide[38,39] accounts for the absence of protein production *in vivo*. This can be checked by pulse labeling of protein or mRNA. Stability of the mRNA depends on its secondary structure and it can be quickly degraded as shown for the *sib* mutant[43] of λ phage. It is, however, more likely that nascent polypeptide is degraded.[38] The Lon$^-$ strain[44] of *E. coli,* which is deficient in one of the intracellular proteases,[39] has been used to reduce proteolysis of foreign gene products in *E. coli.*

Polypeptides smaller than 10 kDa are difficult to produce using the pLcII vector. This is probably because small polypeptides are soluble in the cytoplasm and quickly degraded. This problem can be overcome by inserting a short coding sequence at the *Bam*HI site in pLcII. A small *Bam*HI fragment of the myosin light chain gene encoding a polypeptide 100 amino acids long was used to produce several small polypeptides.[44a]

Large-Scale Protein Synthesis

If a small-scale experiment shows that the fusion protein is expressed at a high level, large-scale synthesis can then be carried out. We routinely produce a large quantity of protein using 2-liter Erlenmeyer flasks.[9]

Method.[9] Ten milliliters of overnight culture is added into a 2-liter flask containing 1 liter 2× TY medium with 50 μg/ml ampicillin and grown at 30° in a shaking incubator. When the optical density of the culture at 600 nm reaches 1.0, the temperature of the culture is quickly raised to 42° by swirling the flask in a 70° water bath. The flask is then immersed in a 42° water bath for 15 min and the culture is continued at 37° for the time period estimated by the isotope-labeling experiment. Once protein synthesis is induced at 42°, it continues at 37°. Cells are usually harvested after 3–4 hr. About 4 g of packed cells is obtained from 1 liter culture in the case of the CIIFXβ-globin expression vector.

[42] J. A. Hautala, C. L. Bassett, N. H. Giles, and S. R. Kushner, *Proc. Natl. Acad. Sci. U.S.A.* **76,** 5774 (1981).

[43] G. Guarneros, C. Montanez, T. Hernandez, and D. Court, *Proc. Natl. Acad. Sci. U.S.A.* **79,** 238 (1982).

[44] S. Gottesman, M. Gottesman, J. E. Shaw, and M. L. Pearson, *Cell* **24,** 225 (1981).

[44a] K. Nagai, unpublished results.

Protein Purification

Most cII fusion proteins are insoluble in the cell and form inclusion bodies.[1] Out of 12 proteins produced using the pLcII(nic$^-$) vector, only the CIIFX-troponin C fusion protein was soluble in *E. coli*.[16] In order to find out whether the fusion protein is soluble or not, a small-scale experiment should be carried out using the following procedure, and both precipitate and supernatant should be run on a polyacrylamide–SDS gel. If the fusion protein is insoluble the following procedure may be used for large-scale protein purification.

Methods.[9] The cells (100 g) are thawed and suspended in 80 ml of lysis buffer [50 m*M* Tris–HCl, pH 8.0, 25% sucrose (w/v), 1 m*M* EDTA]. The cells are lysed by addition of 200 mg lysozyme dissolved in 20 ml of lysis buffer and left to stand for 30 min on ice. It is important to suspend the cells completely in lysis buffer before addition of lysozyme. Otherwise, unlysed cells will be spun down with inclusion bodies and the fusion protein will be contaminated with other cellular proteins. With lysozyme treatment, the solution becomes very viscous. Then, $MgCl_2$, $MnCl_2$, and DNase I (Sigma, D5025) are added to final concentrations of 10 m*M*, 1 m*M*, and 10 μg/ml, respectively. The viscosity of the solution should decrease as soon as DNase I is added. After a 30-min incubation, 200 ml of detergent buffer [0.2 *M* NaCl, 1% deoxycholic acid (w/v), 1% Nonidet P-40 (v/v), 20 m*M* Tris–HCl (pH 7.5), 2 m*M* EDTA] is added to the lysate which is then centrifuged at 5000 *g* for 10 min.[7,45] The supernatant, which is slightly orange colored, is carefully removed and the pellet completely suspended in 0.5% Triton X-100, 1 m*M* EDTA solution and centrifuged at 5000 *g* for 10 min.[46] The milky supernatant is carefully removed and the pellet suspended again in the Triton–EDTA solution and centrifuged. This procedure is repeated until a tight pellet is obtained. The first pellet contains the fusion protein and the membrane fraction with two major membrane proteins. As shown in Fig. 7, the Triton wash procedure solubilizes the membrane fraction completely and leaves the fusion protein in the final pellet.[9]

Further Purification. The fusion protein in the pellet can only be solubilized in a strong denaturant such as 8 *M* urea or 6 *M* guanidine–HCl solution. CIIFXβ-globin,[9] CIIFXα-globin,[24] CIIFX-actin,[14] CIIFX-TMVP

[45] T. M. Gilmer, J. T. Parsons, and R. L. Erikson, *Proc. Natl. Acad. Sci. U.S.A.* **79,** 2152 (1982).

[46] F. A. O. Marston, P. A. Lowe, M. T. Doel, J. M. Shoemaker, S. Whiter, and S. Angal, *Bio/Technology* **2,** 800 (1984).

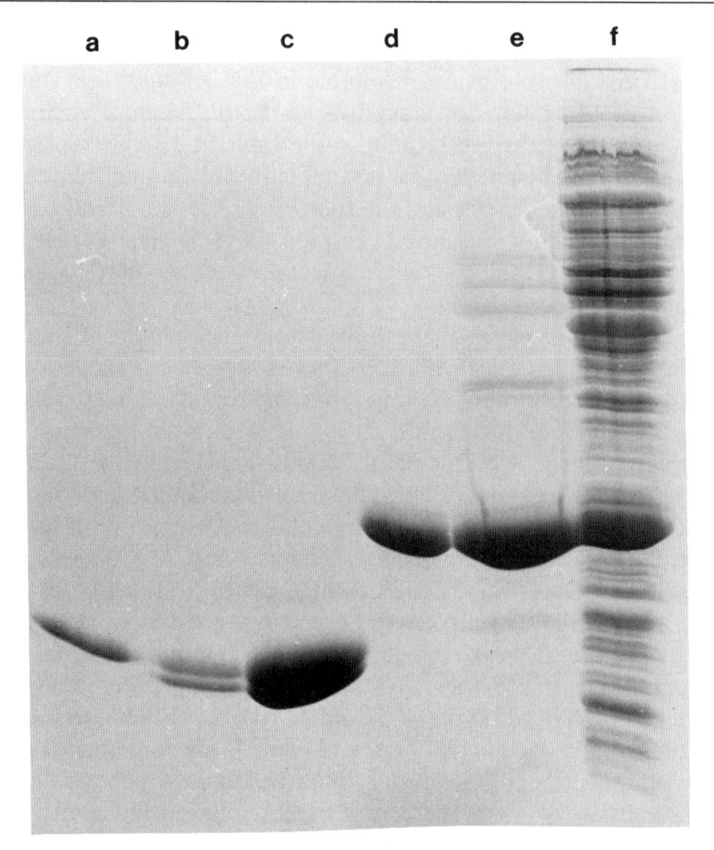

FIG. 7. Polyacrylamide–SDS gel analysis of the cleavable fusion protein CIIFXβ-globin at various stages of purification. (a) β-Globin purified from human blood; (b and c) reconstituted Hb ($\alpha_2\beta_2$); (d) CIIFXβ-globin purified on CM-Sepharose and Sephacryl S-200 columns; (e) pellet after the Triton X-100 washing procedure; (f) total cellular protein.

(tobacco mosaic virus coat protein),[20] and CIIFX-MLC (myosin light chain)[15] have been purified by ion-exchange chromatography on either CM-Sepharose 6B-CL or DEAE-Sepharose 6B-CL (Pharmacia) in the presence of 8 *M* urea. Myosin light chain purified by this procedure was readily folded.[15] However, it is necessary to purify CIIFXα-globin,[24] CIIFX-TMVP,[20] and CIIFXβ-globin[9] further by gel filtration on a Sephacryl S-200 (Pharmacia) column equilibrated with 5 *M* guanidine–HCl, 50 m*M* Tris–HCl, pH 8.0, 1 m*M* dithiothreitol, 1 m*M* EDTA in order to fold these proteins in good yield.

Digestion of CIIFX-Fusion Proteins with Blood Coagulation Factor X_a

Reagents

N-Benzoyl-L-isoleucyl-L-glutamylgylcyl-L-arginine-*p*-nitroanilide hydrochloride (Bz-Ile-Glu-Gly-Arg-pNA) (Kabi Diagnostica, Stockholm, Sweden)
Russell's viper venom (Sigma, V2501)
Cyanogen bromide-activated Sepharose prepared by the procedure of Kohn and Wilchek[47] or from Pharmacia

Isolation of Factor X from Bovine Blood

There is no commercial source of factor X_a suitable for this work. However, the proenzyme can readily be isolated from bovine blood plasma by batch adsorption of the vitamin K-dependent proteins on insoluble barium salts followed by a single chromatographic purification step.

A purification procedure was developed by Esnouf and Williams[48] and has been modified by several groups.[49–51]

Method.[51] Bovine blood is collected directly into plastic bottles containing 0.1 vol of 0.12 *M* sodium citrate buffer, 0.05 *M* benzamidine–HCl (pH 6.9). Pooling of blood prior to separation of the blood plasma can result in hemolysis. The plasma is obtained by low-speed centrifugation (1500 *g*, 4°, 20 min) or by continuous centrifugation for large volumes. Barium chloride is added to a final concentration of 80 m*M* by slow addition of a 1 *M* stock solution. This results in coprecipitation of the vitamin K-dependent plasma proteins with the barium citrate precipitate. After stirring for 30 min (10–15°), the precipitate is collected by centrifugation (1500 *g*, 4°, 15 min) and washed twice with 0.1 plasma volume of 0.15 *M* sodium chloride, 5 m*M* benzamidine–HCl saturated with barium citrate. The proteins are released from the precipitate by reprecipitating the barium ions as the less soluble barium sulfate. The precipitate is resuspended in 0.01 plasma volume of 20 m*M* sodium citrate buffer, 5 m*M* benzamidine–HCl (pH 6.9), and 1 *M* sodium sulfate is slowly added to obtain a slight molar excess of sulfate over barium ions. The reprecipitation is observed as the color of the suspension changes from white to

[47] J. Kohn and M. Wilchek, *Biochem. Biophys. Res. Commun.* **107,** 878 (1982).

[48] M. P. Esnouf and W. J. Williams, *Biochem. J.* **84,** 62 (1962).

[49] C. M. Jackson and D. J. Hanahan, *Biochemistry* **7,** 4506 (1968).

[50] K. Fujikawa, M. E. Legaz, and E. Davie, *Biochemistry* **11,** 4882 (1972).

[51] H. C. Thøgersen, T. E. Petersen, L. Sottrup-Jensen, S. Magnusson, and H. R. Morris, *Biochem. J.* **175,** 613 (1978).

gray. The supernatant is dialyzed into 20 m*M* sodium citrate buffer, 1 m*M* benzamidine–HCl (pH 6.9) in the cold.

Factor X can be completely separated from other proteins in the barium citrate adsorbate by ion-exchange chromatography on DEAE-Sepharose 6B-CL (Pharmacia) at 20°. The column volume should be approximately 1.5% of the initial plasma volume and the column length five times its diameter. The column is eluted at 5 ml cm^{-2} hr^{-1}, first with 20 m*M* sodium citrate buffer, 1 m*M* benzamidine–HCl (pH 6.9) until the unbound material has been eluted, then with a linear gradient to 0.4 *M* sodium citrate buffer, 1 m*M* benzamidine–HCl (pH 6.9) (total of 10 column volumes). The two peaks of factor X, X_1, and X_2, were eluted toward the end of the gradient, at conductivities (25°) of 34 and 38 mmho, respectively. Both variants have the same substrate specificity after activation and should be treated identically in the following experiments. The combined yield is about 5 mg/liter plasma. The material is concentrated to 2–3 mg/ml by ultrafiltration (PM-10 membrane, Amicon) and stored frozen at −20° as the proenzyme or passed through a Sephadex G-50 (fine, Pharmacia) column to exchange the citrate buffer with 50 m*M* Tris–HCl, pH 8.0, in preparation for activation with immobilized Russell's viper venom.

Preparation of Immobilized Russell's Viper Venom

Method. Fifteen milliliters of cyanogen bromide-activated Sepharose-6B (binding capacity ~30 μmol/ml agarose) is washed extensively with water and 0.1 *M* $NaHCO_3$ and 40 mg of Russell's viper venom is allowed to react at 4° overnight with gentle rolling before blocking excess reactive groups with 0.2 *M* Tris–HCl, pH 8.6 (24 hr, 4°). Then the Sepharose is washed with 50 m*M* Tris–HCl, pH 8.0, and stored at 4°.

Assay Procedure for Blood Coagulation Factor X_a

The chromogenic substrate Bz-Ile-Glu-Gly-Arg-pNA is dissolved in water at the concentration of 3 m*M* based on the extinction coefficient ($\varepsilon_{mM}^{316\ nm}$ = 1.3 × 10). One hundred microliters of 3 m*M* Bz-Ile-Glu-Gly-Arg-pNA solution is added to 900 μl of 50 m*M* Tris–HCl, pH 8.0, 100 m*M* NaCl in a cuvette. Five microliters of enzyme solution (OD_{280} ~0.1) is added to the cuvette and an increase in the optical density at 405 nm is recorded.

Activation of Proenzyme Factor[50]

A 1 × 3 cm column is packed with the immobilized Russell's viper venom and a 15-ml solution of factor X (5 mg/ml) in 50 m*M* Tris–HCl, pH

8.0, 1 mM $CaCl_2$ is circulated through the column using a peristaltic pump at room temperature. Ca^{2+} is essential for activation of factor X by Russell's viper venom. Aliquots are withdrawn occasionally and the specific activity is determined. The circulation is continued until no further increase in specific activity is observed. It typically takes 12 hr and a specific activity of 0.4 ΔOD/min/μg (enzyme) is obtained. The activated enzyme solution can be stored frozen in 50 mM Tris–HCl, pH 8.0, at −20° without significant loss of activity for at least 2 years.

Cleavage of CIIFX Fusion Proteins with Factor X_a[9]

CIIFXβ-globin (100 mg) is dissolved in 100 ml of 8 M urea, 50 mM Tris–HCl, pH 8.0, 1 mM DTT, 1 mM EDTA and extensively dialyzed against 50 mM Tris–HCl, pH 8.0, 100 mM NaCl at 4° under an atmosphere of nitrogen. CIIFXβ-globin is extremely unstable at higher temperature and precipitates at room temperature. Therefore, the digestion is carried out on ice at an enzyme to substrate ratio of 1 : 200 (w/w) for 12 hr. As shown in Figs. 3 and 7, the CIIFXβ-globin is completely cleaved at the junction and the authentic β-globin is liberated. CIIFX-actin,[14] CIIFX-troponin C,[16] CIIFX-MLC (myosin light chain),[15] and CIIFXTMVP (tobacco mosaic virus coat protein)[20] are also cleaved at the peptide bond following the tetrapeptide. CIIFX-MLC and CIIFXTMVP were also cleaved internally, but at a lower rate. The rate of cleavage depends on the sequence following the cleavage site and a small-scale digestion should be carried out for each fusion protein.

Comments

Substrate Specificity of Blood Coagulation Factor X_a

The natural substrate of blood coagulation factor X_a, prothrombin, is converted to thrombin by specific cleavage at Arg^{274}-Thr^{275} and Arg^{323}-Ile^{324} in a polypeptide of 582 amino acids.[21–23] Both of these sites are preceded by the same tetrapeptide, Ile-Glu-Gly-Arg. Nagai and Thøgersen showed that the CIIFXβ-globin fusion protein can be cleaved by factor X_a at the peptide bond following the tetrapeptide sequence. There are 6 other Arg and 14 other lysine residues in CIIFXβ-globin but the peptide bonds following these Arg and Lys residues remained uncleaved.[8] In contrast, the homologous protease trypsin cleaved the CIIFXβ-globin at every peptide bond following these residues.[8]

The specificity of X_a has been extensively studied using chromogenic substrates, but the work focused only on the peptide sequence preceding

TABLE I
CIIFX FUSION PROTEINS CLEAVED BY BLOOD COAGULATION FACTOR X_a[a]

FX									
P_4	P_3	P_2	P_1	P'_1	P'_2	P'_3	P'_4	Protein	Reference
Ile-	Glu-	Gly-	Arg=	Val-	His-	Leu-	Thr	β-Globin	Nagai and Thøgersen[8]
Ile-	Glu-	Gly-	Arg=	Ser-	Pro-	Leu-	Thr	β-Globin mutant	Miyazaki and Nagai[53]
Ile-	Glu-	Gly-	Arg=	Val-	Leu-	Ser-	Pro	α-Globin	Pagnier and Nagai[24]
Ile-	Glu-	Gly-	Arg=	Ser-	Tyr-	Ser-	Ile	TMV coat protein	Turner[20]
Ile-	Glu-	Gly-	Arg=	Asp-	Asp-	Asp-	Ile	Actin	Karlsson and Nagai[14]
Ile-	Glu-	Gly-	Arg=	Ala-	Pro-	Lys-	Lys	Myosin light chain	Reinach *et al.*[15]
Ile-	Glu-	Gly-	Arg=	Met-	Ala-	Ser-	Met	Troponin C	Karlsson *et al.*[16]
Ile-	Glu-	Gly-	Arg=	Gly-	Glu-	Lys-	Pro	TFIIIA	Hansen and Thøgersen[20c]
Ile-	Glu-	Gly-	Arg=	Phe-	Thr-	Ser-	Lys	MATα1	Tan and Richmond[20a]
Ile-	Glu-	Gly-	Arg=	Lys-	Glu-	Thr-	Ala	RNase A	Nambiar *et al.*[20d]
Ile-	Glu-	Gly-	Arg=	Ser-	Asp-	Glu-	Lys	Caltrin	Heaphy and Gait[b]

[a] Blood coagulation factor X_a cleaves between P_1 and P'_1 sites.
[b] S. Heaphy and M. Gait, unpublished results.

the cleavage site.[52] The cleavage rate does, however, also depend on the sequence following the cleavage site. When the first fusion protein CIIFXβ-globin was made, only limited data existed on the specificity of factor X_a. Since then, more than 10 CIIFX fusion proteins have been made and treated with factor X_a. Table I summarizes the sequence of the fusion proteins which have been cleaved by factor X_a. As can be seen from Table I, the P'_1 site can be occupied by various amino acid residues. In the case of actin, the tetrapeptide is followed by a row of negatively charged amino acid residues but the site is still readily cleaved by factor X_a.[14]

Some fusion proteins are cleaved internally although the sites do not have the Ile-Glu-Gly-Arg sequence. CIIFXβ-globin is exclusively cleaved at the peptide bond following the tetrapeptide but when the histidine residue at 143 is replaced with Arg, CIIFXβ-globin is also cleaved at the Arg-Lys bond between 143 and 144 in β-globin.[9] CIIFXβ-globin is cleaved more slowly at the Arg-Lys bond than at the tetrapeptide site and intact β-globin[8] can be isolated in good yield. This observation clearly shows that the substrate specificity of factor X_a is not absolutely limited to the tetrapeptide sequence. Myosin light chain[53] and TMV coat protein[20] were also

[52] R. Lottenberg, U. Christensen, C. M. Jackson, and P. L. Coleman, this series, Vol. 80, p. 341.
[53] G. Miyazaki and K. Nagai, unpublished results.

cleaved internally by factor X_a, but the intact authentic protein could be isolated by partial digestion.[15,20] CIIFXα-globin,[24] CIIFX-myoglobin,[13] CIIFX-actin,[14] CIIFX-troponin C,[16] and β-galactosidase-FX-RNase A[20d] were all cleaved uniquely at the tetrapeptide site. Specific inhibitors (anti-thrombin-III[54] for factor X_a and thrombin and hirudin[55] for thrombin) can be used to demonstrate unequivocally whether an observed cleavage is catalyzed by factor X_a or thrombin (most possible contaminant) or other contaminating enzymes.

As a further test of the stringency of the substrate specificity of factor X_a, we incubated the following proteins with factor X_a: bovine serum albumin, human plasminogen, and bovine pancreatic trypsin inhibitor (reduced and carboxymethylated) and chicken erythrocyte histones H2A, H2B, and H5 (acid extracted). Under the condition where β-globin is completely liberated from the CIIFXβ-globin, all or most of each of these proteins remained uncleaved while histone H5 was extensively degraded.[56]

These experiments show that factor X_a is very specific for the Ile-Glu-Gly-Arg tetrapeptide and the method is potentially applicable to production of many eukaryotic proteins.

Other Cleavage Methods

CIIFX-MLC (myosin light chain) was cleaved exclusively at the peptide bond following the tetrapeptide with bovine thrombin.[15] The amino acid sequence around the cleavage site is Ile-Glu-Gly-Arg-Ala-Pro-Lys-Lys in this fusion protein. The -Arg-Ala-Pro-Lys- sequence resembles the cleavage site (-Arg-Gly-Pro-Arg-) in the natural substrate of thrombin, fibrinogen.[57] The Lys residue at the P'_3 does not seem to be essential since thrombin cleaves a mutant of fibrinogen in which Arg (at the P'_3) following Pro is replaced by Ser. The cleavage with thrombin would not be widely applicable because the specificity resides also in the sequence following the cleavage site.

In the case of TMV coat protein, difficulty was encountered in solubilizing the fusion protein to allow stoichiometric cleavage with factor X_a. Turner took advantage of the absence of internal Met residues in this protein and introduced a Met residue immediately preceding the first

[54] E. T. Yin, S. Wessler, and P. J. Stoll, *J. Biol. Chem.* **246,** 36994 (1971).

[55] F. Markwardt, "Blutgerinnungshemmende Wirkstoffe aus Blutsaugenden Tieren." Fischer, Jena, German Democratic Republic, 1963.

[56] H. C. Thøgersen, unpublished results.

[57] M. Blombäck, B. Blombäck, E. F. Mammen, and A. S. Prasad, *Nature* (*London*) **218,** 314 (1968).

amino acid of the coat protein. An authentic TMV coat protein could then be liberated by cyanogen bromide cleavage.[20]

Properties of Eukaryotic Proteins Produced in *E. coli*

β-Globin

β-Globin produced in *E. coli* using the pLcII/factor X_a method was folded and reconstituted with hemin and α chain to form $\alpha_2\beta_2$ tetramers (Fig. 7). The oxygen-binding properties of the reconstituted Hb were essentially the same as those of native human Hb.[9] Nagai *et al.* have constructed more than 10 mutants to date by site-directed mutagenesis. Furthermore, Luisi and Nagai have crystallized three mutants made in *E. coli* and solved their structures by high-resolution X-ray crystallographic analysis. The structure of these mutants differed from that of native Hb only in the vicinity of the mutations, showing that β-globin produced in *E. coli* is structurally and functionally identical to native human Hb.[58]

Myosin Light Chain[15]

Chicken myosin light chain produced in *E. coli* can be bound to scallop myofibril from which endogenous light chains have been removed by treatment with EDTA. Myosin light chain produced in *E. coli* had the same ATPase inhibitory activity as native chicken myosin light chain when reconstituted with scallop myofibril. Three mutants of myosin light chain have been constructed in order to study the mechanism of Ca^{2+} regulation in muscle contraction.[15]

TMV Coat Protein[20]

Native TMV coat protein, released from the fusion protein by cyanogen bromide cleavage, could be readily refolded by dialysis. Reduction in pH then led to the formation of long helical rods characteristic of the authentic protein. This system is currently being exploited to study the structural basis of RNA binding and protein–protein aggregation using site-directed mutagenesis.[20]

Myoglobin[13]

Varadarajan *et al.*[13] made a CIIFX fusion protein containing the human myoglobin polypeptide. CIIFX-Mb was found to be stable and could

[58] B. F. Luisi and K. Nagai, *Nature (London)* **320,** 555 (1986).

be readily folded and reconstituted with hemin with the N-terminal extension of the CII polypeptide which was then cleaved off with trypsin instead of factor X_a. This was possible because myoglobin in the native form is stable against proteolysis. The optical absorption and NMR spectra of human Mb produced in *E. coli* were indistinguishable from those of native human Mb.

Troponin C

The cIIFX-troponin C fusion protein is soluble and properly folded in *E. coli* cells. The fusion protein was cleaved by factor X_a specifically at its recognition site, liberating the authentic troponin C. Several mutants have been constructed and their properties are being studied.[16]

Pancreatic Ribonuclease A

RNase A was produced in *E. coli* as a β-glactosidase fusion protein and the factor X_a cleavage site was inserted at the junction between these functional domains. Authentic RNase A was liberated by factor X_a and folded *in vitro* to give the active form of the enzyme.[20d]

Conclusion

As shown by these experiments, the pLcII(nic⁻) fusion protein expression vector is a straightforward and general method for production of eukaryotic proteins in *E. coli*. Most proteins produced by this method can be readily purified and cleaved with blood coagulation factor X_a, and the authentic proteins refolded into their native conformations. This opens the way for a study of protein function, folding, and structure by site-directed mutagenesis. The pLcII(nic⁻) and M13 mp11FX vectors are available from K. Nagai on request.

Acknowledgments

We thank Drs. S. Brenner, D. Brown, P. M. Esnouf, K. McKenney, E. Remaut, and C. Young for essential materials and help; Drs. M. F. Perutz, A. Klug, S. Brenner, D. R. Turner, S. Anderson, D. Altschuh, K. McKenney, F. Calabi, and B. F. Luisi for helpful discussion and encouragement; Drs. B. F. Luisi, S. MacLeod, J. Kendrick-Jones, F. Reinach, R. Karlsson, J. Pagnier, G. Miyazaki, T. Rabbitts, T. J. Richmond, and M. Waye for allowing us to mention their unpublished results. K.N. thanks Daphne Monet and Johannes Brahms for encouragement. This work is partly supported by National Institute Grant R01 HL 31461-01.

[30] Expression Plasmid Containing the λ P_L Promoter and cI857 Repressor

By GREGORY MILMAN

Introduction

When the leftward promoter of λ phage, P_L, is placed in front of a gene, the gene product may be expressed as 5% or more of soluble bacterial protein. Continuous high levels of expression would be detrimental to bacterial growth, potentially lethal, and strongly selected against. Therefore, cells harboring a plasmid containing P_L also need to contain the λ cI repressor to turn off the promoter. One approach to providing the λ cI repressor is to use a λ lysogen containing the temperature-sensitive cI857 gene.[1] When the repressor protein is inactivated to induce expression by a shift in temperature from 30 to 42°, other λ proteins may be induced as well. An alternative approach is to construct a self-regulated plasmid which contains both the repressor gene and the P_L promoter. An advantage of this latter approach is that the plasmid DNA can be transfected freely into bacterial strains which do not contain a λ repressor. This feature is useful, for example, when transfecting into bacterial strains where expression can be selected genetically. A self-regulated plasmid is also easier to examine in a variety of strains which may give better product yield, for example, those that lack proteases.

Principle

Plasmid pHE6[2] (Fig. 1) is a self-regulated P_L expression plasmid derived from pUC8[3] and λ sequences.[4] It was created by replacing the *Hae*II–*Sma*I region of pUC8 with two adjacent segments of λ phage DNA: that from the *Hae*II site at base 37,061 to the *Bgl*II site at base 38,103 ligated to that from the *Bgl*II site at base 35,711 to the *Hpa*I site at base 35,261. The construction introduced a synthetic DNA fragment,

[1] M. Rosenberg, Y.-S. Ho, and A. Shatzman, this series, Vol. 101, p. 123.

[2] G. Milman, A. L. Scott, M.-S. Cho, S. C. Hartman, D. K. Ades, G. S. Hayward, P.-F. Ki, T. August, and S. D. Hayward, *Proc. Natl. Acad. Sci. U.S.A.* **82,** 6300 (1985).

[3] J. Messing and J. Vieira, *Gene* **19,** 269 (1982).

[4] D. L. Daniels, J. L. Schroeder, W. Szybalski, F. Sanger, A. R. Coulson, G. F. Hong, D. F. Hill, G. B. Petersen, and F. R. Blattner, *in* "Lambda II" (R. W. Hendrix, J. W. Roberts, F. W. Stahl, and R. A. Weisenberg, eds.), pp. 519–676. Cold Spring Harbor Lab., Cold Spring Harbor, New York, 1983.

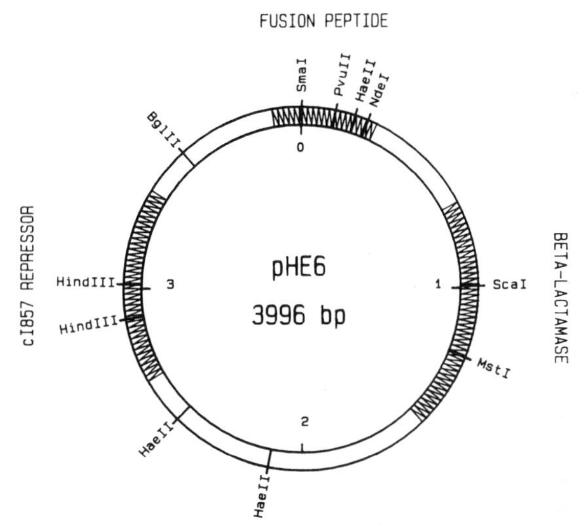

FIG. 1. Plasmid pHE6. Plasmid pHE6 (3996 bp) is illustrated with the center of the unique *Sma*I site at position 1. The direction of transcription is clockwise for the fusion peptide and β-lactamase and counterclockwise for the cI857 repressor. The λ phage P_L promoter is located between the *Bgl*II and *Sma*I sites. The restriction sites in the polylinker cloning region about the *Sma*I site are too close together to be included on this map. A detailed illustration of the polylinker cloning region is presented in Fig. 2.

—CCGGATCCCC—, between the GTT blunt end of the *Hpa*I site of the λ DNA and the GGG blunt end of the *Sma*I site of pUC8. Foreign DNA may be inserted into the unique *Bam*HI, *Sma*I, *Sal*I, *Acc*I, and *Pst*I restriction sites. Blunt-ended DNA inserted into the *Sma*I restriction site automatically gains *Bam*HI linkers and can be excised from the plasmid by digestion with *Bam*HI. Table I lists some useful restriction sites in pHE6.

Plasmid pHE6 produces proteins from inserted foreign DNA in the form of fusion polypeptides which begin with 33 amino acids from the amino terminus of the λ phage N-protein plus the amino acids which are coded for by the linker DNA prior to the insert. To be expressed, the foreign DNA polypeptide must be in the same reading frame as the λ N-protein. The DNA sequence, amino acid reading frame, and restriction enzyme cleavage sites in the immediate vicinity of the *Sma*I–*Hin*dIII sites are illustrated in Fig. 2.

Plasmid pHE6 was modified by deleting a small number of bases between the *Sma*I site and the initiation site of the λ N-protein to allow insertions in the unique restriction sites in all three reading frames. A number of altered reading frame plasmids were identified and the approxi-

TABLE I
SELECTED RESTRICTION SITES IN pHE6[a]

Enzyme	Recognition sequence	Site location
*Aat*II	(GACGT^C)	554
*Acc*I	(GT^JKAC)	11
*Acy*I	(GR^CGYC)	195, 551, 933
*Afl*III	(A^CRYGT)	2361
*Aha*III	(TTT^AAA)	895, 1587, 1606
*Ava*I	(C^YCGRG)	3995
*Ava*II	(G^GLCC)	1109, 1331
*Bam*HI	(G^GATCC)	4, 3990
*Bcl*I	(T^GATCA)	2782
*Bgl*I	(GCCNNNN^NGGC)	185, 1355
*Bgl*II	(A^GATCT)	3533
*Cfr*I	(Y^GGCCR)	41, 1080, 3779
*Dra*II	(RG^GNCCY)	493
*Hae*II	(RGCGC^Y)	198, 2121, 2491
*Hin*dII	(GTY^RAC)	12, 2863, 3419, 3633
*Hin*dIII	(A^AGCTT)	24, 2889, 3014
*Mst*I	(TGC^GCA)	1250
*Nar*I	(GG^CGCC)	195
*Nde*I	(CA^TATG)	247
*Pst*I	(CTGCA^G)	20
*Pvu*I	(CGAT^CG)	156, 1104
*Pvu*II	(CAG^CTG)	125
*Sal*I	(G^TCGAC)	10
*Sca*I	(AGT^ACT)	992
*Sma*I	(CCC^GGG)	1
*Ssp*I	(AAT^ATT)	668
*Taq*I	(T^CGA)	11, 819, 2263, 2598

[a] The site locations in this table were generated using software and facilities of BIONET.

mate size of the deletions determined by restriction analysis. Although the plasmids have not been sequenced to determine the exact number of bases of λ N-protein remaining, the expression of inserted DNA of known sequence indicated that plasmid pHE7 has the reading frame illustrated in Fig. 2.

Materials and Reagents

Bacterial strain DH1 harboring plasmids pHE6 or pHE7: Can be obtained from G. Milman, Department of Biochemistry, The Johns Hopkins University, Baltimore, Maryland

```
                              PLASMID pHE6

33 AA
Lambda
N-protein . Val Pro Asp Pro Arg Gly Ser Val Asp Leu Gln Pro Ser Leu
            GTT CCG GAT CCC CGG GGA TCC GTC GAC CTG CAG CCA AGC TTG
                    BamHI . SmaI  BamHI   SalI    PstI      HindIII
                                          HincII


                              PLASMID pHE7

? AA
Lambda                           Gly Ile Arg Arg Pro Ala Ala Lys Leu
N-protein . . . . . . . .    ??? GGG ATC CGT CGA CCT GCA GCC AAG CTT G
                                   BamHI   SalI    PstI      HindIII
                                           HincII
```

FIG. 2. Cloning sites in pHE6 and pHE7. The DNA sequence, amino acid reading frame, and restriction enzyme cleavage sites are illustrated for plasmids pHE6 and pHE7 in the immediate vicinity of the *Sma*I site shown in Fig. 1. Plasmid pHE7 was created by *Bal*31 deletion of pHE6, and the number of amino acids of λ N-protein deleted is not known at this time.

Molecular Cloning[5] describes the following procedures:
- Preparation of LB growth media and LB agar plates with ampicillin
- Purification of plasmid DNA
- Restriction enzyme digestions
- Agarose gel electrophoresis

Enzymes:
- Calf alkaline phosphatase, 24 U/μl (Boehringer–Mannheim)
- T4 ligase (Collaborative Research)
- Restriction enzymes
- *Bal*31 exonuclease

SDS polyacrylamide gel electrophoresis materials for separation of polypeptides[6]

Lysis solution: 13% sucrose, 250 m*M* NaCl, 10 m*M* EDTA, 50 m*M* Tris–HCl (pH 7.5), 0.25 mg/ml lysozyme, 0.1% Nonidet P-40, and 1 m*M* PMSF added just before mixing with cells

[5] T. Maniatis, E. F. Fritsch, and J. Sambrook, "Molecular Cloning: A Laboratory Manual." Cold Spring Harbor Lab., Cold Spring Harbor, New York, 1982.

[6] U. K. Laemmli, *Nature* (*London*) **227,** 680 (1970).

Methods and Comments

Insertion of Foreign DNA in Expression Plasmid

Best results are obtained when the sequence of the foreign DNA is known and a restriction enzyme cleavage site which inserts the foreign gene open reading frame in phase with the N-protein sequence in the expression plasmid (Fig. 2) can be chosen. Types of DNA which may be inserted are illustrated in Fig. 3. The direction of reading for the polypeptide coded by the DNA fragment is indicated by the top arrow. If the DNA is cleaved by a single restriction enzyme, giving *Bam*HI-compatible overlapping ends as illustrated in Fig. 3A (e.g., by *Bam*HI, *Bgl*II, *Bcl*I, *Mbo*I, *Sau*3A, *Dpn*I, or equivalent), then the DNA can be inserted both in the sense and antisense directions. Cleavage by a second enzyme, e.g., *Pst*I in Fig. 3B, permits insertion in the sense orientation only, and also selects for plasmids containing inserts. If the sequence of the foreign DNA is not known, or if there are no convenient restriction enzyme cleavage sites, then the foreign DNA is cleaved at a restriction site upstream from the coding region, and the DNA is then digested for an appropriate time with the exonuclease *Bal*31 to remove upstream DNA.

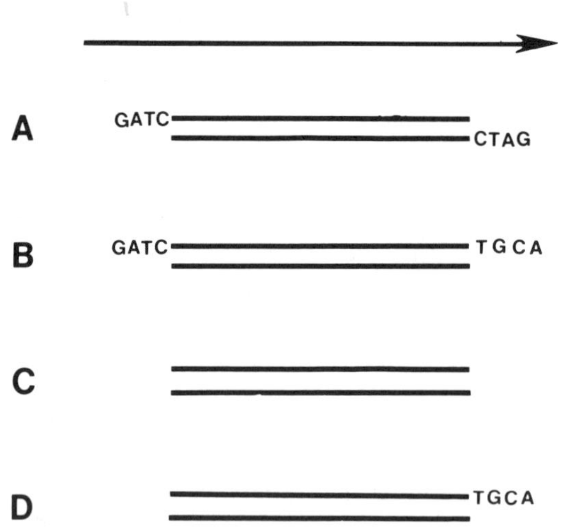

FIG. 3. Examples of DNA fragments for insertion in expression plasmids. The arrow indicates the direction of the polypeptide coding region. (A) DNA cleaved with single restriction enzyme (*Bam*HI in this example); (B) DNA cleaved with two restriction enzymes (*Bam*HI and *Pst*I in this example); (C) DNA digested with *Bal*31 exonuclease; (D) DNA digested with *Bal*31 exonuclease and then with second restriction enzyme (*Pst*I in this example).

The blunt-ended DNA produced by *Bal*31 (Fig. 3C) may be inserted in the *Sma*I site of pHE6. However, if the DNA can also be cleaved at a downstream site (e.g., *Pst*I in Fig. 3D), then analogous to Fig. 3B inserts, the orientation of the DNA in pHE6 is fixed and self-ligation of the plasmid is avoided.

Plasmid DNA (pHE6 or pHE7) is prepared by amplification with chloramphenicol and purification by established methods.[5] Approximately 1 μg of the plasmid is cleaved with a 10-fold excess of the appropriate restriction enzymes to accept the insertion of the foreign DNA. If only a single restriction enzyme is used, calf alkaline phosphatase (1 μl) is added to the restriction digest during the last 10 min of the reaction. Dephosphorylation by alkaline phosphatase decreases self-ligation of the plasmid and increases the relative number of colonies containing inserts.

Foreign DNA is conveniently purified by electrophoresis in low-melting temperature agarose gel. The gel is stained with ethidium bromide and the desired DNA band is identified and excised in the smallest possible volume of agarose (50 μl). The agarose containing the band is liquified by heating to 65° and the DNA is isolated by phenol–chloroform extraction. Alternatively, the DNA–agarose mixture is cooled to 37°, and then appropriately cleaved expression plasmid DNA, T4 ligase, and concentrated T4 ligase buffer are added and the mixture is incubated at 16° for 3 to 18 hr to ligate the foreign DNA into the plasmid. Solidification of the agarose does not prevent ligation. The agarose containing the ligated DNA is again liquified by heating to 65°, cooled to 37°, and the mixture is diluted 10- to 50-fold into buffer used to transfect competent *E. coli*.[5] The transfected bacteria are spread on LB Amp plates and incubated at 30°. Ampicillin-resistant colonies are selected and screened for the presence of the desired plasmids. Rapid lysis minipreps[5] of plasmid DNA are prepared from the colonies and analyzed for proper inserts by size and restriction maps.

Identification of Expressed Proteins

Induction of Expression. Overnight cultures at 30° are prepared from bacteria containing plasmids with inserted foreign DNA, and also from bacteria containing pHE6 without an insert as a negative control. The overnight cultures are diluted 100-fold into 3 ml of 30° L broth in 13-ml test tubes (e.g., Falcon 2057) and grown at 30° with vigorous aeration (300 rpm) for 3 to 4 hr until they reach an OD of 0.3 to 0.5 at 600 nm. Protein expression is induced by transferring the tubes to a 42° water bath while the shaking incubator is also adjusted to 42°. The tubes are replaced in the 42° shaking incubator and incubation is continued for 1 to 4 hr. A 1.5-ml sample of the cells is transferred to a 1.5-ml microcentrifuge tube and the bacteria are sedimented by centrifugation for 10 sec in a microcentrifuge (e.g., an Eppendorf model 5412). The bacteria are washed by suspending in 1 ml 50 m*M* Tris–HCl (pH 7.5) and again sedimented by centrifugation. The bacterial pellets may be frozen at this stage.

SDS–Polyacrylamide Gel Electrophoresis. The bacterial pellets are resuspended in 200 μl SDS sample application buffer for polyacrylamide gel electrophoresis as described by Laemmli[6] and boiled for 10 min. Any insoluble material is removed by centrifugation for 10 min. Samples of 3 to 15 μl are subjected to SDS–polyacrylamide gel electrophoresis and the polypeptides are stained with Coomassie blue. Often, the expressed polypeptide can be identified by comparing the protein staining pattern of bacteria containing the foreign DNA plasmid with that of bacteria containing pHE6. Examples are illustrated in Fig. 4.

Immunological Assays. The availability of antibody which reacts with the expressed foreign DNA polypeptide enables immunological assays for expression. The expressed polypeptide can be detected by immunoblot procedures.[7] Polypeptides from the SDS–polyacrylamide gel are electrophoretically transferred to nitrocellulose and reacted with antibody. The antibody which binds to the expressed protein is detected by ^{35}S-labeled protein A or enzyme-coupled secondary antibody. Examples of immunoscreening are illustrated in Fig. 5.

Antibody produced against the 33-amino acid N-terminal fragment of the lambda N-protein will react with other expressed polypeptides which also share this determinant. The cross-reactivity of this N-terminal fragment can be used to identify fusion polypeptides on immunoblots for which there are no other antibodies. In the example in Fig. 5, the major upper band in lane B is a fusion polypeptide detected by antibody against the λ N-protein determinants of the fusion polypeptide in lane A.

Enzymatic Assays. If the inserted foreign DNA produces an enzyme whose activity can be assayed, then expression can be detected enzymatically. If appropriate bacterial mutants and selective systems are available, then expression may be detected genetically. For example, the expression of bacterial guanine–xanthine phosphoribosyltransferase[8] and herpesvirus (type 1) thymidine kinase[9] could be detected by enzymatic activities.

No Detectable Expression of Foreign Polypeptides. No detectable polypeptides are produced by some foreign DNA inserts whose known sequences should theoretically produce a protein. When no explanation is known for the lack of synthesis, the best recourse is to try another expression plasmid.

[7] H. Towbin, T. Staehelin, and J. Gordon, *Proc. Natl. Acad. Sci. U.S.A.* **76,** 4350 (1979).
[8] S. W. Liu and G. Milman, *J. Biol. Chem.* **258,** 7469 (1983).
[9] A. S. Waldman, E. Haeusslein, and G. Milman, *J. Biol. Chem.* **258,** 11571 (1983).

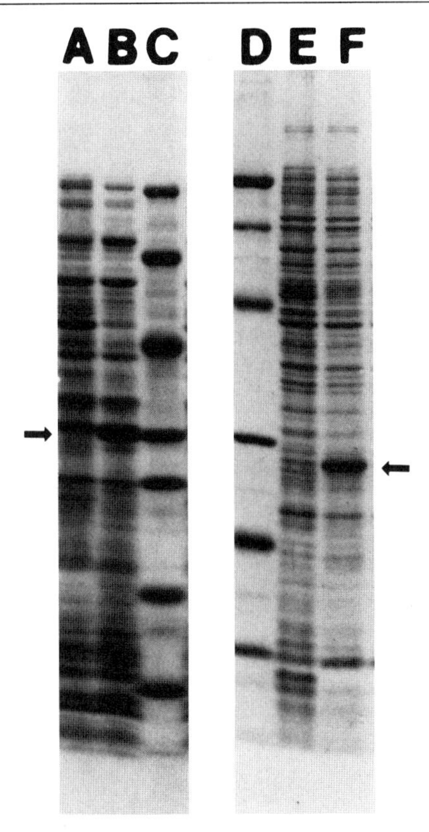

FIG. 4. Examples of expression in pHE6-derived plasmids. Coomassie blue-stained SDS–polyacrylamide gels of crude extracts of bacteria are illustrated. The major bands in lanes B and F (arrows) are the expressed proteins. Lanes A–C were provided by J. D. Roberts and R. L. McMacken (Johns Hopkins University). (A) pHE6 derivative lacking insert; (B) pHE6 derivative expressing lambda O protein; (C and D) molecular weight markers; (E) pHE6 derivative with antisense insert; (F) pHE6 derivative expressing Epstein–Barr virus nuclear antigen.[2]

Conditions for Optimal Expression

The induction time for optimal expression of any polypeptide must be determined empirically by sampling extracts at different times after induction. If the expressed proteins are stable, highest levels of expression are often obtained by overnight incubation. Otherwise, induction is usually limited to 1 to 4 hr.

If an expressed polypeptide is not obvious in a Coomassie blue-stained polyacrylamide gel, the proteins may be pulse labeled with

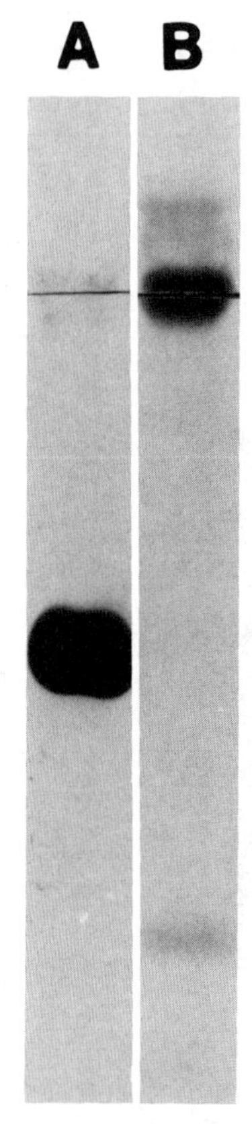

FIG. 5. Cross-reactivity of fusion proteins demonstrated by immunoblots. Crude extracts of induced bacteria containing pHE6 derivatives were subjected to SDS–polyacrylamide gel electrophoresis, transferred to nitrocellulose, probed with antibody to bacterially synthesized Epstein–Barr virus nuclear antigen,[2] and immunostained. (A) Epstein–Barr virus nuclear antigen; (B) Epstein–Barr virus early antigen. The major upper band in B is the fusion early antigen peptide detected by antibody reactivity to the λ N-protein amino terminal determinants.

[^{35}S]methionine to detect synthesis and the stability determined by chase with unlabeled methionine. The methods were described by Rosenberg *et al.*[1]

Purification of Expressed Proteins

Growth and Harvesting of Cells. A 10-ml sample of saturated overnight culture of bacteria grown at 30° is added to 500 ml of LB medium prewarmed to 30° in a 2800-ml culture flask. The culture is grown at 30° in a rotary shaking incubator at 300 rpm. The OD at 600 nm is determined every 30 min and plotted on semilog graph paper. If the expressed polypeptides are unstable, the culture is incubated until a 5-fold dilution of the culture reaches an OD between 0.2 and 0.3. Then, the temperature is rapidly shifted to 42° by the addition of 500 ml of LB medium at 54°. The incubator temperature is adjusted to 42°, and incubation continues for the empirically determined optimal time. If the expressed polypeptides are stable, the temperature is shifted earlier, when the culture reaches an OD of 0.3. The cells are then incubated an additional 16 to 18 hr (overnight) at 42°.

The bacteria are harvested by a 10-min centrifugation at 5000 *g* (6000 rpm in a Sorvall GSA rotor). The supernatant is discarded and the cells are resuspended in 100 ml ice-cold 50 m*M* Tris–HCl (pH 7.5) and again pelleted by centrifugation. The pelleted cells are resuspended in 30 ml of solution containing 50 m*M* Tris–HCl (pH 7.5), 10 m*M* EDTA, and 1 m*M* PMSF and stored frozen at −80°.

Preparation of Crude Cell Extract. The frozen cells (approximately 30 ml) are thawed quickly in a warm water bath with agitation so that the temperature of the solution containing the cells never increases above 5°. As the last ice melts, 90 ml of ice-cold lysis solution is added. The cell suspension is transferred to centrifuge tubes and incubated on ice for 30 min to allow lysis. The lysed cells are centrifuged for 1.5 hr at 100,000 *g* (e.g., 35,000 rpm in a Beckman 50.2 rotor). The supernatant forms the crude cell extract. The high-speed pellets are saved if the expressed proteins are insoluble.

Purification of Expressed Proteins. The purification procedure will depend upon the protein being purified. When the expressed protein represents greater than 1% of the soluble protein, conventional ion exchange, sizing, and adsorption chromatography steps generally yield greater than 99% purity.

If the expressed protein is located in the high-speed pellets, then detergents may be used to solubilize the protein. We have had only limited success in solubilizing pellet proteins in urea and guanidine thiocyanate followed by dialysis against aqueous buffers to renature the proteins.

Acknowledgment

Supported by Grant MV-287 from the American Cancer Society and Grants ESO3131 and GM32950 from the National Institute of Environmental Health Sciences and the National Institute of General Medical Sciences, National Institutes of Health.

[31] Expression and Secretion of Foreign Proteins in *Escherichia coli*

By GUY D. DUFFAUD, PAUL E. MARCH, and MASAYORI INOUYE

Whether the gene for a given protein has been cloned or a DNA fragment is found to contain an open reading frame, the expression of protein, preferably in large amounts, often becomes a main objective. Our laboratory has developed a set of expression and secretion vectors designed to overcome many of the problems encountered when expressing a given gene in *Escherichia coli*. These vectors have been designed for use at specific steps during the cloning and expression of proteins as well as for the particular needs that might come with the nature of the protein to be produced, such as the necessity to secrete products that are toxic for cell growth. We discuss here the different vectors we have constructed and how they can best be utilized for cloning, expression, and secretion.

Cloning and Expression in *E. coli*

Expression Vectors

The first step in achieving expression of a protein is to clone the DNA fragment in generalized expression vectors. These vectors must possess a certain number of essential properties: a multiple restriction cloning site, an upstream promoter region which will promote the expression of the foreign gene, and some way of controlling that expression, particularly if the foreign protein is suspected of being lethal to the cell. Other properties are also important: a high copy number of the vector, a suitable marker for selection, a capacity for cloning large fragments. We have constructed vectors that encompass all of these properties: the pIN-II-A and pIN-III-A series (Table I). The pIN vectors are derived from pBR322, a plasmid produced in roughly 30 copies per bacterial cell.[1] To promote expression, the pIN vectors employ the efficient *lpp* gene promoter (*lpp*^{P}).[2] This gene is responsible for the constitutive expression of lipoprotein, a major outer membrane protein, and has been integrated into these vectors. Downstream of *lpp*^{P}, the *lac*UV5 promoter operator (*lac*^{PO}) fragment has been inserted. Thus, expression will occur only in the presence of a *lac* in-

[1] D. Stueber and H. Bijard, *EMBO J.* **1,** 1399 (1982).
[2] K. Nakamura and M. Inouye, *EMBO J.* **1,** 771 (1982).

TABLE I
EXPRESSION VECTORS BASED ON THE *lpp–lac* PROMOTER SYSTEMS AND CORRESPONDING HOST STRAINS

Plasmid	*E. coli* host	Comments	Reference
pIN-I-A1 pIN-I-A2 pIN-I-A3	W620	Constitutive expression, product in the cytoplasm	3
pIN-II-A1 pIN-II-A2 pIN-II-A3 pIN-II-B1 pIN-II-B2 pIN-II-B3 pIN-II-C1 pIN-II-C2 pIN-II-C3	SB221	Inducible promoter; *lpp*^{P}-*lac*^{PO}. Transformation should be carried out in an *E. coli* strain containing *lacI*^{q}. The pIN-II-B2 plasmid is unstable due to the creation of an open reading frame during construction. Its use is not recommended	3
pIN-III-A1 pIN-III-A2 pIN-III-A3 pIN-III-B1 pIN-III-B2 pIN-III-B3 pIN-III-C1 pIN-III-C2 pIN-III-C3	W620 or SB221	Inducible promoter; *lpp*^{P}-*lac*^{PO}. Plasmid contains *lacI*. The pIN-III-B2 plasmid is unstable as discussed above for the pIN-II-B2 plasmid	3
pIN-III-ompA1 pIN-III-ompA2 pIN-III-ompA3	W620 or SB221	Secretion expression vector. For use when extracytoplasmic localization of the product is desired	10
pIN-III (*lpp*$^{P-5}$)	W620 or SB221	Improved *lpp* promoter; 3- to 4-fold higher production	4
pIC-III	SB4288	Hybrid protein expression vector; cloning sites are *Eco*RI, *Sma*I, *Bam*HI in the *lacZ* coding region	

Strain	Description of *E. coli* strains, phenotype	
W620	K-12 F^{-} *thi-I pyrD*36 *gltA*6 *galK*30 *strA*129 *supE*44 *relA recA*	21
SB221	K-12 F^{-} *hsdR leuB*6 *lacY thi* Δ*trpE*5 *lpp*/F′ *lacI*^{q} *proAG lacZYA*	15
SB4288	K-12 F^{-} *recA thi*-1 *relA mal*-24 *spc*12 *supE*-50 DE5(Δ*lac-proB*)	23

ducer. Downstream of this tandem promoter region a multiple restriction site linker has been inserted (*Eco*RI, *Hin*dIII, *Bam*HI) at three positions determined by the properties of *lpp*: the A, B, and C sites. At each site, the linker has been inserted in such a way that cloning at any restriction site can be accomplished in all three possible reading frames (Fig. 1). This allows the choice of a vector with a reading frame compatible with the fragment to be cloned. The A sites are located right after the initiation codon of lipoprotein, the B sites right after the cleavage site of the lipoprotein precursor, and the C sites nine amino acid into the mature portion of lipoprotein. Constructions employing the B and C sites give rise to hybrid proteins which can use the properties of the lipoprotein signal peptide for localization in the different *E. coli* compartments, as discussed below. Cloning into any of these sites results in the usage of the lipoprotein Shine–Dalgarno sequence, initiation codon, and termination codon, which is convenient if the fragment to be cloned lacks one or more of these essential features. In addition there is a termination codon in each of the reading frames and a ρ-independent efficient transcription termination signal. Detailed reviews on the construction of all these vectors can be found elsewhere.[2,3]

Utilization of the pIN-II and pIN-III Expression Vector

Cloning and Maintenance. Successful utilization of expression vectors requires certain considerations. For cloning it is desirable that the gene have compatible ends with the available restriction sites (six different types of restriction fragments can be cloned). However, if one or two of the sites are not compatible, it is always possible to clone together with commercially available gene linkers in order to make the sites compatible. It should be pointed out that if the gene contains its own ribosome-binding region and initiation codon, a unique *Xba*I site after lac^{PO} can be used as an additional cloning site.

Another important consideration is the choice of host. For the expression to be properly regulated it is necessary to have a host in which the *lac* repressor (*lacI*) is expressed. The pIN-II vectors require a host containing the *lacI* gene, such as *E. coli* strain SB221 (Table I). This will suppress expression of the product during the cloning procedure. If such a strain is not available or cannot be used, the vector of choice becomes one of the pIN-III series. These vectors have the same characteristics as pIN-II except that they have the *lacI* gene inserted at a unique *Sal*I site within the expression vector itself. In this way expression of the gene is blocked

[3] Y. Masui, J. Coleman, and M. Inouye, *in* "Experimental Manipulations of Gene Expression" (M. Inouye, ed.), p. 15. Academic Press, New York, 1983.

until a *lac* inducer such as isopropyl-β-D-thiogalactopyranoside (IPTG) is added to the growth medium. It is important to note that we have found that certain culture media such as yeast extract can contain forms of *lac* activator that induce expression of a cloned gene to significant levels even in the absence of IPTG. Levels of background induction can be high enough to be lethal to the cell if the gene product is toxic per se or if high amounts are toxic. To avoid these problems, it is preferable to use a minimal medium supplemented with casamino acid. For the same reasons, it is also advisable to store the cloned gene in the purified plasmid form rather than to maintain the plasmid in the host cell in a growth medium where background induction may result in mutations in the cloned protein and/or its expression system. If such a mutant plasmid is used, the protein might be altered in its function and/or its level of expression might be lowered in comparison to the original plasmid.

Expression. Upon induction with IPTG, the level of protein production of a cloned gene in a pIN vector might reach higher than 20% of the total cellular protein. However, this level of production depends on the particular protein under study. To compensate for this, we have modified *lpp*^{P} by means of oligonucleotide-directed site-specific mutagenesis. The *lpp*^{P} was systematically altered at the -35 and -10 regions in order to bring them closer to the consensus sequence and create a more effective promoter.[4] The *lpp*^{P} (wild type) sequence is

-35 region -10 region
ATCAAAAAAATATTCTCAACATAAAAAACTTTGTGTAATACTTGTAAC

and it was changed to *lpp*$^{P-5}$, the most effective of the designed promoters, by two substitutions in the -35 region (TTGACA) and one single substitution in the -10 region (TATACT). This new promoter was found to increase the level of expression of β-galactosidase by approximately 4-fold over the already efficient wild-type *lpp*^{P} and can easily be substituted in any pIN vector by simply replacing the *Pst*I–*Xba*I fragment from the vector with the cloned gene with the equivalent fragment from pIN-III-(*lpp*$^{P-5}$). When this is done, the level of expression is expected to increase at least 3- to 4-fold. A word of caution is necessary when using this powerful promoter: overproduction per se might be lethal. If a protein is already expressed at a level of 10–20% of the total cellular protein, attempting to increase this percentage may be counterproductive.

The size of the protein to be expressed is another factor to be considered. It would seem that the protein size would be a limiting factor when using *E. coli* as the host cell. However, a recent report suggests that this

[4] S. Inouye and M. Inouye, *Nucleic Acids Res.* **13,** 3101 (1985).

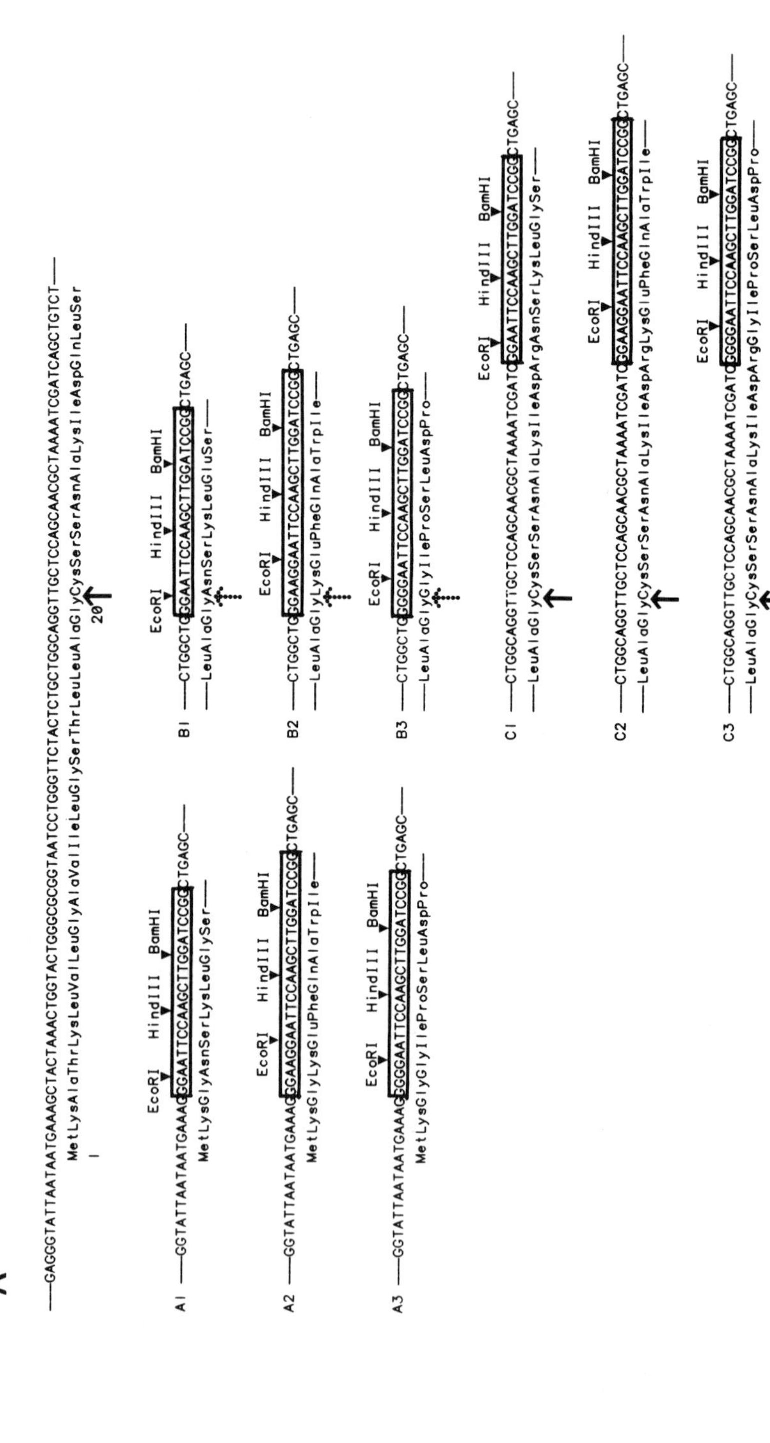

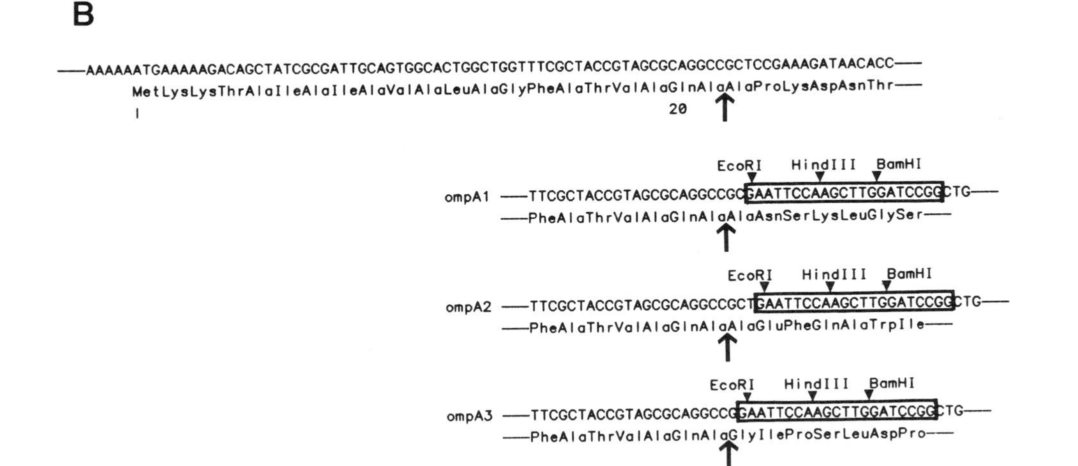

FIG. 1. (A) DNA sequence of the three reading frames of the A, B, and C sites of the pIN vectors. The natural lipoprotein gene is shown at the top and the positions of the A, B, and C sites are aligned below. (B) DNA sequence of the three reading frames of the ompA secretion vectors. The entire ompA signal sequence and cleavage site region is shown at the top and the ompA1, ompA2, and ompA3 reading frames are shown below. The solid arrows denote signal peptide cleavage sites. In the pIN-B series the signal peptide cleavage site is unknown, as indicated by the dotted arrow. The linker DNA is boxed and the positions of restriction enzyme cleavage sites are indicated by arrow heads.

might not be critical. Two to four *lacZ* genes have been fused head to tail, in phase, and have been shown to express the corresponding dimers, trimers, and tetramers of β-galactosidase (up to 460 kDa).[5] All products showed enzymatic activity, which was, however, strongly reduced for the trimer and tetramer forms. Also, the latter were largely degraded to small products (such as a dimer) as could be seen on SDS–PAGE gels. We have also recently been able to express a fused pentameric unit of β-galactosidase using a different strategy: gene fusion was done in such a way that only the tail unit could be active since only this unit possessed the entire *lacZ* coding sequence. This pentamer was expressed with a pIN-III-A3-$lpp^{P\text{-}5}$ vector, showed activity, and was clearly identifiable, with little degradation, upon SDS–PAGE stained with Coomassie blue (K. Tsung and M. Inouye, unpublished results). As was the case for the tetrameric β-galactosidase, foreign proteins might be proteolitically degraded in *E. coli*. In such instances it is advisable to transfer the expression vector with the cloned gene to an *E. coli* strain deficient in proteolytic activity such as a lon^- strain.[6]

Other Expression Vectors and Their Usage

The pIN-I vectors are like the pIN-II vectors except that the expression of a cloned gene is not under lac^{PO} control.[3] Thus they can be used only for constitutive expression of proteins that are known to be at least tolerated by the host cell (or even needed). Their advantage is that they do not require induction and thus have more flexibility in the choise of a host cell and media. Constitutive expression is advantageous because of the potential to make large amounts of product while at the same time minimizing manipulations during cell growth. These vectors could also be suited, for example, to produce proteins meant to be used as antigens for antisera production or even for large-scale production of the antigenic fragment when properly cloned.

pIC-III vectors are different from pIN-III vectors in that they generate a hybrid of a desired protein and β-galactosidase. A linker has been inserted into a pIN-III-*lacZ* vector immediately after the initiation codon in such a way that *lacZ* is out of phase with the initiation codon. A DNA fragment inserted within this linker, in the appropriate reading frame, results in the expression of a β-galactosidase hybrid protein.[3] Since such a hybrid is expected to have a β-galactosidase activity selection of the hybrid is made easy by using agar plates containing β-galactosidase indicator. This vector is best used in the production of hybrid protein for

[5] W. Kuchinke and B. Muller-Hill, *EMBO J.* **4,** 1067 (1985).
[6] S. A. Goff and A. L. Goldbert, *Cell* **41,** 587 (1985).

rapid preparation of antisera, since any *lacZ* hybrid protein is very large and quite easy to purify in a single step via preparative SDS–PAGE.

Secretion Vectors in *E. coli*

There are several advantages in using a secretion vector: (1) the amino terminus can be changed to other than a methionine by adjusting the cleavage site in such a way that, after cleavage of the signal peptide, the desired amino acid is at the amino terminus; (2) the usage of secretion vector is essential for proteins such as nucleases or proteases which are harmful if kept in the cytoplasm after production; (3) proteins secreted into the periplasmic space (the space between the outer membrane and the cytoplasmic or inner membrane) are considered to be more stable because there is a lower proteolytic activity in the periplasmic space; (4) the reducing environment of the cytoplasm does not allow the formation of disulfide bonds which can easily be formed in the oxidizing environment of the periplasm; (5) proteins secreted to the periplasmic space can easily be released from the cell by osmotic shock and readily purified. Based on these considerations we have developed secretion vectors that can accomplish the processes discussed above. We describe below their constructions and methods that can be used with them.

Secretion in E. coli: Requirements for a Secretion Vector

Escherichia coli contains four major subcellular compartments: the cytoplasm, the cytoplasmic or inner membrane, the outer membrane, and the periplasmic space between the inner and outer membranes. Secretory proteins are synthesized first in the cytoplasm as precursor proteins which have a peptide extension of 20–25 amino acids, termed the signal peptide, at their amino termini. This signal peptide is an essential element for directing the precursor protein to the inner membrane and subsequent translocation across the inner membrane. In *E. coli*, outer membrane proteins, periplasmic proteins, and some inner membrane proteins have been shown to be produced as precursors containing a signal peptide. Signal peptides have been found to have unique features that account for their functions. Upon translocation of the precursor protein across the inner membrane, the signal peptide is cleaved by a signal peptidase located in the inner membrane. The mature protein then reaches its final location (outer membrane, periplasmic space, or inner membrane) by information contained within its amino acid sequence.[7] In general, pro-

[7] G. D. Duffaud, S. K. Lehnhardt, P. E. March, and M. Inouye, *Curr. Top. Membr. Transp.* **24,** 65 (1985).

teins normally destined for export, such as hormones and serum proteins, will be easily secreted by *E. coli*. Whether other type of proteins can be secreted across the membrane is not easily predicted.

The signal peptide has been shown to guide translocation of the amino terminus of the mature protein to the outside surface of the inner membrane. The amino terminus of any fused protein should also be translocated in a similar fashion. Whether the rest of the protein is translocated across the membrane depends on the structure of the protein. For example, a protein containing a long stretch of hydrophobic amino acids could likely remain in the inner membrane, since a hydrophobic stretch is not easily translocated across the membrane. Such a protein might still be translocated across the membrane by engineering the hydrophobic stretch as discussed later. Conversely, proteins normally located in the cytoplasm can be guided to localize to the inner membrane by fusing them to a noncleavable signal peptide (pIN-II-B or pIN-III-B). Or a protein might be guided to the outer membrane by fusing it to a cleavable signal peptide of an outer membrane protein (pIN-II-C or pIN-III-C). These vectors have been created in our laboratory and are listed in Table I. The construction and function of these vectors has been reviewed elsewhere and will only be discussed briefly in this chapter.[8,9]

pIN-III-OmpA Secretion Vectors

The pIN-III-OmpA vector has been constructed by replacing the fragment between the unique *Xba*I site and the unique *Eco*RI site of the pIN-III-A3 vector with the DNA fragment carrying the coding region for the signal peptide of the OmpA protein, a major outer membrane protein.[10] The signal peptide of OmpA was chosen for two reasons: (1) OmpA is a major protein of the *E. coli* outer membrane and thus its signal peptide is thought to be efficient in the translocation process and (2) the OmpA signal peptide does not need to be modified before processing, as is the case for the lipoprotein signal peptide, and it is processed efficiently by the major processing pathway that uses signal peptidase I. Thus this vector is the system of choice when the primary objective is the translocation of a protein outside the cytoplasm: it provides all of the advantages of the pIN-III expression vectors and at the same time the necessary information for secretion.

[8] C. A. Lunn, M. Takahara, and M. Inouye, this series, in press.

[9] C. A. Lunn, M. Takahara, and M. Inouye, *Curr. Top. Microbiol. Immunol.* **125,** 59 (1986).

[10] J. Ghrayeb, H. Kimura, M. Takahara, H. Hsiung, Y. Masui, and M. Inouye, *EMBO J.* **3,** 2437 (1984).

Utilization of Secretion Vectors

General cloning strategies and expression protocols follow the same requirements addressed above. The question now is which vector to choose and what to expect. Some precautions must be taken before a secretion vector is utilized. First, as discussed earlier, a protein might not be secreted even if it is fused to a signal peptide: β-galactosidase, a cytoplasmic protein, was not exported to the outer membrane when fused to the signal peptide of *lam*B which codes for an outer membrane protein.[11] Second, it is also important to consider that the fusion of a signal peptide with a foreign protein might not yield a functional cleavage site, thus impairing secretion. This particular problem and ways to solve it are discussed below. Finally, the ultimate localization of the protein will depend on information intrinsic to the sequence itself, not on the origin of the signal peptide.

pIN-III-OmpA has been utilized to secrete both bacterial and eukaryotic proteins to the periplasmic compartment of *E. coli*.[10,12,13] The first protein to be expressed using this vector was β-lactamase.[10] The mature portion of this protein was fused to the OmpA signal peptide by insertion at the *Eco*RI site of pIN-III-OmpA3 (see Fig. 1). Upon induction of gene expression, fully procesed β-lactamase was secreted into the periplasmic space and after a 3-hr induction accumulated to 20% of total cellular protein. The OmpA signal peptide was correctly removed, resulting in the production of β-lactamase with four extra amino acid residues at its amino terminus due to the linker sequence in the vector. As a next step, this linker sequence was deleted by site-specific mutagenesis and the new product had the same amino terminus as authentic β-lactamase. No significant accumulation of precursor was observed.

Since β-lactamase is an *E. coli* protein normally secreted into the periplasmic space, pIN-III-OmpA was tested for the production and secretion of an extracellular protein from a gram-positive bacteria: staphylococcal nuclease A from *Staphylococcus aureus*. This protein was cloned from an *E. coli* vector in which it is poorly expressed to the *Bam*HI site of pIN-III-OmpA.[13] The nuclease was now produced to an extent of 3% of the total cellular protein, was accurately processed at the cleavage site of the OmpA signal peptide (i.e., with a peptide extension originating

[11] F. Moreno, A. V. Fowler, M. Hall, T. J. Silhavy, I. Zabin, and M. Schwartz, *Nature (London)* **286,** 356 (1980).

[12] J. Ghrayeb and M. Inouye, *J. Biol. Chem.* **259,** 463 (1984).

[13] M. Takahara, D. W. Hibler, P. J. Barr, J. A. Gerlt, and M. Inouye, *J. Biol. Chem.* **260,** 2670 (1985).

from the linker and an N-terminal portion of the nuclease gene), and was translocated to the periplasmic space of *E. coli*. The nuclease accounted for 10% of the protein in the periplasm. When the linker portion was deleted via site-specific mutagenesis, nuclease A was produced in its authentic form and accounted for 12% of the periplasmic protein of which one-sixth was in the unprocessed form. To increase the level of expression of the nuclease, the *lpp* promoter was replaced with $lpp^{P\text{-}5}$ as described above. With this vector the production of nuclease was increased almost 4-fold after a 3-hr induction. However, overproduction of this protein resulted in a higher accumulation of the unprocessed precursor (about one-third of the total nuclease protein).[4]

Recently, human growth hormone (hGH) has been successfully expressed and secreted using pIN-III-OmpA3.[14] A DNA fragment carrying the hGH gene was first cloned into the *Eco*RI site out of frame. The linker portion was then removed by site-specific mutagenesis to adjust the reading frame and this new clone was able to express authentic hGH to an amount of 6% of the total cellular protein. This protein was exported to the periplasmic space, where it constituted approximately 30% of the total protein. The protein was purified to homogeneity in a two-step procedure and characterized by trypsin digestion, CD spectroscopy, and amino acid sequencing of the N-terminus. The expressed protein had the same structural characteristics as authentic hGH, including the correct formation of two disulfide bridges. It is interesting to note that in this case expression was obtained without the addition of IPTG and that the addition of IPTG resulted in the accumulation of precursor without a significant increase in the total level of hGH.

These examples underline some of the advantages of the pIN-III-OmpA secretion vector. They can be employed to secrete proteins of different origin (β-lactamase, nuclease A, and hGH). These proteins are identical to their authentic counterpart in terms of activity, conformation, amino terminal residue, and disulfide bond formation. They are secreted in large amounts (even if toxic for the cell, like nuclease A) to the periplasmic space from which they can easily be purified.

Problems Inherent to Secretion Vectors

Secretion of Cytoplasmic Proteins. Cytoplasmic proteins do not contain a signal peptide and remain in the cytoplasm as soluble proteins. Whether they can be secreted across the membrane or not depends on their primary structure, as discussed earlier. Complete translocation

[14] H. M. Hsiung, N. G. Mayne, and G. W. Becker, *Bio/Technology*, in press.

might be inhibited by the presence of hydrophobic sequences, as discussed earlier. The existence of such sequences results in a stop-transfer sequence which blocks the translocation process across the inner membrane or will promote insertion within the inner membrane. It has been shown that a secretory protein can be engineered to contain a stop-transfer signal, making this protein an inner membrane protein.[15] The function of the stop-transfer signal can be destroyed by altering the hydrophobic region to a more polar or hydrophilic sequence, allowing translocation. When designing such mutations care should be taken not to alter the structure or function of the protein.

Structural Requirements for Processing and Translocation of Secretory Proteins; Structural Compatibility between the Signal Peptide and the Secretory Protein. There is a possibility that the signal peptide of the hybrid protein might not be cleaved by an *E. coli* signal peptidase. The properties of the signal peptides should be considered, when creating a hybrid protein, to ensure that it will be processed. Several features are conserved among all signal peptides and are thought to account for their function. The 20- to 25-amino acid prokaryotic signal peptide can be divided into three parts. The amino terminal portion (usually four to five amino residues) is always positively charged due to the presence of up to three basic amino acids. The positive charges arising from these residues are thought to help the initial binding to the nascent peptide chain to negative charges of the inner membranes and the putative secretory components. The amino terminal portion of the signal peptide is followed by a highly hydrophobic segment of about 10 residues, usually punctuated by glycine and/or proline residue. According to the loop model this hydrophobic segment is able to insert in the lipid bilayer by forming a loop structure.[7,16] The cleavage site region follows the hydrophobic core of the signal peptide. It is usually joined to the hydrophobic core by serine and/or threonine residues and it is separated from the cleavage site itself by a short, rather hydrophobic, segment. The C-terminus of the signal peptide is always a small side chain amino acid: glycine, alanine, or serine. The first amino acids of the mature protein are not as well conserved. However, by using the Chou and Fasman parameters to predict the secondary structure at the cleavage site, one observes that precursor proteins might have a conserved β-turn structure in this area.[17] It is possible that the small side chain amino acid at the cleavage site of the signal peptide is a

[15] J. Coleman, M. Inukai, and M. Inouye, *Cell* **43,** 351 (1985).

[16] M. Inouye, R. Pirtle, I. Pirtle, J. Sekizawa, K. Nakamura, J. Di Rienzo, S. Inouye, S. Wang, and S. Halegoua, *Microbiology,* 34 (1979).

[17] P. Y. Chou and G. D. Fasman, *Biochemistry* **47,** 251 (1977).

specific requirement of the signal peptidase and that the conserved β-turn structure at the cleavage site is a main requirement for accurate processing, thus explaining why it is not necessary to conserve the amino acids at the N-terminus of the mature sequence.[18,19] Evidence for the function of the different regions of the signal peptide described above have been provided by a whole set of mutations naturally found or created by site-specific mutagenesis and other methods.[7]

It has been shown that mutation on either side of the cleavage site could hamper processing, and that a combination of mutations could make processing totally defective.[19] Thus when cloning a protein into a secretion vector care should be taken that the newly created cleavage site has the required structural elements, as well as a proper amino acid at the C-terminus of the signal peptide. If these conditions are not met defective processing might result. At best the defect might result in a slow processing rate and at worst in an unexpected cleavage site or a totally defective processing. However, when such a problem arises it should be possible to at least partially remedy it. For example, a linker could be inserted into the construction in such a way that new amino acids are encoded to create the desired structure. More simply, site-specific mutagenesis could be performed on either side of the cleavage site in order to change one or two amino acids, preferably in the signal peptide, so as not to change the amino acid sequence of the secretory protein. We have also found that the signal peptide can have a direct effect on the rate of synthesis of the secretory protein. Pollitt *et al.* have found that the deletion of the glycine residues at positions 9 and 14 of the lipoprotein signal peptide resulted in a 4-fold increase of lipoprotein synthesis *in vivo* as compared to the rate of synthesis of the wild-type lipoprotein (when expressed with the same vector).[20] A significant increase was also observed in the *in vitro* rate of synthesis. However, when the mutant lipoprotein signal peptide was fused to β-lactamase no effect was observed on the rate of lipo-β-lactamase synthesis (C. A. Lunn and M. Inouye, unpublished results). It seems that a certain amount of compatibility between the signal peptide and the secretory protein is required if the signal peptide is to have an effect on the rate of protein synthesis.

Other Secretion Vectors: pIN-II-B, pIN-II-C, pIN-III-B, and pIN-III-C

These vectors have cloning sites right after the lipoprotein signal peptide (B site) or nine amino acid residues after the cleavage site (C site). A

[18] S. Pollitt, S. Inouye, and M. Inouye, *J. Biol. Chem.* **261,** 1835 (1986).
[19] S. Inouye, G. D. Duffaud, and M. Inouye, *J. Biol. Chem.* **261,** 10970 (1986).
[20] S. Pollitt, S. Inouye, and M. Inouye, *J. Biol. Chem.* **260,** 7965 (1985).

protein cloned at one of these sites should be translocated across the cytoplasmic membrane. However, localization of the protein to the outer membrane is possible when cloning into the C sites. The hybrid protein expressed in a pIN-II-C or pIN-III-C vector contains the first nine amino acids of lipoprotein, including the lipid-modified cysteine at position +1. These amino acid residues have been shown to direct the secretion of β-lactamase, a periplasmic protein,[12] and staphyloccoccal nuclease A (M. Takahara and M. Inouye, unpublished results) to the outer membrane.

When a protein is fused to the B site, the hybrid is expected to be exported to the periplasmic space. However, the hybrid may still be bound to the inner membrane by the uncleaved lipoprotein signal peptide. This is the case for staphyloccoccal nuclease A which was found to be localized in the inner membrane fraction when it was cloned into the B site (M. Takahara and M. Inouye, unpublished results).

General Considerations

Cloning, expression, and secretion of any protein is greatly simplified by using specifically designed vectors such as the ones that are described here. Regardless of the strategy used, there is a set of common considerations that apply to expression and/or secretion of a protein when these vectors are utilized.

Expression Levels

The pIN vectors offer a ready solution to many of the problems that might be encountered when one seeks to express a protein. However, in some cases the expression of a given protein might not be as high as one would desire. Any of the vectors described above can be changed to a higher level of expression just by changing its promoter (lpp^{P}) by a still more efficient promoter: $lpp^{P\text{-}5}$. This procedure is made easy by the existence of two unique restriction sites on each side of the promoter region: *Xba*I and *Pst*I.

If the expression is still not high enough there are other factors that should be considered, such as the codon usage, the secondary structure of the mRNA, the protein stability, and the toxicity of the product to the cell. If the cloned fragment contains codons rarely used by *E. coli* then

[21] E. T. Wurtzel, R. N. Movva, F. Ross, and M. Inouye, *J. Mol. Appl. Genet.* **1,** 61 (1981).
[22] P. J. Green and M. Inouye, *J. Mol. Biol.* **176,** 431 (1984).

low levels of expression might result. Alteration of some of the codons by site-specific mutagenesis could improve the overall expression.

Host Strains

The choice of a host strain is also critical. The nature of the vector used determines what type of host have to be used, e.g., a strain containing the *lacI* gene or not. Also it is important to use a lon^- strain or strains having low proteolytic activity if the protein is particularly sensitive to proteolytic activity.

Growth Conditions

Establishing satisfactory growth conditions is an important first step in the successful expression of a protein. Expression of a particular protein might be more successful at the exponential phase than at the stationary phase. With the pIN vector system expression can be induced at any point of the growth curve by addition of IPTG. Another important factor is the medium used for growth. Care should be taken that no *lac* inducer is present when using an inducible vector. The nature of the medium might also have a more indirect role. A growth medium can be altered in its osmolarity, reducing agents, nutrient composition, etc., any of which may affect overall expression levels. Selection of the right components of the growth medium could further improve expression of the protein either by ensuring a more stable secreted product, or by giving the cell optimum growth conditions to produce the protein. Growth temperature can also be an important factor.

Some problems such as the glycosylation of certain eukaryotic proteins are out of the reach of the *E. coli* sysem at present. However, it should be noted that the *E. coli* system is able to provide an environment adequate for the formation of the correct disulfide bonds of hGH in the periplasmic space.

Toxicity

There are several ways to deal with products that are toxic for the cell. One is to have the protein secreted by using a secretion vector. Another way is to induce only tolerable levels of the protein. If the product is particularly stable the cells could even be left to die and the protein harvested later.

Maintenance of Expression Vectors

Whether proteins are toxic to the *E. coli* cell or not, the cloned genes are always subjected to spontaneous mutations. In particular toxic products will be under more pressure to have either their toxicity diminished by a mutation in the structural gene or the expression level lowered by mutations in the promoter region. To avoid such problems the newly obtained clones should be stored in the form of purified DNA form as soon as constructed.

Acknowledgments

This work has been supported by Grants GM11145 and GM19043D from NI General Medical Sciences, Grant NP387M from the American Cancer Society, and a Grant from Eli Lilly Company. Guy Duffaud is recipient of a Fujii Fellowship.

[32] Engineering for Protein Secretion in Gram-Positive Bacteria

By SHING CHANG

Introduction

The cell envelope of gram-positive bacteria contains a single layer of cell membrane separating the cytoplasm from the environment. Secreted proteins are synthesized in the cytoplasm and then transported across the membrane; after completion of this translocation process, certain secreted protein species are anchored or attached to the membrane, while others are excreted into the media. Although the genetics of gram-positive bacteria are less advanced than the well-studied *Escherichia coli,* the ability of the former to excrete proteins into the medium makes them the more favorable candidates as the host organisms for the production of secreted proteins. Some members of this group of bacteria produce a number of economically important exoenzymes, such as cellulases, α-amylase, and proteases, in large quantity.[1,2]

The molecular mechanism of protein secretion in bacteria has been reviewed recently.[3-5] Significant progress has been made in using the *E. coli* model system, and the majority of the conclusions reached also apply

[1] F. G. Priest, *Bacteriol. Rev.* **41,** 711 (1977).

[2] A. G. Glenn, *Annu. Rev. Microbiol.* **30,** 41 (1976).

[3] A. P. Pugsley and M. Schwartz, *FEMS Microbiol. Rev.* **32,** 3 (1985).

[4] D. Oliver, *Annu. Rev. Microbiol.* **39,** 615 (1985).

[5] S. A. Benson, M. N. Hall, and T. J. Silhavy, *Annu. Rev. Biochem.* **54,** 101 (1985).

to the gram-positive bacteria. In general, the precursors to all secreted proteins contain an amino terminal extension of about 20 to 35 residues in length known as the signal peptide. This fragment, although relatively short in length, plays a central role in protein translocation across membranes. In most cases, this signal peptide is removed from the precursor by signal peptidases either co- or posttranslationally, resulting in the formation of the mature protein in the designated extracytoplasmic locations.

Signal peptide can be divided into three distinct regions based on their common structural features: (1) the amino terminal region, which contains one or more positively charged residues, (2) a stretch of 10–20 hydrophobic or neutral residues, and (3) the carboxy end four to seven residues, including the signal peptidase cleavage site sequence. The cleavage site (↓) is frequently located after an alanine residue, with Ala-X-Ala↓ as the most frequently found sequence at the C terminus of the signal peptide.[6,7] In addition, proline or glycine residues are commonly found in the region between positions -4 to -7. These conserved features found in the three regions in the signal peptide likely reflect the specific functions they each perform during the translocation process that directs the secretion of the target protein.

Secreted Proteins from Gram-Positive Bacteria

I researched the literature and compiled a summary (Table I) of the amino terminal sequences of protein secreted from gram-positive bacteria. The general features of signal peptides from gram-positives are not significantly different from those of the gram-negatives. The noted exceptions are that the gram-positive signal peptides are slightly longer, and that they tend to contain more positively charged residues clustered within the amino terminal region. Among the proteins shown in Table I, two special classes of proteins that are significantly different from the majority can be identified.

Precursors to Proteases. All of the protease precursors contain a signal peptide-like amino terminal sequence, followed by a stretch of "pro" sequence ranging from about 70 to 190 residues. This feature appears to be similar to the preproprotein found in certain eukaryotic protease precursors. However, maturation of at least one of these bacterial "prepro-proteases" does not appear to involve the initial cleavage of the signal sequence followed by a second cleavage step that removes the "pro"

[6] D. Perlman and H. O. Halvorson, *J. Mol. Biol.* **167,** 391 (1983).
[7] G. von Heijne, *J. Mol. Biol.* **173,** 243 (1984).

sequence. Instead, removal of the "prepro" sequence from alkaline protease has been shown to involve an autocatalytic processing that converts the precursor to the active form of mature protease.[8]

Precursors to Lipoproteins. A specialized modification and processing pathway is utilized for the biosynthesis of membrane-bound lipoproteins in both gram-positive and gram-negative bacteria.[9] These lipoproteins contain fatty acid and glyceride covalently linked to the cysteine at the amino termini of the mature protein. The most commonly found recognition (lipoprotein consensus) sequence, Leu-Ala-Gly-Cys,[10] is located at the junction between the signal peptide and the mature protein sequences in the precursors of lipoproteins. The lipoprotein signal peptidase cleaves between the Gly and the modified Cys residues, leaving the mature lipoprotein anchored in the membrane via the fatty acids attached to the terminal Cys residue. Certain prelipoprotein precursors can also be processed to form excreted protein by the cleavage of the signal sequence at a different site, irrespective of the presence or absence of a lipoprotein modification signal, thus releasing an unmodified mature protein via the regular secretory pathway.[11]

Directed Secretion of Heterologous Proteins

The general strategy for directing the secretion of a protein of interest in gram-positive bacteria is to construct and express a fusion gene that encodes a chimeric "preprotein," i.e., a secretory precursor consisting of an amino terminal signal peptide from one gene followed by the target (heterologous) protein sequence from a different gene. In general, the following two assumptions are made. (1) The placement of a signal peptide at the amino terminus is sufficient to direct the translocation of the target protein, provided that (2) the target proteins are normally secreted in their native hosts.

A number of examples of expression and directed secretion of heterologous proteins in gram-positive bacteria have been reported in the literature, some of which will be discussed in this chapter. In the cases documented so far, the strategy outlined above was employed. There are several variations in the techniques used for the construction of fusion genes. These include (1) precise fusions between the bacterial signal peptide sequence and the target protein sequence, (2) fusions that contain a

[8] S. D. Power, R. M. Adams, and J. A. Wells, *Proc. Natl. Acad. Sci. U.S.A.* **83,** 3096 (1986).

[9] H. C. Wu, J. S. Lai, S. Hayashi, and C. Z. Giam, *Biophys. J.* **37,** 308 (1982).

[10] J. B. K. Nielsen and J. O. Lampen, *J. Biol. Chem.* **257,** 4490 (1982).

[11] S. Hayashi, S.-Y. Chang, S. Chang, and H. C. Wu, *J. Biol. Chem.* **259,** 10448 (1984).

TABLE I
SIGNAL PEPTIDE SEQUENCES FROM GRAM-POSITIVE BACTERIA

Organism/gene/(strain)	Sequence[a]	References[b]
	1 (pos. 1) ... 10 ... 20 ... 30 ... 40 (scale)	
Bacillus amyloliquefaciens		
α-Amylase	MIQKRKRTVSFRLVLMCTLLFVSLPITKTSA/VNGTLMQYFE	T1, T2
Alkaline protease	MRGKKVWISLLFALALIFTMAFGSTSSAQAAGKS....(70)....HAY/AQSV	T3, T4
Neutral protease (P)	MGLGKKLSVAVAASFMSLTISLPGVQAAENPQLK....(168)...VEH/AATT	T4
Neutral protease (F)	-------S-------------------------....(168)...SGH/AATT	T5
B. cereus		
β-Lactamase (569/H)	MKNKRMLKIGICVGILGLSITSLEAFTGESLQVEAKEKTGQVK/HKNQATHKEF	T6, T7
β-Lactamase (5/B)	----K-----M-------------VT---CA---------------/----------	T8
B. licheniformis		
β-Lactamase (749/C)	MKLWFSTLKLKKAAAVLLFSCVALAG*CANNQTNA/SQPAEKNEKT	T9, T10
β-Lactamase (Ser27)	-------------------------LAGSA/NNQ---/-----------	T11
α-Amylase (FD02)	MKQQKRLYARLLTLLFALIFLLPHSAAAA/ANLNGTLMQY	T12
B. stearothermophilus		
α-Amylase	MLTFHRIIRKGWMFLLAFLLTASLFCPTGQHAKA/AAPFNGTMMQ	T13, T14
Neutral protease	MNKRAMLGAIGLAFGLLAAPIGASAKGESIVWNE....(193)...GQP/VAGA	T15
B. subtilus		
α-Amylase (168)	MFAKRFKTSLLPLFAGFLLLFHLVLAGPAAA∫SA∫ETANKSNE/LTAPSIKSGT	T16
α-Amylase (NA64)	----------------------Y------------------------------	T17
β-Glucanase	MPYLKRVLLLLVTGLFMSLFAVTATA∫SA∫KTGGSFFDPF	T18
Levansucrase	MNIKKFAKQATV(I)LTFTTALLAGGATQAFA/KETNQKPYKE	T19, T20
Alkaline protease	MRSKKLWISLLFALTLIFTMAFSNMSAQAAGKSS....(67)....HEY/AQSV	T21, T22
Neutral protease	MGLGKKLSVRVAASFMSLSISLPGVQAAEGHQLK....(184)...VQH/AAAT	T23
Corynebacterium diphtheriae		
Diphtheria toxin	MVSRKLFASILIGALLGIGAPPSAHA/GADDVVDSSK	T24
Staphylococcus aureus		
Protein A (8325)	MKKKNIYSIRKLGVGIASVTLGTLLISGGVTPAANA/AQHDEAQQNA	T25
Staphylokinase	MLKRSLLFLTVLLLLFSFSSITNEVSA/SSSFDKGKYK	T26
β-Lactamase	MKKLIFLIVIALVLSA*CASNSSHA/KELNDLEKKY	T27
Nuclease	MTEYLLSAGICMAIVSILLIGMAISNVS	
	KGQYAKRFFFFATSCLVLTLVVVSSLSSSANA/SQTDNGVNRS	T28

Organism / protein	Sequence	Ref.
Streptomyces plicatus Endo H	M F T P V R R R V R T A A L A L S A A A A L V L G S T A A S G A S A T P S P A P A P / A P A P V K Q G P T	T29
S. antibioticus ORF438	M P E L T R R R A L G A A A V V A A G V P L V A L P A A R A / D D R G H H T P E V	T30
S. coelicolor Agarase	M V N R R D L I K W S A V A L G A G A G L A G P A P A A H A / A D L E W E Q Y P V	T31
S. hygroscopicus α-Amylase	M Q Q R S R V L G G T L A G I V A A A A A T V A P W P S Q A / T P P G Q K T V T A	T32
S. limosus α-Amylase	M A R R L A T A S L A V L A A A A T A L T A P T P A A A / A P P G A K D V T A	T33
S. lividans β-Galactosidase	M P H S P V S P A E S P A P Q P G R R R P V V S R R R L L E G G A A V L G A L A L S A S P L T A Q A A V R R A A / A D E P P E W N D F	T34
S. avidinii Streptavidin	M R K I V V A A I A V S L T T V S I T A S A S A / D P S K D S K A Q V	T35

[a] The position of most of the initiation methionine in these proteins was deduced from DNA sequence data without verification by protein sequence determination. Abbreviations used are: A, Ala; C, Cys; D, Asp; E, Glu; F, Phe; G, Gly; H, His; I, Ile; K, Lys; L, Leu; M, Met; N, Asn; P, Pro; Q, Gln; R, Arg; S, Ser; T, Thr; V, Val; W, Trp; Y, Tyr. The amino termini of the mature proteins are marked by the "/" sign; the predicted signal peptidase cleavage sites are presented by the authors are indicated by "ſ." Processing sites for lipoproteins are indicated by "*." The numbers shown in parentheses indicate the length of the sequences in the "prepro" region that are not shown. Sequences identical to the ones at the same positions shown in the previous line are indicated by "---."

[b] Key to references: T1, K. Takkinen, R. F. Pettersson, N. Kalkkinen, I. Palva, H. Soderlund, and L. Kaariainen, *J. Biol. Chem.* **258,** 1007 (1983); T2, I. Palva, R. F. Pettersson, N. Kalkkinen, P. Lehtovaara, M. Sarvas, H. Soderlund, K. Takkinen, and L. Kaariainen, *Gene* **15,** 43 (1981); T3, J. A. Wells, E. Ferrari, D. Henner, D. A. Estell, and E. Y. Chen, *Nucleic Acids Res.* **11,** 7911 (1983); T4, N. Vasantha, L. D. Thompson, C. Rhodes, C. Banner, J. Nagle, and D. Filpula, *J. Bacteriol.* **159,** 811 (1984); T5, H. Shimada, M. Honjo, I. Mita, A. Nakayama, A. Akaoka, K. Manaba, and Y. Furutani, *J. Biotechnol.* **2,** 75 (1985); T6, A. Sloma and M. Gross, *Nucleic Acids Res.* **11,** 4997 (1983); T7, P. S. F. Mezes, Y. Q. Yang, M. Hussain, and J. O. Lampen, *FEBS Lett.* **161,** 195 (1983); T8, W. Wang, P. S. F. Mezes, Y. Q. Yang, R. W. Blacher, and J. O. Lampen, *J. Bacteriol.* **163,** 487 (1985); T9, K. Neugebauer, R. Sprengel, and H. Schaller, *Nucleic Acids Res.* **9,** 2577 (1981); T10, J. Kroyer and S. Chang, *Gene* **15,** 343 (1981); T11, S. Hayashi, S.-Y. Chang, S. Chang, and H. C. Wu, *J. Biol. Chem.* **259,** 10448 (1984); T12, M. A. Stephens, S. A. Ortlepp, J. F. Ollington, and D. J. McConnell, *J. Bacteriol.* **158,** 369 (1984); T13, H. Ihara, T. Sasaki, A. Tsuboi, H. Yamagata, N. Tsukagoshi, and S. Udaka, *J. Biochem.* **98,** 95 (1985); T14, R. Nakajima, T. Imanaka, and S. Aiba, *J. Bacteriol.* **163,** 404 (1985); T15, M. Takagi, T. Imanaka, and S. Aiba, *J. Bacteriol.* **163,** 824 (1985); T16, M. Yang, M. Galizzi, and D. Henner, *Nucleic Acids Res.* **11,** 237 (1983); T17, H. Yamazaki, K. Ohmura, A. Nakayama, Y. Takeichi, K. Otozai, M. Yamasaki, G. Tamura, and K. Yamane, *J. Bacteriol.* **156,** 327 (1983); T18, N. Murphy, D. J. McDonnell, and B. A. Cantwell, *Nucleic Acids Res.* **12,** 5355 (1984); T19, A. Fouet, M. Arnaud, A. Klier, and G. Rapoport, *Biochem. Biophys. Res. Commun.* **119,** 795 (1984); T20, M. Steinmetz, D. Le Coq, S. Aymerich, G. Gonzy-Treboul, and P. Gay, *Mol. Gen. Genet.* **200,** 220 (1985); T21, M. L. Stahl and E. Ferrari, *J. Bacteriol.* **158,** 411 (1984); T22, S.-L. Wong, C. W. Price, D. S. Goldfarb, and R. Doi, *Proc. Natl. Acad. Sci. U.S.A.* **81,** 1184 (1984); T23, M. Y. Yang, E. Ferrari, and D. Henner, *J. Bacteriol.* **160,** 15 (1984); T24, L. Greenfield, M. J. Bjorn, G. Horn, D. Fong, G. A. Buck, R. J. Collier, and D. A. Kaplan, *Proc. Natl. Acad. Sci. U.S.A.* **80,** 6853 (1983); T25, M. Uhlen, B. Guss, B. Nilsson, S. Gatenbeck, L. Philipson, and M. Lindberg, *J. Biol. Chem.* **259,** 1695 (1984); T26, T. Sako and N. Tsuchiba, *Nucleic Acids Res.* **11,** 7679 (1983); T27, J. R. McLaughlin, C. L. Murray, and J. C. Rabinowitz, *J. Biol. Chem.* **256,** 11283 (1981); T28, D. Shortle, *Gene* **22,** 181 (1983); T29, P. W. Robbins, R. B. Trimble, D. F. Wirth, C. Hering, F. Maley, G. F. Maley, R. Das, B. W. Gibson, N. Royal, and K. Biemann, *J. Biol. Chem.* **259,** 7577 (1984); T30, V. Bernan, D. Filpula, W. Herber, M. Bibb, and E. Katz, *Gene* **37,** 101 (1985); S.-Y. Chang and S. Chang, unpublished data; T31, M. J. Buttner, I. Fearnley, and M. Bibb, personal communication; T32, S. Hoshiko, O. Makabe, C. Nojiri, K. Katsumata, E. Satoh, and K. Nagaoka, *J. Bacteriol.* **169,** 1029 (1987); T33, M. J. Virolle, C. Long, S.-Y. Chang, S. Chang, and M. Bibb unpublished results; T34, T. Eckhardt, personal communication; T35, C. E. Argarana, I. D. Kuntz, S. Birken, R. Axel, and C. R. Cantor, *Nucleic Acids Res.* **14,** 1871 (1986).

short extraneous sequence, derived either from one or both of the genes, and/or derived from DNA linker sequence used for the fusion construction, and (3) fusions that link the target protein to the carboxy terminal region of a bacterial preprotein.

Most of these variations of fusion techniques have been applied to target proteins of both prokaryotic and eukaryotic origins. Perhaps the most extensively studied secretion vector system is the one derived from the promoter and the signal peptide coding region of the *Bacillus amyloliquefaciens* α-amylase gene. The cloned gene in *B. subtilis* directs high-level synthesis of secreted α-amylase (ca. 3–5 g/liter), which makes it an attractive vector system.[12,13] The *E. coli* TEM β-lactamase fused to the α-amylase signal peptide was expressed and efficiently processed; at the maximum level, it reached slightly more than 10 mg/liter.[14] Proteolysis, instead of poor transcription of the fusion gene, was shown to be the major cause for this lower than expected yield. The similar system has also been used to express human α-interferon[15] and Semliki Forest virus glycoproteins E1 and E2,[13,16] in *B. subtilis*; both of these proteins were expressed to about 1 mg/liter level. The interferon molecules are mostly secreted, as determined by activity assay, whereas the viral proteins remain cell associated.

Other secretion systems have also been used to express heterologous genes. The promoter and signal peptide coding sequence of the *B. subtilis* α-amylase was fused to the coding sequence for the mature protein of *E. coli* β-lactamase, and, under improved laboratory conditions, up to 100 mg/liter of β-lactamase was secreted.[17,18] The coding sequence for the mature protein of *B. subtilis* α-amylase was fused to that of the "prepro" sequence of the neutral protease,[19] and the alkaline phosphatase coding

[12] I. Ulmanen, K. Lundstrom, P. Lehtovaara, M. Sarvas, M. Ruohonen, and I. Palva, *J. Bacteriol.* **162,** 176 (1985).

[13] M. Sibakov, P. Lehtovaara, R. Pettersson, K. Lundstrom, N. Kalkkinen, I. Ulmanen, K. Takkinen, L. Kaariainen, I. Palva, and M. Sarvas, *in* "Genetics and Biotechnology of Bacilli" (J. A. Hoch and A. T. Ganesen, eds.), p. 153. Academic Press, New York, 1984.

[14] I. Palva, M. Sarvas, P. Lehtovaara, M. Sibakov, and L. Kaariainen, *Proc. Natl. Acad. Sci. U.S.A.* **79,** 5582 (1982).

[15] I. Palva, P. Lehtovaara, L. Kaariainen, M. Sibakov, K. Cantell, C. H. Schein, K. Kashiwagi, and C. Weissmann, *Gene* **22,** 229 (1983).

[16] K. Lundstrom, I. Palva, L. Kaariainen, H. Garoff, M. Sarvas, and R. F. Pettersson, *Virus Res.* **2,** 69 (1985).

[17] K. Ohmura, T. Shiroza, K. Nakamura, A. Nakayama, K. Yamane, K. Yoda, M. Yamasaki, and G. Tamura, *J. Biochem.* **95,** 87 (1984).

[18] K. Nakamura, T. Furusato, T. Shirosa, and K. Yamane, *Biochem. Biophys. Res. Commun.* **128,** 601 (1985).

[19] M. Honjo, A. Akaoka, A. Nakayama, H. Shimada, and Y. Furutani, *J. Biotechnol.* **3,** 73 (1986).

sequence of *E. coli* fused to the signal peptide of protein A from *Staphylococcus aureus*[20,21]; these target proteins were synthesized and excreted by the *B. subtilis* and *S. aureus* hosts, respectively.

A number of other eukaryotic genes have also been expressed in *B. subtilis* and *S. aureus*. Excretion of some of the target proteins into the medium was achieved. The signal peptide of the *B. licheniformis* penicillinase (*PenP*) gene was used to direct the expression of human β-interferon, resulting in most of the interferon molecules being cell bound.[22] Similar results were also obtained with human growth hormone. However, when the *PenP* signal peptide containing the $Cys_{27} \rightarrow Ser_{27}$ substitution (at position 27 from the initiation Met codon) was used for the fusion construction, the growth hormone molecules were efficiently excreted.[23] Most likely, the Cys_{27} residue, when present at the amino terminus of these eukaryotic proteins, is modified to form prelipoprotein that directs the processed proteins to the membrane. When Cys is converted to Ser at this position, cleavage by signal peptidase occurred after the downstream Ala residue, resulting in the excretion of the expressed heterologous proteins.

In addition, excretion of the following proteins has been reported in either *S. aureus* or *B. subtilis*: mouse β-interferon directed by the signal peptidase of *B. subtilis* α-amylase[24]; human β-interferon directed by the "prepro" sequence of the *B. amyloliquefaciens* neutral protease[19]; human insulin-like growth factor I (IGF-I) directed by fusion to the precursor of *S. aureus* protein A at the position 271 residues into the mature protein[25]; and human angiotensin I directed by fusion to the pre-α-amylase of *B. subtilis* 477 residues into the mature protein.[26] Yields of these heterologous proteins were variable depending on the systems used. The last three examples are interesting because considerable length of bacterial sequence was included in these chimeric genes. Fusion of either α-amylase or β-interferon to the "prepro" sequence of the neutral protease fragment resulted in synthesis and excretion of enzymatically or biologically active

[20] B. Nilsson, L. Abrahmsén, and M. Uhlén, *EMBO J.* **4,** 1075 (1985).

[21] L. Abrahmsén, T. Moks, B. Nilsson, U. Hellman, and M. Uhlén, *EMBO J.* **4,** 3901 (1986).

[22] S. Chang, D. Ho, O. Gray, S.-Y. Chang, and J. McLaughlin, *in* "Genetics of Industrial Microorganisms" (Y. Ikeda and T. Beppu, eds.), p. 227. Kodansha, Tokyo, 1983.

[23] S.-Y. Chang and S. Chang, unpublished observations.

[24] T. Shirosa, K. Nakazawa, N. Tashiro, K. Yamane, K. Yanagi, M. Yamasaki, G. Tamura, H. Saito, Y. Kawade, and T. Taniguchi, *Gene* **34,** 1 (1985).

[25] B. Nilsson, E. Holmgren, S. Josephson, S. Gatenbeck, L. Philipson, and M. Uhlen, *Nucleic Acids Res.* **13,** 1151 (1985).

[26] K. Yamane, T. Shiroza, T. Furusato, K. Nakamura, K. Nakazawa, K. Yanagi, M. Yamasaki, and G. Tamura, *in* "Molecular Biology of Microbial Differentiation" (J. A. Hoch and P. Setlow, eds.), p. 117. Am. Soc. Microbiol., Washington, D.C., 1985.

target proteins. The excreted α-amylase has a similar molecular weight as that of the mature protein produced from the native host, suggesting correct or nearly correct processing of the chimeric precursor by *B. subtilis*; no data on the processing of the interferon are available yet.

Angiotensin and IGF-I are small peptides of 10 and 70 residues in length, respectively. The corresponding synthetic genes were fused to the carboxy terminal region of the bacterial secretory protein coding sequences with Met and Asp-Pro sequences preceding these two respective heterologous sequences, which allowed subsequent chemical cleavages to remove the bacterial protein portions and to recover the products without extraneous sequences. The yield of the processed amylase–angiotensin fusion protein was difficult to assess from the published data, but the yield of protein A–IGF-I fusion protein is comparable to that of the same protein A fragment. Furthermore, the authors showed that in a protein A hyperproducer strain, the yield of the hybrid protein was increased by 20-fold, to 100 mg/liter. Fusion proteins that contain a portion of protein A sequence also allow efficient purification with IgG affinity column chromatography.[20] It could be a very useful system provided that a suitable method for the subsequent chemical/enzymatical cleavage to separate the target protein from the protein A moiety is available.

From the relatively few examples available at present, it is clear that to successfully direct target proteins for secretion in gram-positive bacteria, proteolysis of the target protein is a major problem to overcome. In several cases, the addition of protease inhibitors to the medium or the use of protease-deficient host strains has improved the stability of the target proteins. Recently, *B. subtilis* strains deficient in the two major extracellular proteases has been constructed.[27,28] They would be useful host strains for expression work, and they could also be used to isolate additional mutations affecting other extracellular proteases.

In setting strategies for the secretion of heterologous proteins in gram-positive bacteria, we should consider additional factors that may play certain roles in secretion. Of particular importance is the sequence in the mature (target) protein and its role in secretion/processing. For example, we converted the first codon of the mature *E. coli* alkaline phosphatase from Arg to Pro. Strains carrying this point mutation accumulated normal levels of alkaline phosphatase upon induction, but a significant fraction of the precursor protein remained unprocessed, and it was not released from the periplasm by osmotic shock.[29] Thus a single amino acid replacement

[27] M. L. Stahl and E. Ferrari, *J. Bacteriol.* **158,** 411 (1984).
[28] F. Kawamura and R. H. Doi, *J. Bacteriol.* **160,** 442 (1984).
[29] S.-Y. Chang and S. Chang, unpublished observations.

in the target protein could significantly alter the processing step. Because the structural requirements for signal peptidase cleavage are only poorly understood, it is not yet possible to predict whether the signal sequence of a given chimeric preprotein can be efficiently processed in bacteria. Another interesting example is the study on the transport of an internal protein A fragment by its own signal peptide.[21] When one region (region B) of protein A was fused directly behind its signal peptide, precursor molecules were made but not processed or transported; the same fragment, when fused behind the signal peptide with an additional amino terminal fragment of the mature protein, resulted in secretion and processing of the fusion protein. These examples illustrate that the assumptions listed at the beginning of this section are not always correct. While the presence of the signal peptide is necessary, it is not sufficient in all cases. The contribution of the sequence in the target protein to the secretion and processing of the preprotein is equally important.

Recently, the *E. coli* β-galactosidase protein (a cytoplasmic protein) has been fused to the signal peptide of the *Streptomyces* β-galactosidase (a secreted protein) protein. This fusion protein was secreted and processed in *Streptomyces*.[30] This result contrasts the observations made in *E. coli* when using β-galactoside fusions (see Refs. 4 and 5 for reviews), and it suggests the possible existence of a different secretion mechanism for the *Streptomyces* protein.

Final Remarks

Using several different chimeric signal peptide–target protein combinations, secretion of heterologous proteins in gram-positive bacteria has been achieved. Certain limiting factors have also been analyzed. In some cases, better host strains or growth conditions that improve the yield of the secreted target proteins have been identified. In general, poor expression and secretion is not necessarily caused by the low level transcription of the gene, but frequently by proteolysis of the target protein and/or, possibly, by the poor transport of the target protein. The last point is of particular importance in attempting to engineer for efficient secretion. If there are certain features in the mature protein sequence that are required for interacting with either the homologous signal peptide or with certain specific host components, these requirements must be satisfied when the target protein is fused to a heterologous signal peptide in a foreign host. Significant future progress in secretion of heterologous protein has to await better understanding of the molecular mechanisms governing pro-

[30] T. Eckhardt, personal communication.

tein secretion in bacteria. Although most of the recent advances in this field were made in *E. coli,* the fundamental underlying principles are likely to be very similar in different species. Many of the gram-positive genes encoding secreted proteins are also expressed in *E. coli*; various *E. coli* strains can be used as the surrogate host for studying the structural requirements of gram-positive secretory proteins. Several genetic loci in *B. subtilis* that affect protein secretion are also known[31,32]; these can be incorporated into the host genetic background to further improve the efficiency of secretion.

The majority of the secretion studies in gram-positive bacteria have been conducted using *B. subtilis*. Many other species, which are morphologically distinct and occupy different ecological niches, can also produce abundant secreted proteins. Among them, streptomycetes constitute the most diverse group of gram-positive bacteria. A number of recently characterized signal peptide sequences from *Streptomyces* are included in Table I. Comparative studies on the secretion of proteins in this group of organisms could provide potentially interesting data and open up new ways to engineer secretory proteins of commercial importance.

Acknowledgment

I am grateful to many of my colleagues for communicating their results prior to publication.

[31] K. Hitotsuyanagi, K. Yamane, and B. Maruo, *Agric. Biol. Chem.* **43,** 2343 (1979).
[32] N. Tomioka, M. Honjo, K. Funahashi, K. Manabe, A. Akaoka, I. Mita, and Y. Furutani, *J. Biotechnol.* **3,** 85 (1985).

[33] Expression and Secretion Vectors for Yeast

By GRANT A. BITTER, KEVIN M. EGAN, RAYMOND A. KOSKI, MATTHEW O. JONES, STEVEN G. ELLIOTT, and JAMES C. GIFFIN

It is now common practice to express heterologous genes in the yeast *Saccharomyces cerevisiae*. This is in large part due to the utility of this organism as a host for the production of commercially relevant proteins.

Vectors designed for efficient expression of heterologous genes have been developed which employ promoter elements derived from the alcohol dehydrogenase,[1] phosphoglycerate kinase,[2] acid phosphatase,[3] glyceraldehyde-3-phosphate dehydrogenase,[4] galactokinase,[5] and mating factor-α[6] genes. These promoters, in general, result in efficient transcription of the heterologous gene. However, the level of accumulation of the heterologous protein varies widely depending on the foreign gene expressed. This is presumably due to different translational efficiencies determined by mRNA secondary structure, untranslated leader sequences, codon utilization, or protein stability. Considerable basic research is being conducted by expressing certain genes in yeast where high-level expression of the given protein is not required or attempted. These studies employ various other promoters and expression strategies. This chapter, however, will focus on methods used to efficiently express and secrete biologically active proteins from *Saccharomyces cerevisiae*.

Standard recombinant DNA methods will not be discussed in this chapter since they have been reviewed in detail elsewhere.[7-9] Methods of introducing recombinant DNA into yeast cells have also been described in detail.[10,11] This chapter will describe yeast expression vectors utilizing episomal vectors. The expression cassettes from these vectors can also be integrated into the yeast chromosome where they will be present at controlled copy number and exhibit a high degree of mitotic stability. Methods of integrating yeast DNA into the chromosome, which will not be discussed in this chapter, have been described in detail.[12,13] Finally, there

[1] R. A. Hitzeman, F. E. Hagie, H. L. Levine, D. V. Goeddel, G. Amerer, and B. D. Hall, *Nature (London)* **293,** 717 (1981).

[2] M. R. Tuite, M. J. Dobson, N. A. Roberts, R. M. King, D. C. Burke, S. M. Kingsman, and A. J. Kingsman, *EMBO J.* **1,** 603 (1982).

[3] A. Miyanohora, A. Toh-E, C. Nozaki, F. Hamada, N. Ohtomo, and K. Matsubara, *Proc. Natl. Acad. Sci. U.S.A.* **80,** 1 (1983).

[4] G. A. Bitter and K. M. Egan, *Gene* **32,** 263 (1984).

[5] C. G. Goff, D. T. Moir, T. Kohro, T. Gravins, R. A. Smith, E. Yamasaki, and A. Taunton-Rigby, *Gene* **27,** 35 (1984).

[6] G. A. Bitter, K. K. Chen, A. R. Banks, and P.-H. Lai, *Proc. Natl. Acad. Sci. U.S.A.* **81,** 5330 (1984).

[7] R. Wu, L. Grossman, and K. Moldave (eds.), this series, Vol. 100.

[8] R. Wu, L. Grossman, and K. Moldave (eds.), this series, Vol. 101.

[9] T. Maniatis, E. F. Fritsch, and J. Sambrook, *in* "Molecular Cloning: A Laboratory Manual." Cold Spring Harbor Lab., Cold Spring Harbor, New York, 1982.

[10] A. Hinnen, J. B. Hicks, and G. R. Fink, *Proc. Natl. Acad. Sci. U.S.A.* **75,** 1929 (1978).

[11] H. Ito, Y. Fukuda, and A. Kimara, *J. Bacteriol.* **153,** 163 (1983).

[12] R. J. Rothstein, this series, Vol. 101, p. 202.

[13] F. Winston, F. Chumley, and G. R. Fink, this series, Vol. 101, p. 211.

are several other recent reviews of yeast cloning and secretion methods which are directly relevant to this chapter.[14–19]

Extrachromosomal Replication Vectors

Selectable Markers

The first yeast genes to be molecularly cloned were those encoding biosynthetic enzymes and these genes are now routinely utilized as selectable markers on recombinant yeast plasmids. The cloned *TRP1, LEU2, HIS3,* and *URA3* genes are commonly used as markers and numerous auxotrophic strains exist with mutations in the appropriate chromosomal gene. Medium lacking tryptophan, leucine, histidine, or uracil is used to select cells containing the plasmid. In addition, several dominant selectable markers for yeast have been developed. These include selection for aminoglycoside G418 resistance[20] [encoded by the *Escherichia coli Tn*601(903) aminoglycoside phosphotransferase gene], copper resistance[21] (encoded by the yeast *CUP1* gene), and methotrexate resistance[22,23] (*E. coli* plasmid R388 or mouse *DHFR* gene). Use of dominant selectable markers greatly increases the range of host strains which may be used with a particular plasmid, including those strains which lack any auxotrophic markers. Additionally, increasing the selective pressure may allow establishment of strains with higher plasmid copy numbers.

DNA Replicators

A number of DNA sequence elements have been isolated which confer the capability of autonomous replication of colinear DNA in yeast. These elements have been termed ARS (for autonomously replicating sequence) and those isolated from yeast DNA presumably are chromosomal origins of replication.[24] ARS-containing vectors replicate once per cell cycle[25] but

[14] A. Hinnen and B. Meyhack, *Curr. Top. Microbiol. Immunol.* **96,** 101 (1982).
[15] C. P. Hollenberg, *Curr. Top. Microbiol. Immunol.* **96,** 119 (1982).
[16] N. Gunge, *Annu. Rev. Microbiol.* **37,** 253 (1983).
[17] K. Struhl, *Nature (London)* **305,** 391 (1983).
[18] R. A. Hitzeman, C. Y. Chen, F. E. Hagie, J. M. Lugovoy, and A. Singh, "Recombinant DNA Products: Insulin, Interferon, and Growth Hormone," p. 47. CRC Press, Boca Raton, Florida, 1984.
[19] G. A. Bitter, *Microbiology,* 330–334 (1986).
[20] A. Jimenez and J. Davies, *Nature (London)* **287,** 869 (1980).
[21] S. Fogel and J. W. Welch, *Proc. Natl. Acad. Sci. U.S.A.* **79,** 5342 (1982).
[22] A. Miyajima, I. Miyajama, K.-I. Arai, and N. Arai, *Mol. Cell. Biol.* **4,** 407 (1984).
[23] J. Zhu, R. Contreras, D. Gheysen, J. Ernst, and W. Fiers, *Bio/Technology* **3,** 451 (1985).
[24] D. H. Williamson, *Yeast* **1,** 1 (1985).
[25] W. L. Fangman, R. H. Hice, and E. Chlebowicz-Sledziewska, *Cell* **32,** 831 (1983).

accumulate to multiple copies per cell due to segregational bias. They are mitotically unstable, exhibiting a segregation frequency of 10–30% per generation under nonselective conditions.[26] In a typical culture grown under selective conditions, only 20–30% of the cells contain plasmid and cells lacking plasmid are capable of undergoing a limited number of additional cell divisions. Such behavior of plasmids is clearly suboptimal for efficient heterologous gene expression.

An alternative to ARS elements is the origin of DNA replication of the yeast 2μ plasmid. This 6318-bp plasmid[27] has no known function and is present at 50–100 copies/cell in most laboratory strains of *Saccharomyces cerevisiae*. The origin of DNA replication of the yeast 2μ plasmid has been localized to a 222-bp segment.[28] A DNA sequence, termed the *REP*3 or *STB* locus, has been localized to a 294-bp segment[29] and directs mitotic equipartitioning of the 2μ plasmid. Thus, a plasmid which includes the 2μ origin of replication and *REP*3 locus will be stably maintained at high copy numbers when the appropriate replication proteins are present in the cell. These may be supplied *in trans* by the endogenous yeast 2μ plasmid.

Figure 1 is a restriction endonuclease map of plasmid pYE,[6] which has been utilized for construction of both expression and secretion vectors. The plasmid includes the *E. coli* plasmid pBR322 with an intact origin of replication and β-lactamase gene (expression of which confers resistance to ampicillin). These features allow propagation in *E. coli* for further recombinant DNA manipulations. The yeast *TRP1* gene allows selection of yeast cells containing the plasmid. Tryptophan prototrophy is a convenient selection for yeast since the selective medium is easily prepared (0.67% yeast nitrogen base without amino acids, 2% dextrose, 0.5% casamino acids). Plasmid pYE has a unique restriction endonuclease recognition site for *Bam*HI located in the pBR322 portion of the plasmid. The 2μ DNA segment contains a *Cla*I restriction site. However, in this sequence context (GATCGATC) it is methylated in *dam*$^+$ strains of *E. coli* (such as HB101). The methylated sequence is not cleaved by *Cla*I. Thus, by propagating the shuttle vector in these hosts, only the *Cla*I site in pBR322 is subject to cleavage and may be used as a unique cloning site.

The entire yeast 2μ plasmid, linearized at the *Eco*RI site in the large unique region, is cloned in pYE. In this construction, the origin of replication, the *REP*3 locus, and the *FLP, REP1,* and *REP2* genes are all intact. The *FLP, REP1,* and *REP2* proteins, expressed from their endoge-

[26] A. W. Murray and J. W. Szostak, *Cell* **34,** 961 (1983).

[27] J. L. Hartley and J. E. Donelson, *Nature (London)* **286,** 860 (1980).

[28] J. R. Broach, Y.-Y. Li, J. Feldman, M. Jayaram, J. Abraham, K. A. Nasmyth, and J. B. Hicks, *Cold Spring Harbor Symp. Quant. Biol.* **47,** 1165 (1982).

[29] M. Jayaram, A. Sutton, and J. R. Broach, *Mol. Cell. Biol.* **5,** 2466 (1985).

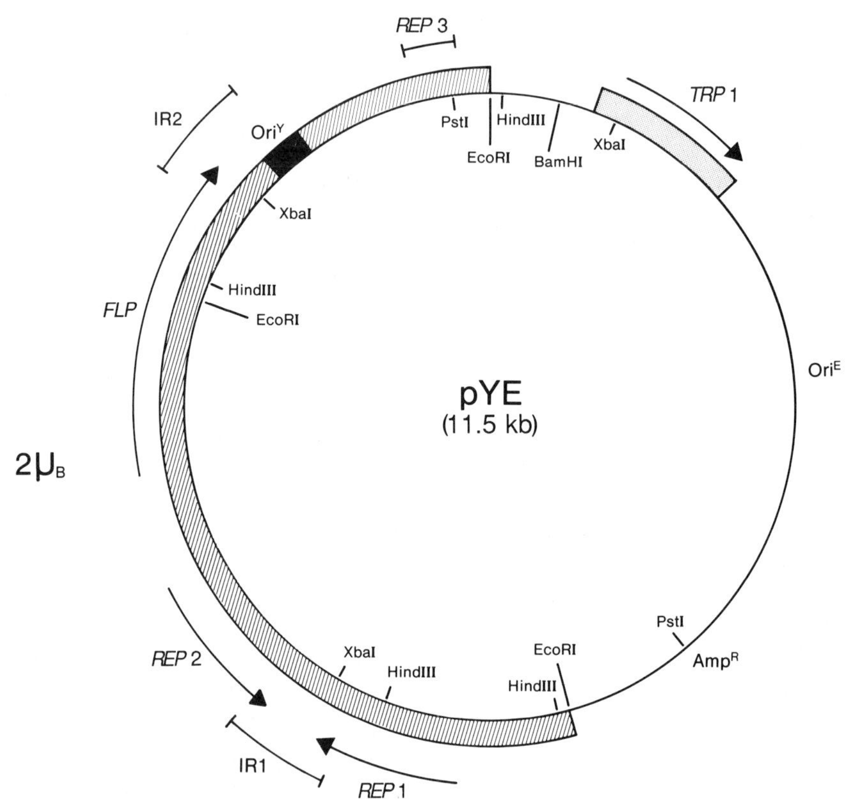

FIG. 1. *Saccharomyces cerevisiae–E. coli* shuttle vector pYE. DNA sequences derived from pBR322 are depicted as a single line. The yeast *TRP1* gene, cloned by blunt-end ligation into the *Sal*I site of pBR322, is depicted by the stippled segment. The yeast 2μ plasmid is indicated by the hatched segment. The yeast and *E. coli* origins of DNA replication are indicated. The location of yeast genes and loci are depicted and the direction of transcription of the yeast genes is indicated by arrows.

nous 2μ promoters on pYE, are involved in amplification of the plasmid to high copy numbers. This has been demonstrated to occur even in strains lacking the endogenous 2μ plasmid (*cir*° strains; see below). It is advantageous to utilize *cir*° hosts for several reasons. Recombination between the endogenous 2μ plasmid and the recombinant plasmid may restructure the expression vector. Because of the expression of the 2μ genes present on pYE, it is capable of self-amplification in the absence of endogenous 2μ plasmid. By propagating the vector in *cir*° strains, high copy numbers may be achieved without the potential complications due to recombination with the endogenous 2μ plasmid. Furthermore, by utilizing *cir*° hosts, the copy number of the recombinant expression vector may be increased since it represents the only 2μ plasmid in the cell.

The 2μ plasmid has two inverted repeats (IR1 and IR2; Fig. 1). Recombination between these two sites is catalyzed by the product of the *FLP* gene.[30] Since the *FLP* gene is expressed on the plasmid pYE in yeast, rapid recombination between IR1 and IR2 will result in two forms of plasmid. This phenomenon is illustrated in Fig. 2. Expression vector pGPD-1(HBs),[4] which includes the entire yeast 2μ plasmid cloned as in pYE, was propagated in a *cir*° yeast strain and then shuttled back into *E. coli*.[31] More than 600 *E. coli* clones were pooled (in order to be representative of the plasmid population present in yeast) and the plasmid amplified and purified. By digesting with a restriction enzyme which cleaves the 2μ DNA on each side of an inverted repeated (e.g., *Hin*dIII), two new restriction fragments are observed in the plasmid preparation recovered from yeast which were not present in the parent plasmid constructed and propagated in *E. coli* (Fig. 2, lanes 1 and 2). The new fragments are due to the second form of the 2μ plasmid generated by recombination in yeast. If individual *E. coli* clones are analyzed, only one form of the plasmid is observed (Fig. 2, lanes 3 and 4). The staining intensity of the restriction fragments in the plasmid population recovered from yeast (Fig. 2, lane 2) indicates an equivalent abundance of each form of the plasmid, which is corroborated by analysis of a large number of individual *E. coli* clones (data not shown).

The above data demonstrate that plasmids such as pYE, which self-amplify in *cir*° yeast hosts, recombine with each other to generate a steady state population of plasmids with two forms of the 2μ insert. We have observed no other types of restructuring of plasmids using this system. However, this theoretical possibility should be borne in mind when evaluating heterologous gene expression systems.

Centromere Plasmids

High-level expression of heterologous genes is generally facilitated by increasing the copy number of the expression vectors. The previous section described vectors capable of stable maintenance at high copy numbers. There may be instances, however, where it is preferable to maintain expression vectors at low copy number. These include cases where expression of the heterologous gene is toxic to yeast and plasmid stability is low. It has been demonstrated that addition of centromere (CEN) sequences to unstable plasmids can reduce segregational loss to 1–3% per

[30] J. R. Broach, V. R. Guarasio, and M. Jayaram, *Cell* **29,** 227 (1982).

[31] G. Miozzari, P. Neiderberger, and R. Huffer, *J. Bacteriol.* **134,** 48 (1978).

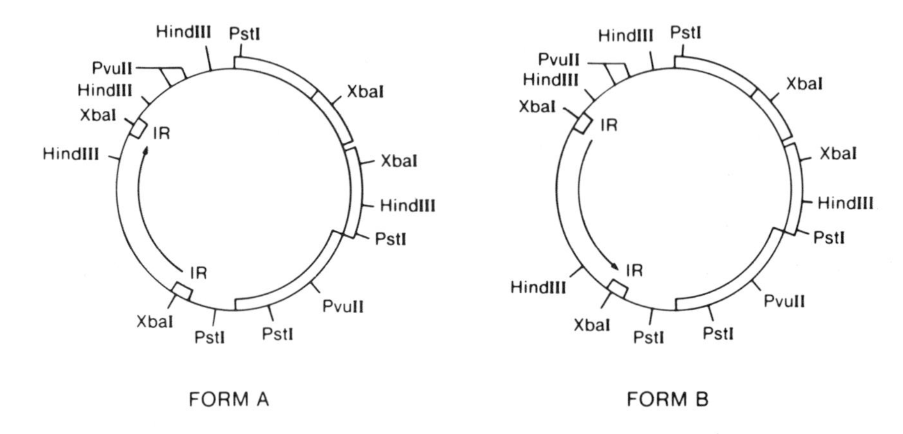

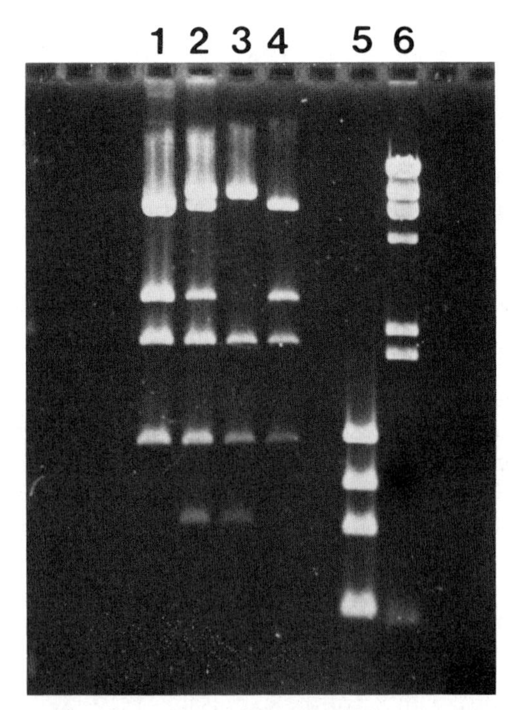

FIG. 2. Interconversion of two forms of 2μ plasmid in yeast. The expression vector pGPD-1(HBs) has been described elsewhere.[4] This plasmid was constructed and propagated as form B in *E. coli* and was expected to interconvert to form A when introduced into yeast. Restriction endonuclease maps of both forms of the plasmid are presented. The plasmid was introduced into *Saccharomyces cerevisiae* RH218 (*Mata gal2 trp*1 *cir*^{o31}) by transformation and shuttled back into *E. coli*. The plamsid was purified from a pool of more than 600 *E. coli* clones as well as from 2 individual *E. coli* clones. The purified DNA samples were restricted with *Hin*dIII and analyzed by agarose gel electrophoresis. Lane 1, pGPD-1(HBs) form B constructed and propagated in *E. coli*. Lane 2, plasmid population recovered from yeast (pool of *E. coli* clones). Lane 3, pGPD-1(HBs) form A recovered from a single *E. coli* clone. Lane 4, pGPD-1(HBs) form B recovered from a single *E. coil* clone. Lane 5, phage ϕX 174 DNA restricted with *Hae*III. Lane 6, phage λ DNA restricted with *Hin*dIII.

generation under nonselective conditions.[32] While the copy number of CEN-containing plasmids (even those including 2μ replicons) is controlled at one to two per haploid cell,[33] the increased stability may be an advantage in particular situations (i.e., in cases where there is strong selection against maintenance of the plasmid). Additionally, it appears that heterologous protein secretion may be more efficient in certain cases when the expression system is stably maintained at low copy numbers[34] (S. Elliott, unpublished results).

Measurement of Plasmid Stability. Plasmid stability of a culture is determined by plating appropriate culture dilutions on nonselective medium, then replica plating the resulting colonies onto medium which is selective for cells containing the plasmid. Plasmid stability is expressed as the percentage of cells containing plasmid, as indicated by the percentage of total colonies on the nonselective plate that grow on the selective plate.

Recombinant yeast plasmids containing the entire 2μ plasmid are lost at a rate of 0.08–4% per generation during nonselective growth.[35] In selective medium, recombinant plasmids reach steady state plasmid stability levels. The steady state level is determined by a number of factors including the plasmid selectable marker, replication efficiency, segregation efficiency, and possible deleterious effects on the cell due to maintaining the plasmid (such as expression of a toxic heterologous protein). For example, we have determined that a recombinant 2μ plasmid which is lost at a rate of 0.3% per generation under nonselective conditions is maintained in 96% of the cells when the *TRP1* gene[36] is used as a selectable marker. In contrast, if a gene encoding human immune interferon (IFN-γ) is constitutively expressed from the 2μ plasmid, the rate of plasmid loss increases to 3% per generation under nonselective conditions and the steady state plasmid stability is 70% under *TRP1* selection (data not shown). Plasmid stability thus varies widely depending on the particular 2μ recombinant vector. As mentioned above, heterologous gene stability may be increased by incorporating CEN sequences into the plasmid, or by integrating the gene into the chromosome. Since these manipulations result in lower gene dosage, such an approach may not yield optimal heterologous gene expression levels.

Measurement of Plasmid Copy Number. It is widely assumed that recombinant plasmids incorporating the 2μ replication/amplification system are maintained at high copy numbers, and a rule of thumb has devel-

[32] L. Clarke and J. Carbon, *Nature* (*London*) **287,** 504 (1980).
[33] G. Tschumper and J. Carbon *Gene* **23,** 221 (1983).
[34] R. A. Smith, M. J. Duncan, and D. T. Moir, *Science* **229,** 1219 (1985).
[35] A. B. Futcher and B. S. Cox, *J. Bacteriol.* **157,** 283 (1984).
[36] G. Tschumper and J. Carbon, *Gene* **10,** 157 (1980).

oped in the literature such that 2μ-based vectors are assumed to be present at an average 20 copies per cell. In actual fact, there is a wide variability in average plasmid copy number per cell for different recombinant vectors. This fact was anticipated from the observed differences in plasmid stability (above). In characterizing a yeast expression system, therefore, it is imperative that plasmid copy number be measured.

High copy number plasmids may be visualized by ethidium bromide staining of restriction enzyme-digested whole-cell DNA which has been size fractionated by agarose gel electrophoresis. This method may be quantitative[37] by comparison of the plasmid band to the staining intensity of the rRNA genes (of defined copy number) but does not detect low copy number plasmids. An attractive alternative, therefore, is hybridization of radioactively labeled specific DNA probes to size-fractionated DNA. In order to accurately quantitate plasmid copy number, a hybridization probe is utilized which is complementary to both the plasmid and a native chromosomal yeast gene. Such DNA probes may derive from either the promoter, transcription terminator, or selectable marker of the expression vector (see below) since these sequences are all also represented in the yeast genome. Whole-cell DNA is isolated as described in Procedures, digested with an appropriate restriction enzyme, and fractionated by size in an agarose gel. The DNA is transferred to nitrocellulose and hybridized to the radioactive probe.[9] By utilizing a restriction enzyme that does not cleave within the DNA fragment used as probe, two of the size-fractionated restriction fragments will hybridize to the probe. Knowledge of the genomic and plasmid restriction map will allow identification of the fragment representing either the chromosomal gene or the recombinant plasmid. The chromosomal fragment in each sample is an internal standard for quantitation and generally represents one (haploid strains) or two (diploid strains) genes per cell. Comparison of the amount of probe hybridized to the plasmid fragment to that hybridized to the chromosomal fragment (of defined copy number) allows quantitation of the plasmid copy number per cell. It has been noted that the transfer/binding efficiency of DNA fragments in Southern blot analyses is size dependent. Therefore, if the sizes of the genomic and plasmid restriction fragments are very different, the copy number should be confirmed by utilizing a different restriction enzyme which generates genomic and plasmid fragments of similar size. Alternatively, a method which employs *in situ* hybridization of the separated DNA fragments in the gel may be employed.[37a] This "unblot" method is independent of transfer or binding efficiencies associated with the Southern blot procedure.

[37] J. R. Broach, this series, Vol. 101, p. 211.

[37a] M. Purrello and I. Balazs, *Anal. Biochem.* **128,** 393 (1983).

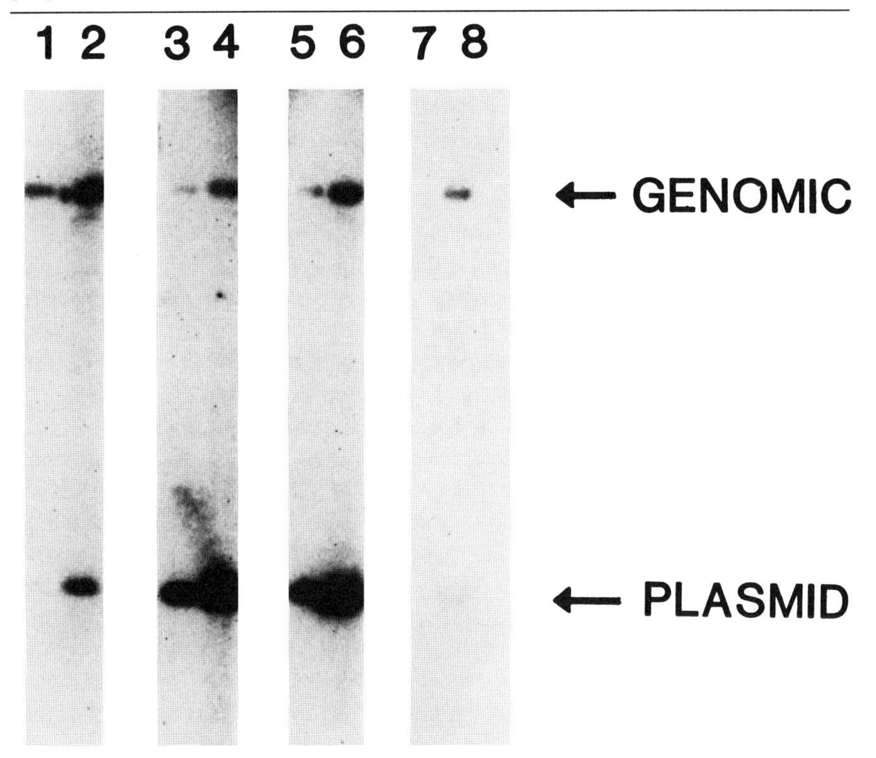

FIG. 3. Determination of plasmid copy numbers. *Saccharomyces cerevisiae* containing expression vectors pGPD(s)γ4 (lanes 1 and 2) or pGPD(G)γ4–9 (lanes 3, 4, 5, 6, 7, and 8) were cultured as described in the text and whole-cell DNA isolated as described in Procedures. The DNA was digested with *Bgl*II, fractionated by agarose gel electrophoresis, transferred to nitrocellulose, and hybridized to a nick-translated yeast *PGK* terminator (*Bgl*II–*Hin*dIII 380-bp segment[39]).

An example of this method of plasmid copy number determination is depicted in Fig. 3. Expression vector pGPD-2 has been described[4] (see below). Expression vector pGPD(s)-2 is identical except for the introduction of a *Sal*I site 240 bp 5′ to the TATA box of the GPD promoter.[38] Expression vector pGPD(G)-2 contains a DNA sequence from the yeast *GAL*1, 10 intergenic region cloned in the *Sal*I site of pGPD(s)-2 which renders the GPD promoter inactive in glucose and induced by galactose.[38] A gene encoding human immune interferon (IFN-γ) was cloned into each vector to generate pGPD(s)γ4 and pGPD(G)γ4-9.[38] Whole-cell DNA from different strains was digested with *Bgl*II, electrophoresed in a

[38] G. A. Bitter and K. M. Egan, manuscript in preparation (1987).

1% agarose gel, transferred to nitrocellulose, and hybridized to a ^{32}P-labeled DNA fragment which includes the 3′ coding region of the yeast *PGK* gene.[39] This DNA fragment will hybridize to both the chromosomal *PGK* gene and the plasmids since both vectors use this DNA sequence as a transcription terminator (Fig. 5, below). The 4-kbp fragment represents the genomic *PGK* gene while the 1.6-kbp *Bgl*II fragment is derived from the plasmid. (This experiment was repeated with *Eco*RI digests which yielded genomic and plasmid fragments of 5 and 6 kb, and corroborated the quantitation obtained with *Bgl*II digests.) For each sample of restricted DNA, two loads of DNA, differing by a factor of 10, were applied to the gel. This facilitates quantitation in cases where the extent of hybridization to the two fragments is significantly different.

The yeast host strain used in the experiment depicted in Fig. 3 is a diploid. Therefore, the genomic *PGK* fragment represents two genes per cell. It is evident that pGPD(s)γ4 is present at a copy number of less than one, consistent with the low plasmid stability of the strain (data not shown). In contrast, pGPD(G)γ4-9 grown in glucose (Fig. 3, lanes 3 and 4) is present at a copy number of 40–60 per cell (the plasmid fragment in the low DNA load has 2–3 times the hybridization of the genomic band in the 10× DNA load). The pGPD(G)γ4-9 strain was grown in galactose (conditions where the promoter is induced and IFN-γ expressed) for 3–4 generations and exhibited a plasmid copy number of 20–30 (Fig. 3, lanes 5 and 6). If the same strain was serially cultured for more than 50 generations in galactose (conditions in which IFN-γ is expressed), the plasmid copy number dropped to less than 1 per cell (Fig. 3, lanes 7 and 8).

The yeast host strain used for the experiment in Fig. 3 was *cir*°. Both expression vectors analyzed contain the entire 2μ plasmid cloned at the *Eco*RI site in the large unique region (as in plasmid pYE). The results demonstrate, therefore, that plasmid copy number may vary between 1 and 60 per cell depending on the vector, growth conditions, and heterologous gene insert (and its expression).

It should be noted that plasmid copy number measured in this way represents average copy number for the cell population. There is likely to be some variability in plasmid copy number per cell but direct methods for measuring this distribution in the population are not currently available. Since some percentage of the cells in the culture do not contain plasmid, however, a more accurate plasmid copy number is obtained by dividing the measured copy number by the fraction of cells containing plasmid (as measured by the plasmid stability test, above).

[39] R. A. Hitzeman, F. E. Hagie, J. S. Hayflick, C. Y. Chen, P. H. Seeburg, and R. Derynck, *Nucleic Acids Res.* **10,** 7791 (1982).

Direct Expression Vectors

In addition to an appropriate shuttle vector, three additional functional components are required for heterologous gene expression: (1) An efficient promoter element is required to support high-level transcription initiation. Thus far, only homologous yeast promoters have been shown to be functional for this purpose; (2) for optimal mRNA accumulation, DNA sequences which impart transcription termination and/or polyadenylation should be included. These sequences can be of either yeast origin or from heterologous organisms[18] and there does not appear to be a quantitative effect on expression level due to use of different yeast terminators[4]; (3) the gene encoding the heterologous protein must be appropriately positioned between the transcription promoter and terminator elements in the shuttle vector. The heterologous gene must include a translation termination codon as well as the ATG initiator codon. The mRNA sequence encoded by the heterologous gene determines the intrinsic translational efficiency. Additional factors, such as mRNA abundance and protein half-life, contribute to the steady state levels of the heterologous protein.

A generic yeast expression cassette is depicted in Fig. 4. Most yeast promoter elements have been isolated as DNA fragments truncated within the untranslated leader of the native yeast gene and include at least 300 bp of upstream DNA. The amount of 5′ DNA required for full promoter activity has not been precisely defined for all yeast promoter elements thus far studied. However, functional yeast promoters, as in higher eukaryotes, are considerably larger than those of prokaryotes. The isolated promoter elements include the yeast transcription start site but not the ATG translation initiation codon. Translation initiation in yeast, as in higher eukaryotes, generally starts at the first AUG in the mRNA. Since no essential consensus sequences have been identified in yeast untranslated leaders, there appears to be no analogy to the bacterial ribosome-binding site. Assembling a yeast expression cassette thus involves posi-

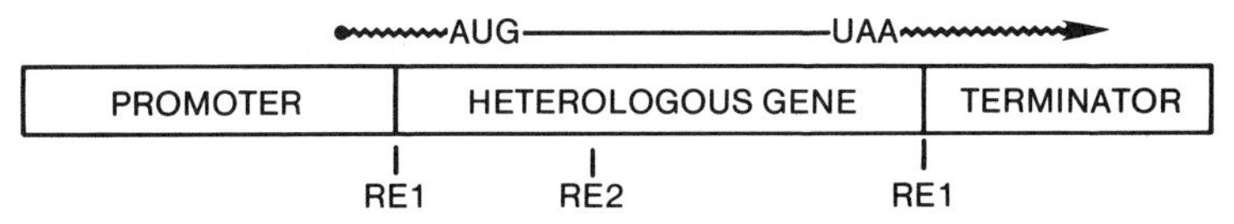

FIG. 4. Generic yeast expression cassette. The three functional components (promoter, heterologous gene, transcription terminator) are indicated. RE1 and RE2 represent restriction enzyme recognition sites in the expression vector. Several of the methods used to assemble expression cassettes (see text) result in loss of the restriction site (RE1) which separates the promoter from transcription terminator of the generalized expression vector (e.g., the *Bam*HI site of pGPD-1 and pGPD-2, Fig. 5).

tioning the promoter element, which includes the transcription start site, upstream of the heterologous gene in such a manner that the ATG translation start codon of the heterologous gene represents the first AUG in the mRNA. If other ATG triplets are present before the intended translation start, translation initiation at the first AUG in the mRNA will either represent a frame shift mutation with no heterologous protein synthesis or, in the case of an in-frame upstream ATG, may result in a heterologous protein fusion. The DNA sequence between the promoter and intended translation initiation codon, including sequences generated in the cloning manipulations, must therefore be carefully examined in constructing expression vectors.

Although there are no consensus sequences in yeast untranslated leader regions, genes which encode highly expressed yeast proteins are markedly A rich and G deficient in this region. Certain heterologous genes will encode untranslated leaders which are clearly suboptimal for translation in yeast. In such instances, it is advisable to resynthesize the 5′ end of the heterologous gene. Thus, using the example in Fig. 4, a DNA segment may be chemically synthesized[40] and used to replace the natural sequence between the leftmost RE1 and RE2. The synthetic segment should incorporate an optimized untranslated leader (e.g., the native sequence from a highly expressed yeast gene) and a sequence downstream of the ATG which encodes the desired amino acid sequence of the foreign protein. It may be preferable to use optimal yeast codons[41] (i.e., those preferentially utilized in highly expressed yeast genes) in this region. Employing such a strategy for expression of the hepatitis B virus surface antigen gene in yeast resulted in more than a 10-fold increase in protein expression level.[4] In designing such synthetic segments, mRNA secondary structure, which may decrease translation efficiency, should be minimized. Additionally, thought should be given to codon selection in the synthetic gene. By chemically synthesizing a complete gene encoding human immune interferon (IFN-γ) which incorporated an optimized untranslated leader and optimal yeast codons, IFN-γ was expressed in yeast as 10% of the total cell protein.[42]

A number of yeast genes contain introns and the splicing mechanism in yeast has been studied in detail. However, when genes containing heterologous introns are expressed in yeast, the splicing appears to be aberrant.[18] At this time, therefore, heterologous gene expression in yeast

[40] M. H. Caruthers, *in* "Chemical and Enzymatic Synthesis of Gene Fragments, a Lab Manual" (H. G. Gassen and A. Lang, eds.), pp. 71–79. Verlag Chemie, Weinheim, Federal Republic of Germany, 1982.

[41] J. L. Bennetzen and B. D. Hall, *J. Biol. Chem.* **257,** 3026 (1982).

[42] J. L. Fieschko, K. M. Egan, T. Ritch, R. A. Koski, M. D. Jones, and G. A. Bittner, *Biotechnol. Bioeng.* **29,** 1113 (1987).

is limited to cDNA, genomic clones which lack introns, or synthetic genes.

As examples of generalized yeast expression vectors, restriction endonuclease maps of two vectors employing the glyceraldehyde-3-phosphate dehydrogenase gene (*GPD*) promoter are depicted in Fig. 5. Both vectors incorporate the entire 2μ plasmid linearized at the *Eco*RI site in the large unique region (see above). The large size of yeast expression vectors limits the available unique restriction sites which may be engineered between the promoter and terminator elements. The expression vectors pGPD-1 and pGPD-2 (Fig. 5) have been constructed[4] with unique *Bam*HI sites between the *GPD* promoter and transcription terminator. In these constructions, cloning strategies were developed which prevented the generation of any additional inverted repeats which might lead to plasmid structural instability.

Both expression vectors employ the yeast *TRP*1 gene for selection. In pGPD-1, the *TRP*1 gene also supplies transcription termination/polyadenylation signals for RNA polymerase II molecules which have initiated transcription at the *GPD* promoter. The transcription terminator region from the yeast *PGK* gene is employed in pGPD-2. No difference in levels of heterologous gene expression were observed with the two transcription terminators.[4]

Assembly of Expression Cassettes

The paucity of unique restriction sites in yeast expression vectors, as well as constraints on sequences surrounding the translation initiation codon, do not allow for inclusion of polylinkers between the promoter and terminator. Since it is unlikely that the heterologous gene has the appro-

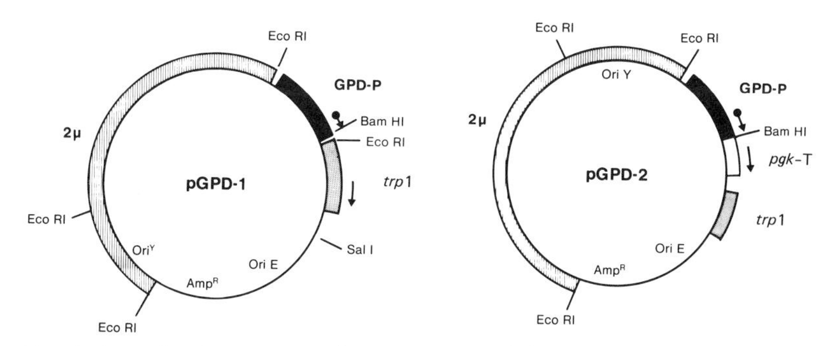

FIG. 5. Generalized yeast expression vectors. Restriction endonuclease maps of expression vectors pGPD-1 and pGPD-2 are presented. DNA derived from the yeast *TRP1* gene, 2μ plasmid, and pBR322 are presented as in Fig. 1. The GPD portable promoter is represented by the black segment and the *PGK* terminator by the open segment. Reproduced from Bitter and Egan[4] with permission of the publisher.

priate flanking restriction sites, the cloning of these genes into the expression vector employs one of three manipulations. Synthetic linkers may be added[9] to the heterologous gene such that the termini are complementary with those of the linearized expression vector (e.g., *Bam*HI cleaved pGPD-1 or pGPD-2). These manipulations will add DNA sequences to the untranslated leader region which must be evaluated (above) for potentially deleterious effects on translation. Alternatively, certain noncomplementary restriction ends may be rendered cohesive by *in vitro* manipulations. These methods have been described in detail elsewhere.[43] For example, a *Bam*HI-restricted vector can be ligated to a *Sal*I restriction fragment by partial end filling of each terminus. This cloning strategy has the advantage that the partially end-filled termini of the vector are no longer complementary. Thus, the frequency of clones containing the inserted gene is greatly increased.

The third method of expression cassette assembly involves blunt-end ligation. Nonhomologous 5′ overhang restriction ends may be end filled with Klenow fragment. This introduces additional DNA sequences in the untranslated leader which should be evaluated. Alternatively, blunt ends may be generated by removal of cohesive termini with either mungbean or S1 nuclease. While blunt-end ligation is the most generally applicable method, it suffers the disadvantage of low ligation frequency. Plasmids containing the heterologous gene insert often need to be screened for by colony hybridization. In order to facilitate screening for such clones, we have used synthetic oligonucleotide probes, the sequence of which spans the junction between the promoter and heterologous gene insert. Under the appropriate hybridization conditions, such probes will detect only *E. coli* colonies which contain the heterologous gene cloned in the correct orientation.

Analysis of Heterologous Gene Transcription

A first step in analyzing expression of the heterologous gene in yeast is analysis of its mRNA. By analyzing the transcript by Northern blots, it is possible to determine whether transcription is initiating and terminating near the expected sites in the promoter and terminator. Precise transcription start sites may be determined by S1 nuclease[44] or primer extension mapping.[45] In addition, mRNA abundance may be measured in a manner similar to plasmid copy number measurement (above). Total cell RNA is extracted, electrophoresed in agarose under denaturing conditions, and subjected to hybridization (with or without transfer to nitrocellulose) as described in Procedures. A hybridization probe is utilized which will de-

[43] M. C. Hung and P. C. Wensink, *Nucleic Acids Res.* **12,** 1863 (1984).
[44] A. J. Berk, and P. A. Sharp, *Cell* **12,** 721 (1977).
[45] S. L. McKnight, E. R. Gravis, R. Kingsbury, and R. Axel, *Cell* **25,** 385 (1981).

tect both the heterologous transcript and a native yeast transcript. For example, for expression vectors derived from pGPD-2 (Fig. 5), the *PGK* terminator DNA fragment will detect both the heterologous transcript and the yeast *PGK* transcript since both mRNAs are homologous to portions of this DNA. An example of such an analysis is presented in Fig. 6. The

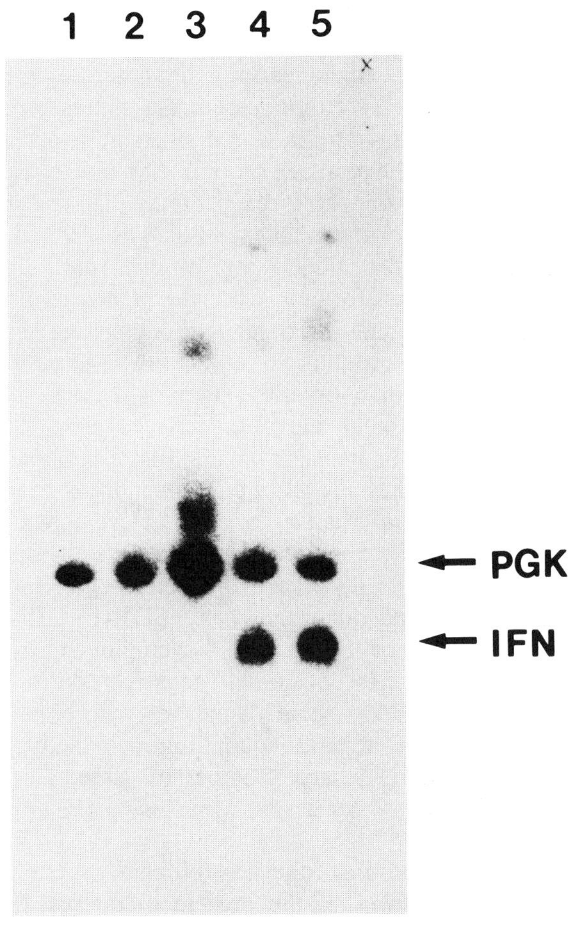

FIG. 6. Measurement of relative transcript abundance. Total RNA was extracted from various *Saccharomyces cerevisiae* strains as described in Procedures, electrophoresed in an agarose gel, and the gel dried. A *PGK* gene terminator fragment (legend to Fig. 3) was hybridized directly to the gel. Lane 1: nontransformed strain. Lane 2: cells containing expression vector pPG73 (M. O. Jones and R. A. Koski, unpublished results) containing the yeast *PGK* promoter and terminator but no heterologous gene insert. Lane 3: cells containing the intact yeast *PGK* gene on a multicopy plasmid. Lanes 4 and 5: cells containing plasmids expressing human immune interferon analogs from vectors employing the *PGK* gene promoter and terminator.

yeast *PGK* transcript serves as an internal standard in each lane for measuring heterologous transcript abundance. In contrast, however, to plasmid copy number determinations (above), the Northern analysis yields a relative transcript abundance. An absolute measurement of mRNA abundance is not obtained. Thus, this method allows determination of relative transcription efficiency of different expression vectors by comparison to a given native yeast mRNA.

Analysis of Heterologous Protein Expression Level. The level of heterologous protein expression will depend on mRNA abundance, translational efficiency (initiation and elongation), and heterologous protein half-life. In order to quantitate heterologous protein expression, a crude cell lysate should be analyzed such that the expression level may be determined as a percentage of the total cell protein. The simplest quantitation is obtained by SDS–PAGE of whole-cell proteins. Typically, 0.5 OD · ml (e.g., 0.5 ml of a culture at an OD_{600} of 1.0) of cells is resuspended in Laemmli sample buffer, boiled for 10 min, and electrophoresed as described.[46] Cells transformed with the same expression vector lacking the heterologous gene insert serve as a negative control. The expressed foreign protein is identified by its size, and by its absence in the negative control. Comparison of the staining intensity of the heterologous protein relative to total stainable material provides an estimate of expression level. This analysis is typically limited to heterologous proteins which are expressed at a level of 0.5% or more of the total cell protein. Confirmation of the identity of the heterologous protein may be obtained by immunoblot analysis[46a] using antisera specific for the heterologous protein. In cases where the heterologous protein expression level is too low for quantitation by SDS–PAGE of whole-cell lysates, sensitive RIA's or EIA's may be employed. In these cases, appropriate controls for false positives (negative control yeast lysate) as well as inhibition (spiking negative extract with bona fide product) should be performed for accurate quantitation. Finally, heterologous protein expression level in a large number of samples may be readily quantitated by immunodot-blot assay, again using the appropriate controls.

In addition to quantitation of the heterologous protein, its biological activity should be measured. If the protein of interest is an enzyme, standard enzyme assays can be used with yeast cell extracts. Hormones, lymphokines, and other biological response modifiers expressed in yeast require *in vitro* biological assays with appropriate target cells. The potency of potential vaccines produced in yeast can be tested by measuring

[46] U. K. Laemmli, *Nature (London)* **227,** 680 (1970).
[46a] W. N. Burnette, *Anal. Biochem.* **112,** 195 (1981).

antibody response of appropriately purified and formulated product in animal systems.

Gentle cell lysis methods are required to prepare samples for biological assays. Small samples can be prepared by hypotonic lysis of spheroplasts, or by vortexing cell suspensions in glass test tubes with 0.5-mm glass beads. French press lysis is convenient for disrupting 1–5 g of cells. Cells should be lysed in buffers that are compatible with the biological activity of the protein of interest. In some cases, addition of urea or a detergent such as Triton X-100 releases the protein of interest from cell debris.

All of the above assays require recovery of intact proteins in yeast cell extracts. Since proteases have historically been a problem with protein purifications from yeast, care should be taken to minimize their activity. Cell extracts should be prepared at 0–4° in the presence of protease inhibitors and assayed as soon as possible. Protease-deficient yeast strains such as 20B-12,[47] available from the Yeast Genetic Stock Center (University of California, Berkeley), can be used to construct appropriate host strains with reduced protease activity. Methods for reducing proteolysis during purification of proteins from yeast have been reviewed.[48] Intracellular turnover of heterologous proteins can be monitored with [^{35}S]methionine pulse–chase experiments. If the foreign protein is much less stable than native yeast proteins, then other expression strategies should be considered. These include regulated (inducible) promoters and secretion systems.

Secretion Vectors

Saccharomyces cerevisiae produces certain of its own proteins at very high levels and efficient expression of heterologous proteins has been obtained using the approaches outlined above. In addition, this organism secretes a restricted number (0.5% of total cell protein) of its own proteins into the culture medium. Initial attempts at secretion of heterologous proteins from *Saccharomyces cerevisiae* were motivated by the anticipation that recovery and purification of a secreted protein would be more efficient. A variety of heterologous signal peptides have been demonstrated to direct protein secretion in *Saccharomyces cerevisiae,* indicating conservation among different organisms in the mechanism of protein translocation.[19] Recent work has focused on homologous yeast leader peptides to program protein secretion. Heterologous protein secretion has been obtained using secretion signals from the yeast

[47] E. W. Jones, *Genetics* **85,** 23 (1977).

[48] J. R. Pringle, *Methods Cell Biol.* **12,** 149 (1975).

α-factor pheromone precursor,[6,49–51] invertase,[52] and acid phosphatase[52] precursors.

There are a number of situations in which it is desirable to produce heterologous proteins in a secretion system. These include foreign proteins that are toxic to yeast or unstable when produced by direct expression in the cytoplasm. Many proteins being considered for use as human therapeutics are themselves secreted proteins which are processed from a larger precursor. These proteins thus have specific NH_2 termini (rarely methionine). Direct expression of these genes results in proteins with NH_2 terminal methionine. Methionine aminopeptidases will remove this residue in certain sequence contexts.[53] However, the methionine will not be removed in other sequence contexts or may be removed from only a fraction of the molecules when the protein is produced at high levels. Under these conditions, the expressed protein represents an analog of the natural material which may have reduced activity or other undesirable properties. Production of such proteins in a secretion system, however, provides a method for the generation of native NH_2 termini. Finally, many proteins of commercial import contain intramolecular disulfide bonds. The cytoplasm of a cell is a reducing environment and few, if any, cytoplasmic proteins contain disulfide bonds. Direct expression of many heterologous proteins thus results in improperly folded (and possibly insoluble) products which must be extracted and subjected to an *in vitro* refolding process. This refolding occurs with variable efficiency and may be impossible for certain complex structures. The formation of accurate disulfide bonds in secreted proteins appears to be catalyzed by protein disulfide isomerase, an enzyme localized to the luminal side of the endoplasmic reticulum in eukaryotic secretory cells.[54] It seems likely, therefore, that heterologous proteins produced in a yeast secretion system will assume the correct tertiary structure. It has recently been demonstrated, by tryptic peptide mapping, that an α-interferon secreted from yeast has the identical disulfide bond structure as the natural human protein.[55]

[49] A. J. Brake, J. P. Merryweather, D. G. Coit, U. A. Heberlein, F. R. Masiarz, G. T. Mullenbach, M. S. Urdea, P. Valenzuela, and P. J. Barr *Proc. Natl. Acad. Sci. U.S.A.* **81,** 4642 (1984).

[50] A. Singh, E. Y. Chen, J. M. Lugovoy, C. N. Chang, R. A. Hitzeman, and P. H. Seeburg, *Nucleic Acids Res.* **11,** 4049 (1983).

[51] A. Miyajima, M. W. Bond, K. Otsu, K. Arai, and N. Arai, *Gene* **37,** 155 (1985).

[52] R. A. Smith, M. J. Ducan, and D. T. Moir, *Science* **229,** 1219 (1985).

[53] F. Sherman and J. W. Stewart, *in* "The Molecular Biology of the Yeast Saccharomyces: Metabolism and Gene Expression" (J. N. Strathern, E. W. Jones, and J. R. Broach, eds.), pp. 301–333. Cold Spring Harbor Lab., Cold Spring Harbor, New York, 1982.

[54] R. B. Freedman, *Trends Biochem. Sci.* **106,** 438 (1984).

[55] K. M. Zsebo, H. S. Lu, J. C. Fieschko, L. Goldstein, J. Davis, K. Duker, S. V. Suggs, P. H. Lai, and G. A. Bitter, *J. Biol. Chem.* **261,** 5858 (1986).

Use of the prepro leader region of the yeast α-factor pheromone precursor, which has proved to be generally useful for secretion of heterologous proteins from yeast, will be described in this section. Although use of other signal peptides[52] will necessitate different secretion vector constructions, the analysis and characterization of the secreted product will be similar. The methods and considerations for direct expression vector construction described above also apply to the development of secretion vectors.

Construction of Prepro-α-Factor Gene Fusions

We have developed two plasmids for construction of prepro-α-factor gene fusions[6,55] (Fig. 7). Plasmids pαC2 and pαC3 contain a 1.7-kb yeast genomic fragment which includes the *MFα1* structural gene as well as its promoter and transcription termination sequences. Each plasmid contains the yeast DNA segment cloned as a *Bam*HI fragment in pBRΔHS (a derivative of pBR322 in which the *Hin*dIII and *Sal*I sites were each deleted by separate restriction, end filling, and blunt-end ligation). Plasmid pαC3 was derived from plasmid pαC2 by site-directed mutagenesis.[55] The serine codon at position 81 of the prepro-α-factor leader was converted to an AGC serine codon which introduces a *Hin*dIII site beginning 10 bp

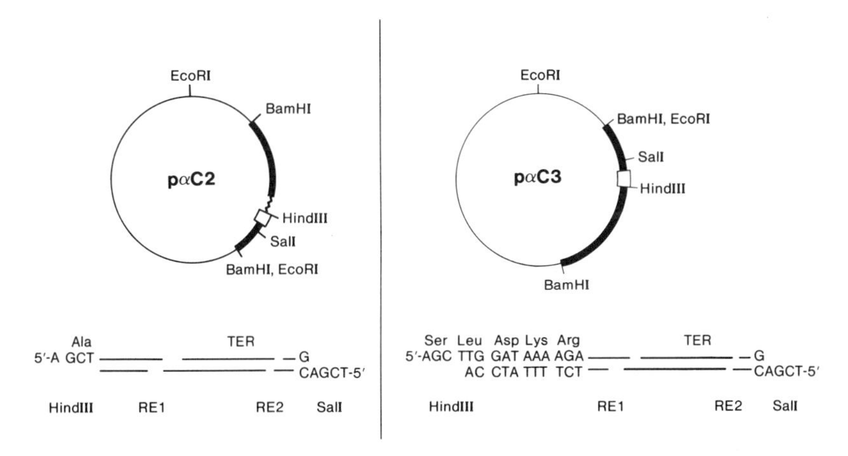

FIG. 7. Plasmids for construction of α-factor gene fusions. Restriction endonuclease maps for plasmids pαC2 and pαC3 are depicted. DNA sequences required at the 5′ termini of the heterologous gene for creating the in-frame gene fusion are depicted. The first codon of the heterologous protein-coding region occurs immediately after the GCT^{Ala} codon in pαC2 and immediately after the AGA^{Arg} codon in pαC3. RE1 and RE2 refer to restriction enzyme sites within the heterologous gene to which synthetic linkers may be fused for assembly of the gene fusion. TER refers to the in-frame termination codon required in the heterologous gene coding region. Refer to the text for a description of construction strategies.

upstream from the start of the Lys codon of the KR endoprotease recognition sequence (see below). In both plasmids, the internal *Hin*dIII repeats of the *MF*α1 structural gene have been deleted.

The strategy for assembly of heterologous gene fusions, as well as the required DNA sequence at the *MF*α1 gene fusion site, are also depicted in Fig. 7. The heterologous gene encoding the protein to be secreted is cloned into pαC2 or pαC3 as a *Hin*dIII to *Sal*I fragment. In the case of constructions using pαC2, the heterologous gene may actually contain a "pseudo-*Hin*dIII" site at the 5′ end. The GCT of the *Hin*dIII recognition site represents the last alanine codon of the spacer peptide. Thus, the first nucleotide of the first codon of the heterologous gene occupies the position of the last T in the *Hin*dIII site of pαC2. For codons which begin with T, the *Hin*dIII site is regenerated. In other cases, however, a *Hin*dIII cohesive terminus is used without regenerating the *Hin*dIII site. Since, for both pαC2 and pαC3, the resulting gene encodes a protein fusion, the cloning manipulations must be very precise. This is most readily accomplished by chemically synthesizing a gene as a *Hin*dIII–*Sal*I fragment. If this is not possible, a synthetic linker from the *Hin*dIII site in pαC2 or pαC3 to a restriction site (RE1; Fig. 7) near the 5′ end of the heterologous gene may be synthesized which encodes the proper protein fusion sequence. If the heterologous gene does not include a *Sal*I site on the 3′ side and close to the translation stop codon, then a synthetic linker fusing a restriction site (RE2; Fig. 7) at the 3′ end of the gene to the *Sal*I site may be employed. Alternatively, the 3′ end of the heterologous gene may be fused to the *Sal*I site by blunt-end ligation. In isolating the heterologous gene fragment, excessive 3′ flanking DNA should be eliminated since it may have deleterious effects on transcription and translation in yeast.

The native α-factor precursor as well as the hybrid precursors encoded by constructions utilizing pαC2 and pαC3 are depicted in Fig. 8. The 13-amino acid α-factor peptide is synthesized as a 165-amino acid prepro-polyprotein precursor containing 4 copies of the α-factor peptide.[56] The native precursor contains a hydrophic 20- to 22-amino acid NH_2 terminus which initiates translocation into the endoplasmic reticulum. There is a 61-amino acid pro segment of unknown function which contains three sites of N-linked glycosylation. The α-factor peptides are separated by spacer peptides of the sequence Lys-Arg(Glu-Ala)$_{2-3}$ or Lys-Arg-Glu-Ala-Asp-Ala-Glu-Ala. The repeating peptide units are excised from the precursor by cleavage on the carboxyl side of arginine by an endoprotease encoded by the yeast *KEX2* gene and termed the KR endoprotease.[57] The Lys-Arg on the COOH terminus of the first three units

[56] J. Kurjan and I. Herskowitz, *Cell* **30,** 933 (1982).

[57] D. Julius, A. Brake, L. Blair, R. Kunbawa, and J. Thorner, *Cell* **37,** 1075 (1984).

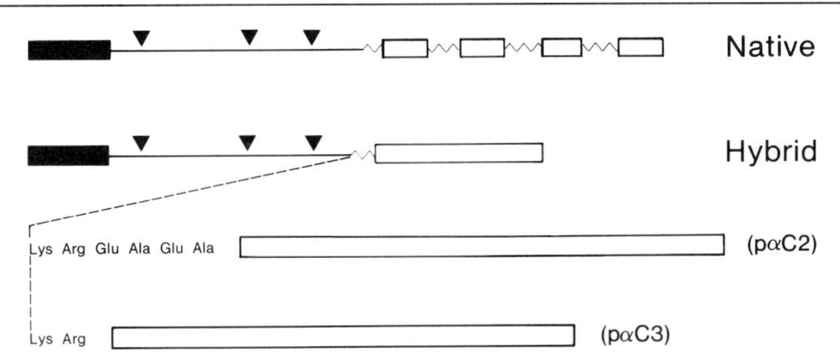

FIG. 8. Structure of native and hybrid α-factor precursors. The hydrophobic NH_2 terminus of the prepro leader region is depicted as a black box while the N-linked glycosylation sites on the pro segment are represented by inverted triangles and the spacer peptides are indicated by squiggles. The α-factor peptides in the native precursor and the heterologous protein in the hybrid precursor are depicted by open boxes. The amino acid sequence of the spacer peptide is indicated for gene fusions constructed in pαC2 (top) or pαC3 (bottom).

excised is presumably removed by a carboxypeptidase B-like activity which may be the YscE enzyme identified *in vitro* in yeast membrane functions.[58] The Glu-Ala or Asp-Ala dipeptides are excised from the NH_2 terminus by dipeptidyl aminopeptidase A which is encoded by the yeast *STE*13 gene.[59] These three proteolytic events yield the fully processed, mature α-factor peptide.

The hybrid precursors (Fig. 8) encoded by gene fusions assembled in pαC2 and pαC3 each include the first 83 amino acids of the native α-factor precursor. This segment includes signals which direct secretion of the hybrid precursor. The fusion constructed in pαC2 includes the spacer peptide Lys-Arg(Glu-Ala)$_2$ while that constructed in pαC3 has the spacer peptide Lys-Arg. The first fusion protein thus contains processing sites for the *KEX*2 and *STE*13 gene-encoded proteases. The gene fusion constructed in pαC3 contains the KR endoprotease recognition sequence but is independent of processing by dipeptidyl aminopeptidase A since the Glu-Ala dipeptides have been eliminated from the precursor. Neither hybrid precursor requires COOH-terminal processing since termination codons are included at the end of the coding region of the heterologous gene.

The gene fusions constructed in pαC2 and pαC3 may be subcloned as a *Bam*HI fragment into a shuttle vector (such as pYE, above) for expression in yeast. The *Bam*HI fragment encoding the gene fusion includes the native yeast *MF*α1 promoter and transcription terminator sequences and is efficiently expressed in yeast.[55] However, it should be noted that alternative promoters and terminators may be employed for expression according to the strategies outlined in Direct Expression Vectors.

[58] J. Achstetter and D. H. Wolf, *EMBO J.* **4,** 173 (1985).

[59] D. Julius, L. Blair, A. Brake, G. Sprague, and J. Thorner, *Cell* **32,** 839 (1983).

Analysis of Heterologous Protein Secretion

The methods utilized to analyze heterologous protein secretion are the same as those used for proteins expressed directly in the cytoplasm (Direct Expression Vectors). The methods of sample preparation, however, differ considerably since proteins secreted into the culture medium are generally more dilute than proteins present in concentrated cell extracts. Secretion offers an advantage over direct cytoplasmic expression in that, within limits, the culture medium conditions may be adjusted to be compatible with retention of biological activity of the secreted product. Thus, the pH, ionic strength, temperature, etc., of the cell culture medium, may be specified appropriately if the optimal parameters have been determined for the specific secreted protein. It should be noted, however, that conditions optimal for protein stability and activity may be suboptimal for efficient fermentation of the yeast strain. Thus, culture conditions should be formulated which optimize both aspects of the production process.

Measurement of Secretion Efficiency. It has been demonstrated that different heterologous proteins are secreted with different efficiencies using the prepro-α-factor leader region.[55] Thus, certain proteins are secreted entirely into the culture medium while others accumulate primarily intracellullarly with only a small percentage secreted into the medium. The mechanisms which determine this partitioning between the cell and medium have not been clearly defined and no simple rules exist which are predictive of the secretion efficiency of a particular protein. In addition to differential partitioning, the total amount of heterologous protein synthesized may vary significantly for different α-factor gene fusions.

The heterologous protein secreted into the culture medium may be quantitated by the methods described in Direct Expression Vectors. If the product concentration is too low, culture supernatants can be rapidly concentrated by ultrafiltration, dialysis against polyethylene glycol, lyophilization, or acetone precipitation. Depending upon the method chosen, samples may be readily desalted or buffer conditions changed during the process of concentration. Alternatively, proteins can be concentrated specifically by immunoprecipitation, lectin columns, or other purification methods specific to the particular protein. The degree to which a sample should be concentrated will depend upon the sensitivity of the technique used for analysis. Nonspecific methods, such as SDS–PAGE, usually require a higher degree of concentration than do immunological methods. If the gel is to be stained with Coomassie blue, it may be necessary to concentrate a sample as much as 100- to 1000-fold, whereas with most silver staining methods the sample may require only 10-fold concentration. For most immunological methods such as RIA, ELISA, and immunoblot analysis, sample concentration is usually unnecessary.

The amount of heterologous protein secreted does not always increase in proportion to the culture cell mass, particularly in high cell density fermentations. Therefore, the product concentration in the culture medium should be correlated to cell mass as a function of growth of the culture. In addition, the amount of heterologous protein which accumulates intracellularly should be measured. In examining secretion efficiency, it is convenient to separate total cell-associated material into intracellular and wall-associated or periplasmic fractions. This can be readily accomplished by harvesting the cells and digesting the cell wall with glusulase or zymolyase (in the presence of osmotic stabilizing agents such as 1.0 *M* sorbitol) and separating the postenzyme supernatant from the spheroplasts. The material present in the spheroplasts is operationally defined as intracellular protein, and that in the supernatants as wall-associated or periplasmic protein. Boiling each fraction in the presence of a detergent, such as SDS, is usually sufficient to solubilize the protein for quantitation by SDS–PAGE or immunological methods. Less severe extraction conditions are generally required for retention of biological activity of the product.

By determining the total amount of heterologous protein produced, the efficiency of the expression system may be evaluated. If these levels are low, then total protein production should be increased by optimizing transcription and/or translation efficiencies (Direct Expression Vectors). Analysis of the heterologous protein partitioning between the cytoplasm, cell wall, and culture medium defines the secretion efficiency. It is highly desirable to achieve 100% secretion of the heterologous protein into the culture medium. Optimization of this process is currently an area of intense research.

Analysis of Hybrid Precursor Processing

The yeast secretory process is a complex series of events associated with membrane-bound subcellular compartments. In addition to the carbohydrate modifications of the pro segment, the native α-factor precursor is processed by three different proteolytic enzymes to generate the mature α-factor peptides. It is important, therefore, to characterize the processing of the hybrid precursor and secreted heterologous protein.

KR Endoprotease Processing. It has been demonstrated that hybrid protein precursors which have not been cleaved by KR endoprotease are secreted into the culture medium.[55] These fusions proteins may have reduced or no biological activity. Their presence, therefore, should be quantitated prior to measuring the specific activity (biological activity units per milligram protein) of the secreted heterologous protein. This quantitation is readily performed by immunoblot analysis[46] using antisera raised against the heterologous protein.

The hybrid precursor will contain N-linked carbohydrate on the three glycosylation sites within the pro segment (Fig. 8). Because of the large size and heterogeneous nature of yeast carbohydrate modifications,[60] the precursors are not readily visualized after SDS–PAGE. However, by treating the secreted proteins with endoglycosidases which remove the N-linked carbohydrate (see Procedures) the heterogeneously sized precursors are reduced to a single band. An example of this analysis is depicted in the Western blot shown in Fig. 9. Lane 1 contains untreated medium proteins from a yeast strain secreting IFN-Con_1.[55] A single major band of apparent molecular weight 18,000 is present. The sample in lane 2 was treated with Endo H prior to gel electrophoresis. A new major band with apparent molecular weight of 28,000 appears which is reactive with the antisera. This protein is the deglycosylated hybrid protein precursor and, in this preparation, represents approximately 90% of the total secreted IFN-Con_1 immunoreactive material.

The secretion of unproteolyzed hybrid precursor may be due to limiting KR endoprotease activity when the precursor is produced at high levels from multicopy vectors. This deficiency might be corrected by incorporating the cloned *KEX2* gene[57] into multicopy plasmids. This strategy, however, may result in internal cleavages at dibasic residues within the heterologous protein. For at least one hybrid precursor, the KR endoprotease processing efficiency appeared greater when Glu-Ala dipeptides were present in the spacer peptide.[55] This indicates that amino acid sequence context may affect the efficiency of cleavage by KR endoprotease.

If conditions can be developed in which complete cleavage by KR endoprotease occurs, then the yield of correctly processed secreted heterologous protein will be optimized. If this is not possible, then the precursor must be separated from the processed heterologous protein. This separation is readily accomplished for heterologous proteins which are not glycosylated since the large mass of carbohydrate on the pro segment of the precursor allows resolution of the two forms by gel exclusion chromatography. In cases where the heterologous protein is glycosylated, however, different procedures may be required to separate the precursor and processed forms of the heterologous protein.

[60] C. E. Ballou, *in* "The Molecular Biology of the Yeast Saccharomyces: Metabolism and Gene Expression" (J. N. Strathern, E. W. Jones, and J. R. Broach, eds.), pp. 335–360. Cold Spring Harbor Lab., Cold Spring Harbor, New York, 1982.

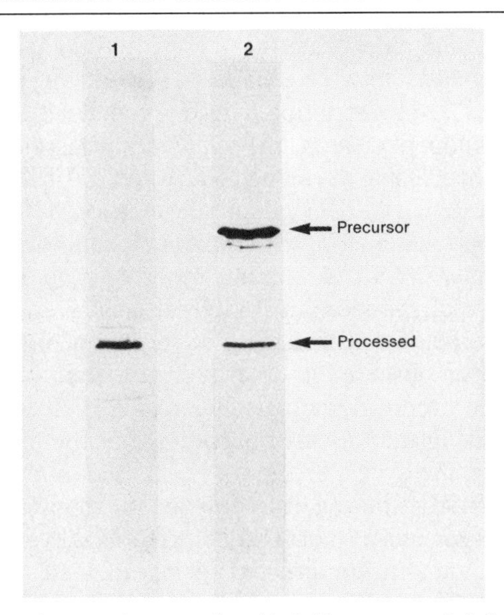

FIG. 9. Detection of secreted unproteolyzed hybrid precursor. Cell-free culture supernatants from a yeast strain secreting IFN-Con_1 were subjected to immunoblot analysis as described.[55] Lane 1, untreated medium proteins. Lane 2, proteins were digested with Endo H as described in Procedures prior to SDS–PAGE.

NH_2 Terminal Analysis of Secreted Heterologous Proteins. One of the advantages of a secretion system is production of heterologous proteins with authentic NH_2 termini. Although size analysis by SDS–PAGE may indicate accurate processing of the heterologous protein, this may be demonstrated only by direct NH_2 terminal amino acid sequence analysis. These methods have been reviewed in detail elsewhere.[61–63] The secreted heterologous protein (processed form) may be purified by conventional means for this purpose. Alternatively, if the secreted product is abundant and well resolved from other secreted yeast proteins, it may be purified by preparative SDS–PAGE and directly subjected to NH_2 terminal analysis.[64]

It has been demonstrated that secreted proteins programmed by gene fusions constructed in pαC2 contain authentic (fully processed) NH_2 termini.[6] However, when these precursors are overproduced, the secreted product contains NH_2 terminal Glu-Ala extensions due to rate-limiting proteolysis by dipeptidyl aminopeptidase A.[6,55] Two strategies exist for

[61] P. Edman and G. Begg, *Eur. J. Biochem.* **1,** 80 (1967).

[62] R. M. Hewick, M. W. Hunkapiller, L. E. Hood, and W. J. Dreyer, *J. Biol. Chem.* **256,** 7990 (1981).

[63] P. H. Lai, *Anal. Chim. Acta* **163,** 243 (1984).

[64] M. W. Hunkapiller, E. Lujan, F. Ostrander, and L. E. Hood, this series, Vol. 91, p. 227.

eliminating these dipeptides from the secreted product. The Glu-Ala codons may be eliminated from the gene fusion by utilizing plasmid pαC3 for construction (Fig. 7). The resulting hybrid precursor is thus independent of dipeptidyl aminopeptidase A. Alternatively, the processing enzyme may be overproduced using the cloned *STE*13 gene.[59] This latter approach might allow more efficient KR endoprotease cleavage[55] as well as complete exoproteolytic removal of the Glu-Ala dipeptides.

Aberrant Proteolysis. In our original report of prepro-α-factor leader-directed secretion of β-endorphin,[6] internal cleavages on the carboxyl side of lysines were noted within the β-endorphin peptide. Such internal cleavages have been observed in other secreted heterologous proteins as well as within the prepro-α-factor leader region.[55] Aberrant proteolysis occurs, in general, in only a small percentage of the total heterologous protein. However, the hybrid precursor is exposed to a number of different proteolytic activities during secretion and the integrity of the product should be monitored. This is most readily performed by immunoblot analysis using polyclonal antisera directed against the heterologous protein.

Glycosylation. The α-factor precursor pro segment contains three sites of N-linked glycosylation (Fig. 8). It has been demonstrated that heterologous proteins containing potential sites of N-linked glycosylation are, in fact, glycosylated during prepro-α-factor-directed secretion from yeast (S. G. Elliott and G. A. Bittner, unpublished results). Carbohydrate addition to glycoproteins is dependent on the accessibility of the glycosylation sites, which is affected by protein folding (both rate and final conformation). Therefore, it cannot be assumed, a priori, that all potential sites of glycosylation in heterologous proteins will be modified. *Saccharomyces cerevisiae* also effects O-linked glycosylation, and this modification may occur on mammalian proteins secreted using the α-factor system.

The methods used for analysis of carbohydrate addition are described in Procedures. It should be noted that the structure of yeast oligosaccharides is markedly different than the structures present on mammalian glycoproteins (reviewed by Ballou[60]). Thus, the effect of yeast carbohydrate addition on the biological activity of these products should be tested in appropriate *in vivo* systems.

Procedures

Mung Bean Nuclease Digestion. This reaction should be done in glass tubes rather than plastic. Suspend the phenol-extracted and ethanol-precipitated restricted DNA in TNE buffer (6 m*M* Tris–HCl, pH 8.0, 6 m*M* NaCl, 0.2 m*M* EDTA) to a concentration of 100 ng/μl. Add one-tenth volume of 10× mung bean buffer (300 m*M* sodium acetate, pH 4.7, 500 m*M* NaCl, 10 m*M* $ZnCl_2$). Add 0.5–1.0 U of mung bean nuclease/1 μg of DNA. (If the nuclease is to be diluted, dilute it into 1× mung bean buffer). Incubate the reaction for 10 min at 37°. To stop the reaction, add 1 μl of 20% SDS/100 μl of reaction mix, and heat at 65° for 5 min. Add one-tenth volume of 500 m*M* Tris, pH 9.5, and one-tenth volume of 8 *M* LiCl; then extract the solution with an equal volume of phenol/chloroform (1 : 1), followed by an extraction with ether. Add 2 vol of cold 100% ethanol to precipitate the DNA.

S1 Nuclease Digestion of Cohesive Termini. Suspend the phenol-extracted and ethanol-precipitated restricted DNA in double-distilled H_2O to a concentration of 100 ng/μl. To this solution, add one-tenth volume of 10× S1 buffer (2.5 *M* NaCl, 10 m*M* $ZnSO_4$, 300 m*M* sodium acetate, pH 4.5). Chill the reaction mix to 14° (in order to minimize strand separation and therefore to prevent excess single-strand cutting by S1 nuclease). Add the S1 nuclease to a concentration of 200 U/pmol DNA. (If the S1 nuclease must be diluted, dilute it into S1 buffer.) Incubate the reaction for 60 min at 14°. Stop the reaction with 25 m*M* EDTA, and bring to pH 8.0 with one-tenth volume of 500 m*M* Tris, pH 9.5. Then extract the solution with an equal volume of phenol/chloroform (1 : 1), followed by an extraction with ether. Add 2 vol of cold 100% ethanol to precipitate the DNA.

End Filling with Klenow. The Klenow end-filling reaction on 5′ overhang termini works well in a variety of buffers including standard restriction enzyme buffers and ligation buffers. To the restricted DNA in the appropriate buffer at a concentration of 100 ng/μl, add one-tenth volume of 1 m*M* deoxynucleoside triphosphates (only those necessary to fill in the recessed 3′ end need be added). Add the Klenow fragment to a concentration of about 1 U/pmol DNA. Incubate the reaction for 10 min at room temperature. Stop the reaction by heating for 10 min at 65°. The solution can now be precipitated with 2 vol of ethanol, or used directly for subsequent ligation or restriction reactions.

Yeast RNA Preparation. Cells from approximately 50 OD · ml of late log phase cultures (e.g., 50 ml of culture at OD_{600} = 1.0) are harvested by centrifugation, resuspended in 10 ml ice-cold H_2O, and centrifuged again. Resuspend the cell pellets in 3 ml RNA extraction buffer (0.15 *M* NaCl, 5 m*M* EDTA; 4% SDS, 50 m*M* Tris–HCl, pH 7.5). Add 3 ml phenol–chloroform–isoamyl alcohol (50 : 50 : 1) and, after 2 min at room temperature, vortex vigorously for 5 min with glass beads. Centrifuge and collect the aqueous phase. Reextract the organic phase with 3 ml of extraction buffer. Pool the aqueous phases and extract once with phenol/chloroform/isoamyl alcohol and once with chloroform. Precipitate the RNA by the addition of 2.5 vol of ethanol, rinse with 70% ethanol, and dry *in vacuo*.

Yeast DNA Preparation. Cells from approximately 100 OD · ml of late log phase cultures are harvested by centrifugation, resuspended in 10 ml H_2O, and centrifuged again. Prepare spheroplasts rapidly by resuspending the cells in 3 ml SCE (1 *M* sorbitol, 0.1 *M* sodium citrate, 0.01 *M* EDTA, pH 5.8) containing 2.5 mg/ml zymolyase 60,000 (Miles), and incubating 3–5 min at 37°. Dilute the spheroplast suspension with 3 vol of ice-cold SCE, and collect the spheroplasts by centrifugation at 4000 *g* for 5 min. The spheroplasts are lysed in 5 ml 25 m*M* EDTA, 0.15 *M* NaCl and extracted with 0.5 vol of equilibrated phenol, and once with 0.5 vol of chloroform. DNA is precipitated from the aqueous phase with 1.5 vol of 2-propanol. The final DNA pellets are washed with 70% 2-propanol and dried *in vacuo*.

Analysis of Carbohydrate Additions to Heterologous Proteins. The presence of oligosaccharide additions to the heterologous protein may be demonstrated by sensitivity to a variety of enzymes. Three commercially available enzymes are effective at removing N-linked carbohydrate from yeast glycoproteins. These include endoglycosidase H (Endo H; Miles, New England Nuclear and Genzyme), endoglycosidase F (Endo F; New England Nuclear), and N-glycanase (Genzyme). Endo H and Endo F treatment of glycoproteins leaves the sugar N-acetyl-galactosamine (GlcNac) bound to the asparagine at the glycosylation site. N-glycanase treatment, on the other hand, results in complete removal of protein carbohydate. Endo F does have an N-glycanase activity and exhaustive digestion can result in removal of the asparagine bound GlcNac. The following reaction conditions will meet most needs but may need to be adjusted for each particular glycoprotein.

Endo H. Boiling the sample in 0.05% SDS prior to deglycosylation results in faster and more complete deglycosylation. SDS concentrations higher than 0.05% may inhibit the enzyme. To the SDS-treated protein sample (100 μg/ml–2 mg/ml) is added a one-tenth volume of 1 *M* sodium citrate, pH 5.5, followed by 10–50 mU/ml final concentration of enzyme. Incubate at 37° until digestion is complete (4–16 hr).

Endo F. Some preparations of endo F have a contaminating protease activity. EDTA is required to minimize protease activity. Reaction conditions are 100 m*M* sodium phosphate, pH 6.1, 50 m*M* EDTA, 1% Nonidet P-40, 0.1% SDS, and 1% 2-mercaptoethanol. The sample containing 0.1–2 mg/ml protein is boiled for 10 min followed by the addition of 10–50 mU/ml Endo F. Incubate at 37° until digestion is complete (4–16 hr).

N-Glycanase. The glycoprotein sample (0.1–2 mg/ml) is boiled in 0.5% SDS and 0.1 *M* 2-mercaptoethanol for 10 min. The sample is then diluted into sodium phosphate buffer. The final concentrations are 0.17% SDS, 33 m*M* 2-mercaptoethanol, 0.2 *M* sodium phosphate, pH 8.6, 1.25% Nonidet P-40. N-Glycanase is added to 2–10 U/ml and incubated 4–16 hr at 37°.

[34] Vaccinia Virus as an Expression Vector

By ANTONIA PICCINI, MARION E. PERKUS, and ENZO PAOLETTI

Introduction

A methodology is presented for using vaccinia virus to express foreign genes. These recombinant vaccinia viruses can be used for a variety of purposes, such as (1) understanding the genetic regulation involved in the expression of endogenous vaccinia functions, (2) determining the fate of foreign gene products in a background free of native influences, (3) producing biologically active molecules, (4) elucidating roles of pertinent antigens in eliciting defined immunological responses, (5) producing live recombinant vaccines, and (6) potentially using in gene replacement therapy.

Biology of Poxviruses

Poxviruses comprise a large group of complex animal viruses whose genetic information is contained within a double-standed DNA molecule and whose replication is confined to the cytoplasm of the infected cell (for recent reviews, see Refs. 1 and 2). Vaccinia virus is the prototype poxvirus. It is a laboratory virus that does not circulate in nature. Vaccinia is one of the largest animal viruses with a DNA genome of approximately 187 kb, sufficient to encode 150–200 gene products. Some of these gene products are included in the enzymatic arsenal localized either within the virion itself or which are induced upon infection. Significantly the naked DNA of poxviruses, unlike that of many other animal virus groups, is not by itself infectious. The virus has a broad host range capable of infecting a variety of animals and replicates on a variety of tissue culture cells.

Principle of the Method

General Protocol for the Insertion of Foreign Genes into Vaccinia Virus

The ability to transfer endogenous genetic markers into infectious vaccinia virus was first demonstrated by Sam and Dumbell[3] using thermo-

[1] B. Moss, *in* "Virology" (B. N. Fields, ed.), p. 685. Raven, New York, 1985.

[2] S. Dales and B. G. T. Pogo, "Biology of Poxviruses" (Virology Monographs 18). Springer-Verlag, New York, 1981.

[3] C. K. Sam and K. R. Dumbell, *Ann. Virol.* (*Inst. Pasteur*) **132E,** 135 (1981).

sensitive markers and by Nakano *et al.*[4] with vaccinia virus deletion mutants. Marker rescue has subsequently been used to localize on the vaccinia genome the genes encoding thymidine kinase,[5,6] DNA polymerase,[7,8] rifampicin resistance,[9] various temperature-sensitive genetic markers,[10–12] and sequences specifying host range.[13] Extensions of these basic protocols have allowed for the insertion of foreign genes into viable vaccinia virus. The basic strategy is outlined in Fig. 1. A foreign genetic element is isolated and inserted at a specific site within the vaccinia DNA. Since the DNA of vaccinia is so large (187 kb), manipulation of cloned subgenomic fragments facilitates these constructions. It is essential that the insertion of the foreign gene occur at a locus of the vaccinia DNA such that the flow of genetic information necessary for virus replication is not disrupted. The foreign insert is flanked by what would normally be contiguous DNA sequences in the vaccinia genome. Additionally, appropriate regulatory transcriptional and translational signals must be engineered immediately flanking the foreign DNA insert to allow for transcription of the foreign gene and its subsequent translation into a protein. The use of vaccinia transcriptional regulatory sequences is required for optimal expression of the foreign gene since vaccinia encodes its own RNA polymerase which recognizes nonvaccinia promoter signals poorly or not at all. The chimeric construct is cloned into a convenient vehicle such as pBR322 so that sufficient quantities may be obtained for subsequent manipulations. This chimeric "donor" plasmid is introduced into tissue culture cells by variations of standard transfection procedures.[14] Since naked poxvirus DNA is not infectious, rescue of the donor DNA sequence is accomplished by concomitant infection of that same cell by infectious "rescuing" vaccinia virus. The infectious rescuing vaccinia virus proceeds through its normal replication cycle. When a DNA molecule replicated from the rescuing virus localizes itself in the cytoplasm in close proximity to the input donor DNA, *in vivo* recombination can occur between the vaccinia DNA sequences flanking the foreign DNA and the homologous DNA sequences on the replicating vaccinia DNA. This event

[4] E. Nakano, D. Panicali, and E. Paoletti, *Proc. Natl. Acad. Sci. U.S.A.* **79,** 1593 (1982).
[5] J. P. Weir, G. Bajszar, and B. Moss, *Proc. Natl. Acad. Sci. U.S.A.* **79,** 1210 (1982).
[6] A. Vassef, F. Ben-Hamida, and G. Beaud, *Anal. Virol. (Inst. Pasteur)* **134E,** 375 (1983).
[7] E. V. Jones and B. Moss, *J. Virol.* **49,** 72 (1984).
[8] P. Traktman, P. Sridhar, R. C. Condit, and B. E. Roberts, *J. Virol.* **49,** 125 (1984).
[9] J. Tartaglia and E. Paoletti, *Virology* **147,** 394 (1985).
[10] R. C. Condit, A. Motyczka, and G. Spizz, *Virology* **128,** 429 (1983).
[11] M. J. Ensinger and M. Rovinsky, *J. Virol.* **48,** 419 (1983).
[12] R. Drillien and D. Spehner, *Virology* **131,** 385 (1983).
[13] S. Gillard, D. Spehner, and R. Drillien, *J. Virol.* **53,** 316 (1985).
[14] F. L. Graham and A. J. van der Eb, *Virology* **52,** 456 (1973).

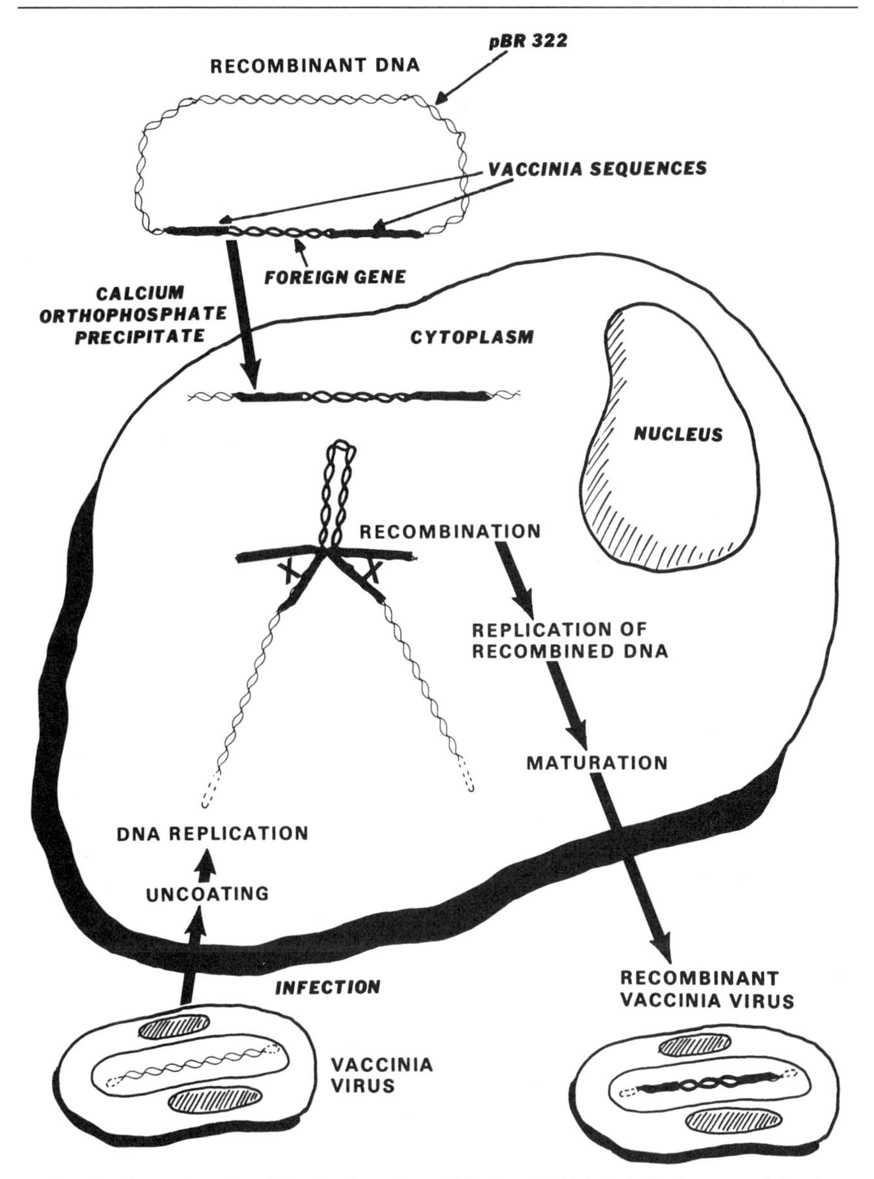

FIG. 1. General protocol for the insertion of foreign DNA into infectious vaccinia virus. Foreign DNA inserted at a locus of a cloned vaccinia subgenomic fragment is introduced into tissue culture cells by transfection procedures. The cell is coinfected with rescuing vaccinia virus. *In vivo* recombination occurs between the vaccinia DNA sequences flanking the foreign insert and homologous DNA sequences on the replicating vaccinia genome, forming a novel recombinant DNA molecule which in turn can be replicated and packaged into infectious recombinant vaccinia virus.

happens with a frequency of approximately 0.1% and results in the generation of a recombinant DNA molecule. This novel vaccinia recombinant genome can itself be replicated and participate in the maturational steps resulting in infectious progeny recombinant virus. Infectious progeny virus containing the foreign insert can be identified by a variety of methods including DNA hybridization techniques, by phenotypic properties such as drug sensitivity and thermal responses, by enzymological methods, or by immunological detection systems. Once identified, the recombinant virus can be prepared as a pure virus stock by repeated plaque cloning. No helper virus is required for its propagation.

Foreign Genes Expressed in Recombinant Vaccinia Viruses

Since the first demonstration in 1982 of the insertion and expression of the herpes simplex virus thymidine kinase gene,[15,16] a large variety of genes have been inserted into vaccinia virus. The expression of foreign genes in viable vaccinia virus recombinants reported to date is presented in Table I. In general, the foreign genes represent those genetic elements which are responsible for the production of immunity to specific pathogens which is related to one of the more exciting uses of vaccinia vectors: live recombinant vaccines. However, the list also includes biochemical markers which can be used in selection, detection, or quantitative procedures. One of the paramount successes thus far has been the finding that when the foreign genes are expressed under vaccinia virus regulation, the expressions both at the RNA and protein level are similar to if not identical with native synthesis. Furthermore, the gene products exhibit authentic secondary modifications such as glycosylation and pertinent antigens encoded by the foreign genes are immunologically presented as they are in a natural infection. For example, the hepatitis B virus surface antigen is secreted from infected cells as a morphologically distinct 22-nm particle. In contrast, the influenza hemagglutinin and the herpes simplex glycoprotein D become associated with the membrane of the infected cell. These considerations are imperative if one wants to use vaccinia virus as an expression vector which exhibits true biological fidelity, in particular, for the production of live recombinant vaccines or biologically active molecules.

Packaging Potential of the Vaccinia Genome

One of the significant advantages of vaccinia virus as a eukaryotic cloning vector is its ability to package considerable quantities of foreign

[15] D. Panicali and E. Paoletti, *Proc. Natl. Acad. Sci. U.S.A.* **79,** 4927 (1982).
[16] M. Mackett, G. L. Smith, and B. Moss, *Proc. Natl. Acad. Sci. U.S.A.* **79,** 7415 (1982).

TABLE I
FOREIGN GENES EXPRESSED IN RECOMBINANT VACCINIA VIRUSES

	Reference
1. Herpes simplex virus thymidine kinase	Panicali and Paoletti[15]; Mackett *et al.*[16]
2. Influenza virus hemagglutinin	Panicali *et al.*[17]; Smith *et al.*[18]
3. Hepatitis B virus surface antigen	Smith *et al.*[19]; Paoletti *et al.*[20]
4. Herpes simplex virus glycoprotein D	Paoletti *et al.*[20]; Cremer *et al.*[21]
5. *Plasmodium knowlesi* sporozoite antigen	Smith *et al.*[22]
6. Chloramphenicol acetyltransferase	Mackett *et al.*[23]
7. Rabies virus glycoprotein	Kieny *et al.*[24]; Wiktor *et al.*[25]
8. Transmissible gastroenteritis virus gp195	Hu *et al.*[26]
9. Vesicular stomatitis virus G protein	Mackett *et al.*[27]
10. Vesicular stomatitis virus N protein	Mackett *et al.*[27]
11. Influenza virus nucleoprotein	Yewdell *et al.*[28]
12. Human factor IX (Christmas factor)	de la Salle *et al.*[29]
13. Neomycin resistance gene	Franke *et al.*[30]
14. Sindbis virus structural proteins	Rice *et al.*[31]
15. β-Galactosidase	Chakrabarti *et al.*[32]
16. Epstein–Barr membrane antigen gp340	Mackett and Arrand[33]
17. Sindbis virus structural proteins	Franke and Hruby[34]
18. Respiratory syncytial virus glycoprotein G	Ball *et al.*[35]
19. Tobacco etch virus proteins	Dougherty *et al.*[36]

genetic material (Table I[15–36]).[17,37] This is especially important for the construction of recombinant viruses which are to function as polyvalent vaccines. As noted above, the protocol for inserting foreign genes into vaccinia virus dictates that the foreign DNA be located at nonessential

[17] D. Panicali, S. W. Davis, R. L. Weinberg, and E. Paoletti, *Proc. Natl. Acad. Sci. U.S.A.* **80,** 5364 (1983).

[18] G. L. Smith, B. R. Murphy, and B. Moss, *Proc. Natl. Acad. Sci. U.S.A.* **80,** 7155 (1983).

[19] G. L. Smith, M. Mackett, and B. Moss, *Nature (London)* **302,** 490 (1983).

[20] E. Paoletti, B. R. Lipinskas, C. Samsonoff, S. Mercer, and D. Panicali, *Proc. Natl. Acad. Sci. U.S.A.* **81,** 193 (1984).

[21] K. J. Cremer, M. Mackett, C. Wohlenberg, A. L. Notkins, and B. Moss, *Science* **228,** 737 (1985).

[22] G. L. Smith, G. N. Godson, V. Nussenzweig, R. S. Nussenzweig, J. Barnwell, and B. Moss, *Science* **224,** 397 (1984).

[23] M. Mackett, G. L. Smith, and B. Moss, *J. Virol.* **49,** 857 (1984).

[24] M.-P. Kieny, R. Lathe, R. Drillien, D. Spehner, S. Skory, D. Schmitt, T. Wiktor, H. Koprowski, and L. P. Lecocq, *Nature (London)* **312,** 163 (1984).

[25] T. J. Wiktor, R. I. Macfarlan, K. J. Reagan, B. Dietzschold, P. J. Curtis, W. H. Wunner, M.-P. Kieny, R. Lathe, J.-P. Lecocq, M. Mackett, B. Moss, and H. Koprowski, *Proc. Natl. Acad. Sci. U.S.A.* **81,** 7194 (1984).

[26] S. Hu, J. Bruszewski, T. Boone, and L. Souza, *in* "Approaches to Vaccines: Molecular and Chemical Basis of Virus Virulence and Immunogenicity" (R. M. Chanock and R. A. Lerner, eds.), p. 219. Cold Spring Harbor Lab., Cold Spring Harbor, New York, 1984.

loci so as not to disrupt the flow of genetic information. Nonessential genes such as thymidine kinase or DNA sequences deleted from the virus[38] provide obvious targets for the insertion of foreign genetic elements. Additional nonessential loci have been identified empirically and their location is illustrated in Fig. 2. More than a dozen nonessential loci are localized within the leftmost 30 kb of the vaccinia genome.

These sites can be used as loci from which additional viable deletion mutants of vaccinia virus can be generated. In addition to the spontaneously occurring deletion of 9.8 kb of DNA toward the left end of the vaccinia genome (S variant),[38] other deletion mutants have been generated in the L variant vaccinia genome using this approach and are shown in Fig. 2. Since the 21.4 kb of DNA deleted is not essential for viral replication, foreign genes can be readily inserted anywhere within this large target. The availability of these viable vaccinia deletion mutants provides additional space for packaging foreign DNA. This has significant advantages with respect to immunization against multiple pathogens (polyvalent vaccines), for the coordinated expression of a family of related gene products, or the production of a large number of biologically active molecules from a single vaccinia infection by the simultaneous expression of multiple foreign genes.

Utilizing the above information, we have generated a vaccinia virus recombinant that contains multiple foreign genes.[39] We will describe the generation of this multiple recombinant, which contains the coding sequences for the influenza virus hemagglutinin, the hepatitis B virus surface antigen, and the herpes simplex virus glycoprotein D. The generation of this triple vaccinia recombinant is an example of the versatility of

[27] M. Mackett, T. Yilma, J. K. Rose, and B. Moss, *Science* **227,** 433 (1985).

[28] J. W. Yewdell, J. R. Bennink, G. L. Smith, and B. Moss, *Proc. Natl. Acad. Sci. U.S.A.* **82,** 1785 (1985).

[29] H. de la Salle, W. Altenburger, R. Elkaim, K. Dott, A. Dieterle, R. Drillien, J.-P. Cazenave, P. Tolstoshev, and J.-P. Lecocg, *Nature (London)* **316,** 268 (1985).

[30] C. A. Franke, C. M. Rice, J. H. Strauss, and D. E. Hruby, *Mol. Cell. Biol.* **5,** 1918 (1985).

[31] C. M. Rice, C. A. Franke, J. H. Strauss, and D. E. Hruby, *J. Virol.* **56,** 227 (1985).

[32] S. Chakrabarti, K. Brechling, and B. Moss, *Mol. Cell. Biol.* **5,** 3403 (1985).

[33] M. Mackett and J. R. Arrand, *EMBO J.* **4,** 3229 (1985).

[34] C. A. Franke and D. E. Hruby, *J. Gen. Virol.* **66,** 2761 (1985).

[35] L. A. Ball, K. K. Y. Young, K. Anderson, P. L. Collins, and G. W. Wertz, *Proc. Natl. Acad. Sci. U.S.A.* **83,** 246 (1986).

[36] W. G. Dougherty, C. A. Franke, and D. E. Hruby, *Virology* **149,** 107 (1986).

[37] G. L. Smith and B. Moss, *Gene* **25,** 21 (1983).

[38] D. Panicali, S. W. Davis, S. R. Mercer, and E. Paoletti, *J. Virol.* **37,** 1000 (1981).

[39] M. E. Perkus, A. Piccini, B. R. Lipinskas, and E. Paoletti, *Science* **229,** 981 (1985).

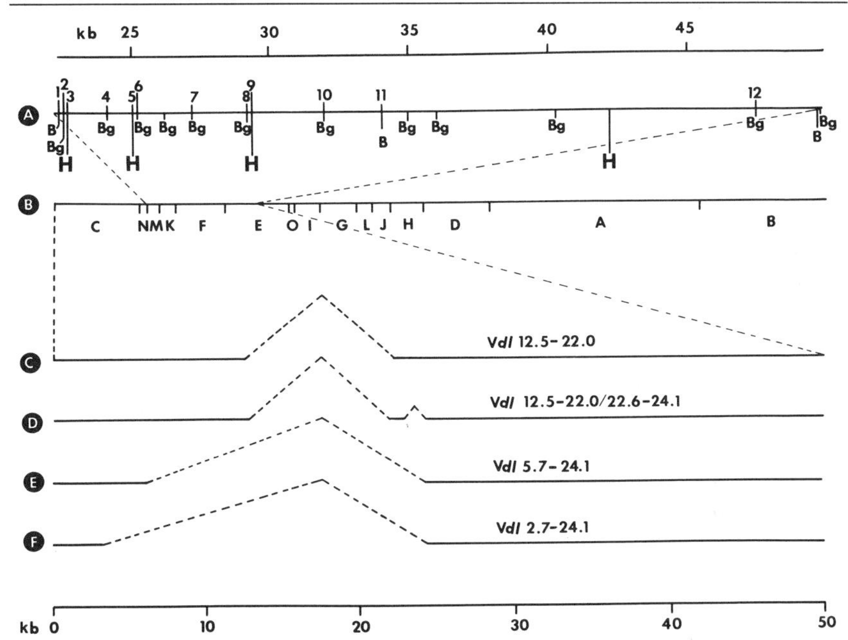

FIG. 2. Insertions and deletions of vaccinia virus. The physical map of the entire L variant vaccinia genome[38] as defined by *Hin*dIII restriction enzyme sites is indicated in (B) and the fragments are referred to as A–O in decreasing size. The bottom line indicates the scale for the left 50 kb in the L- variant vaccinia genome referring to lines (C)–(F). The top line indicates the scale for a blow-up of the region encompassing 25–45 kb from the left terminus and refers to line (A). The location of a dozen viable insertion mutants of vaccinia virus generated by the site-specific insertion of appropriately modified herpes simplex virus thymidine kinase coding sequences is shown in line (A). The numbers 1–12 mark the insertion sites and are identified as B, *Bam*HI; Bgl, *Bgl*II; H, *Hin*dIII sites. Other *Bam*HI, *Bgl*II, and *Hin*dIII enzyme sites are also shown. A series of viable deletion mutants derived from the L variant prototypic vaccinia genome[38] are indicated in lines (C)–(F). The deletion mutants are indicated as Vdl and assigned map coordinates based on kilobases of DNA deleted (indicated by dotted lines) and referenced to the left terminus of the L variant vaccinia genome.

vaccinia virus as an expression vector and exemplifies an approach to the construction of polyvalent vaccines.

Procedures

Donor Plasmid Construction

We will present three representative donor plasmid constructions, designed to permit the insertion of three different foreign genes into three nonessential sites in the vaccinia genome. Vaccinia *Hin*dIII fragments are cloned into the *Hin*dIII site of pBR322 or related plasmids and purified by standard methods.[40] These plasmids are used directly to provide flanking vaccinia arms in chimeric donor plasmid constructions, and as the source of vaccinia DNA for subcloning, promoter mapping and isolation, and subsequent genetic manipulations. Since vaccinia recognizes eukaryotic or prokaryotic promoters poorly, foreign genes are placed under the control of vaccinia promoters for optimal expression. The three constructions detailed below demonstrate the general strategies of (1) inserting a foreign gene proximal to a vaccinia promoter at a nonessential location in the genome, (2) replacing nonessential sequences downstream from a vaccinia promoter with a foreign gene, and (3) translocating a vaccinia promoter and foreign gene to a nonessential site within the vaccinia genome.

1. An early vaccinia promoter is located in the *Hin*dIII F region of vaccinia immediately adjacent to the unique *Bam*HI restriction site (Fig. 2). To construct a plasmid in which the *Bam*HI site in vaccinia is unique, the *Bam*HI site of pBR325 is destroyed by cutting with *Bam*HI endonuclease, followed by treatment with T4 polymerase to generate blunt ends, and self-ligation using T4 DNA ligase. This modified pBR325 plasmid is then cleaved with *Pst*I, and a 5.6-kb vaccinia *Pst*I fragment derived from *Hin*dIII F and containing the early promoter and unique *Bam*HI site is inserted, generating a novel plasmid pRW120. *Bam*HI linkers are ligated onto a 1780-bp DNA fragment containing a cDNA copy of the hemagglutinin gene from influenza virus.[17] The resulting *Bam*HI-ended DNA fragment is inserted into *Bam*HI-cleaved pRW120, in the correct orientation (right to left) relative to the vaccinia promoter, generating plasmid pDP122B.[17]

2. A 1.8-kb fragment derived from a *Hin*dIII–*Sal*I digest of cloned vaccinia *Hin*dIII M is directionally cloned into the *Hin*dIII–*Sal*I region of pBR322, generating pMP62. We had previously mapped an early vaccinia promoter within pMP62, adjacent to a *Rsa*I restriction site by transcription mapping (unpublished). Since both the vaccinia and pBR322 sequences contain multiple *Rsa*I sites, it is not possible to insert a foreign gene directly into the plasmid at the desired location. Instead, an 800-bp *Sph*I–*Bgl*II frament containing the promoter is isolated from pMP62, and subjected to partial *Rsa*I digestion. The 200-bp *Rsa*I (partial)–*Bgl*II fragment containing the promoter is isolated, and *Hin*dIII linkers ligated at the *Rsa*I (blunt) ends. To facilitate subsequent manipulations, the *Sph*I site of pMP62 is also changed to a *Hin*dIII site by linker addition. This

[40] T. Maniatis, E. F. Fritsch, and J. Sambrook, "Molecular Cloning: A Laboratory Manual." Cold Spring Harbor Lab., Cold Spring Harbor, New York, 1982.

modified pMP62 is cleaved with *Hin*dIII and *Bgl*II, and the 800-bp fragment is replaced with the 200-bp *Hin*dIII-linked *Rsa*I (partial)–*Bgl*II fragment, in which the promoter is located immediately upstream from the *Hin*dIII site. The resulting plasmid, pMP62-15, contains an upstream vaccinia arm, containing the vaccinia promoter and approximately 500 bp of rightward vaccinia sequences, and a downstream vaccinia arm, consisting of approximately 800 bp of vaccinia sequences leftward from the vaccinia *Sph*I (now *Hin*dIII) site. Six hundred base pairs of nonessential vaccinia is deleted between the two arms in this construct. A 1330-bp DNA fragment coding for the herpes simplex virus type 1 glycoprotein D[20] is modified by *Hin*dIII linker addition and the resulting *Hin*dIII fragment inserted in the proper orientation right to left into the *Hin*dIII site downstream from the vaccinia promoter in pMP62-15, generating the chimeric donor plasmid, pLP2.[39]

3. To provide flanking vaccinia arms around the nonessential vaccinia site at the *Hin*dIII M–K junction (Fig. 2), a 2.7-kb (partial) *Bgl*II fragment spanning this region is cloned into the *Bam*HI site of pBR322H$^-$, a pBR322 derivative in which the *Hin*dIII site is destroyed by a fill-in reaction using the Klenow fragment of *E. coli* polymerase. Ligation of the vaccinia *Bgl*II fragment into the pBR322 *Bam*HI site destroys both *Bgl*II and *Bam*HI sites. Thus the only *Hin*dIII site in the resulting plasmid, pMP18, is the site at the vaccinia *Hin*dIII M–K junction, and the only *Bgl*II site is located in the vaccinia *Hin*dIII K region, about 100 bp to the right of the *Hin*dIII site. Both the *Hin*dIII and *Bgl*II sites are known to be nonessential for vaccinia virus replication (Fig. 2). A 360-bp *Hin*dIII–*Bam*HI fragment containing a vaccinia promoter derived from vaccinia *Hin*dIII fragment I (Fig. 2) was inserted between the *Hin*dIII and *Bgl*II sites of pMP18, generating pMP18PP. The promoter-containing fragment is constructed such that the promoter is immediately upstream from the *Hin*dIII site. *Hin*dIII linkers are added to a 1090-bp DNA fragment containing the hepatitis B virus surface antigen.[20,41] This *Hin*dIII fragment is inserted in the proper orientation (right to left) downstream from the translocated promoter at the unique *Hin*dIII site of pMP18PP, generating plasmid pPP18sAg.[39]

In Vivo Recombination

The generation of the vaccinia virus recombinant containing the genes encoding the Influenza virus hemagglutinin (InfHA), the herpes simplex virus glycoprotein D (HSVgD), and the hepatitis B virus surface antigen

[41] F. Galibert, E. Mandart, F. Fitoussi, P. Tiollais, and P. Charnay, *Nature* (*London*) **281,** 646 (1979).

(HBsAg) begins with the construction of the chimeric donor plasmids as described above. Each foreign gene is then inserted sequentially into the vaccinia genome via *in vivo* recombination and screened via a replica filter technique described below (Fig. 3).

A number of variations of the calcium phosphate precipitation method for transfecting DNA[14] have been utilized with vaccinia virus with comparable results. A simplified procedure resulting in efficient transfection is

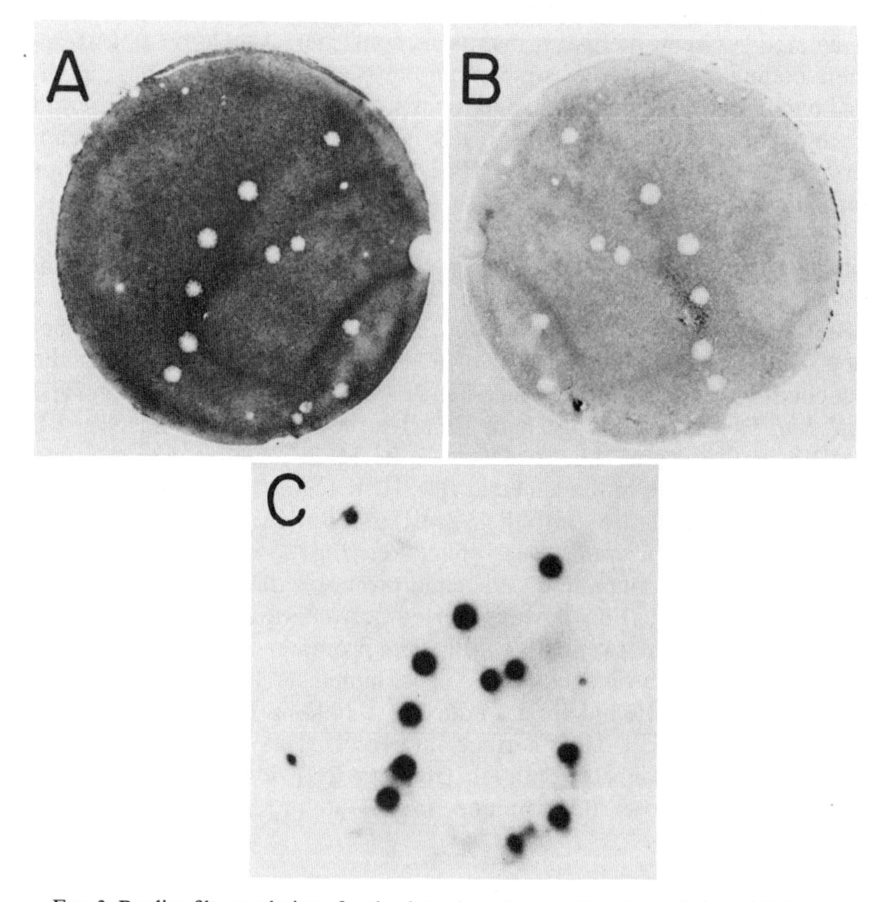

FIG. 3. Replica filter technique for the detection of recombinant poxviruses. (A) Progeny virus from an *in vivo* recombination is plated on a lawn of cells and plaques visualized with neutral red staining and the monolayer is quantitatively imprinted onto a nitrocellulose filter. (B) A mirror-image replica is generated by contacting a second nitrocellulose filter to the primary imprint, transferring a portion of the infected monolayer from the primary filter to the second. (C) One of the filters is prepared for *in situ* hybridization and the plaques generated by recombinant virus containing the foreign DNA sequences are detected with a ^{32}P-labeled probe corresponding to the foreign DNA and autoradiogrammed.

described. To 50 μg of purified donor plasmid DNA in 100 μl of H_2O is added 10 μl of 2.5 M $CaCl_2$ and 110 μl of 2× HeBS (2× HeBS is 0.28 M NaCl, 0.05 M HEPES buffer, and 1.5 mM sodium phosphate, pH 7.0). This mixture is held at room temperature for 30 min, then mixed with 220 μl of rescuing vaccinia virus titered to contain 1–10 pfu/cell. Monolayer cells in a 6-cm Petri dish are infected with the above solution. After a 1-hr adsorption period at 37° 3 ml of liquid nutrient overlay medium is applied. After 24 hr, the cells are harvested and virus released by three cycles of freezing and thawing. This lysate, which contains a mixed population of both wild-type and recombinant vaccinia viruses, is used to infect new cell monolayers. Viral plaques are visualized by neutral red staining and the monolayer is imprinted onto a nitrocellulose filter by contact.[42] A mirror image replica is imprinted onto a second nitrocellulose filter by contact with the primary filter and stored at −20°. The primary nitrocellulose filter is used for *in situ* hybridization.[42] The filters are prepared for hybridization by successive treatment with 0.5 N NaOH, 1 M Tris–HCl, pH 7.6, and 0.1 M Tris–HCl, pH 7.6. The filters are baked at 80° in a vacuum oven for 2 hr. After baking they are hybridized with a ^{32}P-labeled nick-translated fragment as a specific probe to detect viral plaques containing recombinant vaccinia virus which contain the foreign DNA insert. The hybridization conditions begin with a prehybridization time of 4 hr at 65° in which the filters are immersed in a solution containing 4× SSC (20× SSC is 3 M NaCl, 0.3 M sodium citrate, pH 7.0), 5× Denhardt's [10× Denhardt's is 0.2% (w/v) bovine serum albumin, 0.2% (w/v) polyvinylpyrrolidone, 0.2% (w/v) Ficoll 400], 0.1% (w/v) sodium dodecyl sulfate, 0.1% (w/v) tetrasodium pyrophosphate, and 100 μg/ml salmon sperm DNA. The filters are contained in an appropriate vessel such that there is minimum excess volume not in contact with the filters. After 4 hr, the denatured radioactive probe (specific activity of approximately 1×10^8 cpm/μg) at a concentration of 1×10^5–1×10^6 cpm/ml is added directly to the hybridization solution. Hybridization is allowed to continue for at least 18 hr at 65°. After hybridization the filters are washed in 0.2× SSC, 0.1% sodium dodecyl sulfate, 0.1% tetrasodium pyrophosphate at 65° for at least 5 hr. The filters are dried, placed on a support covered with Saran Wrap, and autoradiographed using Kodak XR-2 film exposed overnight at −70° with Du Pont Cronex Lightning Plus intensifying screens.

After autoradiography, positive signals identifying plaques generated by recombinant vaccinia virus are aligned with areas of the replica nitrocellulose filter. Infectious recombinant virus can be recovered by punching the nitrocellulose filter with a cork borer and inoculating monolayer

[42] L. P. Villarreal and P. Berg, *Science* **196,** 183 (1977).

cell cultures. Two or three additional plaque purifications under agar assure homogeneity of the virus preparation. Finally, the recombinant virus is amplified to high titer, and purified on sucrose gradients.

This methodology provides a general approach to identifying vaccinia virus recombinants based solely on the presence of the foreign nucleic acid sequence and independent of its expression into a protein. This scheme is used to identify the triple recombinant containing the InfHA, HSVgD, and HBsAg genes and is outlined below. (1) The vaccinia recombinant vP53 is generated using S variant VTK^{-}79 as rescuing virus and the chimeric donor plasmid pDP122B containing InfHA sequences. (2) The insertion of HSVgD coding sequences is effected with the recombinant vaccinia virus vP53 as rescuing virus and pLP2 as donor plasmid. A recombinant vaccinia vP124 containing both InfHA and HSVgD sequences is identified by replica filter plating and hybridization. (3) This recombinant vP124 is then used as the rescuing virus for the insertion of the HBsAg coding sequences from donor plasmid pPP18sAg. The triple recombinant vP168 is isolated and plaque purified. In all three cases the procedure is that of the general protocol and at specific times each vP is probed with the appropriate ^{32}P-labeled nick-translated sequences. All three foreign coding sequences are inserted at nonessential sites in the vaccinia genome, and transcription of the inserted genes is right to left relative to the vaccinia genome and regulated by vaccinia promoters.

Recombinant viruses are routinely analyzed to verify correct DNA insertion and for mRNA synthesis and protein expression. Additionally, the recombinant viruses are used in animal studies aimed at determining the immunological responses to the foreign antigens.

Expression Analysis

DNA Analysis

The type of analysis that is routinely performed on purified DNA from vaccinia virus recombinants is illustrated in Fig. 4. The DNA from vP168 was extracted[38] and cleaved with *Hin*dIII or *Bam*HI restriction endonucleases. The DNA fragments were separated on agarose, blotted onto GeneScreen Plus (NEN/E. I. Du Pont deNemours and Company) by the Southern method,[43] and probed with ^{32}P-labeled nick-translated[44] DNA sequences homologous to the foreign genes. As shown at the top of Fig. 4, the HSVgD, HBsAg, and InfHa sequences were inserted 23.9, 25.1, and

[43] E. M. Southern, *J. Mol. Biol.* **98,** 503 (1975).

[44] P. W. Rigby, M. Dieckmann, C. Rhodes, and P. J. Berg, *Mol. Biol.* **113,** 237 (1977).

34.1 kb, respectively, from the left terminus of the vaccinia genome as referenced to the L variant vaccinia genotype. The autoradiogram is used to ascertain the genomic organization of the triple recombinant. With the appropriate enzyme, each insert can be released as an original intact fragment or as a larger fragment which is part of the vaccinia genome. This type of analysis verifies the genomic organization of the construct and can detect if any rearrangements of the vaccinia genome have occurred.

RNA Analysis

RNA from vaccinia-infected cells can be isolated and analyzed by standard procedures.[45] Mapping vaccinia mRNA via S1 nuclease analysis[46,47] allowed a number of vaccinia promoters to be identified (unpublished), some of which are used in the construction of vP168.

By using different vaccinia promoters varied levels of RNA expression for a single foreign gene can be obtained. A Northern blot analysis shown in Fig. 5 emphasizes the difference in expression levels with different vaccinia promoters. Steady state RNA levels are shown for HBsAg RNAs (lanes C and D) and InfHA RNAs (lanes G and H). It should be noted that vP142 (lane H) is a double-vaccinia recombinant containing InfHA and HSVgD genes under the regulation of two different promoters. HSVgD RNA expression from vP142 is shown in lane K.[48]

Analysis of Protein Expression

Various assays are available to measure expression of the foreign gene at the protein level. For example, functional herpes simplex virus thymidine kinase can be quantitated by monitoring the phosphorylation of [^{125}I]iododeoxycytidine[15,49] a property not exhibited by eukaryotic or vaccinia-induced thymidine kinase. A gene encoding another enzyme, β-galactosidase, has also been introduced into the vaccinia genome.[32] Detection of expression of this enzyme follows the principles used originally in the prokaryotic system.[50] The substrate 5-bromo-4-chloro-3-indolyl-β-

[45] R. K. Barth, K. W. Gross, L. D. Gremke, and N. D. Hastie, *Proc. Natl. Acad. Sci. U.S.A.* **79,** 500 (1982).

[46] A. J. Berk and P. A. Sharp, *Cell* **12,** 721 (1977).

[47] R. F. Weaver and C. Weissmann, *Nucleic Acids Res.* **7,** 1174 (1979).

[48] E. Paoletti, M. Perkus, A. Piccini, S. Wos, and B. R. Lipinskas, *in* "Medical Virology IV" (L. M. de la Maza and E. M. Peterson, eds.), p. 409. Erlbaum, Hillsdale, New Jersey, 1985.

[49] W. C. Summers and W. P. Summers, *J. Virol.* **24,** 314 (1977).

[50] T. J. Silhavy and J. R. Beckwith, *Microbiol. Rev.* **49,** 398 (1985).

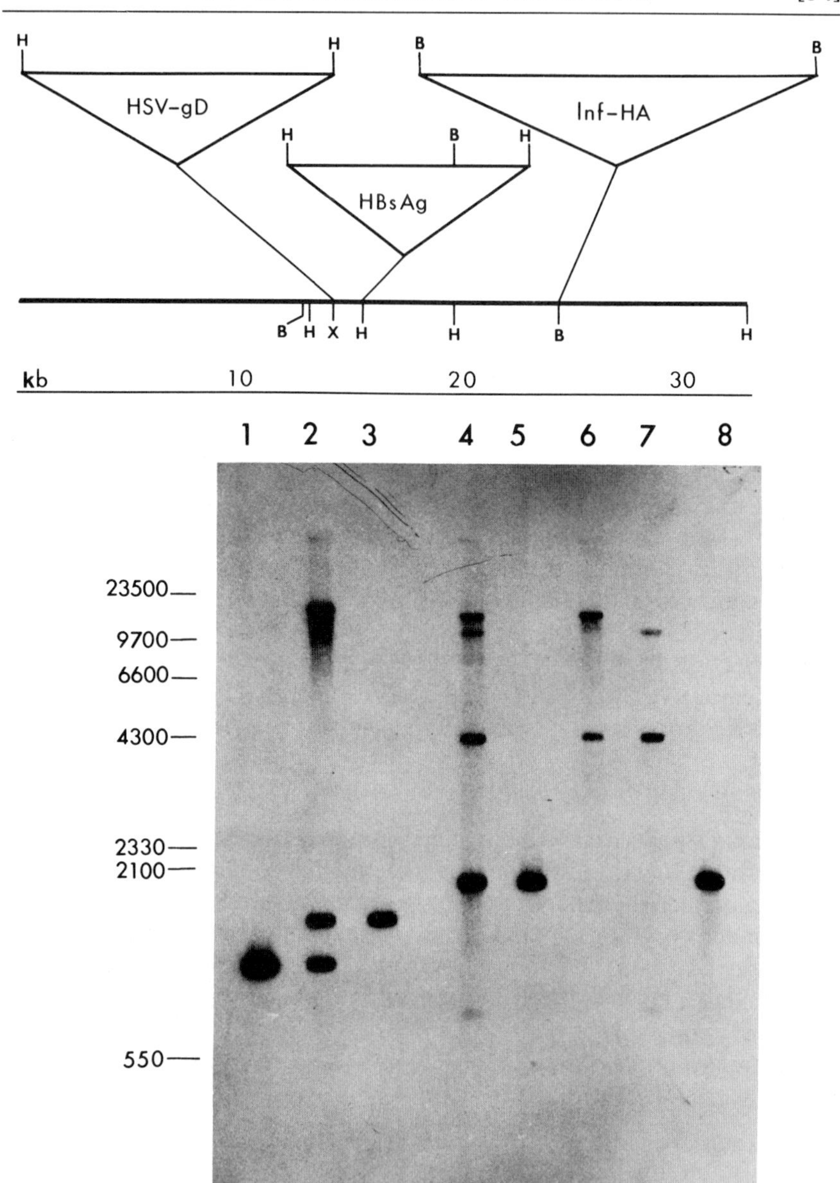

FIG. 4. Structure of the vaccinia virus recombinant harboring the HSVgD, HBsAg, and InfHA coding sequences. The leftmost 30 kb of the vaccinia S variant genome[38] encompass-

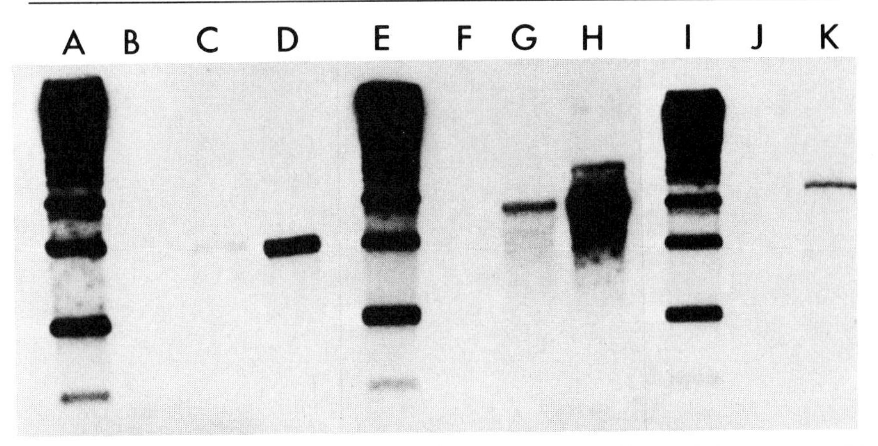

FIG. 5. Messenger RNA analysis of foreign genes expressed in vaccinia virus recombinants. RNA was extracted 3 hr postinfection with vaccinia virus at a multiplicity of 10 pfu/cell in the presence of 40 μg/ml of cytosine arabinoside. Cells were lysed by homogenization and RNA extracted and purified by modifications of the guanidine–HCl method.[45] Twenty micrograms of RNA was electrophoresed on 1.5% agarose gels, blotted onto nitrocellulose paper, and hybridized with nick-translated probes of the respective foreign gene in pBR322. An autoradiogram is shown. Lanes A, E, and I show hybridization of radiolabeled probes with restricted pBR322. Lanes B, F, and J indicate absence of detectable hybridization to RNA isolated from cells infected with wild-type vaccinia virus. Lanes C and D indicate HBsAg mRNA levels in cells infected with the vaccinia recombinant vP59 and vP139, respectively. Lanes G and H show the mRNA levels found in cells infected with the vaccinia recombinants expressing the influenza HA, vP53 and vP142, respectively, while lane K shows the HSVgD mRNA expressed by the double-recombinant vP142 (Paoletti *et al.*[48]).

---

ing the terminal *Hin*dIII F, M, K, and internal F fragments is shown at the top. The restriction sites identified are B (*Bam*HI) and H (*Hin*dIII). The foreign genes were inserted at specific sites as follows: HSVgD coding sequence was inserted at point X, which denotes the deletion of approximately 0.6 kb of vaccinia sequences described in the construction of pMP62-15. HBsAg and InfHA coding sequences were inserted at a *Hin*dIII site and *Bam*HI site, respectively. Additional *Bam*HI and *Hin*dIII sites are also shown. Southern blot hybridization analysis of the vaccinia virus recombinant, vP168, is shown at the bottom. DNA was extracted from purified vaccinia virus, digested with restriction endonucleases, and blotted to GeneScreen Plus after electrophoretic separation of the DNA fragments on agarose. vP168 DNA was digested with *Hin*dIII (lane 2) or *Bam*HI (lane 4 and lanes 6 to 8) and probed with a mixture (lanes 1 to 5) of or single (lanes 6 to 8) ^{32}P-labeled nick-translated HBsAg, HSVgD, and InfHA sequences. Purified DNA fragments containing the coding sequences for HBsAg (lane 1), HSVgD (lane 3), and InfHA (lane 5) were run as standards. Sizes of the fragments were obtained with *Hin*dIII-cleaved λ DNA as markers (Perkus *et al.*[39]).

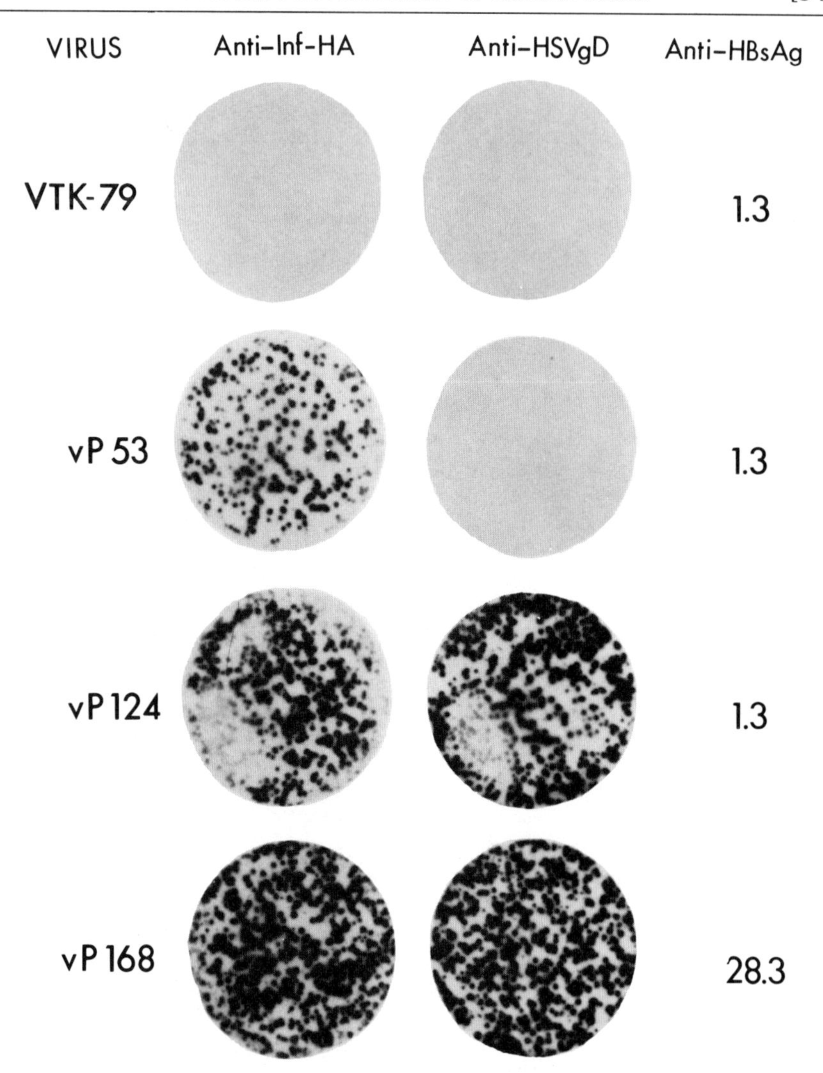

FIG. 6. Detection of InfHA, HSVgD, and HBsAg expression by vaccinia virus recombinants. BSC-40 cells were infected with either wild-type vaccinia virus, VTK^{-}79, or the vaccinia virus recombinants containing coding sequences for the InfHA (vP53), the InfHA and HSVgD (vP124), or the InfHA, HSVgD, and HBsAg (vP168). Two days after infection, the plaques were visualized by neutral red staining. The monolayers were exposed to antiserum to InfHA or to HSV at room temperature for 1 hr. Unreacted antibody was removed by washing in phosphate-buffered saline and ^{125}I-labeled protein A from *Staphylococcus aureus* was added. Unreacted ^{125}I-labeled protein A was removed by washing in saline. The monolayers were imprinted on nitrocellulose filters, dried, and autoradiographed. Film densities indicate presence of antibody-bound ^{125}I-labeled protein A. Unlike the InfHA and the

D-galactoside is added to the culture dishes and β-galactosidase expression is monitored by the appearance of blue plaques.

The *in vitro* expression of a number of genes in recombinant vaccinia virus has relied on immunological assays. These require that the protein product be structurally correct such that it will react with antiserum generated against the native product. Two types of immunological tests are illustrated in Fig. 6. Detection of InfHA, HSVgD, and HBsAg expression in the triple-recombinant vP168 is accomplished using antiserum followed by ^{125}I-labeled protein A for InfHA and HSVgD. A commercially available radioimmunoassay kit AUSRIA II (Abbott Laboratories) is used for HBsAg detection. As described in the legend to Fig. 6, these assays give a qualitative indication of protein expression. Often a quantitative approach in the detection of protein levels is desired. In considering refinement of vaccinia as an expression vector for vaccine applications it is clear that increased levels of expression of the foreign antigen would result in a more potent vaccine. However, taking into consideration the complexities of the immunological system it is reasonable to assume that in a polyvalent vaccine optimal response to a number of antigens may require varied levels of expression. With this in mind, we have set out to define vaccinia virus regulatory elements that allow expression of foreign genes at different levels. As shown in Fig. 7, different levels of expression of the InfHA, as a test antigen, in vaccinia virus vectors have been obtained using a variety of DNA regulatory signals. The detection is performed as an immunodot where infected cell lysates are applied to nitrocellulose. The filter is then processed with anti-influenza HA serum followed by ^{125}I-labeled protein A. The variation in intensities of the signals, as detected by autoradiography,[51] gives a measure of vaccinia-regulated InfHA ex-

[51] E. Paoletti, M. E. Perkus, A. Piccini, S. M. Wos, B. R. Lipinskas, and S. R. Mercer, *in* "Vaccines '85: Molecular and Chemical Basis of Resistance to Parasitic, Bacterial, and Viral Diseases" (R. A. Lerner, R. M. Chanock, and F. Brown, eds.), p. 147. Cold Spring Harbor Lab., Cold Spring Harbor, New York, 1985.

HSVgD, which are localized on the infected cell membrane, the HBsAg synthesized by vaccinia virus is excreted from the infected cells. Therefore, BSC-40-infected cell extracts were prepared by three cycles of freezing and thawing, and the presence of HBsAg was detected with a commercially available radioimmunoassay kit AUSRIA II (Abbott). The numbers represent the positive-to-negative ratio (*P*/*N*) as defined by the manufacturer. The horizontal columns represent BSC-40 cells infected with the wild-type VTK^{-}79 or the vP53, vP124, or vP168 recombinant vaccinia viruses. The vertical columns represent radioimmunoassays of unfixed, infected BSC-40 monolayers in which antiserum to InfHA or to HSVgD was used, or the *P*/*N* ratio as detected with the AUSRIA II kit (Perkus *et al.*[39]).

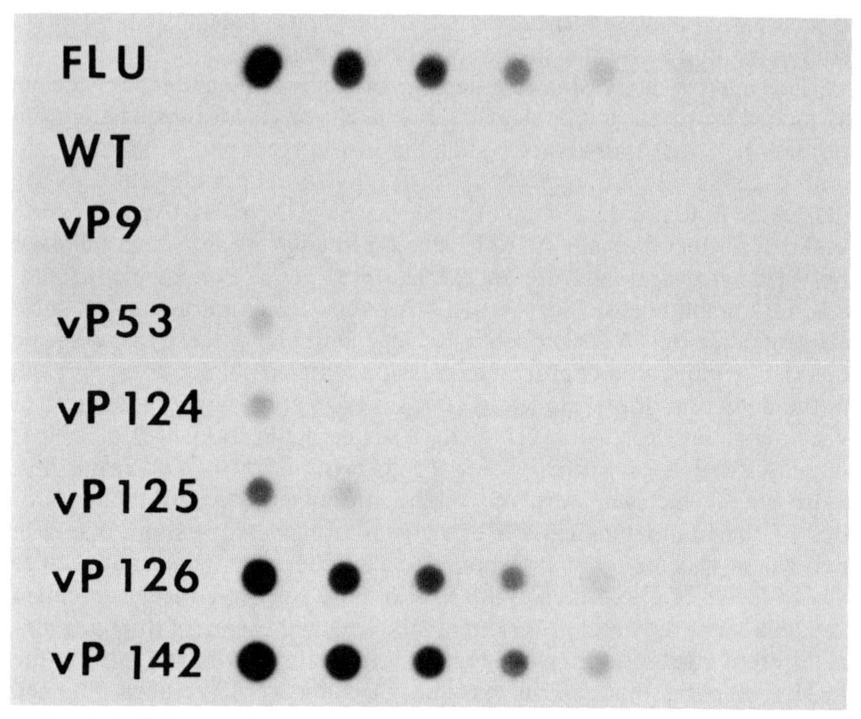

FIG. 7. Radioimmunoblot analysis of influenza HA expression under vaccinia virus regulation. Monolayers of CV-1 cells were infected at 1 pfu/cell with influenza virus, wild-type vaccinia virus, or recombinant vaccinia viruses (vP) expressing the influenza HA gene from different vaccinia promoters. Twenty-four hours post infection, the cultures were frozen and thawed three times to fracture the cells and 2-fold serial dilutions were fixed to a nitrocellulose filter. Specific anti-HA serum was applied to the filter followed, after appropriate washings to remove residual unreacted immunoglobulins, with radiolabeled protein A. An autoradiogram is shown (Paoletti *et al.*[51]).

pression in vaccinia-infected cells. It is of interest to note that the difference of InfHA expression between two vaccinia recombinants using two promoters, vP53 and vP142, is reflected in the different levels of RNA expression between these two vPs as shown in Fig. 5, lanes G and H, respectively.

Animal Studies

In vivo analyses using recombinant vaccinia viruses are a direct assessment of expression level and immunogenicity of the foreign gene products. To this end, studies using laboratory animals are performed. After inoculation, the immune sera can be tested for the presence of antibodies directed against the particular antigen or cellular immune responses measured. InfHA antibody levels have been detected using a hemagglutination inhibition assay.[17] The immune serum is monitored for its ability to inhibit the agglutination of either guinea pig or chicken erythrocytes by authentic InfHA. Specific antibody levels can be detected in a general assay which involves neutralization of viral infectivity. We have used this approach to detect levels of antibodies directed against vaccinia, influenza, and herpes simplex viruses.[17,20,52] HBsAg antibodies can be detected using a commercially available AUSAB radioimmunoassay (RIA) kit (Abbott Laboratories).[20] Protection of vaccinated animals to subsequent challenge with the pathogen has been reported.[18,20,53]

Future Prospects

There has been a renaissance of interest in vaccinia virus. This interest has arisen from the demonstration of using vaccinia virus as a vector for the expression of foreign genes. Vaccinia displays a number of characteristics which make it an ideal expression vector. Some of these are (1) the ability to incorporate large amounts of exogenous DNA, (2) the faithful transcription of the exogenous genes such that the RNAs are translated into protein which resemble the native product in structure, function, and localization, (3) the cytoplasmic site of vaccinia replication, allowing gene expression to proceed without interference by the host genome, and (4) the potentiality as a live recombinant vaccine as a target against human and veterinary pathogens. The data presented here attest to the feasibility of using vaccinia virus as an expression vector with particular attention paid to its potential as a vaccine vector. Future research efforts will focus on the manipulation of the vaccinia genome with respect to defining the genetic parameters which control the expression of both endogenous and exogenous genes. For example, further identification of nonessential vaccinia DNA sequences will establish the minimum amount of genetic material necessary for virus replication. Dissection of the vaccinia genome will serve to attenuate the virus and may influence both host range and virulence. Analysis of regulatory control signals for vaccinia transcription will be aimed at defining the properties of a vaccinia promoter. As this data base accumulates, consensus sequences can be generated to determine the subtle differences in the temporal and quantitative control of vaccinia transcription. Additionally, the translation of exogenous gene transcripts encoding foreign proteins need to be analyzed in terms of RNA utilization and host RNA competition. Fundamentally, the definition of vaccinia molecular mechanisms will lead toward the elucidation of regulatory replication functions involved in vaccinia biogenesis.

[52] E. Paoletti, B. R. Lipinskas, S. Woolhiser, and L. Flaherty, *UCLA Symp. Mol. Cell. Biol., New Series* **21,** 663 (1984).

[53] B. Moss, G. L. Smith, J. L. Gerin, and R. H. Purcell, *Nature (London)* **311,** 67 (1984).

Author Index

Numbers in parentheses are footnote reference numbers and indicate that an author's work is referred to although the name is not cited in the text.

A

Abel, P., 272
Abraham, E. P., 195
Abraham, J., 519, 532(28)
Abrahmsén, L., 513, 514(20), 515(21)
Abrenski, K., 19
Achstetter, J., 537
Adams, C. W., 458
Adams, R. M., 509
Adams, S. P., 278, 293, 301(9), 320
Adams, T. L., 285
Adelman, J., 390, 391(2), 402, 415(11)
Ades, D. K., 482, 489(2), 490(2)
Adhya, S. L., 23
Ahne, F., 373, 374(21)
Aiba, S., 136(*cc*), 141, 162(*z*), 163, 510(T14, T15), 511
Akaoka, A., 510(T5), 511, 512, 513(19), 516
Akiyoshi, D. E., 306
Alberts, B. M., 29
Alborn, W. E., Jr., 168, 169(40), 171(40), 172(40), 178(40), 266
Allen, M. M., 221
Allen, N. E., 168, 169(40), 171(40), 172(40), 178(40), 266
Allet, B., 403, 421, 422(13), 424(13), 427
Alt, F., 268
Altenbuchner, J., 118, 165(19), 167, 460
Altenburger, W., 549(29), 550
Alton, N. K., 58(36), 66, 161(*i*), 162
Amalric, F., 339, 341(15)
Amasino, R. M., 305, 306
Amerer, G., 517
Ames, B. N., 61
Ammirato, P. V., 302
An, G., 257, 282, 293, 294(11), 295(22), 296, 297, 301(11, 20, 21), 302(20, 21)
Anderson, A. R., 299, 310
Anderson, D., 35, 198
Anderson, E. H., 370
Anderson, K., 549(35), 550
Angal, S., 473
Anzai, H., 136(*aa*), 141, 161(*d*), 162
Anziano, P. Q., 369, 373(14), 374(14)
Appleyard, R. K., 421
Arber, W., 77
Arai, K.-I., 518, 534
Arai, N., 518, 534
Argarana, C. E., 511
Armour, S., 266
Armour, S. L., 278
Arnaud, M., 510(T19), 511
Arntzen, C. J., 219, 221(19)
Arrand, J. R., 549, 550
Artz, S., 12, 19(7)
Ash, L., 390
Astell, C. R., 176
Astier, C., 202, 203, 210(50), 217, 232
Auerbach, J. I., 136(*ee*), 141, 161(*k*), 162
Auerswald, E. A., 14, 162(*w*), 163, 236, 275
August, T., 482, 489(2), 490(2)
Ausubel, F. M., 48, 50(42), 219, 223(20), 224, 233, 255
Axel, R., 511, 530
Ayala, E., 421
Aykent, S., 272
Aymerich, S., 510(T20), 511

B

Bachrach, H. L., 415
Backendorf, C., 432
Backman, K., 37, 168(45), 169, 171(45), 433
Bagdasarian, M., 48, 50(43)

Bagdasarian, M. M., 48, 50(43)
Bailey, C. R., 138 (*jj*), 141
Bajszar, G., 546
Baker, B., 284, 314
Balaz, E., 261
Balazs, E., 320
Balazs, I., 524, 541
Baldacci, G., 369, 372(13), 373
Baldari, C., 12, 19(18), 105
Baldwin, J. E., 195
Ball, L. A., 549, 550
Ballard, A., 461
Ballou, C. E., 540, 542
Baltz, R. H., 127, 167, 171(26), 173(26), 176, 177, 178(26), 183(60), 195(26), 198(26)
Banks, A. R., 517, 519(6), 534(6), 535(6), 541(6), 542(6)
Banner, C., 510(T4), 511
Barker, R. F., 273, 285
Barnason, A., 272
Barnes, W. M., 294
Barnett, L., 104, 105(4, 5)
Barnwell, J., 549
Barr, P. J., 501, 534
Barry, G. F., 306
Barth, P. T., 37, 48(24), 236
Barth, R. K., 557, 559(45)
Barton, K., 256, 282, 293
Bassett, C. L., 472
Bates, P. F., 83, 85(4)
Baughman, G., 451
Baumann, G., 286
Baxter, J. D., 392, 406
Beach, D., 373
Beachy, R., 272
Beaud, G., 546
Beck, C. F., 456, 458(6)
Beck, E., 14, 17, 162(*w*), 163, 236, 275
Becker, A., 95(6), 96, 101(6), 103(6), 103, 169
Becker, G. W., 502
Beckers, F., 323
Beckman, R. J., 406
Beckwith, J. R., 557
Bedbrook, J., 257, 266, 289
Beerman, N. E., 168(44), 169, 171(44), 178(44)
Begg, G., 541
Beggs, J. D., 367
Belagaje, R., 195
Belagaje, R. M., 390, 391(4), 402, 403(4), 404(20), 405(20), 407, 411(4), 414(4), 415(4)
Bellard, M., 40
Benavente, A., 461
Ben-Hamida, F., 546
Benigni, R., 166
Bennett, D. R., 294
Bennett, P. M., 283
Bennetzen, J. L., 528
Bennink, J. R., 549(28), 550
Benson, S. A., 507
Benz, R., 323, 338, 349(13)
Benzinger, R., 29
Beppu, T., 131(*g*), 135(*w*), 141, 162(*x*), 163
Berdy, J., 166
Berg, P., 87, 246, 258, 296, 334, 343
Berg, P. J., 555, 556
Berger, S. L., 423
Berk, A. J., 12, 19(8), 60, 530, 557
Berman, M. L., 11, 12, 19(16)
Berman, V., 511
Bernard, H.-U., 418, 421(5), 462
Bernardi, G., 369, 372(13), 373
Bernards, A., 196
Bernan, V., 162
Bertino, J., 268
Betlach, M. C., 35, 37(2), 55, 161(*b*), 162, 223, 417
Better, M., 299
Bevan, M. W., 257, 278, 282, 293, 320
Bibb, M., 511
Bibb, M. J., 116, 117, 118(6, 13), 119(13), 121(7), 127, 129, 133(*p*, *q*), 134(*q*), 138(*kk*), 141, 159(7), 161(*j*), 162(*bb*), 163, 165, 167, 177, 178(66), 196(20)
Bibikova, M., 201, 202(21)
Biemann, K., 511
Biggin, M. D., 467
Birken, I. D., 511
Bijard, H., 492
Birch, A. W., 136(*bb*), 141
Birnboim, H. C., 29, 35, 45(16), 66, 245, 300, 369, 408, 454
Birnstiel, M. L., 197
Birr, E., 131(*k*), 132(*k*), 141
Bitter, G. A., 517, 518, 519(6), 521(4), 522(4), 525(4), 527(4), 528(4), 529(4), 534(6), 535(6, 55), 537(55), 538(55), 539(55), 540(55), 541(6, 55), 542(6, 55)
Bittner, M. L., 278, 293, 301(9), 320
Bjorn, M. J., 510(T24), 511

Bjornsti, M.-A., 198
Blacher, R. W., 510(T8), 511
Blair, L., 536, 537, 540(57), 542(59)
Blanc, H., 368
Blankenship, D. T., 195
Blattner, F. R., 77, 95(7), 96, 106, 107, 482
Blechl, A. E., 77, 95(7), 96
Blin, N., 88
Bloemheuvel, G., 202, 203(49), 211(49)
Blombäck, B., 479
Blombäck, M., 479
Blum, P., 12, 19(7)
Bochner, B. R., 61
Boeke, J. D., 3, 12, 12(11), 23(11)
Bogorad, L., 201, 204(19), 231
Boissy, R., 176
Bolivar, F., 35, 37(2), 55, 161(*b*), 162, 223, 224, 385, 402, 415(1, 2), 417, 463
Bond, M. W., 534
Bonnert, H. J., 335
Boone, T., 549
Borrias, W. E., 200, 201(13), 202(13, 18), 203(18, 49), 204(13, 18, 19, 48), 206(13), 207(18, 39, 48), 210(18), 211(18, 49), 212(13, 39, 48), 216, 224, 233, 234(15), 237, 242(15)
Borst, P., 196, 423
Bos, C. J., 367
Botstein, D., 77, 81(6), 82(6), 111, 176, 196(60a, 62), 243, 416
Boxer, S. G., 402, 414(6), 415(6), 462, 479(13), 480(13)
Boyer, H. W., 35, 37(2), 55, 63, 161(*b*), 162, 207, 223, 402, 415(2), 417, 463
Bradford, M. M., 68
Bradley, D. E., 14, 23
Bradley, S. G., 167
Brady, C., 162(*aa*), 163, 452, 458(1)
Brake, A. J., 534, 536, 540(57), 542(59)
Brammar, W. J., 74
Brand, L., 257, 266(20), 306
Brand, L. A., 278, 293, 301(9), 320
Brasch, M., 50
Braun, A. C., 237, 319
Brawner, M. E., 136(*ee*), 141, 161(*k*), 162
Brechling, K., 549(32), 550, 557(32)
Brenner, S. A., 104, 105(4, 5), 462, 463(20d), 478(20d), 479(20d), 481(20d)
Breznovitz, A., 321
Brisson, N., 314, 335(10)
Broach, J. R., 39, 47(29), 519, 521, 532(28)
Brosius, J., 55, 56(12), 58(12), 59(12), 60, 61(18, 20), 68(18), 434
Brousseau, R., 385
Brown, E. L., 407
Brown, R. M., Jr., 233
Brown, S. E., 48, 50(42)
Brown-Luedi, M., 258, 259(28), 261(28)
Brunel, F., 35, 36(4), 37(21), 39(4, 21), 41, 42(4), 45(4), 46(4), 52(21a)
Bruning, H. J., 168
Brusslan, J., 215, 216(6), 220(6, 11)
Bruszewski, J., 549
Brutlag, D., 15, 17(45)
Bruton, C. J., 116, 117, 121(7), 133(*q*), 134(*q*), 138(*jj*), 139(*ll*), 141, 151, 152, 153(59), 159(7), 160, 163, 164(59), 165(59), 167, 196(20)
Bryant, D. A., 215, 219, 221(18)
Bryson, V., 61
Buchholz, W. G., 306
Buchmann, I., 306
Buchwalder, A., 461
Buck, G. A., 510(T24), 511
Budar, F., 292
Buell, G., 116, 422, 427
Buikema, W. J., 48, 50(42)
Bujard, H., 162(*y*), 163, 460
Bukhari, A. I., 23
Bulawela, L., 462
Bullerjahn, G. S., 236, 242(36)
Burgett, S. G., 161(*m*), 162, 266
Burke, D. C., 517
Burke, J. F., 12, 19(9), 83, 129, 168
Burke, W. F., Jr., 211
Burnette, W. N., 532, 540(46)
Butkowski, R. J., 462, 477(23)
Butler, M. J., 138(*jj*), 141
Buttner, M. J., 119, 133(*p*), 141, 165, 511
Buzash-Pollert, E., 432
Buzby, J. S., 201, 202(22, 23), 213(22, 23), 216, 217, 231, 233
Byers, M. J., 370
Bytebier, B., 279, 284(16)

C

Caboche, M., 304, 336(11)
Calabi, F., 471
Callahan, P. X., 391

Callis, J., 360, 361
Cameron, J., 81, 176, 178(59)
Cantell, K., 512
Cantor, C. R., 105(17), 108, 511
Cantwell, B. A., 510(T18), 511
Caplan, A., 293
Carbon, J., 35, 167, 523
Carere, A., 166
Carle, G. F., 181, 245
Carlton, S., 66
Carr, N. G., 199, 212
Carter, P., 468
Cartwright, N. J., 41, 53(37)
Caruthers, M. H., 392, 528
Casadaban, M. J., 162, 219, 223(21), 224, 233, 421
Casavant, N. C., 191
Cashmore, A. R., 285(30, 31), 286
Castets, A.-M., 202, 204(39), 207(39), 212(39)
Castilho, B. A., 233
Cate, R. L., 60, 61(20)
Cazenave, J.-P., 549(29), 550
Cesareni, G., 12, 19(15, 18), 39, 50(27), 104, 105(4)
Chakrabarti, S., 549, 550, 557(32)
Chambers, A. E., 168(47), 169, 173(47)
Chambon, P., 40
Chan, C. S. M., 246, 249(17, 18), 253(17)
Chang, A., 268
Chang, A. C. Y., 161(*i*), 162(*y*), 163
Chang, C. N., 534
Chang, S., 509, 510(T10, T11), 511, 513, 514
Chang, S. Y., 129, 509, 510(T11), 511, 513, 514
Chapman, J. L., 195
Chapon, C., 432
Charnay, P., 553
Chater, K. F., 116, 117, 121(7), 133(*q*), 134(*q*), 138(*jj*), 139(*ll*), 140(*mm*), 141, 151, 152, 153(59), 157, 159(7), 160, 163, 164(59), 165(58, 59), 166, 167, 196(20)
Chauvat, F., 201, 202(20), 203, 210(50), 213(46), 214(46)
Chen, C. Y., 518, 525(39), 526, 527(18), 529(18)
Chen, E. Y., 61, 441, 510(T3), 511, 534
Chen, H.-Z., 471
Chen, J., 403
Chen, K. K., 517, 518, 519(6), 534(6), 535(6), 541(6), 542(6)
Chérif-Zahar, B., 373
Cheroutre, H., 423, 427(22)
Chevalier, N., 36, 37(21), 39(21), 52(21a)
Chiang, C. C., 293, 301(21), 302(21)
Chilton, M.-D., 256, 278, 282, 292, 293, 294, 310, 320
Chinenova, T. A., 117
Chirgwin, J., 36, 40(17), 62
Chirikjian, J. G., 80
Chiu, P., 35
Chlebowicz-Slediewska, E., 518
Cho, M.-S., 482, 489(2), 490(2)
Choi, C.-H., 296
Chollet, A., 421
Chou, J., 162, 219, 223(21), 224
Chou, P. Y., 503
Christensen, U., 478
Christie, G. E., 444, 446(27)
Chua, N.-H., 258, 263(23), 285
Chumley, F., 517
Chung, C. H., 471, 472(39)
Claeys, H., 462, 477(21)
Clark, A. J., 106, 409
Clark, D. J., 453
Clark, W. G., 234
Clarke, L., 35, 167, 523
Cleary, J. M., 18
Clemans, D., 137(*gg*), 141
Close, T. J., 58, 59(17), 293
Cobley, J., 201, 202(24), 214(24), 242
Cocking, E. C., 314
Coetzee, J. N., 14
Cohen, S., 268
Cohen, S. N., 117, 129, 131(*j*), 135(*u*), 141, 161(*i, j*), 162(*y, bb*), 163, 165, 166(12), 234, 421
Cohen-Bazire, G., 199, 200, 215, 237
Coit, D. G., 534
Coleman, J., 392, 405, 406(21), 409(21), 411(21), 493(3, 15), 494, 498(3), 503, 505
Coleman, P. L., 478
Colkitt, M., 432
Collier, R. J., 35, 510(T24), 511
Collins, F. S., 109
Collins, J., 51, 54, 83, 168, 170(32), 172(30), 205
Collins, P. L., 549(35), 550
Colman, A., 370

Comai, L., 255
Comstock, L. J., 402, 433, 434
Condit, R. C., 546
Conley, P. B., 200
Contreras, R., 423, 518
Cooper, P. K., 204
Corbin, D., 48, 214, 233, 237(23), 270, 280, 281(18), 293, 299(15), 458
Cortese, R., 12, 19(15), 39, 50(27), 109
Coster, H., 338
Coulson, A. R., 12, 19(1), 107, 467, 482
Court, D., 54, 162(*aa*), 163, 452, 458(1), 461, 472, 473(7)
Courtney, M., 461
Covarrubias, L., 35, 223, 224
Covey, S., 261
Cox, B. G., 314
Cox, B. S., 523
Cox, K. L., 183
Crabtree, G. R., 406
Crameri, R., 121, 167
Crawford, D. L., 166
Crawford, I. P., 444, 446(27)
Crawford, R. L., 166
Crayton, M. A., 170
Crea, R., 20, 385, 402, 415(1, 2), 463
Cremer, K. J., 549
Crosa, J. H., 35, 37(2), 55, 161(*b*), 162, 223
Crosa, Y. H., 417
Cross, S. H., 168(46), 169, 173(46), 196(46), 198(46)
Crouse, G. F., 106
Crowl, R., 415
Cryer, D. R., 246
Cullum, J., 118, 136(*bb*), 141, 165(19), 167
Currier, T. C., 232, 235
Curtis, P. J., 549
Curtis, S. E., 215
Czernilofsky, A. P., 278, 314

D

Dahl, H.-H. M., 168
Dahlbeck, D., 285
Dahlberg, A. E., 436
Dale, J. W., 165
Dale, P. J., 330
Dales, S., 545
Dalgarno, L., 402, 432, 461
Damm, D., 285
Daniels, D. L., 107, 482
Danilenko, V. N., 117, 165
Daniell, H., 204
Danna, K. J., 408
Danos, O., 12
Das, H. K., 418, 421(5)
Das, R., 511
Davey, M. R., 314
Davie, E., 475, 476(50)
Davies, J., 161(*n*), 162, 266, 267, 518
Davis, A. R., 415
Davis, J., 534, 535(55), 537(55), 538(55), 539(55), 540(55), 541(55), 542(55)
Davis, L. G., 87
Davis, R. W., 39, 47(28), 74, 77, 81(6), 82(6), 176, 178(59), 233, 253, 357, 377
Davis, S. W., 549, 550, 551(38), 552(17), 556(38), 558(38), 563(17)
Davison, J., 35, 36(4), 37(19, 21), 37(4, 19, 21), 41, 42(4, 18), 44, 45(4, 18, 20, 20a), 46(4, 18), 47(19), 52(21a)
Day, L. A., 24
De, B., 272
Debarbouille, M., 389, 432
De Beuckeleer, M., 292
Deblaere, R., 278, 279, 284(16)
De Block, M., 271, 278, 285(12), 289(12), 291(37), 293, 301(10), 320
Deboeck, F., 279, 284(16), 292
deBoehr, E., 168
de Boer, D., 202, 203(49), 211(49)
de Boer, H. A., 390, 391(2), 402, 415(10, 11), 433, 434, 442
De Cleene, M., 254, 299, 307, 313
Decraemer, H., 282
de Crombrugghe, B., 162(*aa*), 163
de Framond, A., 256
de Framond, A. J., 282
Degrave, W., 423, 427(22)
De Greve, H., 278, 279, 282, 284(16), 292
Delaney, S. F., 199, 212
De Lange, T., 196
de la Salle, H., 549, 550
Delamarter, J., 422, 427
De Lay, J., 254
Deley, J., 299
De Ley, J., 307, 313
Dellaporta, S. L., 289
deLorimier, R., 219, 221(18)

de Louvencourt, L., 367
den-Dulk-Ras, H., 308, 310(15)
Denhardt, D. T., 4, 13, 14(21), 15(21), 17(21), 18(21)
Denniston-Thompson, K., 77, 95(7), 96
Dente, L., 12, 19(15), 39, 50(27), 105
Depicker, A., 273, 275(58), 320
Derbyshire, K., 48
Derom, C., 431
Deruelles, J., 200, 207(15), 216, 232
Derynck, R., 423, 525(39), 526
Deshayes, A., 314, 336(11)
Deshusses, J., 50
D'Eustachio, P., 87
Devenish, R. J., 246, 251(12)
Devereux, J., 256
Devilly, C. I., 202
Devos, R., 403, 421, 422(13), 423, 424(13), 427(22)
de Vries, L., 201, 202(20), 213(46), 214(46)
de Waard, A., 235
de Zamaroczy, M., 369, 372(13)
Dhaese, P., 273, 275(58), 277, 282, 313
Dhawale, S. S., 367
Dieckmann, M., 246, 334, 343, 556
Dieterle, A., 549(29), 550
Dietzschold, B., 549
Dinkelspiel, K., 402, 415(10), 433
Di Rienzo, J., 503
Dirks, R., 316(8), 318, 332
Distler, S., 373, 374(21)
Ditta, G., 214, 233, 237(23), 270, 280, 281(18), 282, 293, 299(15), 458
Ditta, M., 48
Dobson, M. J., 517
Doel, M. T., 473
Doi, R. H., 297, 510(T22), 511, 514
Doly, J., 29, 300, 369, 408
Donalson, J. E., 408
Donelson, J. E., 45, 519
Donn, G., 327
Doolittle, W. F., 199
Doornebos, J., 432
Dott, K., 549(29), 550
Dotto, G. P., 3, 5, 6(12), 12, 15, 17, 18(14, 46), 19(14), 21(14), 22(14)
Dougherty, W. G., 549, 550
Douglas, C., 309
Douglas, C. J., 293
Dowbenko, D., 415
Dowding, J. E., 303
Downing, M. R., 462, 477(23)
Draper, J., 314
Dressler, D., 4, 13, 14(21), 15(21), 17(21), 18(21)
Dretzen, G., 40
Dreyer, W. J., 541
Drillien, R., 546, 549(29), 550
Drummond, M. H., 292
Dudley, R., 259, 260, 261
Dudley, R. G., 320
Duffaud, G. D., 499, 503(7), 504(7)
Dujon, B., 368
Duker, K., 534, 535(55), 537(55), 538(55), 539(55), 540(55), 541(55), 542(55)
Dull, T. J., 55, 56(12), 58(12), 59(12), 434
Dumbell, K. R., 545
Duncan, J., 116
Duncan, M. J., 523, 534, 535(52)
Dunn, J. J., 432
Dunsmuir, P., 289
Dunwell, J. M., 326
Durandt, J., 327
Dutchik, J. E., 69, 74(1), 77(1), 81(1), 82(1)
Duyvesteyn, M. G. C., 235
Dzelzkalns, V. A., 201, 204(19), 231

E

Ecclesshall, R., 246
Ecker, J. R., 357
Eckhardt, T., 131(*h*), 141, 161(*g*), 162, 511, 515
Ebert, P. R., 296
Edgell, M. H., 191
Edman, P., 541
Edstroem, J. E., 114
Egan, K. M., 517, 521(4), 522(4), 525(4), 527(4), 528(4), 529(4)
Egner, C., 170
Eichholtz, D., 255, 256(6), 259(6), 260(6), 270(6), 278, 287(9), 301, 302(37)
Elanskaya, I., 201, 202(21)
Elhai, J., 200, 201(11), 214(11), 215, 231(8), 234, 236(29), 237(29), 240(29), 241(29), 242(29)
Elion, J., 462, 477(23)
Elkaim, R., 549(29), 550

Ellis, L. F., 409
Ellis, S., 391
Emery, S., 36, 37(20, 20a), 45(20, 20a)
Enea, V., 3, 12, 17, 18(14), 19(14), 21(14), 22(14, 68)
Engebrecht, J., 233
Engler, C. R., 394
Enquist, L., 41, 77, 168, 170
Ensinger, M. J., 546
Erdahl, W. S., 15
Erfle, M., 35, 50(8)
Erikson, R. L., 473
Ernst, J., 518
Escanez, S., 116
Esnouf, M. P., 475
Espardellier, F., 202, 217, 232
Esser, K., 373, 374(22), 375(23), 381(23)
Estell, D. A., 510(T3), 511
Ettayebi, M., 442
Evans, D. A., 302

F

Faber, H. E., 77, 95(7), 96
Falkow, S., 35, 37(2), 55, 161(*b*), 162, 223, 417
Fangman, W. L., 518
Fare, L. R., 161(*h*), 162
Fasman, G. D., 503
Faugeron-Fonty, G., 369, 372(13)
Fay, P., 232
Fayerman, J. T., 132(*l*), 141, 167, 168(44), 169, 171(44), 178(44)
Fearnley, I., 511
Feiss, D., 50
Feiss, M., 95, 103, 169, 170
Feix, G., 286
Feldman, J., 519, 532(28)
Ferrari, E., 510(T3, T21, T23), 511, 514
Fiedler, G., 135(*z*), 141
Fiers, W., 403, 414, 417, 418, 421(6), 422(4, 13), 423, 424(4, 13), 425(27), 426(27), 427(22), 431(4), 461, 469(2), 470(2), 518
Fieschko, J. C., 534, 535(55), 537(55), 538(55), 539(55), 540(55), 541(55), 542(55)
Filpula, D., 162, 510(T4), 511
Fink, C., 255, 256(6), 257, 259(6), 260(6), 266(20), 270(6), 278, 293, 301(9), 320
Fink, G. R., 243, 245, 246(1, 8), 349, 367, 375(4), 517
Finlay, D. R., 282
Finnegan, D. J., 408
Finnegan, J., 234
Fischer, R. A., 170
Fischer, S. H., 167
Fischhoff, D., 270
Fisher, S. H., 163, 165
Fishman, S. E., 167
Fishmann, S. E., 118, 165(20)
Fitoussi, F., 553
Flaherty, L., 563
Flavell, R. A., 168
Flavell, R. B., 278, 320
Flick, J., 255, 256(6), 257, 260(6), 266(20), 270(6)
Flick, J. S., 278, 293, 301(9), 320
Fling, M., 273
Flores, E., 201, 214(25), 237, 241(40), 242(40)
Floss, H. G., 116
Fogg, G. E., 232
Fogel, S., 518
Follin, A., 306
Fong, D., 510(T24), 511
Foor, F., 140(*oo*), 141, 164
Fornwald, J. A., 136(*ee*), 141, 161(*k*), 162
Forster, W., 337, 339(3)
Foster, S. G., 117, 163, 167
Fouet, A., 510(T19), 511
Fournier, P., 367
Fowler, A. V., 501
Fraley, R., 255, 256, 257, 258, 259, 260(6, 11), 266(20), 270(6, 11), 272, 278, 287(9), 293, 301(9), 302(37), 306, 310, 320
Franceschini, T., 436, 437(22)
Franck, A., 258
Frank, M., 181
Franke, C. A., 549(31, 36), 550
Franklin, F. C. H., 48, 50(43)
Franklin, N., 418, 421(5)
Fransen, L., 424, 425(27), 426(27)
Fravell, R. B., 293
Freedman, R. B., 534
Freeman, R. J., 117, 162
Frey, J., 48, 50(43)
Friedberg, D., 202
Friedman, A. M., 48, 50(42)

Frischauf, A. M., 104, 110, 168, 171(36), 173(36), 178(36), 197
Fritsch, E. F., 29, 40, 61, 62(28), 63(28), 64(28), 77, 81(9), 82(9), 93, 95(9), 96, 103(9), 181, 182(70), 195(70), 196(70), 207, 222, 223(25), 226(25), 281, 282(20), 283(20), 315, 357, 369, 440, 445(23), 467, 470(29), 485, 487(5), 517, 524(9), 552
Fromm, M., 258, 278, 311, 323, 337, 339(7), 341(7), 351, 357(2), 360, 361(8), 363(2), 364, 365(2), 366
Frontali, C., 167
Frutor, J. S., 391
Fry, J., 270
Fry, J. E., 278, 287(9)
Fry, J. S., 278, 293, 301(9), 302(37), 320
Fujii, I., 116
Fujikawa, K., 475, 476(50)
Fukuda, Y., 373, 375(24), 517
Fukuhara, H., 367
Fulford, W., 15
Funahashi, K., 516
Furlong, L.-A., 77
Furumai, T., 127
Furusato, T., 512, 513
Furutani, Y., 510(T5), 511, 512, 513(19), 516
Futcher, A. B., 523

G

Gabay, J., 389
Gaillard, C., 369, 372(13)
Gaillardin, C., 367
Gait, M. J., 105, 407, 478, 480
Galiba, G., 331
Galibert, F., 553
Galizzi, M., 510(T6), 511
Gallagher, M. L., 211
Galluppi, G. R., 278, 293, 301(9), 320
Ganesan, A. K., 204
Gardner, R., 258, 259(28), 261(28)
Garfinkel, D. J., 293
Garoff, H., 512
Gasser, C., 272
Gatenbeck, S., 510(T25), 511, 513
Gatz, C., 460
Gautier, F., 54
Gay, P., 510(T20), 511
Geider, K., 12, 17, 19(19)
Geiduschek, E. P., 229
Gendel, S., 202, 203, 210(40)
Genetello, C., 255, 278, 306
Gentz, R., 162(*y*), 163
George, H. J., 390, 391(3), 402, 411(12)
George, J., 80
Gerin, J. L., 563
Gerlt, J. A., 501
Gheysen, D., 431, 518
Gheysen, G., 277, 313
Ghrayeb, J., 493(10), 500, 501(10), 505(12)
Giam, C. Z., 509
Gibbons, P., 137(*ii*), 138(*ii*), 141
Gibbons, P. H., 140(*oo*), 141
Gibson, B. W., 511
Gibson, T. J., 467
Giddings, T. H., Jr., 233
Gil, J. A., 134(*s*), 141
Gilbert, W., 17, 35, 40, 54, 385, 392, 393(18), 408, 423
Giles, N. H., 472
Gillard, S., 546
Gillespie, D., 227, 228
Gilmer, T. M., 473
Glassman, D. L., 390, 391(3), 402, 411(12)
Glenn, A. G., 507
Godson, G. N., 61, 64(26), 65(26), 66(26), 67, 549
Goeddel, D. V., 385, 390, 391(2), 402, 415(1, 11), 433, 517
Goedell, P. V., 423
Goff, C. G., 517
Goff, S. A., 498
Gold, L., 403, 432
Gold, M., 95(6), 96, 101(6), 103(6)
Goldberg, A. L., 471, 472(39)
Goldberg, R. B., 286
Goldberg, S. B., 278, 293, 301(9), 320
Goldbert, A. L., 498
Golden, J. W., 200
Golden, S. S., 199, 202, 210, 215, 216(6), 217, 220(6, 10, 11), 221, 223(24), 225(10, 24), 233
Goldfarb, D. S., 297, 510(T22), 511
Goldfarb, M., 108
Goldstein, L., 534, 535(55), 537(55), 538(55), 539(55), 540(55), 541(55), 542(55)
Gonzalez, A., 266
Gonzy-Treboul, G., 510(T20), 511
Goodfellow, M., 166

Goodman, H. M., 36, 40(17), 62, 273, 275(58)
Goosen, T., 367
Gordon, J., 488
Gordon, M., 254, 257
Gordon, M. P., 282, 292, 293, 294(11), 301(11), 305, 306, 309
Gorman, C. M., 67, 303, 361
Goryushin, V. A., 233
Gostimskaya, N. L., 117
Gottesman, M. E., 463, 472(28)
Gottesman, S., 472
Gourse, R. L., 436, 444, 451(28)
Goursot, R., 369, 372(13)
Graham, F. L., 349, 546, 554(14)
Graham, M. Y., 69, 74(1), 77(1), 81(1), 82(1)
Grant, R. A., 15, 18(47)
Gravins, T., 517
Gravis, E. R., 530
Gray, A., 59
Gray, C. P., 17
Gray, G. S., 116, 161(*f*), 162
Gray, H. B., 63
Gray, O., 513
Gray, P. W., 423
Green, P. J., 493(24), 505
Greene, P. J., 35, 37(2), 55, 161(*b*), 162, 223
Greene, P. Y., 417
Greenfield, L., 35, 510(T24), 511
Gregory, R. J., 436
Gremke, L. D., 557, 559(45)
Grigorava, R., 337, 339(3)
Grigorieva, G. A., 201, 202, 207(28), 213(33), 217, 232(11), 233
Grindley, N. D. F., 7, 161(*o*), 162
Gritz, L., 266, 267
Gronenborn, B., 5, 314, 335(10)
Gross, K. W., 557, 559(45)
Gross, M., 510(T6), 511
Grossman, A. R., 200
Grossman, L., 517
Grosveld, F. G., 168, 196
Grubman, M. J., 415
Grundström, T., 161(*c*), 162
Grunstein, M., 35
Grunwald, D. J., 77
Guarasio, V. R., 521
Guarneros, G., 472
Guerineau, M., 177, 178(65), 191(65)
Guglielmi, G., 219, 221(18)
Guild, G. M., 50, 168
Guilley, H., 258, 261, 320
Guiney, D., 282
Gunge, N., 518
Guo, L.-H., 385
Gupta, S. L., 403
Gusella, J. F., 109
Guss, B., 510(T25), 511
Gustafson, G., 266
Gustafson, G. D., 278
Gustafsson, P., 392
Gutell, R. R., 59

H

Haas, D., 283, 284(28)
Haas, R., 12, 19(19)
Haber, J. E., 243
Haeberli, P., 256
Haenlin, M., 168, 169(39), 171(39), 172(39), 173(39), 178(39)
Haeusslein, E., 488
Hagie, F. E., 517, 518, 525(39), 526, 527(18), 529(18)
Hahn, P., 258, 259(28), 261(28)
Hain, R., 278, 314
Halegoua, S., 503
Hall, B. D., 517, 528
Hall, M. N., 389, 501, 507
Hall, T. C., 285, 286, 307
Halperin, W., 309
Halvorson, H. O., 508
Hamada, F., 517
Hanahan, D., 35, 63, 93, 177, 178(63), 188, 195(74), 475
Hanawalt, P. C., 204
Hansen, P. K., 462, 478
Harada, T., 233
Hardon, E. M., 202, 206(45)
Haring, V., 48
Harms, C. T., 316(7), 318, 330
Harris, J. E., 138(*jj*), 141, 152
Harris, T. J. R., 461, 463(1), 473(1)
Harris Cramer, J., 368
Hartley, J. L., 45, 519
Hartman, S. C., 482, 489(2), 490(2)
Hasegawa, A., 385

Hasegawa, M., 167
Haselkorn, R., 199, 200, 215, 216(6), 220(6, 10, 11), 225(10), 229
Hass, M. J., 303
Hastie, N. D., 557, 559(45)
Hatfield, G. W., 458
Hatfull, G., 48
Ha-Thi, V., 36, 37(19, 20–21), 39(19, 21), 41, 45(20, 20a), 47(19)
Haury, J. F., 232
Hautala, J. A., 472
Hawley, D. K., 54
Hayashi, E., 136(*aa*), 141, 161(*d*), 162
Hayashi, S., 509, 510(T11), 511
Hayflick, J., 390, 391(2), 402, 415(10, 11), 433
Hayflick, J. S., 525(39), 526
Hayward, G. S., 482, 489(2), 490(2)
Hayward, S. D., 482, 489(2), 490(2)
Hazlet, M. A., 285
Heaphy, S., 478, 480
Heberle-Bors, E., 326
Heberlein, U. A., 534
Hedges, R. W., 14
Heiter, P., 253
Helinski, D. R., 48, 214, 233, 237(23), 261, 270, 275, 277(34), 280, 281(18), 282, 293, 294, 299(15), 320, 418, 421(5), 458, 462
Hellman, U., 513, 515(21)
Helms, C., 69, 74(1), 77(1), 81(1), 82(1)
Henner, D., 510(T3, T16, T23), 511
Herber, W., 162, 511
Herdman, M., 199, 200(4), 201, 207(15), 211(26), 212(26), 216, 232
Hering, C., 511
Hermann, R., 13, 18(23)
Hernalsteens, J.-P., 277, 278, 285(12), 287(12), 292, 293, 301(10), 307, 320
Hernandez, T., 472
Herrera-Estrella, L., 271, 278, 285(12, 30, 31), 286, 287(12), 291(37), 293, 294(14), 301(10), 314, 320, 336(11)
Herrero, A., 242
Herschfield, M. V., 418, 421(5)
Hershberger, C. L., 118, 132(*m*), 133(*n*), 141, 161(*g*), 162, 165(20), 167, 168, 169(19, 37), 172(19), 178(19, 37), 183(19), 197(19), 391
Herskowitz, I., 536
Heslot, H., 367
Hess, D., 327
Heusterspreute, M., 35, 36(4), 37(19, 20–21), 39(4, 19, 21), 42(4, 18), 44, 45(4, 18, 20, 20a), 46(4, 18), 47(19)
Hewett-Emmett, D., 462, 477(22)
Hewick, R. M., 541
Heyer, W. D., 370
Heyneker, H. L., 35, 37(2), 55, 161(*b*), 162, 223, 385, 390, 391(2), 402, 415(1, 2, 11), 417, 463
Hibler, D. W., 501
Hice, R. H., 518
Hicks, J. B., 243, 246(1), 289, 349, 367, 375(4), 517, 519, 532(28)
Hill, D. F., 14, 15, 21(33), 107, 482
Hill, L. R., 167
Hille, J., 310
Hillen, W., 460
Hindley, J., 403
Hinds, M., 50
Hines, J. C., 18
Hinnen, A., 243, 246(1), 349, 367, 375(4), 517, 518
Hintermann, G., 121, 167
Hirashima, A., 505
Hironaka, C., 272
Hirose, T., 33, 385, 402, 415(1, 2), 463
Hirsch, P., 256, 270(12)
Hirsch, P. R., 279, 282(17), 293
Hirth, L., 258
Hiti, A. L., 415
Hitotsuyanagi, K., 516
Hitzeman, R. A., 35, 517, 518, 525(39), 526, 527(18), 529(18), 534
Ho, D., 513
Ho, K. K., 215
Ho, Y.-S., 461, 463, 482, 490(1)
Hobbs, J. N., Jr., 168, 169(40), 171(40), 172(40), 178(40), 266
Hodnett, J. L., 63
Hoekema, A., 256, 270(12), 279, 282(17), 293
Hoes, R. H., 19
Hofer, S., 116
Hoffmann, N., 255, 256(6), 259(6), 260(6), 270(6), 272
Hoffmann, N. L., 278, 287(9), 293, 301(9), 302(37), 320
Hofker, M., 308, 310(15)

Hofschneider, P. H., 5, 323, 324(32), 337, 349(4)
Hogness, D. S., 50, 168, 408
Hohmeyer, C., 12, 19(19)
Hohn, B., 77, 83, 93, 95(8), 96, 103(5), 104, 168, 170(32), 171(36), 172, 173(36), 178(36), 205, 254, 258, 314, 323(6)
Hohn, T., 95, 103(5), 258, 314, 323(6), 335(10)
Holben, W. E., 436
Hollenberg, C. P., 518
Holmes, D. S., 246
Holmgren, E., 513
Holy, A., 60, 61(18), 68(18)
Holzschu, D., 12, 19(7)
Hong, G. F., 107, 467, 482
Honjo, M., 510(T5), 511, 512, 513(19), 516
Hood, E. E., 310
Hood, L. E., 541
Hooykaas, P. J. J., 256, 270(12), 277, 279, 282(17), 293, 305, 307, 308(13), 310(15)
Hooykaas-Van Slogteren, G. M. S., 277, 307, 308(13)
Hopwood, D. A., 116, 117, 118(6, 13), 119(13), 121, 127, 133(*o, q*), 134(*q, r, t*), 138(*jj*), 141, 151, 159, 161(*e*), 162, 165, 166(15), 167, 168, 170(41), 177, 178(41, 66), 195, 196(20)
Horinouchi, S., 131(*g*), 135(*w*), 141, 162(*x*), 163, 236
Horiuchi, K., 4, 5, 6, 13, 14(22), 15(22), 17(22), 18(22), 21(22), 22(22)
Horn, G., 510(T24), 511
Horowitz, H., 444, 446(27)
Horsch, R., 256, 257, 258, 260(11), 261(14), 262(14), 264, 266(20), 270(11), 271, 272
Horsch, R. B., 278, 287(9), 293, 301(9), 302(37), 320
Hoschek, G., 286
Hoshiko, S., 511
Hotta, Y., 136(*dd*), 141
Houck, C., 255
Houghton, J., 202
Houlker, C., 462
Housman, D., 109
Howard, B. H., 67, 303, 361
Howarth, A., 258, 259(28), 261(28)
Howell, S., 259, 260
Hruby, D. E., 549(30, 31, 36), 550
Hsiung, H., 493(10), 500, 501(10), 502
Hsiung, H. M., 385, 390, 391(4), 392, 402, 403(4), 411(4), 414(4), 415(4)
Hsu, Y. C., 39
Hu, N.-T., 12, 233
Hu, S., 549
Huffer, R., 521
Huang, H. C., 61
Huang, H. V., 108
Hugly, S., 306
Hui, A., 402, 415(10), 433, 442
Hull, R., 261
Hulsebos, T. J. M., 14, 21(32)
Humayun, M., 433
Hung, M. C., 530
Hunkapiller, M. W., 541
Hunter, H., 432
Hunter, I. S., 167, 168(47), 169, 173(47)
Hussain, M., 510(T7), 511
Hutchinson, C. R., 167
Hutchinson, C. A., III, 191
Hütter, R., 121, 167
Huyard, A., 369, 372(13)
Hwang, M., 140(*oo*), 141
Hyman, B. C., 368
Hynes, M. J., 367

I

Idler, K. B., 273
Ihara, H., 510(T13), 511
Iida, S., 235
Ikeda, H., 103, 116, 117, 160, 166(15)
Imanaka, T., 510(T14, T15), 511
Ingolia, T. D., 161(*m*), 162, 195, 266
Inokuchi, Y., 505
Inouye, M., 392, 405, 406(21), 409(21), 411(21), 436, 437(22), 456, 492, 493(3, 4, 10, 15, 22, 24), 494(2), 495, 498(3), 499, 500, 501(10), 502(4), 503(7), 504(7), 505(12)
Inouye, S., 392, 436, 437(22), 493(4), 495, 502(4), 503, 504
Installe, P., 316(9), 318, 332
Inukai, M., 493(15), 503
Inze, D., 293, 306
Iserentant, D., 431
Ish-Horowicz, D., 83, 129, 168
Itakura, K., 33, 385, 402, 415(1, 2), 463
Ito, H., 373, 375(24), 517

Ito, Y., 131(*i*), 141, 161(*a*), 162
Itoh, S., 429
Itoh, T., 7
Itoh, Y., 283, 284(28)
Iyer, V. N., 23, 254, 293, 314, 336(13)
Iyer, R. V., 15
Iyer, V., 15

J

Jackson, C. M., 475, 478
Jacob, H. E., 337, 339(3)
Jacobs, M., 316(8, 9), 318, 332
Jacobsen, H., 420
Jakes, K., 432
Jakes, K. S., 17
Janssen, G. R., 119, 133(*p*), 141
Janssen, M. J. T., 200, 201(14), 202(14), 204(14)
Janvier, M., 199, 200(4)
Jaurin, B., 135(*u*), 141, 161(*c*), 162
Jayaram, M., 519, 521, 532(28)
Jaye, M., 461
Jayne, S., 285
Jeffrey, A., 433
Jen, G., 310
Jhurani, P., 390, 391(2), 402, 415(11)
Jimenez, A., 161(*n*), 162, 266, 518
Johnson, M.-J., 424, 425(27), 426(27)
Johnston, S., 18, 21(60)
Johnson, M. J., 33
Jonard, G., 258, 261, 320
Jones, C., 109
Jones, E. V., 546
Jones, E. W., 533
Jones, G., 271
Jones, G. H., 165, 195
Jones, J. D. G., 289
Jones, M., 528
Jones, M. D., 132(*l*), 141, 167
Jones, M. G. K., 330
Jones, P. T., 468
Joos, H., 234, 255, 278, 280
Joseph, R., 165
Josephson, S., 513
Joset-Espardellier, F., 203, 210(50)
Joyce, C. M., 7, 161(*o*), 162
Julius, D., 536, 537, 540(57), 542(59)

K

Kaariainen, L., 510(T1, T2), 511, 512
Kado, C. I., 293
Kaiser, A. D., 169(49), 170
Kaiser, D., 95
Kalkkinen, N., 510(T1, T2), 511, 512
Kanezawa, A., 385
Kao, F. T., 109
Kao, K. N., 316(5), 318, 322, 343
Kaplan, D. A., 35, 510(T24), 511
Karch, F., 235
Karlsson, R., 462, 463(14, 16), 466(14), 473(16), 477(14, 16), 478, 479(14, 16), 481(16)
Karn, J., 104, 105(4, 5)
Kashiwagi, K., 512
Kassavetis, G. A., 229
Kaster, K. R., 161(*m*), 162, 266
Kathoon, H., 15
Katoh, T., 136(*cc*), 141, 162(*z*), 163
Katsumata, K., 511
Katumata, K., 136(*aa*), 141, 161(*d*), 162
Katz, E., 134(*r*), 141, 162, 511
Kaudewitz, F., 367, 373, 374(21)
Kaufman, R., 268
Kaulen, H., 285
Kawade, Y., 513
Kawai, S., 349
Kawamura, F., 514
Kawashima, E., 33, 421, 424, 425(27), 426(27)
Kayes, L., 310
Kehoe, P., 200, 201(11), 214(11), 215, 231(8), 234, 236(29), 237(29), 240(29), 241(29), 242(29)
Kellems, R., 268
Kelley, M. R., 22
Kelly, J. M., 367
Kemp, J. D., 273, 285
Kempe, T., 7, 12
Kempenaers, W., 423
Kemper, B., 3, 4(3), 12, 19(17)
Kendall, K., 165
Kendrick-Jones, J., 462, 463(15, 16), 473(16), 474(15), 477(15, 16), 478(15, 16), 479(15, 16), 480(15), 481(16)
Kennedy, N., 36, 37(20, 20a), 45(20, 20a)
Kessler, C., 34, 37(1)
Khorana, H. G., 407

Khyen, N. T., 201, 232
Ki, P.-F., 482, 489(2), 490(2)
Kidd, S., 22
Kiefer, D. O., 77
Kieny, M. P., 549
Kieser, H. M., 116, 121(7), 133(*q*), 134(*q*), 141, 159(7), 167, 196(20)
Kieser, T., 116, 117, 118(6, 13), 119(13), 121(7), 125, 133(*o, p, q*), 134(*q*), 141, 159(7), 161(*e*), 162, 164, 165, 167, 177, 178(66), 196(20)
Kim, K., 48
Kim, M. H., 18
Kimara, A., 517
Kimura, A., 373, 375(24)
Kimura, H., 493(10), 500, 501(10)
King, A. A., 140(*m,m*), 141, 163, 165(58), 167
King, R. M., 517
Kingsbury, R., 530
Kingsman, A. J., 517
Kingsman, S. M., 517
Kirst, H., 266
Kirst, H. A., 168, 169(40), 171(40), 172(40), 178(40)
Kishore, G., 272
Klapwijk, P., 254
Klee, H., 254, 257, 261(14), 262(14), 264, 269, 272, 308
Klee, H. J., 282, 293
Kleid, D. G., 385, 402, 415(1)
Klenow, H., 420
Klier, A., 510(T19), 511
Klock, G., 460
Kobayashi, I., 95, 96(1), 97(1), 102(1), 103, 186, 198(72)
Kobayashi, T., 135(*w*), 141, 162(*x*), 163
Koblitz, D., 316(6), 318, 328
Koblitz, H., 316(6), 318, 328
Kohn, J., 475
Kohli, V., 461
Kohro, T., 517
Koller, K.-P., 165
Kolowsky, K. S., 202, 206(44), 217
Konigsberg, W., 15
Konings, R. N. H., 12, 13(20, 20a), 14(24), 15(24, 26, 26a), 17(26, 26a), 18(24, 26, 26a), 19(20, 20a, 26, 26a), 21(26, 26a), 22(20, 20a), 23(20), 24
Kopf, J., 273
Koprowski, H., 549
Kornberg, A., 15, 17(45)
Korsuize, J., 235
Koski, R. A., 528
Koukolikova-Nicola, Z., 254
Kowalski, D., 418
Kozak, M., 257, 296, 432
Koziel, M. G., 285
Kramer, J., 310
Kraszewski, A., 385, 402, 415(1)
Krens, F. A., 278, 314, 321, 323, 343
Kreuzaler, F., 285
Krisch, H. M., 219, 223(22), 224, 427
Kroeker, W. D., 418
Krogman, D. W., 215
Kroyer, J., 510(T10), 511
Krugel, H., 135(*z*), 141
Krzyzek, R. A., 390, 391(3), 402, 411(12)
Kuchinke, W., 498
Kück, U., 373, 375(23), 381(23)
Kuhlemeier, C. J., 200, 201(13, 14), 202(13, 14, 18), 203(18), 204(13, 14, 18, 19), 206(13, 45), 207(18, 39), 210(18), 211(18), 212(13, 39), 216, 233, 234(15), 242
Kuhstross, S., 135(*v*), 141, 161(*l*), 168(43a), 169, 170(43a)
Kunbawa, R., 536, 540(57)
Kunisawa, R., 237
Kuntz, I. D., 511
Kuperstock, Y. M., 320
Kurjan, J., 536
Kurz, C., 14, 17
Kushner, S. R., 472

L

Laemmli, U. K., 61, 289, 393, 409, 446(29), 448, 470, 485, 488, 541
Lai, J. S., 509
Lai, P.-H., 517, 519(6), 534(6), 535(6, 55), 537(55), 538(55), 539(55), 540(55), 541(6, 55), 542(6, 55)
Lambowitz, A. M., 381
Lampen, J. O., 510(T7, T8), 511
Lang, B. F., 367, 369, 373(14), 374(14, 21), 381
Lang-Hinrichs, C., 373, 375(23), 381(23)
Langner, A., 162(*y*), 163

Langridge, P., 286
Langridge, W. H. R., 325, 337, 339(6), 341(6, 9), 343(6), 345(9), 349(6), 350(6)
Larson, J. L., 132(*m*), 133(*n*), 141, 161(*g*), 162, 167, 168, 169(19, 37), 172(19), 178(19, 37), 183(19), 197(19)
Larson, R., 40, 45(32)
Laskey, R. A., 471
Laskowski, M., 418
Lathe, R., 461, 549
Laudenbach, D. E., 203
Laumore, F. S., 471, 472(39)
Lauteuberger, J. A., 461, 473(7)
Lawhorne, L., 29
Lawrence, D. W., 245, 246(8)
Leach, D. R., 111
Lecocq, J.-P., 461
Lecocq, L. P., 549(29), 550
Le Coq, D., 510(T20), 511
Leder, P., 87, 337, 338(5), 339(5), 341(5), 350(5), 351, 355(4), 356(4)
Lee, K., 257, 266
Leemans, J., 255, 278, 279, 284(16)
Leemans, R., 431
Legaz, M. E., 475, 476(50)
Legerski, R. J., 63
Lehman, I. R., 176, 178(59)
Lehnhardt, S. K., 499, 503(7), 504(7)
Lehrach, H., 104, 110, 168, 171(36), 173(36), 178(36), 197
Lehtovaara, P., 510(T2), 511, 512
Leisinger, T., 283, 284(28)
Leitner, E., 373, 374(22)
Lemaux, P. G., 200
Lemke, P., 373, 375(23), 381(23)
Lemmers, M., 292
Lengyel, P., 403
Lerman, L. S., 12, 19(4)
Leung, D. W., 423
Levine, A., 257, 266(20)
Levine, H. L., 517
Levinson, A., 3, 4(5), 109, 268
Levinson, A. D., 423
Li, B. J., 325, 337, 339(6), 341(6), 343(6), 349(6), 350(6)
Li, C. H., 390
Li, S. C., 61, 66(27)
Li, Y.-Y., 519, 532(28)
Liang, X.-W., 282
Lieb, M., 417
Lin, T. C., 15
Lindberg, M., 510(T25), 511
Lindenmaier, W., 54
Linsmaier, E. M., 316(3), 318, 326
Lipinskas, B. R., 549, 550, 553(20, 39), 557, 559(39, 48), 561(39), 562(51), 563(20)
L'Italien, J. J., 390, 391(3), 402, 411(12)
Little, P. F. R., 108, 168(46), 169, 173(46), 196(46), 198(46)
Liu, S. W., 488
Lloyd, A., 272, 287
Lockau, W., 232
Loenen, W. A. M., 74, 106
Loerz, H., 314, 316(7), 318, 330
Logtenberg, T., 200, 201(13), 202(13), 204(13), 206(13), 212(13)
Lomedico, P., 415
Lomonosoff, G., 261
Lomovskaya, N. D., 117, 151
Long, C., 511
Long, G. L., 406
Long, S. R., 48, 50(42)
Lopez, J., 15
Lörz, H., 284
Lottenberg, R., 478
Lowe, P. A., 473
Lu, H. S., 534, 535(55), 537(55), 538(55), 539(55), 540(55), 541(55), 542(55)
Lu, P., 282
Ludwig, G., 162(*w*), 163, 236, 275
Lugovoy, J. M., 518, 527(18), 529(18), 534
Lugtenberg, B., 310
Luisi, B. F., 480
Luiten, R. G. M., 12, 13(20a), 14, 15, 19(20a), 22(20a), 24
Lujan, E., 541
Lundquist, R. C., 293
Lundstrom, K., 512
Lunn, C. A., 500
Lupski, J. R., 61, 64(26), 65(26), 66(26), 67
Lurz, R., 48, 50(43)
Lydiate, D. J., 116, 117, 118(6), 121(7), 133(*q*), 134(*q*, *t*), 141, 159(7), 166(15), 167, 168, 170(41), 178(41), 196(20)
Lyons, A., 370
Lysenko, T. G., 233

M

McAndrew, S., 415
McClure, W. R., 54

McConnell, D. J., 510(T12), 511
McDonald, J. K., 391
McDonnell, D. J., 510(T18), 511
McFadden, B. A., 204
McKenney, K., 54, 452, 458(1)
McKercher, P. D., 415
McKnight, S. L., 530
McLaughlin, J. R., 510(T27), 511, 513
MacLeod, A. R., 462
MacNeil, I., 137(*ii*), 138(*ii*), 141
MacNeil, T., 137(*ii*), 138(*ii*), 141
Maaløe, O., 453
Mabe, J. A., 168(44), 169, 171(44), 178(44)
Macfarlan, R. I., 549
Mackett, M., 548, 549, 550
Maenhaut, R., 285(30), 286
Maes, M., 234, 255, 280
Magnusson, S., 462, 475, 477(21)
Makabe, O., 511
Malcolm, S., 266
Malcolm, S. K., 278
Maley, F., 511
Maley, G. F., 511
Maliga, P., 316(4), 318, 321, 331, 332(27)
Malpartida, F., 116, 134(*t*), 140(*nn*), 141, 151, 161(*n*), 162, 164, 168, 170(41), 178(41), 195
Malyguine, E., 80
Mammen, E. F., 479
Manabe, K., 510(T5), 511, 516
Mandak, V., 258, 314, 323(6)
Mandart, E., 553
Mandel, M., 199, 200(4), 237
Mangin, M., 369, 372(13)
Maniatis, T., 12, 19(4), 29, 40, 61, 62(28), 63(28), 64(28), 77, 81(9), 82(9), 93, 95(9), 96, 103(9), 181, 182(70), 195(70), 196(70), 207, 222, 223(25), 226(25), 281, 282(20), 283(20), 315, 357, 369, 440, 445(23), 467, 470(29), 485, 487(5), 517, 524(9), 552
Manis, J. J., 137(*gg*, *hh*), 141
Mann, C., 253
Mann, D. A., 195
Mann, K. G., 462, 477(23)
Mantei, N., 423
March, P. E., 499, 503(7), 504(7)
Margulis, L., 199
Markwardt, F., 479
Marmenout, A., 414, 424, 425(27), 426(27)
Marotta, G., 369, 372(13)
Marmur, J., 246
Marner, F.-J., 306
Marrs, B. L., 234
Marsh, L., 35, 50(8)
Marston, F. A. O., 473
Martial, J. A., 392, 406
Martin, J. F., 134(*s*), 141
Marton, L., 287, 301, 321
Maruo, B., 516
Marzluf, G. A., 367
Mashiko, H., 135(*y*), 141, 161(*p*), 162
Masiarz, F. R., 534
Masuda, T., 95, 170
Masui, Y., 392, 405, 406(21), 409(21), 411(21), 493(3, 10), 494, 498(3), 500, 501(10), 505
Mathes, H. W. D., 105
Matsubara, K., 170, 517
Matsuhashi, Y., 136(*aa*), 141, 161(*d*), 162
Matsushima, P., 127, 167, 171(26), 173(26), 176, 177, 178(26), 183(60), 195(26), 198(26)
Maundrell, K., 378, 381(26)
Maurer, K. H., 167
Maurer, R., 391, 406, 433
Maxam, A. M., 40, 392, 393(18), 408, 423
Mayne, N. G., 390, 391(4), 402, 403(4), 411(4), 414(4), 415(4), 502
Mazur, B. J., 18
Mead, D., 3, 4(3)
Mead, D. A., 12, 19(17)
Mellando, R. P., 471, 472(38)
Melli, M., 114
Menczel, L., 331
Mendzhul, M. I., 233
Mercer, S. R., 549, 550, 551(38), 553(20), 556(38), 558(38), 561, 562(51), 563(20)
Merchez, M., 35, 36(4), 39(4), 41, 42(4), 45(4), 46(4)
Mergio, A., 255
Merlo, D. J., 285, 292
Merryweather, J. P., 534
Meselson, M., 35, 93, 102, 188, 195(74)
Messens, E., 254, 278, 285(12), 287(12), 293, 301(10), 310, 320
Messing, J., 5, 7, 10, 12, 13(10, 12), 19(10, 13), 20(10, 12, 13, 66), 21(10, 12, 13), 22(13), 23(10, 12), 29, 30, 31(10, 12), 33(10, 12), 35, 37(6), 40, 45(32), 56, 110, 161(*q*), 162, 177, 178(64), 258, 259(28), 261(28), 298, 320, 463, 482

Metrione, R. M., 391
Meyer, B., 433
Meyer, B. J., 391, 406
Meyer, J., 235
Meyer, R., 48, 50
Meyer, T., 12, 17, 19(19)
Meyerowitz, E. M., 50, 168
Meyhack, B., 518
Mezes, P. S. F., 510(T7, T8), 511
Michayluk, M. R., 316(5), 318, 322, 343
Mieschendal, M., 418
Miki, B., 314, 336(13)
Milcarek, C., 188
Miller, B. L., 367
Miller, J., 285
Miller, J. H., 41, 287, 455
Miller, K. Y., 367
Miller, W. L., 392, 406
Mills, A. D., 471
Milman, G., 414, 482, 488, 489(2), 490(2)
Miozzari, G., 521
Mita, I., 510(T5), 511, 516
Mitchell, A. Z., 327
Miwa, T., 170
Miyajima, A., 518, 534
Miyajama, I., 518
Miyake, T., 33
Miyanohora, A., 517
Miyazaki, G., 478
Miyoshi, K., 385
Mizuno, T., 505
Mkrtumian, N. M., 117, 151
Model, P., 15, 18
Moffat, L. F., 67, 303, 361
Mohier, E., 168, 169(39), 171(39), 172(39), 173(39), 178(39)
Moir, D. T., 517, 523, 534, 535(52)
Moks, T., 513, 515(21)
Moldave, K., 517
Molendijk, L., 278, 287, 301, 314, 321(5), 323(5), 343
Molin, S., 392
Montanez, C., 472
Montoya, A. L., 292
Mooraka, Y., 233
Moore, D. D., 77
Moore, D. M., 415
Moore, L., 299
Moore, L. W., 310
Moreno, F., 501
Morgan, D. O., 415
Morgan, E. A., 436, 442
Morin, N., 140(*oo*), 141, 164
Morinaga, Y., 436, 437(22)
Morino, T., 198
Morris, H. R., 475
Morrison, D. A., 246
Moss, B., 545, 546, 548, 549(16, 27, 28, 32, 37), 550, 557(32), 563(18)
Moss, M. T., 165
Motyczka, A., 546
Movva, R. N., 493(22), 505
Mozer, T., 257, 266(20)
Mueller, M., 315, 323(14)
Mullenbach, G. T., 534
Muller, M., 350, 351
Müller, R., 424, 425(27), 426(27)
Müller-Hill, B., 5, 418, 498
Mulligan, R. C., 296
Murai, N., 285
Murakami, T., 136(*aa*), 141, 161(*d*), 162
Murashige, T., 287, 316(2), 318, 331
Murata, K., 373, 375(24)
Murphy, B. R., 549, 563(18)
Murphy, E., 266
Murphy, E. B., 278
Murphy, N., 510(T18), 511
Murray, A. W., 519
Murray, C. L., 510(T27), 511
Murray, K., 77, 170, 471, 472(38)
Murray, M. G., 285
Murray, N., 104, 106
Murray, N. E., 77
Musso, R., 162(*aa*), 163
Muth, W., 131(*k*), 132(*k*), 141
Myers, R. M., 12, 19(4)

N

Nabeshima, S., 136(*dd*), 141
Nagahari, K., 39, 48(26)
Nagai, K., 402, 412(20), 414(5), 415(5), 461, 462(8, 9), 463(8, 9, 14, 15, 16, 20b), 465(8), 466(8, 12, 14, 24), 468(9), 469, 470(8), 472(9), 473(9, 16, 24), 474(9, 15, 24), 477(8, 9, 14–16), 478(8, 9), 479(14, 15, 16, 24), 480(9, 15), 481(16)
Nagaoka, K., 136(*aa*), 141, 161(*d*), 162, 511
Nagata, S., 423

Nagata, T., 316(10), 318, 329
Nagle, J., 510(T4), 511
Nagy, F., 258, 263(30), 285, 316(4), 318, 331
Nagy, J. I., 321, 332(27)
Najarian, R., 423
Nakagawa, S., 385
Nakajima, R., 510(T14), 511
Nakamura, K., 405, 456, 492, 494(2), 503, 512, 513
Nakanishi, N., 131(*i*), 141, 161(*a*), 162
Nakano, E., 546
Nakano, M. M., 135(*x*, *y*), 141, 161(*p*), 162
Nakatsukasa, W. M., 168(44), 169, 171(44), 178(44)
Nakayama, A., 510(T5, T17), 511, 512, 513(19)
Nakazawa, K., 513
Nambiar, K. P., 462, 463(20d), 478, 479(20d), 481(20d)
Narang, S. A., 385, 387(5)
Nasmsyth, K. A., 519, 532(28)
Navashin, S. M., 165
Navins, J. R., 295
Nayak, D. P., 415
Negrutiu, I., 316(8, 9), 318, 331, 332
Neiderberger, P., 521
Nelson, M., 432
Nelson, R., 272
Nesian, M., 67
Nester, E., 254, 257
Nester, E. W., 282, 292, 293, 294(11), 305, 306, 309
Nesterova, N. V., 233
Neugebauer, K., 13, 18(23), 510(T9), 511
Neumann, E., 323, 324(32), 337, 349
Neve, R., 54
Neves, A. G., 391
Newlon, C. S., 243, 246, 249(6), 251(12), 252(6)
Ng, P., 442
Nichols, B. P., 444, 446(27)
Niclen, S., 12, 19(1), 467
Nielsen, J. B. K., 509
Nilsson, B., 510(T25), 511, 513, 514(20), 515(21)
Nishi, T., 429
Nishiguchi, M., 337, 341(9), 345(9)
Nishizawa, M., 349
Nitsch, C., 316(1), 318, 322
Nitsch, J. P., 316(1), 318, 322
Nojiri, C., 136(*aa*), 141, 161(*d*), 162, 511
Noller, H. F., 55, 56(12), 58(12), 59(12), 434
Nomura, M., 444, 451(28)
Nordstrom, K., 392
Normark, S., 161(*c*), 162
Norrander, J., 7, 12
Notkins, A. L., 549
Novick, R. P., 66
Nozaki, C., 517
Nunberg, J., 268
Nussenzweig, R. S., 549
Nussenzweig, V., 549

O

Oberto, J., 36, 37(19), 39(19), 47(19)
O'Callahan, C. H., 67
O'Connell, K., 257, 258, 266(20)
O'Connor, B. H., 423
Odell, J., 258, 259, 260, 263
Odell, J. T., 285
Ogawara, H., 135(*x*, *y*), 141, 161(*p*), 162
Ohmura, K., 510(T17), 511, 512
Ohnuki, T., 136(*cc*), 141, 162(*z*), 163
Ohtomo, N., 517
Okamoto, T., 14
Okanishi, M., 127, 136(*dd*), 141, 180
Okayama, H., 258
Okker, R. J. H., 310
Olfson, P., 233
Oliver, D., 507
Ollington, J. F., 510(T12), 511
Olson, M., 245
Olson, M. V., 69, 74(1), 77(1), 81(1), 82(1), 181
Omer, C. A., 117, 131(*j*), 141, 166(12)
Ōmura, S., 116
Ooms, G., 254
Orr-Weaver, T. L., 243
Ortlepp, S. A., 510(T12), 511
Oshida, T., 131(*i*), 141, 161(*a*), 162
Osiewacz, H. D., 373, 375(23), 381(23)
Osterburg, G., 14
Ostrander, D. A., 63
Ostrander, F., 541
Otsu, K., 534
Otten, L., 292
Otten, L. A. B. M., 311
Overgaard-Hansen, K., 420

Owens, G. C., 201, 204(19)
Ozaki, L. S., 64

P

Padgett, R. A, 81
Pagnier, J., 462, 463, 466(12, 24), 473(24), 474(24), 478, 479(24)
Pakrasi, H. B., 219, 221(19)
Palva, I., 510(T1, T2), 511, 512
Panasenko, S. M., 176, 178(59)
Panem, S., 423
Panicali, D., 546, 548, 549, 550, 551(38), 552(17), 553(20), 556(38), 557(15), 558(38), 563(17, 20)
Paoletti, E., 546, 548, 549, 550, 551(38), 552(17), 553(20, 39), 556(38), 557(15), 558(38), 559(39), 561(39), 562, 563(17, 20)
Papas, T. S., 461, 473(7)
Paradiso, M. J., 140(*oo*), 141
Paschal, J, 266
Paschal, J. W., 168, 169(40), 171(40), 172(40), 178(40)
Parsons, J. T., 473
Pastan, I., 67
Paszkowski, J., 258, 278, 314, 315, 319, 322(17), 323(14), 325, 326, 335(10), 337, 339(8), 345(8), 348, 349(8), 350, 351
Pearce, N., 314
Pearson, M. L., 472
Peden, K., 56
Peeters, B. P. H., 12, 13(20, 20a), 14(24, 25), 15(24–26a), 17(25–26a), 18(24–26a), 19(20, 20a, 25–26a), 21(24–26a), 22(20, 20a, 25), 23(20)
Pennica, D., 423
Penswick, J. R., 314, 335(10)
Perkus, M. E., 550, 553(39), 557, 559(48), 561, 562(51)
Perlman, D., 508
Perlman, P. S., 369, 373(14), 374(14)
Perlmutter, A. P., 60, 61(20)
Pernodel, J.-L., 177, 178(65), 191(65)
Perry, D., 195
Perucho, M., 108
Perutz, M. F., 461, 462(9), 463(9), 468(9), 469(9), 472(9), 473(9), 474(9), 477(9), 478(9), 480(9)
Peters, R., 13, 14(24), 15(24), 18(24), 21(24)
Petersen, G. B., 482
Petersen, G. P., 14, 21(33)
Petersen, T. E., 462, 475, 477(21)
Peterson, G. B., 107
Peterson, R. B., 232
Pettersson, R. F., 510(T1, T2), 511, 512
Petrov, P. A., 166
Petruska, J., 314, 326, 348
Philipson, L., 510(T25), 511, 513
Phillips, S. L., 191
Piccini, A., 550, 553(39), 557, 559(39, 48), 561(39), 562(51)
Pictet, R., 36, 40(17), 62
Pilacinski, W. P., 390, 391(3), 402, 411(12)
Piper, M., 373, 378, 381(26)
Piret, J. M., 152, 160, 163, 167
Pirkl, E., 13, 18(23)
Pirotta, V., 114, 168, 169(38, 39), 171(38, 39), 172(38, 39), 173(38, 39), 178(38, 39)
Pirrotta, V., 433
Pirtle, I. L., 405, 411, 503
Pirtle, R. M., 405, 411, 503
Plaetinck, G., 427
Platt, T., 444, 446(27)
Plutzky, J., 406
Pogo, B. G. T., 545
Polke, C., 17
Pollitt, S., 504
Pometto, A. L., 166
Porter, R. D., 201, 202(22, 23), 213(22, 23), 215, 216, 217, 231(12), 232(10), 233
Poteete, A., 168(45), 169, 171(45)
Potrykus, I., 258, 278, 314, 315, 316(7), 318, 319, 322(17), 323(6, 14), 325, 326, 329, 330, 335(10), 337, 339(8), 345(8), 348, 349, 350, 351
Potter, H., 337, 338(5), 339(5), 341(5), 350(5), 351, 355, 356
Poulis, J., 254
Poustka, A., 104, 168, 171(36), 173(36), 178(36)
Pouwels, P. H., 367
Power, S. D., 509
Poyart, C., 461, 462(9), 463(9), 468(9), 469(9), 472(9), 473(9), 474(9), 477(9), 478(9), 480(9)
Prasad, A. S., 479
Pratt, D., 15
Prentki, P., 219, 223(22), 224, 235

Presnell, S. R., 462, 463(20d), 478(20d), 479(20d), 481(20d)
Prestidge, L. S., 50, 168
Pribnow, D., 54, 403, 432
Price, C. W., 510(T22), 511
Priefer, U., 37
Priest, F. G., 507
Primrose, S. B., 370
Pringle, J. R., 533
Projan, S. R., 66
Prozzi, D., 52
Ptashne, M., 406, 433
Puck, T. T., 109
Pugsley, A. P., 507
Pühler, A., 37, 131(*k*), 132(*k*), 137(*ff*), 141
Pulleyblank, D., 202, 210(40)
Purcell, R. H., 563
Purrello, M., 524, 541
Putterman, D. G., 24

Q

Queen, C., 471
Queener, S. W., 195
Quetier, R., 292
Quigley, M., 246

R

Rabinowitz, J. C., 510(T27), 511
Rabitts, T. H., 462
Rackwitz, H. R., 104, 110, 168, 171(36), 173(36), 178(36), 197
Raker, M. A., 59
Rao, R. N., 50, 135(*v*), 141, 161(*b*, *m*), 162, 167, 168(43a), 169(40), 170(42, 43, 43a), 171(26, 40), 172(40), 173(26), 178(26, 40, 42, 43), 195(26), 198(26), 266
Rapoport, G., 510(T19), 511
Ratzkin, B., 528
Ravensbergen, C. J. C., 432
Ravetch, J. V., 17
Ray, D. S., 4, 13, 14(21), 15(21), 17(21), 18(21), 21(60)
Reagan, K. J., 549
Ream, L. W., 293
Reed, K. C., 195
Reeves, S., 306
Reich, T., 314, 336(13)
Reichelt, B. Y., 199
Reichert, N. A., 285
Reilly, B. E., 198
Reinach, F., 462, 463(15, 16), 473(16), 474(15), 477(15), 478(16), 479(15, 16), 480(15), 481(16)
Reiss, B., 162(*w*), 163, 236, 275, 335
Remaut, E., 403, 414, 417, 418, 421(5, 6), 422(4, 13), 424(4, 13), 427, 431(4), 461, 469(2), 470(2)
Reznikoff, W. S., 320
Rhodes, D., 462, 463(20b)
Rhodes, C., 246, 334, 343, 510(T4), 511, 556
Rice, C. M., 549(30), 550
Richards, C., 273
Richards, J. E., 95(7), 96
Richards, K., 258, 261, 320
Richardson, M. A., 132(*l*), 141, 167, 168(44), 169, 171(44), 178(44)
Richaud, F. V., 219, 223(21), 224
Richmond, M. H., 283
Richmond, T. J., 462, 463(20a), 478
Riedel, G., 168(45), 169, 171(45)
Rigby, P. W. J., 246, 334, 343, 556
Riggs, A. D., 385, 402, 415(1, 2), 463
Rinderknecht, E., 423
Rippka, R., 199, 199(4), 200, 207(15), 216, 232
Roa, M., 432
Robberson, D. L., 63
Robbins, P. W., 511
Roberts, B. E., 546
Roberts, N. A., 517
Roberts, G. P., 140(*oo*), 141
Roberts, J., 266
Roberts, J. L., 278
Roberts, T. M., 168(45), 169, 171(45)
Robertson, B. H., 415
Robinson, C. W., 394
Robinson, S. J., 200, 215, 229
Rodicio, M. R., 139(*ll*), 141, 151, 157, 163, 167
Rodriguez, R. L., 35, 37(2), 54, 55, 58, 59(17), 161(*b*), 162, 223, 297, 417
Rogers, S., 255, 256(6), 257, 258, 259(6), 260(6, 11), 266(20), 270(6, 11), 272
Rogers, S. G., 50, 168, 170(43), 178(43), 278, 287(9), 293, 301(9), 302(37), 306, 320

Rose, J. K., 549(27), 550
Rosenberg, M., 54, 136(*ee*), 141, 161(*k*), 162(*aa*), 163, 452, 458(1), 461, 462, 463, 471, 482, 490
Rosenberg, S., 93
Rosenberg, S. M., 95, 96(1), 97(1), 102(1, 2), 186, 198(72)
Ross, F., 493(22), 505
Ross, G. W., 67
Rosteck, P. R., 118, 165(20), 167, 391
Roth, J. R., 77, 81(6), 82(6)
Rothstein, R. J., 243, 320, 517
Roulland-Dussoix, D., 63, 207
Rovinsky, M., 546
Rownd, R. H., 368
Royal, N., 511
Rubin, R. A., 293
Ruby, C. L., 140(*oo*), 141
Rückert, B., 48, 50(43)
Rudd, B. A. M., 116
Ruiz, A. A., 61, 64(26), 65(26), 66(26)
Ruohonen, M., 512
Russel, M., 22
Rutter, W. J., 36, 40(17), 62
Ruvkun, G. B., 219, 223(20), 224, 233, 255
Ruysschaert, M.-R., 424, 425(27), 426(27)
Ryan, M. J., 407

S

Saiki, R. K., 292
Saito, A., 429
Saito, H., 513
Sakaguchi, K., 39, 48(26)
Sako, T., 510(T26), 511
Sam, C. K., 545
Sambrook, J., 29, 40, 61, 62(28), 63(28), 64(28), 77, 81(9), 82(9), 93, 95(9), 96, 103(9), 181, 182(70), 195(70), 196(70), 207, 222, 223(25), 226(25), 281, 282(20), 283(20), 315, 357, 369, 440, 445(23), 467, 470(29), 485, 487(5), 517, 524(9), 552
Samson, S. M., 195
Samsonoff, C., 549, 553(20), 563(20)
Sanders, J. P. H., 423
Sanders, P., 255, 256(6), 257, 259(6), 260(6), 266(20), 270(6), 272
Sanders, P. R., 278, 287, 293, 301(9), 320
Sanger, F., 12, 19(1), 107, 467, 482
Santerre, R., 266
Santerre, R. F., 406
Sarojini, G., 204
Sarvas, M., 510(T2), 511, 512
Sasaki, T., 510(T13), 511
Sassone-Corsi, P., 40
Sato, H., 198
Sato, M., 429
Satoh, E., 511
Saul, M. W., 258, 278, 314, 315, 323(6, 14), 325, 326, 337, 339(8), 345(8), 348, 349(8), 350, 351
Sauer, R. T., 433
Scalenghe, F., 114
Scarpulla, R., 385
Schaefer, L., 403
Schaeffer-Ridder, M., 323, 324(32), 337, 349(4)
Schafer, F., 106
Schaller, H., 5, 13, 14, 17, 18(23), 162(*w*), 163, 236, 275, 335, 510(T9), 511
Schein, C. H., 512
Schekman, R., 15, 17(45)
Schell, J., 234, 255, 271, 277, 278, 279, 282, 284(16), 285(12, 30, 31), 286, 287(12), 291(37), 292, 293, 306, 307, 313, 314, 320, 335
Scherer, S., 81, 233, 377
Scherzinger, E., 48
Schieder, O., 292
Schieven, G. L., 61
Schilling-Cordaro, C., 255
Schilperoort, R., 254, 256, 270(12)
Schilperoort, R. A., 277, 278, 287, 293, 301, 305, 307, 308(13), 310(15), 311, 314, 321(5), 323(5), 343
Schimke, R., 268
Schindler, D., 266
Schmeissner, U., 452, 458(1)
Schmetterer, G., 242
Schmid, K., 460
Schmidhauser, T., 261, 277(34), 282
Schmidhauser, T. J., 294
Schmidt, R., 266
Schmidt, R. J., 406
Schmitt, D., 549
Schmitt, R., 460
Schneider, K., 456, 458(6)

Schneider, T., 403
Schoenmakers, J. G. G., 12, 13(20), 14(24), 15(24, 26, 26a), 17(26, 26a), 18(24, 26, 26a), 19(20, 26, 26a), 21(24, 26, 26a, 32), 22(20), 23(20), 24
Schöffl, F., 286
Scholz, P., 48
Schoner, B. E., 390, 391(4), 402, 403(4), 404(20), 405(20), 409, 411(4), 414(4), 415(4)
Schoner, R. G., 390, 391(4), 402, 403(4), 404(20), 405(20), 409, 411(4), 414(4), 415(4)
Schottel, J. L., 129, 161(*j*), 162
Schreier, P. H., 335
Schrempf, H., 116, 118, 121(7), 133(*q*), 134(*q*), 141, 159(7), 165(21), 167
Schröder, G. S., 306
Schroder, J., 306
Schroeder, J. L., 107, 482
Schumm, J. W., 77
Schumperli, D., 54
Schwartz, D. C., 105(17), 108
Schwartz, M., 389, 432, 501, 507
Schwarzstein, M., 423
Schweickart, V., 409
Sciaky, D., 292
Scocher, R., 325
Scott, A. L., 482, 489(2), 490(2)
Seamans, C., 415
Seeburg, P., 20, 61
Seeburg, P. H., 390, 391(2), 402, 415(11), 441, 525(39), 526, 534
Seed, B., 3, 4(5), 108, 109
Seegers, W. H., 462, 477(22)
Seftor, E. A., 335
Seijffers, J., 202
Sekizawa, J., 503
Selzer, G., 7, 116
Semon, D., 424, 425(27), 426(27)
Sen, A., 370
Sengupta-Gopalan, C., 285
Seno, E. T., 160, 163
Shaffer, P. W., 232
Shah, D., 272
Shall, S., 373, 378, 381(26)
Shand, R., 12, 19(7)
Shapira, S. K., 219, 223(21), 224
Shapiro, J. A., 23
Shapiro, L., 35
Sharp, P. A., 12, 19(8), 60, 530, 557
Sharp, W. R., 302
Sharpe, G. S., 37, 48(24), 50, 236
Sharrock, R. A., 444, 451(28)
Shatzman, A., 461, 463, 482, 490(1)
Shaw, A., 424, 425(27), 426(27)
Shaw, C. H., 293
Shaw, J. E., 472
Shaw, P., 116
Shaw, W. V., 67, 361
Sheldon, E. L., 77
Shen, S.-H., 402
Shepard, H. M., 433
Shepard, M. G., 402
Shepherd, R., 258, 259(28), 261(28)
Sheppard, R. C., 407
Sherman, F., 35, 245, 246(8), 534
Sherman, L. A., 202, 210, 216, 217, 221, 223(24), 225(24), 233, 236, 242(36)
Sherratt, D., 55, 234, 422, 427(17)
Sherwood, P. J., 423
Shestakov, S. V., 201, 202, 207(28), 213(33), 217, 232(11), 233
Shillito, R., 254, 258
Shillito, R. D., 278, 314, 315, 319, 322(17), 323(6, 14), 325, 326, 337, 339(8), 345(8), 348, 349(8), 350, 351
Shimada, H., 510(T5), 511, 512, 513(19)
Shimatake, H., 162(*aa*), 163, 452, 458(1), 462
Shimizu, K., 108
Shimotsu, H., 135(*w*), 141, 162(*x*), 163
Shindoh, Y., 135(*x*, *y*), 141
Shine, J., 36, 40(17), 62, 402, 432, 461
Shinedling, S., 403
Shinedling, T., 432
Shinozaki, K., 202
Shirahama, T., 127
Shirosa, T., 512, 513
Shivarova, N., 337, 339(3)
Shoemaker, J. M., 473
Shortle, D., 243, 416, 510(T28), 511
Shoyab, M., 370
Sias, S., 390, 391(2), 402, 415(11)
Sibakov, M., 512
Siebenlist, U., 54
Siegel, N., 272
Siegele, D. A., 170

Sigmund, C. D., 442
Signer, E., 106
Silhavy, T. J., 501, 507, 557
Silver, D., 3, 4(5), 109
Silverman, M., 233
Silvestri, L. G., 167
Simoens, C., 306
Simon, M., 233
Simon, R., 37
Simon, R. D., 242
Simonet, J.-M., 167, 177, 178(65), 191(65)
Simons, G., 403, 414, 421, 422(13), 424(13), 427, 431
Simons, G. F. M., 15
Simonsen, C., 268, 423
Simpson, J., 285(31), 286
Simpson, R. B., 54
Singer, B. S., 403, 432
Singh, A., 518, 527(18), 529(18), 534
Skalka, A. M., 35
Skatrud, P. L., 195
Skoog, F., 287, 316(2, 3), 318, 326, 331
Skory, S., 549
Sleeter, D. D., 55, 56(12), 58(12), 59(12), 434
Slightom, J. L., 95(7), 96, 285, 286
Sloma, A., 510(T6), 511
Small, B., 415
Smiley, B. L., 64
Smith, A. R. W., 41, 53(37)
Smith, C., 275
Smith, C. A., 204
Smith, C. P., 116, 121(7), 133(*q*), 134(*q*), 141, 159(7), 163, 165, 167, 196(20)
Smith, E. F., 305
Smith, G., 106
Smith, G. L., 548, 549(16, 28, 37), 550, 563(18)
Smith, H. O., 197
Smith, M., 11, 12, 19(2), 259, 411
Smith, R. A., 517, 523, 534, 534(52)
Smith, R. E., 391
Smithies, O., 77, 95(7), 96, 256
Snyder, L., 140(*oo*), 141
Snyder, M., 253
Soberson, X., 35, 223, 224
Soderlund, H., 510(T1, T2), 511
Sollazzo, M., 109
Som, T., 7
Sommer, R., 14, 17
Sottrup-Jensen, L., 462, 475, 477(21)
Southern, E., 246
Southern, E. M., 291, 333, 556
Southern, P. J., 87
Souza, L., 549
Spaink, H., 310
Spehner, D., 546, 549
Spizz, G., 546
Sprague, G., 537, 542(59)
Sprengel, R., 335, 510(T9), 511
Squires, C., 61, 66(27)
Squires, C. L., 61, 66(27)
Sridhar, P., 546
Stabel, P., 278, 314
Stachel, S., 254, 257, 282, 293, 294(11), 301(11)
Stachel, S. E., 310
Stackhouse, J., 462, 463(20d), 478(20d), 479(20d), 481(20d)
Staechel, S., 273, 275(58)
Staehelin, T., 488
Stafford, D. W., 88
Stahl, F. W., 74, 95, 96(1), 97(1), 102(1), 111, 186, 198(72)
Stahl, M. L., 510(T21), 511, 514
Stahl, M. M., 95, 96(1), 97(1), 102(1), 186, 198(72)
Stahl, U., 373, 374(22), 375(23), 381(23)
Stalker, D., 275
Staneloni, R. J., 293
Stanfield, S., 214, 233, 237(23), 270, 280, 281(18), 293, 299(15), 458
Stanier, R. Y., 199, 200(4), 207(15), 216, 232, 237
Stanisich, V. A., 14, 283
Stanssens, P., 403, 417, 422(4), 424(4), 427, 431(4), 461, 469(2), 470(2)
Stanzak, R., 167, 171(26), 173(26), 177(26), 195(26), 198
Staskawicz, B. J., 285
Staples, D. H., 403
Stark, M. J. R., 436
Stavodubstev, L. I., 165
Stein, D., 131(*h*), 141
Steinbiss, H. H., 278, 314
Steinmetz, M., 510(T20), 511
Steitz, J. A., 403, 432, 436, 461
Steller, H., 168, 169(38, 39), 171(38, 39), 172(38, 39), 173(38, 39), 178(38, 39)
Stephens, M. A., 510(T12), 511
Stepien, P. P., 385

Sternberg, N., 41, 168, 170
Steudle, E., 338
Stevens, S. E., 201, 202(22, 23), 213(22, 23), 217, 219, 221(18), 231(12), 232(10), 233
Stewart, J. W., 534
Stewart, T. A., 87
Stewart, W. D. P., 200, 232
Stiegler, P., 403
Stiles, J. I., 35
Stinchcomb, D. T., 39, 47(28), 377
St. John, T. P., 81
Stock, C. A., 285
Stohl, L. L., 381
Stoll, P. J., 479
Stonesifer, J., 167
Stoorvogel, W., 200, 201(13), 202(13), 204(13), 206(13), 212(13)
Stormo, G., 403
Stotish, R. L., 140(*oo*), 141
Straub, J., 292
Straus, N. A., 202, 203, 210(40)
Strauss, A., 370
Strauss, J. H., 549(30, 31), 550
Streicher, S. L., 140(*oo*), 141
Streuli, M., 423
Struhl, K., 39, 47(28), 377, 518
Stüber, D., 460
Studier, F. W., 432
Stueber, D., 492
Sturm, B., 392
Suarez, J. E., 151, 159
Suggs, S. V., 33, 534, 535(55), 537(55), 538(55), 539(55), 540(55), 541(55), 542(55)
Sugimoto, K., 14
Sugisaki, H., 14
Sugiura, M., 202
Sulzinski, M., 346
Sumi, S. I., 385
Summers, W. C., 557
Summers, W. P., 557
Sun, S. M., 286
Sunderland, N., 326
Sung, W. L., 385, 387(5)
Surosky, R. T., 243, 249(5, 6), 251, 252(5, 6)
Sutcliffe, J. G., 56, 57(13), 58(13), 59(13), 273, 277(57)
Sutton, A., 519
Sutton, D. W., 285
Suzuki, K., 127, 180
Suzuki, M., 385
Swamy, K. H. S., 471, 472(39)
Swanberg, S. L., 168(45), 169, 171(45)
Swift, R. A., 83, 85(4)
Syvanen, M., 170
Szabo, A., 402, 414(6), 415(6), 462, 479(13), 480(13)
Szalay, A. A., 202, 206(43, 44), 216, 217, 237, 325, 337, 339(6), 341(6, 9), 343(6), 345(9), 349(6), 350(6)
Szczesna-Skapura, E., 12, 19(17)
Szostak, J. W., 35, 243, 519
Szybalski, W., 61, 107, 482

T

Tadayuki, I., 136(*cc*), 141, 162(*z*), 163
Takagi, M., 510(T15), 511
Takahara, M., 493(10), 500, 501(10)
Takahashi, H., 198
Takanami, M., 14
Takaoka, C., 429
Takebe, I., 316(10), 318, 329
Takebe, Y., 444, 451(28)
Takeda, K., 131(*i*), 141, 161(*a*), 162
Takeda, T., 127
Takeishi, K., 405
Takeichi, Y., 510(T17), 511
Takizawa, N., 233
Takkinen, K., 510(T1, T2), 511, 512
Talmage, K., 35, 385
Tamura, G., 510(T17), 511, 512, 513
Tan, S., 462, 463(20a), 478
Tandeau de Marsac, N., 200, 202, 204(39), 207(39), 212(39)
Taniguchi, T., 429, 513
Tartaglia, J., 546
Tashiro, N., 513
Taunton-Rigby, A., 517
Tavernier, J., 424, 425(27), 426(27), 427
Taya, Y., 423, 427(22)
Taylor, D. P., 136(*ee*), 141, 161(*h*, *k*), 162, 234
Taylor, E., 170
Taylor, L., 258
Taylor, L. P., 278, 311, 323, 337, 339(7), 341(7), 351, 357(2), 363(2), 364(2), 365(2), 366

Teertstra, R., 202, 203(49)
Teeuwsen, V. J. P., 200, 201(14), 202(14), 204(14), 211(49)
Tessié, J., 338, 339, 341(15)
Tessier, L.-H., 461
Tewfik, E. M., 167
Thatte, V., 23
Thia-Toong, L., 277, 292, 307
Thiel, T., 233
Thierry, F., 12
Thøgersen, H. C., 402, 412(20), 414(5), 415(5), 461, 462(8), 463(8), 465(8), 466(8), 469, 470(8), 475, 477(8), 478(8), 479
Thomas, A. A. M., 201, 202(18), 203(18), 204(18), 207(18), 210(18), 211(18)
Thomas, C., 233, 275
Thomas, D. Y., 385
Thomas, M., 81
Thomas, P. S., 229, 230
Thomas, T. L., 307
Thomashow, L. S., 306
Thomashow, M. F., 306
Thompson, C. J., 116, 127, 133(*o*), 134(*r*), 141, 161(*e*, *f*), 162, 167
Thompson, D. V., 273
Thompson, L. D., 510(T4), 511
Thorner, J., 536, 537, 540(57), 542(59)
Thuriaux, P., 370
Thuring, R. W. J., 423
Thurlow, D. L., 436
Tiemeir, D., 170
Timberlake, W. E., 367
Timko, M. P., 285(30, 31), 286
Timmis, K. N., 48, 50(43)
Tiollais, P., 553
Tischer, E., 36, 40(17), 62
Tizard, R., 424, 425(27), 426(27)
Tobek, I., 117
Tobin, L., 37, 48(24), 236
Toh-E., A., 517
Tolstoshev, P., 461, 549(29), 550
Tomioka, N., 202, 516
Tomizawa, J., 7
Tournis-Gamble, S., 36, 37(20, 20a), 45(20, 20a)
Towbin, H., 488
Townsend, C. O., 305
Townsend, J., 257, 266
Toyama, H., 136(*aa*), 141, 161(*d*), 162
Tracy, S., 73
Traktman, P., 546
Treiber, G., 29
Trimble, R. B., 511
Trnkl, H., 373, 374(21)
Tsao, H., 418, 421(6)
Tschumper, G., 523
Tso, J. Y., 385
Tsong, T. Y., 338
Tsuboi, A., 515(T13), 511
Tsuchiba, N., 510(T26), 511
Tsukagoshi, N., 510(T13), 511
Tucker, P. W., 95(7), 96
Tudar, I., 314
Tudzynski, P., 373, 375(23), 381(23)
Tuite, M. R., 517
Tumer, N., 272
Tumer, N. E., 229
Turco, E., 114
Turner, D. R., 462, 463(20), 474(20), 477(20), 478, 479(20), 480(20)
Twigg, A. J., 55, 422, 427(17)
Tye, B. K., 35, 243, 246, 249(5, 6, 17, 18), 251, 252(5, 6), 253(17)
Tzagoloff, H., 15

U

Udaka, S., 510(T13), 511
Ueda, M., 415
Uhlen, M., 510(T25), 510, 513, 514(20), 515(21)
Uhlin, B. E., 392, 409
Ullrich, A., 36, 40(17), 62
Ullrich, U., 59
Ulmanen, I., 512
Umezawa, H., 127, 180
Unger, B., 460
Unger, R.-C., 106
Uozumi, T., 135(*w*), 141, 162(*x*), 163
Urdea, M. S., 534
Uris, D., 52

V

Valenzuela, P., 534
Valerio, D., 196

van Arkel, G. A., 200, 201(13, 14), 202(13, 14, 18, 20), 203(18, 49), 204(13, 14, 18, 39, 48), 206(13, 45), 207(18, 39, 48), 210(18), 211(18), 212(13, 39, 48), 213(46), 214(46), 216, 224, 233, 234(15), 237, 242(15)
van Bloom, J. H., 15, 432
van Brussel, T. A. N., 310
van Charldrop, R., 432
Van Cleemput, M., 444, 446(27)
Van den Broeck, G., 285(30), 286
van den Broek, H. W. J., 367
van den Elzen, P., 257, 266
van den Hondel, C. A., 15
van den Hondel, C. A. M. J. J., 201, 202(18), 203(18), 204(18, 39, 48), 207(18, 39, 48), 210(18), 211(18), 212(39, 48), 216, 224, 233, 234(15), 237, 242(15), 367
van de Putte, P., 202, 233
Van der Eb, A. J., 349
van der Eb, J., 546, 554(14)
van der Ende, A., 201, 202(18, 20), 203(18), 204(18, 48), 207(18, 48), 210(18), 211(18), 212(48), 213(46), 214(46), 216, 224, 237
Van der Heyden, J., 424, 425(27), 426(27)
Van derleyden, J., 294
van der Plas, J., 202, 203(49), 211(49), 432
Vander Ploeg, L. H. T., 196
Van der Straeten, D., 423
van de Vate, C., 202, 206(45)
van Dijk, M., 202, 203(49), 211(49)
Van Duin, J., 432
Van Frank, R. M., 195
van Gorcom, R. F. M., 367
van Haute, E., 234, 280, 293
Van Hawte, L., 255
van Heugten, H. A. A., 200, 201(13), 202(13), 204(13), 206(13), 212(13)
Van Heuverswyn, H., 423, 427(22)
Van Knippenberg, P. H., 432
van Leen, R. W., 201, 202(18), 203(18), 204(18), 207(18), 210(18), 211(18)
Van Lijsebettens, M., 306
van Montagu, M., 234, 254, 255, 271, 277, 278, 279, 280, 282, 284(16), 285(12, 30, 31), 286, 287(12), 291(37), 292, 293, 294(14), 301(10), 306, 307, 310, 313, 320
Vannier, P., 80
van Straaten, P., 462
Van Vliet, A., 424, 425(27), 426(27)
van Wezenbeek, P. M. G. F., 14, 21(32)
Vapnek, D., 58(36), 66, 161(*i*), 162
Varadarajan, R., 402, 414(6), 415(6), 462, 479(13), 480(13)
Vasantha, N., 510(T4), 511
Vasil, I. K., 330
Vasil, V., 330
Vasquez, D., 266
Vassef, A., 546
Vasser, M., 402, 433, 434
Vedel, F., 203
Veeneman, G. H., 15, 432
Velten, J., 285
Verbeek, S., 202, 204(48), 207(48), 212(48), 216, 224, 237
Verhoeven, E., 23
Verhoeyen, M. E., 468
Vermaas, W. S. J., 219, 221(19)
Vieira, J., 12, 13(12), 19(13), 20(12, 13, 66), 21(12, 13), 22(13), 23(12), 31(12), 33(12), 35, 37(6), 56, 110, 161(*q*), 162, 177, 178(64), 298, 320, 463, 482
Vienken, J., 351
Villarreal, L. P., 555
Virolle, M. J., 511
Visser, J., 367
Vodkin, L. O., 286
Voegeli, P., 116
Vogelstein, B., 227, 228
Vogt, V. M., 418
Vögtli, M., 121
von Heijne, G., 508
Vonshak, A., 200, 201(11), 214(11), 215, 231(8), 234, 235, 236(29), 237(29), 240(29), 241(29), 242(29)

W

Waffenschmidt, S., 306
Walbot, V., 258, 278, 311, 323, 337, 339(7), 341(7), 351, 357(2), 360, 361(8), 363(2), 364(2), 365(2), 366
Waldman, A. S., 488
Waldron, C., 266, 278
Wallace, D. M., 423
Wallace, R. B., 33
Wallis, J., 35

Wallroth, M., 270
Walsby, A. E., 232
Walz, D. A., 462
Wang, J. C., 169(49), 170
Wang, K., 254, 278, 293, 294(14)
Wang, S., 503
Wang, W., 510(T8), 511
Wang, Y., 323, 324(32)
Ward, D. F., 463, 472(28), 477(22)
Ward, J. M., 117, 119, 127, 133(*o*, *p*, *q*), 134(*q*), 141, 161(*e*), 162(*bb*), 163, 165, 167, 196(20)
Warren, G., 234, 255, 280
Waterbury, J. B., 199, 200(4), 207(15), 216, 232
Watson, B., 257
Watson, B. D., 282, 284(28), 293, 294(11), 301(11, 21), 302(21)
Watson, J. M., 283
Watson, M. P., 293
Waye, M., 462, 468
Weaver, R. F., 557
Webster, R. E., 15, 18(47)
Weiler, E. W., 306
Weinberg, R. L., 549, 552(17), 563(17)
Weir, J. P., 546
Weir, L., 337, 338(5), 339(5), 341(5), 350(5), 351, 355(4), 356(4)
Weisbeek, P. J., 202, 203(49), 204(48), 207(48), 211(49), 212(48), 216, 224, 237
Weisblum, B., 236
Weiss, B., 188
Weissman, C., 423
Weissman, S. M., 109, 403
Weissmann, C., 512, 557
Welch, J. W., 518
Wells, J. A., 509, 510(T3), 511
Wenars, K., 367
Wennekes, L. M. J., 367
Wensink, P. C., 408, 530
Wertman, K. F., 176, 196(60a)
Wertz, G. W., 549(35), 550
Wesolowski, M., 367
Wessler, S., 479
West, J., 392
West, R. W., 54
Weyers, A., 15
White, F., 254
Whiter, S., 473
Whitton, B. A., 199
Wickner, W., 15, 17(45)
Widera, G., 54
Widholm, J. M., 357
Wienken, J., 338
Wigler, M., 108
Wiktor, T. J., 549
Wilchek, M., 475
Willett, N. S., 48
Willi, M., 335
Williams, B. G., 77
Williams, J., 202, 203, 210(40)
Williams, J. G. K., 202, 206(43, 44), 216, 217, 219, 221(19), 237
Williams, S. T., 166
Williams, W. J., 475
Williamson, D. H., 518
Willingham, M. C., 67
Willmitzer, L., 278, 292
Winter, G., 467
Winter, J., 272
Winston, F., 517
Wirth, D. F., 511
Wohlenberg, C., 549
Wohlleben, W., 131(*k*), 132(*k*), 137(*ff*), 141
Wojciuch, E., 232
Wolansk, B., 140(*oo*), 141
Wolf, D. H., 537
Wolf, K., 369, 373(14), 374(14, 21), 381
Wolfe, L. B., 111, 176, 196(62)
Wolk, C. P., 200, 201(11), 214(11, 25), 215, 231(8), 232, 233, 234, 235, 236(29), 237(29), 240(29), 241(29, 40), 242(29, 40)
Wong, E., 402, 433
Wong, S.-L., 510(T22), 511
Wong, T. K., 337
Woo, S. C., 278, 293, 301(9), 320
Wood, J., 289
Wood, H. N., 237, 319
Woolhiser, S., 563
Wos, S., 557, 559(48), 561, 562(51)
Woychik, R. P., 87
Wright, A. P. H., 378, 381(26)
Wright, H. M., 117, 118(13), 119(13), 177, 178(66)
Wu, A. M., 444, 446(27)
Wu, H. C., 509, 510(T11), 511
Wu, R., 35, 170, 385, 517
Wulff, D. L., 162(*aa*), 163
Wullems, G. J., 278, 287, 301, 321(5), 323(5), 343

Wunner, W. H., 549
Wurtzel, E. T., 493(22), 505
Wykes, E. J., 35, 50(8)
Wyman, A. R., 111, 176, 196(60a, 62)

Y

Yadav, N., 292, 294
Yagi, S., 385
Yagi, Y., 137(*hh*), 141
Yakobson, E., 282
Yamada, C., 202
Yamada, Y., 136(*aa*), 141, 161(*d*), 162
Yamagata, H., 510(T13), 511
Yamaguchi, T., 131(*i*), 141, 161(*a*), 162
Yamamoto, H., 167
Yamamoto, K. R., 29
Yamane, K., 510(T17), 511, 512, 513, 516
Yamasaki, E., 517
Yamasaki, M., 510(T17), 511, 512, 513
Yamata, Y., 302
Yamazaki, H., 510(T17), 511
Yanagi, K., 513
Yang, M., 510(T16), 511
Yang, M. Y., 510(T23), 511
Yang, Y. Q., 510(T7, T8), 511
Yanisch-Perron, C., 12, 13(12), 20(12), 21(12), 23(12), 31(12), 33(12), 161(*q*), 162, 177, 178(64)
Yano, S., 131(*i*), 141, 161(*a*), 162
Yanofsky, M., 257
Yanofsky, M. F., 282, 305
Yanovsky, C., 418, 521(5), 444, 446(27)
Yansura, D. G., 385, 402, 415(1)
Yao, F.-L., 385, 387(5)
Yelverton, E., 402, 423, 433
Yewdell, J. W., 549, 550
Yi, B.-Y., 296
Yilma, T., 549(27), 550
Yin, E. T., 479
Yoda, K., 512
Yot, P., 80
Young, K. K. Y., 549(35), 550
Young, R. A., 74, 436
Young, Y., 337, 349(4)
Yuan, R., 102

Z

Zabin, I., 501
Zagursky, R. J., 11, 12, 19(16)
Zahab, D. M., 385, 387(5)
Zaitlin, M., 337, 341(9), 345(9), 346
Zalacain, M., 161(*n*), 162
Zambryski, P., 254, 255, 271, 273, 273(58), 278, 287, 291(37), 293, 294(14), 310
Zavala, F., 64
Zehetner, G., 110
Zentgraf, H., 13, 18(23)
Zerbib, D., 339, 341
Zhu, J., 423, 518
Zimmerman, R. A., 436
Zimmermann, U., 323, 336, 337, 338, 349(13), 351
Zinder, N. D., 3, 4, 5, 6(12), 12, 13(11), 14(22), 15, 17(22), 18(14, 22, 46), 19(14), 21(14, 22), 22(14, 22, 68), 23(11)
Zink, B., 14, 15(22)
Zissler, J., 106
Zoller, M., 11, 12, 19(2), 259
Zoller, M. J., 416
Zsebo, K. M., 534, 535(55), 537(55), 538(55), 539(55), 540(55), 541(55), 542(55)
Zuker, M., 403
Zubay, G., 471
Zwieb, C., 436

Subject Index

A

Actinorhodin, 117
Agarase, secretion in gram-positive bacteria, 511
Agmenellum quadruplicatum PR-6
  ATCC number, 217
  genes
    cloning and analysis, 201
    inactivation by recombination with cloned, altered allele, 219
  as host for genetic engineering experiments, 217
  properties, 201
  psbA genes, inactivation, 221
  random *Sau*3AI fragments, ligated to Tn*I*, 231
  transformation, donor DNA for, 216
Agrobacterium
  DNA transfer from *E. coli* to, 299–301
    direct transfer, 300–301
    triparental mating, 299–300
  host range, 277
  induction of crown galls, 305–306
  plant transformation in absence of tumor formation, 306–307
    detection of, 307
  strains, 307
  transformation of *Asparagus*, 307
  transformation of Liliaceae and Amaryllidaceae, 277, 307
  transformation of monocots, method of detection, 308–311
  vector systems, 278
Agrobacterium-mediated gene transfer, 313, 335, 351
Agrobacterium tumefaciens
  3111-SE, containing pMON vector, plant transformation with, use of hygromycin and methotrexate markers for, 270–272
  A208-SE, containing pMON vector, plant transformation with, use of hygromycin and methotrexate markers for, 270–272
  introduction of pMON plasmids into, 269–272
  strain C58C1Rif^R, 279, 283
  Ti plasmid. *See* Ti plasmids
Alkaline phosphatase assay, 456
Alkaline protease, secretion in gram-positive bacteria, 510
Aminoglycoside phosphotransferase II, as marker of plant transformation, 320
α-Amylase, secretion in gram-positive bacteria, 510–513
Anabaena
  colonies, 232
  conjugal transfer of plasmids to, mating procedures, 237–238
  filamentous strains, introduction of DNA into, 214
  properties, 232
  shuttle vectors, 233
    construction, 234
  strains, conjugation of plasmids to, 242
Anabaena M-131
  conjugal transfer of plasmids into, 238, 239, 241
  isolation of plasmids from, 242
Anabaena PCC7120
  conjugal transfer of plasmids into, 238, 239, 241
  exconjugate, antibiotic resistance, 236
  gene cloning and analysis, 201
  isolation of plasmids from, 242
  nitrogen fixation, genetics of, 200
  properties, 201
  psbA genes, inactivation, 218–219, 221
  transformants, RNA isolation, 228–230
Anabaena variabilis, auxotrophic mutants, isolation, 232
Anacystis, conjugation of plasmids to, 242
Anacystis nidulans, properties, 204
Anacystis nidulans R2
  analysis of DNA and RNA from, 221
  chromosomal DNA, isolation, 209
  cotranformation, with two linked markers, 220
  gene cloning and analysis, 200–201

genetic engineering experiments
equipment, 221–222
methods, 223–230
growth, 207, 221
as host for genetic engineering experiments, 217
isolation of plasmid DNA from, preparative scale procedure, 208–209
mutagenesis, 211–213, 221
procedure combining transposon mutagenesis and recombination, 213
plasmid curing, 211
plasmids, 202–203, 207
capable of autonomous replication in, transformation of, 209–210
lacking cyanobacterial origin of replication, transformation, 210
properties, 200–201
psbA genes, inactivation by recombination with cloned, altered allele, 217–219, 221
recombination, 204–206
RNA, Northern analysis, 229–230
Sm^R Met^+ transformants, 206
Sm^R Met^- transformants, 206
source, 206
transformants
analysis of plasmid content, small-scale DNA isolation for, 208
DNA
isolation using glass fines, materials and reagents for, 222
preparation of glass fines for binding, 227–228
DNA miniprep, 225–227
materials and reagents for, 222
gene cassettes and antibiotics used for selection, 223–224
miniprep DNA, Southern analysis, 226–227
RNA isolation, 228–230
materials and reagents for, 222–223
transformation, 201, 223–225
with alleles of unknown phenotype, 220
by chimeric plasmid DNA, 217–219
donor DNA for, 216
frequencies, variation between hybrid plasmids, 204
materials and reagents for, 222
procedure, 209–210
transposon mutagenesis, 212–213
Anacystis nidulans R2-SPc, 204
source, 214
Angiotensin I, secretion in gram-positive bacteria, 513–514
Antibiotic resistance genes, 36
insertional activation, 119
introduction into plant protoplasts by electroporation, 337
restriction sites in, 35
in vector construction, 34–35
Antibiotics, as potential selective agents for development of new markers for plants, 265
Anti-Shine–Dalgarno sequence, 432
of 16 S rRNA gene, mutagenesis, 441
role in translation and ribosome specificity, 433
Aphanocapsa, conjugation of plasmids to, 242
Aphanocapsa PCC6714
ATCC number, 217
as host for genetic engineering experiments, 217
Arabidopsis thaliana
kanamycin resistance, selectability, 264
transformation, *in vitro*, by Ti vector system, 302–303
Aspergillus, nonreplicative vectors, 367
ATG codon
exposed, of expression vector, joining coding region to, 429–430
of expression vector
methods to expose as blunted end, 418–420
tailoring of coding regions for precise fusion to initiation signals, 420
Autonomously replicating sequence, 367, 518

B

Bacillus amyloliquefaciens, secreted proteins, 510, 512
Bacillus cereus, secreted proteins, 510
Bacillus cereus spheroplasts, transformation with plasmid DNA, 337
Bacillus licheniformis, secreted proteins, 510, 513

Bacillus stearothermophilus, secreted proteins, 510
Bacillus subtilis
  deficient in extracellular protease, 514
  secreted proteins, 510, 512, 513
Bacterial spheroplasts, for plant transformation, 314, 337
Bacteriophage. *See also* Helper phage
  f1, 14
    genomes, as cloning vectors, 12–13
    intergenic region, plasmids carrying, 3
    single-stranded DNA, vectors developed from, 3
  fd, 14
    genomes, as cloning vectors, 12–13
    single-stranded DNA, vectors developed from, 3
    transcriptional terminator, 162
  Ff, 14–15, 23
    genome, circular genetic map, 15, 16
    interference-resistant mutants, 22
    intergenic region, as morphogenetic signal and complementary and viral strand replication origin, 15–17
    single-stranded DNA genome, mechanism of replication, 15–17
    viral strand origins, 18
      plus morphogenetic signal, cloning into plasmid, 19
  IKe, 15
    in construction of new cloning vectors, 13
    genome, circular genetic map, 15, 16
    interference-resistant mutants, 22
    intergenic region, as morphogenetic signal and complementary and viral strand replication origin, 15–17
    replication properties, 13
    single-stranded DNA genome, mechanism of replication, 15–17
    viral strand origins, 18
      plus morphogenetic signal, cloning into plasmid, 19
  IKe-9, as helper phage in production of plasmid DNA, 24–25
  IR1
    as helper phage in production of plasmid DNA, 24
    high-titer stocks, preparation, 28–29
    packaging of plasmid DNA strands by, 22
  λ, 151
    clones, DNA analysis, 110
    DNA
      in vitro packaging, 95–103
      purification, 69–82
    growth, 73–76
      amplification on grid plates, 75
      number of host cells, 75–76
      phage inputs, 73–75
      protocol, 76–78
    leftward promoter (P_L), 482
      expression vectors based on, 417
      insertion into ΔPIP2-ASD-HGH, 444–445
      isolation, 444
      repressor, 417
    libraries
      construction, 110–111
      library amplification, 114–115
      ligation, 113–114
      packaging, 114
      plating, 114–115
      preparation of insert DNA, 112–113
      procedures, 111–113
      vector preparation, 111–112
    red and *gam* genes, 74, 105–106
    transcriptional terminators, 162
  M12, genomes, as cloning vectors, 12–13
  M13, 14
    biology, 4–6
    in construction of new cloning vectors, 13
    DNA synthesis, 4
    gene I protein, 4
    gene II protein, 4
    gene III protein, 4
    gene IV protein, 4
    gene V protein, 4
    gene VI protein, 4
    gene VII protein, 4
    gene VIII protein, 4
    gene IX protein. 4
    genome, 4
    intergenic region
      functional regions, 5

plasmids carrying, 3
structure, 5
mutants, compensating for loss of functional domain B, 5–6
polylinker system, 35
replication
phases, 4
properties, 13
single-stranded DNA, vectors developed from, 3
M13KO7, 22
as helper phage in production of plasmid DNA, 24–25
high-titer stocks, preparation, 28–29
packaging of plasmid DNA strands by, 22
Mike
as helper phage in production of plasmid DNA, 24–25
high-titer stocks, preparation, 28–29
packaging of plasmid DNA, 22
*Mike*Δ
chimeric genome, 22
high-titer stocks, preparation, 28–29
packaging of plasmid DNA, 22
ϕC31, 151
attachment site, 161
derivatives, large-scale isolation of DNA from, 157–158
map, 154
mutational cloning with, 164
repressor gene, 161
vectors lacking attP site, and mutational cloning, 160–164
ϕC31KC304, shotgun cloning experiment using, 153
R408
as helper phage in production of plasmid DNA, 24–25
high-titer stocks, preparation, 28–29
packaging of plasmid DNA strands by, 22
TG1, 164
Bacteriophage λ replacement vectors, in library construction and screening, 104
Bacteriophage λ vectors
capacity of, 104
EMBL series, 103–115
Black Mexican Sweet maize suspension culture cells, protoplasts, isolation, 358–359
Bluescribe M13$^+$, genotype/phenotype, 178
Bovine growth hormone
derivatives, expression, 397–398
in *Escherichia coli*, 390–401
expression
analysis of, 393–394
in *Escherichia coli*, 390–391
effect of minor changes in 5′-terminal coding region of bGH gene, 401
expression plasmids
cloned genes in, 5′-end sequence of, 396–397
construction, 393, 395–397
structure, 394
gene
5′-coding region, replacement with synthetic gene fragments, 390–391
fragments, preparation, 392–393
generic first cistron sequence for expression of, 403
phenylalanine form, without amino-terminal methionine, production, 391
two-cistron expression system, 390, 406, 409–414
bacterial strains, 391
chemical synthesis of oligonucleotides for, 392
enzymatic ligations, 392–393
materials, 391
methods, 391
plasmids, 391–392
purification of oligonucleotides for, 393
results, 395–400

C

c2RB, 85–86
c2RBH, 85–87
c2XB, 85–86
c2XB AMP$^-$, 86–87
in genetic selection system for cosmid libraries, 94
c2XBHC, 85–87
Callus
transformed, gene expression in, 366
transformation, *in vitro*, by Ti vector system, 302

Calothrix PCC7601. *See Fremyella diplosiphon*
Carrot protoplasts
  containing TMV coat protein, FITC-antibody staining of, 348
  electroporated, gene expression in, 363, 364
  preparation, 342–344
  TMV-RNA transfected, immunological detection of, 346–347
    methods to increase specificity and sensitivity of, 347–348
  uptake and expression of TMV-RNA in, by electrotransfection, 345–346
Cathepsin C
  in production of natural sequence bovine growth hormone, 391, 399–401
    Met-Asp-bGH or Met-Val-bGH preparation for cleavage, 394–395
  properties, 391
Cauliflower mosaic virus
  19 S-NOS cassette vector, 258–260
  19 S promoter, 258–260
  19 S promoter-NOS 3′ cassette, nucleotide sequence of, 259–260
  35 S-NOS vector, 261–262
  35 S promoter, 258, as functional promoter in maize transient assays, 366
  35 S promoter-NOS 3′ cassette, nucleotide sequence of, 263
  DNA, as marker of plant transformation, 320
  as possible gene vector, 335
CDA, 117
Cell-free transcription-translation system, protease-free, 471–472
Chloramphenicol acetyltransferase
  assay, 66–67, 361–364
    in plant transformants, 304–305
  expression in recombinant vaccinia virus, 549
  as marker for plant transformation, 279
Chromosomal deletions, large, 243
  construction, outline of method, 244–245
Chromosome, circular derivatives of. *See also* Ring chromosome
  generation, 243
Chromosome walking, 198
Cloning vector
  ideal, 36
  number of unique restriction sites, 36
  size, 36
cNEO, 86–87
Complementary DNA, cloned, tailoring to obtain precise fusion of coding sequences to prokaryotic translation signals, 417–420
Complementation, selection of recombinant clones by, 41–42
Corynebacterium diphtheriae, secreted proteins, 510
*cos*4 vector
  characteristics, 173
  genotype/phenotype, 178
Cosmid cloning, principles, 174–175
Cosmid libraries
  construction, 94. *See also cos* site vectors
  genetic selection, 94
  hybridization screening, 93–94
Cosmid recombinants, identification, by complementation, 41–42
Cosmids, 83
  capacity of, 104
  concatamerization during ligation to inserted DNA, 83, 85
  in construction of genomic libraries, problems with, 83
  in vitro packaging, 93
  transduction into bacteria, 93
Cosmid shuttle vector
  for cloning-analysis of *Streptomyces* DNA, 166–198
    materials, 176–179
    materials, storage, 196
    method, 179–196
    obtaining desired clone, 196
    solutions, 177
    strains, 177–178
  cloning sites, used to generate end-labeled probes, 197, 198
  DNA, cleavage, and preparation of two arms, 185
  insert size, and number of clones, 167
  ligation of donor to vector DNA, 186
  size, and copy number, 196
Cosmid vectors
  characteristics, 173
  preparation, 89–90
  refinement, 168
cos sequences, in *in vitro* packaging, 103

cos site vectors
containing one *cos*, 172–175
containing two *cos*, 175–176
double, 83–85
bacterial strains, 87
in construction of cosmid libraries, 84–85, 94
enzymes, 88
insert DNA
dephosphorylation, 91
large-scale partial digest, 90–91
preparation, 90–92
size selection, 91–92
testing partial digest conditions, 90
ligation of vector and insert DNA, 92–93
materials, 87–88
methods, 88–94
plasmids, 87
reagents, 87–88
solutions, 88
structure, 85–87
single, 83
Crown gall disease, 277, 278, 305
Cyanobacteria
akinetes, 232
chemical mutagenesis, 212
chromatic adaptation, 200
cloned genes, inactivation, 223
conjugal transfer of plasmids to, 231–243
age and amount of donor, 238–240
age of recipients, 239, 242
antibiotic concentration used for selection, 240
conjugative plasmids, 239
efficiency of transfer, 241
filters, 239, 241
mating procedures, 236–238
medium for mating, 240, 241
principle, 233–234
purification of exconjugates, 241–242
recovery of plasmid, 241–242
timing of transfer of filters, 239–240
variant protocols, 238–241
ectopic mutagenesis, 231
filamentous strains, photosynthesis-defective mutants, 232
forms, 199
gene cloning in
buffers, 207
media, 207
procedures, 208–213
solutions, 207–208
strains, 206–207
genes, mutational analysis, 199
genetic analysis, 232–233
gene transfer in, 200–201, 215
by conjugation of plasmid DNA from *E. coli*, 231
heterocysts, 200, 232
host–vector systems, 213
using hybrid plasmids, 202
using integrational vectors, 202
integrational vectors, 206
light harvesting, 200, 215
mutagenesis, 211–215, 231
nitrogen fixation, 200, 215, 232
nitrogen metabolism, 199
photosynthesis, 199, 215, 232
phycobilisomes, 200
plasmid curing, 211
plasmids, 202
capable of autonomous replication in, 215–216
properties, 199, 215, 232
recombination, 216
shuttle vectors, 233–234
construction, 234–236
strains, conjugation of plasmids to, 242
transformants
analysis of DNA and RNA from, 216, 220–221
RNA isolation, 228–230
transformation, 201–202
by chromosomal recombination, 216–221
variations, 199
wild-type strains, source, 214
Cyanobacterial chromosome, genetic engineering of, 215–231

D

DEAE–cellulose, purification of λ DNA on, 69–82
advantage of, 82
applications, 81
degraded samples, 81
dilution of lysates, 78
dilution of plate stocks, 71

imperfectly plaque-purified samples, 80
minimization of scale, 82
nonstoichiometric bands, 78–80
overdigestion, 78–80
protocol for, 69–73
RNA-containing samples, 81
scaling up, 81–82
trouble shooting, 78–79
underdigestion, 78–80
unstable clones, 80
Dicots, transformation, 277–278, 313, 351
Dihydrofolate reductase, coding regions, modified and wild-type, partial sequences of, 268
Diphtheria toxin, secretion in gram-positive bacteria, 510
DNA
chromosomal, replicons obtained from, 367
flanking vector segments with *cos* sites, generation, 172–176
genomic, preparation, for construction of cosmid libraries, 88–89
high-molecular-weight donor, for use in cosmid shuttle vector, isolation, 179–181
inserts for shotgun cloning, preparation, 62–63
λ *cos* mapping, 197
linear fragments, transformation, 243
mitochondrial
fragments, cloning of, 372–374
replicons obtained from, 367
of yeast and fungi, isolation, 370–372
replicative form, 13
single-stranded, 3, 12
of (recombinant) plasmids, production of, 19
hybridization probes, preparation, 32–33
isolated from filamentous particles in superinfection of cells harboring recombinant pKUN plasmids, 24–25
mutagenesis, 32–33
nucleotide sequence analysis, 32–33
packaged by IKe and Ff, primers used for sequence analysis, hybridization studies, or site-directed mutagenesis, 21
plasmids for production of
bacteria, 24–26
bacteriophages, 24–26
equipment, 23
reactions, 23
reagents, 23
production, 3, 13
interference by plage with replication of phage, 3–4
vectors for production of
buffers, 26–28
enzymes, 26
media, 26
nutritional supplements, 26–28
preparation of high-titer phage stocks for, 28–29
single-stranded recombinant plasmid
examples of packaging, 24–25, 33–34
rapid dideoxy sequencing, 24–25, 33–34
Streptomyces, partial cleavage and dephosphorylation, 181–183
DNA viruses, as possible gene vectors, 335–336

E

*Eco*K restriction, during *in vitro* packaging, 102
Electroporation
apparatus, 351–356
capacitor charge indicator, 355
capacitors, 351–355
components, 354–356
cuvettes, 355–356
oscilloscope attachment, 355
power supply, 355
RC circuits, 352–354
resistors, 354
sources, 341
switches, 354
of carrot protoplasts, $MgCl_2$ requirement, 349
cautions, 352
cell membrane poration by, 337–338
electrical field requirements, 338–339, 350
conditions for nucleic acid uptake into plant and animal cells, 339, 340
definition, 351

DNA and RNA uptake mediated by, 336–366
effect of nuclear envelope removal by colchicine treatment prior to, 350
efficiency
effect of DNA concentration, 357–358
effect of solution pH, 357
effect of temperature, 357
factors affecting, 349–350
parameters affecting, 356–357
electric pulse generators, types of, 339–341
expression problems after, solutions, 365–366
field strength, effect on efficiency, 49
introduction of nucleic acids into isolated subcellular organelles by, 350
of maize protoplasts, 359–360
method, 341–348
nonionic buffers, for high-frequency transformation of plant protoplasts, 349
number of pulses, effect on efficiency, 349
parameters, 356–358
protocol, 342–348, 356–358
of RNA, 360–361
temperature of medium, effect on efficiency, 349–350
theory, 337–338, 351–356
transformation of *N. plumbaginifolia* protoplasts by, 332
transformation of *N. tabacum* protoplasts by, 323–325
transformation of *P. hybrida* protoplasts by, 329
Electroporation chamber, 338–339
EMBL3, 104–107, 111
cloning efficiency, 104
restriction map of, 107
structure, 105
EMBL3-12, 110
EMBL3A, 107, 111
EMBL3AS, 107
EMBL3S, 107
EMBL4, 104–107, 110, 111
structure, 105
EMBL5, structure, 105
EMBL6, 109
structure, 105
EMBL7, structure, 105
EMBL vectors
amber derivatives, 107–108
derivatives with further cloning sites, 108–109
end fragment cloning, 109–110
genetic selection, 106
*Not*I-linking fragment cloning, 109
structure, 105–106
Endogenous phage, *in vitro* packaging systems, 95–96
assay, 100–101
Endoglycosidase F, removal of N-linked carbohydrate from yeast glycoproteins, 544
Endoglycosidase H
removal of N-linked carbohydrate from yeast glycoproteins, 544
secretion in gram-positive bacteria, 511
5-Enolpyruvylshikimate-3-phosphate synthase, expression cassette vectors expressing coding sequences for, 272
Epstein–Barr membrane antigen gp340, expression in recombinant vaccinia virus, 549
Escherichia coli
1046 (*recA*$^-$, *supE, supF, hsdS*$^-$, *met*$^-$), 87
as alternative host for *Streptomyces* shuttle vectors, 198
BHB2688, 87
BHB2690, 87
cloning and expression in, 492–499
cosmid cloning, identification of specific clones in, 195
DH1, genotype/phenotype, 178
ED8767, 87
expression and secretion of foreign proteins, 492–507
expression levels, 505–506
and growth conditions, 506
and host strain, 506
toxicity, 506
expression of β-globin fusion protein in, 462, 465
expression of unfused protein in
bacterial strains, 421
chemicals, 421
design of expression vector, 417–420
detection of expressed protein, 429

effect of bacterial strain used, 431
effect of nature of ribosome-binding site, 430–431
effect of strong transcription terminator downstream from cloned gene, 431
enzymes, 420–421
expressed protein levels, 427, 429–430
expression plasmid constructions, 423–429
general DNA methodology, 422–423
materials, 420–421
methods, 422–429
plasmids, 421–422
principle, 417–420
reagents, 421
results, 422–429
expression vectors, 492–494
cloning, 494–495
expression, 495–498
maintenance, 494–495, 506–507
filamentous phages, 14
foreign gene expression in, 461
GM48, in restriction site bank construction, 37
HB101, 207
hybrid proteins produced in, 462
isotope labeling of, after heat induction, 470–471
large-scale synthesis, 472
produced in insufficient quantities, 471–472
purification, 473–474
small-scale protein induction, 470
synthesis and sequence-specific proteolysis of, 461–481
infection, by filamentous phages, 13
JE2571, bacteriophages propagated on, 25
JE2571[pCU53], growth, 28
JM101
bacteriophages propagated on, 25
colonies for transformation, 26
growth, 28
JM101[pCU53], colonies for transformation, 26
JM109
genotype/phenotype, 178
yield of single-stranded plasmid DNA, 11
K12-803, 207
MM294, in restriction site bank construction, 37
MV1184
infection with M13KO7, 10
yield of single-stranded plasmid DNA, 11
MV1190, yield of single-stranded plasmid DNA, 11
MV1304, yield of single-stranded plasmid DNA, 11
plasmid, hygromycin B phosphotransferase gene, 161
plasmid DNA from
primary pool analysis, 191–192
rapid isolation, 190–191
promoterless β-lactamase gene, 161
promoterless galactokinase gene, 161
protease, degradation of foreign gene product, 402
random transductants, restriction enzyme analysis of plasmid DNAs from, 188–190
restriction site bank vector pJRD158, 42, 43
restriction site bank vector pJRD184, 45–46
ribosomal RNA mutants, 433
secretion in, requirement for secretion vector, 499–500
secretion vectors, 499–505
problems, 502–504
and secretion of cytoplasmic proteins, 502–503
structural compatibility with signal peptide, 503–504
structural requirements for processing and translocation of, 503–504
utilization of, 501–502
SF8, genotype/phenotype, 178
SMR10, 87, 98, 101–103
single-strain packaging system, 93
specialized ribosome system, 433–434
applications, 450–452
construction, 434–445
induction of hGH synthesis, 446–448
labeling cells *in vivo*, 448–450
as novel system for expression of heterologous proteins, 450–451

probe specificity and induction of mutated rRNA species, 445–446
properties, 445–450
in regulation of rRNA synthesis, 451
specificity, 448–450
transduction, with ligated DNA packaged into λ particles, 187–188
vector DNA from, isolation, 183–184
Escherichia coli–yeast, shuttle restriction site bank vectors, 47–48
Escherichia coli K-12
strain CB454, 453
strain CB806, 453
strain RV308, 391, 406
strain SB221, 493
strain SB4288, 493
strain SF8, 369
strain W620, 493
Eukaryotes
mitochondrial cloning vectors for, 366–382
cloning of mitochondrial DNA fragments for, 372–374
cloning of selection marker, 375
cloning procedures, 369
enzymes, 369
isolation of DNA, 369
isolation of mitochondrial DNA for, 370–372
maintenance of, in host, 379–380
materials, 369
media, 370
prerequisites of establishing, 381
principle of method, 368
procedures for, 370–376
reagents, 369
strains for, 369
test for replication activity of mitochondrial hybrid vectors, 375–377
transformants
molecular analysis, 376–379
whole-cell DNA, preparation, 377–378
molecular cloning in, 366
Eukaryotic genes
5′-terminal sequence, effect on translational efficiency, 402
cloning, 416
expression, 336
in *Escherichia coli*, 414
with synthetic two-cistron system, 401–416
translation of mRNAs containing, 402
two-cistron expression system, 401–416
bacterial strains, 406
chemical synthesis of DNA linkers, 407
cloning of synthetic linkers, 407–408
DNA isolation, 408
enzymes, 406
gel purification of DNA restriction fragments for, 407
linkers with blunt ends, cloning, 407–408
linkers with compatible ends, cloning, 408
methods, 406
PAGE analysis of protein production, 409
plasmid construction, 409–414
plasmids, 406
protein production by transformants, 408–409
reagents, 406
transformation of *E. coli*, 408
Eukaryotic protein
produced in *E. coli*, properties, 480–481
production by engineered bacteria, 416
Expression cassettes
integration into yeast chromosome, 517
for plant transformation, 254
Expression plasmid, containing λ P_L promoter and cI857 repressor, 482–491. *See also* Plasmid pHE6; Plasmid pHE7
conditions for optimal expression, 489–490
enzymatic assays, 488
fusion proteins, cross-reactivity demonstrated by immunoblots, 488, 490
growth and harvesting of cells, 491
identification of expressed proteins, 487–488
immunological assays of expressed proteins, 488, 490
induction of expression, 487–488
insertion of foreign DNA into, 486–487
materials, 484–485
methods, 486–491

F

G

exoenzymes, 507
expression and secretion vectors, 501
protein secretion, 507
effect of sequence of mature (target) protein, 514–515
factors affecting, 514–516
molecular mechanisms, 507, 515–516
requirements, 515–516
secreted proteins from, 508–509
Growth hormone. *See* Bovine growth hormone; Human growth hormone

H

Helper phage
for production of plasmid DNA, 24–25
for production of single-stranded DNA, 21–22, 33
Helper phage M13KO, 74
Helper phage M13KO19, 8
Helper phage M13KO7, 7–8
effect of addition of plasmid origin, 7–8
growth, 9
in production of single-stranded plasmid DNA, 9–11
structure, 7
Hepatitis B virus surface antigen, expression in recombinant vaccinia virus, 548–550, 553
detection, 560–561
Herpes simplex virus glycoprotein D, expression in recombinant vaccinia virus, 548–550, 553
detection, 560–561
Herpes virus thymidine kinase
assay, 557
DNA encoding, uptake into mouse cells, 337
gene, expression in recombinant vaccinia virus, 548, 549
uptake into mouse cells, 341
Human factor IX, expression in recombinant vaccinia virus, 549
Human fibroblast interferon, expression in *E. coli*, effect of bacterial strain used, 431
Human growth hormone, expression vector, 502
Human immune interferon, construction of expression vector for, 423–425
Human interleukin 2
construction of expression vector for, 427–429
expression in *E. coli*, effect of bacterial strain used, 431
Human protein C, two-cistron expression system for, 409–414
Human tumor necrosis factor, construction of expression vector for, 424– 427
Hybrid colonies, screening for, 35
Hygromycin B, 266
Hygromycin phosphotransferase
coding sequences, wild-type and modified, 266–267
as marker for plant transformation, 266, 279

I

Influenza virus hemagglutinin, expression in recombinant vaccinia virus, 548–550, 553
detection, 560–561
Influenza virus nucleoprotein, expression in recombinant vaccinia virus, 549
Insert-directed integration, 160–164
Insertional inactivation, 34
in vector construction, 34–35
Insulin-like growth factor, secretion in gram-positive bacteria, 513–514
Integrative cloning vector, 367
Integrative recombination, in cyanobacteria, 210
α-Interferon
secretion from yeast, 534
secretion in gram-positive bacteria, 512
β-Interferon, secretion in gram-positive bacteria, 513
Interleukin 2. *See* Human interleukin 2
In vitro packaging, 83, 95, 103, 169
of *Bacillus subtilis* phage, 198
complementary two-strain system, 95–96, 103
cos-less system, 100
of cosmids, 93
*Eco*K restriction during, 102
endogenous phage levels, in various systems, 100–101
extract, preparation, 98–100, 114
of λ DNA, 95–103, 170–172
packaging technique, 100

in λ particles, preparation and transduction into *E. coli*, 185–188
of ligated DNA into λ particles, technique, 186–187
method, 95–96
one-strain *cos*-less system, 97
packaging efficiency, 102
of recombinant λ DNA, 168
SMR10 system, 100–103
of *Streptomyces* phage, 198
two-strain complementary system, 100
sonic extract plus freeze–thaw lysate preparation method, 103
In vivo packaging, of λ, 169–170

K

KC304 vector
cloning sites, 138
copy number, 138
map, 154
markers, 138
parent replicon, 138
for shotgun cloning, 152
KC310 vector, 165
KC505 vector
cloning sites, 139
copy number, 139
map, 154
markers, 139
parent replicon, 139
KC515 vector, 160, 164
cloning sites, 139
copy number, 139
map, 155
markers, 139
parent replicon, 139
KC516 vector
cloning sites, 139
copy number, 139
map, 155
markers, 139
parent replicon, 139
KC518 vector, 154, 160
cloning sites, 139
copy number, 139
map, 155
markers, 139
parent replicon, 139
KC604 vector, *lacZ* gene in, 161
KC680 vector, 165
KC684 vector, 163
cloning sites, 140
copy number, 140
map, 155
markers, 140
parent replicon, 140
Kluyveromyces lactis, vector/host systems, 367

L

β-Lactamase
assay, 67–68
expression vector, 501
secretion in gram-positive bacteria, 510, 512
lacZ gene
of *E. coli*, 161
modified, producing α-complementing fragment of β-galactosidase, 161
λ *cos* mapping, 197–198
Lethal zygosis, 119
Lethal zygosis phenotype, 161
Levansucrase, secretion in gram-positive bacteria, 510
Lignin, enzymes degrading, from *Streptomyces*, 166
Lipoproteins, precursors, 509
Liposome fusion, for plant transformation, 336
Liposomes, for plant transformation, 314
Lolium multiflorum, protoplasts, transformation, 329–331
Loric cosmid, characteristics, 173
lpp-lac promoter systems, expression vectors based on, 492–493
host strains, 493
Ltz^+ phenotype, 119
Ltz^+ plasmids, 161
Lysopine dehydrogenase. *See* Octopine synthase

M

Messenger RNA, formation of local secondary structures, 403
Messenger RNA–rRNA duplex formation, 432–433
Methylenomycin, 117

N-Methyl-*N'*-nitro-*N*-nitrodoguanidine, mutagen of cyanobacteria, 212
Microinjection, for plant transformation, 314, 336
Monocots, transformation, 277–278, 313, 351
  via *A. tumefaciens*, detection, 305–313
Morphogenetic signal, of bacteriophage, 17, 18
Mtx^R dihydrofolate reductase, as marker for plant transformation, 279
Myoglobin, produced in *E. coli*, 480–481. *See also* Fusion proteins, CIIFX
Myosin light chain, produced in *E. coli*, 480. *See also* Fusion proteins, CIIFX

N

Neomycin phosphotransferase
  assay, in plant transformants, 304
  as marker for plant transformation, 279
Neomycin phosphotransferase II, as selectable marker of plant transformation, 288, 289
Neomycin resistance gene, expression in recombinant vaccinia virus, 549
Neurospora, nonreplicative vectors, 367
Neurospora crassa, plasmids for construction of replicating vectors similar to mitochondrial genomic DNA, 381
Neutral protease, secretion in gram-positive bacteria, 510, 512
N-glycanase, removal of N-linked carbohydrate from yeast glycoproteins, 544
Nicotiana plumbaginifolia
  protoplasts, transformation, 331–332
  transformed, inheritance of transferred genes in, species dependence of, 291
Nicotiana tabacum
  protoplasts, transformation, 321–327, 331–332
  transformed, inheritance of transferred genes in, effect of antibiotic concentration, 291
Nonreplicative cloning vector, 367
Nopaline dehydrogenase. *See* Nopaline synthase
Nopaline synthase, assays, 311–313
Nostoc
  filamentous strains, introduction of DNA into, 214
  properties, 232
  shuttle vectors, 233
    construction, 234
  strains, conjugation of plasmids to, 242
Nostoc ATCC27896, conjugation of plasmids into, 241
NotEMBL3A, 109
*Not*I-linking fragment clones, 109
NTG. *See* *N*-Methyl-*N'*-nitro-*N*-nitrosoguanidine
Nuclease
  secretion in gram-positive bacteria, 510
  from *Streptomyces*, 166

O

Octopine synthase, assay, 311–313
ompA secretion vectors, reading frames, DNA sequence of, 497
ORF438, secretion in gram-positive bacteria, 511

P

Pancreatic ribonuclease A, produced in *E. coli*, 481. *See also* Fusion proteins, CIIFX
p*cos*2EMBL cosmid
  characteristics, 173
  cloning efficiency, 104
  genotype/phenotype, 178
Penicillinase, secretion in gram-positive bacteria, 513
Penicillium chrysogenum, 369
  auxotrophic mutants, transformation to prototrophy, 375
  map, 374
  mitochondrial DNA
    compilation of hybrid vectors obtained from, 373
    isolation, 371–372
  mitochondrial vectors, construction, 370
  origin of replication, integration into hybrid vectors, 380–381
  transformants, assay for free plasmid molecules in whole-cell DNA of, 378–379

transformation, with hybrid vectors, 375–376
transformation frequencies with mitochondrial vectors, 376–377
whole-cell DNA, preparation, 377
Petunia hybrida
protoplasts, transformation, 327–329, 332
transformation, with *Agrobacterium*-derived vectors, 287
Photosynthesis, cyanobacterial and plant, 199
pIJ610 cosmid, characteristics, 173
Plage, 3
Plant cells
Agrobacterium-derived vectors for cloning in, 277–292
Agrobacterium acceptor strains, 279
binary vectors, 282–283
cointegration vectors, 279–282
expression of chimeric genes in, 283–286
Plant culture media, 315–318
Plant genomic DNA, introduction into plant protoplasts by electroporation, 337
Plant organ transformation, *in vitro*, by Ti vector system, 302–303
Plants
transformants
analysis, 303–305
chloramphenicol acetyltransferase assay, 304–305
extraction of plant tissue, 303–304
neomycin phosphotransferase assay, 304
transformed
analysis of gene expression in, 288–289
callus induction assay, 288–289
DNA microscale preparation, 290
inheritance of transferred genes in, 291–292
neomycin phosphotransferase assay, 289
T-DNA organization in, 289–290
T-DNA structure, molecular analysis, 290–291
Plant-specific promoters, for expression of chimeric genes in plants, 285–286
Plant transformation
advantages, 314
Agrobacterium-derived systems, 254–257, 287–288
binary vectors, 256–257
cointegrating intermediate vectors, 255–256
binary Ti vectors for, 292–305
binary vectors, uses, 283
chimeric genes for, 278
by direct DNA transfer, 351
expression cassette vectors
design of, 257–265
uses of, 272
hybrid marker genes, 319–321
improved vectors for, 253–277
development of new selectable markers for, 264–269
after infection of wounded plant parts, 278
in vitro, by Ti vector system, 301–303
opines as indicators of, 306
systems for, 253–254
T-DNA-mediated
expression of opine synthase gene, 292
number of T-DNA insertion loci, 292
techniques, 277–278
Plant vectors
binary, DNA sequence of multiple cloning sites in, 297
for cloning-promoter expression, media, 299
Plasmid ASD-HGH, 445
Plasmid DNA vectors, phage intergenic region, 6–7
Plasmid-encoded genes, in plant protoplasts after electroporation, 341
Plasmid GM102
cloning sites, 132
copy number, 132
markers, 132
parent replicon, 132
Plasmid N3, 23
Plasmid p175–6, promoter strength and tetracycline resistance, 59
Plasmid p232–8, 55
Plasmid pABD1, 320
Plasmid pACC184, chloramphenicol acetyltransferase gene, 161

Plasmid pACYC184, 23
Plasmid pαC2, 536
  gene fusions constructed in, 536–537, 541
  restriction endonuclease map, 535
Plasmid pαC3, 536
  gene fusions constructed in, 536–537
  restriction endonuclease map, 535
Plasmid pARC1, 165
  cloning sites, 131
  copy number, 131
  map, 142
  markers, 131
  parent replicon, 131
Plasmid pASD-PSDR, 435, 444
Plasmid pAT153, 422
Plasmid pAYC177, kanamycin phosphotransferase gene, 161
Plasmid pBC6
  cloning sites, 131
  copy number, 131
  map, 142
  markers, 131
  parent replicon, 131
Plasmid pBR322, 42, 46, 55–56, 58–59, 234, 279–280, 492
  ampicillin resistance (β-lactamase) gene, 161
  derivatives, mobilization by conjugative plasmids, 234
  failure of replication in heterologous system, 203–204
  palindromic sequences, restriction enzymes recognizing, 38–39
  restriction sites, 37–39
  as vector for propagating cyanobacterial genes in *E. coli* for reintroduction into *A. nidulans* R2, 223
Plasmid pBR328, 235
  in genetic engineering in cyanobacteria, 219
  as vector for propagating cyanobacterial genes in *E. coli* for reintroduction into *A. nidulans* R2, 223
Plasmid pBT37
  cloning sites, 131
  copy number, 131
  map, 142
  markers, 131
  parent replicon, 131
Plasmid pCaMVCAT, DNA, electroporation transfer into protoplasts, 363–364
Plasmid pCAO170, 165–166
  attachment site, 161
  cloning sites, 131
  copy number, 131
  map, 142
  markers, 131
  parent replicon, 131
Plasmid pCB237, 459
Plasmid pCB238, 459
Plasmid pCB267, 456–459
Plasmid pCB302, 456–459
  insertion of promoters in, 460
Plasmid pCB303, 457–459
Plasmid pCB304, 459
Plasmid pcI857, 418, 421
Plasmid pCU1, 23
Plasmid pCU53, 23
Plasmid pCZ11, 412, 413
Plasmid pCZ19, 413
Plasmid pCZ100, 410, 411
Plasmid pCZ104, 396–398
Plasmid pCZ108, 396–398
Plasmid pCZ110, 396–398
Plasmid pCZ115, 396–398
Plasmid pCZ118, 410–412
Plasmid pCZ140, 410, 412
Plasmid pCZ143, 410, 412
Plasmid pCZ144, 410, 412
Plasmid pCZ145, 410, 412
Plasmid pCZ151, 396–398
Plasmid pCZ152, 396–398
  construction, 390, 391
Plasmid pCZ154, 396–398
Plasmid pCZ161, 396–398
Plasmid pCZ451, 413
Plasmid pCZ460, 413
Plasmid pDam1, 376, 377
Plasmid pDP122B, 552
Plasmid pDU1, 234–235
Plasmid pE194, 236
Plasmid pEB11
  cloning sites, 131
  copy number, 131
  map, 143
  markers, 131
  parent replicon, 131
Plasmid pEB102
  cloning sites, 132

copy number, 132
map, 143
markers, 132
parent replicon, 132
Plasmid pEMBL8, 39
Plasmid pFCLV7, 214
Plasmid pFJ103, genotype/phenotype, 178
Plasmid pFJ342
cloning sites, 132
copy number, 132
map, 143
markers, 132
parent replicon, 132
Plasmid pFM111, 373, 377
Plasmid pFM141, 373, 377
Plasmid pGA482
features for plant transformation, 294–295
map of, 295
Plasmid pGA492, 295–296
map of, 296
Plasmid pGA580, map of, 298
Plasmid pGM4
cloning sites, 132
copy number, 132
map, 143
markers, 132
parent replicon, 132
Plasmid pGM102, map, 144
Plasmid pGPD-1, restriction endonuclease map, 529
Plasmid pGPD-1(HBs), 521
Plasmid pGPD-2, 525
restriction endonuclease map, 529
Plasmid pGPD(s)-2, 525
Plasmid pGPD(s)γ4, 525–526
copy number measurement, 525
Plasmid pGPD(s)γ4-9, 525–526
copy number measurement, 525
Plasmid pGSH160, 283–285
map of, 286
Plasmid pGSJ280, 283–285
map of, 286
Plasmid pGV825, 281
Plasmid pGV941, 283
construction, 283, 284
Plasmid pGV1500, 280
map of, 281
T-DNA expression vectors derived from, 283
Plasmid pGV2260, 279–280
introduction of T-DNA vector into, 280
Plasmid pGV3850, 255
Plasmid pHE6
cloning sites in, 485
construction, 482–483
map, 483
selected restriction sites in, 484
Plasmid pHE7, 484
cloning sites in, 485
Plasmid pHGH207-1*, replacement of trp operator and SD sequence with synthetic oligonucleotide fragments, 442–444
Plasmid pHGH-PSDR, insertion of hGH gene into, 444
Plasmid pHJL125, genotype/phenotype, 178
Plasmid pHJL197
cloning sites, 132
copy number, 132
map, 144
markers, 132
parent replicon, 132
Plasmid pHJL210
cloning sites, 133
copy number, 133
map, 144
markers, 133
parent replicon, 133
Plasmid pHJL302
cloning sites, 133
copy number, 133
map, 144
markers, 133
parent replicon, 133
Plasmid pHJL401
cloning sites, 133
copy number, 133
map, 145
markers, 133
parent replicon, 133
Plasmid pHP341, 376, 377
Plasmid pIC, 35
Plasmid pIC-III, 498–499
Plasmid pIJ61
cloning sites, 133
copy number, 133
map, 145

markers, 133
parent replicon, 133
Plasmid pIJ101, 118
Plasmid pIJ486, 119, 164
cloning sites, 133
copy number, 133
markers, 133
parent replicon, 133
restriction map, 120
Plasmid pIJ487
cloning sites, 133
copy number, 133
markers, 133
parent replicon, 133
restriction map of, 120
Plasmid pIJ680
cloning sites, 133
copy number, 133
map, 145
markers, 133
parent replicon, 133
Plasmid pIJ702, 151
cloning sites, 134
copy number, 134
map, 145
markers, 134
parent replicon, 134
temperature-sensitive mutant, 165
Plasmid pIJ860
cloning sites, 134
copy number, 134
map, 146
markers, 134
parent replicon, 134
Plasmid pIJ922, genotype/phenotype, 178
Plasmid pIJ941
cloning sites, 134
copy number, 134
map, 146
markers, 134
parent replicon, 134
Plasmid pIJ943
cloning sites, 134
copy number, 134
map, 146
markers, 134
parent replicon, 134
Plasmid pIMAI, 406, 409–410
Plasmid pIMIA, 392
Plasmid pIN, 492, 505
A, B, C sites, DNA sequence of reading frames of, 496
Plasmid pIN-I, 498
Plasmid pIN-II, utilization, 494–498
Plasmid pIN-II-A, 492–493
Plasmid pIN-II-B, 504–505
Plasmid pIN-II-C, 504–505
Plasmid pIN-III, utilization, 494–498
Plasmid pIN-III-A, 492–493
Plasmid pIN-III-B, 504–505
Plasmid pIN-III-C, 504–505
Plasmid pIN-III-OmpA, 500, 501–502
Plasmid pJAS14
cloning sites, 135
copy number, 135
map, 146
markers, 135
parent replicon, 135
Plasmid pJHL202, genotype/phenotype, 178
Plasmid pJO158, 39
Plasmid pJO184, 47–48
restriction map of, 49
Plasmid pJRD158, 39, 42, 43, 46
deletion of 121-bp Dde1-RsaI fragment of, 44
restriction map of, 43
Plasmid pJRD158b, 46
Plasmid pJRD182, addition of 83-bp segment to, 44–45
Plasmid pJRD184, 45–46, 53
palindromic sequences, restriction enzymes recognizing, 38–39
restriction map of, 46
Plasmid pJRD185, 46–47
Plasmid pJRD203, 39, 41
cleavage at *Bam*HI site, to prevent formation of polycosmids, 40–41
Plasmid pJRD205, 41
Plasmid pJRD215, 41, 48–51, 53
restriction sites, 50–51
wide host range properties, 50
Plasmid pKC31, genotype/phenotype, 178
Plasmid pKC222, genotype/phenotype, 178
Plasmid pKC293
cloning sites, 135
copy number, 135
map, 146
markers, 135
parent replicon, 135

Plasmid pKC420
  characteristics, 173
  structure, 171
Plasmid pKC462a
  characteristics, 173
  genotype/phenotype, 178
Plasmid pKC505, 168
  characteristics, 173
  structure, 169
Plasmid pKC513
  characteristics, 173
  structure, 171
Plasmid pKC531
  characteristics, 173
  structure, 172
Plasmid pKC561
  characteristics, 173
  structure, 170
Plasmid pKC575
  construction, 197
  structure, 197
Plasmid pKK175-6, 55–57, 60
  nucleotide sequence of cloning region, 57
  restriction map of, 56
Plasmid pKK232-8, 57–58, 60
  assay of promoter activity using, 63–68
  nucleotide sequence of cloning region of, 59
  preparation, 61–62
  promoter strength and chloramphenicol resistance or CAT activity, 59
  restriction map of, 58
Plasmid pKK3535, 434
  introduction of spectinomycin mutation, 442
Plasmid pKN402, 409–411
Plasmid pKST2
  cloning sites, 135
  copy number, 135
  map, 147
  markers, 135
  parent replicon, 135
Plasmid pKUC9, multiple cloning sites, 21
Plasmid pKUC19, multiple cloning sites, 21
Plasmid pKUN, 13, 19
  (recombinant) DNA strands, selective packaging, 31–32
  cloning in, 30–31
  genetic organization, 20
  packaging of F strand, 22
  packaging of I strand, 22
  properties, 19
Plasmid pKUN9
  construction, 19–21
  DNA, helper phage for production of, 24–25
  efficiency of cloning in, 33
  isolation, from *E. coli* strains, 29–30
  multiple cloning sites, 21
  restriction enzyme cleavage sites, 21
Plasmid pKUN19
  construction, 19–21
  DNA, 24–25
  efficiency of cloning in, 33
  isolation, from *E. coli* strains, 29–30
  lacZ' gene, nucleotide sequence of 5'-terminal end, 20
  multiple cloning sites, 21
  restriction enzyme cleavage sites, 21
Plasmid pLAFR1, 48
Plasmid pLcII(nic$^-$), 462
  cloning into, 468–470
  map, 469
  use for expression of eukaryotic proteins, 464–474
Plasmid pLcIIFXβ-globin(nic$^-$), 463–464
Plasmid pMCP10
  cloning sites, 135
  copy number, 135
  map, 147
  markers, 135
  parent replicon, 135
Plasmid pMCP28
  cloning sites, 135
  copy number, 135
  map, 147
  markers, 135
  parent replicon, 135
Plasmid pMG312
  cloning sites, 135
  copy number, 135
  map, 147
  markers, 135
  parent replicon, 135
Plasmid pMH158, 39, 47–48
  restriction map of, 47
Plasmid pMON, transfer into *A. tumefaciens*, 269–272

Plasmid pMON200, 255–256, 261–263, 272
assembly, 273–275
endonucleases not cleaving, 273, 274
left inside homology fragment, 255–256
major regions, 273, 274
map of, 256
restriction endonuclease cleavage sites, 273, 274
size, 275
Plasmid pMON237, 258–260
carrying coding sequence for methotrexate resistance, 268–269
map of, 259
Plasmid pMON316, 261–262
mammalian hormone cDNA for α-subunit of human chorionic gonadotropin inserted into, 272
map of, 262
TMV coat protein expressed from cDNA inserted into, 272
Plasmid pMON321, 268–269
map of, 269
Plasmid pMON408, 266
Plasmid pMON410, 266–268
map of, 267
Plasmid pMON505, 257, 261–263, 272
assembly, 275–277
enzymes not cutting, 276
major regions, 275, 276
map of, 264
restriction endonuclease cleavage sites, 275, 276
Plasmid pMON530, 261–265
carrying coding sequence for hygromycin resistance, 266–268
map of, 265
Plasmid pMP18PP, 552
Plasmid pMP62, 552
Plasmid pMP62-15, 553
Plasmid pMS63
cloning sites, 136
copy number, 136
map, 147
markers, 136
parent replicon, 136
Plasmid pMS75
cloning sites, 136
copy number, 136
map, 148
markers, 136
parent replicon, 136
Plasmid pMT660, 165
cloning sites, 136
copy number, 136
markers, 136
parent replicon, 136
Plasmid pNOSCAT, DNA, electroporation transfer into protoplasts, 363–364
Plasmid pNSY, 385–386
Plasmid pNSY-Leu-1, construction, with leader peptides, 386
Plasmid pNSY-Ser-1, construction, with leader peptides, 386
Plasmid pOA154
cloning sites, 136
copy number, 136
map, 148
markers, 136
parent replicon, 136
Plasmid pOS-1, 258
Plasmid pPLAN B2, transformation efficiency, in *Anacystis nidulans* R2, 210–211
Plasmid pPLc236*trp*, 421
Plasmid pPLc245, 422
Plasmid pPLcmu299, 422
Plasmid pPM201, 373, 377
Plasmid pPM217, 373, 377
Plasmid pPUC29, 294
map of, 205
source, 214
Plasmid pRK248cIts, 418
Plasmid pRK290, 48
Plasmid pRL1, 235, 236, 241
Plasmid pRL10, 236
Plasmid pRL2, 235
Plasmid pRL3, 235
Plasmid pRL4, 235
Plasmid pRL5, 236
Plasmid pRL6, 236, 238, 239, 241
Plasmid pRL7, 236
Plasmid pRL8, 236
Plasmid pRL11, 236, 238, 239, 241
Plasmid pRW120, 552
Plasmid pSAM2, 178
Plasmid pSc13, 376
Plasmid pScL3, 377
Plasmid pScP1, 373
Plasmid pScP8, 373
Plasmid pScP9, 373
Plasmid pSG111, transformation efficiency, in *Anacystis nidulans* R2, 210–211

Plasmid pSK21-K3
  cloning sites, 136
  copy number, 136
  map, 148
  markers, 136
  parent replicon, 136
Plasmid pSKO2, 164
  cloning sites, 136
  copy number, 136
  map, 148
  markers, 136
  parent replicon, 136
Plasmid pSKO3, 165
  cloning sites, 137
  copy number, 137
  map, 148
  markers, 137
  parent replicon, 137
Plasmid pSLP114, 165
Plasmid pSLP124, 165
Plasmid pSP325, 376
Plasmid pSP525, 377
Plasmid pSP530, 373, 377
Plasmid pSP533, 373, 377
Plasmid pSP1015, 373
Plasmid pSW1
  cloning sites, 137
  copy number, 137
  map, 149
  markers, 137
  parent replicon, 137
Plasmid pTiB6S3, 279
Plasmid pTiC58, DNA, electroporation with, $MgCl_2$ requirement, 349
Plasmid pU118, 6–7
Plasmid pUC, 35
  efficiency of cloning in, 33
Plasmid pUC 19, with M13 IG region in same location as 119, in opposite orientation, 8
Plasmid pUC118
  with 2.5-kb insert, 8
  with M13KO7 as helper phage, 8
  structure, 6
Plasmid pUC119, 6–7
  with M13KO19 as helper phage, 8
  with M13mp8 phage carrying kanamycin gene, 8
  structure, 6
Plasmid pUC1112
  cloning sites, 137
  copy number, 137
  map, 149
  markers, 137
  parent replicon, 137
Plasmid pUC1120
  cloning sites, 137
  copy number, 137
  map, 149
  markers, 137
  parent replicon, 137
Plasmid pUC303, 204
  map of, 205
  source, 214
  transformation efficiency, in *Anacystis nidulans* R2, 210–211
Plasmid pUCD9k3, uptake by carrot protoplasts, 344–345
Plasmid pUF3, 214
Plasmid pUF12, 214
Plasmid pUG1, 213
Plasmid pUG2, 213
Plasmid pUH24, 202–203, 207
  restriction map, 203, 214–215
  transposon-tagged derivative, 203–204
Plasmid pUH25, 202–203, 207
Plasmid pVE138
  cloning sites, 137
  copy number, 137
  map, 150
  markers, 137
  parent replicon, 137
Plasmid pVE223
  cloning sites, 138
  copy number, 138
  map, 149
  markers, 138
  parent replicon, 138
Plasmid pVE30
  cloning sites, 137
  copy number, 137
  map, 149
  markers, 137
  parent replicon, 137
Plasmid pVS1, 283
Plasmid pVW1, 234–235
Plasmid pVW1C, 235
Plasmid pWOR126
  cloning sites, 138
  copy number, 138
  map, 150

markers, 138
parent replicon, 138
Plasmid pYE, 521
restriction endonuclease map of, 519–520
Plasmid pZ150, M13KO7 as helper phage with, 11
Plasmid R1162, 48
Plasmid R300B, 48–50, 236
Plasmid R702, 239
Plasmid R7K, 239
Plasmid RK2, 282
Plasmid RP4, failure of replication in heterologous system, 203–204
Plasmid RSF1010, 39, 48–50
failure of replication in heterologous system, 203–204
restriction map of, 49
Plasmids, 3
autonomously replicating, 378–379
broad-host-range, 233
chimeric, 233–234
conjugal transfer to cyanobacteria, 231–243
for construction of replicating vectors similar to mitochondrial genomic DNA, 381
DNA, single-stranded
production, 9–11
yield, 3–4
for eukaryotic cloning, 366–367
position and orientation of IG region, effect on packaging of single-stranded DNA, 8, 11
for production of single-stranded DNA, 19
self-regulated, containing P_L promoter and repressor gene, 482
size, and transformation frequency, in cyanobacteria, 204
wide-host-range, 48
Plasmid S-a, 239
Plasmid SCP1, 117, 118
Plasmid SCP2, 117
Plasmid SLP1, 117
Plasmid SLP2, 118
Plasmid SLP3, 118
Plasmid SLP4, 117
Plasmodium knowlesi sporozoite antigen, expression in recombinant vaccinia virus, 549
PM8 vector, 164
cloning sites, 140
copy number, 140
map, 156
markers, 140
parent replicon, 140
Pock formation, 161
as detection system for transformants, 119
Podospora anserina, plasmids for construction of replicating vectors similar to mitochondrial genomic DNA, 381
Poxvirus
biology, 545
DNA, 545
recombinant, replica filter technique for detection of, 554
pPZ74 cosmid, characteristics, 173
Probe vectors, 164
Proinsulin
fused gene, expression in *Escherichia coli*, 385
human, synthesis in *Escherichia coli*, 385–389
Prokaryotes, molecular cloning, 366
Prokaryotic promoters, 54–68
Promoter activity, quantifying, 61
Promoter expression vectors, to create *cat* fusion proteins, 297–298
Promoter fragments, subcloning of, 60
Promoter–probe vectors, 54–55, 116, 164
with galactokinase gene as indicator, 452
transcriptional gene fusions using, 60–61
uses, 460–461
Promoters
cloned, selection of up and down mutations in, 61
composite, cloning of, 60
fusion to *cat* gene, for plant transformation, 296–297
shotgun cloning, 61–63
strong, subcloning of, 60
Promoter strength
assays, using transcriptional gene fusions, 63–68
quantitative assay, 64–68
semiquantitative assay, 64–65
Protease
digestion of hybrid proteins produced in *E. coli*, 471

precursors, 508–509
from *Streptomyces*, 166
Protein A
secretion in gram-positive bacteria, 510, 513
transport of integral fragment by signal peptide, 515
Protocatechuate oxygenase, in *Pseudomonas*, 53
Protoplasts
direct gene transfer to, 313–336
comparison with other gene transfer systems, 335–336
frequency, 314–315
materials, 315–319
media, 315–318
protocols, 319–332
source of hybrid-selectable gene, 319–321
electroporated, gene expression in, 363–364
electroporation, 359–360
electrotransformation, frequency, factors affecting, 348–350
isolation, 358–359
reagents, 358
from leaf mesophyll of *P. hybrida*
preparation, 327–328
source of material, 327
transformants
culture, 329
regeneration of plants from, 329
selection, 329
transformation, 328–329
by electroporation, 329
F medium method, 328
quick method, 328
misinterpretation of apparently resistant colonies as transformants, 332
from sterile shoot culture of *N. plumbaginifolia*
preparation, 331
source of material, 331
transformants
culture, 332
selection, 332
transformation, 331–332
by electroporation, 332
PEG method, 331–332
from sterile shoot culture of *N. tabacum*
cotransformation, 325
culture, 321–322, 325
growth, 321
isolation, 321–322
preparation, 322
source of material, 322
subculture, 322
transformation, 323–325
by electroporation, 323–325
F medium method, 323
quick method without electroporation, 325
transformed
genetic analysis of progeny, 326–327
regeneration of plants from, 325–326
transformed lines, selection, 325–326
from suspension culture of *L. multiflorum*, 329–330
preparation, 330
source of material, 330
transformants
culture, 331
selection, 331
transformation, 330–331
F medium method, 330
transformation, 351
carrier DNA, 321
criteria for confirmation of, 333
DNA preparation for, 320–321
by electroporation, 356
in vitro, by Ti vector system, 301–302
phenotypic change with, 333
transforming DNA, physical form of, 321
transformed
activity assay for product of transforming gene, 334–335
molecular analysis of DNA, 333–334
regenerated plants from, DNA analysis, 333–334
uptake of plasmid DNA and RNA into, 337
Prototechuate oxygenase, in *Pseudomonas*, 41
Pseudomonas
ATCC19151
in restriction site bank construction, 39
vanillate utilization, 53
cosmid gene bank
construction, 51–52
screening, 51–52

gene bank, construction, 40–41
vanillate utilization in, 41

R

Rabies virus glycoprotein, expression in recombinant vaccinia virus, 549
Replicative vector, 367
Respiratory syncytial virus glycoprotein G, expression in recombinant vaccinia virus, 549
Restriction enzymes, 34
Restriction site bank, 36
addition and deletion of small DNA segments, 42–43
construction, 37
bacteria, 37–39
DNA techniques, 40
enzymes, 40
materials, 37–42
plasmids for, 39
techniques, 37–42
definition, 37
Restriction site bank vectors
advantage of, 53
construction of, rationale for, 36–37
Ribosomal RNA
16 S, analysis of mutations in, 451–452
16 S gene, mutagenesis of ASD region of, 441
mutations, 432–452
Ribosomes, direction to single mRNA species, 432–452
Ring chromosome
mitotic stability, 253
of yeast, construction, 251–252
RNA. *See also* Messenger RNA; Ribosomal RNA
electroporation-mediated transfer of, 360–361
Rolling circle mechanism, 4, 17
Roundup, plant tolerance to, 272
rrnB operon, 434
removal of tandem promoters P1 and P2, 436–441
site-directed mutagenesis on double-stranded plasmid DNA, 436–441
Russel's viper venom, immobilized, preparation, 476

S

Saccharomyces cerevisiae
auxotrophic mutants, transformation to prototrophy, 375
cloning vehicles based on plasmid DNA, 367
diploid 168C, source, 245
GRF18, in restriction site bank construction, 39
as host for production of commercially relevant proteins, 516–544
MC333, in restriction site bank construction, 39
mitochondrial autonomously replicating sequences for construction of vectors, 367–368
mitochondrial cloning vectors for, mitotic stability of, 380
mitochondrial origin of replication, 372
mitochondrial vectors, construction, 370
nonreplicative vectors, 367
origin of replication, integration into hybrid vectors, 380–381
protein secretion, 533
strain 320-13B, source, 245
strain 320-16C, source, 245
strain 8534-8C, source, 245
strain AH 22, 369
strain DPT6, chromosome III, analysis of deletion on, 247–249
strain GM119, source, 245
transformants, assay for free plasmid molecules in whole-cell DNA of, 378–379
transformation frequencies with mitochondrial vectors, 376–377
whole-cell DNA, preparation, 377
Schizosaccharomyces pombe
auxotrophic mutants, transformation to prototrophy, 375
cloned mitochondrial DNA fragments, allocation to genetic map of, 373–374
mitochondrial cloning vectors for, mitotic stability of, 380
mitochondrial DNA, compilation of hybrid vectors obtained from, 373
mitochondrial vectors, construction, 370

origin of replication, integration into hybrid vectors, 380–381
restriction and gene map of, 374
strain *leu*1-32 h$^-$, 369
transformants, assay for free plasmid molecules in whole-cell DNA of, 378–379
transformation frequencies with mitochondrial vectors, 376–377
vector/host systems, 367
whole-cell DNA, preparation, 377
SCP2 vector, 165
Self-transfer, 118–119
Semliki forest virus, secretion in gram-positive bacteria, 512
Shine–Dalgarno sequence, 432, 461
effect on mRNA translatability, 433
Short homopeptide leader sequences, for production of human proinsulin in *Escherichia coli*, 385–389
bacterial strain, 385
codon selection for, effect on proinsulin production, 389
cyanogen bromide cleavage of fused protein to give intact proinsulin, 389
expression of fused protein-containing proinsulin with, 388–389
insertion of synthetic oligonucleotide leader sequences, 386–388
materials, 385
methods, 386–389
oligonucleotides encoding, 386–387
proinsulin gene-containing plasmid for, 385–386
Shotgun cloning, 116
with plasmid vector, 119–130
of promoters, 61–63
using phage vector, 152–164
using plasmid vector
buffers, 121–123
media, 121–123
using *Streptomyces* plasmid vector pIJ486
bacterial strains, 123
outline of, 122
Shuttle vectors, for gene cloning in cyanobacteria, 202–204
Signal peptide
from gram-positive bacteria, 510–511
in protein secretion, 533
regions, 508
structural compatibility with secretion vector, 503–504
Signal structure
expression vectors, 452–461
agarose gel electrophoresis of DNA, 454
assays, 453
buffers, 453–454
DNA restriction fragment isolation, 454
filling in of single-stranded 3′ recessed ends, 455
growth media, 453
material, 453
plasmid DNA isolation, 454
procedures, 456–460
reagents, 453
restriction and ligation of DNA, 455
results, 456–460
screening for promoters from uncharacterized DNAs, 458–460
strains, 453
promoter-probe vectors, 456
levels of indicator gene expression, 459
Sindbis virus structural proteins, expression in recombinant vaccinia virus, 549
Site-directed chromosomal rearrangements, in yeast, 243–253
SLP1.2 vector, 165
Staphylococcal nuclease A, expression vector, 501–502
Staphylococcus aureus, protein secretion in, 510, 513
Staphylokinase, secretion in gram-positive bacteria, 510
Streptavidin, secretion in gram-positive bacteria, 511
Streptomyces
absence of promoters, 150
antibiotics, 116–117
β-galactosidase, fusion protein, 515
bifunctional plasmids, 130
cellular differentiation, 166
clone recognition, 150
cloning vectors
compatible, 151
markers, description of, 161–162

commercially useful therapeutic compounds from, 166
DNA
cloned in KC304, screening library of, 158–159
cloned in ϕC31-based vectors, small-scale isolation and analysis, 159–160
mole GC base composition, 166–167
DNA content, 166
gene cloning procedures, 116–166
genome plasticity, 167
host–vector systems, 198
media for, 177–179
phage vectors, 151–152, 167
plasmid cloning vectors, selection, 130–151
plasmids, 118–119, 167
minilysate procedure for analysis of, 129–130
protein secretion, 516
restriction systems, 130
selectable markers, 150
shotgun cloning experiments, 116, 164
transformation systems, 167
vector DNA from, isolation, 184–185
Streptomyces ambofaciens
3212, genotype/phenotype, 178
ATCC 15154, genotype/phenotype, 178
genomic library of, generated using shuttle cosmid vector pKC505, 179–196
analysis of, 188–194
analysis of clones, 195–196
cloning artifact, 195–196
isolation of specific clones, 195
preparation of DNA fragments, 179–185
preparation of recombinants, 185–188
Streptomyces antibioticus
IMRU3720, melanin (tyrosinase) gene, 162
secreted proteins, 511
Streptomyces avidinii, secreted proteins, 511
Streptomyces azureus ATCC 14921, thiostrepton resistance gene, 162
Streptomyces coelicolor
agarase gene, 165
gene isolation, with shotgun cloning using plasmid vector, 119–130
plasmid cloning vectors, 150
protoplast formation, 127
secreted proteins, 511
Streptomyces coelicolor A3(2)
chromosomal DNA, isolation, 123–125
chromosomal linkage map, 117
native plasmids, 117
partially digested DNA, size fractionation, 125
transposable elements, 117
Streptomyces fradiae ATCC 10745, aminoglycoside phosphotransferase gene, 161
high-resistance mutation, 161
Streptomyces griseofuscus
C581, genotype/phenotype, 178
cloned DNA put into, 191–193
DNA isolation from, 183
plasmid DNA from, rapid isolation, 193–194
spiramycin resistance genes from *S. ambofaciens* selected in, 195
transformation, 192–193
Streptomyces hygroscopicus
NRRL 2387, hygromycin phosphotransferase gene, 161
secreted proteins, 511
Streptomyces kanamyceticus
ISP5500, kanamycin resistance (ribosomal) gene, 161
M1164, kanamycin resistance gene, 161
Streptomyces lavendulae 1080, streptothricin acetyltransferase gene, 162
Streptomyces limosus, secreted proteins, 511
Streptomyces litmocidini ISP5164, amikacin resistance gene, 161
Streptomyces lividans
66
DNA amplification, 118
as host for initial cloning experiments, 117–118
plasmid cloning vectors, 130–150
plasmids, 118
promoterless β-galactosidase gene, 161
3104
mycelial pellet, storage, 127
protoplasts, preparation, 127–128
M386, pIJ486 plasmid DNA from, large-scale preparation, 125–127

protoplasts
formation, 127
freezing for storage, 128
selection of clones, 128–129
transfection of, 158
transformation, 128
secreted proteins, 511
TK54, genotype/phenotype, 178
TK64, genotype/phenotype, 178
Streptomyces plicatus, secreted proteins, 511
Streptomyces ribosidificus SF733, ribostamycin resistance gene, 161
Streptomyces rimosus ATCC10970, tetracycline resistance gene, 162
Streptomyces vinaceus NCIB8852, viomycin phosphotransferase gene, 162
Streptomyces viridochromogenes ssp. *sulfomycini* ATCC29776, sulfomycin resistance gene, 162
Superinfection interference, 21–22
Synechococcus
conjugation of plasmids to, 242
PCC6301, transformation, 201–202
with pBR322, 204
PCC7002. *See Agmenellum quadruplicatum* PR-6
host–vector systems, 213
transformation, 201–202, 213
PCC7942. *See Anacystis nidulans* R2
transformation, 216
Synechocystis
conjugation of plasmids to, 242
PCC6308, transformation, 201–202
PCC6714, transformation, 202
PCC6803
ATCC number, 217
cryptic plasmids, 213
gene cloning and analysis, 201
genome, random insertion of *E. coli* plasmids into, 231
host–vector systems, 213
as host for genetic engineering experiments, 217
photosystem II component genes, inactivation by recombination with cloned, altered allele, 219
properties, 201
psbA genes, inactivation, 221
shuttle vectors, 213–214
transformation, 202
transformation, 216

T

TG1 vector
attachment site, 161
cloning system based on, 164
host range of, 164
TG78 vector, 164
cloning sites, 140
copy number, 140
map, 156
markers, 140
parent replicon, 140
Ti plasmid-derived vectors, 254–257
T-DNA, 279
in transfer and expression of foreign genes in plants, 279
Ti plasmids
binary vectors, 256–257, 282–283, 294–295
foreign DNA inserted into, 293
plant species transformed by, 293
for plant transformation and promoter analysis, 292–305
promoter expression vectors based on, 295–296
size, 293
cis elements, 293
introduction into plant protoplasts by electroporation, 337
plant transformation systems based on, 254
regions essential for mobilization and integration of transferred DNA into plant cells, 254
T-DNA border sequences, 254
T-DNA region, 277, 278, 306
genes, 306
transfer of foreign DNA into plant cells using, 277
transformation of monocots using
materials, 307–308
method of detection, 308–311
reagents, 307–308
as transformation vectors, 293
types of, 306
as vectors for genetic manipulation of plants, 306–307
vir region, 254, 279, 282

Tobacco callus transformation, *in vitro*, by Ti vector system, 302
Tobacco cells, dominant selectable marker genes for, 279
Tobacco etch virus proteins, expression in recombinant vaccinia virus, 549
Tobacco leaf fragments, transformation, with *Agrobacterium*-derived vectors, 287–288
Tobacco mosaic virus
  coat protein
    expressed from cDNA inserted into pMON316, 272
    produced in *E. coli*, 480. *See also* Fusion proteins, CIIFX
  RNA
    introduction into plant protoplasts by electroporation, 337
    in plant protoplasts after electroporation, 341
Tobacco protoplasts, β-galactosidase expression *in situ* after electroporation, 365
Transcription promoters, plasmid vectors that select or screen for, 54–55
Transcrition initiation signals, analysis, 68
Transmissable gastroenteritis virus gp195, expression in recombinant vaccinia virus, 549
Transposon Tn*5*
  aminoglycoside phosphotransferase gene, 162
  mutagenesis of *Anacystis nidulans* R2, 212–213
  neomycin phosphotransferase II coding sequence, chimeric gene for expression of, 264
  in transformation of *Anacystis nidulans* R2, 220
Transposon Tn*901*, mutagenesis of *Anacystis nidulans* R2, 212–213
Troponin C, produced in *E. coli*, 481. *See also* Fusion proteins, CIIFX
Tumor-inducing plasmids. *See* Ti plasmids
Tumor necrosis factor. *See* Human tumor necrosis factor
Two-cistron expression system, 401–416
  for bovine growth hormone, 390
  definition, 403
  derived directly from one cistron expression system encoding hybrid proteins, 415
  design, 404–406
  functional first cistron, DNA sequence of, 404
  for Met-bGH, DNA sequences of, 406
  plasmid construction, 409–414
  starting plasmids, 414

U

Undecylprodigiosin, 117
Unfused protein, expression
  in *Escherichia coli*, 416–431
  in two-cistron expression vector, 416

V

Vaccinia virus
  biology, 545
  deletion mutants, generation, 550
  as expression vector, 545–563
    chimeric donor plasmid, 546
    future applications, 563
    principle of method, 545–551
    uses, 545
  gene products, 545
  genome, 545
    packaging potential, 548–551
  host range, 545
  as ideal expression vector, 563
  influenza hemagglutinin expression regulated by, radioimmunoblot analysis, 561–562
  insertion of foreign genes into, protocol for, 545–548
  insertions and deletions, 550–551
  nonessential DNA sequences, identification, 563
  recombinant
    analysis of protein expression, 557–562
    animal studies, 562–563
    containing multiple foreign genes, 550–551
    DNA analysis, 556–558
    foreign genes expressed in, 548
    infectious progeny, 548

preparation as pure virus stock, 548
RNA analysis, 557, 559
rescuing, in rescue of donor DNA sequence, 546
transcriptional regulatory sequences, for expression of foreign genes, 546
triple recombinant
analysis of protein expression, 560–561
construction of chimeric donor plasmid, 554
donor plasmid construction, 551–553
identification, 556
in vivo recombination, 553–556
procedures, 551–556
structure, 558–559
Vectors
one-*cos* containing, 172–175
two-*cos* containing, 175–176
Vesicular stomatitis virus
G protein, expression in recombinant vaccinia virus, 549
N protein, expression in recombinant vaccinia virus, 549
Viral vectors, for plant transformation, 314

Y

Yeast. *See also Saccharomyces cerevisiae*; *Schizosaccharomyces pombe*
2μ plasmid, 519–521
interconversion of two forms, 521–522
origin of DNA replication, 519
α-factor pheromone precursor, prepro leader region
gene fusion construction, 534–537
native and hybrid, structure, 536–537
autonomously replicating sequences
construction, 366–382
vectors containing, 518–519
autonomous replication of colinear DNA in, 518
chromosomal translocations, 253
chromosome III
deletion derivative of, 250–251
ring derivative, construction, 251–252
ring derivatives, construction, 253
direct expression vectors, 527–533
DNA, preparation, 543–544
expression and secretion vectors, 516–544
end filling with Klenow, 543
mung bean nuclease digestion procedure, 542–543
procedures, 542–544
S1 nuclease digestion of cohesive termini, 543
translational efficiency, 517
expression cassettes, 517
assembly, 529–530
expression vectors, using episomal vectors, 517
extrachromosomal replication vectors, 518–526
centromere plasmids, 521–526
copy number measurement, 523–526
stability measurement, 523
DNA replicators, 518–521
selectable markers, 518
fusion of two chromosomes, 253
generalized expression vectors, 529
generic expression cassette, 527–528
heterologous gene transcription in, analysis, 530–532
heterologous proteins
analysis of carbohydrate additions to, 544
biological activity, measurement, 532–533
expression level, analysis, 532
secretion
analysis, 538–539
efficiency, measurement, 538–539
hybrid precursor processing, analysis, 539–540
introns, 528–529
KR endoprotease processing, 539–540
N-linked glycosylation in, 542
promoters, 527
protein purification from, effect of proteases, 533
RNA, preparation, 543
secreted heterologous proteins
aberrant proteolysis, 542
amino-terminal analysis of, 541–542
glycosylation, 542
secretion vectors, 533–542
site-directed chromosomal rearrangements, 243–253

construction of ring chromosome, 251–252
deletions on both arms of chromosome, 249–250
deletions on one arm of chromosome, 246–247
DNA preparation, 245–246
frequency of deletions, 252
materials, 245–246
media, 245
methods, 245–246
physical analysis of deletion chromosomes, 247–249
requirements for, 252
results, 246–247
strains, 245
transformation, 245–246
usefulness, 252–253
untranslated leader regions, 528
vectors for expression of heterologous genes, 517